Palaeozoic Palaeogeography and Biogeography

Palaeozoic Palaeogeography and Biogeography

EDITED BY

W. S. McKERROW

Department of Earth Sciences
University of Oxford

&

C. R. SCOTESE

Shell Development Company
Houston, Texas

Memoir No. 12
1990

Published by
The Geological Society
London

THE GEOLOGICAL SOCIETY

The Geological Society of London was founded in 1807 for the purpose of 'investigating the mineral structures of the earth'. It received its Royal Charter in 1825. The Society promotes all aspects of geological science by means of meetings, special lectures and courses, discussions, specialist groups, publications and library services.

It is expected that candidates for Fellowship will be graduates in geology or another earth science, or have equivalent qualifications or experience. All Fellows are entitled to receive for their subscription one of the Society's three journals: *The Quarterly Journal of Engineering Geology*, the *Journal of the Geological Society* or *Marine and Petroleum Geology*. On payment of an additional sum on the annual subscription, members may obtain copies of another journal.

Membership of the specialist groups is open to all Fellows without additional charge. Enquiries concerning Fellowship of the Society and membership of the specialist groups should be directed to the Executive Secretary, The Geological Society, Burlington House, Piccadilly, London W1V 0JU.

Published by the Geological Society from:
The Geological Society Publishing House
Unit 7
Brassmill Enterprise Centre
Brassmill Lane
Bath
Avon BA1 3JN
UK

(*Orders*: Tel. 0225 445046)

First published 1990
Reprinted 1994

© The Geological Society 1990. All rights reserved.
No reproduction, copy or transmission of this publication may be made without written permission. No paragraph of this publication may be reproduced, copied or transmitted save with the written permission or in accordance with the provisions of the Copyright Act 1956 (as Amended) or under the terms of any licence permitting limited copying issued by the Copyright Licensing Agency, 33–34 Alfred Place, London WC1E 7DP. Users registered with Copyright Clearance Center: this publication is registered with CCC, 27 Congress St., Salem, MA 01970, USA. 0435–4052/90 $03.00.

British Library Cataloguing in Publication Data
Palaeozoic Palaeogeography and Biogeography.
 1. Geological eras
 I. McKerrow, W. S. (William Stuart), *1922*– II. Scotese, C. R. (Christopher Robert) *1959*– III. Geological Society of London IV. Series
 551.7

ISBN 0–9033174–9–4

Contents

Scotese, C. R. & McKerrow, W. S. Revised World Maps and Introduction, 1

Palaeomagnetism and Palaeoclimates

Torsvik, T. H., Smethurst, M. A., Briden, J. C. & Sturt, B. A. A review of palaeomagnetic data from Europe and their palaeogeographical implications, 25

Bachtadse, V. & Briden, J. C. Palaeomagnetic constraints on the position of Gondwana during Ordovician to Devonian times, 43

Kent, D. V. & Van Der Voo, R. Palaeozoic palaeogeography from palaeomagnetism of the Atlantic-bordering continents, 49

Witzke, B. J. Palaeoclimatic constraints for Palaeozoic palaeolatitudes of Laurentia and Euramerica, 57

Scotese, C. R. & Barrett, S. F. Gondwana's movement over the South Pole during the Palaeozoic: evidence from lithological indicators of climate, 75

Van Houten, F. B. & Hou, Hong-Fei. Stratigraphic and palaeogeographic distribution of Palaeozoic oolitic ironstones, 87

Early Palaeozoic Biogeography

Cocks, L. R. M. & Fortey, R. A. Biogeography of Ordovician and Silurian faunas, 97

Bergström, S. M. Relations between conodont provincialism and the changing palaeogeography during the Early Palaeozoic, 105

Finney, S. C. & Chen, Xu. The relationship of Ordovician graptolite provincialism to palaeogeography, 123

Berry, W. B. N. & Wilde, P. Graptolite biogeography: implications for palaeogeography and palaeoceanography, 129

Rickards, R. B., Rigby, S. & Harris, J. H. Graptoloid biogeography: recent progress, future hopes, 139

Crick, R. E. Cambrian–Devonian biogeography of nautiloid cephalopods, 147

Burrett, C., Long, J. & Stait, B. Early–Middle Palaeozoic biogeography of Asian terranes derived from Gondwana, 163

Liao, Wei-Hua. The biogeographic affinities of East Asian corals, 175

Sheehan, P. M. & Coorough, P. J. Brachiopod zoogeography across the Ordovician–Silurian extinction event, 181

Silurian–Devonian Biogeography

Boucot, A. J. Silurian biogeography, 191

Tuckey, M. E. Distributions and extinctions of Silurian Bryozoa, 197

Colbath, G. K. Palaeogeography of Middle Palaeozoic organic-walled phytoplankton, 207

Nestor, H. Biogeography of Silurian stromatoporoids, 215

Berdan, J. M. Silurian and Early Devonian biogeography of ostracodes in North America, 223

Edwards, D. Constraints on Silurian and Early Devonian phytogeographic analysis based on megafossils, 233

Young, G. C. Devonian vertebrate distribution patterns and cladistic analysis of palaeogeographic hypotheses, 243

Stock, C. W. Biogeography of Devonian stromatoporoids, 257

Pedder, A. E. H. & Oliver, W. Rugose coral distribution as a test of Devonian palaeogeographic models, 267

Blodgett, R. B., Rohr, D. M. & Boucot, A. J. Early and Middle Devonian gastropod biogeography, 277

Poncet, J. Biogeography of Devonian algae, 285

Streel, M., Fairon-Demaret, M. & Loboziak, S. Givetian–Frasnian phytogeography of Euramerica and western Gondwana based on miospore distribution, 291

Hou, Hong-Fei & Boucot, A. J. The Balkhash–Mongolia–Okhotsk region of the Old World Realm (Devonian), 297

Carboniferous–Permian Biogeography

Bambach, R. K. Late Palaeozoic provinciality in the marine realm, 307

Kelley, P. H., Raymond, A. & Lutken, C. B. Carboniferous brachiopod migration and latitudinal diversity: a new palaeoclimatic method, 325

Lane, N. G. & Sevastopulo, G. D. Biogeography of Lower Carboniferous crinoids, 333

Waters, J. A. The palaeobiogeography of the Blastoidea (Echinodermata), 339

Ross J. P. R. & Ross, C. A. Late Palaeozoic bryozoan biogeography, 353

Ziegler, A. M. Phytogeographic patterns and continental configurations during the Permian Period, 363

Palaeozoic Geography

Lottes, A. L. & Rowley, D. B. Early and Late Permian reconstructions of Pangaea, 383

Nie, Shangyou, Rowley, D. B. & Ziegler, A. M. Constraints on the location of the Asian microcontinents in Palaeo-Tethys during the Late Palaeozoic, 397

Robardet, M., Paris, F. & Racheboeuf, P. R. Palaeogeographic evolution of southwestern Europe during Early Palaeozoic times, 411

Young, T. P. Ordovician sedimentary facies and faunas of southwest Europe: palaeogeographic and tectonic implications, 421

Isaacson, P. E. & Sablock, P. E. Devonian palaeogeography and palaeobiology of the Central Andes, 431

Revised World maps and introduction

C. R. SCOTESE[1] & W. S. McKERROW[2]

[1] *Bellaire Research Center, Shell Development Company, PO Box 481, Houston TX 77001, USA*
[2] *Department of Earth Sciences, Oxford University, Parks Road, Oxford OX1 3PR, UK*

Abstract: We review the highlights of the 1988 symposium on Palaeozoic Biogeography and Palaeogeography, and present a revised set of 20 Palaeozoic base maps that incorporate much of the new data presented at the symposium. The maps include 5 major innovations: (1) A preliminary attempt has been made to describe the motion of the Cathaysian terranes during the Palaeozoic; (2) a more detailed description of the events surrounding the Iapetus Ocean is presented; (3) an alternative apparent polar wandering path for Gondwana has been constructed using the changing distributions of palaeoclimatically restricted lithofacies; (4) new palaeomagnetic data have been incorporated that places Laurentia and Baltica at more southerly latitudes, and adjacent to Gondwana, during the Early Devonian; Siberia is also placed further south in the light of biogeographic data presented at the symposium; (5) Kazakhstan is treated as a westward extension of Siberia, rather than as a separate palaeocontinent. The relationships between climatic changes, sea level changes, evolutionary radiations and intercontinental migrations are discussed.

A symposium on Palaeozoic Biogeography and Palaeogeography, sponsored by the Geological Society of London, the Palaeontological Association and the International Lithosphere Program, was held at the Department of Earth Sciences, Oxford University in August, 1988. The 40 papers in this volume include new data and new syntheses on palaeomagnetism, palaeoclimatology and the distribution of certain lithologies, but the majority of contributions are on Palaeozoic biogeography. Following the procedure adopted by the previous (and first) global palaeobiogeographical symposium, held at Cambridge University in 1971 (Hughes 1973), experts on major Palaeozoic fossil groups were invited to plot faunal and floral distributions on maps supplied before the symposium. An important goal of this symposium was to produce a revised set of world maps that will provide a common framework, not only for reviewing Palaeozoic biogeography, but as a contribution towards the future overall understanding of Palaeozoic plate motions and tectonics.

The maps provided at the symposium differed from previous ones (Scotese *et al.* 1979, 1985; Scotese 1984, 1986) in the following ways:

(1) a new scenario is presented for the motions of the eastern Asian continents (South China, North China, Indochina and Shan Thai-Malaya);
(2) more detail is shown for the continental margins of the Iapetus Ocean.
(3) a narrow (rather than a wide) ocean is shown between Laurentia (North America) and Gondwana during the Devonian.

During the course of the meeting, it became clear that some maps did a better job of explaining the biogeographic distributions than others. In general, the Ordovician, Silurian, Carboniferous and Permian maps were acceptable to most contributors, but the Late Silurian and Devonian maps did not fare so well. In the light of these comments, and some new data, we have prepared a revised set of Palaeozoic base maps (Figs 2 to 22). After a brief discussion on Palaeozoic biogeography, we first discuss the data from each palaeocontinent used in the constructions of the maps, and we then present a chronological account of the main changes in continental configurations during the Palaeozoic.

Palaeozoic biogeography

The reader will note that the biogeographic terminology used by the authors in this memoir is not consistent from one contribution to another. The editors considered trying to impose greater uniformity in the use of such terms as province, realm, region, etc., but, a consensus on terminology among the symposium participants does not exist. Although one of us (McKerrow & Cocks 1986, p. 185) has made a plea for distinguishing 'Provinces' (separated by barriers) from 'Realms' (climatically controlled), we have decided to allow each contributor to use his own definitions of these terms.

Apart from personal and national prejudices, there is also an underlying scientific reason for the diversity in terminology: different animals and plants have different ecological controls affecting their biogeography. Some are obviously marine and others non-marine, so that continents or oceans which are barriers to some are the channels for migration of others. Moreover, different taxa have different degrees of mobility: terrestial forms are often good indicators of land masses separated by only narrow stretches of sea, as are benthic ostracodes (see **Berdan**, this volume), which have no pelagic larval stage. Forms which are pelagic for part or all of their ontogeny may be more or less reflective of climate or of water masses with limited temperature changes. Thus, different terminologies are used to describe the biogeography of the different groups. The time has not yet arrived when we can attain consistency in terminology, but we hope that this volume, by showing up the different terms still employed, may help to accelerate the development of a unified approach to palaeobiogeography (see **Bambach** & **G. Young**, this volume).

Most contributors have plotted their data on the symposium maps, but some have not. Many contributions also include accounts of the evolution of the various fossil groups. This has resulted in a certain unevenness in presentation, but it has also drawn attention to some significant new conclusions about controls on migration through time.

Evolutionary radiations during times of rising sea level often appear to have been accompanied by migration of marine faunas and floras. Sea level changes in the Arenig and Llanvirn appear to have affected the migration of some groups (**Finney, T. Young**, this volume), but the more prominent changes in the Late Ordovician, known to be related to glacial episodes, have affected the biogeography of nautiloids (**Crick**), blastoids (**Waters**), graptolites (**Berry & Wilde**; **Rickards** *et al.*), stromatoporoids (**Nestor**) and brachiopods (**Sheehan & Coorough**). Similar changes in the Devonian and Carboniferous are also suggested for nautiloids (**Crick**), for stromatoporoids (**Stock**), for blastoids (**Waters**), and for brachiopods (**Kelley** *et al.*). Clearly, the comparison of sea level changes and migration times of different animal groups will reward further study.

Palaeozoic palaeocontinents

The boundaries of the larger continents used in the revised base maps (Fig. 1) are essentially the same as those defined previously (Ziegler *et al.* 1977*a, b*; Scotese *et al.* 1979), but some small terranes are now also included.

In the light of the new data presented in this volume, we have

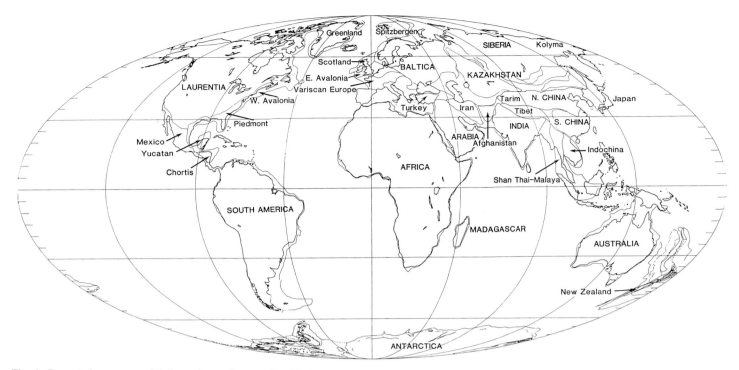

Fig. 1. Present day sutures of Palaeozoic continents, after Ziegler *et al.* (1977*a, b*) and Scotese *et al.* (1979).

prepared a revised set of base maps for the 20 Palaeozoic time intervals used previously (Figs 2 to 21). The data used to construct these maps and the outstanding uncertainties and controversies are discussed next.

Laurentia

The North American palaeocontinent included northwest Ireland, Scotland, Greenland, the North Slope of Alaska and the Chukotsk peninsula of NE USSR. The Barents Sea continent (including Svalbard) was probably attached to Laurentia before the Early Ordovician; it may have collided with NE Greenland in the late Llandovery, so we infer that it was possibly separate from Laurentia in the Late Ordovician and Early Silurian. Mexico, Baja California and the Chortis block of Honduras have been rotated 11° clockwise with respect to cratonic Laurentia along a line parallel to the Torreon–Monterrey shear zone. The rotation parameters used to reassemble Laurentia are given in Table 1.

The Early Palaeozoic margin of Laurentia can be recognized in the Appalachians by the transition from platform carbonates to deeper water slope and offshore deposits (Rodgers 1968; Fortey & Cocks 1986). In the British Isles and East Greenland, the continental margin is marked by very thick Late Precambrian clastic sequences: the Dalradian Supergroup in Scotland (Harris *et al.* 1978) and the Eleonore Bay and Tillite Groups in East Greenland between present latitudes 70° and 76°N (Henriksen 1985), and possibly the Hecla Hoek sequences in Svalbard (Harland 1985).

During the Cambrian and Early Ordovician, the shallow cratonic environments of Laurentia were characterised by endemic faunas (Cocks & Fortey 1982), including bathyurid trilobites and several brachiopod genera (**Cocks & Fortey**, this volume), nautiloids (**Crick**, this volume) and the conodont Midcontinent Faunal Region (**Bergström**, this volume). The deeper water faunas show a greater correlation with climate (**Cocks & Fortey, Finney & Chen, Berry & Wilde, Rickards *et al.***, this volume), and are consistent with Laurentia lying near the equator. By the Silurian, most faunas in Laurentia were cosmopolitan, but the benthic ostracodes (**Berdan**, this volume) and some terrestial taxa were still distinct (Cocks & Fortey 1982). Continental collisions occurred between Laurentia and Baltica in the Late Silurian Scandian Orogeny, and between Avalonia and Laurentia in the Emsian/Eifelian Acadian Orogeny (McKerrow 1988*a, b*), so that the Laurentian faunas lost much of their regional identity during the Silurian and Devonian.

The Cambrian through Devonian orientation of Laurentia is based on the recent syntheses of Van der Voo (1988) and **Kent & Van der Voo** (this volume). The Devonian position of Laurentia was 20° to 30° further south than previously thought (Scotese *et al.* 1979, 1985). This latitudinal shift is based on new palaeomagnetic determinations for the Early Devonian (Miller & Kent 1988; Stearns *et al.* 1989) and the Late Devonian (Miller & Kent 1986). These new data agree well with palaeomagnetic results from Scotland (Briden *et al.* 1984; **Torsvik *et al.***, this volume) and the southerly location is in better agreement with Devonian biogeographic and palaeoclimatic indicators (see **Witzke**, this volume, and discussion above).

The Early Carboniferous orientation of Laurentia is based on revised palaeomagnetic determinations from the Mauch Chunk red beds of the Central Appalachians (Kent & Opdyke 1985) and from Maritime Canada (Scotese *et al.* 1984). By the Late Carboniferous and Permian, Laurentia had collided with Gondwana and had become part of the supercontinent Pangea. The Late Carboniferous and Permian orientation of Laurentia is based on the combined palaeomagnetic poles from all the Pangean continents (Van der Voo *et al.* 1984) and is similar to the Permian maps of **Lottes & Rowley** (this volume).

Baltica

Baltica consists of the major part of northern Europe; it is bounded on the west by the Iapetus suture, on the east by the Ural suture, on the south by the Variscan/Hercynian suture, and on the SW by the suture of the Tornquist Sea (Cocks & Fortey 1982), which closed in the Ashgill, and lies near, but not along, the Tornquist Line.

In the Early Ordovician, Baltica is characterised by a distinctive group of asaphid trilobites (**Cocks & Fortey**, this volume). This

Table 1. *Rotation parameters used to reassemble palaeocontinents*

Plate	Pole of Rotation			Reference Plate
	Latitude	Longitude	Angle	
(1) Laurentia				
Florida	62.20	−15.90	78.80	Africa
Greenland	−50.07	26.29	7.74	Laurentia[1]
N. Alaska	70.11	−128.16	−78.00	Laurentia
Mexico	−48.60	94.10	13.00	Laurentia
Baja California	−46.48	−76.52	7.48	Mexico
Chortis	−39.70	87.90	31.10	Mexico
N. Scotland	−82.30	−25.90	33.50	Laurentia
(2) Gondwana				
Arabia	26.50	21.50	−7.60	Africa[2]
Madagascar	−1.70	−87.80	22.20	Africa
S. America	45.50	−32.20	58.20	Africa[3]
Yucatan	−2.92	97.80	74.50	S. America
India	−22.38	−157.10	55.91	Madagascar
Ceylon	−11.12	−99.73	28.67	India
Variscan Europe	41.85	36.60	13.67	Africa
Iberia	−45.73	−178.30	37.10	Europe
Apulia	21.80	116.80	2.20	Europe
Australia	−1.58	39.02	−31.29	E. Antarctica[3]
E. Antarctica	5.46	−162.90	−92.60	India
Marie Byrdland	62.27	21.84	13.27	E. Antarctica[3]
W. Antarctica	−64.24	−75.64	91.09	South America[3]
Ellsworth	−63.81	−79.40	87.02	South America[3]
N. New Zealand	24.19	−19.91	44.61	Australia[3]
S. New Zealand	65.14	−52.00	62.38	Marie Byrdland[3]
Chatham Rise	41.00	−15.90	7.47	S. New Zealand[3]
(A) Cimmerian terranes				
Shan Thai	5.98	114.75	138.37	Australia
Tibet	−29.90	−84.49	111.20	Shan Thai
Iran	52.37	32.75	32.38	Tibet
Turkey	−17.72	−135.8	11.19	Iran
(B) Cathaysian terranes				
S. China	2.17	161.40	39.30	Australia
N. China	−37.80	−51.10	40.40	S. China
Indochina	9.40	93.80	41.60	S. China

[1] Lawver & Scotese (1989)
[2] McKenzie *et al.* (1970)
[3] Lawver & Scotese (1987)

fauna extends south to the Holy Cross Mountains in southern Poland, where deeper facies are present. The planktonic graptolites and the pelagic trilobites of Baltica show mixtures between low latitude (Laurentian) forms and high latitude (Gondwanan) forms (Cocks & Fortey 1982); their distributions appear to be related to climate (**Cocks & Fortey**, this volume) and suggest that Baltica was at an intermediate latitude in the Early Ordovician.

There are no reliable palaeomagnetic poles from the Cambrian or Ordovician of Baltica (**Torsvik *et al.***, this volume), but the faunal evidence for intermediate latitudes is confirmed by some sedimentary facies. Middle Ordovician detrital limestones provide no evidence for a warm climate, but, in the Late Ordovician, reefs are present and suggest that by this time Baltica, like Laurentia, was close to the equator (Webby 1984; Bruton *et al.* 1985).

After the Late Silurian Scandian Orogeny, some sinistral strike slip movements continued through the Early Devonian. The position of Baltica with respect to Laurentia after this closure of the Iapetus Ocean (**Fig. 12**) is a 'tight' fit, which compensates for the extension that occurred along the margins of the North Atlantic Ocean prior to the Late Cretaceous/Early Tertiary continental rifting. A tight fit of the continents is also required to superimpose Middle and Late Palaeozoic poles from Europe and North America (Frei & Cox 1987; **Lottes & Rowley**, this volume).

Avalonia

The Ardennes of Belgium and northern France, England, Wales, southeastern Ireland, the Avalon Peninsula of eastern Newfoundland, much of Nova Scotia, southern New Brunswick and some coastal parts of New England constitute Avalonia. These areas are characterized by the following.

(1) A basement, which includes Late Precambrian (c. 600 Ma) arc rocks, and appears to have developed on a margin of Gondwana, though the precise location is uncertain.

(2) Cambrian and Early Ordovician shallow water faunas similar to western Gondwana, but with Baltic faunas in the Late Ordovician (**Cocks & Fortey**, this volume). Silurian ostracodes are identical with those of Baltica (**Berdan**, this volume), suggesting a shallow water connection by the end of the Ordovician.

(3) No warm water sediments prior to the Wenlock.

In southern Nova Scotia, the very thick clastic sequence of the Meguma Group extends upwards to include Early Ordovician sediments. No comparable sequence in known in Avalonia, and it may be that the Meguma terrane separated from Gondwana independently from Avalonia. It was, however, linked to the rest of Nova Scotia by the Early Devonian when numerous plutons are present in both terranes (Keppie 1985).

It appears that at some time during the Early Ordovician, Avalonia rifted from Gondwana, perhaps when calc-alkaline igneous activity started in the Ardennes, Wales (Kokelaar *et al.* 1984) and SE Ireland with consumption of the Tornquist Sea oceanic crust. Avalonia probably collided with Baltica in the early Ashgill (McKerrow 1988*a*) with the closure of the Tornquist Sea and the appearance of ostracodes from Baltica. It then collided with Laurentia in the Emsian/Eifelian Acadian Orogeny (McKerrow 1988*b*). The movements of Avalonia (Figs 6–8) by this hypothesis are consistent with the palaeomagnetic data, which places it at temperate latitudes during the Ordovician (Johnson *et al.* 1988). There are few Cambrian and Early Ordovician poles from England, but Late Ordovician through Devonian results (**Torvik** *et al.*, this volume) place England in subtropical latitudes (Figs 7–12).

Gondwana

The continents forming the core of Gondwana are: South America, Africa, Madagascar, India, Antarctica and Australia; the present reconstruction is similar to the fit originally proposed by du Toit (1937), and differs primarily in the 'tightness' of the fit in order to compensate for pre-drift extension. The poles of rotation used here are given in Table 1.

Though the relative positions of the 6 core continents are well constrained, the locations of numerous small continental blocks that bordered Gondwana are less certain. The following have all been adjacent to Gondwana at some time during the Palaeozoic: Yucatan, Florida, Avalonia, central and southern Europe (**Robardet** *et al.* and **T. Young**, this volume) and the Cimmerian terranes of Turkey, Iran, Afghanistan, Tibet and Southeast Asia (Sengör 1984, 1987).

The orientation of Gondwana used here is based on an apparent polar wander (APW) path determined from palaeoclimatic, rather than palaeomagnetic, data (**Scotese & Barrett**, this volume). During the Cambrian and Early Ordovician, the South Pole was located 20–40° to the south of (present day) North Africa (Figs 2–7). The pole moved rapidly across North Africa in the Late Ordovician and reached Brazil by the end of the Silurian (Fig. 12). In the Devonian, the rate of apparent wandering slowed and the trajectory shifted, first south to Argentina in the Late Devonian (Fig. 16) and then east towards South Africa in the Carboniferous (Fig. 18), and by the Early Permian it was near central Antarctica (Fig. 19).

Earlier published APW paths for Gondwana linked Late Ordovician poles with Carboniferous poles by means of a path through central Africa (McElhinny & Embleton 1974; Schmidt & Morris 1977; Brock 1981). The Devonian portion of the Gondwana APW path used here also differs from the APW paths proposed in this volume by **Bachtadse & Briden** and by **Kent & Van der Voo**. Though the Early and Late Palaeozoic portions of the APW path proposed here are similar to the APW path of **Bachtadse & Briden**, the Devonian portion is different. The Devonian APW path proposed by these authors makes a rapid loop from southern Argentina into central Africa, whereas the palaeoclimatically determined path, which we favour, follows a smooth trajectory from Argentina towards South Africa.

Central and southern Europe

The relationships between the Palaeozoic terranes of Iberia, France, Germany and Bohemia is still uncertain. Palaeomagnetic data from western France (Perroud & Van der Voo 1985) and Spain (Perroud 1983; Perroud *et al.* 1984) indicate that these areas were at high southern latitudes during the Early Palaeozoic (see **Torsvik** *et al.*, this volume), which is consistent with a position adjacent to the north African margin of Gondwana. Faunal evidence supports the view that much of central and southern Europe had low diversity temperate shelf faunas in the Silurian, similar to other areas bordering Gondwana (Cocks & Fortey 1988; **Robardet** *et al.*, this volume). The Early Palaeozoic sequences of Morocco, Spain, France and Bohemia are dominated by clastics; the absence of warm water carbonates (prior to the Middle Devonian) is consistent with their being at high southern latitudes.

Siberia and Kazakhstan

The Palaeozoic continent of Siberia is bounded on the west by the northern half of the Urals and the Irtysch Crush zone, on the south by the South Mongolian arc, and on the northeast by the Verkhoyansk Fold Belt. Throughout most of the Palaeozoic, Siberia was oriented 180° from its present alignment, so that its southern, active, Andean-style margin faced to the north.

Kazakhstan now lies southwest of Siberia. Previous reconstructions (e.g. Scotese *et al.* 1979) have treated the region as a separate and independent continent. We now consider Kazakhstan to be an extension of the Palaeozoic Siberian continent; its configuration may have been similar to the present day relationship of the Malay Peninsula, Sumatra and Java to Southeast Asia. The link with Siberia was possibly an extension of the South Mongolian arc, and we consider that Kazakhstan grew during the Palaeozoic by the accretion of volcanic arcs and related trench deposits. The complex and often mixed aspect of the Kazakhstan faunas may be explained, in part, by accretion of exotic and far-travelled terranes.

The orientation of Siberia and Kazakhstan (Figs 2–21) is based on the palaeomagnetic compilations of Khramov *et al.* (1981) and Khramov & Rodionov (1980). Several contributors to this symposium have suggested that these palaeomagnetic poles place Siberia unacceptably far north during the Silurian and Devonian, and we have made some adjustments accordingly. **Nestor** (this volume) shows that stromatoporoids are normally confined to latitudes within 30° of the equator, but that the Tuva region of (present day) south Siberia was shown to be 40–45° N. on the Wenlock and Ludlow symposium maps. Similarly, coral (**Pedder & Oliver**), gastropod (**Blodgett** *et al.*), algae (**Poncet**) and miospore (**Streel** *et al.*) distributions, as well as climatic constraints (**Witzke**), suggest that Siberia was shown too far to the north during the Devonian.

From the Cambrian to the Early Carboniferous, Siberia and Kazakhstan moved northwards. During the mid-Carboniferous they rotated clockwise, colliding with Baltica along the Ural Mountains in the Late Carboniferous/Early Permian.

China and Tarim

It is now recognized that China consisted of at least 3 Palaeozoic continents (**Nie** *et al.*, this volume): North China (Sino-Korea), South China (the Yangtze platform) and Tarim. The Qilian Shan, an Early Palaeozoic arc, separates Tarim from North China. Though we recognize that Tarim and North China were separate palaeocontinents, for simplicity, we have shown them joined together. Tarim appears to have collided with Siberia and Kazakhstan during the Late Carboniferous/Early Permian (**Nie** *et al.*, this volume), whereas North China and South China did not join Asia until the Late Triassic/Early Jurassic Indosinian Orogeny.

Biogeographic affinities with eastern Gondwana (**G. Young, Hou & Boucot, Liao**, this volume) suggest that South China was located near Australia during most of the Palaeozoic. The orientation of South China is based on Cambrian (Lin *et al.* 1985*a*, *b*), Silurian (Opdyke *et al.* 1987), Devonian (Fang & Van der Voo 1989), Carboniferous and Permian (Chan *et al.* 1984; Zhao &

Coe 1987) palaeomagnetic data. Because the polarity of the Early Palaeozoic poles from South China is not well established, it is uncertain whether the NW or SE margin faced Gondwana. Palaeogeographic arguments can be made for either orientation. We have chosen to show the present SE margin facing Gondwana in order to minimise the amount of post-rifting rotation.

Though South China and North China were probably associated with eastern Gondwana during the Early Palaeozoic, they had rifted off by the Late Palaeozoic. This separation is shown by palaeomagnetic data, by biogeographic differences, and by palaeoclimatic evidence which indicates that, while eastern Gondwana was being glaciated, both South and North China enjoyed tropical climates. Relative plate motions suggest that rifting from Gondwana must have begun by the Middle Devonian (Fig. 15).

The location of North China and Tarim during the Early and Middle Palaeozoic is uncertain. There are few reliable Early Palaeozoic palaeomagnetic results from North China, and much of the Middle Palaeozoic stratigraphic record is missing. Cambrian and Early Ordovician faunas from North China are similar to those of Australia, while South China has more affinities with India (Burrett *et al.*, this volume), but in our maps we show North China attached to South China in the vicinity of Korea, though separated from South China by the V-shaped Qin Ling Ocean (Zhao & Coe 1987).

Cimmeria

Turkey, Iran, Tibet (Qiang Tang and Lhasa), Shan Thai–Malaya and Indo-China are shown (Figs 2 to 21) as several small terranes along the margin of Gondwana; these comprise the Cimmerian continent of Sengör (1984, 1987).

During the Early and Middle Palaeozoic, Cimmeria was the site of an active, Andean-style margin. By the Carboniferous, some of the Cimmerian terranes (Qiang Tang and Shan Thai–Malaya) have glacial-related deposits which indicate that they were still located at high southerly latitudes adjacent to Gondwana; this is unlike China, whose equatorial faunas and floras show that it had separated from Gondwana by this time. Palaeomagnetic data confirm a temperate location for the Shan Thai–Malaya terrane during the Late Palaeozoic (McElhinney *et al.* 1981; Fang & Van der Voo 1989). In the Late Palaeozoic, Cimmeria rifted away from Gondwana (Panjal Traps) and crossed the Palaeo-Tethys Ocean, eventually colliding with the southern margin of Asia during the Middle to Late Triassic Cimmerian Orogeny (Sengör 1984, 1987).

Chronological review

Latest Precambrian and Cambrian (Figs 2 to 5)

Many of the passive margins of Laurentia appear to have been formed between 750 and 600 Ma ago, suggesting to several authors that there may have been a Late Precambrian Pangea (Morel & Irving 1978; Bond *et al.* 1984; Piper 1983, 1987). One possible Precambrian supercontinent, composed of Laurentia, Baltica and Siberia, is shown in Fig. 2. The involvement of Gondwana in a Late Precambrian Pangea is problematic. The Pan-African Orogeny, which dates from this time (*c.* 600 Ma), records a series of continental collisions that led to the formation of Gondwana. If this is the case, the Late Precambrian was a time characterized by the break-up of one supercontinent and the assembly of another.

Though the relative positions of Laurentia, Baltica and Siberia are uncertain, their break-up would have led to the formation of new oceans; one of these was the Iapetus Ocean. In Figs 2 to 5, we show the Iapetus widening as Laurentia diverges from Baltica and Siberia during the Cambrian. At the same time, Gondwana moved steadily southwards so that the Cambrian faunas of Laurentia, Baltica and Siberia remained isolated enough to develop some faunal endemism (Palmer 1973; Bergström & Gee 1985, p. 266).

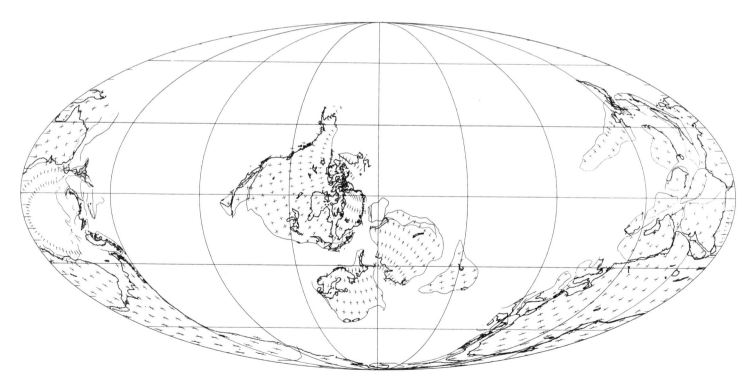

Latest Precambrian

Fig. 2. Latest Precambrian.

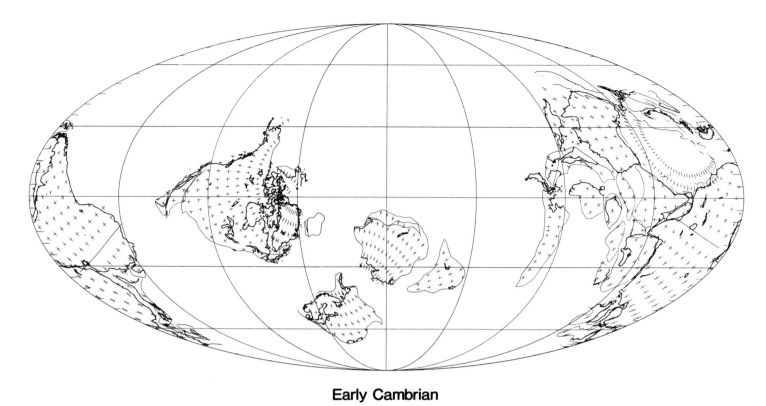

Fig. 3. Early Cambrian.

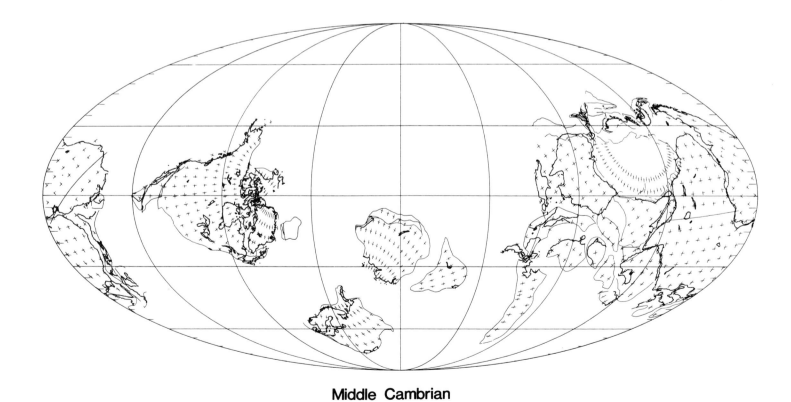

Fig. 4. Middle Cambrian.

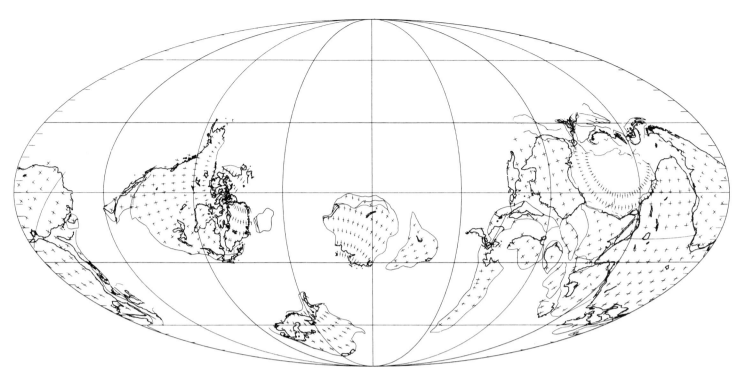

Fig. 5. Late Cambrian.

Ordovician (Figs 6 to 9)

By the Early Ordovician, the Iapetus Ocean and the Tornquist Sea had begun to narrow, with arcs present on the east of Laurentia and in Avalonia (McKerrow 1988a). Apart from pandemic deep water forms, the marine faunas of Laurentia, Baltica and Gondwana remained distinct, suggesting that they were separated by oceans at least 1000 km wide (McKerrow & Cocks 1986). The shallow marine bathyurid trilobites occur in the tropics of Siberia and North China as well as in Laurentia (**Cocks & Fortey**, this volume), suggesting, not only that these three continents were near the Equator, but that they had separations of less than 1000 km.

During the Early Ordovician, (Figs 6 and 7), carbonates are abundant on the cratons of Laurentia, Siberia and the Indo-Australian region of Gondwana, whereas clastic rocks are dominant in Baltica and the North African margin of Gondwana. Warm water carbonates appear in the Late Ordovician of Baltica (Webby 1984; Bruton *et al.* 1985), suggesting a slow northward movement into lower latitudes (Fig. 8).

West Gondwana continued to move southwards during the Ordovician, and it began to cross the South Pole in the Caradoc (Fig. 8). By the Ashgill (Fig. 9), an ice cap covered the whole of North Africa from Morocco to Arabia. Adjacent parts of Gondwana, including Spain and Brittany, have Ashgill sediments which have been interpreted as periglacial (**T. Young**, this volume). There are no Ashgill tilloids in eastern Avalonia, which was probably connected to Baltica, where warm water carbonates were established (see above).

Avalonia probably started to rift from Gondwana in the late Tremadoc (compare Figs 6 and 7), when calc-alkaline rocks appeared in the southern British Isles (Kokelaar *et al.* 1984) and possibly in the Ardennes, and when subduction of the SW margin of the Tornquist Sea originated. In the Middle Ordovician (Fig. 8), the shallower benthic faunas of eastern Avalonia started to lose their affinities with Gondwana and shre links with Baltica, and by the late Caradoc these faunas were identical with Baltica (Cocks & Fortey 1986 and this volume), indicating that shallow benthos with pelagic larval stages could cross the Tornquist Sea. This connection occurs significantly earlier than the Hirnantian stage of the late Ashgill (Fig. 9), when the corresponding faunas became identical across the Iapetus Ocean between Laurentia and Baltica (**Sheehan & Cooragh**, this volume). It is estimated that an ocean around 1000 km, wide would be narrow enough for such complete integration of shallow water benthos with pelagic larvae (McKerrow & Cocks 1986).

The location of the south-western portion of Avalonia is more uncertain. A tillite may exist in the Roxbury Conglomerate of the Boston Bay Group (Bailey *et al.* 1976), but its age is unknown. A tillite is also recorded from the Ordovician of the Meguma terrane of Nova Scotia (Schenk 1972). If the presence of late Ashgill tillites is confirmed in southwestern Avalonia, higher latitudes are indicated.

During the Early Ordovician, several island arcs were present off the eastern margin of Laurentia; they can be readily defined in the Northern Appalachians (Bronson Hill, Tetagouche, Lushs Bight), but are less certain in the British Isles (McKerrow 1988a) and Scandinavia (Stephens & Gee 1985). These arcs appear to have collided with Laurentia progressively, starting in the north during the Early Ordovician and ending with the Caradoc Taconian Orogeny in New England. During the earlier stages of their development, some of these arcs may have been located well out in the Iapetus Ocean. Many of them contain some Early Ordovician brachiopod genera (the 'Celtic Province') which have been considered to be unique to oceanic islands (Neuman 1984); however, more recent records of the same taxa from both Laurentia and Avalonia would suggest that the 'Celtic Province' consists of widely travelled forms (perhaps with a longer than average larval duration) which could colonize volcanic islands, but which are by no means diagnostic of a mid-oceanic setting (McKerrow & Cocks 1986).

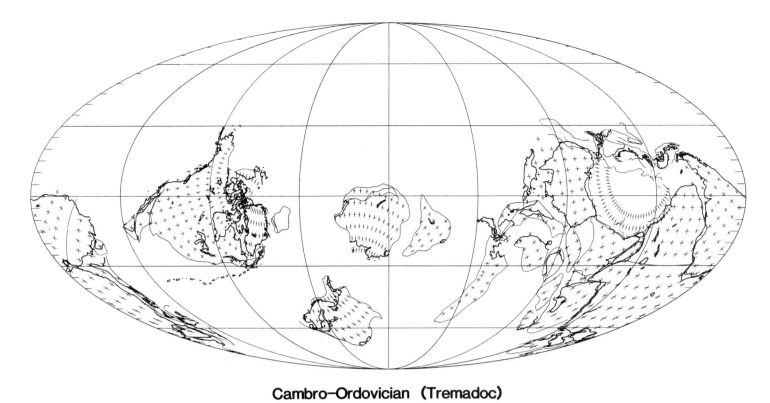

Fig. 6. Basal Ordovician (Tremadoc).

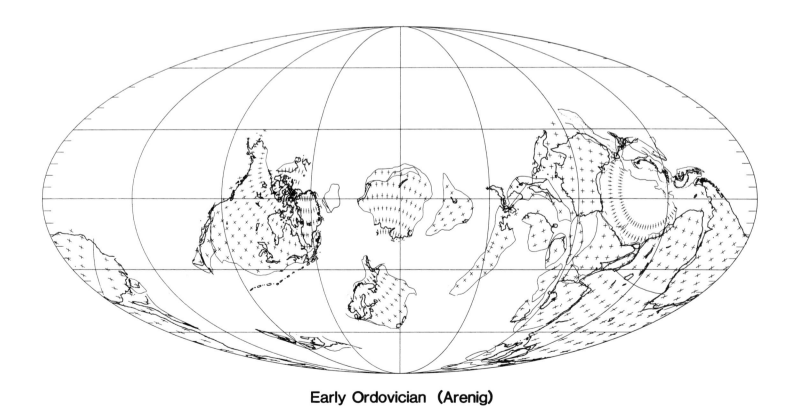

Fig. 7. Early Ordovician (Arenig).

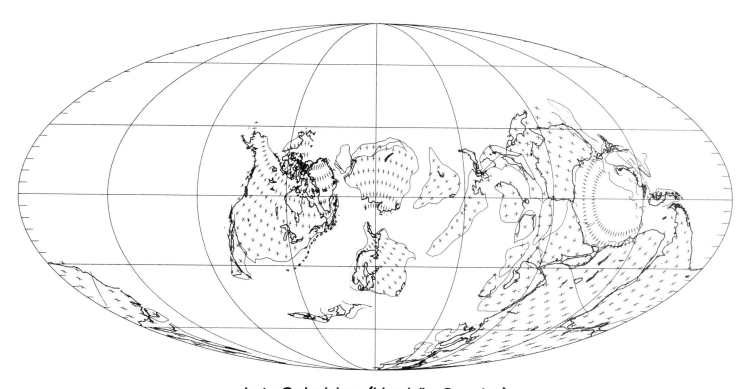

Fig. 8. Middle to Late Ordovician (Llandeilo–Early Caradoc).

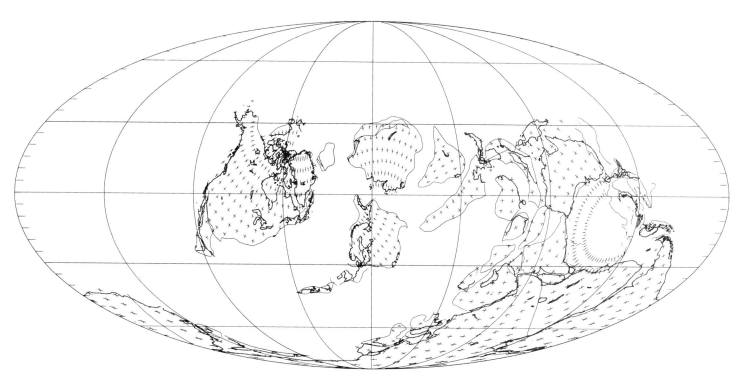

Fig. 9. Latest Ordovician (Ashgill).

Silurian (Figs 10 to 12)

In the late Llandovery (Fig. 10), west-verging nappes were emplaced in north-east Greenland (Hurst *et al.* 1983); this was due perhaps, to the collision of Greenland with Barentsia (the Barents Sea micro-continent including Svalbard). During the Silurian, Laurentia moved southeastwards relative to Baltica, and their collision is marked by the east-verging nappes of the Late Silurian Scandian Orogeny (Stephens & Gee 1985). The collision took place before the completion of all of the strike slip faulting which is so prominent in Scotland and northwest Ireland. The reconstructions (Figs 11 and 12) show Baltica, Avalonia and Barentsia about 10° further south than in most previous North Atlantic reconstructions (Sclater *et al.* 1977). This adjustment allows for the known post-Silurian movements on the Scottish faults (McKerrow & Elders 1989) and it also allows the west-verging Llandovery nappes of northeast Greenland to be situated clear to the north of the east-verging Late Silurian Scandian nappes (McKerrow 1988*a*). It requires approximately 750 km of sinistral strike-slip movement to be of post-Wenlock age.

After the Scandian Orogeny, the southern parts of the Iapetus Ocean still remained open between Avalonia and Laurentia (Figs 11 and 12), with an arc on the Laurentian margin (McKerrow 1988*a*). However, there is increasing evidence for north-westward subduction of Avalonian continental crust in the British Isles during the Late Silurian and Early Devonian (McKerrow & Soper 1989), so the northern parts of Iapetus were not necessarily floored by oceanic crust after the Wenlock.

During the Silurian, carbonates were common in Laurentia, northeastern Avalonia, Baltica, Siberia and equatorial Gondwana. The warm water facies, which had persisted throughout the Early Palaeozoic of Laurentia and appeared in the Caradoc of Baltica, now spread over wider areas of these continents and by the Wenlock, the first coral reefs appeared in England (Fig. 11). The low diversity *Clarkeia* fauna, which is characteristic of the south polar regions, was present in Gondwana, Spain, France and southwestern Avalonia (**Cocks & Fortey**, this volume).

In the latest Silurian, evaporites occur in the United States, consistent with a sub-tropical position (Fig. 12).

The Silurian reconstructions presented here differ from those of **Kent & Van der Voo** (this volume) in the position of Gondwana. **Kent & Van der Voo** (this volume) and **Bachtadse & Briden** (this volume) use an Early Silurian palaeomagnetic pole from an igneous ring complex of Niger (Hargraves *et al.* 1987) to orient Gondwana. This result places the South Pole in southernmost Argentina and requires unusually high rates of plate motion (>15 cm a^{-1}) for Gondwana during the Late Ordovician and Silurian. In the absence of confirming palaeomagnetic results, we have used palaeoclimatic evidence to orient Gondwana (see **Scotese & Barrett**, this volume, and discussion under Gondwana, above).

Devonian (Figs 13 to 16)

By the Early Devonian, most of the movement between Laurentia and Baltica/Avalonia was sinistral strike slip though oblique subduction continued below the Laurentian margin until the Emsian (McKerrow 1988*a*, *b*). Palaeomagnetic data from North America places the Appalachian margin at 40°S during the Early Devonian (Miller & Kent 1988). This result is in good agreement with palaeomagnetic results from Britain, which indicate palaeolatitudes for southern Scotland of 20°–25°S. This more southerly position for Laurentia and Baltica (Euramerica) requires that their contact with Gondwana was earlier than previously supposed. Both the palaeomagnetic and faunal (**G. Young**, this volume) evidence suggests that no wide ocean existed between Euramerica and Gondwana during the Devonian.

The closure of the northern Iapetus in the Scandian Orogeny and the southern Iapetus in the Acadian Orogeny resulted in a high mountain belt between Laurentia and Baltica/Avalonia. These mountains served as a barrier to separate the benthos

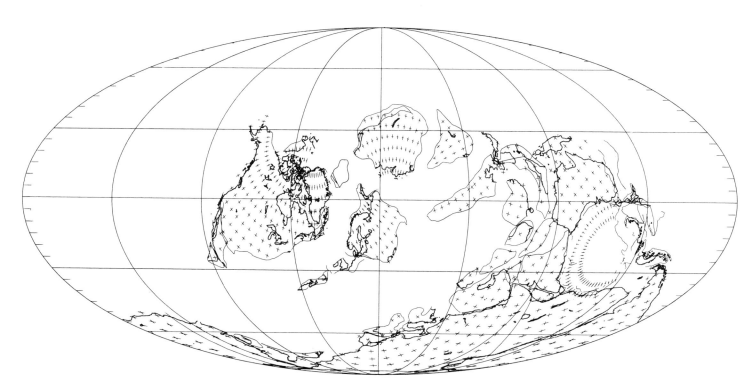

Early Silurian (Llandovery)

Fig. 10. Early Silurian (Llandovery).

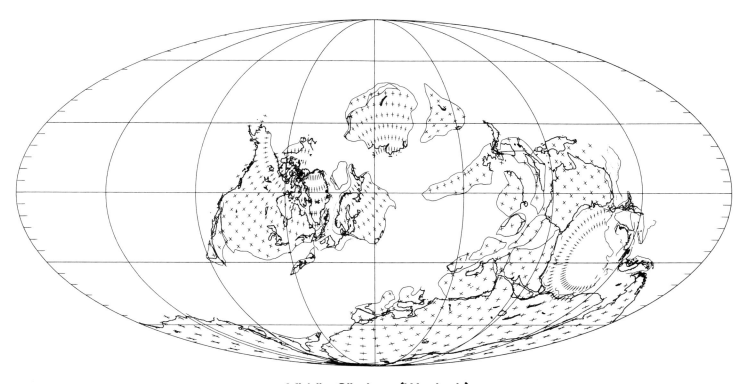

Fig. 11. Middle Silurian (Wenlock).

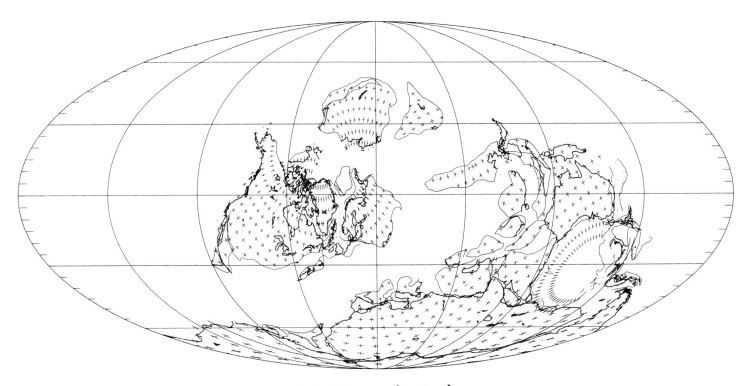

Fig. 12. Late Silurian (Ludlow).

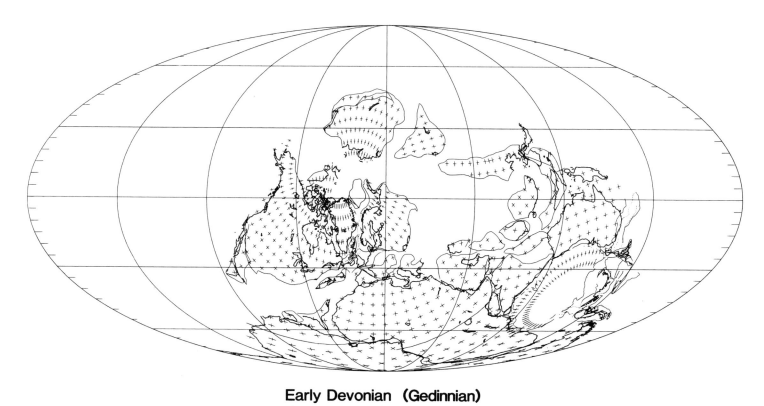

Fig. 13. Early Devonian (Gedinnian).

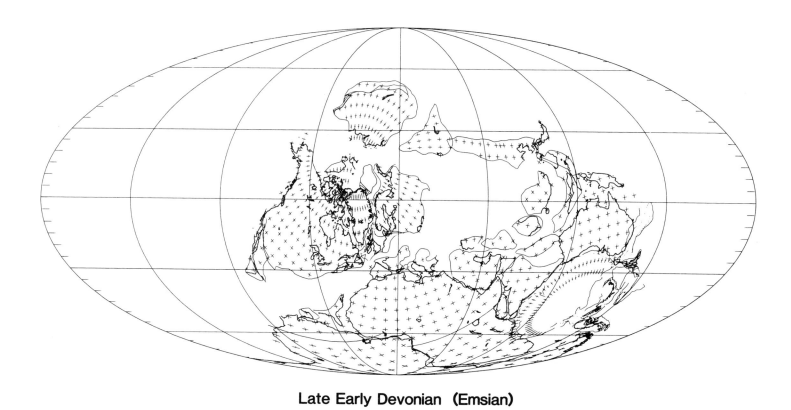

Fig. 14. Late Early Devonian (Emsian).

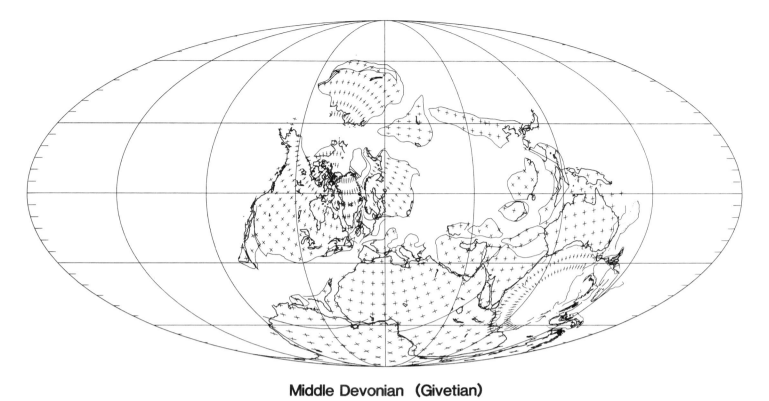

Fig. 15. Middle Devonian (Givetian).

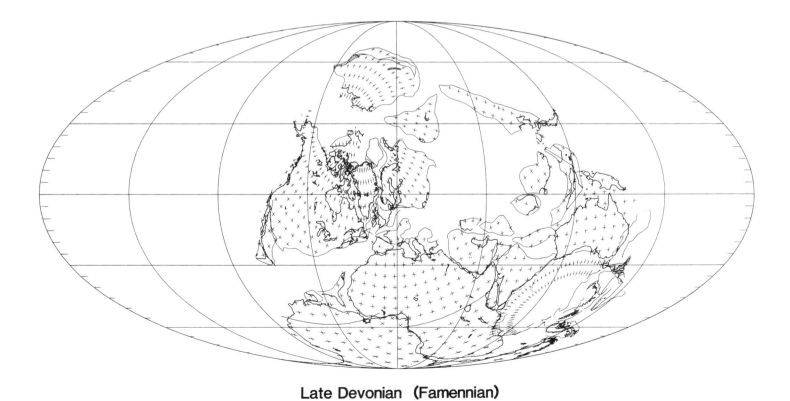

Fig. 16. Late Devonian (Famennian).

of the Appalachian Province of Laurentia from the Rhenish–Bohemian Province (Boucot *et al.* 1969). In western Gondwana, the Appalachian Province occurs in South America (Barrett 1985), while the Rhenish–Bohemian Province is present in North Africa. This separation in both northern and southern continents is hard to explain if there is a wide Devonian ocean north of western Gondwana.

We follow the scenario of Neugebauer (1988), which suggests that, during much of the Devonian, Euramerica and Gondwana were in contact, but not in collision.

Throughout the Devonian, western Gondwana and Euramerica continued to move northward (Figs 13 to 16). This movement is accompanied by the appearance of warm water carbonates in the Middle Devonian of Morocco (Wendt 1985), Brittany and SW England.

Just as the elimination of subduction after the collision of India with Asia resulted in major changes in plate motion during the Tertiary (Scotese *et al.* 1988), so too, the end of subduction on the margins of the Iapetus Ocean can be linked with major changes in plate motion during the Middle and Late Devonian. Most of the effects of this plate reorganisation were seen around Laurentia, where, in addition to the Caledonian/Appalachian mountain belts, new orogens developed along its western (Antler and Caribou Orogenies) and northern (Ellesmerian Orogeny) margins. The Ellesmerian Orogeny may also reflect the collision of Siberia with Laurentia. Elsewhere, rifting occurred on some continental margins: back arc basins developed along the northern margin of Gondwana (Wendt 1985), the Viluy Trough formed in northeastern USSR (Khain 1985), and the Donetz aulocogen originated on the southeastern margin of Baltica. As proposed above, it is also likely that South and North China had started to rift from eastern Gondwana by Early or Middle Devonian times.

The Early Devonian reconstructions presented here (Figs 13 and 14) are similar to those of **Kent & Van der Voo** (this volume). The main difference is in the relative longitudinal positions of Laurentia and Baltica with respect to Gondwana. Palaeomagnetic data predicts approximately 10° of overlap between Laurentia and Gondwana in the Early Devonian. **Kent & Van der Voo** avoid this overlap by shifting Laurentia and Baltica 60° westwards. We, on the other hand, prefer a 'tight fit' that brings the continents as close as possible, and maintains their relative longitudinal positions.

Our Late Devonian reconstruction (Fig. 16) is also different from **Kent & Van der Voo** (this volume) due to the choice of poles for Gondwana. Based on palaeoclimatic evidence, we place the Late Devonian South Pole in north-central Argentina (**Scotese & Barrett**, this volume), whereas Kent & Van der Voo use data (Hurley & Van der Voo 1987) which place the pole in central Africa.

Carboniferous (Figs 17 to 19)

During the Carboniferous and Early Permian, the closure of several oceans resulted in the amalgamation of the western half of Pangea; these included the Rheic Ocean between northern Europe and Gondwana, the Phoibic Ocean between Laurentia and Gondwana, and the Pleionic Ocean between Baltica and Siberia/Kazakhstan (McKerrow & Ziegler 1972).

Though Gondwana and northern Europe were adjacent during most of the Devonian, there appears to have been little deformation until the Late Devonian or Early Carboniferous (Holder & Leveridge 1986). The Hercynian and Variscan Orogenies had their climax in the Westphalian; they record deformation along the southern margins of Baltica and Avalonia, and in the numerous small terranes of southern Europe, some of which were attached to Gondwana. The collisions between Gondwana and northern Europe had a large oblique component related to the clock-wise rotation of Gondwana relative to northern Europe (Neugebauer 1988) (Figs 17 to 19). The small terranes of central and southern Europe were deformed in the transpressive shear zone that resulted

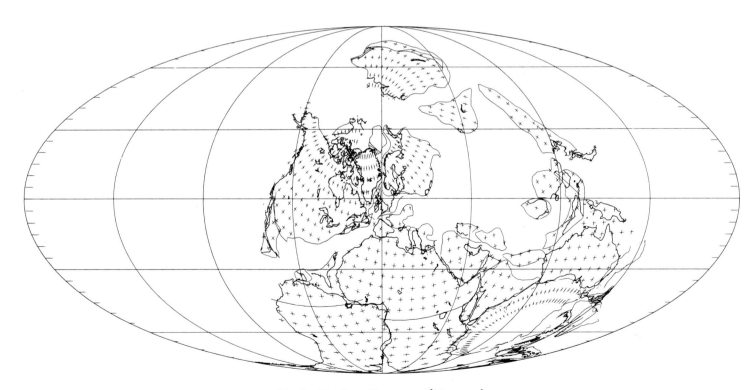

Early Carboniferous (Visean)

Fig. 17. Early Carboniferous (Visean).

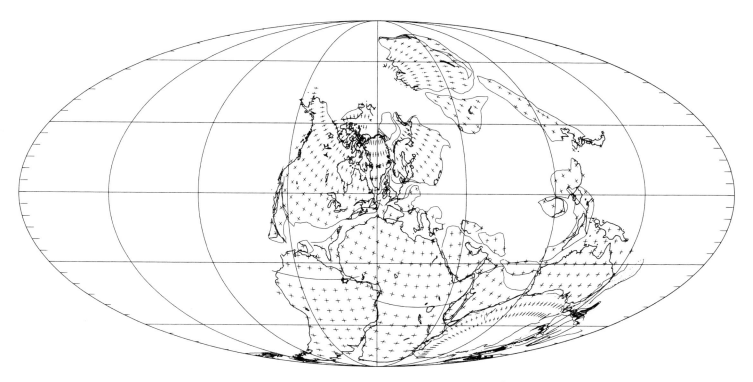

Fig. 18. Early Late Carboniferous (Namurian).

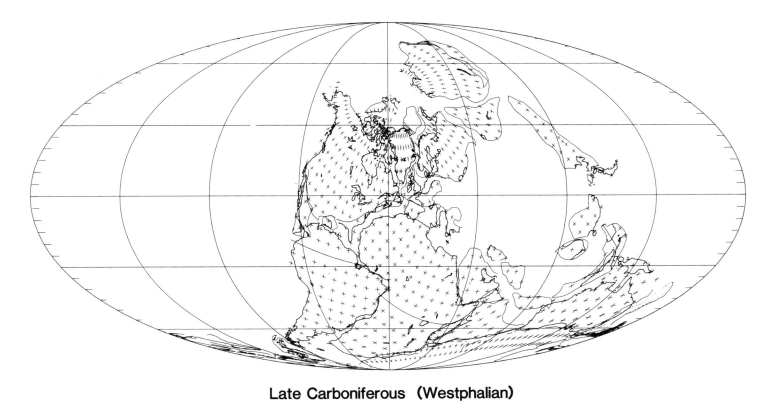

Fig. 19. Late Carboniferous (Westphalian).

from this rotation. The numerous examples of synchronous local compression and tension in adjacent regions in the Variscan/Hercynian Orogen can also be explained by this mechanism.

Due to the clock-wise rotation of Gondwana, the climax of deformation generally progressed from NE to SW along the Hercynian, Variscan, Appalachian/Mauritanide and Ouachita fronts. In eastern Laurentia, the first rumblings of the Alleghanian Orogeny occurred in the Visean (Perry 1978), and the orogen was uplifted and deformed throughout the remainder of the Carboniferous. The youngest folded beds in the Central Appalachians (Monongahela group) are latest Carboniferous in age. On the Allegheny Plateau, the Early Permian Dunkard Group was gently folded during the final phase of deformation.

The Ouachita Mountains of Oklahoma record the final phase of collision between Laurentia and Gondwana. In the Early Carboniferous, Gondwana collided with an island arc on the SE margin of Laurentia; this arc was thrust northward and buried beneath a thick prograding sedimentary wedge during the Late Carboniferous. The final phase of collision in the Early Permian was marked by these deposits being thrust up to form the Ouachita Mountains.

During the Early Carboniferous (Fig. 17), Siberia rifted away from the Arctic margin of Laurentia, opening the Sverdrup Basin (Sweeney 1977). By the mid-Carboniferous, the island arcs of Kazakhstan collided with the SW margin of Siberia, forming the Irtysch and Dzungar fold belts (Fig. 18). Through the remainder of the Carboniferous, Siberia and Kazakhstan rotated clock-wise, and by Late Carboniferous (Fig. 19) were approaching the Uralian margin of Baltica (Nalivkin 1973); this collision occurred in the Early Permian. Recently, it has also been suggested that the Tarim terrane of NW China began to collide with the SW margin of Siberia during the Late Carboniferous (**Nie** *et al.*, this volume).

As these collisions developed, many parts of the continents became emergent, and in the Namurian (mid-Carboniferous) the pole-to-equator temperature gradient began to strengthen (Raymond *et al.* 1989; **Kelley** *et al.*, this volume). The Late Carboniferous was a time of climatic contrasts. Warm, rainy, equatorial swamps extended from the Russian platform across north-central Europe to England and Maritime Canada, and into the mid-continent of North America. Though floristically distinct, similar coal-producing environments occurred in North and South China (**Ziegler**, this volume). Thick evaporite deposits accumulated at sub-tropical latitudes of Hudsons Bay and the Amazon Basin (Ziegler *et al.* 1979; Ronov *et al.* 1984). In the south polar regions, a large ice sheet extended from Argentina across south-central Africa to Madagascar, southern Arabia, India, Antarctica and Australia (Caputo & Crowell 1985).

Permian (Figs 20 and 21)

Many of the continental collisions which began in the Carboniferous were completed in the Permian. The western half of Pangea was assembled, and the new supercontinent, ringed by subduction zones, moved steadily northward. The configuration of the Atlantic-bordering continents is now generally agreed. Our reconstruction (Figs 20 and 21) differs only slightly from that of **Lottes & Rowley** (this volume); both show a tighter fit than previously published maps (e.g. Sclater *et al.* 1977) in order to compensate for the pre-rifting extension of the crust along the continental margins.

The configuration of the blocks that were eventually to form the eastern half of Pangea (China and SE Asia) is still problematic **Nie** *et al.* (this volume) describe no less than 11 terranes in the assembly of Asia. Five of these terranes are shown in Figs 1, 20 and 21: South China (Yangtze), Indochina, Shan Thai–Malaya, Iran and Afghanistan (Helmand). Another five have been combined into composite blocks: Tarim and Sino-Korea are shown as North China; and the West Qiang Tang, East Qiang Tang and Lhasa blocks of **Nie** *et al.* (this volume) are combined in our Tibetan block (Fig. 1). The terranes of northern Manchuria have inadvertantly been omitted here. As **Nie** *et al.* (this volume) show, these terranes were located originally along the southern

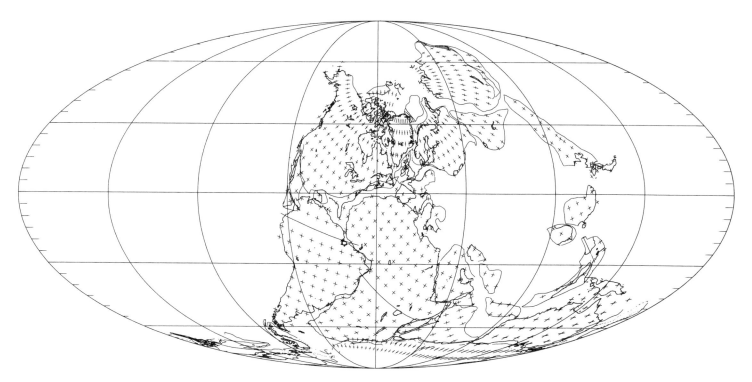

Early Permian (Artinskian)

Fig. 20. Early Permian (Artinskian).

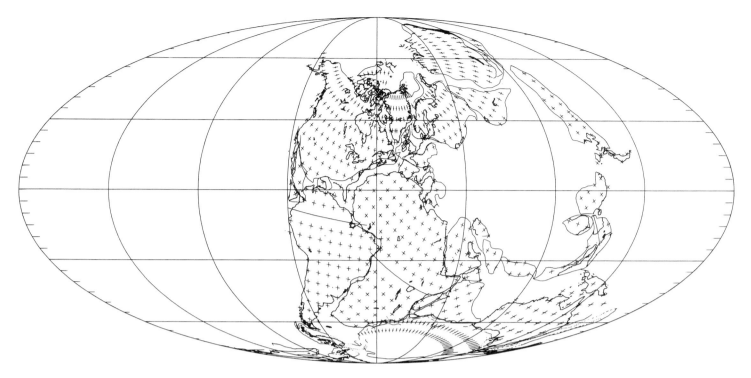

Fig. 21. Late Permian (Kazanian).

margin of Siberia, as an eastward extension of the South Mongolian subduction zone.

In the Permian, the allochthonous terranes of Asia can be divided into two groups: Cathaysian and Cimmerian. The Cathaysian terranes (South China, North China and Indochina) had probably rifted from Gondwana during the Middle Palaeozoic and occupied an equatorial position in the Permian (Figs 20 and 21). During the Early Permian, the western end of North China (Tarim) collided with Siberia (Rowley *et al.* 1985; **Nie *et al*** this volume); and by the Late Permian the NE end (Sino-Korea) had sutured to northern Manchuria (see **Nie *et al.*,** this volume).

South China is shown attached to North China at a hinge near Korea (after Zhao & Coe 1987). The Qinling Ocean, which separated these blocks, closed during the Late Triassic as a result of the scissors-like rotation of these two blocks. Though climatic, biogeographic and palaeomagnetic information constrain the latitude of South China, its longitudinal position in the Late Palaeozoic is more uncertain. **Nie *et al.*** (this volume) show South China in the western part of Proto-Tethys; we on the other hand would place it in the eastern Proto-Tethys, so that its intrusive rocks formed a link between the active margins of the SE China and Indochina and the Tasman arc of Australia (Figs 20 and 21).

The second group of terranes (Turkey, Iran, Tibet, Shan Thai–Malaya) are considered to have formed the elongate continent of Cimmeria by Sengör (1984, 1987). The rifting of Cimmeria from Gondwana in the Permian (compare Figs 20 and 21) is described in more detail by Rowley *et al.* (1985) and **Nie *et al.*** (this volume).

The assembly of western Pangea during the Early Permian had profound climatic and biogeographic effects. In the equatorial regions, the rising Variscan, Appalachian and Mauritanide mountains blocked the wet equatorial easterly winds, casting a rain shadow over much of central Pangea. As northern Europe and the mid-continent of North America moved northward, the coal swamps of the Carboniferous were succeeded by deserts and saline inland seas. In the south temperate and polar regions, the Gondwana ice cap retreated, to be replaced by bogs and peat swamps. The emergence of many continental areas above sea level started the pattern of seasonal monsoons that would become more dominant during the Mesozoic (Crowley *et al.* 1989; Kutzbach & Gallimore 1989).

The emergent Pangea supercontinent, spanning the Earth from pole to pole, provided a wide range of environments suitable for specialised ecological adaptations. **Ziegler** (this volume) describes an actualistic approach to mapping floral distributions. **Bambach** (this volume) suggests that the increase in Late Palaeozoic provinciality in the marine realm was the result of a decrease in the range of regionally distributed genera, rather than any increase in either endemism or the number of recognizable biogeographic units.

Palaeozoic climates

A poster session at the symposium in Oxford was devoted to an informal attempt to characterize climatic changes through the Palaeozoic. Participants were asked to describe the climate in selected palaeogeographic regions as either cold, cool, warm or hot for a succession of time intervals. The results of this exercise are shown in Fig. 22.

It was recognized that the climatic changes for each region result from a combination of secular changes in global climate and the changing latitude of each region through time. Though changes in climate due to latitudinal shift of the continents predominate, a few global climatic events could be recognised, especially the major cooling events in the Late Ordovician (late Ashgill) and the mid-Carboniferous (late Namurian). A less well defined cooling event may have taken place during the Late Devonian (late Frasnian). Prior to these cooling events, faunas were diverse and endemic. During the cooling events, widespread extinctions took place (e.g. **Sheehan & Coorough**, this volume). During the Namurian event, the latitudinal range of floras was compressed towards the Equator (**Kelley *et al.***, this volume). As stated above (in the Palaeozoic Biogeography section), changes in climate are often correlated with sea level changes, and these

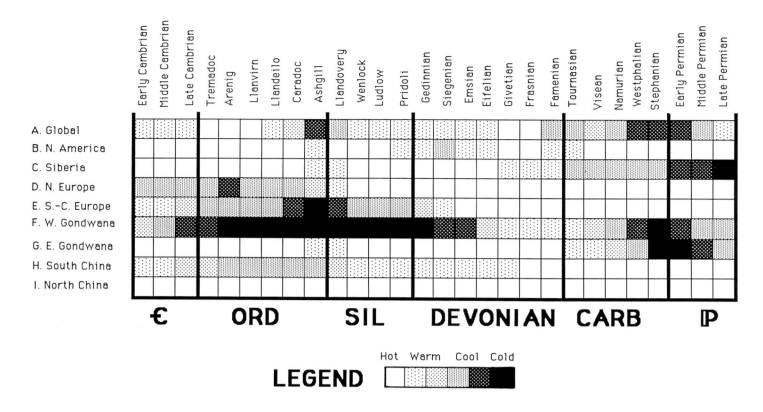

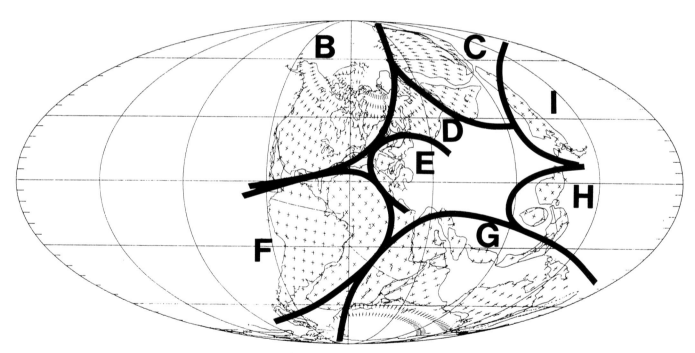

Fig. 22. Climatic changes during the Palaeozoic. The upper diagram shows changes in 9 separate geographic regions (B to I) which are shown in the lower diagram. The density of the stipple in the upper diagram represents temperature determined from faunas and lithofacies.

in turn affect evolutionary radiations and migrations. The correlations between these events deserve much further study.

Conclusions

The symposium provided a opportunity for palaeontologists, palaeobotanists, stratigraphers and palaeomagnetists to meet and exchanges ideas and data on the development of the world during the Palaeozoic Era. One of the messages that came across loud and clear was that *we can no longer rely on a single line of evidence to reconstruct the Palaeozoic world*. We must take an approach that integrates all relevant data from the Earth Sciences; biogeography, tectonics, stratigraphy, palaeomagnetism, sedimentology and palaeoclimatology all have key roles to perform.

The maps presented here have benefitted from using such diverse data. They incorporate five major changes from their predecessors (Scotese *et al.* 1979; Scotese 1984, 1986).

(1) The Cathaysian terranes are shown progressing from north

of East Gondwana to a linear arrangement along an arc postulated to lie between the South Mongolian arc and the east Australian arc.

(2) The terranes bounding the Iapetus Ocean are shown in more detail.

(3) A new polar wandering path for western Gondwana has been constructed using climatically linked lithofacies (**Scotese & Barrett**, this volume).

(4) New palaeomagnetic data has been incorporated (**Kent & Van der Voo**, this volume) that places Laurentia and Baltica further south, adjacent to Gondwana in the Early Devonian; and Siberia is also moved south in the light of biogeographical data.

(5) Kazakhstan is treated as a westward extension of Siberia, rather than as a separate continent.

In many respects, the maps we present here are similar in their precision to the maps of Asia and the New World produced by 16th Century explorers. In the 500 years since the voyages of these early discoverers, we have mapped the Earth 'in space'. We are now embarking on a voyage to map the Earth 'in time'. To do this, we need to assemble systematically a database of palaeogeographic information from the Palaeozoic rocks of every country. To do this well we should take better advantage of recent advances in computer technology, to allow us to manipulate and manage the vast amounts of information. The patterns are becoming too complex and too subtle to unravel using traditional techniques.

The next step, which we are just beginning to take, will be to map the plate boundaries of the Palaeozoic (Zonenshain et al. 1985). At the moment we can recognise some trenches along active margins, but the location of spreading ridges and the motion of the plates (as opposed to motions of the continents) is more difficult to deduce. Plate boundaries, even if preliminary, will provide new insights and important constraints regarding the movements of the continents through time.

Our maps are still provisional. Their accuracy will only be measured by how well they explain stratigraphic, tectonic, climatic and biogeographic patterns through time. The authors welcome comments and criticisms of the maps. Larger format copies are available upon request to Scotese; and a computer program for microcomputers can be provided, which will allow users to replot the maps at any scale and in a variety of projections (Scotese & Denham 1987).

The Symposium on Palaeozoic Biogeography and Palaeogeography was sponsored by the Geological Society of London, the Palaeontological Association and the Working Group 2B of the International Lithosphere Program. We would like to thank AMOCO Production Company, Conoco, Exxon Production Research Company, Shell (USA & UK), Texaco, Mobil Exploration and Production, and the Geological Society of London for their financial support. Special thanks to P. McNiff and C. Elders for their help in the organization of the meeting, to Dung Huynh for his help in preparing the figures, and to A. Hills for expert editorial work. CRS appreciates Shell Development Company's permission to publish.

References

BAILEY, R. H., NEWMAN, W. A. & GENES, A. 1976. Geology of Squantum 'Tillite'. In: CAMERON, B. (ed.) *Geology of Southeastern New England*. New England Intercollegiate Geology Conference, Princeton Press, Princeton, 92–106.

BARRETT, S. F. 1985. Early Devonian Continental Positions and Climate: A framework for paleophytogeography. In: TIFFANY, B. (ed.) *Geological Factors and the Evolution of Plants*. Yale University Press, New Haven, 93–127.

BERGSTRÖM, J. & GEE, D. G. 1985. The Cambrian in Scandinavia. In: GEE, D. G. & STURT, B. A. (eds) *The Caledonide Orogen – Scandinavia and related areas*. Wiley, Chichester, 247–271.

BOND, G. C., NICKESON, P. A. & KOMINZ, M. A. 1984. Breakup of a supercontinent between 625 Ma and 555 Ma: new evidence and implications for continental histories. *Earth and Planetary Science Letters*, **70**, 325–345.

BOUCOT, A. J., JOHNSON, J. G. & TALENT, J. A. 1969. Early Devonian brachiopod zoogeography. *Geological Society of America Special Paper*, **119**.

BRIDEN, J. C., TURNELL, H. B. & WATTS, D. 1984. British paleomagnetism, Iapetus Ocean, and the Great Glen Fault. *Geology*, **12**, 428–431.

BROCK, A. 1981. Paleomagnetism of Africa and Madagascar. In: McELHINNY, M. W. & VALENCIO, D. A. (eds) *Paleoreconstruction of the Continents*. American Geophysical Union Geodynamics Series, **2**, 65–76.

BRUTON, D. L., LINDSTROM, M. & OWEN, A. W. 1985. The Ordovician of Scandinavia. In: GEE, D. G. & STURT, B. A. (eds) *The Caledonide Orogen – Scandinavia and related areas*. Wiley, Chichester, 273–282.

CAPUTO, M. V. & CROWELL, J. C. 1985. Migration of glacial centers across Gondwana during the Paleozoic Era. *Geological Society of American Bulletin*, **96**, 1020–1036.

CHAN, L. S., WANG, C. Y. & WU, X. Y. 1984. Paleomagnetic results from some Permian-Triassic rocks from southwestern China. *Geophysical Research Letters*, **11**, 1157–1160.

COCKS, L. R. M. & FORTEY, R. A. 1982. Faunal evidence for oceanic separations in the Palaeozoic Britain. *Journal of the Geological Society, London*, **139**, 465–478.

—— & —— 1988. Lower Palaeozoic facies and faunas around Gondwana. In: AUDLEY-CHARLES, M. G. & HALLAM, A. (eds) *Gondwana and Tethys*. Geological Society, Special Publication, London, **37**, 183–200.

CROWLEY, T. J., HYDE, W. T. & SHORT, D. A. 1989. Seasonal cycle variations on the supercontinent of Pangaea. *Geology*, **17**, 457–460.

DU TOIT, A. A. 1937. *Our Wandering Continents, an Hypothesis of Continental Drifting*. Oliver and Boyd, Edinburgh.

FANG, W. & VAN DER VOO, R. 1989. Devonian paleomagnetism of Yunnan province across the Shan Thai – S. China suture. *Tectonics* (in press).

FORTEY, R. A. & COCKS, L. R. M. 1986. Marginal faunal belts and their structural implications, with examples from the Lower Palaeozoic. *Journal of the Geological Society, London*, **143**, 151–160.

FREI, L. S. & COX, A. 1987. Relative displacement between Eurasia and North America prior to the formation of oceanic crust in the North Atlantic. *Tectonophysics*, **142**, 111–136.

HARGRAVES, R. B., DAWSON, E. M. & VAN HOUTEN, F. B. 1987. Paleomagnetism and age of Mid-Palaeozoic riung complexes in Niger, West Africa, and tectonic implications. *Geophysical Journal of the Royal Astronomical Society*, **90**, 705–729.

HARLAND, W. B. 1985. Caledonide Svalbard. In: GEE, D. G. & STURT, B. A. (eds) *The Caledonide Orogen – Scandinavia and related areas*. Wiley, Chichester, 999–1016.

HARRIS, A. L., BALDWIN, C. T., BRADBURY, H. D. & SMITH, R. A. 1978. Ensialic basin sedimentation in the Dalradian Supergroup. In: BOWES, D. R. & LEAKE, B. E. (eds) *Crustal evolution in northwestern Britain and adjacent regions*. Geological Journal Special Issue, **10**, 115–38.

HENRIKSEN, N. 1985. The Caledonides of east Greenland 70°–76° In: GEE, D. G. & STURT, B. A. (eds) *The Caledonide Orogen – Scandinavia and related areas*. Wiley, Chichester, 1095–1113.

HOLDER, M. T. & LEVERIDGE, B. E. 1986. Correlation of the Rhenohercynian Variscides. *Journal of the Geological Society, London*, **143**, 141–7.

HUGHES, N. F. (ed.) 1973. Organisms and continents through time. *Special Papers in Palaeontology*, **12**.

HURLEY, N. F. & VAN DER VOO, R. 1987. Paleomagnetism of Upper Devonian reefal limestones, Canning Basin, Western Australia. *Geological Society of American Bulletin*, **98**, 123–137.

HURST, J. M., McKERROW, W. S., SOPER, N. J. & SURLYK, F. 1983. The relationship between Caledonian nappe tectonics and Silurian turbidite deposition in North Greenland. *Journal of the Geological Society, London*, **140**, 123–131.

JOHNSON, R. J. E., VAN DER PLUIJM, B. A. & VAN DER VOO, R. 1988. Paleoreconstruction of the Early Ordovician terranes in the Central Mobile Belt of the Newfoundland Appalachians (abstract). *Geological Society of America, Abstracts with Programs*, **20**, A63.

KENT, D. E. & OPDYKE, N. D. 1985. Multicomponent magnetizations from the Mississipian Mauch Chunck Formation of the Central

Appalachians and their tectonic implications. *Journal of Geophysical Research*, **90**, 5371–5383.

KEPPIE, J. D. 1985. The Appalachian collage. *In*: GEE, D. G. & STURT, B. A. (eds) *The Caledonide Orogen – Scandinavia and related areas*. Wiley, Chichester, 1217–1226.

KHAIN, V. E. 1985. *Geology of the U.S.S.R.*, Gebruder Borntraeger, Berlin.

KHRAMOV, A. N. & RODIONOV, V. P. 1980. Paleomagnetism and reconstruction of paleogeographic positions of the Siberian and Russian plates during the Late Proterozoic and Palaeozoic. *Journal of Geomagnetism and Geoelectricity*, **32**, Supplement III, SIII 23-SIII 37.

——, PETROVA, G. N. & PECHERSKY, D. M. 1981. Paleomagnetism of the Soviet Union, *In*: MCELHINNY M. W. & VALENCIO, D. A. (eds) *Paleoreconstruction of the Continents*. American Geophysical Union, Geodynamics Series, **2**, 177–194.

KOKELAAR, B. P., HOWELLS, M. F., BEVINS, R. E., ROACH, R. A. & DUNKEY, P. N. 1984. The Ordovician marginal basin of Wales, *In*: KOKELAAR, B. P. & HOWELLS, M. F. (eds) *Marginal Basin Geology*. Geological Society, Special Publication, London, **16**, 245–269.

KUTZBACH, J. E. & GALLIMORE, R. G. 1989. Pangaean climates: megamonsoons of the megacontinent. *Journal of Geophysical Research (in press)*.

LAWVER, L. A. & SCOTESE, C. R. 1987. A revised reconstruction of Gondwanaland. *In*: MCKENZIE, G. (ed.) *Gondwana Six: Structure, Tectonics, and Geophysics*, Geophysics Monograph, **40**, American Geophysical Union, Washington D.C., 17–24.

LAWVER, L. A. & SCOTESE, C. R. 1989. A Review of Tectonics Models for the Evolution of the Canada Basin, *In*: *Decade of North American Geology* (in Press).

LIN, J. L., FULLER, M. & ZHANG, W. Y. 1985a. Paleogeography of the North and South China blocks during the Cambrian. *Journal of Geodynamics*, **2**, 91–114.

——, ——, & —— 1985b. Preliminary Phanerozoic polar wander paths for the North and South China Blocks. *Nature*, **313**, 444–449.

MCELHINNY, M. W. & EMBLETON, B. J. J. 1974. Australian paleomagnetism and the Phanerozoic plate tectonics of eastern Gondwana. *Tectonophysics*, **22**, 1–29.

——, —— MA, X. H. & ZHANG, Z. K. 1981. Fragmentation of Asia during the Permian. *Nature*, **293**, 212–216.

MCKENZIE, D. P., MOLNAR, P. & DAVIES, D. 1970. Plate tectonics of the Red Sea and East Anglia. *Nature*, **226**, 243–248.

MCKERROW, W. S. 1988a. The development of the Iapetus Ocean from the Arenig to the Wenlock. *In*: HARRIS, A. L. & FETTES, D. J. (eds) *The Caledonian–Appalachian Orogen*. Geological Society, Special Publication, London, **38**, 405–412.

—— 1988b. Wenlock to Givetian deformation in the British Isles and the Canadian Appalachians. *In*: HARRIS, A. L. & FETTES, D. H. (eds) *The Caledonian–Appalachian Orogen*. Geological Society, Special Publication, London, **38**, 437–448.

—— & COCKS, L. R. M. 1986. Oceans, island arcs and olistostromes: the use of fossils in distinguishing sutures, terranes and environments around the Iapetus Ocean. *Journal of the Geological Society, London*, **143**, 185–91.

—— & ELDERS, C. F. 1989. Movements on the Southern Upland Fault. *Journal of the Geological Society, London*, **146**, 393–5.

—— & SOPER, N. J. 1989. The Iapetus suture in the British Isles. *Geological Magazine*, **126**, 1–8.

—— & ZIEGLER, A. M. 1972. Palaeozoic Oceans. *Nature, Physical Sciences*, **240**, 92–4.

MILLER, J. D. & KENT, D. V. 1986. Paleomagnetism of the Upper Devonian Catskill formation from the southern limb of the Pennsylvania salient. *Geophysical Research Letters*, **13**, 1173–1176.

—— & —— 1988. Paleomagnetism of the Siluro-Devonian Andreas red beds: evidence for an Early Devonian supercontinent? *Geology*, **16**, 195–198.

MOREL, P. & IRVING, E. C. 1978. Tentative paleocontinental maps for the early Phanerozoic and Proterozoic. *Journal of Geology*, **86**, 5, 535–561.

NALIVKIN, D. V. 1973. *Geology of the U.S.S.R.* University of Toronto Press, Toronto.

NEUGEBAUER, J. 1988. The Variscan plate tectonic evolution: an improved "Iapetus model". *Schweiz. Mineral. Petrogr. Mitt..*, **68**, 313–333.

NEUMAN, R. B. 1984. Geology and paleobiology of islands in the Ordovician Iapetus Ocean: review and implications. *Geological Society of American Bulletin*, **95**, 1188–1201.

OPDYKE, N. D., HUANG, K., XU, G., ZHANG, W. Y. & KENT, D. V. 1987. Paleomagnetic results from the Silurian of the Yangtze paraplatform. *Tectonophysics.*, **139**, 123–132.

PALMER, A. R. 1973. Cambrian Trilobites. *In*: HALLAM, A. (ed) *Atlas of palaeobiogeography*. Elsevier, Amsterdam, 3–11.

PERROUD, H. 1983. Palaeomagnetism of Palaeozoic rocks from the Cabo de Penas, Asturia, Spain. *Geophysical Journal of the Royal Astronomical Society*, **75**, 201–215.

—— & VAN DER VOO, R. 1985. Paleomagnetism of the late Ordovician Thouars massif, Vendee province, France. *Journal of Geophysical Research*, **90**, 4611–4625.

——, —— & BONHOMMET, N. 1984. Palaeozoic evolution of the Armorica plate on the basis of paleomagnetic data. *Geology*, **12**, 579–582.

PERRY, W. J. 1978. Sequential deformation of the Central Appalachians. *American Journal of Science*, **278**, 518–542.

PIPER, J. D. A. 1983. Proterozoic paleomagnetism and single continent plate tectonics. *Geophysical Journal of the Royal Astronomical Society*, **74**, 163–197.

—— 1987. *Paleomagnetism and the continental crust*. J. Wiley & Sons, New York.

RAYMOND, A., KELLEY, P. H. & LUTKEN, C. B. 1989. Polar glaciers and life at the equator: the history of Dinantian and Namurian (Carboniferous) climate. *Geology*, **17**, 408–411.

RODGERS, J. 1968. The eastern edge of the North American continent during the Cambrian and Early Ordovician, *In*: E-AN ZEN et al. (eds) *Studies of Appalachian Geology: Northern and Maritime*. Interscience, New York, 141–149.

RONOV, A. KHAIN, V. & SESLAVINKSY, K. 1984. *Atlas of Lithologic-Paleogeographic Maps of the World*, Leningrad.

ROWLEY, D. B. RAYMOND, A., PARRISH, J. T., LOTTES, A. L., SCOTESE, C. R. & ZIEGLER, A. M. 1985. Carboniferous paleogeographic, phytogeographic and paleoclimatic reconstructions. *International Journal of Coal Geology*, **5**, 7–42.

SCHENK, P. E. 1972. Possible Late Ordovician glaciation in Nova Scotia. *Canadian Journal of Earth Sciences*, **9**, 95–107.

SCHMIDT, P. W. & MORRIS, W. A. 1977. An alternative view of the Gondwana Paleozoic apparent polar wander path. *Canadian Journal of Earth Sciences*, **14**, 2674–2678.

SCLATER, J. G., HELLINGER, S. & TAPSCOTT, C. 1977. The paleobathymetry of the Atlantic Ocean from the Jurassic to the present. *Journal of Geology*, **85**, 509–552.

SCOTESE, C. R. 1984. An introduction to this volume: Palaeozoic Paleomagnetism and the Assembly of Pangea, *In*: VAN DER VOO, R., SCOTESE, C. R. & BONHOMMET, N. (eds) *Plate Reconstruction from Paleozoic Paleomagnetism*. American Geophysical Union, Geodynamics Series, **12**, 1–10.

—— 1986. *Phanerozoic reconstructions: A new look at the assembly of Asia*. University of Texas Institute for Geophysics Technical Report No. 66.

—— & DENHAM, C. R. 1987. *User's Guide to Terra Mobilis: A Plate Tectonics program for the Macintosh*. Earth in Motion Technologies, Houston.

—— BAMBACH, R. K. BARTON, C., VAN DER VOO, R. & ZIEGLER, A. M. 1979. Palaeozoic base maps. *Journal of Geology*, **87**, 217–227.

SCOTESE, C. R., BARRETT, S. F. & VAN DER VOO, R. 1985. Silurian and Devonian base maps. *Philosophical Transactions of the Royal Society, London*, **B 309**, 57–77.

——, GAHAGAN, L. M. & LARSON, R. L. 1988. Plate reconstructions of the Cretaceous and Cenozoic ocean basins. *Tectonophysics*, **155**, 27–48.

——, VAN DER VOO, R., JOHNSON, R. E. & GILES, P. S. 1984. Paleomagnetic results from the Carboniferous of Nova Scotia, *In*: VAN DER VOO, R., SCOTESE, C. R. & BONHOMMET N. (eds) *Plate Reconstruction from Palaeozoic Paleomagnetism*, American Geophysical Union, Geodynamics series, **12**, 63–81.

SENGÖR, A. M. C. 1984. The Cimmeride orogenic system and the tectonics of Eurasia. *Geological Society of America, Special Paper*, **195**.

—— 1987. Tectonics of the Tethysides: orogenic collage development in a collisional setting. *Annual Reviews of Earth and Planetary Science*,

15, 213–244.

STEARNS, C. VAN DER VOO, R. & ABRAHAMSEN, N. 1989. A new Siluro-Devonian paleopole from early Paleozoic rocks of the Franklinian Basin, North Greenland fold belt. *Journal of Geophysical Research*. (in press).

STEPHENS, M. B. & GEE, D. G. 1985. A tectonic model for the evolution of the eugeoclinal terranes in the central Scandinavian Caledonides. *In*: GEE, D. G. & S*turt*, B. A. (eds) *The Caledonide Orogen–Scandinavia and related areas*. Wiley, Chichester, 953 or 78.

SWEENEY, J. F. 1977. Subsidence of the Sverdrup basin, Canadian Arctic Islands. *Geological Society of America Bulletin*, **88**, 41–48.

VAN DER VOO, R. 1988. Palaeozoic paleogeography of North America, Gondwana and intervening displaced terranes: comparisons of paleomagnetism with paleoclimatology and biogeographical patterns. *Geological Society of America Bulletin*, **100**, 311–24.

——, PEINADO, J. & SCOTESE, C. R. 1984. A paleomagnetic reevaluation of Pangea reconstructions, *In*: VAN DER VOO, R., SCOTESE, C. R. & BONHOMMET, N. (eds) *Plate Reconstruction from Paleozoic Paleomagnetism*. American Geophysical Union, Geodynamics Series, **12** 11–16.

WEBBY, B. D. 1984. Ordovician reefs and climate: a review. *In*: BRUTON, D. L. (ed.) Aspects of the Ordovician System. *Paleontological Contributions from the University of Oslo*, **295**, 89–100.

WENDT, J. 1985. Distribution of the Continental Margin of Northwestern Gondwana: Late Devonian of the eastern Anti-Atlas (Morocco). *Geology*, **13**, 815–818.

ZHAO, X. & COE, R. S. 1987. Palaeomagnetic constraints on the collision and rotation of North and South China. *Nature*, **327**, 141–144.

ZIEGLER, A. M., HANSEN, K. S., JOHNSON, M. E., KELLY, M. A., SCOTESE, C. R., & VAN DER VOO, 1977a. Silurian continental distribution, paleogeography, climatology, and biogeography, *Tectonophysics*, **40**, 13–51.

—— SCOTESE, C. R., MCKERROW, W. S., JOHNSON, M. E. & BAMBACH, R. K. 1977b. Palaeozoic biogeography of continents bordering the Iapetus (Pre-Caledonian) and Rheic (pre-Hercynian) Oceans, *In*: WEST, R. M. (ed.) *Paleontology and Plate Tectonics*. Milwaukee Public Museum, Special Publication in Biology and Geology, **2**, 1–22.

——, ——, ——, —— & —— 1979. Paleozoic paleogeography. *Annual Reviews of Earth Science*, **7**, 473–502.

ZONENSKHAIN, L. P., KUZMIN, M. I. & KONONOV, M. V. 1985. Absolute reconstructions of the Paleozoic oceans. *Earth and Planetary Science Letters*, **74**, 103–116.

Palaeomagnetism and Palaeoclimates

A review of Palaeozoic palaeomagnetic data from Europe and their palaeogeographical implications

TROND H. TORSVIK,[1] MARK A. SMETHURST,[1] JAMES C. BRIDEN[1] & BRYAN A. STURT[2]

[1] *Department of Earth Sciences, University of Oxford, Oxford OX1 3PR, UK*
[2] *Norwegian Geological Survey, P.B. 3006 Lade N-7002 Trondheim, Norway*

Abstract: Recent palaeomagnetic studies on Devonian and Carboniferous rocks have resulted in a time re-calibration of the Apparent Polar Wander Path (APWP) for Europe, and revision of the shape of the APWP for North America. Differences between previously published versions of these paths are now much reduced. The APWP for southern Britain is different from those for North America and Armorica, thus southern Britain is believed to have been an isolated block within the pre-Hercynian ocean. New continental reconstructions are presented to take account of these conclusions. A lack of sufficient reliable palaeomagnetic data from Baltica make its position on the map uncertain, and hence the significance of the Tornquist Sea between Baltica and Palaeo-Europe remains incompletely understood.

Given the assumption of a geocentric dipole field, a plate's apparent polar wander path (APWP) forms a reference system with respect to the earth's spin axis and to other plates, thus providing important palaeogeographic information. APWPs, however, should not be considered as permanent frames of reference (Van der Voo 1988), and in particular better constraints on magnetic age by means of detailed field-tests have led to substantial changes in the time-calibration of APWPs.

A large number of Ordovician to Permian palaeomagnetic pole positions have been reported from Europe (North of the Alpine Orogenic Belt), and in order to overcome subjective data-selection various grading schemes have been proposed (e.g. McElhinny 1973; Briden & Duff 1981; Van der Voo 1988). The most important attributes include tectonic and magnetic age control by means of field-tests, and the application of stepwise demagnetization and modern analytical techniques. Nevertheless, it is still necessary to make some subjective judgments in data-selection. In this compilation we have attempted to include all 'reliable' Ordovician to Permian palaeomagnetic data from Northern Europe (cf. geographic locations in Fig. 1). The selected palaeomagnetic data (Tables 1–6) essentially follows compilations given in Briden *et al.* (1984, 1988), Torsvik *et al.* (1989b); Perroud *et al.* (1984a, b) and Kramhov *et al.* (1981), with the inclusion of some new data.

In orogenic zones, rotations on both vertical and horizontal axes occur, and terrane rotations have been demonstrated within the British/Norwegian Caledonides (Smethurst & Briden 1988; Robertson 1988; Abrahamsen *et al.* 1979; Torsvik *et al.* 1989a), and the European Hercynides (e.g. McClelland Brown 1983; Bachtadse & Van der Voo 1986). In this review, however, we will address ourselves primarily to assumed 'non-rotated' palaeomagnetic poles.

To aid the definition of APW trends within each tectonic unit, and compare trends between tectonic units we have fitted a smooth path to the data. One advantage of path fitting is that it also constitutes *provisional* data extrapolation. A number of numerical methods for fitting smooth paths to palaeomagnetic poles have been offered in the literature (Gould 1969; Parker & Denham 1979; Thompson & Clark 1981, 1982; Clark & Thompson, 1984; Jupp & Kent 1987). In the present account we have used the method of Jupp & Kent (1987) because most of the previously proposed methods can produce distortion if the data are spread over a large portion of the sphere, and moreover some of the solutions are not invariant to changes in co-ordinate system. The method of Jupp & Kent (1987) aims to fit 'spherical smoothed splines' to a given data-set on the sphere with known ages. The palaeomagnetic pole ages listed in Tables 1–6 represent either approximate magnetic ages quoted in the original studies, or re-interpreted ages by the authors. Magnetic age reinterpretation has been made in cases where there are no independent magnetic age constraints and the palaeomagnetic pole falls on a younger part of the APWP, between well-dated poles. The method we have applied allows various levels of smoothing, as well as weighting of individual data points. We have weighted the data according to their α_{95}, but in the case of some well-dated key poles (e.g. the well-established Upper Silurian/Lower Devonian and Permian poles from Britain) these have been assigned a low α_{95}, value (= 1) to anchor the path. Smoothing methods have certain limitations especially when there are abrupt trend changes in the data, and in such instances it is then necessary to use low smoothing parameters.

Apparent polar wander paths

The British Isles is a key area for palaeomagnetic study of the Lower–Middle Palaeozoic rocks of Europe, and the Ordovician to Permian Apparent Polar Wandering (APW) path from the SE margin of the Laurentian plate, i.e. Northern Ireland and Scotland, is known with some confidence (see e.g. Briden *et al.* 1984, 1988). The Middle–Upper Devonian pole position, however, is not well known, and a number of proposed Devonian poles lie between well-dated Lower Carboniferous and Permian results (Torsvik *et al.*, 1988b). Therefore we regard the majority of reported mid–late Devonian poles (mostly derived from sediments) for which there is no magnetic age control as Carboniferous overprints (*c*. 320 Ma; see Tables 1–3). Conversely tectonic models based on the existing 'Devonian' data-base should be considered with great caution.

For convenience we have divided the British Caledonides into three major units (Fig. 1b): (1) the area north of the Great Glen Fault (GGF); (2) the area between the GGF and the Iapetus Suture; (3) the area south of the Iapetus Suture.

Ordovician to Permian palaeomagnetic south pole positions from these units are listed in Tables 1–3 (Fig. 2). The hallmark of the British APW paths is one of pronounced westerly movement through Ordovician and Silurian times. The Upper Silurian/Lower Devonian 'corner' (Fig. 3) in the path is widely accepted, but the apparent backtracking of the APW path during the Siluro-Devonian poses serious problems in discriminating Siluro-Devonian and Lower Carboniferous poles.

Models have been proposed involving major transcurrent motion along the GGF (e.g. Van der Voo & Scotese 1981; Storetvedt 1987), and as such it has been utilized in certain plate models as a prima-facie candidate for a major crustal megashear. The APW paths for the two northerly structural units, i.e. those on either side of the GGF, are virtually identical (Fig. 3), thus precluding large-scale movements (>500 km) along the GGF (and/or the Highland Boundary/Southern Upland Faults). Conse-

From MCKERROW, W. S. & SCOTESE, C. R. (eds), 1990, *Palaeozoic Palaeogeography and Biogeography*, Geological Society Memoir No. 12, pp. 25–41.

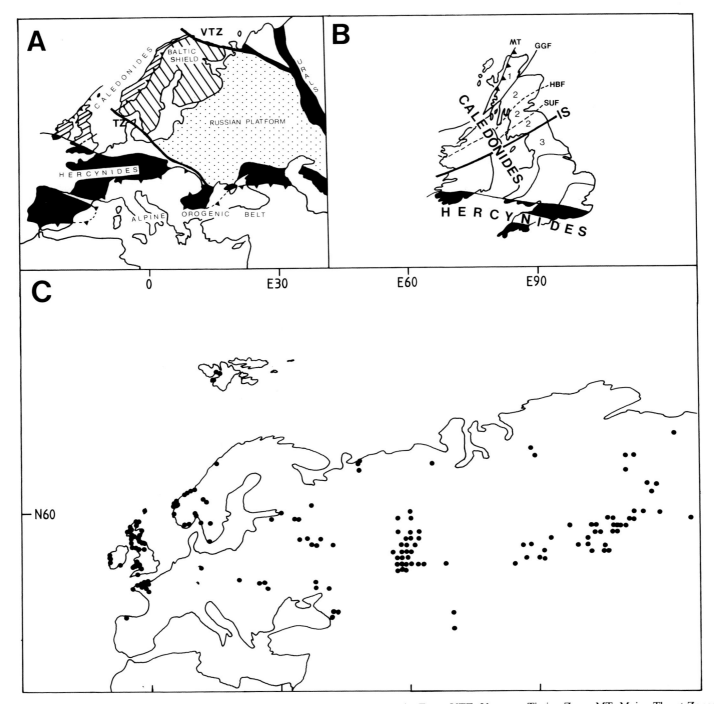

Fig. 1. Principal tectonic units (**a**) of Europe and (**b**) of the British Isles; TZ, Tornquist Zone; VTZ, Varanger Timian Zone; MT, Moine Thrust Zone; GGF, Great Glen Fault; HBF, Highland Boundary Fault; SUF, Southern Upland Fault; IS, Iapetus Suture: 1–3 denote tectonic units discussed in the text. (**c**) the geographical sampling locations for palaeomagnetic poles listed in Tables 1–6.

quently we have combined the palaeomagnetic data from these two units to produce a single APWP for Northern Britain (Fig. 4).

Few new palaeomagnetic data have been reported from southern Britain in recent years, but as previously pointed out by Briden *et al.* (1984, 1988), the majority of Ordovician palaeomagnetic data from southern Britain define an APWP with a more northerly (equatorial) polar trend compared with that recorded north of the suture. This is clearly shown by our fitted paths (Figs 3 & 4) for Ordovician to Middle Silurian times. The post-Lower Devonian paths for northern and southern Britain are essentially similar.

In Northern/Central Europe, overprinting related to Hercynian deformation conceals the original Palaeozoic magnetic signatures. Large discrepancies in declination or azimuth, but fairly coherent inclinations, also point to considerable tectonic/palaeomagnetic rotations of Hercynian age (Bachtadse & Van der Voo 1986). Palaeomagnetic data from the Armorican Massif, however, have shed some interesting light on Palaeozoic reconstructions (Perroud *et al.* 1984*a*). Palaeozoic palaeomagnetic data from the Armorican Massif consist almost entirely of Ordovician and Carboniferous (Hercynian) poles (Fig. 5a). The Silurian and Devonian section of the fitted path is essentially extrapolated between the two comparatively well-defined end-points. The APWP from the

Table 1. *Britain north of GGF.*

Rock-unit	Code	Age	Lat	Long	Reference
Rackwick lavas	(RL)	320 M*	−23	326	Storetvedt & Meland (1985)
Esha Ness ignimbrites	(EN)	320 M*	−20	315	Storetvedt & Torsvik (1985)
Caithness sandstone	(CS)	320 M*	−27	329	Storetvedt & Torsvik (1983)
Argyllshire dykes	(AD)	320 R	−35	355	Esang & Piper (1984a)
John O'Groats Sandstone	(JG)	320 M*	−24	325	Storetvedt & Carmichael (1979)
Shetland Sst. & Lavas	(SL2)	320 M	−24	340	Torsvik et al. (1989d)
Hoy lavas	(HL)	370 R	−14	334	Storetvedt & Meland (1985)
Shetland lavas	(SL1)	370 M	−2	340	Torsvik et al. (1989d)
Eday Sandstone	(ED)	375 R	−8	346	Robinson (1985)
Kishorn-Moine metased.	(KM2)	400 M	−14	320	Torsvik & Sturt (1988)
Sarclet L.ORS	(SA)	400 R	−9	326	Storhaug & Storetvedt (1985)
Borrolan Syenite, Loch Ailsh & dykes[1]	(BS)	408 M	−17	326	Turnell & Briden (1983)
Moine metased. (Ratagen)	(MR)	408 M	1	324	Turnell (1985)
Moine metasediments (IB)[1]	(MMI)	408 M	−1	309	Watts (1982)
Moine metasediments (HB)[1]	(MMH)	408 M	−6	313	Watts (1982)
Ratagen Complex	(RC)	415 R	−15	347	Turnell (1985)
Helmsdale granite	(HG)	420 R	−31	355	Torsvik et al. (1983)
Strontian granite	(SG)	430 R	−21	344	Torsvik (1984)
Borrolan Ledmorite, Pseudo-leucite, Loch Loyal[1]	(BL)	430 R	−13	2	Turnell & Briden (1983)
Caledonian dolerites	(CD)	435 M	−14	347	Esang & Piper (1984b)
Caledonian microdiorites	(CM)	435 M	−16	346	Esang & Piper (1984b)
Kishorn−Moine metased.	(KM1)	450 M	−14	42	Torsvik & Sturt (1988)
Caledonian Dolerites	(CA)	450 M	−5	56	Esang & Piper (1984b)

Palaeomagnetic South Poles are given; LAT, Latitude in degrees;
LONG, Longitude in degrees east;
R, Rock-age; M, magnetic age if not rock-age (*revised age)
[1] Combined pole for reasons of clarity (IB, Intermediate blocking; HB, High blocking)

Table 2. *Britain south of GGF, north of Iapetus*

Rock-unit	Code	Age	Lat	Long	Reference
Peterhead dyke	(PD)	260 R	−41	342	Torsvik (1985b)
Dykes & remag. ORS	(DR)	270 RM	−43	343	Torsvik et al. (1989b)
Queensferry Sill	(QF)	280 R	−38	354	Torsvik et al. (1989b)
Salrock Fm. overprint	(SFO)	320 M*	−28	331	Smethurst & Briden (1988)
Claire Island overprint	(CO)	320 M*	−35	331	Smethurst & Briden (1988)
Tourmakeady & Glensaul	(TG)	320 M*	−31	349	Deutsch & Storetvedt (1988)
Foyers sandstone	(FS)	320 M*	−30	326	Kneen (1974)
Jedburgh Upper ORS	(JB)	320 M*	−32	338	Nairn (1960)
Burntisland & Kinghorn	(BK)	350 R	−14	332	Torsvik et al. (1989b)
Clyde Lava & remag. ORS	(CL)	350 RM	−14	322	Torsvik et al. (1989b)
Cheviot[1]	(CV)	398 R	4	323	Thorning (1974)
Garabal Hill−Glen Fyne	(GH)	404 R	−5	326	Briden (1970)
Lorne Plateau lavas	(LP)	405 R	2	321	Latham & Briden (1975)
ORS lavas & sediments	(MV)	408 R	−4	320	Sallomy & Piper (1973)
Strathmore lavas	(SL)	408 R	2	318	Torsvik (1985a)
Lower ORS	(LW)	408 R	−11	307	Douglass (1987)
Comrie complex	(CM)	408 R	−6	287	Turnell (1985)
Peterhead granite	(PG)	415 R	−20	357	Torsvik (1985b)
Arrochar complex	(AC)	418 R	−8	324	Briden (1970)
Salrock Formation	(SF)	420 R	−2	288	Smethurst & Briden (1988)
Foyers granite	(FG)	420 R	−27	346	Torsvik (1984)
Aberdeenshire gabbros 3[1]	(AG3)	428 M	−6	331	Watts & Briden (1984)
Aberdeenshire gabbros 2[1]	(AG2)	448 M	−5	360	Watts & Briden (1984)
Ballantrae gabbros	(BG)	450 M	−10	26	Piper (1978a)
Ballantrae serpentinites	(BS)	450 M	−12	27	Piper (1978a)
Barrovian zone	(BZ)	450 M	−3	22	Watts (1985a)
Aberdeenshire gabbros 1[1]	(AG1)	468 M	−14	28	Watts & Briden (1984)
Mweelrea Ignimbrites	(MW)	470 R	−11	38	Morris et al. (1973)

Legend as Table 1
ORS = Old Red Sandstone

Table 3. *Britain south of Iapetus*

Rock-unit	Code	Age	Lat	Long	Reference
Exeter lavas	(EL)	280 R	−46	345	Cornwell (1967)
Whin sill	(WS)	281 R	−44	339	Storetvedt & Gidskehaug (1969)
Wackerfield dyke	(WD)	303 R	−49	349	Tarling & Mitchell (1973)
Bristol Upper ORS	(BS)	320 M*	−32	338	Morris et al. (1973)
Hendre & Blodwell intrusive rocks	(HB)	320 M*	−32	346	Piper (1978b)
Lower ORS Wales	(ORS)	398 R	3	298	Chamalaun & Creer (1964)
Lavas Somerset & Gloucester	(SG)	400 R	8	309	Piper (1975)
Shelve volcanic rocks	(SH)	440 R	−5	78	Piper (1978b)
Builth intrusive rocks	(BI)	446 R	2	2	Piper & Briden (1973)
Carrock Fell gabbro	(CA)	448 R	−19	4	Briden & Morris (1973)
Builth volcanic rocks	(BU)	450 R	−3	5	Briden & Mullan (1984)
Breidden Hills	(BR)	453 R	1	17	Piper & Stearn (1975)
Eycott Group	(EG)	460 R	−7	357	Briden & Morris (1973)
Tramore volcanic rocks	(TV)	460 R	11	342	Deutsch (1980)
Borrowdale Volcanic Group	(BV)	460 R	0	23	Faller et al. (1977)

Legend as Table 1

Table 4. *Baltic Shield (Scandinavia)*

Rock-unit	Code	Age	Lat	Long	Reference
Ny−Hellesund dykes		255 R	−38	340	Halvorsen (1970)
Arendal (1)		255 R	−44	341	Halvorsen (1972)
Arendal (2)		255 R	−39	333	Halvorsen (1972)
Ytterøy lamprophyre		256 R*	−43	324	Torsvik et al. (1989c)
Bohuslan dykes (mean)		260 R	−46	345	Abrahamsen et al. (1979)
Oslo Igneous rocks (B)		270 R	−40	340	Storetvedt et al. (1978)
Oslo Graben Lavas		270 R	−45	338	Douglass (1988)
Sunnhordaland dykes		275 R	−43	342	Løvlie (1981)
Sarna body		287 R	−38	347	Bylund & Patchett (1977)
E−Vastergotland sill		287 R	−31	354	Mulder (1971)
Stabben sill (HB)		297 R	−32	354	Sturt & Torsvik (1987)
Scania dolerites		300 R	−37	354	Mulder (1971)
Scania dolerites		300 R	−39	349	Bylund (1974)
Kvamshesten Sandstone	(KH)	360 M	−21	324	Torsvik et al. (1986)
Hornelen Sandstone (A)	(HO)	360 M	−12	327	Torsvik et al. (1988)
Håsteinen (A)	(HA)	360 M	−16	335	Torsvik et al. (1987)
Hitra Sandstone	(HS)	360 M	−8	350	Bøe et al. (1989)
Fongen−Hyllingen P2	(FH)	360 M*	−11	312	Abrahamsen et al. (1979)
Røragen Sandstone	(RO)	360 M*	−19	340	Storetvedt & Gjellestad (1966)
Askøy Pluton (B1)	(AP1)	360 M	−16	340	Rother et al. (1987)
Smøla ORS/substrate	(SM)	365 M	−13	346	Torsvik et al. (1989a)
Ringerike Sandstone	(RS)	415 R	−19	344	Douglass (1988)
Gotland Follingbo limest.	(GF)	425 R	−21	344	Claeson (1979)
Gotland limestone	(GL)	425 R	−19	349	Claeson (1979)
Gotland Medby Limestone	(GM)	425 R	−23	351	Claeson (1979)
Skaane limestone	(SL)	425 R	−22	341	Claeson (1979)
Sulitjelma gabbro	(SG)	443 R	−14	0	Piper (1974)
Askøy Pluton (A1)[†]	(AP2)	474 ?	−25	53	Rother et al. (1987)

Legend as Table 1

[†] The Askøy Pluton (A1) could be rotated since it is situated within a Scandian (Late Silurian) allochthonous unit.

Armorican Massif is in this account considered as representative for the Armorican plate, defined by Van der Voo (1979) to embrace the Armorican and Bohemian Massifs, the Avalon block and southern Britain. The fitted path for Armorica, however, shows a clear discordance with the fitted path for southern Britain from Ordovician to Silurian times (Fig. 4), thus favouring the view that Southern Britain was marginal to Baltica rather than Armorica (Livermore et al. 1985; Briden et al. 1988). It should be noted, however, that a subordinate number of Ordovician palaeomagnetic poles from southern Britain (cf. Thomas & Briden 1976; McCabe 1988) have been argued as having better correspondence with Armorican palaeomagnetic poles (see later).

Palaeomagnetic data from the Baltic Shield (Scandinavia) indicate inconsiderable APW through Middle Palaeozoic time (Fig. 5b). The majority of poles of published pre-Devonian age, for which there is poor or no magnetic age control, plot close to Upper Devonian/Lower Carboniferous poles. Accruing evidence suggests that the latest Caledonian tectonism, including terrane

Table 5. *Northern Europe (Armorican Massif)*

Rock-unit	Code	Age	Lat	Long	Reference
Rozel B		300 M	−37	341	Perroud et al. (1982)
Tregastel−Ploumanac'h		300 M	−34	332	Duff (1979)
Jersey dolerite (A)		300 M	−31	336	Duff (1980)
Crozon dolerites remag.		300 M	−23	322	Perroud et al. (1983)
Montmartin Red Beds		300 M	−38	325	Perroud et al. (1984b)
Laval syncline		300 M	−33	309	Edel & Coulon (1984)
Cambro−Ord. Red beds		300 M	−33	325	Duff (1979)
Cap Frehel		300 M	−39	338	Jones et al. (1979)
Zone Bocaine		300 M	−33	332	Jones et al. (1979)
Paimpol−Brehec		300 M	−26	320	Jones et al. (1979)
San Pedro red beds[2]		320 M*	−23	324	Perroud & Bonhommet (1984)
Carteret (B)		320 M	−18	322	Perroud et al. (1982)
Flamanville		320 M	−30	332	Van der Voo & Klootwijk (1972)
Thouars Massif remag.		320 M	−23	316	Perroud & Van der Voo (1985)
Moulin de Chat remag.		320 M	−21	320	Perroud et al. (1986a)
St Malo dykes		330 M	−30	328	Perroud et al. (1986b)
Jersey dolerite (B)	(JD)	408 M	−1	339	Duff (1980)
Jersey lampr. dykes	(JL)	408 M	16	322	Duff (1980)
Crozon dolerites	(CD)	440 R	3	358	Perroud et al. (1983)
Thouars Massif	(TM)	444 R	34	5	Perroud & Van der Voo (1985)
Cabo de Penas[2]	(CP)	450 R	30	330	Perroud (1983)
Moulin de Chat Fm.	(MC)	480 R	34	343	Perroud et al. (1986a)
Erquy Spilite series	(ES)	493 M	35	344	Duff (1979)

[2] Corrected for the opening of Bay of Biscay
Legend as Table 1

Table 6. *Spitsbergen*

Rock-unit	Code	Age	Lat	Long	Reference
Permo−Carboniferous sst	(PC)	280 R	−36	321	Vincenz & Jelenska (1985)
Billefjorden sandstone	(BF)	350 R	−23	332	Watts (1985b)
Mimer Valley Sandstone	(MV)	360 M	−24	325	Torsvik et al. (1985)
Wood Bay Sst. Combined	(WB)	408 R	−3	322	Torsvik et al. (1985), Jelenska & Lewandowski (1986) and Douglass (1987)

Legend as Table 1

docking (Torsvik et al. 1986, 1987, 1988, 1989a; Sturt & Roberts 1987), occurred in Late Devonian time. This episode is generally known as the Solundian/Svalbardian orogenic phase (Sturt 1983). The apparent lack of APW for the Baltic Shield during Ordovician and Silurian times could, therefore, be explained by widespread Late Devonian remagnetization associated with the Solundian/Svalbardian orogenic phase. The Lower−Middle Palaeozoic APWP for the Baltic Shield should therefore be viewed with considerable caution (see also Pesonen et al. 1989). As yet, only the Middle Silurian Ringerike Sandstone for which there is a positive fold-test, and a stratigraphically related polarity pattern (Douglass 1988), can be regarded as a reliable pre-Devonian pole position from the Baltic Shield.

Few Palaeozoic palaeomagnetic results have been reported from Spitsbergen (Table 6, Fig. 5c). The fitted APWP for Spitsbergen is only based on four Lower Devonian to Permian poles, and is therefore weakly constrained (Fig. 6). Although the Lower Devonian pole is uncertain (see Løvlie et al. 1985 and Douglass 1987), the fitted path follows a pronounced southerly track which corresponds with the post Lower Devonian path of the two British paths (see also Torsvik et al. 1985; Watts 1985a, b; Douglass 1987). Note the discordance between the Spitsbergen and the Baltic Shield path in Fig. 6.

Ordovician to Permian data from the USSR have been divided into three parts, i.e. those data from the Russian Platform taken to be part of Baltica, those from the Urals region and those from the Siberian Platform. We have used the data compiled by Khramov et al. (1981) which are shown in Fig. 7. A number of the Lower−Middle Palaeozoic poles from the Russian Platform plot close to Permian poles from the Platform and Western Europe. Consequently, extensive Late Palaeozoic overprinting may also account for much of the Early to Middle Palaeozoic data from the Urals, and one can hardly distinguish palaeomagnetic poles astride the Urals. Thus, we have combined all data in Fig. 7b, and naturally, a fitted path can not be derived from such an analysis.

Time-calibration of some parts of the paths for the USSR (Russian and Siberian Platforms) is as yet impossible, and our interpretation of these data relies on a trend analysis which is only weakly time constrained. The Ordovician poles from the Siberian Platform, however, are well-grouped around 30°N−310°E (Fig. 7a) and do not conform to Permian poles which fall at intermediate southerly latitudes. The first Palaeozoic palaeomagnetic results from the Russian and Siberian Platforms were obtained at the beginning of the 1960s (see Krahmov & Sholpo 1967). It was shown that the data might be consistent with relative

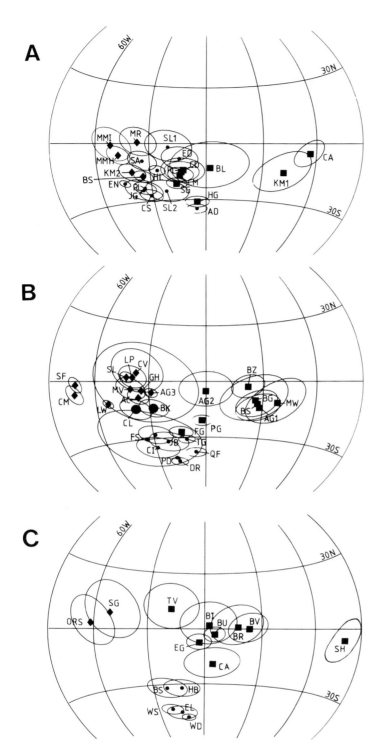

Fig. 2. Ordovician to Permian palaeomagnetic poles from the British Isles (**a**) North of the Great Glen Fault (GGF), (**b**) South of the GGF-North of the Iapetus Suture and (**c**) South of the Iapetus Suture. Cf. Tables 1–3 for pole labels. Square symbols, Ordovician/Silurian poles; Triangles, Upper Silurian/Lower Devonian poles; circles, Devonian to Permian poles. All poles are shown with the oval of 95% confidence around the mean pole (semi-axes dp/dm). Equal-area projection.

rotation of the two platforms, at least during the Early Palaeozoic. An alternative explanation has since been put forward which is that divergence in the Early Palaeozoic poles from the two platforms is due to widespread Upper Palaeozoic remagnetization on the Russian Platform, which is also seen throughout Western Europe. It is now generally accepted that extensive remagnetiz-

ation of Lower–Middle Palaeozoic rocks of the Russian Platform has occurred, probably in Permo-Carboniferous time. However, this does not preclude significant Lower Palaeozoic relative movements/rotation of the Siberian and Russian platforms.

Palaeogeographic reconstructions

Throughout this study we use spherical splines as fitted APWPs (Table 7) from which we generate palaeo-reconstructions. This enables us to animate a plate-tectonic scenario for any given time. No error confidences have been listed for the APWPs, but it should be noted that the uncertainty is notably high and tentative in the extrapolated section of the paths. We emphasize that the fitted APWPs should not be considered absolute and some of the palaeogeographic enigmas are outlined below. All reconstructions are only quantitative in terms of palaeo-latitude, and palaeo-longitude is unconstrained by the paths. As time-calibration of much of the Russian and Siberian Platform APWPs is at present difficult, Baltica is positioned according to data from the Baltic Shield. Siberia is not included in the reconstructions because its Middle–Late Palaeozoic APWP is poorly constrained, and only Ordovician poles show a consistent grouping which deviates systematically from Late Palaeozoic (Permian) palaeomagnetic poles for Europe. For comparison, the coastline of North America is displayed and positioned according to Northern British data (southern margin of Laurentia) in a Bullard *et al.* (1964) fit.

There are only small differences between the paths in Lower Devonian times, and virtually no difference at all between the Permian part of the paths (Fig. 8). In post Lower Devonian time the paths move southward from near the equator, a feature also indicated in the Russian and Siberian Platform APWPs. The Baltic Shield APWP, however, deviates from this pattern, but the bend in its Middle Palaeozoic part of the path probably has no physical significance.

Ordovician

On the balance of both palaeomagnetic and faunal data it is evident that Armorica was situated in high southerly latitudes (probably together with Gondwana) in Middle Ordovician time (450 Ma; Fig. 9). Similarly, palaeomagnetic results from Siberia suggest high latitudes during Ordovician time. Perroud *et al.* (1984*a*) used the Ordovician palaeomagnetic data from Armorica to confirm the existence of a Medio-European Ocean (Whittington & Hughes 1972) which formed part of an implied triple-junction configuration with the Iapetus Ocean. Such an triple-junction pattern has also been suggested by Cocks & Fortey (1982). Based on faunal evidence Cocks & Fortey (1982) argue for a major oceanic separation between Gondwana and Baltica in Early Ordovician time (Tornquist Sea). The Tornquist Sea probably compares with the Medio-European Ocean of Whittington & Hughes (1972). Cocks & Fortey (1982) contend that the Tornquist Sea was essentially closed in Upper Ordovician time.

If we were to accept some 'anomalous' Ordovician palaeomagnetic poles (McCabe 1988) as being representative for southern Britain, it would carry the implication that southern Britain could have been positioned marginal to Armorica (see also Fig. 3a of Kent & Van der Voo 1990). This has obvious implications concerning the width of the Iapetus Ocean during the Ordovician. This southerly position of Southern Britain is based on a few 'anomalous' directions described in the literature together with a new palaeomagnetic result from the Llanvirn aged (*c.* 470 Ma) Stapeley Volcanic Formation (Welsh Borderlands; McCabe 1988). McCabe reports a pre-folding dual-polarity magnetization which implies high southerly latitudes (*c.* 50–55° S) for southern Britain (along with Armorica) during Ordovician times. However, this argument is not certain, and the best palaeomagnetic data (e.g. Builth volcanic rocks; Briden & Mullan 1984) suggest that

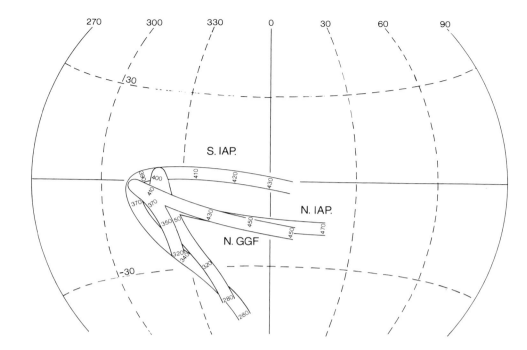

Fig. 3. Spherically smoothed APW paths for the palaeomagnetic poles shown in Fig. 2. S.IAP, South of Iapetus Suture; N.IAP, North of the Iapetus Suture/South of the Great Glen Fault; N.GGF, North of the Great Glen Fault. Numbers along the path represent ages in millions of years. Equal area projection.

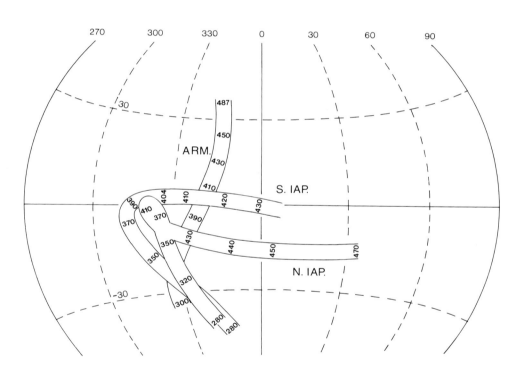

Fig. 4. Spherically smoothed APW paths for Armorica (ARM), South of Iapetus (S.IAP, British Isles) and North of Iapetus (N.IAP). See Table 7 for details. Convention as Fig. 3.

southern Britain was located in mid-latitudes (30–35°S), some 1000–1500 km south of Northern Britain at that time (Figs 9 & 10).

Throughout the Lower–Middle Palaeozoic, Baltica was effectively stationary at low latitudes (see e.g. Fig. 10). This, however, could be an artifact of extensive Devonian/Lower Carboniferous remagnetization. An equatorial position for Baltica is not consistent with faunal evidence which indicates temperate latitudes for Baltica in Ordovician times. This is why a number of published Ordovician reconstructions show Baltica at southerly latitudes around 30–45° (see e.g. Cocks & Fortey 1982; Fortey & Cocks 1988; Livermore et al. 1985). The reconstructions based on Ordovician palaeontological evidence, however, make little allowance for oceanic circulation models, and in particular major current patterns.

Silurian

It has commonly been argued that Armorica (and Gondwana) drifted northwards during latest Ordovician and Silurian times. By Middle–Silurian time (Fig. 11) Armorica appears marginal to southern Britain. In an alternative model, however, Edel (1987b) postulates that Armorica consisting of Central Europe was situated at high latitudes until Middle–Upper Devonian time. Northern and southern Britain appear to have remained fairly stationary at temperate/tropical palaeo-latitudes during the Early Palaeozoic.

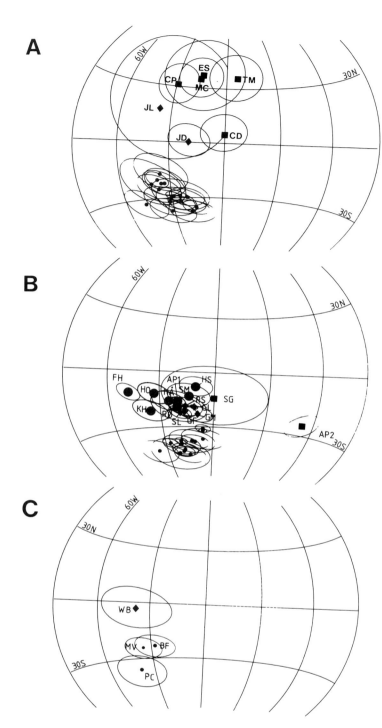

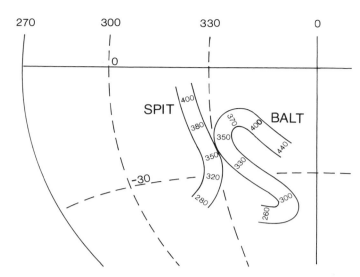

Fig. 6. Spherically smoothed APW paths for Spitsbergen (SPIT) and the Baltic Shield (BALT). Conventions as Fig. 3.

Kent 1988) and Greenland (Stearns *et al.* Pers. Comm.), however, imply a more southerly position of the North American craton in Upper Silurian/Lower Devonian time than had been previously been assumed. Thus, taken as a face value, the APWPs from northern Britain and the North American craton (Laurentia) now converge to a common path in a classical Bullard *et al.* (1964) fit.

The most striking difference between recent palaeoreconstructions shown by Livermore *et al.* (1985) and those presented here, is our equatorial positioning of Baltica throughout Ordovician and Silurian time (Figs 9 & 11). Livermore *et al.* (1985) always show Baltica marginal to southern Britain. The relationship between southern Britain and Armorica also differ somewhat in our reconstructions, and for example during the Silurian we suggest a closer relationship between Armorica and southern Britain (Fig. 11) than that presented by Livermore *et al.* (1985; see their fig. 5). Furthermore, Livermore *et al.* (1985) map the Avalon Platform marginal to Southern Britain, whereas we have plotted the Avalon Platform at high southerly latitudes as part of the Armorican plate.

Early–Middle Devonian

By Early Devonian times the APW paths for northern and southern Britain converge (Figs 10 and 12), implying effective closure of the Iapetus Ocean in Britain. Recent studies in the Scandinavian Caledonides, however, have attempted to establish how much Caledonian terrane accretion can be ascribed to a Late Devonian (Solundian) orogenic event rather than Late Silurian (Scandian) crustal imbrication. The metamorphic signature of folded Devonian sediments in Norway and Western Shetland, and the occurrence of Lower–Middle Devonian island-arc type calc-alkaline lavas (Thirlwall 1981, 1988) cut by a calc-alkaline batholith in Western Shetland suggests final closure of parts of the Iapetus Ocean system in as late as Middle–Late Devonian time (Thirlwall 1988; Torsvik *et al.* 1988*a*).

Baltica and Laurentia most likely collided in the earliest Devonian to form Euramerica. In Middle–Upper Silurian time Laurentia was drifting southward, and during the collisional event Laurentia attained its most southerly Palaeozoic latitudinal position. The southerly latitudinal shift of Laurentia during Silurian times, previously only evident from palaeomagnetic data from northern Britain, is confirmed by (1) new Upper Silurian/Lower Devonian palaeomagnetic data from North America/Greenland (see Kent & Van der Voo 1990) and (2) Late Ordovician/Silurian expansion of evaporites across central North America (Witzke 1988).

Fig. 5. Ordovician to Permian (Carboniferous) from (**a**) the Armorican Massif, (**b**) Baltic Shield and (**c**) Central Spitsbergen. See Tables 4–6 for details. Permo-Carboniferous poles in (a) and (b) are unlabelled. Symbol code in (a,c) as Fig. 2. In (b) large circular symbols denote poles known to represent secondary Solundian/Svalbardian remagnetization.

Conversely the width of the intervening Iapetus Ocean is probably as wide as in Ordovician time. Considerable anti-clockwise rotation of Northern Britain during Ordovician and Silurian time took place (compare Figs 9 & 11; see also Fig. 10), a feature, also characteristic of the North American craton (Van der Voo 1988). The latitudinal position of northern Britain, during Ordovician and Silurian times, however, is somewhat more southerly than latitudes predicted from the North American data-base. This has recently been amplified by Briden *et al.* (1988) and Van der Voo (1988). New palaeomagnetic data from North America (Miller &

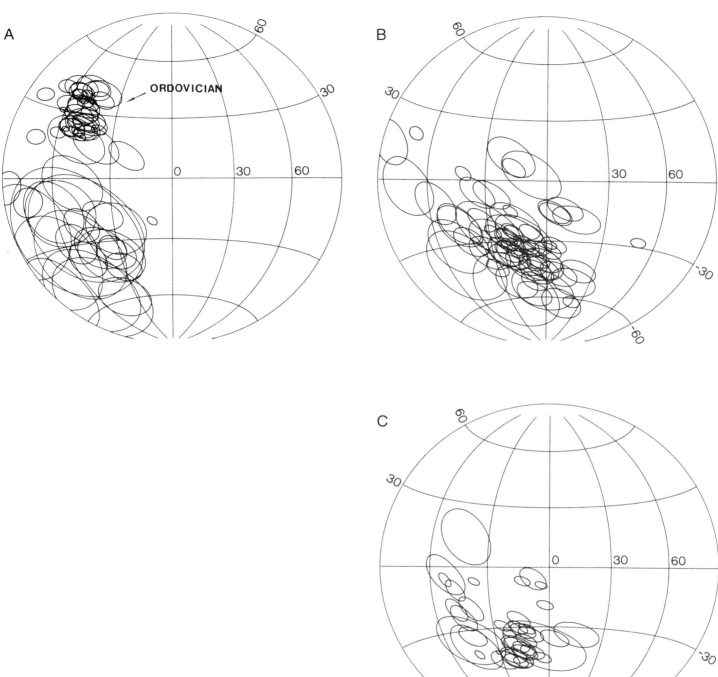

Fig. 7. Ordovician to Permian palaeomagnetic poles (portrayed with oval of 95% confidences) from (**a**) Siberian Platform, (**b**) the 'Urals' and (**c**) Russian Platform. Note in (**a**) that Ordovician palaeomagnetic poles cluster around N30 & E 310.

The precise timing of the docking of Armorica against Laurentia/Baltica is uncertain. Van der Voo (1979) originally suggested that collision was related to the Taconian Orogeny. This view was later revised, and Perroud *et al.* (1984*a*, *b*) and Van der Voo & Johnson (1987) proposed collision in the Early–Middle Devonian, marked by the Acadian Orogeny, to form the Old Red Sandstone Continent. This issue is still not clear due to the sparsity of Siluro-Devonian palaeomagnetic data, and the only Late Silurian–Lower Devonian poles claimed in the literature as being representative for Armorica are from the Jersey dolerite and lamprophyre dykes, for which there is no magnetic age control. The fitted paths for southern Britain and Armorica, however, suggest that the two were at similar latitudes during Early Devonian time.

Late Devonian/Early Carboniferous

The Ordovician–Devonian APWP discordance between Baltica, and Britain and Armorica is eliminated in Carboniferous times (cf. Figs 13 & 14). If the relatively large pre-Carboniferous discordance in the paths between Baltica and southern/northern Britain is real, Baltica and Britain can be assumed to have moved into their present juxtaposition in Late Devonian to Lower Carboniferous time through megashearing along the Tornquist

Table 7. *Palaeozoic Apparent Polar Wander Paths for N. Europe in 10 million year intervals*

Age	N Britain lat	N Britain long	S Britain lat	S Britain long	Armorica lat	Armorica long	'Baltica' lat	'Baltica' long	Spitsbergen lat	Spitsbergen long
260	−42	343					−42	344		
270	−42	342					−42	348		
280	−41	341	−46	342			−41	351	−36	322
290	−39	340	−48	345			−38	352	−35	323
300	−35	337	−48	347	−33	328	−37	351	−34	325
310	−31	333	−45	346	−31	328	−34	347	−32	326
320	−28	332	−38	341	−28	327	−31	342	−30	328
330	−24	330	−32	335	−26	327	−27	337	−28	328
340	−18	328	−25	327	−23	328	−23	333	−27	328
350	−14	327	−20	322	−20	330	−19	331	−25	328
360	−9	326	−12	315	−17	331	−16	333	−22	328
370	−5	325	−8	313	−12	333	−14	337	−20	326
380	−2	324	−3	312	−8	336	−14	339	−16	325
390	+1	322	+1	315	−4	337	−14	341	−13	324
400	0	320	+2	322	0	339	−16	342	−9	323
410	−3	320	+3	334	+6	341	−18	344	−5	322
420	−8	325	+2	348	+10	343	−19	346		
430	−12	335	0	359	+15	345	−21	348		
440	−15	352	−2	006	+20	346	−23	350		
450	−16	004	−3	007	+25	346				
460	−16	019	−3	003	+29	346				
470	−15	036			+33	345				
480					+36	345				
490					+39	344				

Legend as Table 1

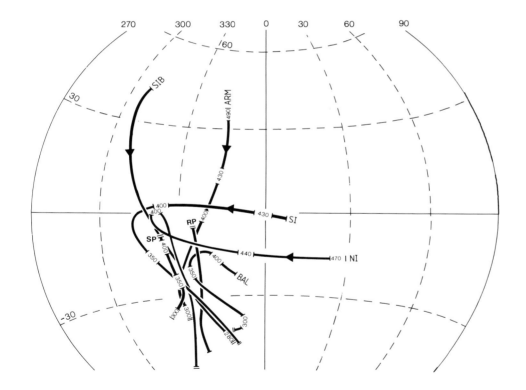

Fig. 8. Compilation of all spherically smoothed paths. SIB, Siberian Platform; ARM, Armorica; SI, South of Iapetus (British Isles); NI, North of Iapetus; BAL, Baltic Shield; SP, Spitsbergen; RP, Russian Platform. Note that SIB and RP are only distributional trend paths.

Zone since there is no evidence for an ocean. Central Spitsbergen is generally considered part of Baltica in Devonian times. However, the palaeo-latitudinal trend for Spitsbergen is consistent with Britain. Consequently, if we were to accept dextral mega-shearing along the Tornquist Zone (Fig. 1a) as for example postulated by Douglass (1988), it would also have been accompanied by sinistral movements along the north-eastern border of Baltica, i.e. the Varanger–Timian Zone (cf. geographic location in Fig. 1a). Devonian or younger movements along the Varanger–Timian Zone, however, are precluded by the continuity of Caledonian Nappes across the zone. We would stress again, however, that the Lower–Middle Palaeozoic data from Scandinavia are of uncertain age and/or tectonic significance (see e.g. Pesonen *et al.* 1989), excepting the Ringerike Sandstone results of Douglass (1988). Therefore reconstructions of the Baltic Shield based on Lower–Middle Palaeozoic palaeomagnetic data are at best speculative. The position of Baltica has therefore traditionally been heavily reliant on the assumption

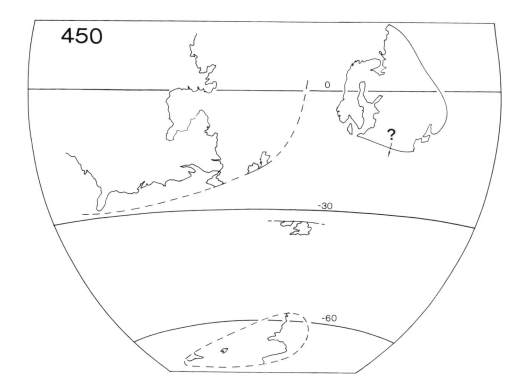

Fig. 9 Ordovician (450 Ma) reconstruction based on palaeomagnetic data. Optional longitude. Laurentia (N. America) positioned according to Northern Britain palaeomagnetic data. Equal-area projection.

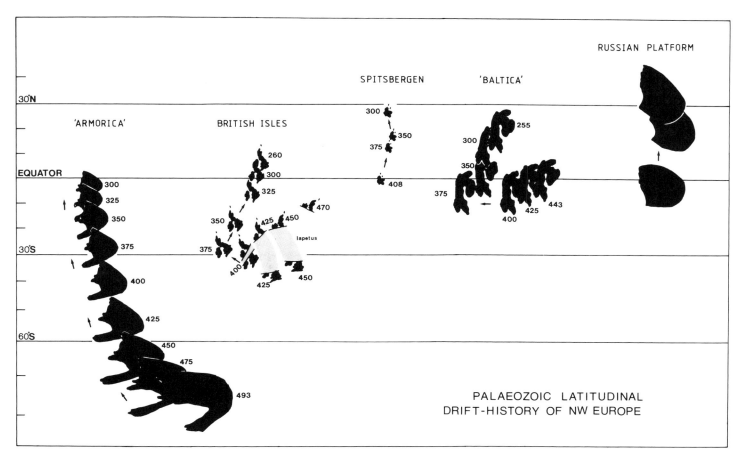

Fig. 10. Palaeozoic latitudinal drift-history of 'Armorica', the British Isles, Spitsbergen, 'Baltica' and the Russian Platform. The latter is a trend plot, i.e. an expression of the latitudinal spread of the palaeo-magnetic data. Numbers represent million years (approximately 25 million years interval). Galls projection.

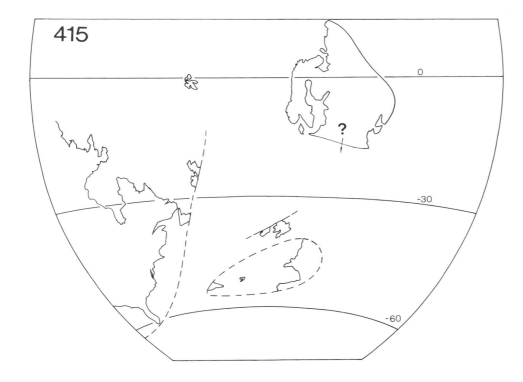

Fig. 11. Silurian (415 Ma) palaeogeographic reconstruction.

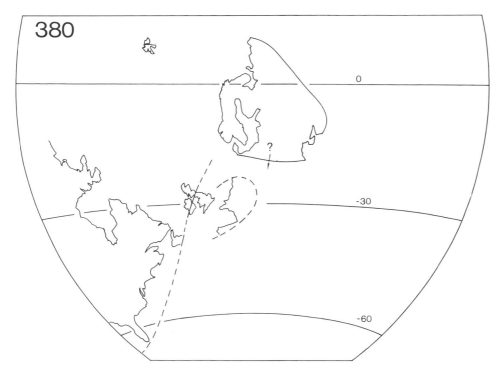

Fig. 12. Early Devonian (380 Ma) palaeogeographic reconstruction.

that southern Britain was marginal to Baltica through Palaeozoic times (e.g. Livermore *et al.* 1987; Briden *et al.* 1988).

Palaeomagnetic arguments for megashearing along the Tornquist Zone (Douglass 1988) are based on a palaeomagnetic discordance between the Ringerike Sandstone pole and Lower Old Red Sandstone poles from Britain. This analysis, however, is of uncertain validity since the Middle Silurian Ringerike Sandstone has been compared with poles of Upper Silurian to Lower Devonian age.

A popular view on the Devonian/Lower Carboniferous tectonic evolution of the Caledonian–Appalachian orogen has been one of megashearing and continental re-arrangement, including the shear zones of Spitsbergen (Billefjorden Fault), northern Britain (Great Glen Fault) and Newfoundland/New England (Cabot Fault). This long-held mobilistic view was is part based on palaeomagnetic data (e.g. Kent & Opdyke 1978, 1979; Van der Voo & Scotese 1981; Storetvedt 1987). New palaeomagnetic and geological information, however, has eliminated the need for such extreme tectonic interpretations (e.g. Irving & Strong 1984, 1985; Kent & Opdyke 1985; Briden *et al.* 1984; Torsvik *et al.* 1985).

Carboniferous–Permian

The Carboniferous is characterized by collisional docking of Gondwana with Euramerica leading to the Hercynian Orogeny. Northward drift of Euroamerica took place during the

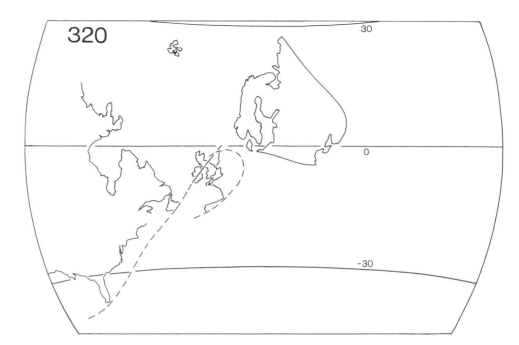

Fig. 13. Carboniferous (320 Ma) palaeogeographic reconstruction.

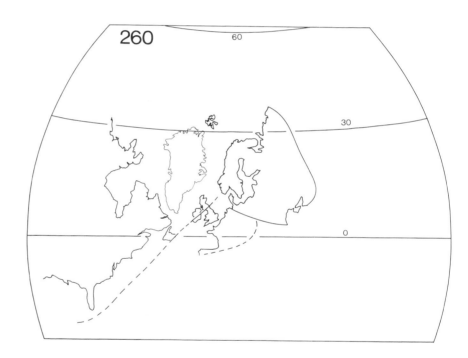

Fig. 14. Permian (260 Ma) palaeogeographic reconstruction. Armorica positioned according to its present position relative to the British Isles. Greenland located according to its present position relative North America after adjusting for opening of the Labrador Sea.

Carboniferous, and the final 'Pangaean' continental assembly was probably achieved in Late Carboniferous/Early Permian time (Fig. 14). Northward-drift of Euroamerica ($c.$ 2–4 cm a^{-1}) is also reflected in the rapid changes in palaeoclimatic and lithological patterns (Steel & Worsley 1984; Witzke 1988). A minor overlap between Britain and Armorica is noted in the 320 Ma reconstruction (Fig. 13), but this we relate to minor inaccuracies in the paths. There are, however, indications that Hercynian palaeomagnetic overprints from Armorica (Armorican Massif) differ in polar longitude compared with contemporaneous data from the British Isles and Baltica (compare Figs 5a & b and 2b & c), which may indicate relative rotations during Middle–Upper Carboniferous time. In the Pangaean continent configuration of Fig. 14 Armorica is located according to its current position relative Britain. Similarly, Greenland is located according to its current position relative North America after adjusting for opening of the Labrador Sea.

Conclusions

The Caledonian and Hercynian assembly of Pangaea entails collisional docking of Laurentia, Baltica, Gondwana and Siberia, and destruction of the intervening Iapetus and Tornquist Oceans. Additionally, a number of minor plates or continental fragments such as Armorica, including the Armorican Massif and Avalon Terranes, each having their separate crustal histories during the Palaeozoic have been postulated. In this review we have addressed ourself primarily to palaeomagnetic data from the SE margin of Laurentia, Baltica and Armorica, and some major points are outlined below.

(1) Armorica together with Gondwana was situated in high southerly latitudes (>60) during the Ordovician, at which time Laurentia was situated in equatorial to temperate southerly latitudes. A problem with respect to Armorica, however, concerns the various models as to which elements comprised the Armorican

plate (cf. review in Young 1987). The time at which Armorica rifted off from Gondwana, if it did at all, and when it eventually docked against Euramerica is unclear. However, assuming a uniform northward-drift ($c.$ 5 cm a^{-1}) in Siluro-Devonian time (Fig. 10) Armorica would appear to have collided with Euramerica during the Lower Devonian (Acadian) to form the Old Red Sandstone Continent.

(2) There is a clear discordance between the latitudinal position of Northern and Southern Britain in Ordovician and Silurian time, suggesting that the intervening Iapetus Ocean was at least 1000–1500 km wide in Middle Ordovician times (Fig. 10). During collision of Laurentia and Baltica, Laurentia reached its most southerly latitudinal position. The convergence vector for Laurentia was probably near SE (present day co-ordinates). The majority of palaeomagnetic data from Southern Britain suggest temperate southerly latitudes during the Ordovician, although some new data, previously regarded as 'anomalous', imply high southerly latitudes (McCabe 1988).

(3) Palaeomagnetic data from Spitsbergen suggest northerly movement harmonious with drift of the British Isles in post-Lower Devonian time (Fig. 10). This is also indicated by data from the Russian Platform, although time calibration is complicated by extensive late Palaeozoic remagnetization. The palaeolatitudinal position of the Baltic Shield (Baltica) during the Lower–Middle Palaeozoic is highly uncertain. The palaeomagnetic data suggest tropical latitudes in Ordovician to Devonian times (Fig. 10). It is suspected, however, that this apparent equatorial position is an artifact of Late Devonian magnetic resetting. Accordingly, palaeomagnetic data from the Baltic Shield are as yet insufficient to shed any light on postulated oceanic separations (Tornquist Sea) between Baltica and Gondwana/Armorica in Lower Ordovician time.

(4) The northern Britain APWP is broadly similar to the North American APWP (cf. Kent & Van der Voo 1990), but there is still a minor latitudinal difference during Ordovician and Silurian time (on the Bullard *et al.* (1965) reconstruction). A number of tectonic models for the assemblage of Euramerica that have been postulated over the last two decades are now clearly invalid and were due to the failure to recognize extensive Late Palaeozoic magnetic overprinting. These remagnetization features were strongly diachronous. In the Appalachians of North America, Permian or Kiaman overprinting is far-reaching, whereas an Upper Devonian/Lower Carboniferous (Svalbardian) remagnetization episode has been identified on the Baltic Shield in Scandinavia. On the other hand, Carboniferous and Early Mesozoic overprinting are frequently observed in the British Isles. In Central–Northern Europe, Carboniferous or Hercynian magnetic overprinting is widespread. Congruent pre-Devonian palaeomagnetic data have only been derived from rocks of the Armorican Massif. Similarly, a hiatus in the APW for the Russian Platform probably reflects Late Palaeozoic/Early Mesozoic remagnetization.

In conclusion it is proposed that future work should be addressed to carefully designed palaeomagnetic studies from southern Britain and Baltica, including the Russian Platform, in order to establish the position of Baltica during the Early–Middle Palaeozoic, and to determine whether southern Britain was part of Armorica in Ordovician time. We submit that the palaeogeographical position of Laurentia and Armorica is fairly well constrained for parts of the Palaeozoic. The position portrayed for Baltica, however, is speculative at best, and new studies of early–mid Palaeozoic rocks are of vital importance in order to provide more detailed insight into Palaeozoic reconstructions.

The Norwegian Research Council for the Humanities and Science (NAVF), the Natural Environment Research Council (NERC) and the Norwegian Geological Survey are acknowledged for financial support. Valuable discussions with R. Van der Voo, D. V. Kent, N. D. Opdyke and C. Scotese are acknowledged. Norwegian Lithosphere Contribution (46).

References

ABRAHAMSEN, N., WILSON, J. R., THY, P., OLSEN, N. O. & EESBENSEN, K. H. 1979. Palaeomagnetism of the Fongen–Hyllingen gabbro complex, southern Scandinavian Caledonides; plate rotation or polar shift. *Geophysical Journal of the Royal Astronomical Society*, **59**, 231–248.

BACHTADSE, V. & VAN DER VOO, R. 1986. Palaeomagnetic evidence for crustal and thin-skinned rotations in the European Hercynides. *Geophysical Research Letters*, **13**, 1, 161–164.

BRIDEN, J. C. 1970. Palaeomagnetic results from the Arrochar and Garabal Hill-Glen Fyne igneous complexes, Scotland. *Geophysical Journal of the Royal Astronomical Society*, **21**, 457–470.

—— & DUFF, B. A. 1981. Pre-Carboniferous palaeomagnetism of Europe north of the Alpine Orogenic Belt. *In*: MCELHINNY, M. W. & VALENCIO, D. A. (eds) *Palaeoreconstruction of the Continents*. Geo-dynamic Series, **2**, American Geophysical Union, 137–149.

—— & MORRIS, W. A. 1973. Palaeomagnetic studies in the British Caledonides-III. Igneous rocks of the Northern Lake District, England. *Geophysical Journal of the Royal Astronomical Society*, **34**, 27–46.

—— & MULLAN, A. J. 1984. Superimposed Recent, Permo-Carboniferous and Ordovician palaeomagnetic remanence in the Builth Volcanic Series, Wales. *Earth Planetary Science Letters*, **69**, 413–421.

——, TURNELL, H. B. & WATTS, D. 1984. British paleo-magnetism, Iapetus Ocean, and the Great Glen fault. *Geology*, **12**, 428–431.

——, KENT, D. V., LAPOINTE, P. L., LIVERMORE, R. A., ROY, J. L., SEGUIN, M. K., SMITH, A. G., VAN DER VOO, R. & WATTS, D. 1988. Palaeomagnetic constraints on the evolution of the Caledonian–Appalachian orogen. *In*: HARRIS A. L. & FETTES, D. J. (eds) *The Caledonian–Appalachian Orogen*. Geological Society, London, Special Publication, **38**, 35–48.

BULLARD, E. C., EVERETT, J. E. & SMITH, A. G. 1965. The fit of the continents around the Atlantic. *Royal Society of London, Philosophical Transections Series A*, **258**, 41–51.

BYLUND, G. 1974. Palaeomagnetism of dykes along the southern margin of the Baltic Shield. *Geologiske Foreningen Stockholm Forhandlinger*, **96**, 231–235.

—— & PATCHETT, P. J., 1977. Palaeomagnetic and Rb-Sr isotopic evidence for the age of the Sarna alkaline complex, western central Sweden. *Lithos*, **10**, 73–59.

BØE, R., ATAKAN, K. & STURT, B. A. 1989. The style of deformation of the Devonian rocks of Hitra and Smøla. *Norsk Geologisk Tidskrift* (in press).

CHAMALAUN, F. H. & CREER, K. M. 1964. Thermal demagnetization studies on the Old Red Sandstone of the Anglo–Welsh Cuvette. *Journal of Geophysical Research*, **68**, 8, 1607–1616.

CLAESSON, C. 1979. Early Palaeozoic geomagnetism of Gotland. *Geologisk Forening Stocholm Forhandlinger*, **101**, 149–155.

CLARK, R. M. & THOMPSON, R. 1984. Statistical comparison of palaeomagnetic directional records from lake sediments. *Geophysical Journal of the Royal Astronomical Society*, **76**, 337–368.

COCKS, L. R. M. & FORTEY, R. A. 1982. Faunal evidence for oceanic separations in the Palaeozoic of Britain. *Journal of the Geological Society, London*, **139**, 465–478.

CORNWELL, J. D. 1967. Palaeomagnetism of the Exeter Lavas, Devonshire. *Geophysical Journal of the Royal Astronomical Society*, **12**, 181–196.

DEUTSCH, E. R. 1980. Magnetism of the mid-Ordovician Tramore volcanics, SE Ireland, and the question of a wide Proto–Atlantic Ocean. *Journal of Geomagnetism and Geoelectricity*, **32**, 77–98.

—— & STORETVEDT, K. M. 1988. Magnetism of igneous rocks from the Tourmakeady and Glensaul inliers, W. Ireland: mode of emplacement and aspects of the Ordovician field pattern. *Geophysical Journal*, **92**, 223–234.

DOUGLASS, D. N. 1987. *Geology and palaeomagnetics of three Old Red Sandstone Basins: Spitsbergen, Norway and Scotland*. PhD thesis, Dartmouth College, Hanover, New Hampshire, USA.

—— 1988. Palaeomagnetics of Ringerike Old Red Sandstone and related rocks, southern Norway: implications for pre-Carboniferous separation of Baltica and British terranes. *Tectonophysics*, **148**, 11–27.

DUFF, B. A. 1979. The palaeomagnetism of Cambro-Ordovician red beds, the Erquy spilite series, and the Tregastel–Ploumanach granite complex, Armorican Massif (France and the Channel Islands). *Geophysical Journal of the Royal Astronomical Society*, **59**, 345–365.

—— 1980. The palaeomagnetism of Jersey volcanics and dykes and the lower Palaeozoic apparent polar wander path for Europe. *Geophysical Journal of the Royal Astronomical Society*, **60**, 355–375.

EDEL, J. B. 1987b. Palaeopositions of the western European hercynides during the Late Carboniferous deduced from palaeomagnetic data: consequences for "stable Europe". *Tectonophysics*, **139**, 31–41.

—— & COULON, M. 1984. Late Hercynian remagnetization of Tournaisian Series from the Laval Syncline, Armorican Massif, France. *Earth and Planetary Science Letters*, **68**, 343–350.

ESANG, C. B. & PIPER, J. D. A. 1984a. Palaeomagnetism of the Carboniferous E-W dyke swarm in Argyllshire. *Scottish Journal of Geology*, **20**, 3, 309–314.

—— & —— 1984b. Palaeomagnetism of Caledonian Intrusive Suites in the Northern Highlands of Scotland: Constraints to tectonic movements within the Caledonian orogenic belt. *Tectonophysics*, **104**, 1–34.

FALLER, A. M., BRIDEN, J. C. & MORRIS, W. A. 1977. Palaeomagnetic results from the Borrowdale Volcanic Group, English Lake District. *Geophysical Journal of the Royal Astronomical Society*, **48**, 111–121.

FORTEY, R. A. & COCKS, L. R. M. 1988. Arenig to Llandovery faunal distribution in the Caledonides. *In*: HARRIS, A. L. & FETTES, D. J. (eds) *The Caledonian-Appalachian Orogen*. Geological Society, London, Special Publication, **38**, 233–246.

GOULD, A. L. 1969. A regression technique for angular variates. *Biometrics*, **25**, 683–700.

HALVORSEN, E. 1970. Palaeomagnetism and the age of the younger diabases in the Ny-Hellesund areas, S. Norway. *Norsk Geologisk Tidskrift*, **50**, 2, 157–166.

—— 1972. On the palaeomagnetism of the Arendal diabases. *Norsk Geologisk Tidskrift*, **52**, 3, 217–228.

IRVING, E & STRONG, D. F. 1984. Evidence against large-scale Carboniferous strike-slip faulting in the Appalachian–Caledonian orogen. *Nature*, **310**, 762–764.

—— & —— 1985. Palaeomagnetism of rocks from Burin Peninsula, Newfoundland: Hypothesis of Late Palaeozoic displacement of Acadia criticized. *Journal of Geophysical Research*, **90**, B2, 1949–1962.

JELENSKA, M. & LEWANDOWSKI, M. 1986. A paleomagnetic study of Devonian sandstone from Central Spitsbergen. *Geophysical Journal of the Royal Astronomical Society*, **87**, 617–632.

JONES, M., VAN DER VOO, R. & BONHOMMET, N. 1979. Late Devonian to early Carboniferous palaeomagnetic poles from the Armorican Massif (France). *Geophysical Journal of the Royal Astronomical Society*, **58**, 287–308.

JUPP, P. E. & KENT, J. T. 1987. Fitting smooth paths to spherical data. *Applied Statistics*, **36**, 1, 34–46.

KENT, D. V. & OPDYKE, N. 1978. Palaeomagnetism of the Devonian Catskill Red Beds: evidence for motion of the coastal New England relative to Cratonic North America. *Journal of Geophysical Research*, **83**, 4441–4450.

—— & —— 1979. The early Carboniferous palaeo-magnetic field of North America and its bearing on tectonics of the Northern Appalachians. *Earth and Planetary Science Letters*, **44**, 365–372.

—— & —— 1985. Multicomponent magnetizations from the Mississippian Mauch Chunk Formation of the Central Appalachians and their tectonic implications. *Journal of Geophysical Research*, **90**, B7, 5371–5383.

—— & VAN DER VOO, R. 1990. Palaeozoic palaeogeography from the palaeomagnetism of the Atlantic-bordering continents. *In*: MCKERROW, W. S. & SCOTESE, C. R. (eds) *Palaeozoic Palaeogeography and Biogeography*. Geological Society, London, Memoir, **12**, 49–56.

KNEEN, S. 1973. The palaeomagnetism of the Foyers Plutonic Complex, Inverness-shire. *Geophysical Journal of the Royal Astronomical Society*, **32**, 53–63.

KHRAMOV, A. N. & SHOLPO, L. E. 1967. *Palaeomagnetism. Principles, methods and geological applications of palaeomagnetology*. Nedra, Leningrad.

——, PETROVA, G. N. & PECHERSKY, D. M. 1981. Palaeomagnetism of the Soviet Union. *In*: MCELHINNY, A. W. & VALENCIO, D. A. (eds) *Palaeoreconstruction of the Continents*. Goedynamic Series, **2**, American Geophysical Union, 177–194.

LATHAM, A. G. & BRIDEN, J. C. 1975. Palaeomagnetic field directions in Siluro-Devonian lavas of the Lorne Plateau, Scotland, and their regional significance. *Geophysical Journal of the Royal Astronomical Society*, **43**, 243–252.

LIVERMORE, R. A., SMITH, A. G. & BRIDEN, J. C. 1985. Palaeomagnetic constraints on the distribution of continents in the late Silurian and early Devonian. *Philosophical Transactions of the Royal Society London*, B **309**, 29–56.

LØVLIE, R. 1981. Palaeomagnetism of coast-parallel alkaline dykes from western Norway; ages of magmatism and evidence for crustal uplift and collapse. *Geophysical Journal of the Royal Astronomical Society*, **66**, 417–426.

MCCABE, C. 1988. *Palaeomagnetism of the Ordovician Stapeley Volcanic Formation, Welsh Borderlands*. Palaeozoic Biogeography and Palaeogeographic symposium, Oxford (abstract).

MCCLELLAND BROWN, E. 1983. Palaeomagnetic studies of fold development in the Pembrokshire Old Red Sandstone. *Tectonophysics*, **98**, 131–139.

MCELHINNY, M. W., 1973. *Palaeomagnetism and Plate Tectonics*. Cambride University Press, Cambridge.

MILLER, J. D. & KENT, D. V. 1988. Palaeomagnetism of the Silurian–Devonian Andreas redbeds: Evidence for an Early Devonian supercontinent? *Geology*, **16**, 195–198.

MORRIS, W. A., BRIDEN, J. C., PIPER, J. D. A. & SALLOMY, J. T. 1973. Palaeomagnetic studies in the British Caledonides- V. Miscellaneous new Data. *Geophysical Journal of the Royal Astronomical Society*, **34**, 69–106.

MULDER, F. G. 1971. Palaeomagnetic research in some parts of Central and Southern Sweden. *Sveriges Geologiske Undersokelse*, **64**, 10, 1–56.

NAIRN, A. E. M. 1960. Palaeomagnetic results from Europe. *Journal of Geology*, **68**, 285–306.

PARKER, R. L. & DENHAM, C. R. 1979. Interpolation of unit vectors. *Geophysical Journal of the Royal Astronomical Society*, **58**, 685–687.

PERROUD, H. 1983. Palaeomagnetism of Palaeozoic rocks from the Cabo de Penas, Asturia, Spain. *Geophysical Journal of the Royal Astronomical Society*, **75**, 201–215.

—— & —— 1984. A Devonian palaeomagnetic pole for Armorica. *Geophysical Journal of the Royal Astronomical Society*, **77**, 839–845.

—— & VAN DER VOO, R. 1985. Palaeomagnetism of the late Ordovician Thouars Massif, Vendee Province, France. *Journal of Geophysical Research*, **90**, B6, 4611–4625.

——, BONHOMMET, N. and ROBARDET, M. 1982. Comment on "A palaeomagnetic study of Cambrian volcanics and intrusives from the Armorican Massif, France" by W. A. MORRIS. *Geophysical Journal of the Royal Astronomical Society*, **69**, 573–578.

——, —— & VAN DER VOO, R. 1983. Palaeomagnetism of Crozon peninsula (France). *Geophysical Journal of the Royal Astronomical Society*, **72**, 307–319.

——, —— & THEBAULT, J. P. 1986a. Palaeomagnetism of the Ordovician Moulin de Chateaupanne formation, Vendee, western France. *Geophysical Journal of the Royal Astronomical Society*, **85**, 573–582.

——, Van der Voo, R. & BONHOMMET, N. 1984a. Palaeozoic evolution of the Armorica plate on the basis of palaeomagnetic data. *Geology*, **12**, 579–582.

——, AUVRAY, B., BONHOMMET, N., MACE, J. & VAN DER VOO, R. 1986b. Palaeomagnetism and K-Ar dating of Lower Carboniferous dolerite dykes from northern Brittany. *Geophysical Journal of the Royal Astronomical Society*, **87**, 143–154.

——, ROBARDET, M., VAN DER VOO, R., BONHOMMET, N. & PARIS, F. 1984b. Revision of the age of magnetization of the Montmartin red beds, Normandy, France. *Geophysical Journal of the Royal Astronomical Society*, **80**, 541–549.

PESONEN, L. J., TORSVIK, T. H., BYLUND, G. & ELMING, S. A. 1989. Crustal evolution of Fennoscandia — Palaeomagnetic constraints. *Tectonophysics* (in press).

PIPER, J. D. A. 1974. Sulitjelma Gabbro, Northern Norway: A palaeomagnetic result. *Earth and Planetary Science Letters*, **21**, 383–388.

—— 1975. Palaeomagnetism of Silurian lavas, of Somerset and Gloucestershire, England. *Earth and Planetary Science Letters*, **25**, 355–360.

—— 1978a. Palaeomagnetism and palaeogeography of the Southern Uplands in Ordovician times. *Scottish Journal of Geology*, **14**, 93–107.

—— 1978b. Palaeomagnetic survey of the (Palaeozoic) Shelve inlier and Berwyn Hills, Welsh Borderland. *Geophysical Journal of the Royal Astronomical Society*, **53**, 355–371.

—— & BRIDEN, J. C. 1973. Palaeomagnetic studies in the British Caledonides-I. Igneous rocks of the Builth Wells–Llandrindod Wells Ordovician inlier, Radnorshire, Wales. *Geophysical Journal of the Royal Astronomical Society*, **34**, 1–12.

—— —— & STEARN, J. E. F. 1975. Palaeomagnetism of the Breidden Hill (Palaeozoic) inlier, Welsh Borderlands. *Geophysical Journal*, **43**, 1013–1016.

ROBERTSON, D. J. 1988. Palaeomagnetism of the Connemara Gabbro, Western Ireland. *Geophysical Journal of the Royal Astronomical Society*, **94**, 1, 51–64.

ROBINSON, M. A. 1985. Palaeomagnetism of volcanics and sediments of the Eday Group, Southern Orkney. *Scottish Journal of Geology*, **21**, 285–300.

ROTHER, K., FLUGE, P. R. & STORETVEDT, K. M. 1987. Palaeomagnetism of the Askøy mafic pluton (late Precambrian), West Norway; events of Caledonian metamorphic remagnetization. *Physics of the Earth and Planetary Interiors*, **45**, 85–96.

SALLOMY, J. T. and PIPER, J. D. A. 1973. Palaeomagnetic studies in the British Caledonides IV. Lower Devonian lavas of the Strathmore Region, Scotland. *Geophysical Journal of the Royal Astronomical Society*, **34**, 47–68.

SMETHURST, M. A. & BRIDEN, J. C. 1988. Palaeomagnetism of Silurian sediments in W. Ireland: evidence for block-rotation in the Caledonides. *Geophysical Journal*, **95**, 327–346.

STEEL, R. J. & WORSLEY, D. 1984. Svalbard's post-Caledonian Strata — an atlas of sedimentological patterns and palaeogeographic evolution. *Journal of petroleum geology of the North European Margin*, Graham Trotman, 59–69.

STORETVEDT, K. M. 1987. Major late Caledonian and Hercynian shear movements on the Great Glen Fault. *Tectonophysics*, **143**, 253–267.

—— & CARMICHAEL, C. M. 1979. Resolution of superimposed magnetizations in the Devonian John O'Groats Sandstone, N. Scotland. *Geophysical Journal of the Royal Astronomical Society*, **58**, 769–784.

—— & GIDSKEHAUG, A. 1969. The magnetization of the Great Whin Sill, Northern England. *Physics of the Earth and Planetary Interiors*, **2**, 105–111.

—— & GJELLESTAD, G. 1966. Palaeomagnetic investigation of an Old Red Sandstone formation of Southern Norway. *Nature*, **212**, 59–61.

STORETVEDT, K. M. & MELAND, A. H. 1985. Geological interpretation of palaeomagnetic results from Devonian rocks of Hoy, Scotland. *Scottish Journal of Geology*, **21**, 337–352.

—— & TORSVIK, T. H. 1983. Palaeomagnetic re-examination of the basal Caithness Old Red Sandstone; aspects of local and regional tectonics. *Tectonophysics*, **98**, 151–164.

—— & —— 1985. Palaeomagnetism of the Middle–Upper Devonian Esha Ness ignimbrite, W. Shetland. *Physics of the Earth ands Planetary Interiors*, **37**, 169–173.

——, PEDERSEN, S., LØVLIE, R. & HALVORSEN, E. 1978. Palaeomagnetism in the Oslo rift. *In*: RAMBERG, J. B. & NEUMANN, E. R. (eds) *Tectonics and Geophysics of Continental Rifts*, D. Reidel Publishing Company, Holland, 289–296.

STORHAUG, K. & STORETVEDT, K. M. 1985. Palaeomagnetism of the Sarchlet Sandstone (Orcadian Basin); age perspectives. *Scottish Journal of Geology*, **20**, 3, 275–284.

STURT, B. A. 1983. *Late Caledonian and possible Variscan stages in the Orogenic evolution of the Scandinavian Caledonides.* The Caledonide Orogen-IGCP Project 27, symposium de Rabat, Morocco (abstract).

—— & ROBERTS, D. 1987. *Terrane linkage in the final stages of the Caledonian orogeny in Scandinavia.* Extended abstract IGCP Project 233 symposium, Nouakchott, Mauritania, 193–196.

—— & TORSVIK, T. H. 1987. A late Carboniferous palaeomagnetic pole recorded from a syenite sill, Stabben, Central Norway. *Physics of the Earth and Planetary Interiors*, **49**, 350–359.

TARLING, D. H. 1985. Palaeomagnetic studies of the Orcadian Basin. *Scottish Journal of Geology*, **21**, (3), 261–273.

—— & MITCHELL, J. G. 1973. A palaeomagnetic and isotopic age for the Wackerfield dyke of Northern England. *Earth and Planetary Science Letters*, **18**, 427–432.

THIRLWALL, M. F. 1981. Implications for Caledonian plate tectonic models of chemical data from volcanic rocks of the British Old Red Sandstone. *Journal Geological Society, London*, **138**, 123–138.

—— 1988. Wenlock to mid-Devonian volcanism of the Caledonian–Appalachian orogen. *In*: HARRIS, A. L. & FETTES, D. J. (eds) *The Caledonian–Appalachian Orogen.* Geological Society, London, Special Publication, **38**, 415–428.

THORNING, L. 1974. Palaeomagnetic results from Lower Devonian rocks of the Cheviot Hills, Northern England. *Geophysical Journal of the Royal Astronomical Society*, **36**, 487–496.

THOMAS, C. & BRIDEN, J. C. 1976. Anomalous geomagnetic field during the late Ordovician. *Nature*, **259**, 5542, 380–382.

THOMPSON, R. & CLARK, R. M. 1981. Fitted polar wander paths. *Physics of the Earth and Planetary Interiors*, **27**, 1–7.

—— & —— 1982. A robust least-square Gondwanan apparent polar wander path and the question of palaeomagnetic assessment of Gondwanan reconstructions. *Earth and Planetary Science Letters*, **57**, 152–158.

TORSVIK, T. H. 1984. Palaeomagnetism of the Foyers and Strontian granites, Scotland. *Physics of the Earth and Planetary Interiors*, **36**, 163–177.

—— 1985a. Magnetic properties of the Lower Old Red Sandstone Lavas in the Midland Valley, Scotland; palaeomagnetic and tectonic considerations. *Physics of the Earth and Planetary Interiors*, **39**, 194–207.

—— 1985b. Palaeomagnetic results from the Peterhead granite, Scotland; implication for regional late Caledonian magnetic overprinting. Physics of the Earth and Planetary Interiors, **39**, 108–117.

—— & STURT, B. A. 1988. Multiphase magnetic overprints in the Moine Thrust Zone. *Geological Magazine*, **125**, 1, 63–82.

——, LØVLIE, R. & STORETVEDT, K. M. 1983. Multicomponent magnetization in the Helmsdale granite, N. Scotland; geotectonic implications. *In*: MCCLELLAND, B. & VANDENBERG, J. (eds) *Palaeomagnetism of orogenic belts*. Tectonophysics, **98**, 111–129.

——, —— & STURT, B. A. 1985. Palaeomagnetic argument for a stationary Spitsbergen relative to the British Isles (Western Europe) since late Devonian and its bearing on North Atlantic reconstruction. *Earth and Planetary Science Letters*, **75**, 278–288.

——, STURT, B. A. & RAMSAY, D. M. 1989d. On the origin and the tectonic implications of magnetic overprinting of the Old Red Sandstone, Shetland. *Geophysical Journal International*, (in press).

——, LYSE, O., ATTERÅS, G. & BLUCK, B. 1989b. Palaeozoic palaeomagnetic results from Scotland and their bearing on the British Apparent polar wander path. *Physics of the Earth and Planetary Interiors*, **55**, 93–105.

——, STURT, B. A., RAMSAY, D. M. & VETTI, V. 1987. The tectonomagnetic signature of the Old Red Sandstone and pre-Devonian strata in the Håsteinen area, Western Norway, and implications for the later stages of the Caledonian Orogeny. *Tectonics*, **6**, 3, 305–322.

——, GRØNLIE, A. & RAMSAY, D. M. 1989c. Palaeomagnetic data bearing on the age of the Ytterøy Dyke, Central Norway. *Physics of the Earth and Planetary Interiors*, **54**, 156–162.

——, ——, RAMSAY, D. M., KISCH, H. J. & BERING, D. 1986. The tectonic implications of Solundian (Upper Devonian) magnetization of Devonian rocks of Kvamshesten, western Norway. *Earth and Planetary Science Letters*, **80**, 337–347.

——, ——, ——, BERING, D. & FLUGE, P. R. 1988. Palaeomagnetism, magnetic fabrics and the structural style of the Hornelen Old Red Sandstone, Western Norway. *Journal Geological Society, London*, **145**, 413–430.

——, ——, ——, GRØNLIE, A., ROBERTS, D., SMETHURST, M., ATAKAN, K., BØE, R. & WALDERHAUG, H. J. 1989a. Palaeomagnetic constraints on the early history of the Møre–Trøndelag Fault Zone, Central Norway: *In*: KISSEL, C. & LAJ, C. (eds) *Palaeomagnetic rotations and continental deformation*. Kluwer Academic Publishers, 431–457.

TURNELL, H. B. 1985. Palaeomagnetism and Rb-Sr ages of the Ratagen and Comrie intrusions. *Geophysical Journal of the Royal Astronomical Society*, **83**, 363–378.

—— & BRIDEN, J. C. 1983. Palaeomagnetism of NW Scotland syenites in relation to local and regional tectonics. *Geophysical Journal of the Royal Astronomical Society*, **75**, 217–234.

VAN DER VOO, R. 1979. Palaeozoic assembly of Pangea: a new plate tectonic model for the Taconic, Caledonian and Hercynian orogenies. *Eos Transactions AGU*, **60**, 241.

—— 1988. Palaeozoic paleogeography of North America, Gondwana, and intervening displaced terranes: Comparison of palaeomagnetism with paleoclimatology and biogeographical patterns. *Geological Society of America Bulletin*, **100**, 311–234.

—— & JOHNSON, R. J. E. 1985. Palaeomagnetism of the Dunn Point Formation (Nova Scotia): High palaeolatitudes for the Avalon terrane in the late Ordovician. *Geophysical Research Letters*, **12**, 6, 337–340.

—— & KLOOTWIJK, C. T. 1972. Palaeomagnetic reconnaissance study of the Flamanville granite, with special reference to the anisotropy of its susceptibility. *Geological Mijnbouw*, **51**, 609–617.

—— & SCOTESE, C. 1981. Palaeomagnetic evidence for a large (ca. 2000 km) sinistral offset along the Great Glen Fault during Carboniferous time. *Geology*, **9**, 583–589.

VINCENZ, S. A. & JELENSKA, M. 1985. Palaeomagnetic investigations of Mesozoic and Palaeozoic rocks from Svalbard. *In*: HUSEBY, E. S. & KRISTOFFERSEN, Y. (eds) *Geophysics of the Polar Regions*. Tectonophysics, **114**, 163–180.

WHITTINGTON, H. B. & HUGHES, C. P. 1972. Ordovician geography and faunal provinces deduced from Trilobite distribution. *Philosophical Transactions of the Royal Society of London*, B **263**, 235–278.

WATTS, D. 1982. A multicomponent, dual-polarity palaeomagnetic regional overprint from the Moine of northwest Scotland. *Earth and Planetary Science Letters*, **61**, 190–198.

—— 1985*a*. Palaeomagnetic resetting in the Barrovian Zones of Scotland and its relationship to the late structural history. *Earth and Planetary Science Letters*, **75**, 258–264.

—— 1985*b*. Palaeomagnetism of the Lower Carboniferous Billefjorden Group, Spitsbergen. *Geological Magazine*, **4**, 383–388.

—— & BRIDEN, J. C. 1984. Palaeomagnetic signature of slow post-orogenic cooling of the north-east Highlands of Scotland recorded in the Newer Gabbros of Aberdeenshire. *Geophysical Journal of the Royal Astronomical Society*, **77**, 775–788.

WITZKE, B. J. 1988. *Constraints for Palaeozoic Palaeolatitudes of Euramerica*. Palaeozoic Biogeography and Palaeogeographic symposium, Oxford (abstract).

YOUNG, G. C. 1987. Devonian palaeontological data and the Armorica problem. *Palaeogeography, Palaeoclimatology, Palaeoecology*, **60**, 283–304.

Palaeomagnetic constraints on the position of Gondwana during Ordovician to Devonian times

V. BACHTADSE[1] & J. C. BRIDEN[1,2]

[1] *Department of Earth Sciences, Oxford University, Parks Road, Oxford OX1 3PR, UK*
[2] *Natural Environment Research Council, Polaris House, North Star Avenue, Swindon SN1 1EU, UK*

Abstract: The currently available palaeomagnetic data base for Gondwana is reviewed and using revised rotation parameters for the fit of the southern continents an attempt has been made to construct an apparent polar wander path for Gondwana during Ordovician to Permo-Carboniferous times using a cubic spline fitting technique. Although the density of the data set is still rather sparse and the quality of the data is variable, our approach seems to be justified when tying the apparent polar wander path to selected palaeopoles of high quality. The palaeogeographic scenario based on our results is rather complex. Rapid northward shift of Gondwana during the Ordovician to the Early Silurian and subsequent collision with Laurentia is followed by divergence and the formation of a wide intervening ocean during the Devonian. The final closure of this ocean did not begin before the Late Devonian and was completed by the Late Carboniferous. If the apparent polar wander path presented in this paper is correct, then extremely high drift rates of about 23 cm a^{-1} have to be postulated for Gondwana during the Late Ordovician–Silurian.

Palaeoclimatic and palaeogeographic data, such as the occurrence of climatically sensitive sediments or the pattern of faunal distributions, cannot, on their own, provide absolute estimates of change in latitude during geological time of a given locality. Since oceanic and atmospheric circulation patterns as well as climatic gradients are, at least in part, controlled by the distribution, configuration and physiography of the continents, independent information on those controlling factors is essential for any meaningful palaeogeographic interpretation. Hence, under the premise that the time averaged palaeomagnetic field is both axial and dipolar (Merrill & McElhinny 1983 and references therein) for at least the last 600 Ma palaeomagnetism still is the only discipline in Earth Sciences which allows a quantitative approach to palaeogeographical problems such as the determination of palaeolatitudes.

Precise knowledge of the position of Gondwana with respect to the northern continents is of crucial importance for the understanding of the geodynamic evolution of Europe and northern America during the time in question and consequently Palaeozoic rocks of the southern continents have been a prime target of palaeomagnetic research since the early days of palaeomagnetism and the results of about fifty studies have been published so far (see reviews by Brock 1981; Embleton 1981; Vilas 1981). However, close inspection of the dataset reveals that the apparent polar wander path (APWP) for Gondwana is still far from being completely known. Large time gaps in the data set as well as conflicting results from different parts of Gondwana yield ambiguous and sometimes conflicting palaeogeographical models for some intervals, including critical periods such as the Siluro-Devonian. The reasons for this incoherence of the data are many. The majority of the older data published in the 1960s and the early 1970s do not fulfil some of the basic reliability criteria such as defined by various authors (Van der Voo 1988, Briden & Duff 1981, McElhinny & Embleton 1976) largely because of the lack of adequate demagnetization experiments in those days. Consequently the multivectorial character of the natural remanent magnetization (NRM) or its nature as a complete secondary overprint was not identified in a number of studies. Further palaeomagnetic work on undeformed continental sediments, abundant on the southern continents and one of the prime targets of palaeomagnetic research, is a rather difficult task due to the intrinsic sparsity of stratigraphically significant fossils: precise stratigraphic ages are rather difficult to establish and, in the absence of geological controls the age of magnetization is often unknown. It does therefore not come as a surprise that less than 20 of the palaeopole positions for the Palaeozoic of Gondwana, as compiled by various authors (Vilas 1981; Brock 1981; Embleton 1981) and additional data published subsequently, fulfil minimal reliability criteria (Table 1 and Fig. 1).

The palaeomagnetic database

The majority of reliable palaeomagnetic results published for Gondwana during the last two decades originate from Australia (e.g. review by Embleton 1981). Unfortunately, however, the significance of the bulk of the data depends strongly on the interpretation of the tectonic evolution of eastern Australia (Lachlan Fold Belt). The variety of Apparent Polar Wander Paths for Gondwana (McElhinny & Embleton 1974; Schmidt & Morris 1977; Morel & Irving 1978; Goleby 1980) reflects the various interpretations of the tectonic setting of the Lachlan Fold Belt and whether eastern Australia can be considered as being an integral part of Australia during the whole Palaeozoic or must be interpreted as an independent terrane docking sometime during the Late Palaeozoic. The answer to this question is of vital importance concerning the geodynamic interpretation of recently published palaeopoles for the Devonian of Australia such as the ones from the Comerong Volcanics (CV; Schmidt *et al.* 1986) and the Snowy River Volcanics (SV; Schmidt *et al.* 1987). The poles which we have used in this analysis are listed in Table 1. The abbreviations referring to individual poles are plotted on Fig. 1 and listed in Table 1 and used in the following text to identify the poles.

The Ordovician

Although they are of rather varying quality, Ordovician palaeopoles have been available for the southern continents (Table 1 and Fig. 1) since the 1960s. However disagreements between the Morel & Irving (1978) APWP, advocating an Ordovician palaeopole off the coast of Libya and a significantly different palaeopole position west of western northern Africa remain conspicuous. A recent palaeomagnetic study of the well dated mid-Ordovician (460 Ma Rb/Sr) Salala ring complex (SA) from eastern Sudan (Bachtadse & Briden 1989), as well as a radiometric age reinvestigation of the Ntonya ring complex (NT), Malawi, (Briden 1968; J. C. Briden & D. C. Rex, unpublished) seem to support the positioning of the Ordovician palaeopole close to the coast of northwestern Africa. Transformation of (western) Australia and (east) Antarctica into a pre-Mesozoic Gondwana configuration

Table 1. *Selected Ordovician to Permo-Carboniferous palaeopole positions from Africa, Australia and Antarctica in a Gondwana reassembly*

Formation	Symbol	age	Lat. [°N]	Long. [°E]	k	α_{95} (dp, dm)	Reference
Africa							
Damara granites	DG	458±8[2]	27	351	–	5	Corner & Henthorn (1978)
Hook intrusive rocks	HK	517±18[2]	14	336	13	36	Brock (1967)
Ntonya Ring complex	NT	485±[3]	28	345	–	(1, 2)	Briden (1968)
Graafwater Formation	GR	Ol[4]	28	14	–	(7, 12)	Bachtadse et al. (1987)
Blaubeker Formation	BL	471±26[2]	51	352	3	(11, 12)	Kröner et al. (1980)
Salala Ring Complex component B	SA	c.460[2]	40	330	–	(8, 12)	Bachtadse & Briden (1989)
Pakhuis/Cedarberg Formations	PAC	Ou/Sl[4]	25	343	–	(12, 21)	Bachtadse et al.(1987)
Aïr intrusive rocks	AIR	435[3]	−43	9	50	6	Hargraves et al. (1987)
Bokkeveld Group	BV	Dm[4]	10	15	–	(9, 11)	Bachtadse et al. (1987)
Dwyka varves	DV	Cu[4]	−26	26	52	10	McElhinny & Opdyke (1968)
Tanzanian redbeds	TR	P−C[4]	−40	64	11	16	McElhinny et al. (1968)
Australia							
Jinduckin Formation	JF	Ol[4]	36	359	10	11	Luck (1972)
Stairway Sandstone	SS	Om[4]	51	34	25	9	Embleton (1972)
Mereenie Sandstone	MS	D(?)[4]	10	17	15	11	Embleton (1972)
Canning Basin	CB	Du[4]	2	16	62	8	Hurley & Van der Voo (1987)
Hervey Group[1]	HG	Du[4]	−5	8	–	(8, 16)	Li et al. (1988)
Mugga Mugga por.[1]	MP	414±8[2]	−16	2	–	(6, 9)	Briden (1966); revised by Goleby (1980)
Snowy river volc. rocks[1]	SV	Dl[4]	−54	15	–	(11, 14)	Schmidt et al. (1986)
Comerong volcanic rocks[1]	CV	Dm[4]	−34	2	42	7	Schmidt et al. (1987)
Eastern Antarctica							
Sør Rondane intrusives	SR	c.469[2]	11	8	–	(5, 6)	Zijderveld (1968)
Taylor valley dykes	TV	460±7[2]	36	9	–	(6, 11)	Manzoni & Nanni (1977)

Palaeomagnetic poles from Australia and Antarctica have been transferred into African coordinates using the rotation parameters given by Lawver & Scotese (1987). Whenever necessary isotopic ages have been recalculated using the appropriate decay constants (Steiger & Jäger 1977). α_{95}: semi angle of the cone of confidence about the palaeopole position at the 95% probability level. dp, dm: semi axis of the oval of 95% confidence about the mean pole. Lat., Long.: latitude, longitude of the palaeomagnetic pole. k is the Fisherian precision parameter associated with the means.
O: Ordovician; S: Silurian; D: Devonian; P−C: Permo−Carboniferous; 1, m, u: lower, middle, upper.
[1] palaeopole positions from the (allochthonous?) eastern Australian Lachlan Fold Belt.
[2] Rb/Sr age
[3] K−Ar age
[4] stratigraphic age

using revised finite rotation parameters (Lawver & Scotese 1987) significantly reduces the gap between Australia and India as compared with Smith & Hallam's (1970) reconstruction and consequently leads to a westward (15°) shift of the western Gondwana palaeopoles (Table 1). A better grouping of the Ordovician palaeopole positions in western North Africa results, yielding an Ordovician mean pole offshore Morocco (Fig. 2), which is virtually identical with the one proposed originally by Briden (1967a) and McElhinny et al. (1968).

The Silurian

Silurian data (e.g. from the Mugga Mugga Porphyries (MP; Briden 1966; Goleby 1980) of Late Silurian age) from the structurally rather complex eastern Australian (Lachlan) Fold Belt, yielded a palaeopole off the southern Argentinean coast (in Pangea configuration and African coordinates). Being at odds with various apparent polar wander paths, such as the one proposed by McElhinny et al. (1968) or Morel & Irving's (1978) path X, which both connect an area of Ordovician palaeopoles in northern Africa with the well established Carboniferous VGPs off southern Africa in a rather straight line, those poles have been interpreted in terms of an exotic terrane model for the Lachlan Fold Belt and their significance for cratonic Australia and by implication Gondwana has been questioned (McElhinny & Embleton 1974). Recently Hargraves et al. (1987) reported palaeomagnetic results from 11 individual igneous ring complexes from the Aïr mountains (AIR), Niger, which yielded a pole position to the west of the coast of southern Africa, close to the area of Palaeozoic pole positions from eastern Australia. In addition to the published age data (Rb/Sr) for those ring complexes, which range from 480 Ma in the north to 400 Ma in the south (Bowden et al. 1987), Hargraves et al. (1987) report ^{40}Ar/^{39}Ar ages from three more centrally positioned intrusions, which yielded an average age of 435 Ma (latest Ordovician/earliest Silurian). Although the significance of those age data for the whole suite of complexes is debatable, the authors claim this age to be representative for their entire palaeomagnetic collection and it is used to constrain the timing of acquisition of the NRM.

Devonian

Not only is the Devonian segment of the Gondwana APWP of crucial importance for the reconstruction of the geodynamic setting of the Acadian, Caledonian and Hercynian orogenies in Europe and North America but it is also the most controversial one.

Fig. 1. Ordovician through Permo-Carboniferous palaeopole positions from Africa (squares), Australia (triangles) and eastern Antarctica (diamonds) plotted on a modified Gondwana reassembly. Pole positions from Australia and Antarctica have been transferred into African co-ordinates using rotation parameters as given by Lawver & Scotese (1987). Labels as in Table 1.

Conflicting data have been emerging from Australia during the last couple of years, yielding both a Late Devonian pole position in Central Africa from Late Devonian carbonates from the western Australian Canning Basin (CB; Hurley & Van der Voo 1987) and a more southerly and westerly palaeopole west of southern Argentina for the early to mid-Devonian Comerong volcanic rocks (CV; Schmidt et al. 1987) and the Snowy River volcanic rocks (SV; Schmidt et al. 1986), both from the Lachlan Fold Belt. A third group of Devonian palaeopoles from the early to mid-Devonian Tasmanian Housetop granite (Briden 1976b) and Embleton's (1977) data from the southeastern Australian Mulga Downs Group of Upper Devonian age (not plotted on Fig. 1) fall conspicuously close to the postfolding (secondary) palaeopoles from the Buchan Cave Limestones and the Snowy River volcanic rocks (Schmidt et al. 1987). This observation and the fact, that Tertiary and Cenozoic palaeopole positions from Australia tend to fall very close to the Permo-Carboniferous VGP for Gondwana makes severe contamination by secondary (Permo-Carboniferous and/or Tertiary–Cenozoic) overprints very likely and this last group will therefore not be considered any further.

In his review of Palaeozoic palaeomagnetic data, Van der Voo (1988) argued convincingly in favour of the Canning Basin pole as being representative for the Late Devonian of cratonic Australia, rather than the Comerong pole from the structurally more complex Lachlan Fold Belt. Supporting that line of argument is a pole position recently published from Late Devonian sediments of the Hervey Group (HG) from eastern Australian (Li et al. 1988), which is in close agreement with the Canning Basin result. Although Li et al. (1988) cannot demonstrate a positive fold test for the rocks studied, they cite evidence supporting their argument that the data from the Hervey Group are representative of the Earth's field during or shortly after deposition of those rocks (i.e. the latest Devonian). This result also allows us tentatively to assume a rather similar age for the (re)magnetization of the Mereenie Sandstone (MS; Embleton 1972), which has long

been considered as being crucial to the interpretation of the Australian APWP. The quality of the Mereenie result has been subject to discussion (Schmidt et al. 1987); for instance the direction of NRM in situ is very close to the direction of the present day (or Tertiary) field direction at the sampling locality. Similarities between the Mereenie and the Canning Basin and Hervey Group poles could therefore be coincidental (Schmidt & Embleton 1987).

The long-standing Devonian palaeopole position in Central Africa, derived from the Msissi Norite in Morocco (Hailwood 1974) has recently been shown to be more complex than originally assumed. Age determinations as well as a reinvestigation of the palaeomagnetism (Salmon et al. 1986) provide evidence for the conclusion that the Msissi result can no longer be thought of as representative of the direction of the Earth's magnetic field during the Devonian. Moreover, results from Devonian sediments from the Gneiguira Formation in Mauritania (Kent et al. 1984), which have often been used as an alternative to the Msissi pole (see Scotese et al. 1985), are located very close to the area of Late Palaeozoic palaeopole positions and overprints southeast of southern Africa and therefore, in the absence of field tests, are more likely to represent a Late Palaeozoic palaeomagnetic overprint, than to be representative for the direction of the geomagnetic field during the Devonian. Nor is the result from the mid-Devonian Bokkeveld Group (BV; Bachtadse et al. 1987) constrained by geological field tests; given the rather complicated rock-magnetic behaviour of these rocks, this result does not fulfil the criteria required for a key pole. Nevertheless in the light of the recently published palaeomagnetic data from Australia (poles CB, HG) it seems now that the position of the Late Devonian palaeopole lay in Central Africa, close to the Bokkeveld pole.

The apparent polar wander path for Gondwana

Although the palaeomagnetic database for the Palaeozoic of Gondwana, and especially for the Silurian to Late Devonian/Early Carboniferous, is still far from complete, an attempt is made to construct an APWP for the southern continents based

Fig. 2. Fitted apparent polar wander path for Gondwana for Ordovician through Permo-Carboniferous times. O, Ordovician; S, Silurian; D, Devonian; C, Carboniferous; P-C, Permo–Carboniferous; l, lower; m, mid; u, upper.

Table 2. *Interpolated palaeopoles for Gondwana from Ordovician through Permo–Carboniferous times (African coordinates)*

Age (Ma)		Lat. [°N]	Long. [°E]
Om	(460)	31	349
Ou	(445)	−15	359
Sl	(435)	−38	8
Su	(422)	−49	23
Dl	(409)	−42	31
Dm	(380)	−6	27
Du	(366)	8	22
Cl	(353)	13	16
Cm	(323)	5	11
Cu	(300)	−27	30
Pl	(286)	−42	61

Notations as in Table 1.

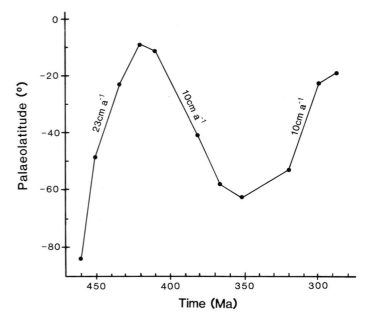

Fig. 3 Palaeolatitude versus time for a given location on the northern margin of Gondwana (Tindouf, 27.83° N, 8.07° W). Also shown are the minimal drift rates during the mid Ordovician to early Silurian, the Early Silurian to Late Devonian and the Late Devonian to the Permo-Carboniferous.

on the data listed in Table 1 and plotted in Fig. 1. The Jupp & Kent (1987) algorithm used to fit a smooth path to ordered data has been chosen since it is provides a 'distortion-free' spherical smoothed spline even if the data set is spread over a large area of the sphere. Where no absolute ages of the magnetization have been given in the original paper, an estimate has been made using the stratigraphic information available and the absolute time scale as proposed by Harland *et al.* (1982). Poles AIR and CB have been used as key poles in order to anchor the path whereas the Ordovician part of the path is controlled by a mean pole (labelled Om in Fig. 2 and Table 2) for the Ordovician. At the younger (Late Palaeozoic) end, the path has been constrained by the Late Carboniferous Dwyka pole (DV; McElhinny & Opdyke 1968) and the combined results for the Permo-Carboniferous (TR; McElhinny 1973). The resulting fitted path is shown in Fig. 2 and the corresponding palaeopole positions for selected time intervals are listed in Table 2. It should be noted that the mean poles of Table 2 that are not based on key poles, as discussed above (e.g. Om, Sl, Du, Cu, Pl), are simply interpolations based on uniform rates of polar shift.

Conclusion and open questions

The use of the revised rotation parameters as proposed by Lawver & Scotese (1987), reduces the differences between the palaeomagnetic data sets for the Ordovician from Africa on one hand and central Australia and eastern Antarctica on the other and therefore in an apparent improvement over previous Gondwana reconstructions. The apparent polar wander path presented in this paper is rather similar to Morel & Irving's (1978) path Y, as well as to the APW paths as proposed by Van der Voo (1988) and Schmidt *et al.* (1987). The Silurian hairpin as shown by Morel & Irving (1978) however was entirely based on eastern Australian data and on the assumption that the Lachlan Fold Belt did not act as an independent terrane during the time under question. The results of this review differ insofar that only data from cratonic Australia, Africa and Antarctica have been used for the APWP construction and the Silurian hairpin is now based on the recent result from the Aïr Massif (Table 1). Thus the palaeomagnetic data at present do not provide any conclusive proof whether eastern Australia was part of stable Australia during the Early to Middle Palaeozoic although general agreement of the East Australian data with the proposed APWP for Gondwana suggests that the parts of Australia were close together if not completely welded. Palaeomagnetic data for the Middle and Late Devonian from western and central Australia (HG, CB) as well as from southern Africa (BV) suggest that the Late Devonian palaeopole was situated in Central Africa.

The validity of the APWP as presented in this paper depends crucially on the Aïr pole (Hargraves *et al.* 1987) and on the correct determination of its age. If our underlying assumptions are correct, then a rather complex palaeogeographical scenario emerges for Gondwana during the Palaeozoic (Fig. 3). Rather rapid northward shift of Gondwana during the Ordovician to Silurian (23 cm a^{-1}), was followed by southward drift until the early Carboniferous (10 cm a^{-1}) and subsequent continuous northward drift with similar minimum drift for the period from the early Carboniferous to the Permo-Carboniferous (see also Fig. 3). As a consequence of these rather high drift rates (especially in the Early to Middle Palaeozoic) and abrupt changes in drift direction in the Late Silurian and the Early Carboniferous (Figs 2 and 3), the Palaeozoic palaeogeographical scenario is controlled by the recurrent collision and divergence of Laurentia and Gondwana and the repeated destruction and formation of intervening oceans (Van der Voo 1988). In this setting, the first collision between Laurentia and Gondwana is achieved during the latest Ordovician to Late Silurian. Divergent motion of Laurentia and Gondwana during Devonian times led to the formation of a substantial ocean between the two supercontinents achieving its maximum width of about 2800 km in the Late Devonian. Convergent motion of Laurentia and Gondwana prevails through the Carboniferous and complete suturing is achieved during the Late Carboniferous.

However open questions still remain to be answered. For example are the extremely high drift rates as implied by the Siluro/Devonian hairpin (Fig. 3) reasonable and can the two rather drastic changes in drift direction during the Silurian and Devonian be related to geological events? How can the obvious discrepancies between the almost coeval Pakhuis/Cedarberg (PAC) and the Aïr palaeopole positions be explained? Is a Central African pole position during the Devonian, rather close to relics of a Late Devonian glaciation in Niger (Caputo & Crowell 1985) but also close to abundant contemporaneous carbonate occurrence in Morocco and Algeria compatible with palaeogeographic models? Providing answers to those questions remains beyond the scope of this review.

This study was supported by a British Petroleum Extramural Research Award to JCB. Our thanks are extended to M. A. Smethurst and T. H. Torsvik for their assistance. This paper benefitted substantially from comments by R. Van der Voo, C. R. Scotese and one anonymous reviewer.

References

BACHTADSE, V. & BRIDEN, J. C. 1989. Palaeomagnetism of the early to mid Ordovician Salala igneous ring complex, Red Sea Hills, Sudan. *Geophysical Journal International* (in press).

——, VAN DER VOO, R. & HAELBICH, I. W. 1987. Palaeomagnetism of the Western Cape Fold Belt, South Africa, and its bearing on the Palaeozoic apparent polar wander path for Gondwana. *Earth and Planetary Science Letters*, **84**, 487–499.

BOWDEN, P., BLACK, R., MARTIN, R. F., IKE, E. C., KINNAIRD, J. A. & BATCHELOR, R. A. 1987. Niger–Nigerian alkaline ring complexes: a classic example of African Phanerozoic anorogenic mid-plate magmatism. *In*: FITTON, J. G. & UPTON, B. G. J. (eds), *Alkaline Igneous Rocks*, Geological Society, London, Special Publication, **30**, 357–379.

BRIDEN, J. C. 1966. Estimates of directions and intensity of the palaeomagnetic field from the Mugga Mugga Porphyry, Australia. *Geophysical Journal of the Royal Astronomical Society*, **11**, 267–278.

—— 1967a. Recurrent continental drift of Gondwanaland. *Nature*, **212**, 246–247.

—— 1967b. Secondary magnetisation of some Palaeozoic rocks from Tasmania. *Papers and Proceedings of the Royal Astronomical Society of Tasmania*, **101**, 43–48.

—— 1968. Palaeomagnetism of the Ntonya ring structure, Malawi. *Geophysical Journal of the Royal Astronomical Society*, **73**, 725–733.

—— & DUFF, B. A. 1981. Pre-Carboniferous paleomagnetism of Europe North of the Alpine Orogenic Belt. *In*: McELHINNY, M. W. & VALENCIO, D. A. (eds) *Paleoreconstruction of the continents*, American Geophysical Union, Geodynamic Series, **2**, 137–149.

BROCK, A. 1967. Palaeomagnetic result from the Hook intrusives, Zambia. *Nature*, **216**, 359–360.

—— 1981. Paleomagnetism of Africa and Madagascar. *In*: McELHINNY, M. W. & VALENCIO, D. A. (eds) *Paleoreconstruction of the Continents*, American Geophysical Union Geodynamic Series, **2**, 65–76.

CAPUTO, M. V. & CROWELL, J. C. 1985. Migration of glacial centers across Gondwana during the Palaeozoic Era. *Bulletin of the Geological Society of America*, **96**, 1020–1036.

CORNER, B. & HENTHORN, D. I. 1978. Results of a palaeomagnetic survey undertaken in the Damara Mobile Belt, South West Africa, with special reference to the magnetisation of the uraniferous pegmatitic granites. *Atomic Energy Board Report* **PEL-260**, Pretoria.

EMBLETON, B. J. J. 1972. The palaeomagnetism of some Palaeozoic sediments from central Australia. *Journal and Proceedings of the Royal Society of New South Wales*, **105**, 86–93.

—— 1977. A Late Devonian palaeomagnetic pole for the Mulga Downs Group, western New South Wales. *Journal and Proceedings of the Royal Society of New South Wales*, **110**, 25–27.

—— 1981. A review of the Paleomagnetism of Australia and Antarctica. *In*: McELHINNY, M. W. & VALENCIO, D. A. (eds) *Paleoreconstructions of the continents*. Geodynamics Series, American Geophysical Union, Geodynamic Series, **2**, 77–91.

GOLEBY, B. R. 1980. Early Palaeozoic paleomagnetism in South East Australia. *Journal of Geomagnetism and Geoelectricity*, **32 SIII**, SIII 11–SIII 21.

HAILWOOD, E. A. 1974. Palaeomagnetism of the Msissi Norite (Morocco) and the Palaeozoic reconstruction of Gondwanaland. *Earth and Planetary Science Letters*, **23**, 376–386.

HARGRAVES, R. B., DAWSON, E. M. & VAN HOUTEN, F. B. 1987. Palaeomagnetism and age of mid Palaeozoic ring complexes in Niger, West Africa, and tectonic implications. *Geophysical Journal of the Royal Astronomical Society*, **90**, 705–729.

HARLAND, W. B., COX, A. V., LLEWELLYN, P. G., PICKTON, C. A. G., SMITH, A. G. & WALTERS, R. 1982. *A geologic time scale*. Cambridge University Press, Cambridge.

HURLEY, N. F. & VAN DER VOO, R. 1987. Paleomagnetism of Upper Devonian reefal limestones, Canning Basin, Western Australia. *Bulletin of the Geological Society of America*, **98**, 138–146.

JUPP, P. E. & KENT, J. T. 1987. Fitting smooth paths to spherical data. *Applied Statistics*, **36**, 34–36.

KENT, D. V., DIA, O. & SOUGY, J. M. A. 1984. Paleomagnetism of Lower–Middle Devonian and Upper Proterozoic–Cambrian(?) rocks from Mejeria (Mauritania, West Africa), *In*: VAN DER VOO, R. & SCOTESE, C. R. (eds) *Plate reconstructions from Palaeozoic Paleomagnetism*. American Geophysical Union, Geodynamic Series, **12**, 99–115.

KRÖNER, A., McWILLIAMS, M. O., GERMS, G. J. B., REID, A. B. & SCHALK, K. E. L. 1980. Paleomagnetism of late Precambrian to early Palaeozoic Mixtite bearing formations in Namibia (South West Africa): The Nama Group and the Blaubeker Formation. *American Journal of Earth Sciences*, **280**, 942–968.

LAWVER, L. A. & SCOTESE, C. R. 1987. A revised reconstruction of Gondwanaland. *In*: McKENZIE, G. D. (ed.) *Gondwana Six: Structure, Tectonics and Geophysics*. American Geophysical Union, Monograph, **40**, 17–23.

LI, Z. X., SCHMIDT, P. W. & EMBLETON, B. J. J. 1988. Paleomagnetism of the Hervey Group, central New South Wales, and its tectonic implications. *Tectonics*, **7**, 351–367.

LUCK, G. R. 1972. Palaeomagnetic results from Palaeozoic sediments of northern Australia. *Geophysical Journal of the Royal Astronomical Society*, **28**, 475–487.

MANZONI, M. & NANNI, T. 1977. Palaeomagnetism of Ordovician lamprophyres from Taylor Valley, Victoria Land, Antarctica. *Pure and Applied Geophysics*, **115**, 961–977.

McELHINNY, M. W. 1973. *Paleomagnetism and Plate Tectonics*. Cambridge University Press, Cambridge.

—— & EMBLETON, B. J. J. 1974. Australian palaeomagnetism and the Phanerozoic plate tectonics of eastern Gondwana. *Tectonophysics*, **22**, 1–29.

—— & —— 1976. Precambrian and early Palaeozoic Palaeomagnetism in Australia. *Philosophical Transactions of the Royal Society, London*, **A280**, 417–432.

—— & OPDYKE, N. D. 1968. The palaeomagnetism of some Carboniferous glacial varves from Central Africa. *Journal of Geophysical Research*, **73**, 689–696.

——, BRIDEN, J. C., JONES, D. L. & BROCK, A. B. 1968. Geological and geophysical implications of palaeomagnetic results from Africa. *Reviews of Geophysics*, **6**, 201–238.

MERRILL, A. T. & McELHINNY, M. W. 1983. *The Earth's magnetic field*, Academic Press, London.

MOREL, P. & IRVING, E. 1978. Tentative paleocontinental maps for the early Phanerozoic and Proterozoic. *Journal of Geology*, **86**, 535–561.

SALMON, E., MONTIGNY, R., EDEL, J. B., PIQUE, A., THUIZAT, R. & WESTPHAL, M. 1986. The Msissi Norite revisited: K/Ar dating, Petrography and Paleomagnetism. *Geophysical Research Letters*, **13**, 741–743.

SCOTESE, C. R., VAN DER VOO, R. & BARRET, S. F. 1985. Silurian and Devonian base maps. *Philosophical Transactions of the Royal Society*, **B309**, 57–77.

SCHMIDT, P. W. & EMBLETON, B. J. J. 1987. A critique of paleomagnetic results from Australian Palaeozoic fold belts and displaced terranes. *In*: LEITCH, E. C. & SCHEIBNER, E. (eds), *Terrane Accretion and Orogenic Belts*. American Geophysical Union, Geodynamic Series **19**, 21–30.

—— & MORRIS, W. A. 1977. An alternative view of the Gondwana Palaeozoic apparent polar wander path. *Canadian Journal of Earth Sciences*, **14**, 2674–2678.

——, EMBLETON, B. J. J., CUDAHY, T. J. & POWELL, C. McA. 1986. Prefolding and premagakinking magnetizations from the Devonian Comerong volcanics, New South Wales, Australia, and their bearing on the Gondwana Pole Path. *Tectonics*, **5**, 135–150.

——, —— & PALMER, H. C. 1987. Pre- and postfolding magnetizations from the early Devonian Snowy River Volcanics and Buchan Caves Limestone, Victoria, *Geophysical Journal of the Royal Astronomical Society*, **91**, 155–177.

SMITH, A. G. & HALLAM, A. 1970. The fit of the southern continents. *Nature*, **225**, 139–144.

STEIGER, R. H. & JÄGER, E. 1977. Subcommission on Geochronology: convention on the use of decay constants in geo- and cosmochronology. *Earth and Planetary Science Letters*, **36**, 359–362.

VAN DER VOO, R. 1988. Paleogeography of North America, Gondwana, and intervening displaced terranes: Comparisons of paleomagnetism with paleoclimatology and biogeographical patterns, *Bulletin of the Geological Society of America*, **100**, 311–324.

VILAS, J. F. A. 1981. Palaeomagnetism of South American rocks and the dynamic processes related with the fragmentation of western Gondwana. *In*: MCELHINNY, M. W. & VALENCIO, D. A. (eds), *Paleoreconstruction of the continents*. American Geophysical Union, Geodynamic Series, **2**, 106–114.

ZIJDERVELD, J. D. A. 1968. Natural remanent magnetisations of some intrusive rocks from the Sør Rondane Mountains, Queen Maud Land, Antarctica. *Geophysical Journal of the Royal Astronomical Society*, **73**, 3773–3785.

Palaeozoic palaeogeography from palaeomagnetism of the Atlantic-bordering continents

DENNIS V. KENT

Lamont-Doherty Geological Observatory and Department of Geological Sciences, Columbia University, Palisades, NY 10964, USA

ROBERT VAN DER VOO

Department of Geological Sciences, University of Michigan, Ann Arbor, MI 48109, USA

Abstract: Revised palaeogeographic reconstructions of the Atlantic-bordering continents and intervening terranes are presented for the Siluro-Devonian boundary, Early Devonian, and Late Devonian, based on incorporation of new palaeomagnetic results that have become available from Laurentia and Gondwana. The key features of the palaeogeographic model are the transpressive collision between the eastern margin of Laurentia and the northwest South American margin of Gondwana in the Siluro-Devonian, and the subsequent Devonian retreat of the north African margin and the development of a wide ocean between Africa and Europe by the Late Devonian. Although rather complex and involving rapid motions especially of Gondwana, the revised Devonian plate tectonic evolution as indicated by palaeomagnetism is not inconsistent with biogeographic and palaeoclimatological evidence.

Background and methodology

The palaeomagnetic, biogeographic, and palaeoclimatological evidence for Palaeozoic palaeogeography of the Atlantic-bordering continents and intervening displaced terranes has been recently reviewed by Van der Voo (1988). However, the continued generation of relevant palaeomagnetic data requires further and substantive modification in palaeogeographic reconstructions for the Siluro-Devonian to Late Devonian time interval. In particular, new results from Laurentia (Miller & Kent 1988; Stearns et al. 1989) place this continental block in higher southern palaeolatitudes in the Siluro-Devonian than previous palaeomagnetic data indicated, whereas additional palaeomagnetic evidence from Australia (Li et al. 1988) supports a central African location for the Late Devonian palaeopole for Gondwana and the retreat of the north African margin to mid-southern latitudes in the Late Devonian.

We present an updated list of selected Ordovician to Devonian palaeomagnetic poles for Laurentia, Gondwana, and the intervening displaced terranes; key Devonian poles from eastern Australia, Baltica, and from Iran are also included (Table 1). Each palaeopole is assigned a quality parameter using the criteria of Van der Voo (1988, 1989). Mean palaeopoles (Table 2), based on what we regard as reliable data, are used to make palaeogeographic reconstructions for three critical time intervals: Siluro-Devonian boundary, Early Devonian, and Late Devonian, providing a basis for comparison to biogeographic and palaeoclimatological evidence. The previous published palaeogeographic maps for Middle/Late Ordovician and the Middle Silurian (Van der Voo 1988) remain unchanged.

Apparent polar wander paths

Our interpretation of the apparent polar wander (APW) paths for Laurentia and Gondwana, based on the palaeopoles listed in Table 1, implies considerable movement of the respective landmasses over the Ordovician to Devonian time interval.

A significant new feature in the APW path for Laurentia is a broad loop (counterclockwise in South Poles) from the Middle and Late Ordovician to the Late Devonian (Fig. 1). There is now a group of palaeopoles that falls at low latitude (South Pole closest to Laurentia) in the Siluro-Devonian, and it is these (AN, PE, and KS; Table 1) that largely motivate our reappraisal of the Middle Palaeozoic palaeogeography of Laurentia. It is noteworthy that similar palaeopoles from Scotland, when fitted against Laurentia in the reconstruction of Bullard et al. (1965), have long been known for the low latitude location. The new and reinterpreted poles (AN, PE, and KS) show convincingly for the first time that Scotland was an integral part of Laurentia in the Middle Palaeozoic (and probably since the Early Proterozoic).

The APW path for Gondwana implies large and surprisingly rapid motions of this large landmass: from the well-documented palaeoposition of the South Pole in north Africa in the Ordovician, the APW path shows a swing to a location near southern Chile during the Silurian, and then loops back to central Africa by the Late Devonian (Fig. 2). The essential features of this APW path are documented by high quality albeit few results from cratonic Gondwana, especially from the Silurian Air ring dike complexes of Niger [pole AR] and the Late Devonian Canning Basin sediments of western Australia [pole CB]. There is however supporting evidence for the Early and Late Devonian positions of the Gondwana APW path from palaeopoles from eastern Australia and from Iran (Fig. 2; Table 1), suspect terranes which are thought to have been adjacent to the main body of Gondwana by the Early Devonian.

Palaeogeographic reconstructions

Ordovician setting

Ordovician palaeogeographic models for Laurentia and Gondwana remain relatively unchanged in the last decade and we use the recently published reconstruction of Van der Voo (1988) for this time interval (Fig. 3). The key elements of the Ordovician model are Laurentia in equatorial palaeolatitudes and the north facing margin of Gondwana in high southern palaeolatitudes, with the South Pole in northwest Africa. Ordovician palaeomagnetic data for Armorica, Avalon and at least parts of the Appalachian Piedmont indicate that these intervening terranes occupied high southern palaeolatitude, peri-Gondwanide locations. Southern Britain was also in intermediate to high southern palaeolatitudes according to some interpretations of the Ordovician palaeomagnetic data (McCabe & Channell 1988) but was adjacent to northern Europe in alternative interpretations (Briden et al. 1984). New Ordovician palaeomagnetic data from the Central Mobile Belt of Newfoundland suggest an intra-oceanic island setting for some of the tectonic elements now located between the Laurentia continental margin and Avalon (Johnson et al. 1988).

Table 1. *Selected palaeomagnetic poles for the Ordovician through Devonian periods for Laurentia, Gondwana and intervening displaced terranes*

Rock Unit, Location	Symbol	Age	Pole Position Lat, Long	k	α95	Reliability 1 2 3 4 5 6 7	Q	Reference
North American Craton and Cratonic Margins (south poles)								
Catskill Red beds, south, PA	CS	Du	26S, 304E	16	16	x x x x	4	Miller & Kent (1986)
Catskill Red beds, north, PA	CN	Du	33S, 296E*	165	7	x x x x x	5	Miller & Kent (1986)
Peel Sound MDL pole, NWT, Canada	PS	Dl?	25S, 279E	66	9	x x x x	4	Dankers (1982)
Peel Sound E pole, NWT, Canada	PE	Su/Dl	1S, 271E	18	10	x x x x	4	Dankers (1982)
Andreas Red beds, PA	AN	Su/Dl	11S, 305E*	13	9	x x x x	4	Miller & Kent (1988)
Kap Stanton, N. Greenland	KS	S/D?	1S, 292E	16	11	x x x x	4	Stearns et al. (1989)
Bloomsburg Fm., south, PA-VA		Su	31S, 297E	36	9	x x x x x	5	Kent (1988)
Bloomsburg Fm., north, PA		Su	29S, 309E*	57	7	x x x x	4	Kent (1988)
Rose Hill Fm., PA-VA		Sm	19S, 309E	18	6	x x x x x x	6	French & Van der Voo (1979)
Wabash Reef Ls., IN		Sm	17S, 305E	74	5	x x x x x x x	7	McCabe et al. (1985)
Ringgold Gap sediments, GA		Sl/Ou	28S, 322E	62	7	x x x x x	5	Morrison & Ellwood (1986)
Cordova secondary, Ontario		446	31S, 282E	18	11	x x x x x	5	Dunlop & Stirling (1985)
Juniata Fm., north, PA		Ou	17S, 306E*	29	13	x x x x x	5	Miller & Kent (1989)
Juniata Fm., south, PA		Ou	19S, 308E	14	21	x x x x	4	Miller & Kent (1989)
Juniata Fm., south, PA-VA		Ou	32S, 294E	53	5	x x x x x	5	Van der Voo & French (1977)
Steel Mntn. Secondary, W. Newf.		451	23S, 319E	22	13	x x x x	4	Murthy & Rao (1976)
Chapman Ridge Fm., TN		Om	27S, 292E	38	15	x x x	3	Watts & Van der Voo (1979)
Moccasin-Bays Fms., TN		Om	33S, 327E	135	6	x x x x x x	6	Watts & Van der Voo (1979)
St George Fm., Newfoundland		Ol	26S, 306E	202	7	x x x x	4	Beales et al. (1974)
Oneota Dolomite, IA, MN, WI		Ol	10S, 346E	18	12	x x x x x x	6	Jackson & Van der Voo (1985)
Gondwana Craton (African coordinates; south poles)								
Canning Basin Reef Ls., Austr.	CB	Du	8N, 23E	62	8	x x x x x x	6	Hurley & Van der Voo (1987)
Bokkeveld Grp., S. Africa		Dm	10N, 15E	33	7	x x x	3	Bachtadse et al. (1987)
Mereenie Ss., Central Austr.		D?	15N, 25E	15	11	x x x x	4	Embleton (1972)
Air Ring Complexes, Niger		435	43S, 9E	50	6	x x x x x	5	Hargraves et al. (1987)
Pakhuis/Cedarberg, S. Africa		Sl/Ou	25N, 343E	9	18	x x x x x	5	Bachtadse et al. (1987)
Stairway Ss., Central Austr.		Om	43N, 31E	25	9	x x x x	4	Embleton (1972)
Taylor Valley dikes, Antarct.		470	36N, 15E	75	11	x x x x	4	Manzoni & Nanni (1977)
Sør Rondane Intr., Antarctica		480	12N, 14E	—	5	x x x	3	Zijderveld (1968)
Jinduckin Fm., N. Australia		Ol	31N, 1E	10	11	x x x x x x	6	Luck (1972)
Graafwater Fm., S. Africa		Ol	28N, 14E	25	9	x x x x x x	6	Bachtadse et al. (1987)
Ntonya Ring Str., Malawi		474	28N, 345E	1054	2	x x x x x	5	Briden (1968)
Hook Intrusives, Zambia		500	14N, 336E	13	36	x x x	3	Brock (1967)
Key Devonian poles from eastern Australia (African coordinates, south poles)								
Hervey Group, N.S.W.		Du/Cl	2N, 14E	14	15	x x x x x	5	Li et al. (1988)
Comerong Volcanics, N.S.W.		Dm/Du	26S, 3E	42	7	x x x x x x	6	Schmidt et al. (1986)
Snowy River Volcs., Victoria	SR	Dl	48S, 14E	26	10	x x x x x x	6	Schmidt et al. (1987)
Devonian poles from Iran (south poles)								
Geirud Basalts	GB	Du/Cl	0S, 32E	380	4	x x x x x	5	Wensink et al. (1978)
Central Iran Red beds	IR	Dl	51S, 16E	18	10	x x x x x x	6	Wensink (1983)
Central Mobile Belt of Newfoundland, New Brunswick, Maine (south poles)								
Compton Fm., Quebec	CF	Du	28S, 257E	30	7	x x x x x x	6	Seguin et al. (1982)
Traveler Felsite, ME	TF	Dl	29S, 262E	16	11	x x x x x	5	Spariosu & Kent (1983)
Dalhousie Volcanics, New Br.	DV	Su/Dl	1S, 268E	236	7	x x x	3	Seguin & Gahé (1985)
Botwood Volc., C. Newfoundland		Sm	13S, 305E	177	9	x x x x	4	Lapointe (1979)
Wigwam red beds, C. Newf.		Sm	25S, 280E	9	14	x x x x	4	Lapointe (1979)
Avalon Basement Terranes of Nova Scotia, New Brunswick, Boston Basin (south poles?)								
Metamorphics, Massachusetts	MM	Du	23S, 306E	11	10	x x x	3	Schutts et al. (1976)
Eastport Fm., ME	EF	Dl	24S, 294E	19	9	x x x x x	5	Kent & Opdyke (1980)
Hershey Fm., ME	HF	Dl	20S, 309E	36	6	x x x x x	5	Kent & Opdyke (1980)
Mascarene Fm., C, New Brunsw.		Su?	28S, 265E	666	4	x x x x	4	Roy & Anderson (1981)
Mascarene Fm, A-B, New Brunsw.	MA	Su	5N, 267E	33	7	x x x x x	5	Roy & Anderson (1981)
Dunn Point Fm., Nova Scotia		Sl/Ou	2N, 316E	79	4	x x x x x x	6	Van der Voo & Johnson (1985)
Nahant Gabbro, Massachusetts		O	34N, 282E	—	—	x x x x	4	Weisse et al. (1985)
Delaware Piedmont (south poles?)								
Metamorphic rocks, DE		Ou	48N, 288E	15	16	x x x x	4	Brown & Van der Voo (1983)
Arden Pluton, DE		Ou	16N, 303E	32	14	x x x x	4	Rao & Van der Voo (1980)
Baltic shield								
Ringerike Old Red, Norway	RO	Su/Dl	19S, 344E	15	9	x x x x x x	6	Douglass (1988)
Southern Ireland, Southern England and Wales (south poles)								
Old Red Sandst., Wales	OR	Su/Dl	3N, 298E	4	13	x x x x x	5	Chamalaun & Creer (1964)
N. Wales Intrusives		O	68N, 288E	17	5	x x x x	4	Thomas & Briden (1976)

(Table 1. continued)

Rock Unit, Location	Symbol	Age	Pole Position Lat, Long	k	α95	Reliability 1 2 3 4 5 6 7	Q	Reference
Builth Inlier Intr., Wales		Ou	2N, 2E	37	11	x x x x x	5	Piper & Briden (1973)
Tramore Volcanics, Ireland		Om	11N, 18E	—	—	x x x x x x	6	Deutsch (1984)
Builth Volcanics, Wales		Ol	3S, 355E	22	4	x x x x x x	6	Briden & Mullan (1984)
Central and Southern Europe (south poles)								
Sediments, Volcanics, Germany	SV	Du	30S, 9E	—	—	x x x x x x	6	Bachtadse et al. (1983)
San Pedro Red beds, Spain*	SP	Dl	22S, 319E*	30	10	x x x x x x	6	Perroud & Bonhommet (1984)
Cabo de Penas, Spain		Ou	30N, 330E	50	6	x x x x	4	Perroud & Bonhommet (1981)
Bucaco, Portugal*		Ou	25N, 335E*	83	6	x x x x	4	Perroud & Bonhommet (1981)
Thouars Massif, France		Ou	34N, 5E	27	5	x x x x x	5	Perroud & Van der Voo (1985)
M. de Chateaupanne Fm., France		Ou	34N, 343E	65	6	x x x x x	5	Perroud et al. (1986)
Erquy Volcanics, France		479	35N, 344E	47	11	x x x	3	Duff (1979)
Arenig Sandst., Poland		Ol	22N, 9E	—	10	x x x x	4	Lewandowski (1987)

Explanation: Symbols denote those poles used in Table 2 for the determination of mean poles and the positioning of the tectonic elements in the figures; age abbreviations are: C, Carboniferous; D, Devonian; S, Silurian; O, Ordovician; u, upper; m, middle; l, lower. k and α95 are the statistical parameters associated with the means. Q, quality factor (maximum 7), comprised of the following reliability criteria: 1, well-determined age for the rocks and the age of magnetization is not demonstrably different from the rock age; 2, sufficient number of samples (greater than 25) and high enough precision (k greater than 10, α95 lower than 16°); 3, demagnetization published in sufficient detail; 4, positive field (fold-, conglomerate-, contact-) tests; 5, tectonic coherence with block or craton and sufficient structural control; 6, presence of antipodal reversals; 7, lack of similarity with poles for younger times (see Van der Voo 1989).

* Poles recalculated after restoration of rotations: for the Pennsylvania salient (Catskill north, Andreas, Juniata north, and Bloomsburg north poles) this involves a 23° clockwise rotation of the northern limb of the salient during thrusting (Miller & Kent 1988; Kent 1988); for Spain and Portugal (San Pedro and Bucaco poles) this involves a 80° counterclockwise rotation during Hercynian deformation of the Ibero–Armorican arc (Perroud & Bonhommet 1981).

Poles have been selected for quality (Q); the poles for the southern British Isles have been taken from Briden & Duff (1981); poles with their quality classification of C or D have not been used.

Table 2. *Mean palaeopoles used to position the major continents and some terranes in figures*

Continental Element	Su/Dl Mean Pole	Dl Mean Pole	Du Mean Pole
Laurentia	5S, 289E (KS,AN,PE)	25S, 279E (PS)	29S, 300E (CS,CN)
Gondwana	45S, 11E #	48S, 14E (SR)	8N, 23E (CB)
Baltica	19S, 344E (RO)	—, —	—, —
Scotland	5S, 317E ##	—, —	—, —
Central Mobile Belt	1S, 268E (DV)	29S, 262E (TF)	28S, 257E (CF)
Avalon	5N, 267E (MA)	22S, 301E (HF,EF)	23S, 306E (MM)
S. Europe (Armorica)	—, —	22S, 319E (SP)	30S, 9E (SV)
S. Britain	3N, 298E (OR)	—, —	—, —
Iran	—, —	51S, 16E (IR)	0S, 32E (GB)

Explanation: Period abbreviations are the same as in Table 1. Means have been calculated from the poles indicated in parentheses and listed by their symbols in Table 1; if no poles are available, then sometimes a pole was calculated by interpolation, as indicated by #. The mean Su/Dl pole for Scotland (##) has been taken from Torsvik (1988) and is based on the mean of his poles CM, LW, SL, MV, LP, GH, CV, MMB, MMI, MR, BS, SA, and KM2.

Poles are given in the coordinates of the individual blocks (African coordinates for Gondwana, based on the reconstruction parameters of Smith & Hallam 1970); in the reconstructions, Iran is left in the same position with respect to Africa as it occupies today, so the Iranian pole coordinates can be directly compared to those for Gondwana.

Siluro-Devonian boundary

The wide palaeolatitudinal separation between Laurentia and Gondwana in the Ordovician is no longer apparent by the Siluro-Devonian boundary (Fig. 4).

Laurentia occupies a more southerly position in the Siluro-Devonian than previously thought, according to recently available palaeomagnetic results. Pre-folding magnetizations from the latest Silurian to earliest Devonian Andreas redbeds from the Appalachians may be oroclinally rotated but clearly indicate a southerly palaeolatitude of about 36°S (Miller & Kent 1988). This result (AN) supports the reality of the PE pole from the Early Devonian Peel Sound Formation of the Canadian Arctic, and suggests that the better described and more usually quoted (PS) pole obtained in the Peel Sound study (Dankers 1982) refers to a magnetization acquired somewhat later in the Early Devonian. Indirect but strong confirmation of the placement of Laurentia almost entirely in the southern hemisphere at around Siluro-Devonian time also comes from secondary magnetizations from the Cambro-Ordovician Kap Stanton unit of north Greenland (KS; Stearns et al. 1989). Since the Peel Sound PE pole, the Andreas AN pole, as well as the Kap Stanton secondary magnetization KS pole do not resemble any younger (or for that matter, older Palaeozoic) palaeopole positions documented for Laurentia, they are thought to provide a reliable mean Siluro-Devonian pole with respect to this continent (Table 2).

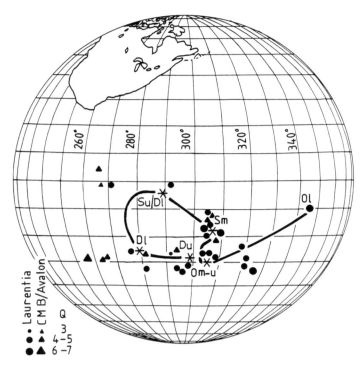

Fig. 1. Ordovician through Late Devonian apparent polar wander path for cratonic Laurentia (North America and Greenland; dots). Palaeopoles for Middle Silurian and Devonian time are added as triangles for the Central Mobile Belt (CMB) and Avalon terranes of the northern Appalachians. The symbols for individual palaeopoles are plotted with different sizes according to the quality factor (Q, as listed in Table 1); the mean palaeopoles (asterisks) used to make the palaeogeographic maps of Figs 4–6 are listed in Table 2. O, Ordovician; S, Silurian; D, Devonian; l, lower; m, middle: u, upper. Note the agreement between Laurentian and CMB/Avalon palaeopoles for Middle Silurian through Devonian time, indicating that these areas were together during that interval.

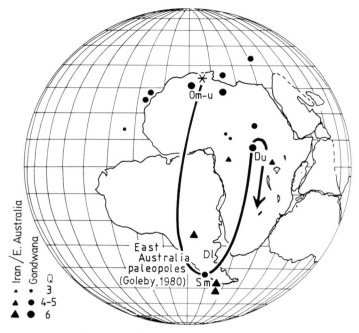

Fig. 2. Ordovician through Late Devonian apparent polar wander path for cratonic Gondwana (dots). Palaeopoles for Devonian time are added as triangles for Iranian and eastern Australian terranes. Symbols for palaeopoles (from Table 1), and abbreviations are explained in Fig. 1. Note the agreement between cratonic Gondwana, Iran and eastern Australia palaeopoles for Devonian time, indicating that these areas were together during that interval.

At about the same time, the north African margin of Gondwana drifted northward into tropical palaeolatitudes, as indicated by the Silurian palaeopole from the Air ring complexes of Niger (Hargraves *et al.* 1987) with supporting evidence from the Early Devonian Snowy River palaeopole from eastern Australia (Schmidt et al. 1987). We interpolate between the Air (AR and Snowy River (SR) poles to derive a mean Siluro-Devonian pole representative for Gondwana (Table 2).

Positioning Laurentia and Gondwana according to the poles discussed above leads to an assembly where northwest South America is opposite the eastern margin of Laurentia at around the Siluro-Devonian boundary (Fig. 4). Siluro-Devonian palaeopoles from the intervening terranes, e.g., Avalon (MA), southern Britain (OR), and the Central Mobile Belt of the northern Appalachians (DV), are consistent with this fit, suggesting that these terranes had docked during the late Caledonian and Acadian orogenies in the Siluro-Devonian. Baltica can also be regarded as part of the supercontinent assembly, judging by the recent palaeomagnetic results with dual polarity and a positive fold test from the Late Silurian (to earliest Devonian?) Ringerike Old Red Sandstone of southern Norway (Douglass 1988).

Early Devonian

The Siluro-Devonian supercontinent assembly does not appear to have been long lived as some of the major continental elements drifted apart over the Devonian. Initial stages of this disruption apparently occurred already in the Early Devonian (Fig. 5): palaeolatitudes for Laurentia based on the PS pole are more northerly by 10° to 20° compared to the Siluro-Devonian. The similarity of palaeolatitudes that can be inferred from Early Devonian palaeopoles from the intervening terranes like the Central Mobile Belt (TF), Avalon (HF, EF), and Armorica (SP), to those predicted from Laurentia, suggest that these terranes stayed fixed with respect to Laurentia and henceforth shared a common drift history.

Gondwana in the meantime did not move far (SR pole) compared to its Siluro-Devonian position. Although the Snowy River pole (SR) comes from the Tasman orogen of eastern Australia, the good agreement of Early Devonian palaeomagnetic results from Iran (IR; Wensink 1983) with the palaeolatitudinal framework for Gondwana suggested by the SR pole provides supportive evidence for the palaeogeography shown here (Fig. 5).

Late Devonian

Later in the Devonian, the action shifts to Gondwana. Early Devonian (PS) and Late Devonian (CN, CS) palaeopoles for Laurentia are not very different, hence Laurentia experienced little palaeolatitudinal change over most of the Devonian. Late Devonian palaeolatitudes from the Central Mobile Belt (CF), Avalon (MM) and Armorica (SV) are also consistent with those of a relatively stationary Laurentia. This is not however the case for Gondwana. The palaeomagnetic results from the Late Devonian of western Australia yields a pole (CB) in central Africa (Hurley & Van der Voo 1987). Supported by new palaeomagnetic data from the Late Devonian (to earliest Carboniferous) Hervey Group of southeastern Australia (Li *et al.* 1988), the central African location for the Late Devonian palaeopole for Gondwana implies a retreat of the north African margin, from subtropical palaeolatitudes in the Early Devonian (SR) to about 50°S palaeolatitude in the Late Devonian (CB) as shown in Fig. 6. We note that a Late Devonian palaeopole from Iran (GB) fits well with the position of Gondwana as indicated by the results

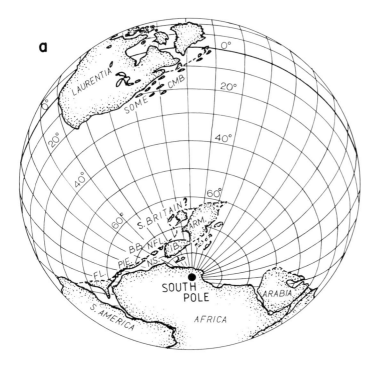

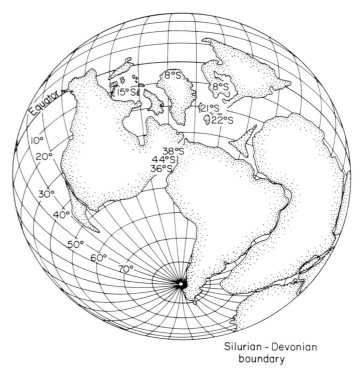

Fig. 4. Palaeogeographic reconstruction for the Silurian–Devonian boundary; for explanation see Fig. 3. The continental elements are positioned according to the mean poles of Table 2.

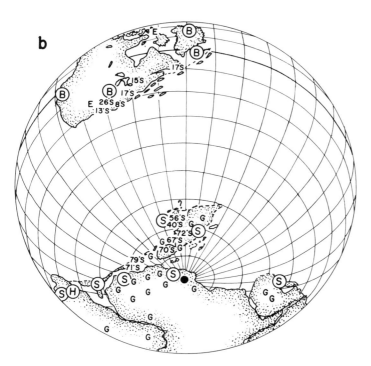

Fig. 3. Middle to Late Ordovician palaeogeographic reconstruction (from Van der Voo 1988). (a). Map showing the names of the separate tectonic elements: CMB, Central Mobile Belt; ARM, Central Europe, and IB, Iberian Peninsula, together constituting Armorica; NFL, eastern Newfoundland; NS, Nova Scotia; BB, Boston Basin; PIE, Avalonian part of the Piedmont Province and FL, northern Florida. (b) The same map as in (a) but with palaeomagnetic palaeolatitudes (numbers), biogeographic (encircled letters), and palaeoclimatological indicators (letters). B, Bathyurid fauna; S, *Selenopeltis* fauna; H, Hungaiid–Calymenid trilobite fauna. G, glacial relicts; E, evaporite occurrences.

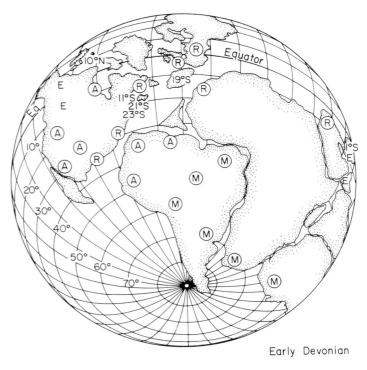

Fig. 5. Palaeogeographic reconstruction for the Early Devonian. For explanation see Fig. 3. A, Eastern Americas realm; R, Rhenish–Bohemian subprovince of the Old World realm; M, Malvinokaffric realm of brachiopod provincialities. The continental elements are positioned according to the mean poles of Table 2.

from Australia, just as was the case for the Early Devonian. Although it is possible that the palaeopole for Gondwana was in central Africa into the Early Carboniferous (Li *et al.* 1988), the main effect of this would be to reduce the time available (hence, increase the drift rate) for Gondwana to move into a Pangaea configuration by about the end Visean.

Discussion

The key elements of the palaeogeographic scenario outlined above involve the transpressive collision of the eastern margin of Laurentia with the northwest South American margin of Gondwana by about the time of the Siluro-Devonian boundary, followed by the opening of a wide ocean between Africa and Europe by the Late Devonian. The suggested supercontinent assembly in the Siluro-Devonian is reminiscent of the model of McKerrow & Ziegler (1972) based primarily on tectonostratigraphic grounds; it is also called for by the more southerly position of Laurentia as indicated by new palaeomagnetic data (Miller & Kent 1988; Stearns *et al.* 1989) which now agree well with the distribution of lithic palaeoclimatic indicators (Heckel & Witzke 1979). A more conventional Pangaea-like supercontinental assembly for the Siluro-Devonian (e.g., Van der Voo 1988), with northwest Africa against eastern Laurentia, cannot be easily accommodated because Laurentia is now seen to occupy a palaeolatitudinal range that overlaps with that of northern South America and Africa.

Palaeomagnetism of course does not provide constraints on the palaeolongitudinal position of the continents and considerable east–west relative motion between the continents may have taken place. Nevertheless, we have chosen to minimize relative motions, hence the longitudinal separations, between Laurentia and Gondwana over the Devonian. This allows the northwest South American margin to remain in close proximity to Laurentia even while a wide ocean develops between Africa and Europe in the Late Devonian (Fig. 6). We believe that this model is still compatible with some important biogeographic constraints; for example, the dispersal of early land vertebrates between Gondwana and the northern continents (Young 1990) may have occurred through the South America–Laurentia land connection over the Devonian.

We wish to thank C. R. Scotese and W. S. McKerrow for organizing and inviting us to participate in the Symposium on Palaeozoic Biogeography and Palaeogeography at Oxford. This research was supported by the National Science Foundation, Earth Sciences Division grants EAR86–12469 (RVdV) and EAR88–03814 (DVK). Lamont–Doherty Geological Observatory Contribution #4431.

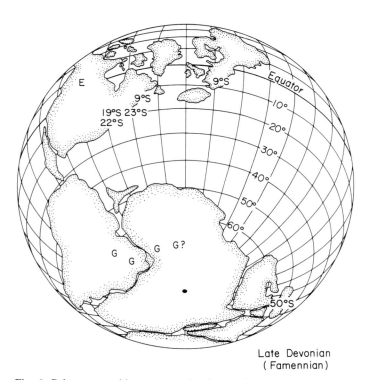

Fig. 6. Palaeogeographic reconstruction for the Late Devonian; for explanation see Fig. 3. The continental elements are positioned according to the mean poles of Table 2.

References

BACHTADSE, V., HELLER, F. & KROENER, A. 1983. Paleomagnetic investigations in the Hercynian mountain belt of western Europe. *Tectonophysics*, **91**, 285–299.

BACHTADSE, V., VAN DER VOO, R. & HAELBICH, I. W. 1987. Paleomagnetism of the western Cape Fold Belt, South Africa, and its bearing on the Paleozoic apparent polar wander path for Gondwana. *Earth and Planetary Science Letters*, **84**, 487–499.

BEALES, F. W., CARRACEDO, J. C. & STRANGWAY, D. W. 1974. Paleomagnetism and the origin of Mississippi-Valley type ore deposits. *Canadian Journal of Earth Sciences*, **11**, 211–223.

BRIDEN, J. C. 1968. Paleomagnetism of the Ntonya Ring Structure, Malawi. *Journal of Geophysical Research*, **73**, 725–733.

—— & DUFF, B. A. 1981. Pre-Carboniferous paleomagnetism of Europe north of the Alpine orogenic belt, *In*: MCELHINNY, M. W. & VALENCIO, D. A. (eds) *Paleoreconstruction of the Continents*. American Geological Union and Geological Society of America Geodynamics Series, **2**, 137–149.

—— & MULLAN, A. J. 1984. Superimposed Recent, Permo–Carboniferous and Ordovician paleomagnetic remanence in the Builth Volcanic Series, Wales. *Earth and Planetary Science Letters*, **69**, 413–421.

——, TURNELL, H. B. & WATTS, D. R. 1984. British paleomagnetism, Iapetus Ocean, and the Great Glen Fault. *Geology*, **12**, 428–431.

BROCK, A. 1967. Paleomagnetic result from the Hook intrusives, Zambia. *Nature*, **216**, 359–360.

BROWN, P. M. & VAN DER VOO, R. 1983. A paleomagnetic study of Piedmont metamorphic rocks in northern Delaware. *Geological Society of America Bulletin*, **94**, 815–822.

BULLARD, E. C., EVERETT, J. E. & SMITH, A. G. 1965. A symposium on continental drift — IV. The fit of the continents around the Atlantic. *Philosophical Transactions of the Royal Society*, **258**, 41–51.

CHAMALAUN, F. H. & CREER, K. M. 1964. Thermal demagnetization studies of the Old Red Sandstone of the Anglo-Welsh cuvette. *Journal of Geophysical Research*, **69**, 1607–1616.

DANKERS, P. 1982. Implications of Early Devonian poles from the Canadian Arctic archipelago for the North American apparent polar wander path: *Canadian Journal of Earth Sciences*, **19**, 1802–1809.

DEUTSCH, E. R. 1984. Mid-Ordovician paleomagnetism and the Proto-Atlantic Ocean in Ireland, *In*: VAN DER VOO, R., SCOTESE, C. R. & BONHOMMET, N. (eds) *Plate Reconstructions from Paleozoic Paleomagnetism*. American Geophysical Union Geodynamics Series, **12**, 116–119.

DOUGLASS, D. N. 1988. Paleomagnetism of Ringerike Old Red Sandstone and related rocks, southern Norway: Implications for preCarboniferous separation of Baltica and British terranes. *Tectonophysics*, **148**, 11–27.

DUFF, B. A. 1979. The paleomagnetism of Cambro–Ordovician redbeds, the Erquy Spilite Series, and the Trégastel–Ploumanac'h granite complex, Armorican Massif (France and the Channel Islands). *Geophysical Journal of the Royal Astronomical Society*, **59**, 345–365.

DUNLOP, D. J. & STIRLING, J. M. 1985. Post-tectonic magnetizations from the Cordova gabbro, Ontario and Palaeozoic reactivation in the Grenville province. *Geophysical Journal of the Royal Astronomical Society*, **91**, 521–550.

EMBLETON, B. J. J. 1972. The palaeomagnetism of some Palaeozoic sediments from central Australia. *Journal and Proceedings of the Royal Society of New South Wales*, **105**, 86–93.

FRENCH, A. N. & VAN DER VOO, R. 1979. The magnetization of the Rose Hill Formation at the classical site of Graham's fold test. *Journal of*

Geophysical Research, **84**, 7688–7696.

HARGRAVES, R. B., DAWSON, E. M. & VAN HOUTEN, F. B. 1987. Paleomagnetism and age of Mid-Paleozoic ring complexes in Niger, West Africa, and tectonic implications. *Geophysical Journal of the Royal Astronomical Society*, **90**, 705–729.

HECKEL, P. H. & WITZKE, B. J. 1979. Devonian world palaeogeography determined from distribution of carbonates and related lithic palaeoclimatic indicators. *Special Papers in Palaeontology of the Palaeontological Association*, **23**, 99–123.

HURLEY, N. F. & VAN DER VOO, R. 1987. Paleomagnetism of Upper Devonian reefal limestones, Canning Basin, Western Australia: *Geological Society of America Bulletin*, **98**, 123–137.

JACKSON, M. J. & VAN DER VOO, R. 1985. A Lower Ordovician paleomagnetic pole from the Oneota Dolomite, Upper Mississippi Valley. *Journal of Geophysical Research*, **90**, 10449–10461.

JOHNSON, R. J. E., VAN DER PLUIJM, B. A. & VAN DER VOO, R. 1988. Paleoreconstruction of Early Ordovician terranes in the Central Mobile Belt of the Newfoundland Appalachians (abstract). *Geological Society of America Abstracts with Programs*, **20**, A63.

KENT, D. V. 1988. Further paleomagnetic evidence for oroclinal rotation in the central folded Appalachians from the Bloomsburg and the Mauch Chunk formations. *Tectonics*, **4**, 749–759.

—— & OPDYKE, N. D. 1980. Paleomagnetism of Siluro-Devonian rocks from eastern Maine. *Canadian Journal Earth Sciences*, **17**, 1653–1665.

LAPOINTE, P. 1979. Paleomagnetism and orogenic history of the Botwood Group and Mount Peyton batholith, Central Mobile Belt, Newfoundland. *Canadian Journal of Earth Sciences*, **16**, 866–876.

LEWANDOWSKI, M. 1987. Results of the preliminary paleomagnetic investigations of some Lower Paleozoic rocks from the Holy Cross Mountains (Poland). *Kwartalnik Geologicne*, in press.

LI, Z. X., SCHMIDT, P. W. & EMBLETON, B. J. J. 1988. Paleomagnetism of the Hervey Group, central New South Wales and its tectonic implications. *Tectonics*, **7**, 351–367.

LUCK, G. R. 1972. Palaeomagnetic results from Palaeozoic sediments of northern Australia. *Geophysical Journal of the Royal Astronomical Society*, **28**, 475–487.

McCABE, C. & CHANNELL, J. E. T. 1988. Paleomagnetism of the Middle Ordovician Stapely volcanic Group, Welsh Borderlands. *EOS*, **69**, 337.

——, VAN DER VOO, R., WILKINSON, B. H. & DEVANEY, K. 1985. A Middle/Late Silurian paleomagnetic pole from limestone reefs of the Wabash Formation (Indiana, U.S.A.). *Journal of Geophysical Research*, **90**, 2959–2965.

McKERROW, W. S. & ZIEGLER, A. M. 1972. Palaeozoic oceans. *Nature Physical Science*, **240**, 92–94.

MANZONI, M. & NANNI, T. 1977. Paleomagnetism of Ordovician lamprophyres from Taylor Valley, Victoria Land, Antarctica. *Pageoph*, **115**, 961–977.

MILLER, J. D. & KENT, D. V. 1986. Paleomagnetism of the Upper Devonian Catskill Formation from the southern limb of the Pennsylvania salient. *Geophysical Research Letters*, **13**, 1173–1176.

—— & —— 1988. Paleomagnetism of the Siluro-Devonian Andreas red beds: evidence for an Early Devonian supercontinent? *Geology*, **16**, 195–198.

—— & —— 1989. Paleomagnetism of the Upper Ordovician Juniata Formation of the central Appalachians revisited again. *Journal of Geophysical Research*, **94**, 1843–1849.

MORRISON, J. & ELLWOOD, B. B. 1986. Paleomagnetism of Silurian-Ordovician sediments from the Valley and Ridge province, northwest Georgia. *Geophysical Research Letters*, **13**, 189–192.

MURTHY, G. S. & RAO, K. V. 1976. Paleomagnetism of Steel Mountain and Indian Head anorthosites from western Newfoundland. *Canadian Journal of Earth Sciences*, **13**, 75–83.

PERROUD, H. & BONHOMMET, N. 1981. Paleomagnetism of the Ibero-Armorican arc and the Hercynian orogeny in western Europe. *Nature*, **292**, 445–448.

—— & —— 1984. A Devonian pole for Armorica. *Geophysical Journal of the Royal Astronomical Society*, **77**, 839–845.

—— & ——, VAN DER VOO, R. 1985. Paleomagnetism of the Late Ordovician Thouars Massif, Vendée province, France. *Journal of Geophysical Research*, **90**, 4611–4625.

——, BONHOMMET, N. & THÉBAULT, J. P. 1986. Palaeomagnetism of the Ordovician Moulin de Chateaupanne formation, Vendée, western France. *Geophysical Journal of the Royal Astronomical Society*, **85**, 573–582.

PIPER, J. D. A. & BRIDEN, J. C. 1973. Palaeomagnetic studies in the British Caledonides — Part I, Igneous rocks of the Builth Wells — Llandridnod Wells Ordovician inlier, Radnorshire, Wales. *Geophysical Journal of the Royal Astronomical Society*, **34**, 1–12.

RAO, K. V. & VAN DER VOO, R. 1980. Paleomagnetism of a Paleozoic anorthosite from the Appalachian Piedmont, northern Delaware: possible tectonic implications. *Earth and Planetary Science Letters*, **47**, 113–120.

ROY, J. L. & ANDERSON, P. 1981. An investigation of the remanence characteristics of three sedimentary units of the Silurian Mascarene Group of New Brunswick. *Journal of Geophysical Research*, **86**, 6351–6368.

SCHMIDT, P. W., EMBLETON, B. J. J. & PALMER, H. C. 1987. Pre- and post-folding magnetizations from the Early Devonian Snowy River Volcanics and Buchan Caves limestone, Victoria. *Geophysical Journal of the Royal Astronomical Society*, **91**, 155–170.

——, ——, CUDAHY, T. J. & POWELL, C. McA. 1986. Prefolding and premegakinking magnetizations from the Devonian Comerong Volcanics, New South Wales, Australia and their bearing on the Gondwana pole path. *Tectonics*, **5**, 135–150.

SCHUTTS, L. D., BRECHER, A., HURLEY, P. M., MONTGOMERY, C. W. & KRUEGER, H. W. 1976. A case study of the time and nature of paleomagnetic resetting in a mafic complex in New England. *Canadian Journal of Earth Sciences*, **13**, 898–907.

SEGUIN, M. K. & GAHÉ, E. 1985. Paleomagnetism of Lower Devonian volcanics and Devonian dykes of northcentral New Brunswick, Canada. *Physics of the Earth and Planetary Interiors*, **38**, 262–276.

—— RAO, K. V. & PINEAULT, R. 1982. Paleomagnetic study of Devonian rocks from Ste. Cécile-St. Sebastien region, Quebec Appalachians. *Journal of Geophysical Research*, **87**, 7853–7864.

SMITH, A. G., and HALLAM, A. 1970. The fit of the southern continents: Nature, v. 225, p. 139–144.

SPARIOSU, D. J. & KENT, D. V. 1983. Paleomagnetism of the Lower Devonian Traveler Felsite and the Acadian orogeny in the New England Appalachians. *Geological Society of America Bulletin*, **94**, 1319–1328.

STEARNS, C., VAN DER VOO, R. & ABRAHAMSEN, N. 1989. A new Siluro-Devonian paleopole from early Paleozoic rocks of the Franklinian Basin, North Greenland fold belt. *Journal of Geophysical Research*, in press.

THOMAS, C. & BRIDEN, J. C. 1976. Anomalous geomagnetic field during the Late Ordovician. *Nature*, **259**, 380–382.

TORSVIK, T. H. 1988. A review of Palaeozoic Palaeomagnetic data from Europe (*abstract*). Proceedings of Palaeozoic Biogeography and Palaeogeography Symposium, Oxford, UK, August 1988, 2–4.

VAN DER VOO, R. 1988. Paleozoic paleogeography of North America, Gondwana, and intervening displaced terranes: comparisons of paleomagnetism with paleoclimatology and biogeographical patterns. *Geological Society of America Bulletin*, **100**, 311–324.

—— 1989. Paleomagnetism of North America; the craton, its margins and the Appalachian Belt, *In*: Pakiser, L. C. & Mooney, W. D. (eds) *Geophysical Framework of the Continental United States*, Geological Society of America Memoir, **172**, in press.

—— & FRENCH, R. B. 1977. Paleomagnetism of the Late Ordovician Juniata Formation and the remagnetization hypothesis. *Journal of Geophysical Research*, **82**, 5796–5802.

—— & JOHNSON, R. J. 1985. Paleomagnetism of the Dunn Point Formation (Nova Scotia): high paleolatitudes for the Avalon terrane in the Late Ordovician. *Geophysical Research Letters*, **12**, 337–340.

WATTS, D. R. & VAN DER VOO, R. 1979. Paleomagnetic results from the Ordovician Moccasin, Bays, and Chapman Ridge formations of the Valley and Ridge province, eastern Tennessee. *Journal of Geophysical Research*, **84**, 645–655.

WEISSE, P. A., HAGGERTY, S. E. & BROWN, L. L. 1985. Paleomagnetism and magnetic mineralogy of the Nahant Gabbro and Tonalite, eastern Massachusetts. *Canadian Journal of Earth Sciences*, **22**, 1425–1435.

WENSINK, H. 1983. Paleomagnetism of red beds of Early Devonian age from central Iran. *Earth and Planetary Science Letters*, **63**, 325–334.

—, ZIJDERVELD, J. D. A. & VAREKAMP, J. C. 1978. Paleomagnetism and ore mineralogy of some basalts of the Geirud Formation of Late Devonian–Early Carboniferous age from the southern Alborz, Iran. *Earth and Planetary Science Letters*, **41**, 441–450.

YOUNG, G. C. 1990. Devonian vertebrate distribution patterns, and cladistic analysis of palaeogeographic hypotheses. *In*: MCKERROW & SCOTESE, C. R. (eds) *Palaeozoic Palaeogeography and Biogeography*. Geological Society, London, Memoir, **12**, 243–255.

ZIJDERVELD, J. D. A. 1968. Natural remanent magnetizations of some intrusive rocks from the Sør Rondane Mountains, Queen Maud Land, Antarctica. *Journal of Geophysical Research*, **73**, 3773–3785.

Palaeoclimatic constraints for Palaeozoic Palaeolatitudes of Laurentia and Euramerica

BRIAN J. WITZKE

Iowa Department of Natural Resources, Geological Survey Bureau, and Department of Geology, University of Iowa, Iowa City, Iowa 52242 USA

Abstract: Palaeozoic lithic palaeoclimatic data were plotted at the series or stage level for Laurentia (North America) and Euramerica. Their distribution was used to infer the extent of humid and arid climatic belts across the continent for specified time intervals. Climatic data are potentially useful for constraining palaeolatitudes for times when zonal atmospheric circulation remains the dominant pattern. Arid lithic indicators (evaporites, carbonate oolite) and humid indicators (coal, bauxite) were constrained to provide the best fit of these data into a simple zonal climatic pattern delineating subtropical arid belts and equatorial and/or temperate humid belts, respectively. An attempt was made to minimize occurrences of arid indicators in interpreted equatorial regions to maximize a zonal fit. Climatic belts can be arranged in a temporally consistent zonal climatic scheme for much of the Palaeozoic, but monsoonal and orographic effects apparently disrupted the general zonal pattern during portions of the Late Palaeozoic. The interpreted climatic patterns suggest progressive southward movement of Laurentia during the Cambrian through Early Devonian followed by general northward drift of Euramerica during the Middle Devonian through Permian. Lithic palaeoclimatic data provide an independent means to test palaeolatitudes interpreted from palaeomagnetic data. There is general agreement between palaeoclimatic and palaeomagnetic interpretations, but some discrepancies, particularly for the Devonian and Early Carboniferous, warrant further attention.

Palaeolatitudinal inferences for individual continental blocks through the Palaeozoic can be drawn from four independent lines of evidence: (1) palaeomagnetic, (2) palaeoclimatic, (3) palaeobiogeographic, and (4) tectonic. Although the use of palaeomagnetic data has been pre-eminent in recent years for interpreting palaeolatitudes on most Palaeozoic global reconstructions, the need for independent latitudinal tests of proposed geographies is imperative. The accumulation of Palaeozoic palaeomagnetic data is progressing, and the resolution of many outstanding palaeogeographic problems will undoubtedly be forthcoming as such studies continue. Nevertheless, evaluation of Palaeozoic '... palaeomagnetic data indicates that reliable data are few and that magnetic overprinting is commoner than normally thought. This means that the palaeomagnetic data for any particular continental block at any one period are unreliable for detailed analysis' (Tarling 1980, p. 11). Therefore, palaeomagnetically-based Palaeozoic reconstructions must be '... tested and modified by other palaeontological and palaeoclimatic evidence' (ibid.) in order to achieve a higher degree of resolution.

Changing climatic patterns through time result from the complex interplay of motions in the atmosphere, hydrosphere, and lithosphere (Robinson 1973). General palaeoclimatic patterns for individual continental blocks through time can be inferred from lithic evidence contained in the stratigraphic record. Such evidence may be useful for evaluating proposed geographies provided that there is some general and potentially predictable relationship between climate and latitude.

Climatic considerations

Global climatic patterns are influenced by the zonal atmospheric circulation system, and the interaction of the basic circulation with orographic barriers and land/ocean masses producing orographic, continental, and monsoonal effects. Because the zonal circulation essentially parallels latitude, it is the zonal pattern that offers the most straight-forward method for using palaeoclimatic data to constrain palaeolatitudes. However, orographic, continental, and monsoonal effects tend to modify zonal patterns and make the relationship between climate and latitude more difficult to evaluate (Parrish & Barron 1986). The relative influence of these modifying effects on disrupting the zonal pattern is, therefore, of paramount importance in using palaeoclimatic data to constrain palaeolatitudes. Resolving the influence of such effects for individual continents requires additional palaeogeographic evidence concerning (1) the location of mountain belts, (2) the distribution of land and sea on the continent, and (3) the relative proximity and latitudinal position of adjacent or nearby continental masses and oceans.

Orographic effects are relatively simple to understand; in general, mountainous belts will be wetter on their windward sides and drier on their lee sides. Therefore, orographic factors should be evident from palaeoclimatic data and are helpful for constraining the direction of prevailing winds (along with other palaeowind and palaeocurrent indicators). Patterns produced by monsoonal effects are more difficult to predict, since knowledge of global land/sea distributions is needed. It is the monsoonal patterns that introduce the greatest level of predictive uncertainty for palaeolatitudinal positioning of various palaeoclimatic indicators. Monsoonal circulation is developed in response to seasonal temperature contrasts over large mid-latitude continental masses, producing strong seasonal variations in wind direction and rainfall that disrupt or eliminate the zonal pattern. Monsoonal effects should be most pronounced during times of low global sea-level stand, since widespread epicontinental seas should have an ameliorating effect on continental climates. Rowley *et al.* (1985) also suggested that the development of monsoonal systems is related to the relative symmetry of continental masses about the equator and the presence or absence of equatorial mountains.

Although monsoons dramatically modify zonal patterns in specific geographic settings, it is the zonal circulation pattern that remains the dominant climatic influence in the modern world despite the many longitudinally-oriented continents that disrupt it (Parrish & Curtis 1982, p. 49). Mesozoic–Cenozoic lithic palaeoclimatic data also tend to show the dominance of zonal patterns for the post-Palaeozoic world, although consideration of monsoonal effects provide a higher level of predictive capability than the zonal model alone (Parrish *et al.* 1982). The influence of monsoons for times during the Palaeozoic certainly cannot be discounted, but the general dominance of zonal patterns will form the central assumption for a first-order evaluation of latitudinal constraints for specific climatic indicators during the Palaeozoic.

Lithic palaeoclimatic indicators

The distribution of palaeoclimatically-sensitive rock types provides the primary evidence for ancient palaeoclimatic patterns. Koppen & Wegener (1924) pioneered the use of such palaeoclimatic data in palaeogeographic analysis, and used it as a powerful argument for the existence of a Late Palaeozoic supercontinent (Pangaea).

The climatic significance of various rock types is discussed in numerous publications (e.g., Frakes 1979; Habicht 1979; Boucot & Gray 1979; Heckel & Witzke 1979; Witzke & Heckel 1989), and only general considerations are reiterated here concerning potential latitudinal constraints of specific sedimentary rock types.

The distribution of carbonate sediment types provides broad latitudinal constraints. Foramol carbonate sediments can occur at any latitude, but Bahamian-style carbonates (including chlorozoan and chloralgal skeletal associations) are limited to tropical-subtropical climates at latitudes between about 35°N and S in the modern world (Lees 1975). The distribution of hermatypic reefs, indurated pellets, and shallow-marine carbonate muds are constrained similarly. Ziegler et al. (1984) found that most shallow shelf carbonates during the Mesozoic–Cenozoic formed between 5 and 35°N and S and that 99% of the occurrences were within 45° of the equator. They concluded that light refraction (a direct function of latitude) is the controlling factor for carbonate production, and, as such, Bahamian-type carbonates should provide extremely important palaeolatitudinal constraints. Therefore, all occurrences of Bahamian-type shallow marine carbonate sediments should be confined to a broad 'carbonate belt' between latitudes of about 45°N and S. Carbonate ooids form in the modern world where salinity is consistently high (Lees 1975), and, hence, marine oolites are considered as potential indicators of generally drier climatic conditions within the confines of the larger 'carbonate belt.'

Evaporites provide unequivocal evidence of arid climates during deposition, and, where associated with carbonates, evaporites are expected to be confined largely to the tropical-subtropical arid belts. Modern marine-derived evaporites are known in local settings at latitudes ranging from about 8–45°N and 5–27°S, and modern nonmarine evaporites range to about 50° latitude. Mesozoic–Cenozoic data (Parrish et al. 1982) show evaporites confined between about 50°N and S, with the bulk between 10 and 40°N and S. Marine-derived carbonate-evaporite associations in the Palaeozoic are constrained in this report to latitudes no higher than about 45°. Evaporites should not be found in equatorial settings at times when zonal atmospheric patterns are dominant, because of consistently high rainfall. However, equatorial evaporites are potentially possible when monsoonal patterns become dominant.

Additional lithic indicators of arid or semiarid climatic conditions have been proposed and include calcareous palaeosols, aeolian (dune) sands, and redbeds. Soils tend to accumulate calcium carbonate in arid or semiarid conditions producing calcretes and caliche. Modern red soils are confined generally between 40°N and S (Habicht 1979), and primary redbeds may be indicators of generally warm climates. Although many redbeds seem to occur in arid palaeoclimatic settings, there is no inherent restriction of ferric oxide deposition to arid climates. In this report, an attempt is made to generally constrain redbeds to warm tropical and subtropical climatic belts, although in the absence of associated evaporitic indicators they are not necessarily excluded from wet belts. The distribution of redbed occurrences on the accompanying maps is far from complete, and additions will need to be made.

Humid climatic conditions are inferred for areas that lack evaporites but display bauxites or thick and widespread coals. Bauxites are limited to areas of high rainfall in tropical and subtropical settings (Frakes 1979, p. 102). Coal is derived from peat, which is deposited in saturated terrains. Localized occurrences of coal do not necessarily imply wet climates, but merely the presence of a locally high water table. However, areas with extensive coal that lack evaporites are most consistent with wet climates. Coal potentially can form at any latitude, but coals should be best developed when humid climates prevail, that is in equatorial and temperate wet belts and orographic or monsoonal rain areas. In general, 'rainfall consistently through the year is essential for forest growth and peat formation, particularly in the tropics where evaporation rates are high' (Ziegler et al. 1987). The presence of thick sequences of terrigenous siliciclastic rocks indicates probable humid conditions in the source areas, and the absence of arid indicators and the presence of coals, bauxites, or kaolinitic latosols further suggest probable humid conditions during deposition.

Marine phosphorites and phosphatic sediments generally delimit areas of marine upwelling. Upwelling is induced generally by divergence of surface currents, and most modern phosphorites are associated with vigorous coastal upwelling along the western margins of continents in subtropical to tropical latitudes. Some phosphatic sediments are associated with dynamic, monsoonal, or zonal upwelling which may occur along east-facing or poleward-facing coasts in specific settings (Parrish 1982). The distribution of ancient phosphorites is potentially useful for testing specific palaeogeographies, but will not in itself necessarily constrain unique geographies. Using the modern distribution of phosphorites as a guideline, Palaeozoic phosphorites would be predicted to be most abundant in west-facing marginal seaways, a prediction that holds true on the Palaeozoic reconstructions of Euramerica presented here.

Glacigenic sediments in Gondwana provide evidence of continental glaciation during the Palaeozoic (Hambrey & Harland 1981; Caputo & Crowell 1985). Although modern continental ice sheets are restricted to latitudes of 60 to 90°, Pleistocene glaciation reached latitudes as low as 38°N. If Palaeozoic glaciations are comparable to those of the Quaternary, similar latitudinal constraints may be expected. Mountain glaciation can occur at any latitude if sufficient elevations are present, but evidence of such glaciations are rare in the pre-Quaternary record due to extensive erosion of mountainous areas.

The distribution of oolitic ironstones is plotted for Palaeozoic Laurentia–Euramerica, but, as Van Houten & Bhattacharyya (1982) noted for Phanerozoic global occurrences, the distribution displays wide latitudinal scatter across differing climatic regimes. Most oolitic ironstones were deposited during the waning phases of detrital influx in relatively shallow and low-energy environments, commonly along an embayed detrital coastline (ibid.). Their occurrences are of limited climatic significance, but may be useful from a palaeogeographic perspective for delimiting proximity to shoreline.

Palaeoclimatically-sensitive lithic data have been compiled at the series or stage level for the Early Palaeozoic of Laurentia and the mid to Late Palaeozoic of Euramerica. The palaeogeographic approach used in this report first attempts to constrain the lithic data into temporally consistent zonal-orographic palaeoclimatic patterns. As a first-order approach, arid climatic indicators (evaporites, carbonate oolite, etc.) are constrained to appropriate zonal arid belts (5–10° to 35–45°N and S) or orographic rain shadows. Occurrences of equatorial evaporites are minimized. Likewise, humid indicators (bauxites, related palaeosols, thick or widespread coals, thick siliciclastic sequences lacking evaporites, etc.) are positioned for 'best fit' with zonal (equatorial or temperate storm belts) and orographic wet belts. For times when zonal patterns appear to be dominant, this approach should offer an effective means for independently evaluating proposed palaeolatitudinal reconstructions. When the lithic data does not reasonably constrain a zonal-orographic fit, other factors, including monsoonal effects, need to be considered.

The apparent width of arid and wet belts through time may also be influenced by non-zonal effects and equator-to-pole temperature gradients. Although the widths of zonal belts may vary slightly through time (perhaps 0–10°), the respective arrangement of wet and dry belts will be maintained as a general consequence of Earth rotation. The temporal distribution of lithic palaeoclimatic indicators on large continental blocks provides an empirical basis for constraining the widths of specific climatic belts through time, which may, in turn, provide indirect evidence of changes in equator-to-pole temperature gradients.

North American and Euramerican reconstructions

Lithic palaeoclimatic data for North America and Euramerica have been compiled from a variety of sources, as outlined by Witzke & Heckel (1989). Most Canadian data was derived from publications of the Geological Survey of Canada; most Russian data is from Vinogradov (1969) and Zharkov (1984); American occurrences are primarily taken from publications of the US Geological Survey, various state geological surveys, and other regional stratigraphic studies; western European data is after Ziegler (1982) and other sources.

The lithic data base used in this report has two inherent problems. First, compilation at the series or stage level provides relatively crude temporal constraints (various maps cover intervals of time ranging from about 8 to 28 million years in length, averaging about 15 Ma). Because continents may drift considerable distances during such periods of time, resolution of general climatic boundaries will need to be time-averaged. Fortunately, North America–Euramerica appears to have been relatively slow in its latitudinal drift for much of the Palaeozoic, but for some episodes of relatively rapid latitudinal movement (especially the Visean–Namurian) an apparent mix of arid and humid lithic indicators may result. Such mixes do not necessarily reflect actual short-term climatic variations, but may result from drift of continental localities across multiple climatic belts during a particular time span. The need for finer temporal resolution becomes especially critical during periods of relatively rapid continental drift, but such a refined lithic compilation requires a higher level of chronostratigraphic resolution than is generally available for entire continental blocks during the Palaeozoic.

This last point brings up the second problem with the data compilation used here; in many areas available biostratigraphic control is sparse to absent, necessitating dependence on relatively coarse levels of correlation. Evaluation of all published correlations was beyond the scope of this report, but it should be noted that disparate correlations exist for some areas where biostratigraphic studies are incomplete or inconclusive. Precise correlations between contrasting biofacies have not yet been achieved in many areas necessitating the erection of local biostratigraphic and chronostratigraphic schemes. In many cases, such schemes can be related to the evolving global chronostratigraphic framework in only general terms. Some lithofacies (e.g., many evaporitic facies) lack reliable biostratigraphic control, and other means of correlation may be needed (such as sea-level event stratigraphy).

The accompanying maps (Figs 1–8, 10–11) are generalized sketches that are approximately equal-area but are non-standard projections. The location and geography of continental margin areas is highly schematic. The potential placement of continental margins is discussed elsewhere, and the reader is referred to discussions and references concerning marginal areas of the Caledonian margin, mid Palaeozoic margins, and the Cordilleran margin found in Barker & Gayer (1985), Witzke & Heckel (1989), and Monger & Ross (1984). Although continental margin areas are of considerable palaeotectonic importance, lithic data from such areas make up only a small portion of the data base. Lithic data from the relatively stable cratonic areas make up the bulk of the data set. The cratonic areas display few large-scale tectonic complications, and lithic data in these areas can be positioned with a higher degree of confidence than the more tectonically complex marginal areas where large-scale lateral displacements are common (thrusting, strike-slip motions, accretion of allochthonous blocks). Thrust faults have not been palinspastically restored (200 km is the desired resolution of the maps, making such restorations of limited significance). The union of Laurentia with Baltica to form Euramerica may have involved strike-slip motions during the Devonian or Early Carboniferous, although such movements are not demonstrable from palaeomagnetic data (Miller & Kent 1986). A slightly modified Bullard fit of Laurentia–Baltica is used in this report for the Devonian through Permian.

An attempt is made on the accompanying maps (Figs 1–8, 10–11) to schematically outline the maximum extent of epicontinental seaways for the given time intervals. Interpretations of seaway extent for each interval are inferred from the extent of preserved marine strata (including outliers), the distribution of shoreline and nonmarine facies, the relative proximity to siliciclastic source areas during maximum transgressive phases, and published palaeogeographic syntheses. When subtidal open-marine facies occur at the zero-edge of preserved strata, seaway extent was extrapolated beyond that edge to reflect maximum inundation. Sea levels undoubtedly fluctuated during the represented time intervals, and tectonic factors modified seaway extent in some areas over shorter time spans than represented by the accompanying maps. Therefore, the illustrated interpretations are generalized and schematic, and do not necessarily reflect marine inundation for any specific moment in time, merely an approximation of maximum seaway extent averaged over the entire epoch or age.

Cambrian Laurentia

All of Cambrian Laurentia must be positioned within the 'carbonate belt', as Bahamian-type carbonates are found across the continent. Potential arid indicators (carbonate oolite) are widespread, and evaporites occur in the eastern and central US and Arctic Canada (Fig. 1). No matter how Laurentia is positioned, arid indicators span 40 to 50° of latitude; if all occurrences are positioned within a single arid belt, the belt would be exceptionally wide. This suggests that both the north and south subtropical arid belts crossed Laurentia. If zonal patterns dominate, a humid equatorial belt lacking evaporites should be identifiable. Potential locations of the equatorial belt are illustrated on Fig. 1.

Cambrian sedimentary patterns within the proposed equatorial belts, especially areas along the Transcontinental Arch ('continental backbone'), contrast with adjacent regions. Regions along the Arch lack carbonate oolites and evaporites, and contain relatively coarse siliciclastic-dominated sequences that are in part very glauconitic. The Mount Simon Sandstone (Dresbachian in upper part) reaches thicknesses to 790 m in Illinois (Willman et al. 1975), implying significant weathering of adjacent cratonic source terranes during the Middle Cambrian. The abundance of quartzarenites in areas paralleling the Transcontinental Arch, likewise, suggests that significant sediment recycling and deep weathering may have characterized the cratonic source areas. It is, therefore, suggested that the proposed equatorial regions are consistent with humid climatic conditions.

The 'best fit' of lithic palaeoclimatic indicators into a zonal pattern suggests a position of Laurentia in low latitudes straddling the equator. The palaeoclimatic approach offers a general verification of palaeomagnetic data, which also place Laurentia across the equator at low latitudes (Scotese et al. 1979; Scotese 1986). Scotese's (1986) Cambrian latitudinal reconstructions are similar to the latitudinal fit suggested here, but are rotated slightly clockwise placing the evaporites of Arctic Canada on the equator. Is the zonal pattern reasonable for Cambrian Laurentia? Most global reconstructions for the Cambrian (e.g., Scotese 1986) show a decided asymmetry of continents about the equator, and Laurentia is not positioned close to any mid-latitude continental mass. If true, monsoonal circulation should be minimal, and a zonal pattern should dominate across Laurentia during the Cambrian.

Ordovician Laurentia

The distribution of Early and Middle Ordovician lithic palaeoclimatic indicators across Laurentia is similar to that of the Cambrian, but a general northward expansion of carbonate oolite-evaporite facies in the central US in the Early Ordovician suggests slight southward movement (Fig. 2). The continent is constrained

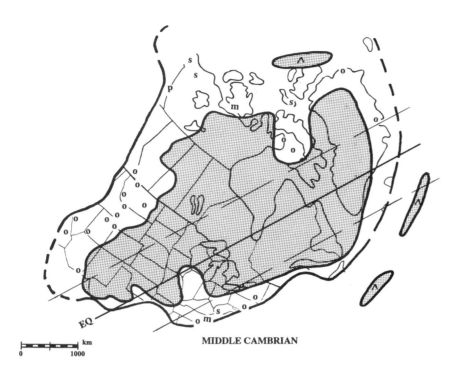

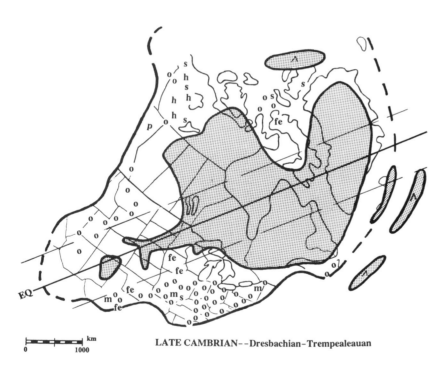

Fig. 1. Interpreted palaeogeography for Middle and Late Cambrian Laurentia; Late Cambrian includes Dresbachian, Franconian, Trempealeauan. Continental margins and distribution of land areas (shaded area) during maximum marine onlap are schematic. Lithic palaeoclimatic indicators are plotted on the base map and constrained to inferred zonal climatic belts. Abbreviations and symbols for this and subsequent maps: o, carbonate ooids and coated grains; p, phosphorite or phosphatic sediments; m, evaporite crystal molds; s, sulfate evaporites (primarily gypsum, anhydrite); h, halite; k, potash salts (most soluble evaporites noted for each evaporite occurrence); fe, oolitic ironstones; r, redbeds; al, bauxite; rectangles, coal; g, glacigenic sediments; ∧, generalized mountainous terrains; EQ, proposed palaeoequator. Dashed lines that parallel equator mark approximate division between proposed humid and arid climatic conditions.

within the 'carbonate belt' throughout the Ordovician. Lower Ordovician strata on the northern side of the Transcontinental Arch are siliciclastic-dominated and lack arid indicators. Middle Ordovician strata (which include the early Caradoc in American chronostratigraphy) are also dominated by siliciclastics (with minor oolitic ironstones) in the same area, as well as along the southern fringe of the Transcontinental Arch. Kaolinitic mudstones and quartzarenites characterize much of this region (Witzke 1980), which is interpreted as the equatorial humid belt. Oolitic carbonate and evaporite facies in the eastern and central US and Arctic Canada lie within the north and south arid belts, respectively.

A significant change in the distribution of arid lithic indicators is noted for the Late Ordovician. Carbonate oolites and evaporites apparently are lacking from Upper Ordovician strata of Arctic Canada, but evaporitic facies are widespread in the central continent (Fig. 3), an area that lacks indicators of arid deposition for the Cambrian through Middle Ordovician. If zonal patterns were maintained, the simplest explanation of the apparent shift is continuing southward drift of Laurentia during the Ordovician, placing the bulk of the continent in the southern arid belt by the Late Ordovician. Migration of the equatorial humid belt into Arctic Canada may also account for the apparent absence of arid indicators in that region. Uplift of Taconic mountains along the eastern seaboard of the US is evidenced by detrital and volcanic influx from orographic sources, which culminated in deltaic progradation (including extensive redbeds) during the Ashgill.

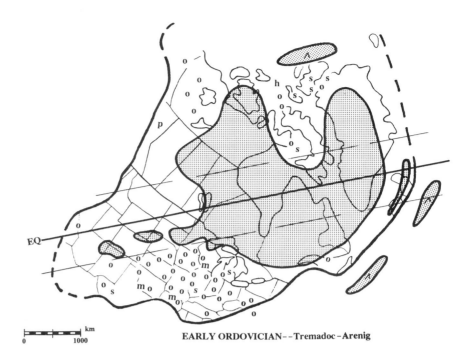

EARLY ORDOVICIAN--Tremadoc-Arenig

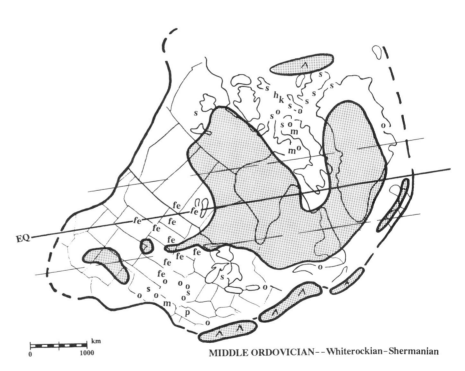

Fig. 2. Interpreted palaeogeography for Early and Middle Ordovician Laurentia; Early Ordovician includes Canadian, Tremadoc, Arenig; Middle Ordovician includes Whiterockian, Chazyan, Black Riveran, Trentonian, Rocklandian, Kirkfieldian, Shermanian, Llanvirn, Llandeilo, early Caradoc. Symbols as in Fig. 1.

MIDDLE ORDOVICIAN--Whiterockian-Shermanian

Laurentian Ordovician latitudes proposed here (Figs 2 & 3) are generally consistent with recent palaeomagnetically-based reconstructions (Scotese 1986; Van der Voo 1988), and differ by no more than about 10° latitude. Minor discrepancies include placement of the Early Ordovician evaporites of the Canadian Arctic and Late Ordovician evaporites of the Williston and Hudson Bay basins on the equator. Available global Ordovician reconstructions (e.g., Scotese 1986) maintain a marked asymmetry of continents about the equator, and Laurentia is isolated some distance from any mid-latitude continental mass. Such a palaeogeographic setting would not be conducive to development of monsoonal circulation across Laurentia, where zonal palaeoclimatic patterns presumably would dominate.

Silurian Laurentia and Baltica

Silurian lithofacies across Laurentia (Fig. 4) are grossly similar to those of the Late Ordovician, and the entire continent must be positioned within the 'carbonate belt'. However, unlike the Late Ordovician, evaporite facies occur in the Canadian Arctic. Arid indicators apparently are absent from northwest Canada, where Late Ordovician carbonate oolite is noted. If zonal patterns are maintained, the simplest explanation for a slight northward shift of evaporitic facies is continuing southward movement of Laurentia during the Silurian. A potential zonal constraint on arid indicators places the equator along the northern fringe of Laurentia (Fig. 4). This construction places the halites of the

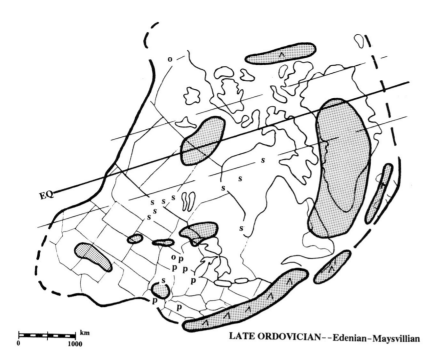

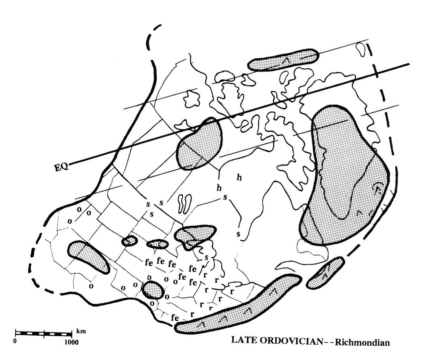

Fig. 3. Interpreted palaeogeography for Late Ordovician Laurentia; Edenian–Maysvillian approximately equivalent to late Caradoc–early Ashgill; Richmondian plot includes mid- to late Ashgill, Hirnantian, Gamachian. Symbols as in Fig. 1.

eastern US near the southern limit of the arid southern carbonate-evaporite belt, that is about 35–45° S.

Baltica joined with Laurentia by the latest Silurian or earliest Devonian forming Euramerica. Therefore, Baltica is placed in close proximity to Laurentia on Figure 4. Silurian Baltica lay entirely within the 'carbonate belt', and evaporites occur in the Timan and Baltic areas. Occurrences of carbonates and evaporites in Baltica are consistent with placement in the southern arid belt adjacent to Laurentia. Many Silurian reconstructions show Baltica colliding with Laurentia slightly southward from a Bullard-fit position (e.g. Ziegler *et al.* 1977; Scotese 1986), and climatic data are also consistent with such a position. Tarling's (1985*a*) northward position of Baltica places the northeast European evaporites near the equator. Varying palaeomagnetic interpretations of Silurian palaeolatitudes of Laurentia have been published, and all generally agree on placing the bulk of the continent in southern latitudes. Equatorial evaporites occur in Hudson Bay and the Boothia–northern Baffin Island area on the reconstructions of Scotese (1986) and Van der Voo (1988), respectively, which is not consistent with zonal climatic patterns. The reconstructions of Turner & Tarling (1982) and Tarling (1985*a*) place the equator along the northern fringe of Laurentia in a position closely similar to that suggested here (Fig. 4). A global asymmetry of continents about the equator is illustrated on all recent palaeomagnetic reconstructions, and the northernmost continent, Siberia (Angara), occurs within the vast northern ocean well within the carbonate-evaporite belt (Witzke & Heckel 1989). The global distribution of continents would not be conducive for the development of vigorous monsoons over northern Laurentia, and a zonal climatic fit seems reasonable.

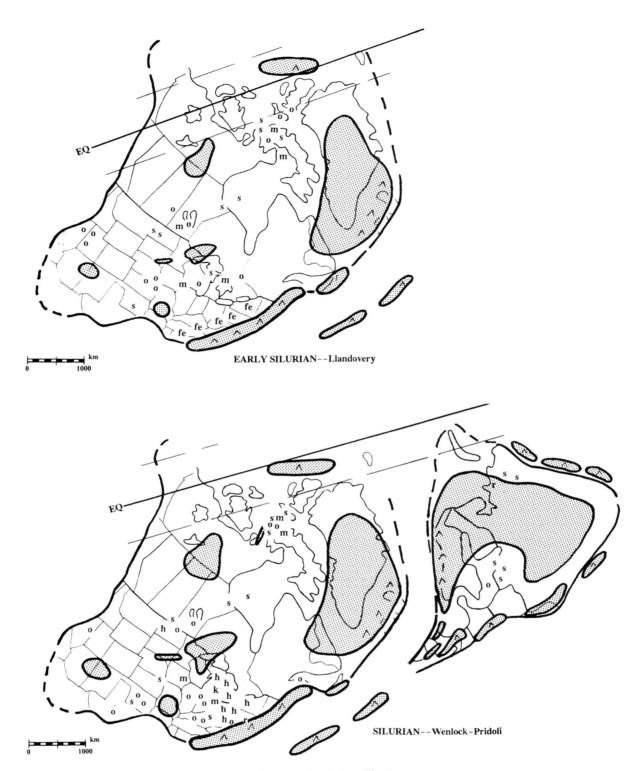

Fig. 4. Interpreted palaeogeography for Silurian Laurentia and Baltica. Symbols as Fig. 1.

Devonian Euramerica

A climatic interpretation of Devonian palaeolatitudes of Euramerica was presented by Heckel & Witzke (1979) and Witzke & Heckel (1989), and only general considerations are presented here. Devonian evaporites are widespread in Canada, but there is an apparent southward shift of the northern limits of these evaporites from Early to Late Devonian time (Figs 5 & 6). Coincident with this shift, siliciclastic sequences with coals become well developed across Arctic Canada and Svalbard. The East European (Russian) Platform also contains extensive evaporites, but coals and bauxites are noted in northern Russia. Redbed deposition was extensive on Euramerica ('Old Red Sandstone Continent'), primarily along the Caledonian and Acadian mountains and around the Baltic Shield.

The general distribution of humid and arid indicators on Euramerica can be constrained into a relatively simple zonal scheme. The continent occupies a position within the 'carbonate belt', and widespread arid indicators are constrained to the southern arid belt. The appearance of humid indicators along the

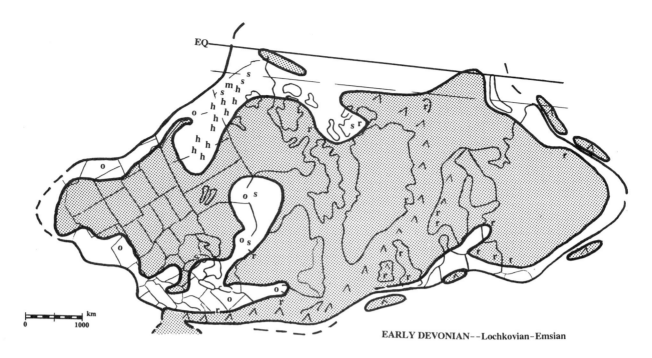

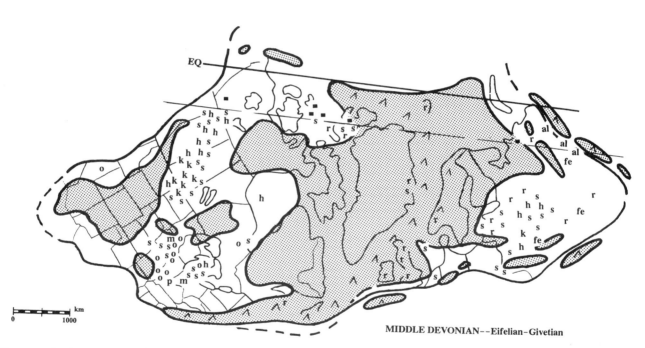

Fig. 5. Interpreted palaeogeography for Early and Middle Devonian Euramerica. Early Devonian includes Emsian, Pragian, Lochkovian, Gedinnian, Siegenian; Middle Devonian includes Eifelian, Givetian, Erian. Symbols as in Fig. 1.

northern fringe of Euramerica is interpreted to correspond to northward movement into the humid equatorial belt. If correct, the Devonian marked a significant reversal in the general drift of Laurentia, which is interpreted to have been primarily southward during the Early Palaeozoic.

Varying interpretations of palaeomagnetic data have resulted in widely disparate Devonian palaeolatitudinal reconstructions of Euramerica. Some reconstructions position Devonian Euramerica in northern latitudes (Ziegler *et al*. 1979) or straddling the equator (Scotese 1986) and differ from those presented here by 30–40° latitude. These reconstructions place arid indicators in equatorial latitudes, necessitating disruption of zonal patterns. Van der Voo (1988) and Miller & Kent (1988) position Late Devonian North America in a more southerly position, but place Canadian evaporite occurrences on the equator and coaly strata at about 20–25° N, possibly indicating potential orographic influences. Early Devonian palaeomagnetic reconstructions by Tarling (1980, 1985*a*) and Miller & Kent (1988) and Late Devonian reconstructions by Turner & Tarling (1982) differ the least (0–10° latitude) from those presented here, and place northern Euramerica in equatorial latitudes. Streel's (1986) equatorial positions for the Late Devonian are based on phytogeographic interpretations and differ by only 5–10° from that shown on Fig. 6.

The proximity of Devonian Euramerica to large continental

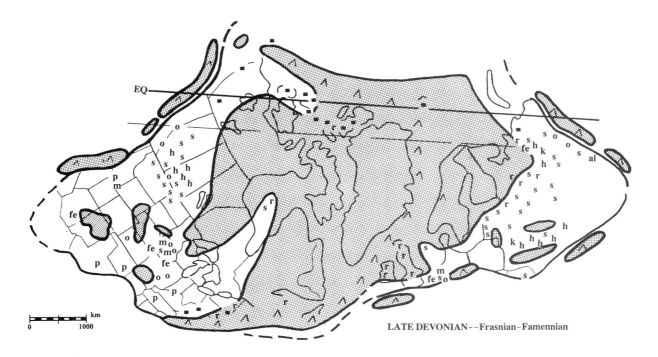

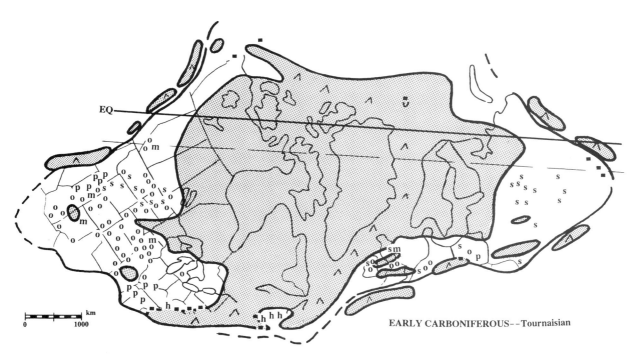

Fig. 6. Interpreted palaeogeography of Late Devonian and Tournaisian for Euramerica. Late Devonian includes Frasnian, Famennian, Senecan, Chautauquan; Early Carboniferous map includes Tournaisian, early Dinantian, Kinderhookian, early Osagean. Symbols as in Fig. 1.

masses, especially Gondwana, is important for interpreting the potential for monsoonal circulation over Euramerica. Varying palaeomagnetic interpretations are possible (e.g., Scotese 1984), but many recent reconstructions place Gondwana and Euramerica in contact or close proximity, at least for the Early Devonian (e.g., Van der Voo 1988; Scotese 1986; Miller & Kent 1988; Witzke & Heckel 1989). The potential zonal climatic patterns proposed here (Figs 5 & 6) seem temporally consistent, although near-equatorial Late Devonian evaporites in the Timan area may indicate the possibility of zonal disruption in northeastern Euramerica. Witzke & Heckel (1989) suggested, however, that monsoonal patterns were not dominant across Euramerica. The Devonian atmospheric circulation model of Parrish (1982), likewise, is not consistent with large-scale monsoonal patterns over North America.

All cited Devonian global reconstructions place Siberia (Angara) in close proximity to Euramerica along its northern margin. Most palaeomagnetically-based reconstructions place Siberia at mid to high latitudes. However, Devonian Siberia was positioned within the northern arid belt by Witzke & Heckel (1989) to be consistent with the extensive carbonate-evaporite facies identified there. Placement of Devonian Euramerica pri-

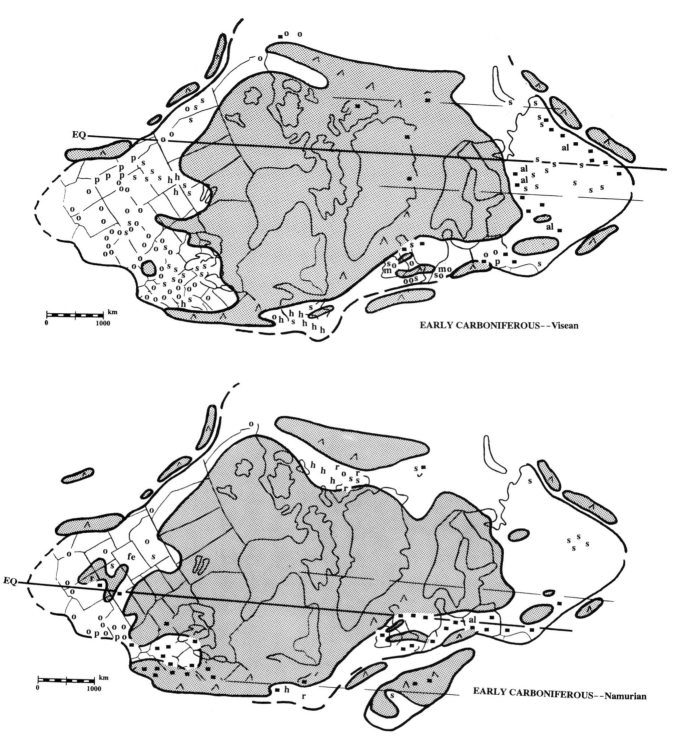

Fig. 7. Interpreted palaeogeography for Early Carboniferous of Euramerica; Hercynian margin and South European block highly schematic. Visean map includes late Osagean, Meramecian, mid- to late Valmeyeran, early Chesterian, late Dinantian; Namurian map includes late Chesterian, early Morrowan, Serpukhovian, earliest Bashkirian. Symbols as in Fig. 1.

marily in the southern hemisphere allows room for Siberian lithofacies to be positioned north of Euramerica within the 'carbonate belt' at latitudes less than 45° N.

Early Carboniferous Euramerica

The apparent northern limit of Tournaisian evaporites in Euramerica shows a slight southward shift compared to that of the Late Devonian, and coals are also shifted southward along the Urals (Fig. 6). Evaporites and oolitic carbonates are widely scattered across the US, western Europe, and the central Russian Platform (Fig. 6), presumably the southern arid belt. If zonal patterns are maximized, continuing and gradual northward movement of Euramerica is inferred for the Tournaisian (Witzke & Heckel 1989).

Visean patterns show some noteworthy changes in the distribution of arid and humid indicators (Fig. 7): (1) carbonate oolite and evaporites once again occur along the northern margin of Euramerica (western Canada, Alaska, Pechora-Novaya Zemlya), and (2) coals and bauxites are widely distributed in a more

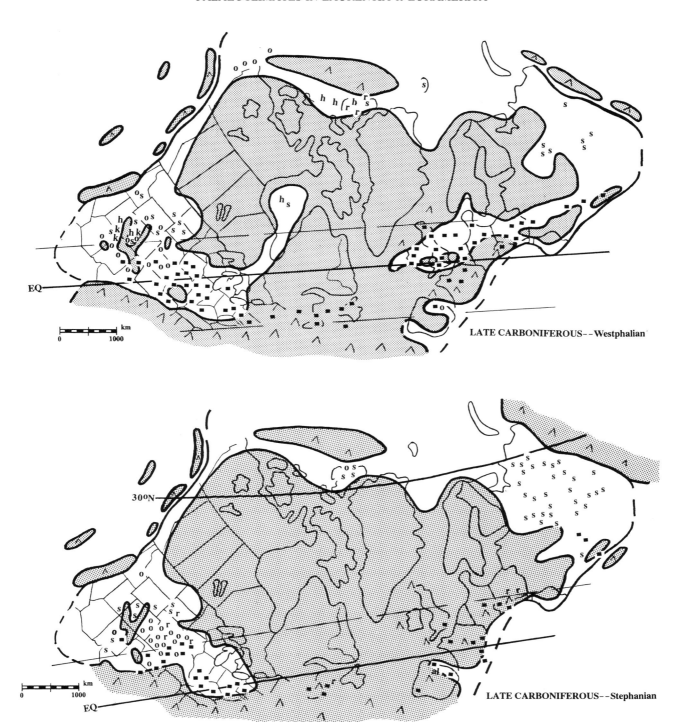

Fig. 8. Interpreted palaeogeography for Late Carboniferous of Euramerica; Gondwana is schematically joined to southern margin. Westphalian map includes late Morrowan, Atokan, Desmoinesian, mid- to late Bashkirian, Moscovian; Stephanian map includes Missourian, Virgilian, Gzelian, Kasimovian, Zhigulevanian, Orenburgian. Symbols as in Fig. 1.

southerly position (especially east Greenland, central and southern Urals, western Russia). Continuing southward migration of coal deposition during the Namurian is marked by a belt of extensive coal (and some bauxite) across southern Euramerica, with evaporites in Arctic Canada and the Russian Platform (Fig. 7). These patterns suggest that northern Euramerica migrated into an arid climate during the Visean–Namurian, and wet climates spread southward coincident with general northward movement of Euramerica across the equator during the Early Carboniferous.

Even though general northward drift is interpreted here, the Visean map (Fig. 7) displays a seemingly chaotic mixture of humid and arid indicators in areas of Europe, and evidence for humid deposition in western Euramerica has not been recognized. Therefore, a simple zonal model that minimizes Visean evaporites on the equator is not possible at the portrayed level of temporal resolution. There are two general possibilities that may explain the non-zonal component. (1) Monsoonal circulation may have begun to influence climatic patterns during the Early Carboniferous as the assembly of Pangaea progressed and continental symmetry about the equator was approached (Rowley *et al*. 1985). Although not portrayed on Fig. 7, union of Gondwana–South Europe and Euramerica along the Hercynian, Appalachian, and

Mauritanian margins was probably achieved before the end of the Namurian. The influence of developing monsoonal circulation over Euramerica during the Visean–Namurian probably accounts for partial loss of clear-cut zonal patterns over areas of the continent. (2) The northward drift of Euramerica during the Early Carboniferous is interpreted to have been the most rapid latitudinal movement of the continent during the Palaeozoic. As such, refined temporal control becomes of critical importance in resolving whether coal/bauxite and evaporite deposition was contemporaneous in local areas of Europe during portions of the Visean, or whether evaporite and coal deposition succeeded each other in a zonally-consistent temporal pattern. The Visean is a lengthy period of time (20 Ma), and considerable climatic changes are possible over such an interval if latitudinal drift was significant. To complicate matters, disparate Visean biostratigraphic correlations, especially between forams and conodonts, complicate resolution in portions of Euramerica.

A palaeomagnetically-based Early Carboniferous equator of Tarling (1980, 1985b) is indistinguishable from the Visean position illustrated on Fig. 7, and differs from the Visean equator of Scotese (1986) and Smith et al. (1981) by about 15–20° and 5–10°, respectively. The Namurian equator of Scotese (1986) is closely similar to that of Fig. 7, being only about 5° apart. Tournaisian and Visean 'palaeoclimatic equators' of Van der Zwan et al. (1985), based on climatic and phytogeographic inferences, are closely similar to those suggested here (Figs 6 & 7). The relative proximity of Euramerica and Gondwana is indicated on virtually all current global reconstructions for the Early Carboniferous, and general climatic patterns between the two continental masses and with Siberia are discussed by Witzke & Heckel (1989). Northward movement of Siberia, apparently coupled with northward drift of Euramerica, is evidenced by migration out of the northern carbonate-evaporite belt into the humid north temperate belt (with abundant coals) during the span of the Late Devonian through late Namurian/early Westphalian (ibid.).

Late Carboniferous Euramerica (northern Pangaea)

Euramerica was joined to Gondwana in the Late Carboniferous (collision was initiated during the Early Carboniferous), and formed the northwestern portion of Pangaea. Kazakhstan apparently collided with Euramerica along the Uralian suture during the Late Carboniferous. The southern areas of Euramerica display exceptional development of coal-bearing facies, especially during the Westphalian. The southern coal belt is clearly separated from an expansive arid belt across the remainder of Euramerica; evaporites occur in the western US, Hudson Bay, Arctic Canada, Svalbard, and the Russian Platform. (Fig. 8). Carbonate oolite is present in coal-bearing sequences in the central US, possibly suggesting climatic variations during the Late Carboniferous in that area.

A simple zonal climatic fit adequately constrains all occurrences of coal to humid equatorial regions and evaporite occurrences to the northern arid belt. The zonal fit is somewhat surprising, since monsoonal circulation patterns might be expected to develop over the giant Pangaean continental mass. It seems possible that the zonal arrangement of equatorial mountains along southern Euramerica-northern Gondwana (Appalachian–Ouachita, Mauritanides) may have countered the effects of the developing monsoon, essentially pinning the equatorial low-pressure cell to that region (Rowley et al. 1985).

The equatorial placement for the Late Carboniferous of Euramerica interpreted from zonal climatic patterns (Fig. 8) is in excellent agreement with available palaeomagnetic data. Equatorial positions used here for the Westphalian–Stephanian are identical to Late Carboniferous positions shown by Scotese et al. (1979), Scotese (1986), and Tarling (1980, 1985b), and offer independent verification of these reconstructions.

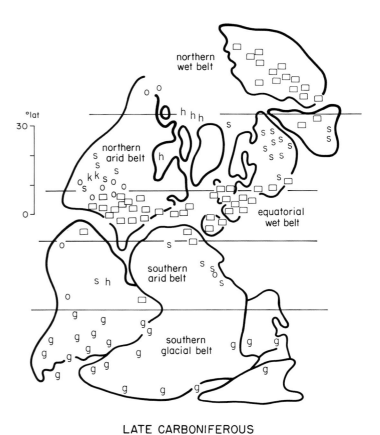

Fig. 9. Plot of Late Carboniferous lithic palaeoclimatic indicators for western Pangaea and Siberia on generalized base map of Scotese (1986). General climatic belts are delimited. Symbols as in Fig. 1.

A general zonal arrangement of Late Carboniferous climatic indicators across Pangaea and Siberia is portrayed in Fig. 9, using the base map of Scotese (1986). A zonal fit of lithic palaeoclimatic indicators is remarkably good. In general, a relatively broad equatorial wet belt (in part due to time-averaged effects of the lithic data) is flanked by arid belts with evaporites to the north and south. Extensive glaciation in southern Gondwana is consistent with a mid to high latitude placement. Siberia displays abundant coals and an absence of Bahamian-type carbonates consistent with placement in the northern wet belt.

Permian Euramerica (northern Pangaea)

Several significant changes characterize Permian lithic patterns across Euramerica from those noted in the Late Carboniferous. (1) The expansive Late Carboniferous coal belt across southern Euramerica is not recognizable during the Permian, although a few minor coals are present in that area for the Wolfcampian (Asselian–Sakmarian) interval (Fig. 10). (2) Late Carboniferous coal-bearing sequences of southern Euramerica are succeeded by Permian evaporitic facies in the south-central US, western Europe, and the southern Russian Platform (East European Basin). (3) Late Carboniferous–Wolfcampian evaporite facies of northern Euramerica are succeeded by mid to Late Permian coal-bearing sequences in Arctic Canada and the Pechora area of northern Russia (Figs 10 & 11). In general, Euramerica was characterized by widespread arid climatic conditions across most of its extent during the Permian, as evidenced by extensive evaporites. Evaporites are particularly well developed in the central and western US, the East European Basin, and the Rotliegendes–Zechstein sequence of western Europe.

The Appalachian–Ouachita–Mauritanide mountains of southern Euramerica remained as a significant palaeotopographic

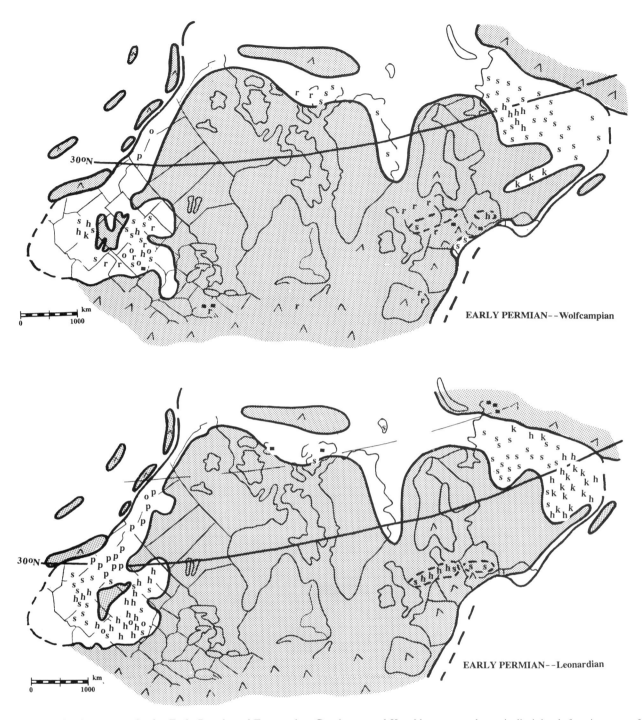

Fig. 10. Interpreted palaeogeography for Early Permian of Euramerica; Gondwana and Kazakhstan are schematically joined, forming part of northern Pangaea. Wolfcampian map includes Asselian, Sakmarian, lower Rotliegendes; Leonardian map includes Artinskian, Kungurian, upper Rotliegendes. Symbols as in Fig. 1.

feature during the Permian, and should presumably have served to pin equatorial low pressure to that region if that area remained at identical latitudes to that of the Late Carboniferous. The general absence of Permian humid indicators in that area, therefore, might suggest that the mountain belt migrated slightly out of equatorial latitudes during the Permian, or, alternatively, that other climatic factors related to monsoonal or orographic effects became operant. Northward drift of Euramerica during the Permian is suggested by the migration of the northern fringe of Euramerica out of the evaporite belt into a presumed mid-latitude wet belt during the mid to Late Permian. Although monsoonal circulation undoubtedly influenced climatic patterns across Pangaea, a simplistic zonal model can constrain arid indicators of Euramerica to the northern arid belt, with progressive drift of the northern fringe into wet temperate climates during the Permian.

Latitudinal placements of Permian Euramerica proposed here (Figs 10 & 11) are in essential agreement with the reconstructions of Scotese (1986), but differ slightly from those of Tarling (1980). In order to place Permian Euramerica in broader palaeogeographic perspective, lithic data has been plotted for the Early and Late Permian (Figs 12 & 13) on base maps of Scotese (1986). Northern and southern arid belts are identified for the Early Permian (Fig. 12) with Siberian coals positioned in the northern

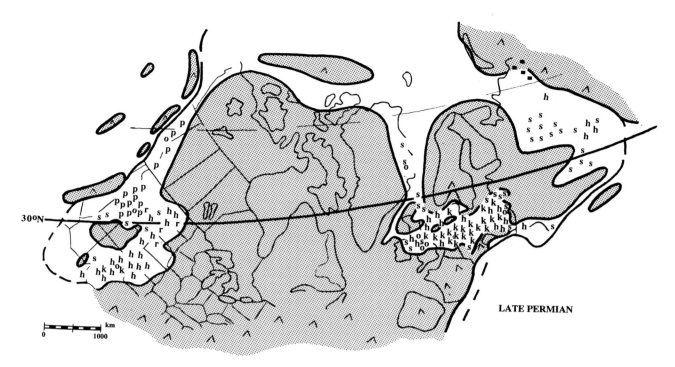

Fig. 11. Interpreted palaeogeography for Late Permian of Euramerica (northern Pangaea). Map includes Ufimian, Kazanian, Tatarian, Dzhulfian, Guadalupian, Ochoan, Zechstein. Symbols as in Fig. 1.

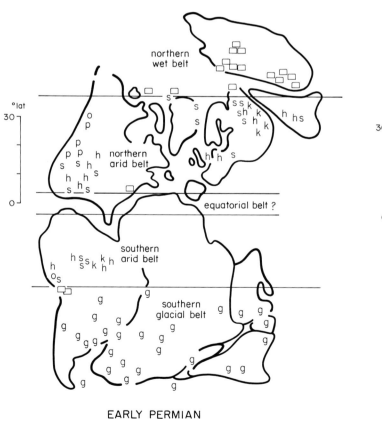

Fig. 12. Plot of Early Permian lithic palaeoclimatic indicators for western Pangaea and Siberia on generalized base map of Scotese (1986). General climatic belts are delimited. Symbols as in Fig. 1.

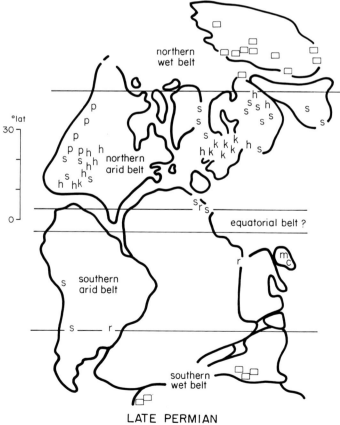

Fig. 13. Plot of Late Permian lithic palaeoclimatic indicators for western Pangaea and Siberia on generalized base map of Scotese (1986). General climatic belts are delimited. Symbols as in Fig. 1.

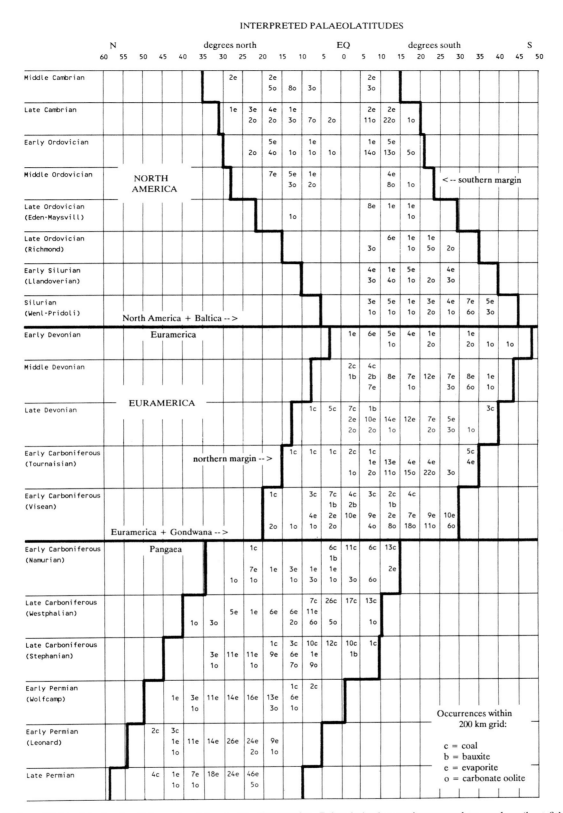

Fig. 14. Compilation of lithic data for the Palaeozoic of Laurentia–Euramerica. Palaeolatitudes are interpreted to produce 'best fit' of lithic data in a zonal-climatic scheme, as portrayed on Figs. 1–8, 10 & 11.

wet belt and glacial deposits of southern Gondwana at appropriate latitudes, in a manner similar to that of the Late Carboniferous. However, the decided asymmetry of northern and southern arid belts, and the general absence of well developed humid indicators in the equatorial region, suggest that non-zonal climatic factors strongly influenced climatic patterns across Pangaea. The Late Permian distribution of palaeoclimatic indicators (Fig. 13) is similar to that of the Early Permian, but the general absence of southern glaciation and the apparent expansion of the southern arid belt possibly suggest that equator-to-pole temperature contrasts became reduced during the Late Permian. The zonal pattern alone does not adequately explain Permian palaeolatitudes, and, as suggested by Rowley et al. (1985), full monsoonal conditions apparently were developed by the Permian.

Conclusions

An attempt has been made to constrain lithic palaeoclimatic indicators for Palaeozoic Euramerica into a zonal-orographic scheme, and latitudinal positionings of the lithic data have been further arranged in a temporally consistent pattern of migration across zonal belts. A tabulation of lithic data on the interpreted palaeolatitudinal base is presented in Fig. 14, which portrays a general southward drift of Euramerica for the Cambrian through Early Devonian and a progressive northward drift for the Middle Devonian through Permian. The placement of lithic data into this framework followed some basic assumptions to maximize a zonal pattern: (1) arid indicators were constrained to produce the 'best fit' of data into appropriate zonal belts, that is from about 5–10° to 35–45° N and S; (2) occurrences of arid indicators, especially evaporites, were minimized from interpreted humid zonal belts, that is in equatorial (0 to 5–10° N and S) and wet temperate (> 45°) belts; (3) occurrences of humid indicators, especially bauxites and widespread coals, were constrained to appropriate humid equatorial or temperate zonal belts or to areas of potential orographic rainfall. These underlying assumptions are, of course, simplistic, but the resulting palaeolatitudinal fit maintains a reasonable temporal consistency.

Palaeozoic lithic data for Euramerica were compiled to produce histograms (Fig. 15) that maximize the distribution of climatic indicators to appropriate latitudes, using the assumptions listed above. In general, the maps in this report (Figs 1–13) were constructed to produce the patterns illustrated on the histograms, so it is no surprise that the bulk of the evaporites, carbonate oolites, coals, and bauxites occur within predicted zonal belts. Although 'best fit' of Palaeozoic lithic data into a zonal scheme was achieved by inspection, the statistical method proposed by Scotese & Barrett (1990) might also be used.

In general, Palaeozoic palaeolatitudes presented in this report provide independent verification of palaeomagnetically based latitudinal reconstructions of Euramerica, and differ by only 0 to 10° for much of the Palaeozoic. The greatest discrepancies are noted for the Devonian and Early Carboniferous, where latitudinal placements on some reconstructions differ by as much 15 to 40°. However, other palaeomagnetically-based reconstructions for the Devonian–Early Carboniferous of Euramerica are virtually identical to the zonal palaeogeographic interpretations of this report, and additional tests (especially biogeographic) are needed to resolve discrepancies in interpretations. The reliability of some palaeomagnetic data may need to be re-evaluated, which seems especially critical for the mid Palaeozoic of Siberia where exceptional climatic anomalies are apparent on the palaeomagnetic reconstructions.

The assumption of the dominance of zonal climatic patterns for the Palaeozoic of Euramerica was central to the interpretations presented here, but this assumption is, of course, not necessarily accurate for the entire span of the Palaeozoic. It is hoped that the distribution of lithic climatic indicators can be more accurately constrained within a more sophisticated zonal–orographic–monsoonal model. This report is merely a first-order attempt to see if a simple zonal–orographic model can be used to constrain the lithic data into a general climatic–latitudinal framework. This approach provided a temporally consistent arrangement that constrained the bulk of the data to predicted latitudes in a generally satisfactory manner. These preliminary results may suggest that zonal patterns dominated Euramerican climates for most of the Early and Middle Palaeozoic. Zonal patterns seem to be disrupted for intervals of the Visean to Late Permian, and the influence of monsoonal circulation over Pangaea is probably the most reasonable explanation. However, even with the development of monsoonal patterns during the Late Palaeozoic, zonal patterns are still evident by the presence of global wet and arid belts at appropriate palaeomagnetically-based latitudes.

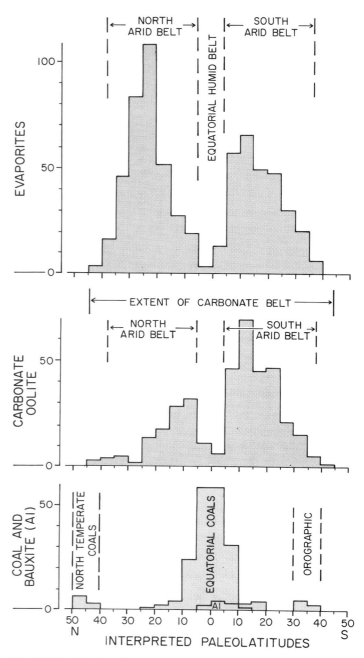

Fig. 15. Distribution of Palaeozoic lithic palaeoclimatic indicators for Laurentia–Euramerica on an interpreted palaeolatitudinal base (see Fig. 14). Evaporites and carbonate oolite are largely restricted to interpreted arid climatic belts and coal–bauxite to humid climatic belts reflecting assumptions used in the palaeogeographic reconstructions.

The efforts of many people have contributed, in one form or another, to the completion of this report. I especially wish to acknowledge C. Scotese, B. Bunker, P. Heckel, G. Ludvigson, R. Anderson, P. Lohmann, J. Day, F. Woodson, and K. Witzke for their assistance and encouragements.

References

BARKER, A. J. & GAYER, R. A. 1985. Caledonide–Appalachian tectonic analysis and evolution of related oceans. *In*: GAYER, R. A. (ed.) *The Tectonic Evolution of the Caledonide–Appalachian Orogen*. Friedr. Vieweg & Son, Braunschweig/Wiesbaden, 126–165.

BOUCOT, A. J. & GRAY, J. 1979. A Cambro–Permian Pangaeic model consistent with lithofacies and biogeographic data. *In*: STRANGWAY, D. W. (ed.) *The Continental Crust and Its Mineral Deposits*. Geological

Association of Canada, Special Paper, **20**, 389–419.

CAPUTO, M. V. & CROWELL, J. C. 1985. Migration of glacial centers across Gondwana during Paleozoic Era. *Geological Society of America Bulletin*, **96**, 1020–1036.

FRAKES, L. A. 1979. *Climates Throughout Geologic Time*. Elsevier, Amsterdam.

HABICHT, J. K. A. 1979. Paleoclimate, Paleomagnetism, and Continental Drift. *American Association of Petroleum Geologists, Studies in Geology*, **9**.

HAMBREY, M. J. & HARLAND, W. B. 1981. Earth's pre-Pleistocene glacial record. *International Geological Correlation Programme Project 38: Pre-Pleistocene Tillites*. Cambridge University Press.

HECKEL, P. H. & WITZKE, B. J. 1979. Devonian world paleogeography determined from distribution of carbonates and related lithic palaeoclimatic indicators. *In*: HOUSE, M. R., SCRUTTON, C. T. & BASSETT, M. G. (eds) *The Devonian System*. Palaeontological Association, London, Special Papers in Palaeontology, **23**, 99–123.

KOPPEN, W. & WEGENER, A. 1924. *Die Klimate der geologischen Vorzeit*. Verlag von Gebruder Borntraeger, Berlin.

LEES, A. 1975. Possible influence of salinity and temperature on modern shelf carbonate sedimentation. *Marine Geology*, **19**, 159–198.

LIVERMORE, R. A., SMITH, A. G. & BRIDEN, J. C. 1985. Paleomagnetic constraints on the distribution of continents in the late Silurian and early Devonian. *Philosophical Transactions of the Royal Society of London, Series B*, **309**, 29–56.

MILLER, J. D. & KENT, D. V. 1986. Synfolding and prefolding magnetizations in the Upper Devonian Catskill Formation of eastern Pennsylvania: implications for the tectonic history of Acadia. *Journal of Geophysical Research*, **91**, 12, 791–12, 803.

—— & —— 1988. Paleomagnetism of the Silurian-Devonian Andreas redbeds: Evidence for an Early Devonian supercontinent? *Geology*, **16**, 195–198.

MONGER, J. W. H. & ROSS, C. A. 1984. Upper Paleozoic volcanosedimentary assemblages of the western North American Cordillera. *9th Congress International de Stratigraphie et de Geologie du Carbonifere*, **3**, 219–228.

PARRISH, J. T. 1982. Upwelling and petroleum source beds, with reference to Paleozoic. *American Association of Petroleum Geologists Bulletin*, **66**, 750–774.

—— & BARRON, E. J. 1986. *Paleoclimates and Economic Geology*. Society of Economic Geologists and Mineralogists, Short Course No. 18.

—— & CURTIS, R. L. 1982. Atmospheric circulation, upwelling, and organic-rich rocks in the Mesozoic and Cenozoic eras. *Palaeogeography, Palaeoclimatology, Palaeoecology*, **40**, 33–66.

——, ZIEGLER, A. M. & SCOTESE, C. R. 1982. Rainfall patterns and the distribution of coals and evaporites in the Mesozoic and Cenozoic. *Palaeogeography, Palaeoclimatology, Palaeoecology*, **36**, 67–101.

ROBINSON, P. L. 1973. Palaeoclimatology and continental drift. *In*: TARLING, D. H. & RUNCORN, S. K. (eds) *Implications of continental drift to the earth sciences*. New York, Academic Press, 451–476.

ROWLEY, D. B., RAYMOND, A., PARRISH, J. T., LOTTES, A. L., SCOTESE, C. R. & ZIEGLER, A. M. 1985. Carboniferous paleogeographic, phytogeographic, and paleoclimatic reconstructions. *International Journal of Coal Geology*, **5**, 7–42.

SCOTESE, C. R. 1984. Paleozoic paleomagnetism and the assembly of Pangea. *In*: VAN DER VOO, R., SCOTESE, C. R. & BONHOMMET, N. (eds) *Plate Reconstruction from Paleozoic Paleomagnetism*. American Geophysican Union, Geodynamics Series, **12**, 1–10.

—— 1986. Phanerozoic Reconstructions: A new look at the assembly of Asia. *University of Texas, Institute for Geophysics, Technical Report*, **66**.

—— & BARRETT, S. F. 1990. Gondwana's movement over the South Pole during the Palaeozoic: evidence for lithological indicators of climate. *In*: MCKERROW, W. S. & SCOTESE, C. R. (eds) *Palaeozoic Palaeogeography and Biogeography*. Geological Society, London, Memoir, **12**, 75–85.

—— BAMBACH, R. K., BARTON, C., VAN DER VOO, R. & ZIEGLER, A. M. 1979. Paleozoic Basemaps. *Journal of Geology*, **87**, 217–277.

SMITH, A. G., HURLEY, A. M. & BRIDEN, J. C. 1981. *Phanerozoic Palaeocontinental World Maps*. University Press, Cambridge.

STREEL, M. 1986. Miospore contribution to the upper Famennian-Strunian event stratigraphy. *In*: BLESS, M. J. M. & STREEL, M. (eds) Late Devonian Events around the Old Red Continent: Annals Societe Geologique Belgique, **109**, 75–92.

TARLING, D. H. 1980. Upper Palaeozoic continental distributions based on palaeomagnetic studies. *In*: PANCHEN, A. L. (ed.) *The Terrestrial Environment and the Origin of Land Vertebrates*. Academic Press, London, 11–37.

—— 1985a. Siluro-Devonian palaeogeographies based on palaeomagnetic observations. *Philosophical Transactions of the Royal Society London, Series B.*, **309**, 81–83.

—— 1985b. Carboniferous reconstructions based on paleomagnetism. *10th Congress International de Stratigraphie et de Geologie du Carbonifere*, 153–168.

TURNER, S. & TARLING, D. H. 1982. Thelodont and other agnathan distributions as tests of Lower Paleozoic continental reconstructions. *Palaeogeography, Palaeoclimatology, Palaeoecology*, **39**, 295–311.

VAN HOUTEN, F. B. & BHATTACHARYYA, D. P. 1982. Phanerozoic oolitic ironstones — Geologic record and facies model. *Annual Review of Earth and Planetary Sciences*, **10**, 441–457.

VAN DER VOO, R. 1988. Paleozoic paleogeography of North America, Gondwana, and intervening terranes: Comparisons of paleomagnetism with paleoclimatology and biogeographical patterns. *Geological Society of America Bulletin*, **100**, 311–324.

VAN DER ZWAN, C. J., BOULTER, M. C. & HUBBARD, R. N. L. B. 1985. Climatic change during the Lower Carboniferous in Euramerica, based on multivariate statistical analysis of palynological data. *Palaeogeography, Palaeoclimatology, Palaeoecology*, **52**, 1–20.

VINOGRADOV, A. P. (ed.) 1969. *Atlas of the Lithological–Paleogeographical Maps of the USSR*. Ministry of Geology, Academy of Science USSR, Moscow, **2**.

WILLMAN, H. B., ATHERTON, E., BUSCHBACH, T. C., COLLINSON, C., FRYE, J. C., HOPKINS, M. E., LINEBACK, J. A. & SIMON, J. A. 1975. *Handbook of Illinois Stratigraphy*. Illinois State Geological Survey, Bulletin 95.

WITZKE, B. J. 1980. Middle and Upper Ordovician Paleography of the region bordering the Transcontinental Arch. *In*: FOUCH, T. D. & MAGATHAN, E. R. (eds) *Paleozoic paleogeography of west-central United States, Rocky Mountain Section*. Society of Economic Paleontologists and Mineralogists, 1–18.

—— & HECKEL, P. H. 1989. Paleoclimatic indicators and inferred Devonian paleolatitudes of Euramerica. *In*: MCMILLAN, N. J., EMBRY, A. F. & GLASS, D. J. (eds) *Devonian of the World*. Proceedings of the Second International Symposium on the Devonian System, Calgary, Canada, **1**, 49–63.

ZIEGLER, A. M., HANSEN, K. S., JOHNSON, M. E., KELLY, M. A., SCOTESE, C. R. & VAN DER VOO, R. 1977. Silurian continental distributions, paleogeography, climatology, and biogeography. *Tectonophysics*, **40**, 13–51.

——, HULVER, M. L., LOTTES, A. L. & SCHMACHTENBERG, W. F. 1984. Uniformitarianism and paleoclimates: inferences from the distribution of carbonate rocks. *In*: BRENCHLEY, P. (ed.) *Fossils and Climate*. John Wiley and Sons Limited, New York, p. 3–25.

——, RAYMOND, A. L., GIERLOWSKI, T. C., HORRELL, M. A., ROWLEY, D. B. & LOTTES, A. L. 1987. Coal, climate and terrestrial productivity: the present and early Cretaceous compared. *In*: SCOTT, A. C. (ed.) *Coal and Coal-bearing Strata: Recent Advances*. Geological Society, Special Publication, **32**, 25–49.

——, SCOTESE, C. R., MCKERROW, W. S., JOHNSON, M. E. & BAMBACH, R. K. 1979. Paleozoic paleogeography. *Annual Review of Earth and Planetary Science*, **7**, 473–502.

ZIEGLER, P. 1982. *Geological Atlas of Western and Central Europe*. Shell Internationale Petroleum, Maatschappij B. V.

ZHARKOV, M. A. 1984. *Paleozoic Salt Bearing Formations of the World*. Springer-Verlag, Berlin.

Gondwana's movement over the South Pole during the Palaeozoic: evidence from lithological indicators of climate

CHRISTOPHER R. SCOTESE[1] & STEPHEN F. BARRETT[2]

[1] *Shell Development Company, Bellaire Research Center, P.O. Box 481, Houston Texas, 77001, USA*
[2] *Amoco Production Co., P.O. Box 3092, Houston, Texas, 77253, USA*

Abstract: A statistical technique is described that uses the geographical distribution of lithological indicators of climate (carbonates, evaporites, coals and tillites) to estimate the past position of the geographic pole. This technique was used to estimate the movement of the South Pole across the supercontinent of Gondwana during the Palaeozoic. Our results indicate that during the Cambrian and Early Ordovician the South Pole was located adjacent to northwestern Africa. The pole moved into the Amazon Basin during the Late Ordovician and into south-central Argentina during the Silurian. Throughout the Devonian and Early Carboniferous the pole moved slowly from a location in southern Argentina to a position near the south coast of Africa. From the Late Carboniferous and into the Permian the South Pole swung eastward across central Antarctica. The Early Palaeozoic and Late Palaeozoic portions of the palaeoclimatically determined APW path are in good agreement with available palaeomagnetic data. The Middle Palaeozoic portion of the palaeoclimatically determined APW path agrees better with the palaeomagnetic data that places the South Pole in southern Argentina, than with the palaeomagnetic results that place the Devonian pole in central Africa.

The anomalous occurrences of ancient tillites in present-day subtropical latitudes or ancient coral reefs near the geographical pole have long been used as evidence for the mobility of continents (Wegener 1915; Koeppen & Wegener 1924; Du Toit 1939). Some lithofacies, such as carbonates, evaporites, coals and tillites are deposited in climatically restricted belts and their changing geographic distribution has been used to infer the latitudinal motion of the continents. For example, Volkheimer (1969) described the latitudinal movement of South America by mapping the changing distribution of climatically restricted lithofacies. Similar descriptions of continental movement across climatic belts have been presented for North America (Heckel & Witzke 1979; Witzke 1990; Witzke & Heckel 1989), Kazakhstan (Ziegler *et al.* 1977), and Gondwana (Caputo & Crowell 1985; Veevers & Powell 1987). In this paper, we describe a statistical technique that uses the geographical distribution of climatically restricted lithofacies to estimate the position of the geographical pole.

An apparent polar wander (APW) path determined using palaeoclimatic data can serve as an independent check of APW paths determined by palaeomagnetic methods. Using the changing latitudinal distribution of palaeoclimatically restricted lithofacies (tillites, coals, evaporites, and carbonates) from Gondwana, we have mapped the apparent polar wander path of the South Pole during the Palaeozoic (Scotese & Barrett 1985). In the final section of the paper we compare this climatically determined APW path for Gondwana with predictions made by palaeomagnetism.

Methods

Climatic principles: the zonal model

We make the assumption that the present-day pattern of Hadley cell circulation (Fig. 1) has been a stable feature of atmospheric circulation for the past 600 million years. As illustrated in Fig. 1, warm air rises at the equator (A, equatorial low pressure zone). As the air rises, it cools and moisture condenses and rains back to earth (equatorial rainy belt). This high-level air mass cools as it moves poleward and descends to earth (B, subtropical high pressure zone) devoid of most of its moisture (subtropical dry belt). The air mass continues to move poleward until it meets cold air escaping from the polar region (C, polar front; temperate rainy belt).

The zonal pattern of atmospheric circulation described above is primarily the result of the differential heating at the equator and cooling at the poles, and the deflection of air masses due to the Coriolis effect. This zonal pattern could be modified by a dramatic increase in either the thermal gradient between the pole and equator or the speed of the Earth's rotation. However, large changes in either of these parameters are unlikely (Parrish 1982; Parrish & Curtis 1982). Modest changes in the pole to equator temperature gradient, on the other hand, probably have been the rule rather than the exception. Climatic belts will expand poleward during times of warming and contract towards the equator during times of cooling. In either case, even during times of dramatic climatic change, the pattern of atmospheric circulation will remain predominantly zonal. Though the zonal pattern has expanded and contracted through time, it is unlikely that climatic changes would have been sufficient to eliminate or add new circulation cells (Parrish 1982; Parrish & Curtis 1982).

Though the pattern of zonal circulation is a basic aspect of the earth's climatic system, there are important effects that disrupt the zonal pattern. These effects include: (1) mountain belts (>2000 m elevation) that restrict and deflect winds, (2) large, east-west oriented landmasses, across which zonal patterns can not be maintained (for example, present-day Eurasia), and (3) large landmasses at low latitudes, that create strongly seasonal monsoonal circulation systems (Parrish *et al.* 1982; Crowley *et al.* 1986, 1987; Kutzbach & Gallimore 1989). While recognizing the importance of these nonzonal effects, in this paper we concentrate on the zonal aspects of the Earth's climate, leaving the nonzonal components to be integrated into a second iteration of the modelling process.

Lithological indicators of climate

Certain rocks types tend to be deposited under restricted climatic conditions. Peat, which may eventually become coal, forms where it is persistently wet (equatorial and temperate rainy belts). Conversely, marine evaporites (halite, anhydrite, and gypsum) occur where it is dry (subtropics). Carbonates, in particular those of Bahamian type, occur where it is warm and where there is adequate sunlight penetration (equatorial, subtropical, and warm temperate regions). Tillites, and other evidence of cold climates and glacial action (dropstone conglomerates, varves, glendonite, striated pavements) occur primarily at temperate and polar latitudes.

Given the relationship of climate and certain lithologies, we have attempted to characterize the latitudinal distribution of several lithologic indicators of climate (carbonates, evaporites, coals, and tillites). The latitudinal distribution of palaeoclimatically restricted lithofacies might be estimated in several ways. A theor-

From McKerrow, W. S. & Scotese, C. R. (eds), 1990, *Palaeozoic Palaeogeography and Biogeography*, Geological Society Memoir No. 12, pp. 75–85.

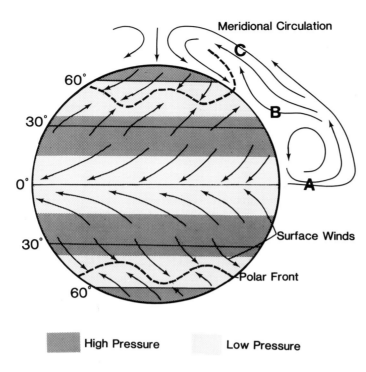

Fig. 1. Zonal pattern of atmospheric circulation (modified after Parrish 1982).

etical approach might attempt to predict the pattern directly from the climate dynamics. Computer simulation of wetness, dryness, and temperature have been made using variants of the General Circulation Model (GCM), (Barron & Washington 1982), however, these simulations are not yet sophisticated enough to take into account all of the factors responsible for the occurrence of coals, evaporites, carbonates, and tillites. In addition, the GCM appears to be 'tuned' to the subtleties of the present-day climate, and may not be a suitable model for the past. Another approach might be to use the present-day distribution of peats (coals), evaporites, carbonates and tillites as a model for the past. This approach, however, also suffers from the problem that, in detail, the present-day climate may not be typical of past climates.

The method we have chosen to characterize the latitudinal distribution of climatically sensitive lithofacies has been to plot the palaeolatitudinal occurrence of Mesozoic and Cenozoic coals, evaporites, carbonates, and tillites in the form of pole-to-pole histograms (Figs. 2a–2d). The frequency of these indicators was summed over 5 degree latitudinal intervals, and the data from the northern and southern hemisphere were combined to form a symmetrical, composite pattern (Fig. 2).

This historical method has the advantages that: sample sizes are large and therefore the observed patterns are probably statistically significant, the palaeolatitudinal positions of the major continents are fairly well constrained, and because the data are sampled over a long time interval (220 million years) secular variations in climate tend to be averaged out. The Mesozoic and Cenozoic plate tectonic reconstructions of Ziegler *et al.* (1983) were used to determine the palaeolatitudes the palaeoclimatic indicators. As noted by pioneering studies of palaeomagnetists eager to confirm the accuracy of their results (Opdyke 1959, 1962; Blackett, 1961; Irving 1964), the observed palaeolatitudes of climatically sensitive lithofacies are in good agreement with their expected latitudinal range. Subsequent studies have confirmed this relationship (Irving & Briden 1962; Briden & Irving 1964; Briden 1968, 1970; Drewry *et al.* 1974; Gordon 1975; Habicht 1975; Ziegler *et al.* 1979, 1981). This agreement between the expected and observed latitudinal range of palaeoclimatically restricted lithofacies both confirms the general accuracy of palaeomagnetism and indicates that the zonal pattern of atmospheric circulation has been a dominant feature of the Earth's climatic system for the past 250 million years.

The latitudinal frequency of carbonates (Fig. 2a) is based on the work of Ziegler *et al.* (1984), who suggested that the penetration of light, rather than water temperature, is the major factor that limits the latitudinal distribution of carbonates. The latitudinal frequency of evaporites (Fig. 2b) and coals (Fig. 3c) is based on

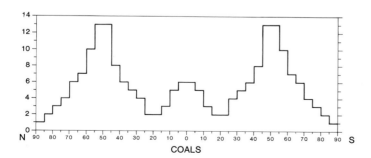

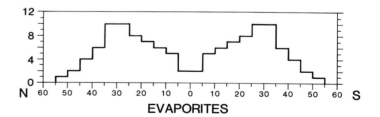

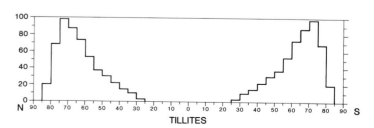

Fig. 2. Histograms illustrating latitudinal frequency of climatically sensitive lithofacies, (**a**), Carbonate, (**b**) Evaporites, (**c**) Coals, (**d**) Tillites.

the work of Parrish et al. (1982). The latitudinal distribution of tillites (Fig. 3d) is more problematical. The frequency distribution is based solely on the area of ice cover during the Pleistocene (McIntyre et al. 1976). Though the sample size is small, the pattern is consistent with the observed occurrence of dropstones and glendonite localities reported from the Mesozoic (Frakes & Francis 1988).

Estimating the position of the geographic pole from the distribution of climatically sensitive lithologies

The histograms describing the latitudinal frequency of climatically sensitive lithofacies (Fig. 2) can be used to estimate the probability that a given lithofacies will occur at a particular latitude. For instance, according to the histogram 2b, 20 of 61 evaporite localities occur between 25 and 35 degrees latitude. A single evaporite locality would therefore have a 0.33 probability of occurring between 25 and 35 degrees. Similarly, it is twice as likely that an evaporite locality will occur at a latitude of 25 degrees than at either a latitude of 5 degrees or 45 degrees.

The probability that a given lithofacies will occur at a particular latitude has also been determined for carbonates, tillites, and coals. Carbonates have a maximum likelihood of occurring between 15° and 25° latitude, the maximum for tillites is between 70° and 80°, and coals have two peaks, one between 45° and 55°, and a subsidiary peak at the equator. The complete range of latitudinally dependent probabilities (P_d) is given in Table 1.

The position of the geographical pole can be estimated by solving the inverse problem. Rather than determine the probability that a given lithofacies will occur at a particular latitude, the frequency distributions can be used to determine the probability that the geographical pole lies at a certain distance from a given lithofacies site. For instance, evaporites have a maximum probability of occurring at a latitude of 30° (Fig. 2b), which is the same as saying that the geographical pole has a maximum probability of occurring at a distance of 60° from any given evaporite locality. In the case of a single evaporite site, the best estimate for the location of the geographical pole would be a doughnut-shaped ring with a radius of 60°, centered about the evaporite locality (Fig. 3a).

As illustrated in Fig. 3b, the probability distributions for two widely separated evaporite localities would resemble in two intersecting rings. Because the probabilities are additive, the best estimate for the location of the geographical pole would be in the areas where the two rings overlap. When 3 or more localities are used the probability distributions begin to intersect in a single region, forming a well-defined maximum (Fig. 3c).

A computer program, 'Pole-Finder', was written to determine the most likely location of the geographical pole. The input data were the latitude and longitude of each lithofacies site and the equator to pole probability distributions described above. The latitudes and longitudes were measured in a Gondwanan reference frame (Lawver & Scotese 1987) in which Africa was held fixed. The probabilities were weighted to take into account the fact that latitudinal sectors near the poles have less area than latitudinal sectors near the equator (Table 1). The program then divided the globe into 2592 cells, each cell 5 degrees on a side, and calculated the distances from a given lithofacies site to the centre of each 5 degree cell. Using the latitudinal distribution for that lithofacies, a probability (P_d) was assigned to each cell based on distance from the lithofacies site and the cell. This last step was then repeated for each lithofacies site and an 'average' probability ($\overline{P}_d$) was calculated for each cell. In the final step, the maximum value of the averaged probabilities (P_{max}) was determined, and all other values were normalized with respect to this maximum value. To display these results, the global grid of normalized probabilities was contoured at intervals representing 10% of the maximum probability value.

This maximum represents the most likely location of the geographical pole. $\overline{P}_{max}$ is the maximum value of all $\overline{P}_d$, where $\overline{P}_d$ is the average probability of each cell, expressed as:

$$\overline{P}_d = \frac{\sum_{i=1}^{N} P_{di}}{N}$$

$\overline{P}_d$ is the probability that the geographic pole lies at given distance from a given lithofacies site, and N is the number of sites.

If the lithofacies data are distributed in a zonal pattern, then as the number of sites increases, the absolute value of the calculated

Table 1. *Calculation of probability functions from latitudinal frequency of palaeoclimatic indicators*

Latitudinal range	Relative area	Weighting factor	Carbonates			Evaporites			Coal			Tillites		
			#	%	P_d	#	%	P_d	#	%	P_d	#	%	P_d
0°– 5°	1.0000	1.000	12	0.09	0.08	2	0.03	0.03	6	0.06	0.04	0	0	0
5°–10°	0.9924	1.0008	14	0.09	0.09	5	0.08	0.07	5	0.05	0.03	0	0	0
10°–15°	0.9772	1.023	19	0.14	0.13	6	0.10	0.09	3	0.04	0.02	0	0	0
15°–20°	0.9546	1.048	20	0.15	0.14	7	0.11	0.11	2	0.02	0.01	0	0	0
20°–25°	0.9248	1.081	20	0.15	0.14	8	0.13	0.12	2	0.02	0.01	0	0	0
25°–30°	0.8879	1.126	19	0.14	0.14	10	0.16	0.16	4	0.05	0.03	3	0.01	0
30°–35°	0.8442	1.185	13	0.09	0.10	10	0.16	0.17	5	0.05	0.03	10	0.02	0.01
35°–40°	0.7941	1.259	9	0.07	0.07	6	0.10	0.11	6	0.06	0.04	15	0.03	0.01
40°–45°	0.7380	1.355	5	0.04	0.04	4	0.07	0.08	8	0.08	0.06	22	0.04	0.02
45°–50°	0.6762	1.479	3	0.02	0.03	2	0.03	0.04	13	0.13	0.10	30	0.06	0.03
50°–55°	0.6093	1.641	2	0.01	0.02	1	0.02	0.02	13	0.13	0.12	37	0.07	0.04
55°–60°	0.5378	1.859	1	0.01	0.01	0	0	0	10	0.10	0.10	53	0.10	0.07
60°–65°	0.4622	2.164	0	0	0	0	0	0	6	0.06	0.07	73	0.14	0.11
65°–70°	0.3830	2.611	0	0	0	0	0	0	6	0.06	0.08	87	0.17	0.16
70°–75°	0.3010	3.322	0	0	0	0	0	0	4	0.04	0.07	98	0.19	0.22
75°–80°	0.2166	4.616	0	0	0	0	0	0	3	0.04	0.08	68	0.13	0.22
80°–85°	0.1307	7.654	0	0	0	0	0	0	2	0.02	0.09	20	0.04	0.11
85°–90°	0.0044	22.900	0	0	0	0	0	0	0	0	0	0	0	0
		Totals	137	100	100	61	100	100	86	100	100	516	100	100

Weighting Factor, 1/Relative Area; #, number of occurrences in each latitudinal sector %; percent of total occurrences; P_d, percent of total occurrences weighted by relative area of sector.

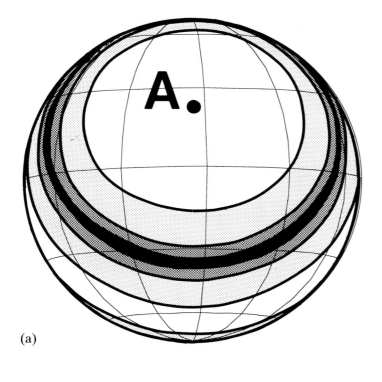
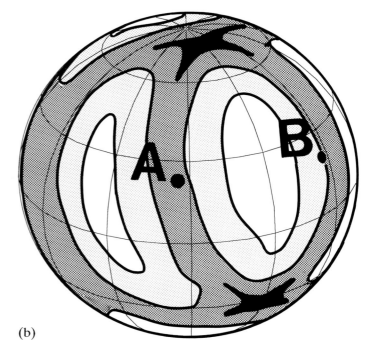
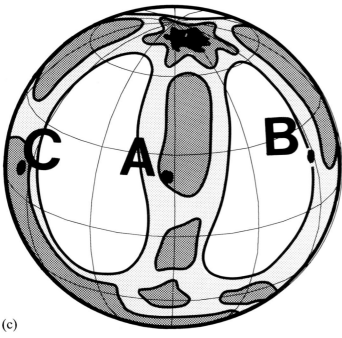

Fig. 3. Illustration showing the progressive improvement in the estimation of geographical pole using, (**a**) one site, (**b**) two sites, and (**c**) three sites. Dense stipple indicates regions of high probability, light stipple indicates regions of low probalility.

maximum probability should also increase. The shape of the area of maximum probability will be determined by the geographical distribution of the data. The best results will be obtained when the lithofacies sites are both widely and uniformly distributed.

A test of the method: estimating the location of the present-day geographical pole from the distribution of modern evaporites, carbonates and peats and tillites

To test the accuracy of this method, the location of the geographical pole was estimated using the present-day geographical distribution of carbonates (Bahamian type), evaporites, and peats (tropical and temperate), as well as tillites. The present-day distribution of evaporites and peats is taken from Ziegler *et al.* 1987. The present-day distribution of carbonates and tillites is based on the lithological database compiled by the University of Chicago Palaeogeographic Atlas Project. Each sample locality represents data from a 5 degree square sample area.

The location of geographical poles estimated from the geographical distribution of coals, evaporites, carbonates, and tillites is given by the symbols in Fig. 4. The contours about the pole, represent the cumulative probability normalized to 100%. Remarkably, the predictions from carbonates (square), coals (circle) and tillites (cross) are all within 10° of the North Pole. The only outlier is the pole predicted from the distribution of evaporites. The region of maximum likelihood based on evaporites is an elongate region stretching from the pole to the tropics along the 170° meridian. The major peak is located at 25° N, 170° E. A

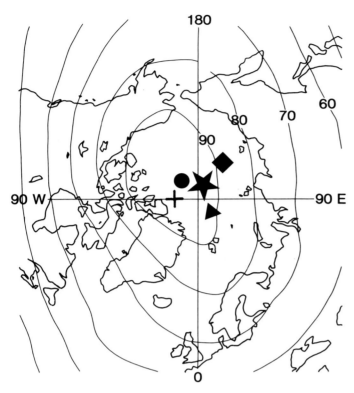

Fig. 4. Predictions of North Pole from the present-day geographical distribution of carbonates (square), evaporites (triangle), peats (circle), and tillites (cross). The composite result is indicated by the large star.

Table 2. *Estimation of present-day geographical pole*

Lithology	Pole Latitude/Longitude		N	K_p
1. Carbonates (Bahamian type)	80°N	145°E	111	0.85
2. Evaporites				
a. (all sites)	25°N	170°E	67	0.75
b. (minus Asia)	85°N	50°E	46	0.78
3. Peat (coal)	85°N	140°W	85	0.78
4. Tillites	85°N	90°W	44	0.80
5. All combined	85°N	140°W	307	0.76
Mean pole using poles 1, 2b, 3 and 4	88°N	171°E	\multicolumn{2}{l}{4, $K = 127.1$, $A_{95} = 8.18$}

subsidiary peak is located at 80°N, 120°W (Fig. 4, triangle). The aberrant evaporite pole is due entirely to the occurrence of evaporites at high latitudes in central Asia. These continental evaporites occur in the rain-shadow of the Himalayan mountains and Tibetan plateau. Because nearly all ancient evaporites are of marine origin, these continental evaporites should be excluded from the analysis. If these localities are removed, the result is a single maximum at 85°N, 50°E. The star in Fig. 4 represents the combined results for carbonates, tillites, evaporites, and coals. The composite pole is located at 85°N, 140°W.

Because each lithofacies is an independent estimator of the position of the geographical pole, a mean pole can be calculated using the same statistical methods employed in palaeomagnetic studies. This technique gives equal weight to each pole. The mean pole based on all four lithofacies is 88°N, 171°E, with confidence limits ($A_{95} = 8.2°$) and estimates of precision ($K = 127.1$) that reflect the tight grouping of the individual poles.

At first glance, it may appear that this test is circular, however, this is not the case. The probabilities used in this test are based on the palaeolatitudinal distribution of Mesozoic and Cenozoic lithologies. No present-day information was used to construct the histograms (Fig. 2) that are the basis of the probability functions (Table 1). The fact that present-day geographical pole can be accurately located using Mesozoic and Cenozoic data suggests that the zonal signal is a robust feature of the climate system.

Table 2 lists the number of localities that were used in each determination, the location of the pole, and an estimate of the precision of the prediction. The precision parameter, K_p, indicates how well the pole has been estimated using the assumption that the lithologic indicators of climate are distributed in a zonal pattern. The maximum value for K_p is 1.0.

$$K_p = \frac{\bar{P}_{max}}{P_{theoretical}}$$

where $\bar{P}_{max}$ is the maximum observed probability, and $P_{theoretical}$ is maximum theoretical probability. A high K_p value (>0.85) indicates that the zonal model is a good predictor of the observed distribution of lithological indicators. A K_p value between 0.70 and 0.85 indicates that the zonal model is a fair predictor, but that there are important nonzonal components. A K_p value less than 0.70 indicates that the distribution of climatic indicators are strongly modified by nonzonal effects.

It is interesting to note that the K_p values for the present-day distribution of lithological indicators are all similar. Similar values are expected because each lithofacies experiences the same, 'instantaneous' climate. For the geological record the assumption of instantaneous sampling will not be valid. The K_p value will decrease in proportion to the length of the geological interval sampled, or if the climate changes rapidly during the sampled interval, such as during glacial and interglacial times. The intermediate K_p value for the present indicates that both zonal and nonzonal effects are important.

Results

An apparent polar wander path for Gondwana determined using palaeoclimatic data

The changing geographic distribution of evaporites, carbonates, coals and tillites from Gondwana has been used to estimate the Palaeozoic apparent polar wander path of the South Pole. Gondwana serves as a good test case to demonstrate this technique because it was a large continent, crossing several climatic zones and for much of the Palaeozoic both the South Pole and equator were located within its borders. The location of the South Pole was determined for 8 time intervals: Early and Middle Cambrian, Late Cambrian–Middle Ordovician, Late Ordovician–Early Silurian, Late Silurian–Early Devonian, Middle and Late Devonian, Early Carboniferous, Late Carboniferous, and Early Permian. Carbonate, evaporite, coal and tillite localities were taken primarily from the compilation of Ronov *et al.* (1984), supplemented by the work of Ziegler *et al.* (1979), Caputo (1985) and Caputo & Crowell (1985) for tillites.

Because of the varying reliability lithofacies descriptions, an index was used to selectively weight the data. Lithofacies occurrences were graded at 3 levels: poor, good, and excellent. An excellent rating was reserved for only those sites at which precise lithological identification and age assignments were available. Generally, descriptions from these sites have been corroborated by numerous studies (e.g. Late Ordovician tillites in North Africa), or have been directly studied by the authors. Good ratings were reserved for sites that were considered to be accurately identified and dated. Poor ratings were used for localities that are known from only a few reports and for which there may be some uncertainty concerning either the lithofacies identification or the age (e.g. Late Devonian tillites of Brazil). A rating of poor was also used to identify those localities that may reside on allochthonous terranes. Only sites with good or excellent ratings have been plotted in Figs 5–12 and were used to estimate the location of the geographical pole.

Review of results by time interval

The location the South Pole determined from the geographical distribution of carbonates, evaporites, coals, and tillites is plotted as a star on Figures 5–12. Lines of latitude, spaced 30° degrees apart, are indicated by dashed lines. The contours, which represent 10% increments of the maximum value, map out the shape of the probability distribution. In addition to the location of the maximum values (geographical pole), the location of the minimum values (equatorial region) should also be noted. The squares represent carbonate localities, the triangles are evaporite localities, the circles represent coal deposits, and the crosses are tillites. Where evaporites and carbonates occur together, a open triangle is plotted on top of a square.

It should be noted that for three time intervals (Early and Middle Cambrian, Late Cambrian–Middle Ordovician, Late Silurian–Early Devonian) the predictions of the location of the South Pole are based only on carbonates and evaporites. For the Late Ordovician–Early Silurian and the Middle–Late Devonian time intervals, carbonates, evaporites and tillites were used. Only for the Late Palaeozoic (Early Carboniferous, Late Carboniferous, and Early Permian) was a full suite of climatic indicators available for analysis.

Early–Middle Cambrian (Fig. 5). Early and Middle Cambrian carbonates and evaporites are reported from nearly all of the margins of Gondwana. Carbonates occur in Bolivia, Argentina, along the Trans-Antarctic Mountain Range, into eastern and northern Australia, and from northern India into the Mideast (Afghanistan, Iran, Turkey, and Arabia). Carbonates of Early and Middle Cambrian age have also been reported from Spain and North Africa. Evaporites occur in northern Argentina and eastern Australia.

The best estimate for the location of the South Pole during the Early and Middle Cambrian is within an arcuate region that extends from North Africa to a location near the present-day North Pole. The location of the pole shown in Fig. 5, (55°N, 7.5°) has been chosen so that the palaeo-equator passes through the area of minimum probability centered in eastern Antarctica.

Late Cambrian–Middle Ordovician (Fig. 6). The distribution of carbonates is more restricted during the Late Cambrian and Ordovician with localities in Argentina, Australia, northern India, and the Mideast. Evaporites occur in central Australia during the Late Cambrian and in western Australia during the Early Ordovician. The westward migration of evaporites across Australia during the Cambrian and Ordovician may reflect the movement of Australia northward across the northern subtropical dry belt.

The best estimate for the location of the South Pole during the Late Cambrian and Middle Ordovician (35°N, 10°W) lies just off the coast of North Africa. During this time the equator crossed the southern tip of South America, bisected eastern Antarctica, and exited Gondwana between India and Australia.

Late Ordovician–Early Silurian (Fig. 7). Carbonates are present in southern South America during the Late Ordovician, and are replaced by a clastic facies during the Early Silurian. Carbonates also occur in Australia, northern India, Tibet; and Late Ordovician carbonates and evaporites have been reported from the Arabian platform. The Late Ordovician and Early Silurian was a time of climatic cooling (Sheehan & Coorough 1990). Late Ordovician tillites are reported from North Africa, Spain, France, and Arabia. Early Silurian tillites are described from Brazil (Caputo 1985), Bolivia, Argentina, and South Africa (Caputo & Crowell 1985). The best estimate for the location of the South Pole during the Late Ordovician and Early Silurian (5°N, 20°W) lies within a broad region that includes the northern part of the Amazon Basin, but also extends into northwestern Africa.

Late Silurian–Early Devonian (Fig. 8). Late Silurian carbonate platforms covered the northeast margin of Gondwana stretching from Australia to the Mideast (Iran, Turkey). By the Early Devonian carbonates extended into Arabia and had spread westward into northernmost Africa. Evaporites of this age have been reported from western Australia, with problematic occurrences described from the Mideast.

During the Late Silurian and Early Devonian, the South Pole moved from northeastern South America to a location in central

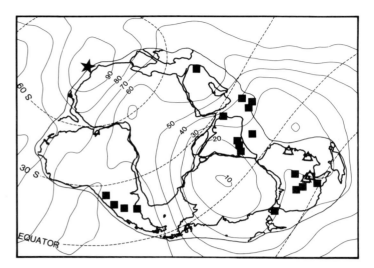

Fig. 6. Late Cambrian–Middle Ordovician. Prediction of South Pole from distribution of climatically sensitive lithofacies: squares, carbonates; triangles, evaporites. Contours represent 10% increments in the value of the maximum probability (best estimate of pole shown as star). The dashed lines represent 30° increments of latitude.

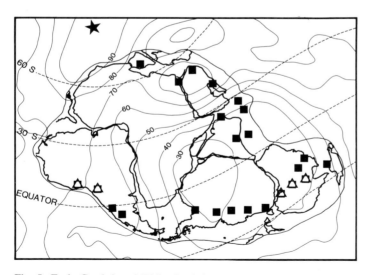

Fig. 5. Early Cambrian–Middle Cambrian. Prediction of South Pole from distribution of climatically sensitive lithofacies; squares, carbonates; triangles, evaporites. Contours represent 10% increments in the value of the maximum probability (best estimate of pole shown as star). The dashed lines represent 30° increments of latitude.

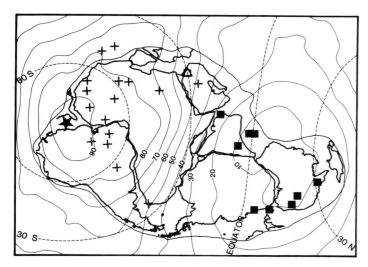

Fig. 7. Late Ordovician–Early Silurian. Prediction of South Pole from distribution of climatically sensitive lithofacies: squares, carbonates; triangles, evaporites; tillites, crosses. Contours represent 10% increments in the value of the maximum probability (best estimate of pole shown as star). The dashed lines represent 30° increments of latitude.

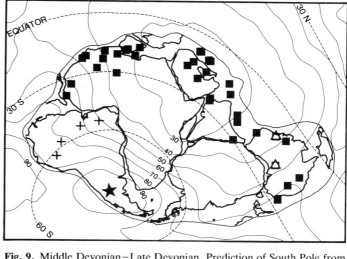

Fig. 9. Middle Devonian–Late Devonian. Prediction of South Pole from distribution of climatically sensitive lithofacies: squares, carbonates; triangles, evaporites; tillites, crosses. Contours represent 10% increments in the value of the maximum probability (best estimate of pole shown as star). The dashed lines represent 30° increments of latitude.

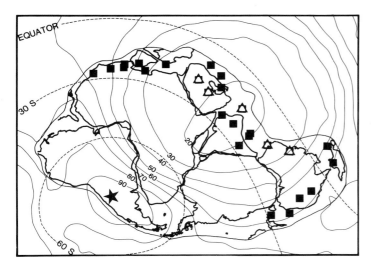

Fig. 8. Late Silurian–Early Devonian. Prediction of South Pole from distribution of climatically sensitive lithofacies: squares, carbonates; triangles, evaporites. Contours represent 10% increments in the value of the maximum probability (best estimate of pole shown as star). The dashed lines represent 30° increments of latitude.

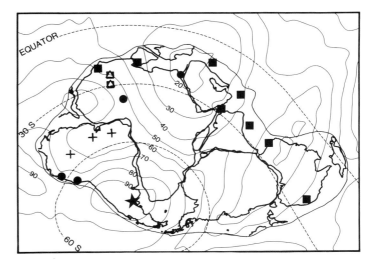

Fig. 10. Early Carboniferous. Prediction of South Pole from distribution of climatically sensitive lithofacies: squares, carbonates; triangles, evaporites; tillites, crosses; coals, circles. Contours represent 10% increments in the value of the maximum probability (best estimate of pole shown as star). The dashed lines represent 30° increments of latitude.

Argentina (30° S, 0° E)). The reappearance of carbonate facies in North Africa reflects the northward movement of that part of Gondwana from high polar latitudes to warm temperate latitudes.

Middle–Late Devonian (Fig. 9). The lithofacies patterns of the Middle and Late Devonian are similar to those of the Early Devonian. Carbonates occur along the entire northern margin of Gondwana, extending southward into north-central and western Africa. Evaporites are present in western Australia; and Late Devonian tillites have been reported from Brazil (Caputo 1985). The best estimate for the location of the South Pole during the Middle and Late Devonian is in south-central Argentina (35° S, 5° E).

Early Carboniferous (Fig. 10). Though there is sparse lithofacies information, the Early Carboniferous is the first time interval for which a full suite of palaeoclimatic indicators is available. Carbonates occur along the northern and northeastern margin of Gondwana. Early Carboniferous evaporites are found in northwest Africa; coals are reported from Peru and north-central Africa. Tillites, are described from Brazil (Caputo 1985) and northwestern Argentina. Despite the relative paucity of data for the Early Carboniferous ($N = 16$), predicted location for the South Pole is fairly well constrained (35° S, 10° E; $K_p = 0.78$).

Late Carboniferous (Fig. 11). 67 sites, comprising all four palaeoclimatic indicators, were used to determine the location of the South Pole during the Late Carboniferous. The most notable change between the Early and Late Carboniferous is the dramatic southward shift of carbonates from the northeastern margin of Gondwana to the northwestern margin of Gondwana. On the Late Carboniferous reconstruction, carbonates extend from the

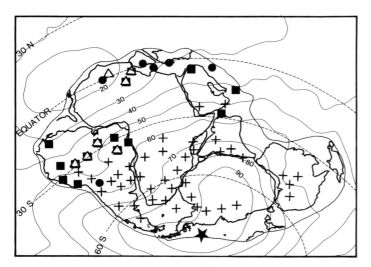

Fig. 11. Late Carboniferous. Prediction of South Pole from distribution of climatically sensitive lithofacies: squares, carbonates; triangles, evaporites; tillites, crosses; coals, circles. Contours represent 10% increments in the value of the maximum probability (best estimate of pole shown as star). The dashed lines represent 30° increments of latitude.

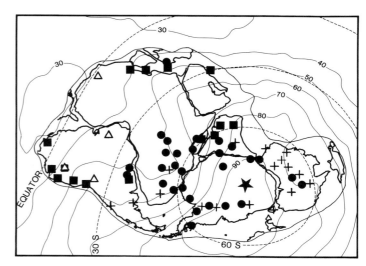

Fig. 12. Early Permian. Prediction of South Pole from distribution of climatically sensitive lithofacies: squares, carbonates; triangles, evaporites; tillites, crosses; coals, circles. Contours represent 10% increments in the value of the maximum probability (best estimate of pole shown as star). The dashed lines represent 30° increments of latitude.

Mideast, across northern Africa, and into northern South America. Evaporites occur in North Africa and the Amazon Basin of Brazil. Temperate coals are reported from northwestern Argentina, and tropical coals are present along the northernmost fringes of Gondwana. Tillites were widespread during the Late Carboniferous, covering much of southern South America, South Africa, Antarctica, Madagascar, India, and Australia (Caputo & Crowell 1985).

Despite the large number of sample sites, the predicted location of the South Pole (50° S, 45° E) based on the combined results of all lithological indicators is not well constrained. However, because of the large sample size and the diversity of lithofacies, it is possible to calculate separate poles for each climatic indicator. As indicated in Table 2, there is fairly good agreement between the poles based on tillites (T) and coals (P). The carbonate pole (C) is divergent and results in the southward extension of the region of maximum probability. The evaporite pole (E) is aberrant due to the small sample size ($N = 7$).

Early Permian (Fig. 12). Like the Late Carboniferous, the Permian pole is also based on a complete suite of lithofacies. Carbonates occur across the northern half of Gondwana with localities in northwestern India, Turkey, North Africa, Colombia, Peru, Bolivia, and eastern Argentina. Evaporites are reported from Morocco and the Amazon Basin. Widespread temperate coals replace tillites as the dominant palaeoclimatic indicator occurring in eastern Argentina, through much of South Africa, and in East Antarctica, Madagascar, India, and Australia. Lingering tillites are reported from southern Argentina, South Africa, East Antarctica, and Australia. The best estimate for the location of the South Pole during the Early Permian is at 25° S, 65° E.

The relative importance of zonal versus non-zonal circulation systems during the Palaeozoic

The K_p values listed in Table 3 for the Palaeozoic reflect the relative importance of zonal versus nonzonal effects. High K_p values correspond to times when the atmospheric circulation can be characterized as 'zonal'. Low K_p values may indicate times when atmospheric circulation was strongly modified by nonzonal effects (e.g. moonsoons). Quick inspection reveals that the time intervals can be divided into two groups. One group consisting of four time intervals (Early Cambrian–Middle Cambrian, Late Cambrian–Middle Ordovician, Late Silurian–Early Devonian, and Middle Devonian–Late Devonian) have high K_p values (≥ 0.84) indicating that the zonal model is a good predictor of the geographical distribution of lithological indicators. At the other extreme are the times with K_p values ≤ 0.69 (Late Ordovician–Early Silurian, Late Carboniferous, and Early Permian). During these time intervals non-zonal aspects of the climate system appear to have dominated. The Early Carboniferous has an intermediate K_p value (0.78) that is similar to the present-day, indicating that both zonal and nonzonal aspects of the climate system were important.

The gradual increase in the K_p values suggest that there was an increase in the importance of the nonzonal components of atmospheric circulation from the Middle to the Late Palaeozoic. This trend is consistent with the expectation that as Pangaea gradually assembled, the climate became dominated by a more monsoonal pattern of atmospheric circulation (Kutzbach & Gallimore 1989). It is also notable that the time intervals with the lowest K_p values were times of extensive glaciation. The low K_p values might therefore also reflect 'climate averaging' resulting from mixing of lithofacies boundaries due to the rapid transition between glacial and inter-glacial climates.

Comparison of the palaeoclimatically determined APW for Gondwana with an APW path based on palaeomagnetic data

The best estimates for the location of the South Pole illustrated in Figures 5–12, have been combined in Fig. 13 to illustrate the apparent movement of the South Pole across Gondwana during the Palaeozoic. The shaded areas represent the best estimate of the location of the South Pole from Early Cambrian to Early Permian times. These areas often overlap and follow one another in a progressive, time dependent sequence (Fig. 13).

We have connected the shaded areas by a smooth curve that passes through the centres of these regions (Fig. 13). This solid line is our interpreted apparent polar wander (APW) path for Gondwana. The dashed line, also plotted in Fig. 13, is the APW path for Gondwana based on a palaeomagnetic data (Bachtadse & Briden 1990). It is remarkable how well portions of the palaeoclimatically determined APW path match the palaeomagnetically

Table 3. *Gondwana poles*

Time interval		Pole Lat./Long.		N	K_p	Lithologies
1. Early Cambrian–Middle Cambrian		45°N	10°W	42	0.84	carbonates, evaporites
2. Late Cambrian–Middle Ordovician		35°N	10°W	36	0.93	carbonates, evaporites
3. Late Ordovician–Early Silurian		5°N	20°W	30	0.69	carbonates, evaporites & tillites
4. Late Silurian–Early Devonian		30°S	0°W	38	0.90	carbonates, evaporites
5. Middle Devonian–Late Devonian		35°S	5°E	43	0.87	carbonates, evaporites & tillites
6. Early Carboniferous		35°S	10°E	19	0.78	carbonates, evaporites, tillites, & coals
7. Late Carboniferous	a.	50°S	45°E	67	0.59	carbonates, evaporites, tillites, & coals
	b.	20°S	55°E	61	0.63	carbonates, evaporites, & tillites
	c.	70°S	65°E	15	0.95	carbonates only
	d.	20°S	115°E	7	0.93	evaporites only
	e.	20°S	35°E	39	0.62	tillites only
	f.	50°S	45°E	6	0.97	coals only
Mean pole of c. d. e. f		45°S	66°E			
8. Early Permian	a.	25°S	65°E	64	0.64	carbonates, evaporites, tillites & coals
	b.	70°S	85°E	14	0.96	carbonates only
	c.	15°S	115°E	4	0.90	evaporites only
	d.	20°S	65°E	17	0.76	tillites only
	e.	50°S	50°E	29	0.74	coals only
Mean pole of b, c, d, e		41°S	79°E			

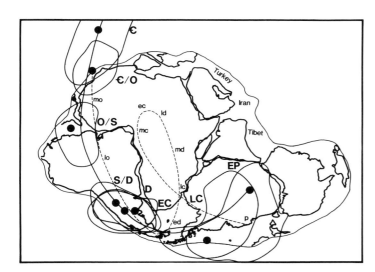

Fig. 13. Gondwana APW. Comparison of palaeoclimatically determined APW path (solid line), with paleomagnetically determined APW path (dashed line, after Bachtadse & Briden 1990). Black circles mark the location of the South Pole from Figs 5–12, (P>90%). C, Early–Middle Cambrian; C/O, Late Cambrian–Early Ordovician; O/S, Late Ordovician–Early Silurian; S/D, Late Silurian–Early Devonian; D, Middle–Late Devonian; EC, Early Carboniferous; LC, Late Carboniferous; EP, Early Permian.

determined APW path. The palaeomagnetic APW path, for most of its length, lies adjacent to the palaeoclimatic APW. The notable exception, however, is the Middle Devonian through Middle Carboniferous 'loop' in the palaeomagnetic APW path.

The Ordovician through Permian portion of both paths begin with the South Pole located in the vicinity of northwestern Africa. The South Pole then moves south across central South America during the Late Ordovician, and into southern South America by the Silurian. The paths diverge in the Devonian and Early Carboniferous. During this interval the palaeoclimatic APW path moves slowly southeastwards towards South Africa, whereas the palaeomagnetic APW path makes a rapid excursion northward into central Africa and quickly loops back towards southern Africa during the middle Carboniferous.

The excursion in the palaeomagnetic APW requires Gondwana to move at rates of 8 to 12 cm per year. These rates are unusually high for plates comprising large areas of continental crust. In comparison, Eurasia has moved at an average rate of 3.8 cm per year during the last 250 million years (Denham & Scotese 1987). Though the palaeoclimatic APW path also requires Gondanwa to move relatively quickly (6.3 cm per yr), this rate is still within the expected range for plates with a high ratio of continental to oceanic crust (3–7 cm per yr).

It is interesting to note that both proposed APW paths differ significantly from previously published palaeomagnetically determined APW paths for Gondwana (McElhinny & Embleton

1974; Schmidt & Morris 1977; Brock 1981). Previous APW paths for Gondwana usually have connected the Ordovician poles in northwest Africa with the Permian poles in central Antarctica by means of route through central Africa. The palaeomagnetic summary of Bachtadse & Briden (1990) and palaeoclimatic results presented here both suggest that the Ordovician through Silurian portion of the path should be displaced westward through South America.

Summary and conclusions

This paper presents a simple, statistical technique that uses lithological indicators of climate to estimate the past position of the geographical pole. This technique provides an important check on apparent polar wander paths determined using palaeomagnetic methods and can provide estimates of the position of the geographical pole for time intervals when palaeomagnetic data is not available.

The technique described here can be adapted for use with other kinds of palaeoclimatic indicators (e.g. oolites, faunal or floral distributions, storm deposits). The principle requirement is that it must be possible to construct an equator-to-pole histogram tha characterizes the latitudinal frequency of the given climatic indicator (e.g. Fig. 2). This raw data must then be converted to area-normalized, latitudinally-dependent probabilities (P_d, Table 1).

This statistical technique was used to estimate the movement of the South Pole across the supercontinent of Gondwana during the Palaeozoic. During the Cambrian and Early Ordovician the South Pole was located adjacent to northwestern Africa (Figs 5 and 6). The pole moved into the Amazon Basin during the Late Ordovician (Fig. 7) and crossed central South America during the Silurian (Fig. 8). Throughout the Devonian and Early Carboniferous the pole moved slowly from a location in southern Argentina to a position near the south coast of Africa (Figs 9 and 10). From the Late Carboniferous and into the Early Permian the South Pole swung eastward across central Antarctica (Figs 11 and 12).

The palaeoclimatically determined APW path of Gondwana is similar to the APW path determined using palaeomagnetic data. The major difference is that there is no palaeoclimatic support for the proposed excursion of the South Pole into Central Africa during the Devonian and Early Carboniferous (Van der Voo 1988; Bachtadse & Briden 1990; Kent & Van der Voo 1990).

In order to understand the palaeogeographical development of the Earth, evidence from several disciplines must be brought to bear: climatology, biogeography, geology and tectonics, and geophysics (palaeomagnetism). Eventually, the story of Earth's history told by each of these disciplines must come into agreement. The challenge in the future will be to erect a system of independent tests and checks that will help bring these dissonant voices together. In this paper we have outlined a method by which lithological indicators of climate can be used to quantitatively evaluate palaeomagnetic data and test palaeogeographical models. Using these tools, we hope to begin painting a more vivid picture of the history of the earth.

This research was begun while the author (CRS) was at the University of Texas, and was supported by the sponsors of the Palaeoceanographic Mapping Project (POMP). Special thanks are given to A. M. Ziegler (University of Chicago) for permission to use unpublished portions of the Palaeogeographic Atlas Project lithologic database and to Malcolm Ross (Rice University) for his help writing the program, 'Pole-Finder'. Copies of the computer program are available from the authors. We would also like to thank Nie Shang-Yu, J. Parrish and B. Witzke for their thoughtful reviews, and Dung Huynh for his help in preparing the final illustrations.

References

BACHTADSE, V. & BRIDEN, J. C. 1990. Palaeomagnetic constraints on the position of Gondwana during Ordovician to Devonian times, *In*: MCKERROW, W. S. & SCOTESE, C. R. (eds) *Palaeozoic Palaeogeography and Biogeography*. Geological Society, London, Memoir, **12**, 43–48.

BARRON, E. J. & WASHINGTON, W. M. 1982. Cretaceous climate: A comparison of atmospheric simulations with the geologic record. *Paleogeography, Paleoclimatology and Paleoecology*, **40**, 103–133.

BLACKETT, P. M. S. 1961. Comparisons of ancient climate with the ancient latitude deduced from rock magnetic measurements. *Proceedings of the Royal Society of London*, **A263**, 1–30.

BRIDEN, J. C. 1968. Paleoclimatic evidence of a geocentric axial dipole field. *In*: PHINNEY, R. A. (ed.) *The History of the Earth's Crust, A Symposium*. Princeton University Press, Princeton, N.J. 178–194.

BRIDEN, J. C. 1970. Palaeolatitude distribution of precipitated sediment. *In*: RUNCORN, S. K. (ed.) *Palaeogeophysics*. Academic Press, New York, N. Y. 437–444.

—— & IRVING, E. 1964. Paleolatitude spectra of sedimentary paleoclimatic indicators. *In*: NAIRN, A. E. M. (ed.) *Problems in Palaeoclimatology*. Interscience, New York, N.Y. 199–224.

BROCK, A. 1981. Paleomagnetism of Africa and Madagascar. *In*: MCELHINNY, M. W. & VALENCIO, D. A. (eds) *Paleoreconstruction of the Continents*, American Geophysical Union Geodynamics Series, **2**, 65–76.

CAPUTO, M. V. 1985. Late Devonian glaciation in South America. *Paleogeography Paleoclimatology and Paleoecology*, **51**, 291–317.

—— & CROWELL, J. C. 1985. Migration of glacial centers across Gondwana during the Paleozoic Era. *Geological Society of America Bulletin*, **96**, 1020–1036.

CROWLEY, T. J., SHORT, D. A., MENGEL, J. G. & NORTH, G. R. 1986. Role of seasonality in the evolution of climate during the last 100 million years. *Science*, **231**, 579–584.

——, MENGEL, J. G., & SHORT, D. A. 1987. Gondwanaland's seasonal cycle. *Nature*, **239**, 803–807.

DENHAM, C. R. & SCOTESE, C. R. 1987. *Terra Mobilis: A Plate Tectonics program for the Macintosh*. Earth in Motion Technologies, Houston, Tx.

DREWRY, G. E. RAMSAY, A. T. S. & SMITH, A. G. 1974. Climatically controlled sediments, the geomagnetic field, and trade wind belts in Phanerozoic time. *Journal Geology*, **82**, 531–553.

DU TOIT, A. L. 1939. *Our Wandering Continents*, Oliver and Boyd.

FRAKES, L. A. & FRANCIS, J. E. 1988. A guide to Phanerozoic cold polar climates from high-latitude ice-rafting in the Cretaceous. *Nature*, **333**, 547–549.

GORDON, W. A. 1975. Distribution by latitude of Phanerozoic evaporite deposits. *Journal of Geology*, **83**, 671–684.

HABICHT, J. K. A. 1975. *Paleoclimate, Paleomagnetism, and Continental Drift*. American Association of Petroleum Geologist, Studies in Geology, No. 9.

HECKEL, P. H. & WITZKE, B. J. 1979. Devonian world paleogeography determined from the distribution of carbonates and related lithic paleoclimatic indicators, *In*: HOUSE, M. R., SCRUTTON, C. T. & BASSETT, M. G. (eds) *The Devonian System*, Palaeontological Association London, Special Papers in Palaeontology, **23**, 99–123.

IRVING, E. 1964. *Paleomagnetism and its Application to Geological and Geophysical Problems*. John Wiley, New York, N.Y.

——, & BRIDEN, J. C. 1962. Paleolatitude of evaporite deposits. *Nature*, **196**, 425–428.

KENT, D. E. & VAN DER VOO, R. 1990. Palaeozoic paleogeography from paleomagnetism of the Atlantic-bordering continents. *In*: MCKERROW, W. S. & SCOTESE, C. R. (eds) *Palaeozoic Palaeogeography and Biogeography*. Geological Society, London, Memoir, **12**, 49–56.

KOEPPEN, W. & WEGENER, A. 1924. *Die Klimate der geologischen Vorzeit*, Verlag von Gebruder Borntraeger, Berlin.

KUTZBACH, J. E. & GALLIMORE, R. G. 1989. Pangaean climates: megamonsoons of the megacontinent. *Journal Geophysical Research*, (in press).

LAWVER, L. A. & SCOTESE, C. R. 1987. A revised reconstruction of Gondwanaland. *In*: MCKENZIE, G. (ed.) *Gondwana Six: Structure, Tectonics, and Geophysics*, Geophysical Monograph **40**, American

Geophysical Union, Washington, D.C. 17–24.

McElhinny, M. W. & Embleton, B. J. J. 1974. Australian paleomagnetism and the Phanerozoic plate tectonics of eastern Gondwana, *Tectonophysics*, **22**, 1–29

McIntyre, A., Moore, T. C., Andersen, B., Balsam, W., Be, A., Brunner, C., Cooley, J., Crowley, T., Denton, G., Gardner, J., Geitzenauer, Hays, J. D., Hutson, W., Imbrie, J., et al. 1976. The surface of ice-age earth. *Science*, **191**, 1131–1137.

Opdyke, N. D. 1959. The impact of paleomagnetism on paleoclimatic studies. *International Journal of Bioclimatology and Biometeorology*, **3 (4A)**, 1–11.

Opdyke, N. D. 1962. Paleoclimatology and continental drift. *In*: Runcorn, S. K. (ed.) *Continental Drift*. Academic Press, New York, New York, pp. 41–65.

Parrish, J. T. 1982. Upwelling and petroleum source beds, with reference to the Palaeozoic. *American Association of Petroleum Geologists Bulletin*, **66**, 750–774.

—— & Curtis, R. L. 1982. Atmospheric circulation, upwelling, and organic-rich rocks in the Mesozoic and Cenozoic Eras. *Paleogeography, Paleoclimatology and Paleoecology*, **40**, 31–66.

——, Ziegler, A. M. & Scotese, C. R. 1982. Rainfall patterns and the distribution of coals and evaporites in the Mesozoic and Cenozoic. *Paleogeography, Paleoclimatology and Paleoecology*, **40**, 67–101.

Ronov, A., Khain, V. & Seslavinsky, K. 1984. *Atlas of Lithologic-Paleogeographic Maps of the World*, Leningrad.

Schmidt, P. W. & Morris, W. A. 1977. An alternative view of the Gondwana Paleozoic apparent polar wander path, *Canadian Journal of Earth Sciences*, **14**, 2674–2678.

Scotese, C. R. & Barrett, S. F. 1985. Paleoclimatic constraints on the motion of Gondwana during the Palaeozoic, (abstract). *The Sixth International Gondwana Symposium, Ohio State University, Columbus, Ohio, August 19–23, 1985*.

Sheehan, P. M. & Coorough, P. J. 1990. Brachiopod zoogeography across the Ordovician-Silurian extinction event, *In*: McKerrow, W. S. & Scotese, C. R. (eds) *Palaeozoic Palaeogeography and Biogeography*, Geological Society, London, Memoir, **12**, 181–187.

Van der Voo, R. 1988. Palaeozoic paleogeography of North America, Gondwana, and the intervening displaced terranes: comparisons of paleomagnetism with paleoclimatology and biogeographical patterns. *Geological Society of America Bulletin*, **100**, 311–324.

Van der Voo, R., Peinado, J. & Scotese, C. R. 1984. A paleomagnetic reevaluation of Pangaea reconstructions, *In*: Van der Voo, R., Scotese, C. R. & Bonhommet, N. (eds), *Plate Reconstruction from Paleozoic Paleomagnetism*, American Geophysical Union, Geodynamics Series, **12**, 11–16.

Veevers, J. J. & Powell, C. Mc. 1987. Late Paleozoic glacial episodes in Gondwanaland reflected in transgressive-regressive depositional sequences in Euramerica. *Geological Society of America Bulletin*, **98**, 475–487.

Volkheimer, W. 1969. Paleoclimatic evolution in Argentina and relations with other regions of Gondwana. *In*: *Gondwana Stratigraphy*. International Union of Geological Sciences, UNESCO, 551–588.

Wegener, A. 1915. *Die Entstehung der Kontinente und Ozeane*. Bruanschwieg, Druck und Verlag von Freidr. Vieweg & Sohn.

Witzke, B. J. 1990. Paleoclimatic constraints for Palaeozoic palaeolatitudes of Laurentia and Euramerica, *In*: McKerrow, W. S. & Scotese, C. R. (eds) *Palaeozoic Palaeogeography and Biogeography*. Geological Society, London, Memoir, **12**, 57–73.

—— & Heckel, P. H. 1989. Paleoclimatic indicators and the inferred Devonian paleolatitudes of Euramerica. *2nd International Symposium on the Devonian System* (in press).

Ziegler, A. M., Barrett, S. F. & Scotese, C. R. 1981. Paleoclimate, sedimentation, and continental accretion. *In*: Moorbath, S. & Windley, B. F. (eds), *The Origin and Evolution of the Earth's Crust. Philosophical Transactions of the Royal Society*, **A 301**, 253–264.

——, Hansen, K. S., Johnson, M. E., Kelly, M. A., Scotese, C. R. & Van der Voo, R. 1977. Silurian continental distributions, paleogeography, climatology, and biogeography, *Tectonophysics*, **40**, 13–51.

——, Hulver, M. L., Lottes, A. L. & Schmachtenberg, W. F. 1984. Uniformitarianism and paleoclimates: inferences from the distribution of carbonate Rocks, *In*: Benchley, P. J. (ed.), *Fossils and Climate*, John Wiley and Sons, Chichester, 3–25.

——, Raymond, A. L., Gierlowski, T. C., Horrell, M. A., Rowley, D. B. & Lottes, A. L. 1987. Coal, climate and terrestrial productivity: the present and early Cretaceous compared. *In*: Scott, A. C. (ed.), *Coal and Coal-bearing Strata: Recent Advances*, Geological Society Special Publication, **32**, 25–49.

——, Scotese, C. R. & Barrett, S. F. 1983. Mesozoic and Cenozoic paleogeographic maps, *In*: Brosche, P. & Sundermann, J. (eds), *Tidal Friction and the Earth's Rotation II*, Springer-Verlag, Berlin. 240–252.

——, ——, McKerrow, W. S., Johnson, M. E. & Bambach, R. K. 1979. Paleozoic paleogeography, *Annual Review of Earth Sciences*, **7**, 473–502.

Stratigraphic and palaeogeographic distribution of Palaeozoic oolitic ironstones

F. B. VAN HOUTEN[1] & HOU HONG-FEI[2]

[1] *Department of Geological and Geophysical Sciences, Princeton University, Princeton, NJ 08544, USA*
[2] *Institute of Geology, Chinese Academy of Geological Sciences, Baiwanzhuang Road, Beijing, PR China*

Abstract: The record of Palaeozoic oolitic ironstones reflects second-order Phanerozoic sea-level changes. Middle Cambrian–earliest Ordovician high stand: scattered ironstones developed on low-latitude Laurentia and locally on high southern-latitude NW Africa and Nova Scotia. Early–early Late Ordovician high stand: major and scattered minor ironstones were widespread on high southern-latitude N Africa and Peri-Gondwanan blocks. Minor ones on C Laurentia, NE and SW Baltica, SW Kazakhstan, and NW and SW Siberia, NW Malaysia, and C Australia accumulated in middle and low latitudes, as did all later Palaeozoic ironstones. Latest Ordovician–earliest Silurian low stand: during southern-latitude glaciation minor ironstones were limited to WC Algeria and low-latitude C and SE Laurentia. Early–early Late Silurian high stand: minor ironstones accumulated on NC, WC, and SE South America, N Africa, and Peri-Gondwanan blocks, while major ones developed on NW Africa and CE Laurentia. Latest Silurian–Early Devonian low stand: major ironstones developed on WC Algeria; minor ones on CE Laurentia, several Peri-Gondwanan blocks, N and W Africa, NC South America, and South China. Middle–early Late Devonian high stand: major ironstones accumulated on N Africa, SE Baltica, NW Iberia, and South China; minor ones on SW and CE Laurentia, and NW and SE Middle Europe. Latest Devonian–earliest Carboniferous low stand: well-developed latest Devonian ironstones were limited to N Africa and NW Middle Europe, minor ones to SE Middle Europe and CE Laurentia; minor earliest Carboniferous ones to NW Middle Europe and NC Africa.

Palaeozoic oolitic ironstones are distinctive sedimentary deposits containing more than 10–15 per cent ooids composed principally of ferric oxide and iron-rich clay minerals. They accumulated in several different kinds of lithostratigraphic sequences and tectonic frameworks, and they varied in abundance in space and time. In order to recognize similarities and evaluate differences among these ironstones we first review information derived from the Phanerozoic record of more than 175 examples.

Principal patterns

Most of the Phanerozoic ironstones developed along passive cratonic margins and, less comonly, in intracratonic and foreland basins, as well as on drifting and docked microplates. They accumulated during times of quiescence, extension, and subdued orogeny, at times when normal sedimentation had waned.

Development of ironstones flourished in the 150–175 Ma Ordovician–Devonian and Jurassic–Palaeogene phases of first-order Phanerozoic cycles (Van Houten & Bhattacharyya 1982). In these favourable phases ironstones were especially abundant in Ordovician time, mostly in high southern latitudes, and in Devonian and Jurassic times, mostly in low and middle latitudes.

More than half of the Phanerozoic ironstones are restricted to ten major sedimentary basins where they developed repeatedly during second-order eustatic high stands (Van Houten & Arthur 1989) which lasted for several to many tens of million years. Sites of repeated ironstones persisted longest on cratons and along their margins. In these successions as many as a dozen ironstones recurred at intervals estimated at a few to several million years.

Lithostratigraphic differences among Phanerozoic ironstones

Most of the oolitic ironstones are associated with shallow marine detrital deposits. These accumulated in both high and low latitudes in several different tectonic frameworks. The relatively uncommon ironstones in mixed carbonate-detrital facies are mostly minor ones that accumulated on major cratons in low and middle latitudes.

As a marked exception to the well-documented marine origin of most ironstones some mid-Cenozoic ones, principally those in the Turgay Basin and northern Aral lowland of western Kazakhstan, were closely associated with coastal plain (paludal-lacustrine and alluvial) and estuarine facies (Zitzmann 1977, p. 364–367). It is not clear, however, to what extent these 'nonmarine' ironstones were related to repeated flooding of the coastal plain by the Uralian Sea (Bronevoi *et al.* 1967).

Beds containing ooids and subordinate peloids have several rather consistent modes (see Guerrak 1987). Thin (less than 20 cm) and lean (less than 30 %) muddy oolites probably accumulated on prograding mudflats at the site of origin of the ooids and peloids, thus providing information about local conditions under which the granules were formed (Bhattacharyya 1989). Thick (1–5 m) oolites containing as much as 80 per cent ooids commonly contain features pointing to repeated winnowing (Bayer *et al.* 1985) and some suggest concentration on subtidal sandwaves (Teyssen 1984). Persistent thin, muddy to well-sorted ironstones display features suggestive of high energy events, such as storm deposits, and are commonly associated with a mixed carbonate-detrital facies (Cotter 1988; Dreesen 1982).

Palaeozoic record

The distribution of Palaeozoic oolitic ironstones (Fig. 1; Table 1) has been plotted on the continental drift reconstructions (Fig. 2) of Scotese (1988). In our Discussion we refer to somewhat different Silurian and Devonian latitudinal positions of Western Gondwana proposed by Hargraves *et al.* (1987) and by Van der Voo & Kent (1988).

Middle Cambrian–earliest Ordovician time (Fig. 2.1)

During this earliest major Phanerozoic high stand of sea level Laurentia, Kazakhstan, Siberia, South China, Southeast Asia, and Australia lay in low latitudes, whereas Baltica, Africa, South America, and Peri-Gondwanan microplates (Table 1) were in higher southern latitudes. The oldest recorded ironstone accumulated in Middle Cambrian time on northeast Nova Scotia and along the Moroccan margin (High Atlas) of North Africa (Destombes *et al.* 1985, p. 322) where marine flooding had prevailed since latest Proterozoic time. Scattered Late Cambrian ironstones were limited to the widely flooded central Laurentia (midcontinent USA). Throughout most of North Africa and on the Peri-Gondwanan blocks a Late Cambrian sedimentary record is missing. With renewed deposition an earliest Ordovician ironstone accumulated on west-central Algeria. In contrast to such a meagre production of ironstones, this episode was marked by one of the most extensive developments of glauconitic deposits

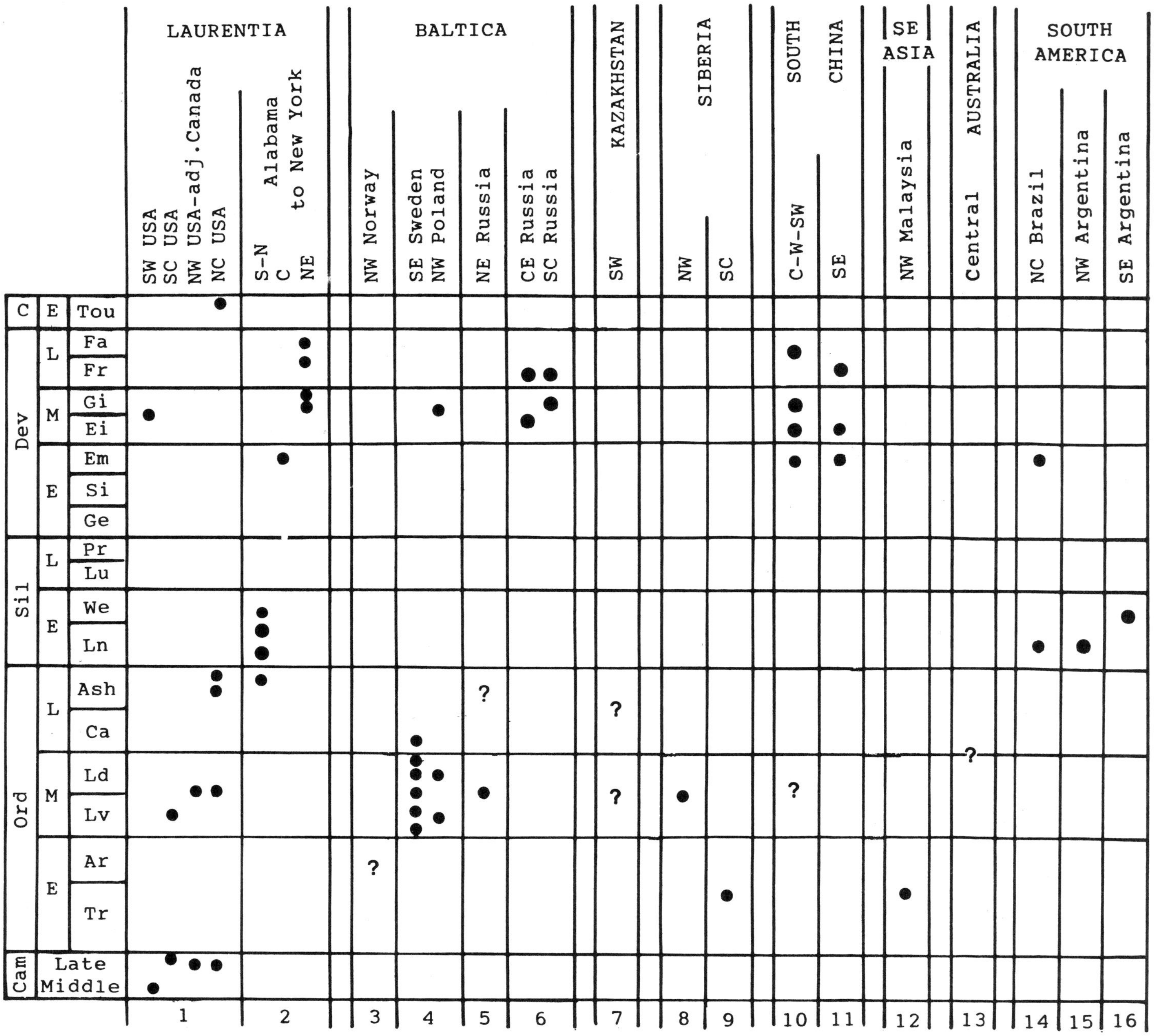

Fig. 1. Stratigraphic distribution of Palaeozoic oolitic ironstones on major cratons and Peri-Gondwanan microplates. Localities numbered as in Table 1.

Table 1. Tectonic framework of Palaeozoic oolitic ironstones. Localities numbered as in Fig. 1.

LAURENTIA
(1) *Cratonic interior.* SW USA, C Arizona; SC USA, New Mexico, Texas, Oklahoma; NW USA–adjacent Canada, Wyoming, S. Dakota, N. Dakota, SE Saskatchewan, S Manitoba; NC USA, Kansas, Nebraska, Iowa, Indiana, Wisconsin
(2) *Taconian and Acadian foreland basin.* S–N, Alabama to New York; C, C Virginia to SC Pennsylvania; NE, C Pennsylvania to SC New York

BALTICA
(3) *Caledonides.* NW Norway, Dunderland (?)
(4) *Southwest margin.* SE Sweden; NW Poland
(5) *Northeast margin.* NE Russia, Pechora Basin.
(6) *Southeast margin.* CE Russia; SC Russia

KAZAKHSTAN
(7) *Southwest margin.* SW Kazakhstan, Syr Darya lowland

SIBERIA
(8) *Northwest interior.* NW Siberia, Tunguska Basin
(9) *Southwest margin.* SC Siberia, Sayan Mountains

SOUTH CHINA
(10) *Yangtze craton.* C–W–SW South China, Hubei, Sichwan, Guizhou, Yunnan
(11) *Huanan block.* SE South China, Hunan, Jiangxi, Guangxi

SOUTHEAST ASIA
(12) *Sibumasu block.* NW Malaysia

AUSTRALIA
(13) *Central interior.* Amadeus Basin

SOUTH AMERICA
(14) *North-central interior.* NC Brazil, Amazon Basin
(15) *West central margin.* NW Argentina, Precordilleran belt
(16) *Southeast margin.* SE Argentina, Sierra Grande

PERI-GONDWANAN Blocks:
- (17) *Avalonia and exotic Appalachian terranes.* CE USA, E Virginia; E NFLD and GB, Avalon Peninsula and Grand Banks; NE Nova Scotia, Antigonish; W Nova Scotia (Mcguma terrane), Torbrook
- (18) *London-Brabant Massif.* N Wales; SW England; SE Ireland; SE Belgium—NE France, Ardennes Shelf
- (19) *Rhenish Massif.* WC FR Germany, Eifel, Hunsruck, Sauerland
- (20) *Armorican Massif.* NW France, Brittany, Normandy
- (21) *Iberian Massif.* C Spain, Celtiberian chains; NW Spain, Cantabrica, Asturias—Leon, Galicia; N Portugal; SW Spain, Osso—Morena, Estremadura
- (22) *Bohemian Massif.* SW DR Germany—SE FR Germany, Thuringia; W Czechoslovakia, Barrandian Basin; C Czechoslovakia, Moravia
- (23) *Holy Cross Mountains.* SC Poland
- (24) *Rhodope Massif.* SE Yugoslavia
- (25) *Anatolia-Iranian Massif.* NW Turkey
- (26) *Italian terranes.* NE Italy, Carnic Alps; NW Sardinia
- (27) *Moroccan mesetas.* C, High Atlas; WC, massifs; N, Rif

AFRICA
- (28) *Northeast Africa.* NC Saudi Arabia (?)
- (29) *North central interior.* SE and SW Libya, Kufra and Murzuk basins; NW and C Libya, Tripolitania and Gargaf; S Tunisia; SE Algeria, N and S flanks of Hoggar
- (30) *Northwest Margin.* CW Algeria—S Morocco, Tindouf Basin and Anti-Atlas
- (31) *Central West Margin.* Mauritania, Adrar; Guinea, Bové Basin

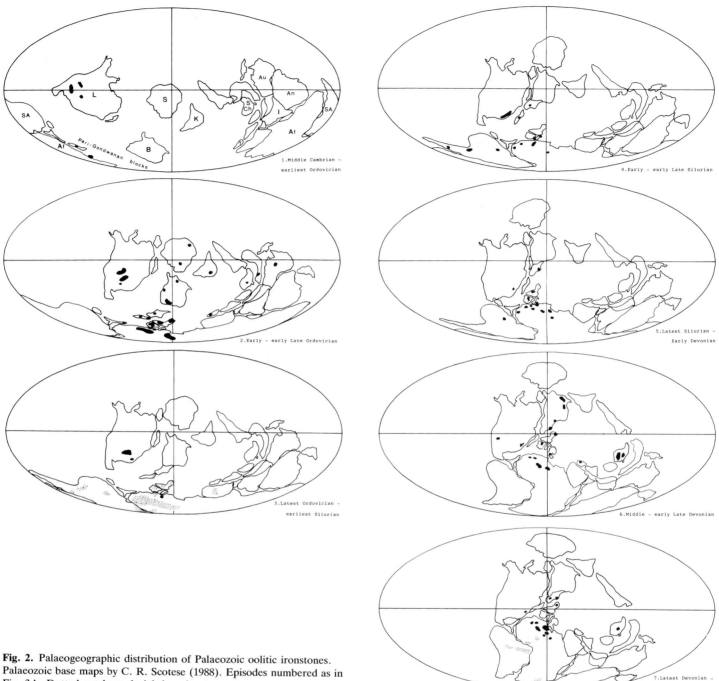

Fig. 2. Palaeogeographic distribution of Palaeozoic oolitic ironstones. Palaeozoic base maps by C. R. Scotese (1988). Episodes numbered as in Fig. 3A. Dotted patches: glacial deposits.

during the entire Palaeozoic Era. In fact, some were closely associated with the few ironstone-bearing sequences that did develop.

Early–early Late Ordovician time (Fig. 2.2)

During the most widespread Palaeozoic flooding of the cratons Africa and the Peri-Gondwanan blocks drifted into higher southern latitudes and Baltica drifted northward. In this episode ironstones developed more widely than at any time during the Phanerozoic Eon. Major ironstones accumulated along a 1400 km belt on marginal and interior northwest Africa (west-central Algeria, southern Morocco), and on Peri-Gondwanan blocks that included eastern Newfoundland, north Wales, northwest Spain and adjacent Portugal, Bohemia-Thuringia, and central Morocco. Scattered minor ones accumulated on high-latitude northwest Libya, north-central Saudi Arabia, and northwest Turkey, as well as on the northeast and southwest margins of mid-latitude Baltica, and on low-latitude southwest Kazakhstan, northwest and south–central Siberia, southwest Southeast Asia, central Australia, and central Laurentia. During the latter part of this episode the Taconian orogeny dominated the northeast margin of Laurentia.

Latest Ordovician–earliest Silurian time (Fig. 2.3)

Development of ironstones was curtailed by southern hemisphere glaciation and widespread lowering of sea level. As North Africa and Peri-Gondwanan blocks drifted rapidly northward a minor latest Ordovician ironstone accumulated on northwest Africa (Guerrak 1987) as well as on low-latitude central and southeast Laurentia.

Early–early Late Silurian time (Fig. 2.4)

A major rise in sea level was accompanied by renewed development of ironstones. Well-developed low-latitude Early Silurian ones were deposited on west-central Algeria, and at the southern and northern ends of a 1500 km tract of ironstones in a foreland basin on eastern Laurentia during a lull between the Taconian and Acadian orogenies. Scattered minor low-latitude ironstones developed on north-central, northwest, and central-west Africa, on Peri-Gondwanan blocks that included northeast Nova Scotia, southwest England, northeast Italy and Sardinia, as well as middle- to high-latitude north-central Brazil, and central-west and southeast Argentina.

Latest Silurian–Early Devonian time (Fig. 2.5)

During a long low stand of sea level, northeast Laurentia, south and west Baltica, and many of the Peri-Gondwanan blocks were involved in low-latitude early Acadian–late Caledonian orogeny. Scattered, mostly minor ironstones developed in middle to low latitudes on central-east Laurentia, on a few Peri-Gondwanan blocks, on north-central, northwest, and central-west Africa, on north-central Brazil, and on South China.

Middle–early Late Devonian time (Fig. 2.6)

A high stand of sea level was accompanied by late Acadian orogeny along central-east Laurentia and an extensional regime in Middle Europe (docked Peri-Gondwanan blocks). This setting induced the second major development of Palaeozoic oolitic ironstones. Well-developed ones accumulated along a 1800 km belt on the low-latitude central-east and southeast Russian Platform (southeast Baltica), and more locally in northwest (Belgium and adjacent France and Germany), east (northwest and south central Poland), southeast (Yugoslavia and Turkey), and southwest (northwest Spain) Middle Europe, as well as in 1200 km tract on middle-latitude north-central and northwest Africa (northwest Libya and southeast Algeria). Ironstones were also widespread throughout South China (Hou & Wang 1984). Very minor ones were deposited on low-latitude southwest and central-east Laurentia.

Latest Devonian–earliest Carboniferous time (Fig. 2.7)

Africa and South America lay in middle and high southern latitudes and were glaciated locally. Well-developed latest Devonian ironstones were limited to a 1200 km tract on middle-latitude north-central and northwest Africa (central Libya, southeast and west-central Algeria) and to low-latitude northwest Middle Europe (Belgium). These deposits developed along the margin of a seaway that was subsequently closed by renewed northward drift of Africa. Minor low-latitude latest Devonian ironstones accumulated on southeast Middle Europe and on central-east Laurentia. In earliest Carboniferous time a single minor ironstone developed on southeast Ireland and a few others on north-central Africa (central Libya).

Precambrian–earliest Palaeozoic precursors

Some early to middle Proterozoic (2300–1500 Ma) iron formations are arenitic (Simonson 1985) types containing aggregated particles including ferriferous ooids (Dimroth 1976). Significantly, however, this type does not occur among the several late Proterozoic (900–570 Ma) iron formations, each of which is lutitic and 'banded'. Moreover, only a few oolitic ironstones have been found in Early and Middle Cambrian deposits in which glauconitic peloids are abundant.

Late Palaeozoic demise

Development of oolitic ironstones declined drastically as Pangaean cratons converged in late Palaeozoic time. The few that did accumulate were associated with marine members of coal measures. These occur in Early Carboniferous deposits of southeast South China, in Middle and early Late Carboniferous deposits of central England (Chowns 1966; Dunham 1960), and in very late Carboniferous (or earliest Permian) deposits of central North China (Shanxi) and of western South China (Fouling). In Middle Permian time, shortly after waning of Gondwanan glaciation, a single well-developed ironstone accumulated in a marine embayment in north-central Australia (Edwards 1958).

Discussion

Palaeozoic ironstones have been found on all of the major cratons except Antarctica, India, and those now lying between South China and Siberia. Their known distribution is, in part, an artefact of geological exploration. Nevertheless, the record of well-explored North America and Australia is meagre whereas that of less extensively explored North Africa is prolific along a broad belt reaching 3000 km from Mauritania to central Libya (Guerrak 1987). Such a marked contrast is reassuring evidence that the present available data are reasonably representative.

Most of the oolitic ironstones, including those in the longest successions of repeated development (Table 2, Nos 1,3,8 & 10), accumulated in shallow marine detrital sequences in both high and low latitudes. The few well-developed exceptions in mixed carbonate-detrital deposits are the low-latitude Early Silurian ironstones in eastern Laurentia (Table 2, Nos 6 & 7), the Middle and Late Devonian ones in South China (Table 2, No. 2), and Late Devonian ones in southeastern Baltica (Table 2, No. 5). In addition, successions of relatively minor ironstones in mixed carbonate-detrital deposits accumulated on middle-latitude southwest Baltica (Table 2, No. 4) as it drifted equatorward in Middle Ordovician time, and on low-latitude northwest Middle Europe (Table 2, No. 9) in Late Devonian time. Even in these carbonate-detrital associations the ironstones occur most commonly in the nearshore detrital facies.

During the most extensive Palaeozoic flooding of the continents, in Ordovician time, ironstones accumulated on most of the major cratons and microplates, and especially those in high southern latitudes (Fig. 2.2). Yet none developed on southeast Baltica, South America, or South China where ironstones did accumulate later in the Era. Some of the numerous sites of Ordovician ironstones on North Africa and the Peri-Gondwanan blocks persisted as favourable places during the Silurian and Devonian periods (Fig. 1). On several of the Peri-Gondwanan blocks, however, there is no Silurian ironstone between an Ordovician ironstone and a Devonian one that accumulated after the blocks had drifted toward the equator and docked in Middle Europe during late Caledonian orogeny (Ziegler 1978). In contrast to this record of successive development, Palaeozoic ironstones apparently never accumulated again on low-latitude south Laurentia, northeast and southwest Baltica, Kazakhstan, Siberia, Southeast Asia, or Australia (Permian only).

Deposition of ironstones diminished considerably during the Silurian Period (Fig. 2.4) as southern-latitude cratonic blocks drifted northward (Hargraves *et al.* 1987). The ones that developed in new tectonic frameworks were mostly on eastern Laurentia, South America, and West Africa. Those in Guinea and the adjacent Amazon Basin of north-central Brazil, and the major deposit in the extensive foreland basin of eastern Laurentia, were succeeded by Devonian ironstones. In contrast, the minor ones in Sardinia and northeast Italy, in northwest and southeast Argentina, and in Mauritania, had no Devonian successors.

Widespread accumulation of ironstones was renewed in the Devonian Period (Figs 2.5, 6 & 7) in low and middle latitudes.

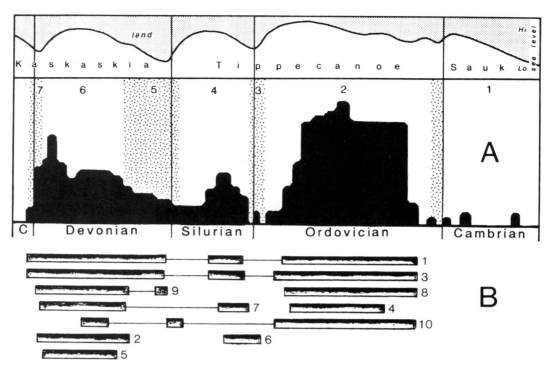

Fig. 3. Temporal patterns of global abundance of Palaeozoic oolitic ironstones. (**A**) Histogram of all known ironstones. Episodes numbered as in Figure 2; names after Sloss (1963). Stippled columns — major low stands of sea level and interregional unconformities. (**B**) Major successions of repeated ironstones, numbered and identified in Table 2.

In addition to development on many of the Ordovician sites major Devonian ironstones were deposited in new settings on southeast Baltica and South China; minor ones accumulated on southwest Laurentia and the microplates of Rhodope and Meguma.

Plotted on palaeocontinental reconstructions the two major episodes of Palaeozoic ironstone accumulation, in the Ordovician and the Devonian periods, are associated with significantly different arrangements of cratonic blocks. The role of their specific global distribution is difficult to assess, however, because longitudinal position cannot be established by palaeomagnetic data alone.

During Ordovician time most of the ironstones developed along the northern margin of Africa and on nearby Peri-Gondwanan blocks, when they had drifted into middle and high southern latitudes. To the north a seaway separated these major sites from middle- to low-latitude Laurentia and Baltica. In middle Palaeozoic time many of the new places of accumulation were in post-orogenic regimes along belts of 'Caledonian' convergence. After the Peri-Gondwanan blocks had docked along eastern Laurentia and southern Baltica by Early Devonian time Africa drifted southward (Hargraves *et al*. 1987; Van der Voo & Kent 1988), opening a southern middle- to low-latitude seaway (Theic ocean). Significant deposits of oolitic ironstones developed along both its margins, in Middle Europe and North Africa, as well as along the margins of its eastward Tethyan expansion onto southeastern Baltica and South China.

Many of the data assembled here are from Laznicka (1985) and Zitzmann (1977). Additional information was supplied by J. C. Garcia-Ramos, S. Guerrak, M. M. Kimberley, Liao S-f, and J. Petranek. This paper is a contribution of IGCP Project 277.

Table 2. *Palaeozoic successions of repeated oolitic ironstones*

Fig. 3B Tectonic Framework	Age	Number	Duration (Ma)
Stable Craton			
(1) Central North Africa C−NW Libya	E Ord− E Carb	10	125−130
(2) South China C−W South China	E−L Dev	7	20−25
Cratonic Margin			
(3) Northwest Africa S Morocco, S−W Algeria	E Ord− E Carb	18	125−130
(4) Southwest Baltica Southern Sweden	M−L Ord	5	22
(5) East-southeast Baltica E−SE Russian Platform	M−L Dev	?	15−20
Foreland Basin			
(6) Southern Appalachians Alabama−Tennessee	L Ord− E Sil	5	20
(7) Central Appalachians Virginia-New York	E Sil− L Dev	8−9	65
Peri-Gondwanan Microplates			
(8) Bohemian Massif W Czecho., SE Germany	E−L Ord	12	45
(9) Ardennes Shelf S Belgium, W FR Germany	E−L Dev	7−8	45
(10) Iberian Massif C−NW Spain, N Portugal	E Ord− M Dev	7	100

References

BAYER, U., ALTHEIMER, E. & DEUTSCHLE, W. 1985. Environmental evolution in shallow epicontinental seas: sedimentary cycles and bed formation. *In*: BAYER, U. & SEILACHER, A. (eds) *Sedimentary and Evolutionary Cycles*. Springer-Verlag, Berlin, 347−381.

BHATTACHARYYA, D. P. 1989. Concentrated and lean oolites: examples from the Nubia Formation at Aswan, Egypt, and significance of the oolite types in ironstone genesis. *In*: YOUNG, T. P. & TAYLOR, W. E. G. (eds) *Phanerozoic Ironstones*. Geological Society, London, Special Publication **46**, 93−103.

BRONEVOI, V. A., GORETSKII, R. G. & KIRYUKLIN, L. G. 1967. New area of middle Oligocene iron ore deposit in the north Aral region and in northeastern Ustyurt. *Lithology and Mineral Resources*, **4**, 436–445.

CHOWNS, T. M. 1966. Depositional environment of the Cleveland ironstone series. *Nature*, **211**, 1286–1287.

COTTER, E. 1988. Heirarchy of sea-level cycles in the medial Silurian siliciclastic succession of Pennsylvania. *Geology*, **16**, 242–245.

DESTOMBES, J., HOLLARD, H. & WILLEFERT, S. 1985. Lower Palaeozoic rocks of Morocco. *In*: HOLLAND, C. H. (ed.) *Lower Palaeozoic of North-Western and West-Central Africa*, John Wiley & Sons, New York, 91–336.

DIMROTH, E. 1976. Aspects of the sedimentary petrology of cherty iron-formation. *In*: WOLF, K. H. (ed.) *Handbook of Strata-Bound and Stratiform Ore Deposits, II*, Elsevier Publishing Company, New York, 203-254.

DREESEN, R. 1982. Storm-generated oolitic ironstones of the Famennian (Fa1b–Fa2a) in the Vendre and Dinant Synclinoria (Upper Devonian, Belgium). *Annales de la Societe Geologique de Belgique*, **105**, 105–129.

DUNHAM, K. C. 1960. Syngenetic and diagenetic mineralization in Yorkshire. *Proceedings of the Yorkshire Geological Society*, **32**, 229–284.

EDWARDS, A. B. 1958). Oolitic iron formations in northern Australia. *Geologisches Rundschau*, **47**, 668–682.

GUERRAK, S. 1987. Palaeozoic oolitic ironstones of the Algerian Sahara: a review. *Journal of African Earth Sciences*, **6**, 1–8.

HARGRAVES, R. B., DAWSON, E. M. and VAN HOUTEN, F. B. 1987. Palaeomagnetism and the age of the mid-Palaeozoic ring complexes in Niger, West Africa, and tectonic implications. *Geophysical Journal of the Royal Astronomical Society*, **90**, 709–729.

HOU, H-f, and WANG, S-t. 1984. Devonian paleogeography of China. *Acta Palaeontologica Sinica*, **24**, 194–197.

LAZNICKA, P. 1985. Strata-related ore deposits: classified by metals, lithologic association, and some quantitative relationships. *In*: WOLF, K. H. (ed.) *Handbook of Strata-bound and Stratiform Ore Deposits, IV*, 12, Elsevier Publishing Company, New York, 1–107.

SCOTESE, C. R. (coordinator) 1988. Phanerozoic Plate Tectonic Reconstructions, 1987. Paleoceanographic Mapping Project, Institute of Geophysics, University of Texas, *Technical Report*, **90**.

SIMONSON, B. M. 1985. Sedimentological constraints on the origin of Precambrian iron-formations. *Geological Society of America Bulletin*, **96**, 244–252.

SLOSS, L. L. 1963. Sequences in the cratonic interior of North America. *Geological Society of America Bulletin*, **74**, 93–113.

TEYSSEN, T. A. L. 1984. Sedimentology of the Minette oolitic ironstones of Luxembourg and Lorraine: a Jurassic subtidal sandwave complex. *Sedimentology*, **31**, 195–211.

VAN DER VOO, R. & KENT, D. V. 1988. Paleozoic paleogeography of Laurentia, Gondwana and southern Europe. 1988 Centenial Celebration, *Geological Society of America, Abstracts with Programs*, A349.

VAN HOUTEN, F. B. & ARTHUR, M. A. 1989. Temporal patterns among Phanerozoic oolitic ironstones and oceanic anoxia. *In*: YOUNG, T. P. & TAYLOR, W. E. G. (eds) *Phanerozoic Ironstones*. Geological Society, London, Special Publication **46**, 33–49.

—— & BHATTACHARYYA, D. P. 1982. Phanerozoic oolitic ironstones — geologic record and facies model. *Annual Review of Earth and Planetary Sciences*, **10**, 441–457.

ZIEGLER, P. A. 1978. North-western Europe: tectonics and basin development. *Geologie en Mijnbouw*, **57**, 589–626.

ZITZMANN, A. 1977. The iron ore deposits of the western U.S.S.R. *In*: ZITZMANN, A. (ed.) *The Iron Ore Deposits of Europe and adjacent areas*. Federal Institute for Geosciences and Natural Resources, **1**, 326–391.

Early Palaeozoic Biogeography

Biogeography of Ordovician and Silurian faunas

L. R. M. COCKS & R. A. FORTEY

Department of Palaeontology, British Museum (Natural History), Cromwell Road, London SW7 5BD, UK

Abstract: The new reconstructions generated for this volume fit well with the majority of Ordovician and Silurian faunal data and are a great improvement on previous attempts. The distribution of selected trilobites, brachiopods and graptolites are plotted on the new maps and confirm the importance of palaeolatitude in controlling the faunal distributions, particularly of the old cratons, which are shown for the Early Ordovician and Late Silurian. Two contrasting patterns of cratonic faunas are (a) disjunct or (b) gradational across a large palaeocontinent, e.g. Gondwana. Marginal and deeper-water biofacies show different patterns, which help to define the edges of palaeocontinents, but which are not so constrained in their palaeolatitudinal distributions. In contrast pelagic trilobites do not help to define palaeocontinents, but were sensitive to palaeotemperature and palaeolatitude. Specific case histories are considered, in particular the closing of Iapetus and the contemporary widening of the Rheic Ocean and the positioning of Avalonia, and the ancestry of the various associated trilobite and brachiopod genera during the later Ordovician. The Ashgill deeper-water *Foliomena* fauna is also plotted.

In contrast to Mesozoic to Recent palaeogeographical maps, in which the positions of the continental plates may be determined from the magnetic reversals preserved in ocean-floor deposits, reconstructions of the Palaeozoic have little objective basis, apart from the need of the sequence of successive maps through geological time to be internally consistent. Palaeomagnetic measurements from continental deposits provide some help from the Precambrian onwards, but the evidence for many of the smaller plates often seems poorly founded for the Ordovician and Silurian, and thus our faunal criteria appear to give the most plausible results, particularly in regard to assessing palaeolatitudes.

Starting with faunal and spatial relationships between North America (Laurentia), Baltica and northwestern Gondwana (North Africa and southern Europe) we have postulated a series of reconstructions (Cocks & Fortey 1982) which we have enlarged to a global scale for the deeper water isograptid facies (Fortey & Cocks 1986) and a peri-Gondwanan scale for several types of biota (Cocks & Fortey 1988).

The new maps generated for this volume can be tested in various ways using fossils. Failure of the maps to pass one or another of these tests either requires a particular *ad hoc* explanation to account for the anomaly, or suggests that some modification to the map is required. If palaeogeographic maps are to be consistent with faunal distributions they should reveal the following.

(1) Broad climatic control on benthic cratonic faunas. If a single continental entity such as Gondwana spans more than one climatic zone, we should expect to discover different kinds of faunas approximating to major palaeolatitudinal belts. There should be a climatic and faunal cline across a single palaeocontinent.

(2) Latitudinal control on epiplanktic faunas without obvious relationship to palaeocontinent distribution. Like living planktic foraminiferans, distribution of Lower Palaeozoic plankton living high in the water column ought to be related to surface water temperature and hence to palaeolatitudinal belts. Although there may be some water mass specificity, oceanic tracts will not, in general, be a major factor in limiting distribution.

(3) Deep water benthics will be found in marginal sites around palaeocontinents. These may show a wider palaeolatitudinal spread than the cratonic faunas, and need not be related to past continental configurations. Palaeogeographies based on such faunas alone may, in the past, have led to misleading reconstructions, by inferring geographical proximity directly from faunal similarity.

With regard to the new maps produced by C. R. Scotese for this volume we note the following.

Arenig–Llanvirn cratonic faunas

This time period is recognized, both on faunal and palaeomagnetic grounds, as one of high continental dispersal relative to palaeolatitude. Cratonic faunas may be expected to show climatic zonation and close correspondence with continental configuration. This is born out well by the reconstructions (Fig. 1). Southern high latitude faunas are Gondwanan. They have highly characteristic endemic trilobite and brachiopod faunas in inshore sites. The brachiopods are dominated by large inarticulate genera such as *Lingulobolus*, *Pseudobolus*, *Ectenoglossa* and *Lingulepis* (illustrated in Cocks & Fortey 1988 fig. 3), and these are joined in Llanvirn times by early dalmanellaceans, and others (e.g. Williams *et al.* 1981). The trilobites are typified by an array belonging to the superfamilies Calymenacea (for example, the genera *Neseuretus*, *Calymenella*, *Plaesiacomia*, *Brongniartella*) and Dalmanitacea (for example, the genera *Ormathops*, *Kloucekia*, *Zeliskella*, *Eudolatites*). Virtually all other relatively inshore trilobites, belonging to such families as Asaphidae, Trinucleidae, and Illaenidae, in this part of Gondwana bear the same signature. This area was clearly recognized on a statistical analysis of faunas by Whittington & Hughes (1972) and was termed the '*Selenopeltis* Province' after another endemic. We prefer the label 'Calymenacean-dalmanitacean' because *Selenopeltis* may be a pelagic trilobite (Fortey & Owens 1987), and has a distribution which extends beyond the confines of shallow water palaeoenvironments. Note that these typical Gondwanan trilobites extend into Avalonia as far as the Lake District, which lay at the margin of early Ordovician Gondwana (Fortey *et al.* 1989). The tropical platforms of Laurentia have long been known to be characterised by trilobites of the family Bathyuridae, and there are many more animals endemic to the same areas, including molluscs and brachiopods. Bathyurid occurrences extend eastwards into the NE Siberian platform, where lithologies are apparently identical, and into North China (Zhou & Fortey 1986). At intermediate latitudes (and separated both from Laurentia and from Gondwana) the shallow-water deposits of the Baltic–Russian platform have another suite of endemics, such as asaphid trilobites of megalaspid, ptychopygine, and pseudobasilicine type, long ago termed the 'Asaphid province' by Whittington (1963). The fourth suite of endemics (*Asaphopsis* province of Whittington & Hughes 1972) are low latitude Gondwanan; these faunas include a variety of dikelokephalinid trilobites, and such distinctive and restricted genera as the blind 'dalmanitacean' *Prosopiscus*. These platform faunas are now known from numerous stations in South China and Australia, and 'spot occurrences' in Bolivia and the Himalaya. As would

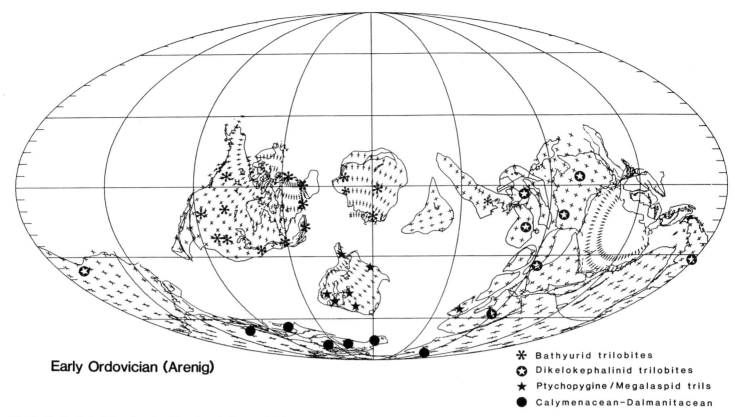

Fig. 1. Early Ordovician Arenig–Llanvirn platform trilobite assemblages showing close relationship to palaeolatitude and palaeogeography. Gondwanan inshore faunas comprise a cline with 'mixed' faunas at intermediate latitudes e.g. Middle East (Cocks & Fortey 1988), where some Baltic forms may also appear.

be expected with a large Gondwana continent extending from palaeoequator to south pole, there is a latitudinal and faunal cline between warm water eastern Gondwanan shelf faunas and cold water western Gondwanan shelf faunas, but no evidence of the disjunct distribution which might indicate a vanished oceanic barrier. Mixed faunas between eastern and western Gondwanan occur in such areas as Iran and Saudi Arabia, where they might be expected (El-Khayal & Romano 1985). Interestingly, some of the Baltic genera also appear in Turkey at what was an appropriate temperate palaeolatitude (Dean 1973).

This new palaeogeographical reconstruction fits well with the data from inshore biofacies in the Early Ordovician. An interesting question is posed by the differences between tropical Gondwanan, and Laurentian/Siberian faunas, although in some groups this difference is hardly apparent, since, for example, it includes the extent of the Midcontinent Conodont Province. Lithological differences are not an adequate explanation: similar tropical calcareous facies to those of Laurentia are known from Australia. Why did bathyurid trilobites extend to North China, but no further East? Could it be that North China was further separated from Gondwana than indicated on the maps? Certainly the South China and Australian faunas are similar to each other, but North China is different and more similar to Laurentia in both faunas and lithofacies.

Early Ordovician pelagic trilobites

Primary climatic control, with less direct relationship to continental configuration, is predicted for such animals, and when plotted on the early Ordovician map (Fig. 2), this is shown very well. Unlike the benthic trilobites and brachiopods discussed above, the epipelagic trilobites *Carolinites* and *Opipeuter* have a pan-equatorial distribution. They are found in on-craton as well as peripheral cratonic sites (central Australia and Siberia, as well as western Ireland, and Spitsbergen). Their high-latitude equivalents were cyclopygids, which are known from a number of marginal Gondwanan sites in the Arenig–Llanvirn, but confined to deeper water biofacies, and are therefore regarded as having been mesopelagic rather than epipelagic (Fortey 1985). They do extend into marginal sites around Baltica. As would be anticipated, the only areas where the cylopygids and equatorial pelagics can co-occur are at former temperate latitudes, for example *Carolinites* is a rare fossil in Baltica and in Turkey (Fig. 2). Thus there are no anomalies between the fossil distributions and the map reconstructions.

Early Ordovician exterior biofacies

Two examples of exterior biofacies have been chosen to illustrate the properties of this kind of faunal assemblage (Fig. 3). The Isograptid graptolite biofacies (Fortey & Cocks 1986) and the olenid biofacies (Fortey 1975; Cook & Taylor 1975). The latter is chiefly dominated by the eponymous trilobite family, but we have also included other occurrences of olenids as an element in larger faunas to indicate the distribution of the family as a whole. Both isograptid biofacies and olenids are independent of the continental configuration, as should be the case. Their palaeo-latitudinal spread is from the equator to high boreal, and this may be associated with uniform oceanic conditions beneath the thermocline.

The olenids lie chiefly where they should, on sites marginal to palaeocontinents, however, one interesting exception to this is their western Australian occurrence in the Canning Basin (Legg 1976). Is it possible that this site may have been more marginal than indicated on the reconstruction? Could this be related to the biofacies patterns in southern China, where the deepest biofacies

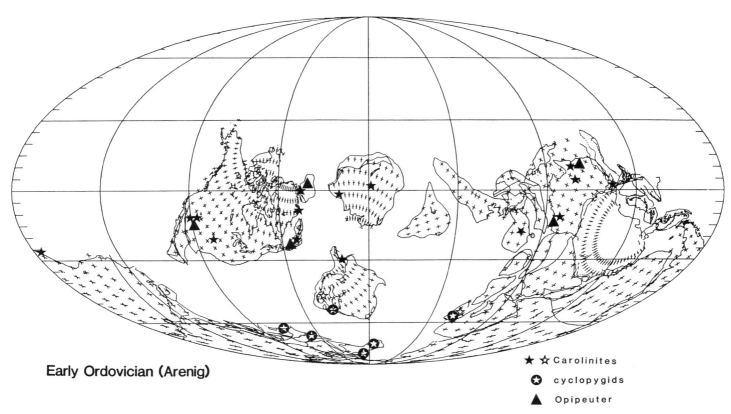

Fig. 2. Early Ordovician Arenig–Llanvirn pelagic trilobites showing close latitudinal control, with overlap between tropical and high latitude pelagics only at temperate latitudes. Open 'stars' show outliers of the genus *Carolinites* known from one or two specimens only, at extreme southward limit of range.

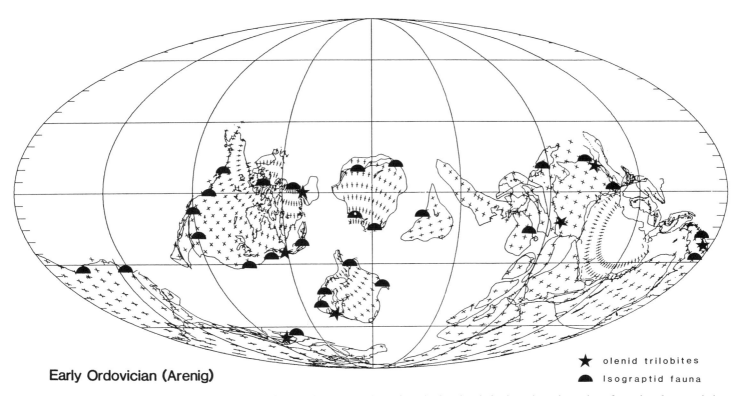

Fig. 3. Early Ordovician marginal or deep water biofacies showing independence from both palaeolatitude and continental configuration, but restriction to sites peripheral to continental masses. Isograptid graptolite biofacies defined by Fortey & Cocks (1986) includes an assemblage of 'oceanic' genera not found in epicratonic graptolitic rocks.

lie to the southeast (Lu *et al.* 1976), or perhaps the orientation or even the location of South China is shown incorrectly on the new reconstructions (cf. Cocks & Fortey 1988, fig. 2)?

The movement of Avalonia

As stated above in the section on Arenig–Llanvirn cratonic faunas, the fossil remains found in the Lake District, South Wales and elsewhere in Avalonia can be unequivocally assigned to the calymenacean–dalmanitacean province so typical of western Gondwana (Cocks & Fortey 1982, 1988). At that time Avalonia was clearly quite separate from Baltica and Laurentia, which had temperate and tropical cratonic faunas respectively. However, by the latest Ordovician and Early Silurian time, Eastern Avalonia had become faunistically indistinguishable from Baltica, and by the Early Silurian, Avalonian–Baltic faunas themselves merged with those of Laurentia and elsewhere to form an extensive cosmopolitan fauna which inhabited most of that world (see section below). There is also independent tectonic and sedimentary evidence to show that, commencing in Middle Silurian and finishing in Middle Devonian time, the Iapetus Ocean which had separated Laurentia from Avalonia–Baltica had closed and that the three continents had merged to form the Old Red Sandstone palaeocontinent. Although the vast mass of Gondwana had also drifted over the South Pole during this period, culminating, for example in the formation of warm-water limestones in Bohemia in the latest Silurian, nevertheless it seems increasingly probable that the Avalonian microcontinent had separated from Gondwana and migrated northward to firstly unite with Baltica and then in turn with Laurentia. Small marginal biotas from the early Ordovician of Avalonia and Laurentia, many well described by Neuman (e.g. 1976), have had their faunal affinities re-evaluated (Fortey 1988) and can be shown to be of mixed parentage. We here attempt to evaluate the larger question of the position of Avalonia as a whole, in an attempt to assess the time at which it separated from Gondwana. In the Arenig, the new reconstructions show Avalonia welded to Gondwana, and the faunal evidence very strongly supports that union, as it does for the succeeding Llanvirn, but for the succeeding 'Llandeilo–Caradoc' map (458 Ma, actually Middle Llandeilo time), Avalonia is shown well separated from Gondwana (Fig. 4). What evidence is there to support this rift so early in the Ordovician? The southern margin of Avalonia (eastern Newfoundland, southern Britain and Belgium) and the northern margin of Gondwana (in this case western Iberia, Armorica and Bohemia) do not show any deeperwater facies or faunas marginal to the cratonic facies in both continents, although the Llanvirn–Llandeilo period is well known for its enhanced volcanic activity. This vulcanism may represent the rifting phase of the separation of Avalonia from Gondwana.

We have plotted Caradoc faunal data for both brachiopods and trilobites on the polar projection for the 'Llandeilo–Caradoc' (Fig. 4), and assigned to each genus present at the various sites an ancestry depending on whether the earliest known record is from Gondwana, Baltica or Laurentia. As can be seen, by Caradoc time the Avalonian faunas had had their original Gondwanan stock diluted considerably by arrivals from other palaeocontinents, both Baltica and Laurentia, and this is also confirmed by the ostracode data (Vannier *et al.* 1989). It is instructive to compare (Fig. 4) contemporary data from Bohemia, which lay to the south of Avalonia, and also Turkey, which was still part of Gondwana, but at the same postulated palaeolatitude as Avalonia. It can be

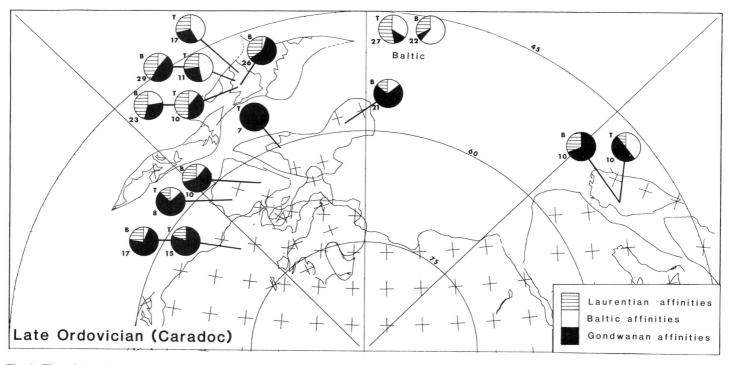

Fig. 4. The origins of western Gondwanan and Avalonian faunas of Caradoc age, plotted on a new 458 Ma polar projection constructed for this volume. The various brachiopod (B) and trilobite (T) faunas have been analysed by their generic origins (the number of genera present are shown for each locality, e.g. 17 trilobites for Northern England). Two Baltic faunas are included at the top of the diagram for comparison. Localities are Northern England (Dean 1962 for Cross Fell trilobites), Bala (Williams 1963 for Gelli–Grin brachiopods; Whittington 1968 for Lower Bala trilobites), Shelve (Williams 1974 for Whittery Shales brachiopods; Whittard 1966 Aldress to Whittery Shales trilobites);
Shropshire (Hurst 1979 Ragdon Member brachiopods); Oslo (Owen & Bruton 1980 Solvang trilobites); Estonia (Rõõmusoks 1970 Johvi brachiopods); Bohemia (Havlíček & Vaněk 1966 Zahorany brachiopods); Armorica (Henry 1980 trilobites); Spain (Villas 1985 Fombuena brachiopods, Hammann 1976 trilobites); Morocco (Destombes 1985 Ouaouglout trilobites, Havlíček 1971 Ktaoua brachiopods); Turkey (new brachiopod data from Bedinan Formation, Dean 1967 Bedinan trilobites).

seen that the Avalonian biotas can be explained by the more northerly position of western Gondwana into more temperate latitudes alone rather than having to postulate a rift between Gondwana and Avalonia in the earlier Ordovician, a rift which is not confirmed by the marginal biofacies and lithofacies which would be expected. Thus we see no reason on faunal evidence alone to support any rift between Gondwana and Avalonia before late Llanvirn or Llandeilo time.

Division of southern hemisphere distributions later in the Ordovician

The closer encroachment of the Gondwana and Baltic continents shown on the reconstructions introduces an intriguing possibility of producing disjunct distributions of at least some taxa by separation of the capacity for dispersal between shelves of the separated Iapetus Ocean to the west and 'proto-Tethyan' Ocean to the east. There is some evidence for this. Some outer shelf trilobites, for example, have distributions encompassing the margins of the 'proto-Tethyan' Ocean, but do not occur in Laurentia at the same palaeolatitudes. Conversely, other taxa are found on both sides of Iapetus, but not in China, at comparable palaeolatitudes. The examples shown on the map (Fig. 5) are distinctive and widespread taxa to either side of the Siberian–Baltic–North Gondwana divide. Another late Ordovician genus with a 'Tethyan' distribution is *Ovalocephalus* (more usually known as *Hammatocnemis*), which occurs widely around the Tethyan margin as in Kazakhstan, China, Iran, Turkey, Poland, and possibly Norway. Assuming the operation of coriolis forces comparable to those of today, we have postulated the positions of two separate gyres (cf. Ross 1975), which might explain the otherwise surprising distribution of *Panderia* in Poland, Uzbekistan, Kazakhstan, Turkistan and France. Another cause of these separations might have been in the erroneous reconstruction of the Iberian–Armorican land masses, which, as T. P. Young shows elsewhere in this volume, may have been much more elongated and might have been the basis of a land barrier stretching from North Africa to Baltica.

In contrast to these patterns of some outer-shelf trilobites in the Caradoc and Ashgill is the distribution of the latest Caradoc and early to middle Ashgill *Foliomena* fauna of Sheehan (1973) which has recently been revised (Cocks & Rong 1988). This distinctive assemblage, which consists chiefly of thin-shelled and small brachiopods, clearly colonized outer-margin sites, and is known from the Laurentian, Baltic–Avalonian, north Gondwanan, south Chinese and Shan–Thai palaeocontinents (Fig. 6), but its absence from intervening and apparently suitable localities is a mystery, although it may have been overlooked since the fossils are small. Once again, the deeper-water position of the fauna means that it would have colonized a broad spectrum of palaeolatitudes, in contrast to the contemporary shallower-water biota.

During the very end of the Ordovician the ice sheet, for which there is some evidence in west Gondwana regions as early as Caradoc times, expanded during the Hirnantian glacial episode, and this is considered to be the chief cause of the breakdown and redistribution of the animal communities and provincialization at the Ordovician–Silurian boundary. Prior to that time, in the late Caradoc and earlier Ashgill, the Richmondian shelly faunas of the Laurentian craton were very distinct from those found in Avalonia-Baltica and north-west Gondwana (for example Bohemia and Iberia), and the latest Ordovician *Hirnantia* faunas succeed them in many areas (for a recent review see Rong & Harper 1988).

Silurian faunal distributions

Following the end-Ordovician glaciation, which resulted in widespread unconformities at many localities, the biotas were slow to recover. There were widespread transgressions and flooding

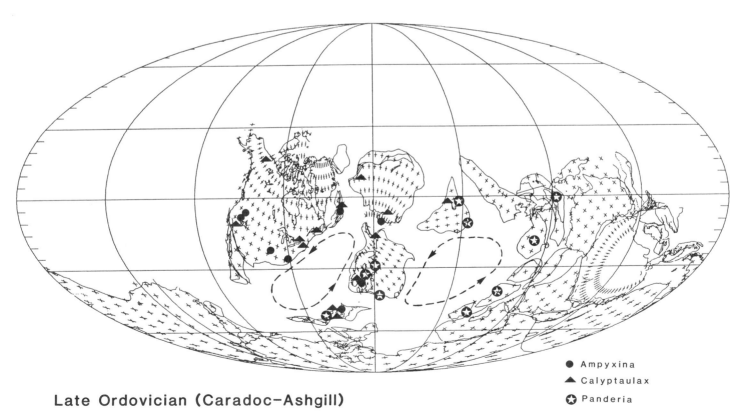

Late Ordovician (Caradoc–Ashgill)

● Ampyxina
▲ Calyptaulax
✪ Panderia

Fig. 5. Separation into eastern and western distributions of certain late Caradoc and Ashgill trilobites, which could have been caused by the gyre patterns shown which would have influenced the dispersal of larvae.

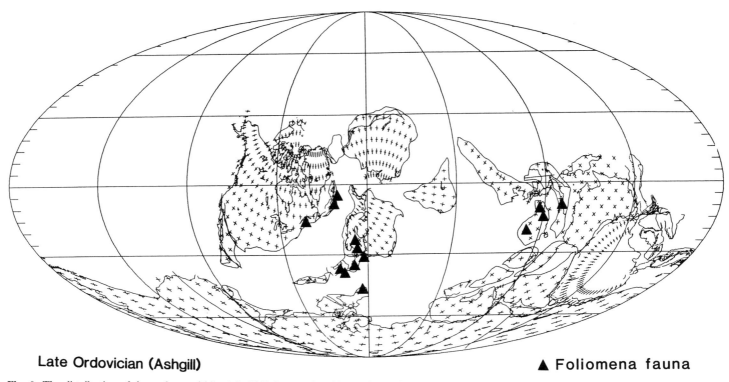

Fig. 6. The distribution of the early to middle Ashgill *Foliomena* brachiopod fauna (data from Cocks & Rong 1988) plotted on the Ashgill map generated for this volume.

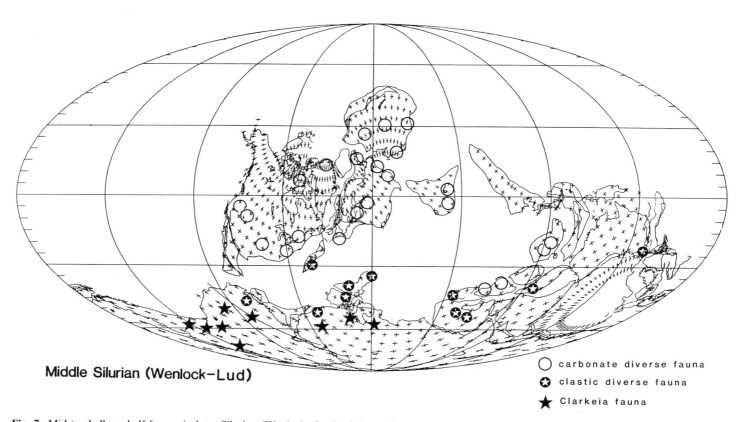

Fig. 7. Mid to shallow shelf faunas in later Silurian (Wenlock–Ludlow) time, plotted on a new map. The *Clarkeia* fauna is the low-diversity polar precursor of the Lower Devonian Malvinofaffric fauna and always occurs in clastic rocks. The more cosmopolitan faunas of the rest of the world (the brachiopod *Salopina* is shown) have been divided between their two principal lithofacies.

of the shelf areas and increased anoxia at the seafloor meant that graptolites are often the only fossils recovered, this being the case for the whole of northwestern Gondwana (Armorica, Bohemia, the Iberian Peninsula, Italy and North Africa) for most of the Llandovery period. Because the chief continents were relatively narrowly separated, faunal communications in the tropical and temperate areas appear to have been very good, and as a result the late Llandovery and early Wenlock benthic shelly faunas are amongst the most cosmopolitan in the whole geological record.

We have plotted the distribution of the Late Silurian *Clarkeia* fauna, which is only found from polar and sub-polar sites (Cocks 1972), and this distribution fits well with the new map generated for this volume (Fig. 7). Also plotted are some, but by no means all of the diverse shelly faunas to be found in late Wenlock and Ludlow times, exemplified in this diagram by the distribution of the mid to shallow shelf brachiopod *Salopina*, which is known from most of the available sites, although some areas, for example North China, have no known Late Silurian deposits of any sort. By dividing their occurrences between those occurring in clastic rocks and those found in limestones (many with bioherms) a further palaeolatitudinal division becomes evident. As in the Ordovician, there is a cline of Silurian faunas along the Gondwanan margin from polar to equatorial palaeolatitudes.

Although the southern part of the Iapetus Ocean is correctly shown as open during this period (the accretionary prism in the Southern Uplands of Scotland was still operating under deepwater graptolitic bearing conditions of sedimentation until the late Wenlock *lundgreni* Zone), the northern collision between Baltica and Laurentia may have taken place earlier, since 'Old Red Sandstone' conditions are known from the early Wenlock of parts of the Midland Valley of Scotland (e.g. Girvan) and also from the middle Wenlock of the Oslo Region, Norway. It is, however, hard to link the subsequent Old Red Sandstone faunas of eurypterids and vertebrates with any particular palaeolatitude.

Thus, apart from the cold-water *Clarkeia* fauna already mentioned, there are few Silurian faunas which are of much use in delimiting palaeocontinental margins and consequently endorsing or disproving the continental configurations shown in the new reconstructions. Berry & Boucot (1972) postulated that Silurian graptolites showed a depth zonation, but so far no worker has been able to differentiate any deepest-water graptolite faunas which might be used to identify marginal sites in a comparable way to the early Ordovician isograptid fauna.

The chief faunal checks on proving or disproving the majority of the suggested palaeogeographic reconstructions come therefore not so much from the Silurian but from the preceding Ordovician period and from the subsequent Devonian. However, in many areas, e.g. Wales (Ziegler *et al.* 1968) and South China (e.g. Wang *et al.* 1987) the succession of benthic animal communities from shallow to deep water are invaluable in reconstructing local palaeogeographies and thus establishing the margins and sutures of contemporary palaeocontinents.

References

BERRY, W. B. N. & BOUCOT, A. J. 1972. Silurian graptolite depth zonation. *24th International Geological Congress*, Montreal, 7, 59–65.

COCKS, L. R. M. 1972. The origin of the Silurian *Clarkeia* shelly fauna of South America and its extension to West Africa. *Palaeontology*, 15, 623–30, pl. 121.

—— & FORTEY, R. A. 1982. Faunal evidence for oceanic separations in the Palaeozoic of Britain. *Journal of the Geological Society, London*, 139, 465–478.

—— & —— 1988. Lower Palaeozoic facies and faunas around Gondwana. In: AUDLEY-CHARLES, M. G. & HALLAM, A. (eds). *Gondwana and Tethys*. Geological Society, Special Publication, 37, 183–200.

—— & RONG, J.-y. 1988. A review of the late Ordovician *Foliomena* brachiopod fauna with new data from China, Wales and Poland. *Palaeontology*, 31, 53–67, pls 8, 9.

COOK, H. E. & TAYLOR, M. E. 1975. Early Palaeozoic continental margin sedimentation, trilobite biofacies, and the thermocline, western United States. *Geology*, 3, 559–562.

DEAN, W. T. 1962. The trilobites of the Caradoc Series in the Cross Fell Inlier of Northern England. *Bulletin of the British Museum (Natural History) Geology*, 7, 65–134, pls 6–18.

—— 1967. The correlation and trilobite fauna of the Bedinan Formation (Ordovician) in southeastern Turkey. *Bulletin of the British Museum (Natural History) Geology*, 15, 81–123.

—— 1973. The Lower Palaeozoic stratigraphy of the Taurus Mountains near Beyşehir, Turkey. III The trilobites of the Sobova Formation (Lower Ordovician). *Bulletin of the British Museum (Natural History) Geology*, 24, 279–348.

DESTOMBES, J. 1985. Ordovician, pp 184–238 In: HOLLAND, C. H. (ed.) *Lower Palaeozoic of North-western and West-central Africa*. Wiley, Chichester.

EL-KHAYAL, A. A. & ROMANO, M. 1985. Lower Ordovician trilobites from the Hanadir Shale of Saudi Arabia. *Palaeontology*, 28, 401–12.

FORTEY, R. A. 1975. Early Ordovician trilobite communities. *Fossils and Strata*, 4, 331–352.

—— 1985. Pelagic trilobites as an example of deducing the life habits of extinct arthropods. *Transactions of the Royal Society of Edinburgh*, 76, 219–230.

—— 1988. Terranes around Ordovician Gondwana: biogeography and stratigraphic implications. *Abstracts for the Vth Ordovician Symposium, St John's, Newfoundland*, 32.

—— & COCKS, L. R. M. 1986. Marginal faunal belts and their structural implications, with examples from the Lower Palaeozoic. *Journal of the Geological Society, London*, 143, 151–160.

—— & OWENS, R. M. 1987. The Arenig series in south Wales. *Bulletin of the British Museum (Natural History) Geology*, 41, 69–307.

——, —— & RUSHTON, A. W. A. 1989. The palaeogeographic position of the Lake District in the earlier Ordovician. *Geological Magazine*, 126, 9–17.

HAMMANN, W. 1976. The Ordovician of the Iberian Peninsula — a review, pp 387–409 In: BASSETT, M. G. (ed.) *The Ordovician System*. University of Wales Press, Cardiff.

HAVLÍČEK, V. 1971. Brachiopodes de l'Ordovicien du Maroc. *Notes et Memoires du Service Géologique, Rabat*, 230, 1–135, pls 1–26.

—— & VANEK, J. 1966. The biostratigraphy of the Ordovician of Bohemia. *Sborník Geologických Ved paleontologie*, 8, 7–69, pls 1–16.

HENRY, J.-L. 1980. Trilobites ordoviciens du Massif armoricain. *Memoires de la Societé géologique et minéralogique de Bretagne*, 22, 1–250, pls 1–48.

HURST, J. M. 1979. The stratigraphy and brachiopods of the upper part of the type Caradoc of south Salop. *Bulletin of the British Museum (Natural History) Geology*, 32, 183–304.

LEGG, D. P. 1976. Ordovician trilobites and graptolites from the Canning Basin, western Australia. *Geologica et Palaeontologica*, 10, 1–58.

LU, Y.-h. and others. 1976. Ordovician biostratigraphy and palaeozoogeography of China. *Memoir of the Institute of Palaeontology, Nanjing*, 7, 1–83, pls 1–14, (in Chinese).

NEUMAN, R. B. 1976. Early Ordovician (Late Arenig) brachiopods from Virgin Arm, New World Island, Newfoundland. *Canadian Geological Survey Bulletin*, 261, 11–61, pls 1–8.

OWEN, A. W. & BRUTON, D. L. 1980. Late Caradoc — early Ashgill trilobites of the central Oslo region, Norway. *Paleontological Contributions from the University of Oslo*, 245, 1–62, pls 1–10.

RONG, JIA-YU & HARPER, D. A. T. 1988. A global synthesis of the latest Ordovician Hirnantian brachiopod faunas. *Transactions of the Royal Society of Edinburgh: Earth Sciences*, 79, 383–402.

RÕÕMUSOKS, A. 1970. *Stratigraphy of the Viruan Series (Middle Ordovician) in Northern Estonia*. Valgus, Tallinn.

ROSS, R. J. 1975. Early Paleozoic trilobites, sedimentary facies, lithospheric plates, and ocean currents. *Fossils and Strata*, 4, 307–329.

SHEEHAN, P. M. 1973. Brachiopods from the Jerrestad Mudstone (early Ashgillian, Ordovician) from a boring in Southern Sweden. *Geologica et Palaeontologica*, 7, 59–76, pls 1–3.

VANNIER, J. M. C., SIVETER, D. J. & SCHALLREUTER, R. E. L. 1989. The composition and palaeogeographical significance of the Ordovician ostracode faunas of southern Britain, Baltoscandia and Ibero–Armorica. *Palaeontology*, 33, 163–322, pls 24–30.

VILLAS, E. 1985. Braquiopodos del Ordovicico medio y superior de Las Cadenas Ibericas Orientales. *Memorias del Museo Paleontologico de La Universidad de Zaragosa*, **1**, 1–223, pls 1–34.

WANG, Y., BOUCOT, A. J., RONG J.-y. & YANG, X.-c. 1987. Community Paleoecology as a geologic tool: the Chinese Ashgillian-Eifelian (latest Ordovician through early Middle Devonian) as an example. *Geological Society of America Special Paper*, **211**, 1–100, pls 1–20.

WHITTARD, W. F. 1966. The Ordovician trilobites of the Shelve Inlier, west Shropshire. Part VIII. *Palaeontographical Society Monograph*, 265–306, pls 46–50.

WHITTINGTON, H. B. 1963. Middle Ordovician trilobites from Lower Head, western Newfoundland. *Bulletin of the Museum of Comparative Zoology, Harvard*, **129**.

—— 1968. A monograph of the Ordovician trilobites of the Bala area, Merioneth. Part IV. *Palaeontographical Society Monograph*, 93–138, pls 29–32.

—— & HUGHES, C. P. 1972. Ordovician geography and faunal provinces deduced from trilobite distribution. *Philosophical Transactions of the Royal Society of London*, **B263**, 235–278.

WILLIAMS, A. 1963. The Caradocian brachiopod faunas of the Bala District, Merionethshire. *Bulletin of the British Museum (Natural History) Geology*, **8**, 327–471, pls 1–16.

—— 1974. Ordovician Brachiopods from the Shelve District, Shropshire. *Bulletin of the British Museum (Natural History) Geology Supplement*, **11**, 1–163, pls 1–28.

——, LOCKLEY, M. G. & HURST, J. M. 1981. Benthic palaeocommunities represented in the Ffairfach Group and coeval Ordovician successions of Wales. *Palaeontology*, **24**, 661–694.

ZHOU, Z-Y. & FORTEY, R. A. 1986. Ordovician trilobites from North and Northeast China. *Palaeontographica*, Abt A, **192**, 157–210, 12 pls.

ZIEGLER, A. M., COCKS, L. R. M. & MCKERROW, W. S. 1968. The Llandovery transgression of the Welsh Borderland. *Palaeontology*, **11**, 736–82.

Relations between conodont provincialism and the changing palaeogeography during the Early Palaeozoic

STIG M. BERGSTRÖM

Department of Geology and Mineralogy, The Ohio State University, Columbus, Ohio 43210, USA

Abstract: World-wide distribution patterns of conodont species during the Cambrian, Ordovician, and Silurian are assessed quantitatively using a Coefficient of Similarity (CS) formula. The validity of the idea is examined that there is a correlation between CS values and the geographic distance between the sample points, and that changes in CS values may, at least in part, reflect plate motions. To minimize effects of local palaeoecological control and the varying degree of exploration of the faunas, well known faunas are used from comparable depositional environments on the several continental plates.

Conodont provincialism can be traced back to Upper Cambrian time, when there was an initial differentiation into low-latitude, warm-water (Midcontinent Faunal Region) and high-latitude, colder-water (Atlantic Faunal Region) faunas. The Lower Ordovician is characterized by great taxonomic diversification as well as striking provincial differentiation. Based on regional distribution patterns, it is possible to recognize a North American Interior Province, a Mediterranean Province, a North Chinese Province, a Siberian Province, a Baltic Province, and an Australian Province during Arenig time. A similar lateral differentiation prevailed during Llanvirn to lower Caradoc time. It is difficult to recognize a close correlation between CS values and inferred distances of plate separations in the Cambrian and Lower and Middle Ordovician.

Unlike some megafossil groups that markedly decreased in provincialism during the Upper Ordovician, the conodont faunas continued to exhibit pronounced lateral differentiation despite the fact that the Baltic, Siberian, and North American plates were then relatively closely together in the equatorial zone. Near the Ordovician–Silurian boundary (probably in the uppermost Ordovician) there was a global, quite conspicuous turn-over in the conodont faunas. During that time, the taxa characteristics of the Atlantic Faunal Region disappeared and apparently, the ancestors of virtually all Silurian taxa are to be found in the Midcontinent Faunal Region.

The Silurian conodont faunas have a cosmopolitan character although there are minor local differences that may be attributed to local environmental control. During Silurian time, the plates that have produced the best conodont faunas (North America, Baltica, Siberia) were all in the equatorial zone, and a general uniformity in environmental conditions may be one of several reasons behind the cosmopolitan character of the Silurian conodont faunas.

Many Lower Palaeozoic conodont species were capable of crossing water bodies of oceanic dimensions, and the distribution of many taxa seems to have been more closely controlled by water temperature and other ecological parameters than by potential migration barriers related to size of water bodies and emerged continental blocks. Nevertheless, because of their pronounced latitudinal differentiation during the Ordovician, the conodont faunas may be used as tools to decipher some of the positions of continental plates.

In the Lower Palaeozoic, several invertebrate groups exhibit some of the most striking cases of provincialism known in the fossil record such as the trilobites in the Cambrian (Palmer 1973, 1979) and the Ordovician (Jaanusson 1979; Whittington & Hughes 1972), and the brachiopods (Williams 1976; Jaanusson 1979) and graptolites (Skevington 1973, 1974; Berry 1979) in the Ordovician. Also some microfossil groups, such as the Ordovician chitinozoans (Laufeld 1979; Grahn & Bergström 1984) and acritarchs (Vavrdova 1974) display conspicuous, but less well known, provincialism. The conodonts also show provincialism in parts of the Lower Palaeozoic, and their most pronounced provincialism is in the Ordovician, a period with striking provincialism in most other fossil groups.

Since Early Palaeozoic conodont provincialism was recognized in the late 1950s (Sweet *et al.* 1959), several papers have summarized aspects of it (e.g., Bergström 1971, 1973; Barnes *et al.* 1973; Sweet & Bergström 1974, 1984; Barnes & Fåhraeus 1975; Fåhraeus 1976; Lindström 1976; Charpentier 1984; Dzik 1983; Miller 1984). However, the latest general review (Charpentier 1984) did not consider post-1979 papers and was based largely on outdated form taxonomy. A vast amount of new data is now available, including much information from previously little studied areas such as eastern Asia. The present study is based, whenever possible, on modern multielement taxonomy and includes data published up to 1987. It differs from previous studies, except Charpentier (1984) and Sweet & Bergström (1984), in using a quantitative rather than qualitative approach.

Materials and methods

Data base

Among the more than 500 papers in the Lower Palaeozoic conodont literature most deal with Ordovician collections. This reflects the fact that conodonts reached their peak of diversity (Sweet 1985) and have great stratigraphic utility in that system (Sweet & Bergström 1971). Cambrian conodonts are less diverse and have been studied in less detail, but those of the Silurian, although morphologically less diverse than the Ordovician ones, are relatively well known. Globally, the coverage is quite uneven; most described faunas are from the northern hemisphere and the information at hand from South America, Africa, Australia, and Antarctica is fragmentary and deficient in several respects. Conodonts share this problem with several other major fossil groups, but at least in the Ordovician, the conodont data now available are as good as, if not better than, those of any other fossil group. One problem specifically affecting the conodont data base is the outdated form taxonomy used in most pre-1970 papers. This has now been largely replaced by a biologically sounder multielement taxonomy, although many taxa remain unrevised. The presence of two rather different systems of classification makes it virtually impossible to use the Lower Palaeozoic conodont literature for regional biogeographical assessments without experience of the classification(s) used. The uncritical use of published form species records (Charpentier 1984) inevitably produces misleading figures of endemics and species diversity. In the present study, the faunal lists have been screened and when appropriate, reinterpreted in terms of modern multielement taxonomy. Taxa not identified to species have not been considered except in the case of highly characteristic species. Although some bias in unavoidable in this procedure, the revised species lists should be more reliable and representative than the original ones.

Methods of analysis

Many now recognized conodont multielement genera, especially those composed of morphologically closely similar species with ramiform elements in the apparatus, appear to represent biologically sound entities. Others, for instance some of those based on species having exclusively coniform (simple cone) elements, can be suspected to be polyphyletic and biologically artificial groupings. In view of this, along with the fact that many conodont genera are characterized by wide regional range, it is preferable to assess the provincialism at the species, rather than genus, level.

Among the several different formulas available for assessing similarity between faunas (Savage *et al.* 1979), I use the Coefficient of Similarity (CS) employed by Clark & Hartleberg (1983):

$$CS = \frac{2v}{a+b}$$

where v is the number of species in common between the two areas, and a and b represent the total number of species in each area.

It is obvious that if the number of species is very different in the two faunas compared, the value of CS will be low, even in a case where all species in the smaller sample are present in the larger one. However, most conodont faunas dealt with in the present study contain a similar number of species. Experience from the present study indicates that CS values below 0.20 reflect low or very low similarity, those between 0.20 and 0.55 moderate to relatively high similarity, and those above 0.55 very high similarity. Values above 0.75 are rare, even between faunas from the same province, and are obtained only from comparison of nearly identical faunas.

Data selection and provincial and stratigraphic classifications

A biogeographical study like the present one should ideally have adequate coverage of the whole spectrum of depositional environments. This goal is currently not attainable because most conodont collections come from shallow-water carbonates, and the faunas from, for instance, dark shales and deeper-water carbonates are less well known. An important bias may be introduced in faunal similarity assessments if one simply compares lists of *all* species known from each region without consideration of potential differences in the coverage of different depositional environments. To minimize this problem, I have tried to select for comparison well-known faunas representing at least broadly comparable depositional environments. This is justified also by the fact that a particular association of conodont species tends to have a considerable lateral distribution, in many cases a major part of a continental plate, and a comprehensive local fauna may therefore be representative of the faunas of an entire region.

There is little agreement in the palaeontological literature regarding the hierachial classification of biogeographic units. For instance, some authors use the term 'realm' more or less as a synonym of 'province' (Savage *et al.* 1979) whereas others prefer to recognize several provinces within a single realm. In the present contribution, I follow Berry (1979) in using 'faunal region' as the largest unit and this is subdivided into 'provinces'. This arbitrary classification follows many of those used previously in Lower Palaeozoic conodont biogeography.

A global stage and series terminology is now accepted for most of the Silurian. No such uniformity exists as yet in the Cambrian and the Ordovician and terms like the 'Lower Ordovician' has been given vastly different stratigraphic scope by different authors. The informal terms Lower, Middle, and Upper Cambrian are precise enough for the present contribution. For the Ordovician, I use the British series terminology while recognizing the many problems associated with the definition and correlation of these units (Whittington *et al.* 1984).

Distribution patterns

Cambrian

Cambrian conodonts remain less studied than those of any other system. One reason for this may be that the morphologically little varied Lower and Middle Cambrian species do not appear to have much biostratigraphic significance (Miller *et al.* 1981). The biostratigraphic usefulness of the group starts in the Upper Cambrian where conodont zonal schemes have been established in several parts of the world (An 1982; Miller 1984).

The records of pre-Upper Cambrian conodonts are too few and too scattered (cf. Müller 1971) to serve as a basis of assessment whether or not there existed a notable degree of provincialism. Most protoconodont and paraconodont species of that age seem to be widely distributed and apparently tolerated a wide range of environmental conditions (Miller 1984), suggesting a pelagic mode of life. Early Cambrian 'small shelly fossils' including some primitive conodonts may exhibit some provincial differentiation (Nowlan *et al.* 1985) but published information does not yet permit a regional assessment.

The Late Cambrian is marked by an important period of conodont evolution that continued into the Early Ordovician. A variety of paraconodonts appeared as well as the first euconodonts; the latter is significant because this group includes most post-Cambrian conodonts. Late Cambrian faunas are recorded from several areas in North America, northeastern China, northern Europe, Australia, and scattered localities in USSR, Iran, India, and a few other areas. The North American, Chinese, and north European late Late Cambrian faunas are now well known, and an excellent up-to-date summary of these is provided by Miller (1984).

There is a close similarity between the faunas described by Miller (1969, 1980, 1984), Miller *et al.* (1982), and others from western and southern USA and those recorded by An (1982) and Wang (1984) from northeastern and northern China. All these occur in shallow-water strata in the vast platform regions that occupied a low-latitude position during the Late Cambrian (Fig. 1). The Late Cambrian conodonts in the Baltic region are found in dark-colored organic-rich shales and limestones which most authors interpret as having been laid down in relatively shallow water at high latitudes. These faunas, which are present in a large region in northern Europe and can be traced into some marginal areas of eastern Canada (Landing 1980), differ conspicuously from those of the North American Interior and China.

In the present study, I selected for comparison faunas from the *Proconodontus* Zone in Texas-Oklahoma (Miller 1984; Miller *et al.* 1982), northeastern China (An 1982; see also Wang 1984), Kazakhstan (Apollonov & Chugaeva 1982; Dubinina Pers. Comm.) and Scandinavia (Müller 1959). The diverse fauna from Kazakhstan is not yet described in full detail but the other faunas are well known and each occurs in a vast region.

As shown in Fig. 1, the Coefficient of Similarity (CS) between the faunas of Texas–Oklahoma and those of China is much higher (0.69) than that between the former and the Baltic area (0.25), which is the same as that between the latter area and China. The fauna of Kazakhstan shares about 50% of the species recorded from the Baltic area (CS = 0.29). The CS between the

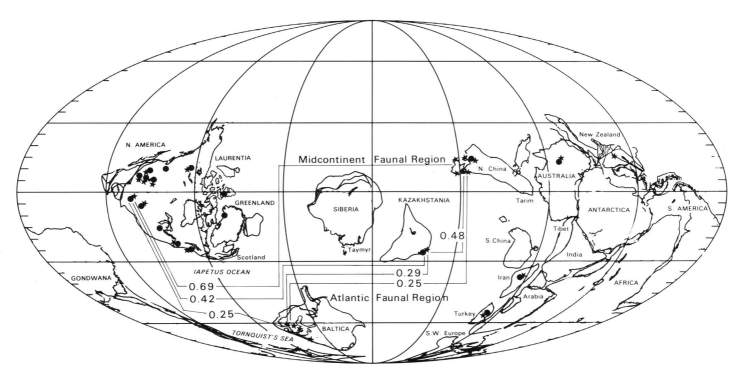

Fig. 1. Late Cambrian conodont distribution patterns. As in Figs 2–7, numerical values are Coefficient of Similarity (CS) between selected faunas. Note that the CS is much higher between Midcontinent Faunal Region faunas in south-central North America, Kazakhstan, and North China than between those of the latter areas and those of the Atlantic Faunal Region in the Baltic area, which indicates the beginning of the provincial differentiation of the conodont faunas. Stars indicate location of faunas with paraconodonts and *Phakelodus*, which have a virtually cosmopolitan distributions; black dots refer to occurrences of *Proconodontus*, which is particularly common and widespread in the Midcontinent Faunal Region. In many cases, one symbol marks several occurrences and no attempt has been made to indicate every single record although it is believed that the regional coverage is reasonably complete; the same applies to Figs 2–7. Map base from Scotese *et al.* (1979) modified by Scotese (unpubl.).

Kazakhstan fauna and those of China and North America is 0.48 and 0.42, respectively. This shows that the North American and Chinese faunas are closely related but they are both distinct from the Baltic one, and the Kazakhstan fauna is intermediate.

These data support Miller's (1984) conclusion that the striking provincialism in the Ordovician conodont faunas can be traced back to late Late Cambrian time, despite the fact that virtually all Cambrian conodont species became extinct in latest Cambrian to earliest Ordovician time. The Late Cambrian horizontal differentiation is similar to that in the Ordovician. In the past, the area of the low-latitude, warm-water Ordovician faunas has been referred to as the 'Midcontinent Province' and that of the high-latitude, cold-water faunas as the 'North Atlantic Province' (see, for instance, Sweet & Bergström 1974). Here, 'faunal region' is used as a replacement of 'province' (cf. Sweet & Bergström 1984) but the well-established terms 'Midcontinent' and 'Atlantic' are retained; hence I recognize a *Midcontinent Faunal Region* and an *Atlantic Faunal Region* in the Late Cambrian. The evidence is insufficient to distinguish provinces within each of these faunal regions where endemism is low, even at the species level. This is shown by the fact that the North American zonal scheme with minor modification is readily applicable to the North Chinese succession, and the Atlantic faunas are closely similar across that faunal region. The faunas of Kazakhstan and Iran (Müller 1973), differ in some respects from those of the two major faunal regions but the biogeographic significance of these faunas requires further study.

Ordovician

One of the most important events in the evolutionary history of conodonts took place during latest Cambrian and earliest Ordovician time when the Cambrian protoconodont–paraconodont faunas were replaced by faunas dominated by euconodonts. Details about this are still incompletely known (Miller 1984), but shortly after their appearance, the euconodonts underwent an explosive radiation resulting in a greater diversity than during any other time period in the history of the phylum (Sweet 1985). Although Lower Ordovician conodonts have been studied extensively, many species remain unrevised, or even undescribed, which complicates biogeographic assessments. The classification of Middle and Upper Ordovician conodonts has reached a more mature stage. The marked provincialism that is characteristic of Ordovician conodonts will be illustrated by examples from the Arenig, late Llanvirn–Llandeilo, and late Caradoc–early Ashgill.

Arenig. Representative Arenig conodont faunas have been described from parts of the USA, for instance west Texas–New Mexico (Repetski 1982), Utah (Ethington & Clark 1982), and New York (Landing 1976); and from the Baltic area (Lindström 1955; Löfgren 1978), Siberia (Abaimova 1972; 1975), China (An *et al.* 1983; An 1987), and central Australia (Cooper 1981). The faunas from the North American Interior, Siberia, North China, and Australia probably inhabited warm waters in the tropical zone whereas the Baltic ones are known mainly from presumably colder waters at mid- to high latitude (Fig. 2). Accordingly, it is no surprise that the former faunas show general similarity at the generic level but differ strikingly from the Baltic ones. As in the Late Cambrian, one may distinguish a *Midcontinent Faunal Region* and an *Atlantic Faunal Region* in the Arenig. Although it is difficult, and probably not very meaningful, to use a certain CS value to define the extent of these regions, their faunas have quite a different aspect with relatively few taxa in common.

This is illustrated by the minimal similarity between the rich Arenig faunas in the El Paso Group of Texas–New Mexico

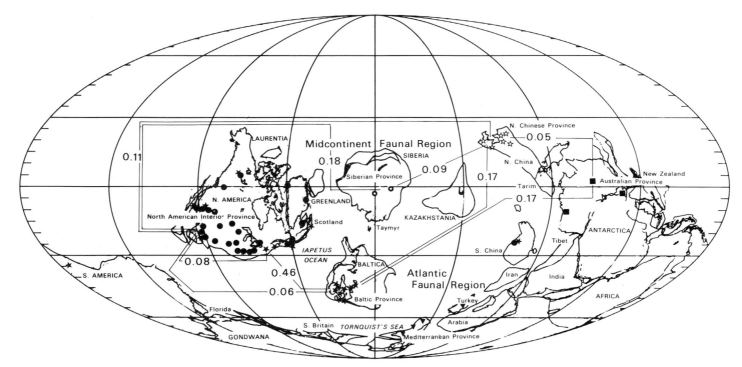

Fig. 2. Arenig (Early Ordovician) conodont faunal distribution patterns. Note the marked provincial differentiation indicated by very low CS numbers. Two faunal regions and six provinces are recognized. Black stars, faunas with characteristic Baltic Province elements; black dots, faunas of North American Interior Province type; black squares, faunas of Australian Province type; open stars, faunas of North China Province type; dots with white star, faunas of Siberian Province type; black triangles, faunas of Mediterranean Province type.

(Repetski 1982), and those of the Baltic area (Atlantic Faunal Region) (Löfgren 1978; Lindström 1955) as indicated by the very low CS of 0.06. Only 2% of the El Paso species are shared with the faunas of the Baltic area, and these are rather cosmopolitan taxa. A clear difference at the generic level is shown by the fact that about 40% of the Baltic genera are unknown in the El Paso and at least 35% of the El Paso genera are unrecorded from the Baltic area. Characteristic El Paso genera not known from the Baltic area include *Acanthodus*, *Cristodus*, *Clavohamulus*, *Reutterodus*, and *Ulrichodina* whereas the common Baltic genera *Baltoniodus*, *Belodella*, *Cornuodus*, *Paroistodus*, *Periodon*, *Scalpellodus*, *Stolodus*, *Strachanognathus*, and *Walliserodus* are unknown in the El Paso. Faunas similar to those of the El Paso are widespread across the North American craton and in the western Appalachians, an area here referred to as the *North American Interior Province*. Deeper-water strata along the continental margins, especially in eastern New York (e.g. the Deepkill Shale; Landing 1976), southwestern Texas (Marathon area; Bergström unpubl.), and western Great Basin (Ethington 1972) contain a considerable number of species present in the Baltic faunas. For instance, a comparison between the Deepkill fauna (Landing 1976) and coeval Baltic ones (Löfgren 1978) gives CS = 0.46 whereas the CS between the faunas of the Deepkill and El Paso is only 0.08. However, the Deepkill contains some characteristic taxa not recorded from the Baltic platform such as *Bergstroemognathus*, *Fryxellodontus*?, *Juanognathus*, and '*Cordylodus*' *horridus*, which are also known in other faunas from the margins of the North American continent.

Little is currently known about Arenig conodonts in Europe outside the Baltic area. The faunas from deep-water strata in South Scotland recorded by Lamont & Lindström (1957) are clearly of Baltic Province type. The conodonts of the shallow-water Durness Limestone of northwesternmost Scotland are quite different but similar to those of the North American Interior Province (Bergström & Orchard 1985). This supports the idea that North Scotland was a part of the Lower Palaeozoic North American plate, or constituted a microplate located on the American side of the Iapetus (Fig. 2).

A few occurrences of Arenig conodonts are known from central Europe (Lindström 1976; Dzik 1984) but most of these faunas are poorly known and not useful for quantitative comparisons. These faunas contain several distinctive, and probably endemic, species (see Sannemann 1955) and, as noted by Lindström (1976), it may be justified to recognize a separate conodont faunal province, here referred to as the *Mediterranean Province*, in this region during Arenig time.

The comprehensive Arenig faunas described from North China by An et al. (1983) show a general similarity, especially in the dominance of coniform taxa such as *Scolopodus*, to coeval cratonic faunas from North America. Yet there are considerable differences at the species level as shown by the low CS of 0.11 between the North Chinese and the El Paso faunas. Only 16% of the Chinese species are shared with the El Paso, and the presence of representatives of several characteristic, and apparently endemic, genera such as *Aurilobodus*, *Paraserratognathus*?, and *Serratognathus* makes these Chinese faunas quite distinct. This makes it justified to recognize a separate *North Chinese Faunal Province* within the Midcontinent Faunal Region during the Early Ordovician. The horizontal extent of this province outside northern China is still unclear but the Arenig faunas in the Yangtze Gorges region in south-central China are quite different and strikingly similar to those of the Baltic area (Zeng et al. 1983; An 1987).

A comparison between the faunas from northern China and the coeval ones from the Baltic area shows minimal similarity (CS = 0.17) and most of the species in common between the two areas are widespread coniform taxa. Also, 45% of the Chinese genera, including several highly distinctive ones such as *Aurilobodus*, *Bergstroemognathus*, *Loxodus*, *Paraserratognathus*?, *Polycaulodus*, *Rhipidognathus*, and *Tangshanodus* are entirely unknown from the Baltic area. Accordingly, there are good reasons to refer these faunas to different faunal regions.

Lower Ordovician conodonts from Siberia have been described

in several papers (see, e.g., Abaimova 1972, 1975), but the available data are difficult to use because the faunas have not been dealt with in terms of modern taxonomy. Nevertheless, Abaimova's (1975) list of species from her Complexes IV and V should be representative of the Siberian Arenig. Significantly, a comparison with the faunas from northern China gives a CS of only 0.09 although 45% of the Siberian genera are recorded in the North Chinese faunas. As may be expected, there is little similarity between the Siberian and the Baltic faunas (CS = 0.26), and the few species in common are wide-ranging taxa. The CS between the Siberian faunas and those of the El Paso is no more than 0.18 but at least 13 out of 42 named species and nine out of 11 genera listed by Abaimova (1975) are recorded from the El Paso, and the general aspect of these faunas is similar. Although the Siberian faunas are referable to the Midcontinent Faunal Region, their distinctiveness justifies the recognition of a separate province, the *Siberian Province*, during Arenig time.

Ordovician conodonts are still incompletely known in Australia and no complete Arenig conodont succession has been described. Cooper's (1981) study of the lower–middle Arenig Horn Valley Siltstone of central Australia is used here for general comparison. The similarity to the faunas from northern China (CS = 0.05) and to those of the El Paso (CS = 0.02) is minimal. Some identifications of Baltic species in the Horn Valley are not accepted by the present author, and following this interpretation, the CS between the Baltic faunas and that of the latter unit is only 0.17. These figures support the suggestion (Bergström 1971) that central Australia can be recognized as a separate province, the *Australian Province*, during Early and Middle Ordovician time. It should be noted that oldest Ordovician conodont faunas from Queensland (Druce & Jones 1971) contain many Midcontinent Faunal Region taxa.

Late Llanvirn–Llandeilo. Conodonts of this age have been studied extensively in North America (see, for instance, Sweet & Bergström 1971, 1976; Harris *et al.* 1979; Tipnis *et al.* 1978), Great Britain (Bergström *et al.* 1986), the Baltic area (Dzik 1976, 1978, 1983; Löfgren 1978; Hamar 1966), Siberia (Moskalenko 1983), and China (An *et al.* 1983; Hao 1981; An 1981). There are also records from scattered localities in western continental Europe, South America, and Australia, but faunas from these areas are incompletely known. The general pattern of lateral faunal differentiation is similar to that of the Arenig. As shown in Fig. 3, the southern parts of the British Isles as well as Baltoscandia were at somewhat lower latitudes than in the Early Ordovician but details of the palaeogeography of this part of the Ordovician are still very uncertain. For instance, the precise location of Scotland and northern Ireland remains unknown although available data suggest that these areas were situated on the North American side of the Iapetus.

North America, Siberia, and North China apparently occupied low latitude positions (Fig. 3) and the conodont faunas from the North American Interior and central Siberia show a general similarity, particularly at the generic level. They are here interpreted as representing the Midcontinent Faunal Region. The faunas of the Baltic area and south-central China are similar to each other, but strikingly different from those of the Midcontinent Faunal Region, and are here interpreted as representing the Atlantic Faunal Region. Faunas having some Atlantic species mixed with some Midcontinent species are known from a few low latitude areas, especially along the margins of the North American plate, where they typically occur in deeper-water and/or offshore strata. In a series of maps Sweet & Bergström (1974) contrasted the distribution of some genera characteristic of either the Midcontinent or the Atlantic Faunal Region. Although much additional data are now available, the illustrated general distribution patterns are still valid, including the location of the border region between the two faunal regions in the Appalachians.

A representative conodont succession from the North American Interior has recently been described by Bauer (1987) from the

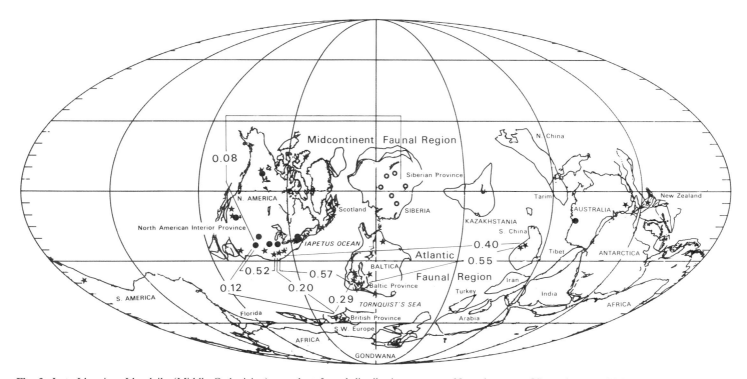

Fig. 3. Late Llanvirn–Llandeilo (Middle Ordovician) conodont faunal distribution patterns. Note that most CS numbers are higher than corresponding Arenig ones, suggesting somewhat decreased degree of provincialism. Also note the close similarity between faunas from the Baltic area and the Appalachian region in eastern North America, which indicates that the Iapetus Ocean was not a migration barrier for many species. Black dots, faunas of the North American Interior Province type; black stars, faunas of Baltic Province type; black triangles, faunas of British Province type; dots with white star, faunas of Siberian Province type.

McLish, Tulip Creek, and lower Bromide formations of Oklahoma. A comparison between those faunas and coeval ones of the Baltic area (Dzik 1978; Löfgren 1978; Bergström unpubl.) shows a CS of no more than 0.20, and only nine out of 39 Oklahoma species are recorded in the Baltic area. The difference at the generic level is equally striking in that 'Bryantodina', Erismodus, Leptochirognathus, Neomultioistodus, Paraprioniodus, Phragmodus, and Plectodina, which are all well represented in Oklahoma, are either completely unknown in coeval Baltic strata or, in the case of Phragmodus, represented by a single rare species. The Oklahoma faunas come from a region occupying a somewhat marginal position on the North American plate, and faunas from sites closer to its centre, such as that of the Dutchtown Formation of Missouri, have virtually nothing in common with those of the Baltic area. On the other hand, the coeval faunas from the Southern and Central Appalachians (Bergström & Carnes 1976; Bergström 1973, and unpubl.) are clearly related both to those of Oklahoma (CS = 0.52) and to those of the Baltic area (CS = 0.57). These Appalachian faunas share about 53% of the species with the latter area but they differ from the Baltic ones in having representatives of most of the genera just listed from Oklahoma. This is especially the case in the faunas of the western thrust belts in the Appalachians (Tazewell and Lee Confacies Belts of Jaanusson & Bergström 1980) whereas some faunas from the eastern thrust belts are strikingly similar to coeval ones from the Baltic area (Bergström et al. 1974, p. 1652). However, these faunas also tend to have a subordinate component of taxa best known from the Continental Interior.

The type Llandeilo Series and associated strata in Wales (Bergström et al. 1986) have faunas of low diversity, which are similar to the fauna described by Lindström et al. (1974) from coeval strata in Brittany, northwestern France. Some affinity to the Baltic faunas is shown by the fact that 75% of the Welsh genera occur in the taxonomically much more varied Baltic faunas, but the CS between the two areas is no more than 0.29 and few species are in common. These Welsh faunas show even less similarity to those of Oklahoma and the Southern and Central Appalachians (CS = 0.12 and 0.20, respectively). The separate British Province recognized in the Caradoc by Sweet & Bergström (1984) can be traced back to Llandeilo and slightly older strata (Ffairfach Group). Nothing is known about older Ordovician conodonts in Wales and England (Bergström & Orchard 1985), and the time of origin of the British Province cannot be established.

The conodont faunas described by Moskalenko (1973, 1983) and Kanagin et al. (1977) from the Vihorevian through Kirenskian–Kudrinian stages of the Siberian platform show close similarity at the generic level to those of the American Midcontinent. Although the Vihorevian and the Dutchtown Formation share some species (Moskalenko 1983), the similarity at the species level between the Siberian and North American faunas is insignificant; for instance, a comparison with the Oklahoma faunas just mentioned show a CS of no more than 0.08. Although some degree of similarity might be obscured by different taxonomic practice, these Siberian faunas appear to have their own special character, as is the case also in the Lower Ordovician. This is reinforced by the fact that they show virtually no similarity, even at the generic level, to coeval faunas from the Baltic area, or to those of south-central China. Because only a few, biogeographically insignificant, taxa have as yet been recorded from corresponding strata in northern China (An et al. 1983), the relationship between the Siberian and North Chinese faunas remains unclear.

Taxonomically varied and biogeographically interesting faunas occur in the Miapo Shale and slightly older strata in the Yangtze Gorges and adjacent regions in south-central China (Hao 1981; An 1987; Bergström unpubl.). No less than 81% of the species known to be present are shared with the Baltic area, and the CS is as high as 0.55. The only notable difference to coeval Baltic faunas is the occurrence of probably endemic species of Belodella, Eoplacognathus, and Polyplacognathus? in the Chinese faunas. This similarity provides a striking example of the very wide distribution of many conodont species in this time interval. Interestingly, at least 65% of the Chinese species are present also in the Southern and Central Appalachians (CS = 0.40).

Late Caradoc–early Ashgill. During this time interval, North America and Siberia continued to occupy an equatorial position, the Mediterranean area was still at high latitudes, and northwestern Europe was moving toward the equator and might have been at about 30° latitude (Fig. 4). As in earlier parts of the Ordovician, the palaeogeography of eastern Asia, and particularly China, remains unclear.

A comprehensive review of the conodont biogeography of this interval was recently published by Sweet & Bergström (1984) and reference is made to that study for more detailed data than can be presented within the scope of the present paper. Based on R- and Q-mode cluster analysis, Sweet & Bergström (1984) recognized a Red River and an Ohio Valley Province within the Midcontinent Faunal Region which occupy large areas in the Continental Interior of North America (Fig. 4). Coeval conodont faunas from the marginal regions of the continent are less well known but information is available from, for instance, Texas (Marathon area; Bergström 1978), the Great Basin (Harris et al. 1979; Ross et al. 1979) and eastern Canada (Nowlan 1981, 1983). Although dominated by species best known from the Midcontinent Faunal Region, these faunas also include some species that are widely distributed in northwestern Europe but uncommon in North America, where most of their occurrences are in deeper-water deposits along the continental margins.

Out of the 43 species recorded from the North American Interior by Sweet & Bergström (1984, table 1), 10 are known also from the Baltic area (CS = 0.30). Interestingly, the CS between the Baltic faunas and the Midcontinent ones from localities interpreted to have been 1–2° from the equator (Sweet & Bergström's locs. FH, S, BI) is 0.24 whereas that between the Baltic faunas and those of the localities at 18–20° latitude (Sweet & Bergström's locs. CO, T, C, BV, CI, MB) is considerably higher (0.44) as might be expected based on palaeogeographic considerations. The clear difference between the North American and Baltic faunas is also evident at the generic level, although 12 out of 21 genera recorded in the North American faunas are represented also in the Baltic area. Yet, several widespread North American genera such as *Pseudobelodina*, *Plegagnathus*, *Culumbodina*, *Oulodus*, and *Rhipidognathus* are unknown from the Baltic area, and the common Baltic genera *Hamarodus*, *Scabbardella*, and *Strachanognathus* have not been found in the Continental Interior of North America. The difference between the Midcontinent and the Atlantic Faunal Regions is as marked as in older parts of the Ordovician, and the rapid decrease in provinciality noted in some megafossil groups in late Middle and early Late Ordovician time is not obvious in the case of the conodonts.

Coeval British faunas have 13 out of 21 genera (62%) in common with the faunas of the Continental Interior of North America (Sweet 1979). This figure is closely similar to that obtained at a comparison between Baltic and North American faunas. About 40% of the species in the British faunas (Sweet & Bergström 1984) have been recorded also in North America, but the many endemic taxa in the latter area depress the CS to only 0.30. This happens to be the same as that obtained between the Baltic and the North American faunas, but it is much smaller than that between the former and the British faunas (CS = 0.68). Clearly, the Baltic and the British faunas are similar in several respects but, as noted by Sweet & Bergström (1984), there are several rather striking differences that justifies separation into a Baltic and a British Province.

At several localities in southern and western Europe and northern Africa the lower to middle Ashgill has produced distinc-

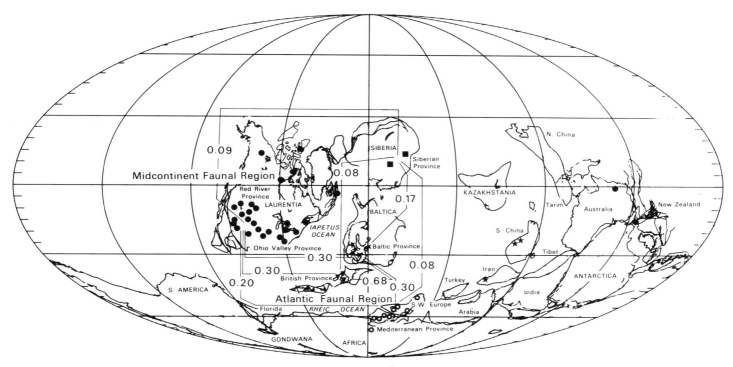

Fig. 4. Late Caradoc–early Ashgill (Late Ordovician) conodont faunal distribution patterns. Note that low CS values indicate considerable provincial differentiation between continental plates. Also note that the Iapetus Ocean has narrowed and the Rheic Ocean has widened. Black dots, faunas of the Red River and Ohio Valley provinces; black stars, faunas of Baltic Province type; black triangles, faunas of British Province type; black dots with white star, faunas of Mediterranean Province type; black squares, faunas of Siberian Province type.

tive conodonts (Knüpfer 1967), including the genera *Istorinus*, *Nordiodus*, and *Sagittodontina*, which may be endemic for this region. The presence of these taxa and the absence of several of the genera commonly represented in northwestern Europe led Sweet & Bergström (1984) and Bergström & Massa (in press) to recognize a separate *Mediterranean Province*. Based on the list of species in Bergström & Massa (in press), the CS between the faunas of the Baltic and Mediterranean provinces is 0.30, and that between the Mediterranean Province and the Red River and Ohio Valley provinces only 0.20. Both these figures provide good support for the distinctiveness of the Mediterranean Province faunas. As indicated by Bergström & Massa (in press), there are indications that the conodont faunas of this province were distinctive already in Early Ordovician time.

Siberia is likely to have been in the equatorial zone during this time interval. Although the long-distance correlations of the Siberian platform strata are somewhat tenuous, it seems likely that the Dolborian and Nirundian conodont faunas described from the Siberian platform by Moskalenko (1973, 1983) are late Caradoc and early Ashgill in age. These Siberian faunas exhibit similarity at the generic level to those of the North American Continental Interior (Moskalenko 1983), but at the species level, these faunas are quite different as indicated by a CS of no more than 0.09. A highly characteristic feature in these Siberian faunas is the presence of several species, representing several genera, that have elements with denticulated cusp edges. These genera, which include *Acanthodus*, *Acanthodina*, and *Acanthocordylodus*, are known elsewhere in the world mainly from Alaska (Harris & Repetski 1987). The species present of the latter two genera appear to represent a distinctive element in these Siberian faunas, which include also other taxa that may be endemic. The Siberian faunas show minimal similarity to coeval Baltic and British faunas (CS = 0.17 and 0.08, respectively), and there appear to be good reasons to recognize a *Siberian Province*, which may extend into Alaska, also in the late Caradoc and early Ashgill.

Few data are currently available on conodonts of this age from other parts of the world. One important occurrence is in the Baota (Pagoda) Limestone of south-central China from which An (1981, 1987) and Chen (1983) recorded, a.o., *Amorphognathus superbus*, *Hamarodus europaeus*, *Scabbardella altipes*, *Protopanderodus insculptus*, and *P. liripipus*, together with probably endemic species of *Belodella* and *Icriodella*. Sweet & Bergström (1984, p. 83) noted that, apart from the presence of *Belodella*, the Baota fauna is closely similar to that in coeval Baltic strata and represents the *Hamarodus europaeus*–*Dapsilodus mutatus*–*Scabbardella altipes* biofacies recognized in many successions in northwestern Europe. This fauna appears to be distributed rather widely in south-central China but its total range is still unknown. Strata of late Middle and Late Ordovician age are missing in northern China (Sheng 1980).

Silurian

There was a remarkable turnover globally in the conodont faunas near the Ordovician–Silurian boundary. The high-diversity Ordovician faunas are replaced by relatively low-diversity faunas that, with relatively minor changes, persist throughout the Silurian. As noted by several authors (Sweet & Bergström 1974; Sweet 1985), the faunas of the Atlantic Faunal Region vanished during the Middle and Late Ordovician and left few descendants in the Silurian. Most of the Silurian stocks appear to have their origins in those of the Midcontinent Faunal Region. The faunal replacement was by no means a catastrophic event; rather, there was a gradual extinction, and some evolution, through the Late Ordovician (Barnes & Bergström 1988). Most Ordovician stocks had disappeared in the latest Ordovician when typical Silurian taxa appeared. This is one of the most severe extinction periods in the history of conodonts, and the very small number of stocks that survived gave rise to the hundreds, if not thousands, of Silurian and younger species. Interestingly, the timing and magnitude of this turnover is quite similar to that experienced by trilobites, brachiopods, graptolites, chitinozoans, and acritarchs.

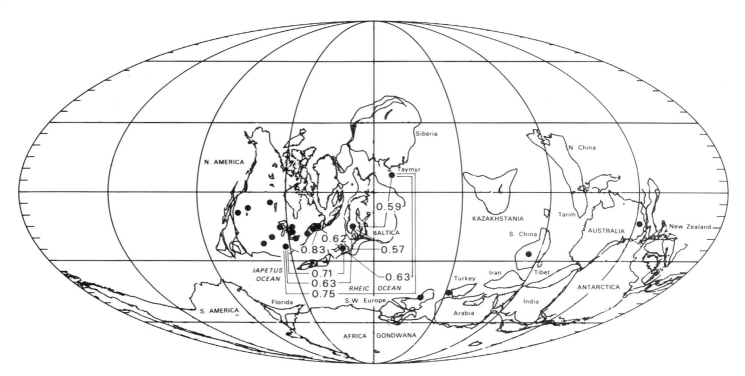

Fig. 5. Early Llandovery (early Early Silurian) conodont faunal distribution patterns. Note that high CS numbers show minimal provincialism between faunas from different continental plates.

The biogeographic effect of this turnover was dramatic on a global scale, and marks a change from strongly provincial to virtually cosmopolitan faunas. Three time intervals (early Llandovery, Wenlock, and Pridoli) have been selected to illustrate the Silurian conodont biogeography.

Early Llandovery. In the reconstruction of Early Silurian palaeogeography used here (Fig. 5), North America, northwestern Europe, and Siberia are placed close to each other in the equatorial zone. As pointed out by Sweet (1985), virtually all our data on post-Ordovician conodonts are from localities between 40°S and 40°N palaeolatitude, and high-latitude faunas are almost unknown. This applies to the Llandovery faunas, and their apparent regional uniformity may be a 'sampling effect', especially as most records come from a belt only 90 degrees wide longitudinally. However, there is no suggestion of latitudinal differentiation from the equator to 40° latitude similar to that present in the Ordovician faunas.

Early Llandovery (*Icriodella discreta–I. deflecta* Zone; the part of the *Distomodus kentuckyensis* Zone that is older than the level of appearance of *Hadrognathus staurognathoides*) conodont faunas are best known from the North American Midcontinent, for instance, the Brassfield Limestone of Ohio (Cooper, 1974, 1975, 1977, 1980); Anticosti Island in eastern Canada (McCracken & Barnes 1981; Fåhraeus & Barnes 1981), Great Britain (Aldridge 1972), the Oslo region, Norway (Aldridge & Mohamed 1982), the USSR Arctic (Männik 1983), and China (Lin 1983). Unfortunately, complete lists of species are not yet available from the three latter areas, and also various differences in taxonomic practice complicate close comparisons. Nevertheless, although the assessment presented here is obviously preliminary, the general trend is quite clear.

The Brassfield Limestone fauna (Cooper 1974, 1975, 1977) is very similar to that of the Becscie and Gun River formations of Anticosti Island (Fåhraeus & Barnes 1981) as shown by the very high CS of 0.83 (Fig. 5). There is a high CS also between the Brassfield fauna and that of the Lower Llandovery of Britain (Aldridge 1972), and between the latter and the Anticosti Island formations just mentioned (CS = 0.71 and 0.62, respectively). Virtually all genera and most species are in common between these faunas. Only non-coniform species are currently recorded from the coeval faunas in the Oslo region (Solvik and Saelabonn formations; Aldridge & Mohamed 1982) and in the USSR (Vodopad Formation of Severnaya Zemlya; Männik 1983). However, these faunas are also strikingly similar to those of North America and Britain at the species level. Based on non-coniform species, the CS between the Norwegian and Russian faunas is 0.59; between the Norwegian and the British faunas 0.57; between the Brassfield and the Norwegian faunas 0.63; betweeen the Russian and the British faunas 0.63; and between the Brassfield and the Russian faunas 0.75. These faunas are so similar to each other that it is highly justified to refer all to the same faunal province. Also younger Llandovery faunas from North America and Europe differ relatively little from each other, and the differences that do exist may be due to local environmental control.

Little is currently known about early Llandovery conodonts from other parts of the world. The succession of Silurian conodont zones in China discussed by Lin (1983) exhibits considerable similarity to those elsewhere in the world. Several well-known and widespread species are recorded from the pre-*Celloni* Zone portion of this succession but there are also some species that are not known from other areas. Regrettably, Lin (1983) gives no complete species lists, and the use of outdated form taxonomy further complicates comparisons. Also, the reported occurrences of species of *Aphelognathus*, *Cyrtoniodus*, and *Microcoelodus* require confirmation because all these genera are known only from Ordovician rocks elsewhere in the world. Additional information is clearly needed before the biogeographic significance of these faunas can be assessed, but presently available data do not suggest that the Chinese species associations are distinctive enough to justify recognition of a separate province. From southeastern Australia, Bischoff (1986) described diverse early Llandovery conodont faunas 'different from those recorded from

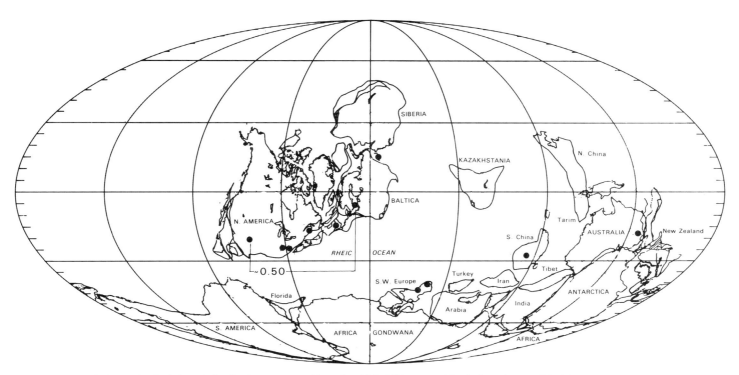

Fig. 6. Wenlock (Middle Silurian) faunal distribution patterns. Note that high CS between North American and Baltic faunas indicates continued minimal provincialism. Also note that the Iapetus Ocean is essentially closed between northern Europe and eastern North America.

North America'. These faunas appear to be quite distinctive but further studies are needed to clarify whether or not their distribution justifies the recognition of a separate faunal province.

Wenlock. In the interpretation used herein, the Wenlock palaeogeography is not markedly different from that of the Llandovery (Fig. 6). The equator was located across North America, the Baltic area, and northern China. England–Wales and much of Siberia was in the equatorial zone and Continental Europe was situated at 35–40°S latitude. This distribution of continents seems to be in general agreement with sedimentological and faunal evidence.

The conodonts were affected by another global extinction event, albeit of far less magnitude than the Late Ordovician one, in latest Llandovery and earliest Wenlock time (Jeppsson 1987). This is the most striking change in the conodont faunas during the entire Silurian and involves the replacement of the *P. amorphognathoides* Zone high-diversity fauna with its several morphologically highly elaborate taxa by a relatively low-diversity fauna composed mainly of morphologically relatively conservative species, most of which are of long-ranging stocks. Of particular significance was the disappearance of several characteristic platform conodonts such as *Apsidognathus*, *Aulacognathus*, *Icriodella*, *Johnognathus*, *Pterospathodus*, and *Tuxekania*, which are entirely unknown in younger strata. It is probably more than a coincidence that the extinction that marks the top of the *P. amorphognathoides* Zone at many places is associated with evidence of a general drop in sea-level from a high-stand during the late Llandovery. Interestingly, Ordovician platform conodonts, although present in strata representing a wide range of depositional environments, are most common and diversified in off-shore, relatively deeper-water shelf deposits. Jeppsson (1987) attributes the extinction event to a general change in the chemistry of the sea water.

The Wenlock faunas discussed herein are not from the entire series but only from the interval corresponding to the *Ozarkodina sagitta* and *Kockelella patula* zones in the European standard succession. Wenlock conodonts have been described from several regions in North America, northern and central Europe, and scattered localities elsewhere in the world but as a whole, they are not as well known as those of the Llandovery, and useful data are available only from a few areas. The fauna of the Clarita Formation of Oklahoma (Barrick & Klapper 1976; Barrick 1977) is perhaps the best known Wenlock fauna from North America, and that from the Island of Gotland, eastern Sweden (Jeppsson 1983, 1984) has been studied in equally great detail. There are still some unsolved taxonomic problems in both these faunas, and the non-*Panderodus* coniforms are still not identified to species in the Gotland collections, but these faunas are clearly strikingly similar. They share the same 10 genera, and a majority of the species are the same, or very closely related. The only notable difference seems to be the presence of two additional species of *Kockelella*, and possibly additional species of *Dapsilodus*, in the Oklahoma fauna; some differences in the case of species of *Ozarkodina* (*Hindeodella* in Jeppsson) and *Ligonodina* (*Delotaxis* in Barrick & Klapper) might reflect differences in taxonomic practice. At any rate, the CS between these faunas is about 0.50 and it may well be higher when taxonomic relationships have been fully clarified.

Although not yet described in full detail, the conodont faunas from the British Wenlock (Aldridge 1975, 1985) appear to be quite similar to those from Oklahoma and Gotland but somewhat less varied taxonomically. Characteristic endemic taxa have not been reported, and the same applies to the faunas of the classic Cellon succession in the Carnic Alps near the border between Italy and Austria (Walliser 1964). Faunas from other parts of the world are still incompletely known but there is no evidence of provincial differentiation (Jeppsson 1983) although some taxa, such as *Belodella* and *Dapsilodus*, are more abundantly represented in some areas than in others, probably because of local environmental control. Also the Ludlow faunas do not show provincial differentiation.

Pridoli. In the reconstruction of latest Silurian (Pridoli) palaeogeography used herein, the location of the main continental plates is very similar to that of the Wenlock (Fig. 7). North

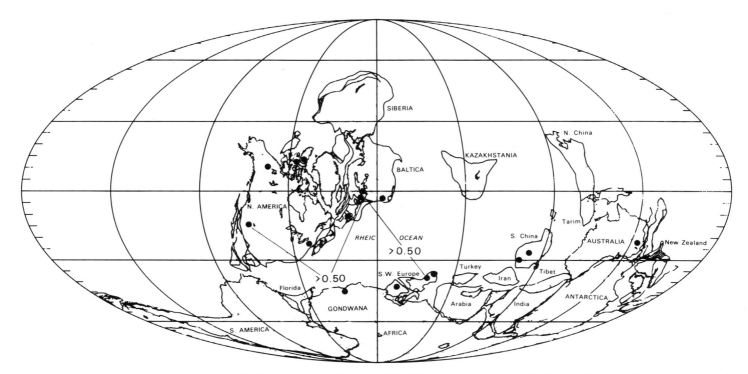

Fig. 7. Pridoli (Late Silurian) conodont faunal distribution patterns. Note that high CS between faunas from North America, the Baltic area, and central Europe shows minimal provincialism.

America, northwest Europe, Kazakhstan, and North China straddled the equator, and the Mediterranean area, northern Africa, South China, Siberia, and Australia were between 0° and 60° latitude. If this reconstruction is reasonably correct, all Pridoli conodont data available are from regions within 45° of the equator.

Pridoli conodonts are known from scattered localities on most continents (Fig. 7), but there are very few modern studies of the entire species associations in these faunas. For instance, the basic work by Walliser (1964) on the Carnic Alps Cellon section does not consider coniform conodonts at all, and also some other extensive studies, such as that of Klapper & Murphy (1974) on the Roberts Mountains Formation of central Nevada, and that of Jeppsson (1975, 1981, 1984) on the succession in Scania and Gotland, South Sweden, provide little data on coniform conodonts. Furthermore, the taxonomy of the Pridoli conodonts is particularly troublesome and not yet stabilized. This complicates quantitative comparisons at the faunas. Because the total number of species is small, the CS values obtained may vary a great deal depending on the interpretation of the morphologic scope and identity of just a few taxa. It should therefore be stressed that the CS values recorded here are highly preliminary and subject to revision when species nomenclature has stabilized, and the entire faunas are described.

For of the present study, I have selected the Roberts Mountains fauna of central Nevada (Klapper & Murphy 1974) as representative for North America, and particularly western USA. In Europe, the British Pridoli succession is largely developed in lithologies unsuitable for conodont extraction and the few records published (Aldridge 1975, 1985) serve mainly to show that the faunas are similar to those elsewhere in Europe. In Sweden Jeppsson (1975, 1981, 1984) has done extensive work on most of the Silurian faunas, but the Pridoli part of the succession is not well exposed, and samples from surface exposures in Scania have not yielded diverse faunas. From Pridoli erratic boulders (Beyrichienkalk), Jeppsson (1981) recorded a small fauna, including *Ligonodina elegans detorta*. In south-central Europe, much conodont work has been carried out since Walliser's basic studies (1964), and in the present paper, I use data from Jeppsson's (1988) study of the Pridoli of the Silurian–Devonian boundary stratotype at Klonk, Czechoslovakia.

All the Pridoli faunas recorded to date are of low diversity (Sweet 1985, Fig. 7); the most varied ones known, which are from the Carnic Alps and Czechoslovakia, include only about 12–15 species distributed among some five to eight genera. Because these are from many collections from a wide range of facies, it seems unlikely that the number of species will increase substantially by additional study. The faunas from Nevada and Czechoslovakia show striking similarity at the generic level in that they contain representatives of *Ozarkodina*, *Ligonodina* (*Delotaxis*) and *Belodella* along with coniform species of *Panderodus* and *Pseudooneotodus*. The Nevada faunas also contain *Pelekysgnathus* and *Coelocerodontus*. Five out of the seven genera (71%) present in Nevada are present in Czechoslovia, and the Carnic Alps faunas are also closely similar. It is difficult to calculate a meaningful CS value for these faunas but it seems safe to suggest that it is >0.50 between Nevada and Czechoslovia, and a comparable value is likely to be obtainable between Nevada and the Carnic Alps.

Jeppsson (1975, p. 13–15) discussed some horizontal distribution patterns in Late Silurian conodont faunas and noted that, by and large, there is little evidence of provincialism. He described several cases of different relative frequencies of some taxa in coeval strata in different regions, and drew attention to diachroneity in the regional appearance and disappearance of some species. However, because information is currently available only from a small number of scattered localities, it is difficult to recognize regional biogeographic patterns as well as the probably important effect of local environmental control. The overall close similarity between the faunas from North America, Europe, and China (Lin 1983) supports the conclusion that the Pridoli was a time interval with minimal provincialism in conodont faunas. Interestingly, although acknowledging the existence of some endemism at the specific, but not generic, level in the Lower Devonian faunas, Klapper & Johnson (1980) concluded that this endemism was not of the extent that it justified the recognition of formal provinces. Hence the pattern of weak, if any, conodont provincialism established in the Early Silurian can be followed into the Early Devonian.

Comparison with Other Groups

The changes in degree of provincialism shown by the conodonts, from minimal in most of the Cambrian to conspicuous throughout the Ordovician to minimal provincialism again in the Silurian, are similar to, but not exactly the same, as those shown by several other major fossil groups, including trilobites, brachiopods, graptolites, chitinozoans, and acritarchs. The provincialism of each of these groups will be briefly discussed below and contrasted with that of the conodonts.

Cambrian. The spatial differentiation of Cambrian trilobites has often been cited as a classical example of provincialism in the fossil record. Trilobite provincialism was distinctive already in the Early Cambrian, and in the Middle and Late Cambrian, it became quite striking (Palmer 1973, 1979). Palmer (1979) distinguished seven provinces in the latter interval, two in North and South America, two in Europe, and three in Asia, at least two of which extended to Australia. No similar provincialism is evident in the distribution of conodonts through most of the Cambrian, and the provincial differentiation recognized in latest Cambrian time is still too poorly known to permit comparison with that exhibited by the trilobites.

Ordovician. Conspicuous provincialism is shown by virtually all widespread and morphologically diverse groups in the Ordovician (Spjeldnaes 1961; Burrett 1973) such as trilobites, brachiopods, graptolites, and conodonts as well as by corals, ostracodes, chitinozoans and acritarchs. Interestingly, although there are many similarities in the areal distribution of provincial units of different groups, the number of provincial units recognized, and also their lateral and temporal extent, tend to differ from group to group. This makes the Ordovician provincialism exceedingly complex, and only some major features can be noted within the limited scope of the present paper. For an authoritative review, see Jaanusson (1979).

In the case of shelly fossils, the North American Midcontinent (including the Canadian Arctic and Greenland) had a distinct provincial fauna during virtually the entire period that shows close similarity to that of Siberia; however, the latter contains many endemic taxa and may represent one or more provinces or subprovinces. Lithic, palaeomagnetic, and faunal evidence suggests that the North American shelly fauna, and its counterparts in Siberia and northern China, inhabited the warm waters of the equatorial zone. Although there are some differences in their spatial distributions, these shelly faunas are, by and large, found in the same areas as those occupied by the conodont Midcontinent Faunal Region. In Europe, a distinct shelly fauna with numerous endemic elements was distributed across a wide area from Baltoscandia to the Moscow Basin and the northern Ural Mountains, forming a separate major provincial unit, the Baltic Province of Williams (1973). Although this shelly fauna became less distinctive in the late Middle and Late Ordovician (Jaanusson 1979), the Baltic Province can be recognized through most of the period. Generally speaking, this province occupies the same area as the Baltic Conodont Province and at least in pre-Late Ordovician time, it occurred in the temperate mid-latitude zone. Conodont faunas of general Baltic type but commonly having some North American Midcontinent stocks are known from marginal areas of the North American plate, especially in the eastern Appalachians. In the Middle Ordovician, the same strata contain a distinctive megafossil assemblage, the Scoto-Appalachian fauna of Whittington & Williams (1955; also see Jaanusson & Bergström 1980), which includes some taxa known from the Baltoscandic area. Especially at the species level, this shelly fauna shows less close similarity to those of the Baltic Province than do coeval conodont faunas. Interestingly, the strata with Scoto-Appalachian faunas in the Girvan area, southwest Scotland, which many interpret to represent a part of the North American plate, have a conodont fauna that is virtually indistinguishable from the coeval ones in the Appalachians, particularly in eastern Tennessee (Bergström 1986a).

Based on megafossils, several biogeographic units can be distinguished in the Mediterranean–Anglo–Welsh region through the Ordovician (Williams 1973; Dean 1967; Whittington & Hughes 1972; Havlíček 1974; Cocks & Fortey 1982). Little is currently known about Early Ordovician conodonts from this large region but the few low-diversity faunas recorded appear to differ from those of the Baltic Province. The Middle Ordovician Anglo–Welsh conodont species associations is different from both coeval North American and Baltic ones, and provincial differences remained to the end of the Ordovician. Regrettably, available data from both shelly fossils and conodonts are still too incomplete to permit a close comparison of their horizontal differentiation. The same is the case in east Asia and Australia, where both the megafossil and conodont faunas contain provincial elements.

For a long time, it has been common practice to recognize two main faunal regions (by some authors referred to as provinces or realms) in the global distribution of Ordovician graptolites, the Pacific (American–Australian) and Atlantic (European) faunal regions (Skevington 1969, 1974; Berry 1979). The latter is best developed in the Arenig–Llanvirn of Europe (except western Ireland, southwestern Scotland, and west-central Norway) and North Africa, but faunas of similar type are also known from Peru–Bolivia, northern Argentina, and southwestern China. The Pacific Faunal Region occurs mainly in North America, Australia, north-central China, Siberia, and Kazakhstan. Berry (1979) recognized a British, a Bohemian, and a Baltic Province within the Atlantic Faunal Region. The Pacific Faunal Region faunas, which are more diverse than those of the Atlantic Faunal Region, are considered to represent the equatorial belt (Skevington 1974). They show some spatial differentiation (Berry 1979) but there are no widely accepted named provinces. Generic-level graptolite provincialism faded away during late Middle and Late Ordovician time but some spatial differentiation at the species level is still recognizable in the Ashgill.

The spatial distribution of graptolites shows similarity to that of the conodonts in the differentiation into two main faunal regions, but it is important to note that these faunal regions do not coincide precisely in lateral range. For instance, in the lower Middle Ordovician of the eastern Appalachians, the graptolites represent the Pacific Faunal Region but the conodonts are of Baltic Province type with some addition of Midcontinent taxa (Jaanusson & Bergström 1980). Because graptolites are best known from areas with shaly successions from which few conodonts have been collected (Bergström 1986b), it is difficult to compare the provincial differentiation of these two groups in detail. Yet there are clearly similarities not only at the faunal region level but also at the provincial level, for instance in the case of the Baltic province.

Laufeld (1976) drew attention to the marked difference between the chitonozoans of the Baltic area, North Africa, and North America, and subsequent work has added much new data on this spatial differentiation (Paris 1981). Some similarity to the provincial pattern of conodonts have been noted (Grahn & Bergström 1984), but the chitinozoan data are still too incomplete to permit a meaningful comparison on a global scale. Based on acritarchs, Vavrdova (1974) recognized a Baltic and a Mediterranean Province in Europe, and the latter was recognized also in eastern Newfoundland by Martin (1982). Yet, also in the case of acritarchs, too little is known about the spatial distribution for a close comparison with that of the conodonts.

Silurian. The Silurian shelly faunas show far less provincial differentiation than those of the Ordovician but two major biogeographic units are commonly recognized, the North Silurian and the Malvinokaffric realms (Boucot 1979, fig. 1). The latter, which covers most of South America, Africa south of the equator, and Australia, is characterized by low-diversity faunas with many endemics. The North Silurian Realm is found in North America, Europe, North Africa, and Asia, that is, areas from which most Silurian shelly fossils and conodonts have been described. At the generic level, brachiopods of this realm are relatively cosmopolitan although some limited differentiation has been recognized in the Middle and Late Silurian (Boucot 1979). The trilobites show a similar distribution picture although details are still poorly known. As noted by Berry (1979), Silurian graptolites show very little endemism and the faunas are essentially cosmopolitan. As far as known, this applies also to the chitinozoans (Laufeld 1979). The lateral differentiation recognized in acritarchs in Africa, North America, and Europe was interpreted as biofacies differences by Cramer & Diez de Cramer (1972). Hence, the general cosmopolitan distribution shown by Silurian conodonts has its counterparts in other important fossil groups.

Causes of Provincialism

The causes behind Recent provincial differentiation are frequently complex and in some cases not well understood, but, as noted by Berry (1979), they include both ecological and historical aspects as well as the phylogeny of particular groups. Because we have currently only incomplete knowledge about many conditions in Early Palaeozoic marine environments, assessment of the relative significance of potential distribution controls is difficult. In the case of conodonts, a further complication is the fact that there is still no agreement whether or not they were primarily benthic, nekto-benthic, nektic, or pelagic (Klapper & Barrick 1978). The recent discovery of conodont animals (Briggs *et al.* 1983; Aldridge 1987; Smith *et al.* 1987) has not added substantial new data to the solution of this problem. However, the presence of similar provincialism in many groups during the Ordovician, and its virtual absence in the Silurian, indicate that the controlling parameters were of such type(s) that similar provincial distribution patterns were established in morphologically quite different groups.

Conventionally, two factors are recognized as being of major importance for the large-scale spatial differentiation of marine organisms, namely water temperature and physical barriers. Different latitudinal zones tend to have faunas of different composition, especially in the planktic and shelf faunas, whereas some deep-water faunas may be remarkably similar over large regions (Cook & Taylor 1977; Cocks & Fortey 1982). Physical barriers include not only emerged areas such as the Isthmus of Panama but also ocean currents, and, in the case of shallow-water benthic organisms with pelagic larval stages, deep-water oceanic areas too wide for successful dispersal during this stage of the life cycle. Also, areas with unsuitable salinity or water temperature may form efficient migration barriers.

A major feature in the Ordovician biogeography is the equatorial belt with its Midcontinent Conodont Faunal Region, Pacific Graptolite Faunal Region, and Midcontinent Shelly Fossil Region. In this case, it appears that the main controlling parameter was water temperature, and possibly water chemistry conditions related to temperature. The broad similarity between various tropical faunas of different plates indicates that interplate dispersal took place to some extent, but the presence of some provincialism suggests that species exchange was selective, and that the oceanic areas apparently served as filters for migration. As far as the conodonts are concerned, many species exhibit a very wide distribution within a particular province but are rare, or unknown, outside that particular area. Good examples of this are several species of *Belodina* and *Pseudobelodina* in the Red River Province and *Acanthodina* and *Acanthocordylodus* in the Siberian Province.

The higher latitude faunas were, in general, conspicuously different from those of the equatorial belt during most of the Ordovician, but some more or less cosmopolitan species were shared, especially in conodonts and graptolites. Occasional dispersal of faunal elements from the tropical zone to high latitudes took place during time intervals with warm-water carbonate sedimentation in the latter region (Jaanusson & Bergström 1980), again suggesting that water temperature was a critical controlling factor. Significantly, some Atlantic elements also occur locally in the Midcontinent Faunal Region but typically, they are found in deeper-water deposits. These may have been laid down in environments with bottom temperatures comparable to those in shallow waters at high latitudes. Bergström & Carnes (1976) described a case of apparent 'submergence' of conodont species from one side of the Iapetus to the other which, in principle, follows the dispersal model proposed by Taylor (1977, text-fig. 14). The common occurrence of Baltic Province elements in the Appalachians along the eastern margin of the North American plate shows that the Iapetus Ocean was not a physical barrier capable of preventing dispersal of these taxa. This is also indicated by the fact that faunas strikingly similar to those known in coeval Middle Ordovician strata in the Baltic Province are recorded from such remote places as southern Tasmania (Burrett *et al.* 1983) and south-central China (Hubei Province; see above). McKerrow & Cocks (1976; see also Cocks & Fortey 1982) have argued that there was a gradual decrease in faunal separation between the North European and American plates during Early Palaeozoic time, and that this closely reflected the narrowing and ultimate closing of the Iapetus Ocean. In the case of conodonts, this decrease of provinciality cannot be recognized because Baltic Province conodont faunas are known from marginal areas of the north American plate in strata as old as Arenig (Bergström *et al.* 1972). Most of these taxa, as well as those in faunas from similar sites of Middle and Late Ordovician age, are, however, completely unknown from the interior regions of the North American and Siberian plates despite abundant evidence of physical continuity between the depositional environments of the marginal shelf and the continental interior. The boundary between the Midcontinent and Atlantic faunal regions recognized along the central part of the Appalachians (Bergström 1971) is distinctly expressed also in the shelly faunas (Jaanusson & Bergström 1980). Clearly, environmental parameters, rather than geographical barriers, prevented the successful dispersal of these mega- and microfossil faunas into the continental interior.

There is little doubt that ocean currents played a significant role in the regional distribution of many conodont species. A comparison between the model of ocean current circulation in the Early Ordovician (Arenig) shown in Fig. 8 and the distribution of coeval provincial conodont faunas (Fig. 2) provides ready explanation of several puzzling distribution patterns in both faunal regions, e.g. the presence of Baltic Province-type conodonts in eastern North America and South China as well as in South America.

A final, and most important, feature in the evolving Ordovician conodont biogeography was the disappearance of the marked provincial differentiation near the end of the period. As noted by Sweet & Bergström (1984) and Barnes & Bergström (1988), virtually all species, and all but eight genera, of the taxonomically quite diverse Late Ordovician conodont faunas became extinct prior to the beginning of the Silurian. The few surviving coniform taxa represent morphologically generalized, widespread, and probably eurythermal species. One of the three surviving genera with compound elements in the apparatus, *Oulodus*, is characteristic of the Middle and Late Ordovician Midcontinent faunas; another, *Ozarkodina*, has a global distribution in Silurian and younger faunas but is known from the Late Ordovician only as a rare element in Atlantic faunas; and a third, *Icriodella*, is essentially

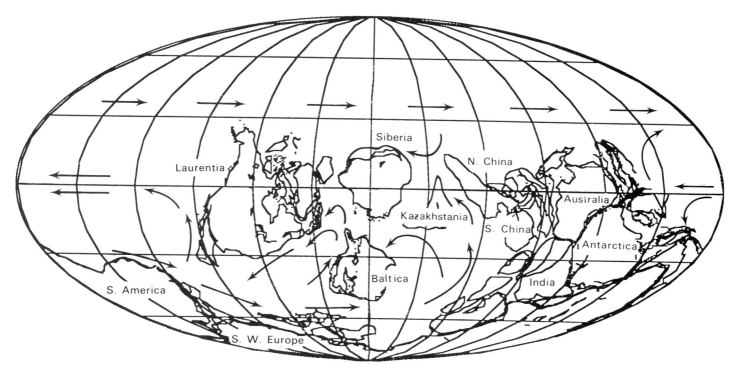

Fig. 8. Ocean current circulation pattern during the Early Ordovician.

cosmopolitan in the Middle and Upper Ordovician but has its oldest occurrences in the Mediterranean Province of the Atlantic Faunal Region.

The disappearance of virtually the entire Midcontinent Faunal Region conodont fauna may have been caused by lowering of the water temperature in the tropical zone during the Late Ordovician glaciation. As noted by Stanley (1984), tropical species tend to be adopted to a relatively narrow temperature interval and during a global cooling event, there is no region with suitable warm temperatures to which these species can migrate. A possible indication of a lowering of the water temperature in the equatorial region in North America is the appearance, in the latest Ordovician of the Mississippi Valley region, of brachiopods related to those of the widespread *Hirnantia* fauna (Amsden & Barrick 1986) which are associated with a low-diversity conodont fauna characterized by *Noixodontus*, a genus closely similar to, if not identical with, *Sagittodontina* of the high-latitude Mediterranean Province. Although it is distributed quite widely geographically, the *Hirnantia* fauna has commonly been interpreted to be of colder-water type. Yet, as noted by Amsden (*in* Amsden & Barrick 1986), brachiopods related to those of the *Hirnantia* fauna are in places present in warm-water (bahamitic) limestones containing a variety of megafossils best known from low-latitude faunas. This suggests that even if the tropical water temperature was lowered, it was still above 20°C. However, it is well known that a cooling of a few degrees C may have a devasting effect on tropical marine faunas.

The factors causing the disappearance of the Atlantic Faunal Region fauna are more puzzling, especially as this fauna included several long-established, morphologically diverse, and geographically widespread stocks. It is significant that this was not an abrupt event; the diversity of the Atlantic fauna was on the decline already in the Middle Ordovician, but the distinctiveness of the fauna remained to near the end of the period. Stanley (1984) drew attention to the fact that during a global cooling event, higher-latitude faunas can be expected to show a much higher species survival rate than low-latitude ones, not only because they include more eurythermal taxa but also because their species should be able to migrate toward the equator following

equatorialward shift in the latitudinal climate belts. However, there is little evidence of such survival and migrations in the Late Ordovician conodont faunas. Sweet & Bergström (1984) and Barnes & Bergström (1988) suggested that the disappearance of the Atlantic Faunal Region as a biogeographic unit was due to the fact that its key areas, especially North Europe, had moved into the tropical zone by latest Ordovician time, and suitable cold-water environments had become greatly restricted, if at all available, in these areas. The fact that the Hirnantian strata in North Europe, although locally quite fossiliferous, are almost barren of conodonts certainly suggests that the environmental conditions were very unfavorable for this group. It is also conceivable that another palaeogeographic factor was significant for the demise of the Atlantic conodont fauna and prevented its survival in shallow waters at mid-latitudes; as shown in Fig. 5, these latitudes were mostly occupied by the Rheic Ocean, and areas above sea level. The submerged parts of southern Europe and northern Africa were positioned close to the South Pole in latest Ordovician time, and conditions there may have been too severe at times for the survival of conodonts adopted to non-arctic water temperatures.

It is possible that the basic cause of the Ordovician provincial differentiation was related to the latitudinally scattered positions of the continental plates during a time period when glaciations created steep temperature gradients across much of the world's oceans. Jaanusson (1979) drew attention to the fact that because the Ordovician North pole appears to have been located far off continents in the northern Pacific, there was probably no ice cap in that polar region. If so, one can expect that there was only very limited differentiation into climatic zones on the northern hemisphere. However, in the palaeogeographic reconstruction shown in Fig. 4, the major plates do not extend much outside the tropical belt north of the equator during Late Ordovician time, and the absence of a northern ice cap may not have strongly influenced the composition of the marine faunas in the vast epicontinental seas. With the North American–Baltic–Siberian continental plates that have produced the vast majority of our fossils positioned in the equatorial zone, and the global climate again being more uniform after the Ordovician glacial episodes,

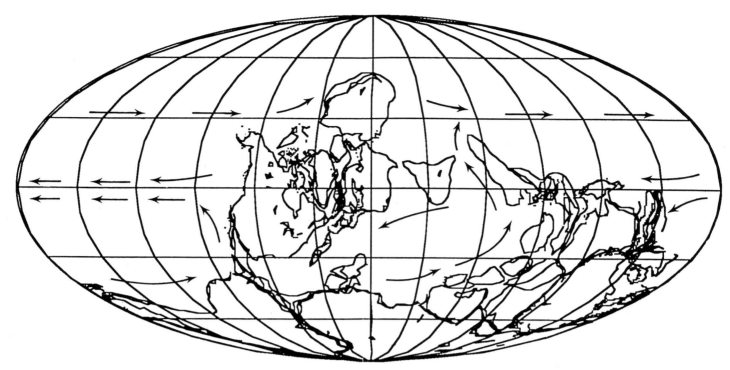

Fig. 9. Ocean current circulation pattern during the Silurian.

it is not surprising that in the Early Silurian, the marked provincialism was replaced by regional faunal uniformity in the low to mid-latitude regions of Laurentia–Eurasia whereas the high-latitude areas of Gondwana continued to have some endemism in the megafossil, and possibly also in the conodont, faunas. Regrettably, because virtually nothing is currently known about high-latitude Silurian conodont faunas, it is difficult to assess whether or not they exhibited a latitudinal differentiation regionally. A contributing factor to the apparent uniformity in the Silurian conodont faunas may have been the relatively simple oceanic circulation pattern (Fig. 9) that appears to have been capable of providing migration routes between most of the continental plates.

I am indebted to V. Jaanusson, C. E. Mitchell, and G. S. Nowlan for reading the manuscript and offering constructive comments, and to H. Hayes and K. Tyler for invaluable assistance.

References

ABAIMOVA, G. P. 1972. Subdivision and correlation of Ordovician deposits from the south-eastern part of the Siberian Platform based on conodonts. *Novosibirsk, Trudy Sibirskogo Nauchno-Issledovatel' skogo Inst. Geol. Geofis. i Minerl. Syr'ya (SNIGGIMS)* **146**, 65–67. (in Russian)

—— 1975. Early Ordovician conodonts from the middle reaches of the Lena River. *Trudy Sibirskogo Nauchno-Issledovatel'skogo Inst. Geol. Geofis. i Mineral. Syra'ya (SNIIGGIMS)* **207**, 1–129. (in Russian)

ALDRIDGE, R. J. 1972. Llandovery conodonts from the Welsh Borderland. *Bulletin of the British Museum (Natural History) Geology*, **22**, 125–231.

—— 1975. The stratigraphic distribution of conodonts in the British Silurian. *Journal of the Geological Society, London*, **131**, 607–618.

—— 1985. Conodonts of the Silurian System from the British Isles. *In*: HIGGINS, A. C. & AUSTIN, R. L. (eds) *A stratigraphical index of conodonts*, Ellis Horwood Ltd, 68–92.

—— 1987. Conodont palaeobiology: a historical review. *In*: ALDRIDGE, R. J. (ed). *Palaeobiology of conodonts*. Ellis Horwood Ltd., Chichester, 11–34.

—— & MOHAMED, J. 1982. Conodont biostratigraphy of the early Silurian of the Oslo region. *In*: WORSLEY, D. (ed.), *IUGS Subcommission on Silurian Stratigraphy. Field Meeting Oslo Region 1982*. Palaeontological Contributions, University of Oslo, **278**, 109–120.

AMSDEN, T. W. & BARRICK, J. E. 1986. Late Ordovician – Early Silurian strata in the central United States and the Hirnantian Stage. *Oklahoma Geological Survey Bulletin*, **139**, 1–95.

AN TAI-XIANG 1981. Recent progress in Cambrian and Ordovician conodont biostratigraphy of China. *In*: TEICHERT, C., LIN LU & CHEN, PEI-JI (eds) *Paleontology in China*. Geological Society of America Special Paper, **187**, 209–235.

—— 1982. Study on the Cambrian conodonts from North and Northeast China. *Science Reports of the Institute of Geoscience, University of Tsukuba*, **B, 3**, 113–159.

—— 1987. *The Lower Paleozoic conodonts of South China*. Beijing. (in Chinese).

——, ZHANG FANG, XIANG WEIDA, ZHANG YOUQIU, XU WENHAO, ZHANG HUIJUAN, JIANG DEBIAO, YANG CHANGSHENG, LIN LIANDI, CUI ZHANTANG, & YANG XINCHANG 1983. *The conodonts of North China and the adjacent regions*. Beijing. (in Chinese).

APOLLONOV, M. K. & CHUGAEVA, M. N. 1982. [Batyrbaisai section of the Cambrian and Ordovician at Malyi Karatau (south Kazakhstan).] *Izvestiya Akadamii Nauk SSR, Seriya Geologicheskaya*, **1982 (4)**, 36–46. (in Russian)

BARNES, C. R. & BERGSTRÖM, S. M. 1988. Conodont biostratigraphy of the uppermost Ordovician and lowermost Silurian. *In*: COCKS, L. R. M. (ed.). *The Ordovician–Silurian Boundary*. Bulletin of the British Museum (Natural History) Geology, **43**, 325–343.

—— & FÅHRAEUS, L. E. 1975. Provinces, communities, and the proposed nectobenthic habit of Ordovician conodontophorids. *Lethaia*, **8**, 133–149.

——, REXROAD, V. B., & MILLER, J. F. 1973. Lower Paleozoic conodont provincialism. *Geological Society of America Special Paper*, **141**, 157–190.

BARRICK, J. E. 1977. Multielement simple-cone conodonts from the Clarita Formation (Silurian), Arbuckle Mountains, Oklahoma. *Geologica et Palaeontologica*, **11**, 47–68.

BARRICK, J. E. & KLAPPER, G. 1976. Multielement Silurian (Late Llandoverian-Wenlockian) conodonts of the Clarita Formation, Arbuckle Mountains, Oklahoma and the phylogeny of *Kockelella*.

Geologica et Paleontologica, **10**, 59–99.

BAUER, J. A. 1987. *Conodont biostratigraphy, correlation, and depositional environments of Middle Ordovician rocks in Oklahoma*. PhD dissertation, The Ohio State University.

BERGSTRÖM, S. M. 1971. Conodont biostratigraphy of the Middle and Upper Ordovician of Europe and eastern North America. *Geological Society of America Memoir*, **127**, 83–157.

BERGSTRÖM, S. M. 1973. Biostratigraphy and facies relations in the lower Middle Ordovician of easternmost Tennessee. *American Journal of Science*, 273-A, 261–293.

—— 1978. Middle and Upper Ordovician conodont and graptolite biostratigraphy of the Marathon, Texas graptolite zone reference standard. *Palaeontology*, **21**, 723–758.

—— 1986a. Significance of conodonts in the Ordovician of the Girvan area — a suspect terrane in the Scottish Caledonides. *Geological Society of America Abstracts with Programs*, **18**, 212.

—— 1986b. Biostratigraphic integration of Ordovician graptolite and conodont zones — a regional review. *In*: HUGHES, C. P. & RICKARDS, R. B. (eds). *Palaeoecology and Biostratigraphy of Graptolites*. Geological Society Special Publication, **20**, 61–78.

—— & CARNES, J. B. 1976. Conodont biostratigraphy and paleoecology of the Holston Formation (Middle Ordovician) and associated strata in eastern Tennessee. *Geological Association of Canada Special Paper*, **15**, 27–57.

—— & MASSA, D. (in press). Stratigraphic and biogeographic significance of Upper Ordovician conodonts from north-western Libya. *In*: *The Geology of Libya*, **3**, Academic Press, London.

—— & ORCHARD, M. J. 1985. Conodonts of the Cambrian and Ordovician Systems from the British Isles. *In*: HIGGINS, A. C. & AUSTIN, R. L. (eds). *A stratigraphical index of conodonts*. Ellis Horwood, Ltd, 32–67.

——, EPSTEIN, A. G., & EPSTEIN, J. B. 1972. Early Ordovician North Atlantic Province conodonts in eastern Pennsylvania. *United States Geological Survey Professional Paper*, **800–D**, D37–D44.

——, RHODES, F. H. T., & LINDSTRÖM, M. 1986. Conodont biostratigraphy of the Llanvirn–Llandeilo and Llandeilo–Caradoc Series boundaries in the Ordovician System of Wales and the Welsh Borderland. *In*: AUSTIN, R. (ed.). *Conodonts: Investigative Techniques and Applications*. Ellis Horwood Ltd, 297–317.

——, RIVA, J. & KAY, M. 1974. Significance of conodonts, graptolites, and shelly faunas from the Ordovician of Western and North-Central Newfoundland. *Canadian Journal of Earth Sciences*, **11**, 1625–1660.

BERRY, W. B. N. 1979. Graptolite biogeography: A biogeography of some Lower Paleozoic plankton. *In*: GRAY, J. & BOUCOT, A. J. (eds). *Historical biogeography, plate tectonics, and the changing environment*. Oregon State University Press, Corvallis, 105–115.

BISCHOFF, G. C. O. 1986. Early and Middle Silurian conodonts from midwestern New South Wales. *Courier Forschungsinstitut Senckenberg*, **89**, 1–337.

BOUCOT, A. J. 1979. Silurian. *In*: ROBISON, R. A. & TEICHERT, C. (eds). *Treatise on Invertebrate Paleontology. Part A*. Geological Society of America and the University of Kansas. Boulder and Lawrence, A167–A182.

BRIGGS, D. E. G., CLARKSON, E. N. K. & ALDRIDGE, R. J. 1983. The conodont animal. *Lethaia*, **16**, 1–14.

BURRETT, C. 1973. Ordovician biogeography and continental drift. *Palaeogeography, Palaeoclimatology, Palaeoecology*, **13**, 161–201.

——, STRAIT, B., & LAURIE, J. 1983. Trilobites and microfossils from the Middle Ordovician of Surprise Bay, southern Tasmania, Australia. *Memoirs of the Association of Australasian Paleontologists*, **1**, 177–193.

CHARPENTIER, R. R. 1984. Conodonts through time and space: Studies in conodont provincialism. *Geological Society of America Special Paper*, **196**, 11–32.

CHEN YING-HUA 1983. The conodonts found in the Pagoda Formation of Guanyinchao, Qijiang, Sichuan Province. *Comprehensive Geological Brigade, Chinese Academy of Geological Sciences Bulletin*, **4**, 137–139.

CLARK, D. L. & HATLEBERG, E. W. 1983. Paleoenvironmental factors and the distribution of conodonts in the Lower Triassic of Svalbard and Nepal. *Fossils and Strata*, **15**, 171–175.

COCKS, L. R. M. & FORTEY, R. A. 1982. Faunal evidence for oceanic separations in the Palaeozoic of Britain. *Journal of the Geological Society, London*, **139**, 465–478.

COOK, H. E. & TAYLOR, M. E. 1975. Early Paleozoic continental margin sedimentation, trilobite biofacies, and the thermocline, western United States. *Geology*, **3**, 559–562.

COOPER, B. J. 1974. *Studies of multielement Silurian conodonts*. PhD dissertation. The Ohio State University.

COOPER, B. J. 1975. Multielement conodonts from the Brassfield Limestone (Silurian) of southern Ohio. *Journal of Paleontology*, **49**, 984–1008.

—— 1977. Toward a familial classification of Silurian conodonts. *Journal of Paleontology*, **51**, 1057–1071.

—— 1980. Toward an improved Silurian conodont biostratigraphy. *Lethaia*, **13**, 209–227.

—— 1981. Early Ordovician conodonts from the Horn Valley Siltstone, central Australia. *Palaeontology*, **24**, 147–183.

CRAMER, F. H. & DIEZ DE CRAMER, M. D. C. R. 1972. North American Silurian palynofacies and their spatial arrangement: Acritarchs. *Palaeontographica (Abteilung B)*, **138**, 107–179.

DEAN, W. T. 1967. The distribution of Ordovician shelly faunas in the Tethyan region. *In*: ADAMS, C. G. & AGER, D. V. (eds). *Aspects of Tethyan biogeography*. Systematics Association Publication, **7**, 11–44.

DRUCE, E. C. & JONES, P. J. 1971. Cambro-Ordovician conodonts from the Burke River structural belt, Queensland. *Australia Bureau of Mines and Mineral Resources, Geology and Geophysics Bulletin*, **110**.

DZIK, J. 1976. Remarks on the evolution of Ordovician conodonts. *Acta Palaeontologica Polonica*, **21**, 395–455.

—— 1978. Conodont biostratigraphy and paleogeographical relations of the Ordovician Mojcza Limestone (Holy Cross Mts., Poland). *Acta Palaeontologica Polonica*, **23**, 51–72.

—— 1983. Relationship between Ordovician Baltic and North American Midcontinent faunas. *Fossils and Strata*, **15**, 59–85.

—— 1984. Early Ordovician conodonts from the Barrandian and Bohemian–Baltic faunal relationships. *Acta Palaeontologica Polonica*, **28**, 327–368.

ETHINGTON, R. L. 1972. Lower Ordovician (Arenigian) conodonts from Pogonip Group, Central Nevada. *Geologica et Palaeontologica*, **Sb 1**, 17–28.

—— & CLARK, D. L. 1982. Lower and Middle Ordovician conodonts from the Ibex area, western Millard County, Utah. *Brigham Young University Geology Studies*, **28(2)**, 1–160.

FÅHRAEUS, L. E. 1976. Conodontophorid ecology and evolution related to global tectonics. *In*: BARNES, C. R. (ed.) *Conodont paleoecology*. Geological Association of Canada Special Paper, **15**, 11–26.

—— & BARNES, C. R. 1981. Conodonts from the Becscie and Gun River Formations (Lower Silurian) of Anticosti Island, Quebec. *In*: LESPÉRANCE, P. J. (ed.). *Stratigraphy and Paleontology*. IUGS Subcommision on Stratigraphy and Ordovician-Silurian Boundary Working Group. Field Meeting, Anticosti-Gaspé, Québec 1981, **II**, 165–172.

GRAHN, Y. & BERGSTRÖM, S. M. 1984. Lower Middle Ordovician chitinozoa from the Southern Appalachians, United States. *Review of Palaeobotany and Palynology*, **43**, 89–122.

HAMAR, G. 1966. The Middle Ordovician of the Oslo region, Norway. 22. Preliminary report on conodonts from the Oslo-Asker and Ringerike districts. *Norsk Geologisk Tidsskrift*, **46**, 27–83.

HAO, SHOUGANG 1981. Study of the biostratigraphy and conodont fauna from the Middle Ordovician Datian Ba and Miapo formations from Nanjing, Zhigui and Yidu areas. *MSc thesis, Beijing University*. (in Chinese)

HARRIS, A. G. & REPETSKI, J. E. 1987. Ordovician conodonts from northern Alaska. *Geological Society of America Abstracts with Programs*, **19**, 169.

——, BERGSTRÖM, S. M., ETHINGTON, R. L. & ROSS, R. J., Jr. 1979. Aspects of Middle and Upper Ordovician conodont biostratigraphy of carbonate facies in Nevada and Southeast California and comparison with some Appalachian successions. *Brigham Young University Geology Studies*, **26(3)**, 7–33.

HAVLÍČEK, V. 1974. Some problems of the Ordovician in the Mediterranean region. *Věstník Ústředního Ústavu Geologického*, **49**, 343–348.

JAANUSSON, V. 1979. Ordovician. *In*: ROBISON, R. A. & TEICHERT, C.

(eds). *Treatise on Invertebrate Paleontology. Part A. Introduction*. Geological Society of America and the University of Kansas. Boulder and Lawrence, A136–A166.

—— & Bergström, S. M. 1980. Middle Ordovician faunal spatial differentiation in Baltoscandia and the Appalachians. *Alcheringa*, 4, 89–110.

Jeppsson, L. 1975. Aspects of Late Silurian conodonts. *Fossils and Strata*, 6, 1–54.

—— 1981. The conodont faunas in the Beyrichienkalk. *In*: Laufeld, S. (ed.). *Proceedings of Project Ecostratigraphy Plenary Meeting, Gotland 1981*. Geological Survey of Sweden, Rapporter och Meddelanden, 25, 13–14.

—— 1983. Silurian conodont faunas from Gotland. *Fossils and Strata*, 15, 121–144.

—— 1984. Sudden appearances of Silurian conodont lineages: Provincialism or special biofacies. *Geological Society of America Special Paper*, 196, 103–112.

—— 1987. Lithological and conodont distributional evidence for episodes of anomalous oceanic conditions during the Silurian. *In*: Aldridge, R. J. (ed.). *Palaeobiology of conodonts*. Ellis Horwood Ltd., Chichester, 129–145.

—— 1988. Conodont biostratigraphy of the Silurian-Devonian boundary stratotype at Klonk, Czechoslovakia. *Geologica et Palaeontologica*, 22, 21–31.

Kanagin, A. V., Moskalenko, T. A., Yadrenkina, A. G. & Semenova, V. S. 1977. About stratigraphical section and correlation of the Middle Ordovician of the Siberian Platform. *Transactions of the Institute of Geology and Geophysics, Academy of Science USSR, Siberian Branch*, 372, "Nauka" Novosibirsk. (in Russian).

Klapper, G. & Barrick, J. E., 1978. Conodont ecology: pelagic versus benthic. *Lethaia*, 11, 15–23.

—— & Johnson, J. G. 1980. Endemism and dispersal of Devonian conodonts. *Journal of Paleontology*, 54, 400–455.

—— & Murphy M. A. 1974. Silurian-Lower Devonian conodont sequence in the Roberts Mountains Formation of central Nevada. *University of California Publications in Geological Sciences*, 111, 1–62.

Knüpfer, J. 1967. Zur Fauna und Biostratigraphie des Ordoviziums (Gräfenthaler Schichten) in Thüringen. *Freiberger Forschungshefte*, C220, 1–119.

Lamont, A. & Lindström, M. 1957. Arenigian and Llandeilian cherts identified in the Southern Uplands of Scotland by means of conodonts, etc. *Transactions of the Edinburgh Geological Society*, 17, 60–70.

Landing, E. 1976. Early Ordovician (Arenigian) conodont and graptolite biostratigraphy of the Taconic allochthon, eastern New York. *Journal of Paleontology*, 50, 614–646.

—— 1980. Late Cambrian–Early Ordovician macrofaunas and phosphatic microfaunas, St. John Group, New Brunswick. *Journal of Paleontology*, 54, 752–761.

Laufeld, S. 1979. Biogeography of Ordovician, Silurian, and Devonian chitinozoans. *In*: Gray, J. & Boucot, A. J. (eds). *Historical biogeography, plate tectonics, and the changing environment*. Oregon State Univ. Press Corvallis, 75–90.

Lin Bao-Yu 1983. New developments in conodont biostratigraphy of the Silurian of China. *Fossils and Strata*, 15, 145–147.

Lindström, M. 1955. Conodonts from the lowermost Ordovician strata of south-central Sweden. *Geologiska Föreningens i Stockholm Förhandlingar*, 76, 517–604.

—— 1976. Conodont palaeogeography of the Ordovician. *In*: Bassett, M. G. (ed.). *The Ordovician System: proceedings of a Palaeontological Association Symposium, Birmingham, September 1974*. University of Wales Press and National Museum of Wales, Cardiff, 501–522.

——, Racheboeuf, P. R. & Henry, J.-L. 1974. Ordovician conodonts from the Postolonnec Formation (Crozon Peninsula, Massif Armoricain) and their stratigraphic significance. *Geologica et Palaeontologica*, 8, 15–28.

Löfgren, A. 1978. Arenigian and Llanvirnian conodonts from Jämtland, northern Sweden. *Fossils and Strata*, 13, 1–129.

Männik, P. 1983. Silurian conodonts from Severnaya Zemlya. *Fossils and Strata*, 15, 111–119.

Martin, F. 1982. Some aspects of late Cambrian and early Ordovician acritarchs. *In*: Bassett, M. G. & Dean, W. T. (eds). *The Cambrian-Ordovician boundary; sections, fossil distributions, and correlations*. National Museum of Wales, Geological Series, 3, Cardiff, 29–40.

McCracken, A. D. & Barnes, C. R. 1981. Conodont biostratigraphy and paleoecology of the Ellis Bay Formation, Anticosti Island, Quebec, with special reference to Late Ordovician – Early Silurian chronostratigraphy and the system boundary. *Geological Survey of Canada Bulletin*, 329, 51–134.

McKerrow, W. S. & Cocks, L. R. M. 1976. Progressive faunal migration across the Iapetus Ocean. *Nature*, 263, 304–306.

Miller, J. F. 1969. Conodont fauna of the Notch Peak Limestone (Cambro-Ordovician), House Range, Utah. *Journal of Paleontology*, 43, 413–439.

—— 1980. Taxonomic revisions of some Upper Cambrian and Lower Ordovician conodonts with comments on their evolution. *University of Kansas Paleontological Contributions Paper*, 99, 1–44.

—— 1984. Cambrian and earliest Ordovician conodont evolution, biofacies, and provincialism. *Geological Society of America Special Paper*, 196, 43–68.

——, Taylor, M. E., Stitt, J. H., Ethington, R. L., Hintze, L. F., & Taylor, J. F. 1982. Potential Cambrian–Ordovician boundary stratotype sections in the western United States. *In*: Bassett, M. G. & Dean, W. T. (eds). *The Cambrian–Ordovician boundary: Sections, fossil distributions, and correlations*. National Museum of Wales Geological Series, 3. Cardiff, 155–180.

Miller, R. H., Cooper, J. D., & Sandberg, F. A. 1981. Upper Cambrian faunal distribution in southeastern California and southern Nevada. *United States Geological Survey Open-File Report* 81–743, 138–142.

Moskalenko, T. A. 1973. Conodonts of the Middle and Upper Ordovician of the Siberian Platform. *Transactions of the Institute of Geology and Geophysics, Academy of Science USSR, Siberian Branch*, 137, 1–144 (in Russian).

—— 1983. Conodonts and biostratigraphy in the Ordovician of the Siberian Platform. *Fossils and Strata*, 15, 87–94.

Müller, K. J. 1959. Kambrische conodonten. *Zeitschrift der Deutschen Geologischen Gesellschaft*, 111, 434–485.

—— 1971. Cambrian conodont faunas. *Geological Society of America Memoir*, 127, 5–20.

—— 1973. Late Cambrian and Early Ordovician conodonts from northern Iran. *Geological Survey of Iran, Report*, 30, 1–77.

Nowlan, G. S. 1981. Late Ordovician–Early Silurian conodont biostratigraphy of the Gaspe Peninsula – a preliminary report. *In*: Lespérance, P. J. (ed.) *Stratigraphy and Paleontology*, IUGS Subcommission on stratigraphy and Ordovician–Silurian Boundary Working Group. Field Meeting, Anticosti-Gaspé, Québec 1981. II, 257–291.

—— 1983. Biostratigraphic, paleogeographic, and tectonic implications of Late Ordovician conodonts from the Grog Brook Group, northwestern New Brunswick. *Canadian Journal of Earth Sciences*, 20, 651–671.

Nowlan, G. S., Narbonne, G. M. & Fritz, W. H. 1985. Small shelly fossils and trace fossils near the Precambrian–Cambrian boundary in the Yukon Territory, Canada. *Lethaia*, 18, 233–256.

Palmer, A. R. 1973. Cambrian trilobites. *In*: Hallam, A. (ed.). *Atlas of Palaeobiogeography*, Elsevier Science Publishing Co., London, 3–11.

—— 1979. Cambrian. *In*: Robison, R. A. & Teichert, C. (eds). *Treatise on Invertebrate Paleontology. Part A. Introduction*. Geological Society of America and the University of Kansas. Boulder and Lawrence, A119–A135.

Paris, F. 1981. Les Chitinozoaires dans le Paleozoique du sud-ouest de l'Europe. *Mémoires de la Societé géologique et minéralogique de Bretagne*, 26, 1–412.

Repetski, J. E. 1982. Conodonts from the El Paso Group (Lower Ordovician) of westernmost Texas and southern New Mexico. *New Mexico Bureau of Mines and Mineral Resources Memoir*, 40, 1–121.

Ross, R. J., Jr., Nolan, T. B., & Harris, A. G. 1979. The Upper Ordovician and Silurian Hanson Creek Formation of Central Nevada. *United States Geological Survey Professional Paper*, 1126–C, C1–C22.

Sannemann, D. 1955. Ordovizium und Oberdevon der bayerischen Fazies des Frankenwaldes nach Conodontenfunden. *Neues Jahrbuch Geologie und Paläontologie, Abhandlungen*, 102, 1–36.

Savage, N. M., Perry, D. G. & Boucot, A. J. 1979. A quantitative analysis of Lower Devonian brachiopod distribution, 160–200. *In*:

GRAY, J. & BOUCOT, A. J. (eds) *Historical biogeography, plate tectonics, and the changing environment*. Oregon State Univ. Press, Corvallis.

SCOTESE, C. R., BAMBACH, R. K., BARTON, C., VAN DER VOO, R. & ZIEGLER, A. M. 1979. Paleozoic base maps. *Journal of Geology*, **87**, 217–277.

SHENG SHEN-FU 1980. The Ordovician System in China. Correlation chart and explanatory notes. *International Union of Geological Sciences Publication*, **1**, 1–7.

SKEVINGTON, D. 1969. Graptolite faunal provinces in Ordovician of northwest Europe. *American Association of Petroleum Geologists Memoir*, **12**, 557–562.

—— 1973. Ordovician graptolites. *In*: HALLAM, A. (ed.). *Atlas of Paleobiogeography*. Elsevier Science Publishing Co., London, 27–35.

—— 1974. Controls influencing the composition and distribution of Ordovician graptolite faunal provinces. *Palaeontological Association London Special Papers in Palaeontology*, **13**, 59–73.

SMITH, M. P., BRIGGS, D. E. G. & ALDRIDGE, R. J. 1987. A conodont animal from the lower Silurian of Wisconsin, USA, and the apparatus architecture of panderodontid conodonts. *In*: ALDRIDGE, R. J. (ed.). *Palaeobiology of conodonts*. Ellis Horwood Ltd. Chichester, 91–104.

SPJELDNAES, N. 1961. Ordovician climatic zones. *Norsk Geologisk Tidsskrift*, 41, 45–77.

STANLEY, S. M. 1984. Marine extinctions: A dominant role for temperature. *In*: NITECKI, M. H. (ed.). *Extinctions*. The University of Chicago Press, Chicago and London, 69–117.

SWEET, W. C. 1979. Conodonts and conodont biostratigraphy of post-Tyrone Ordovician rocks of the Cincinnati region. *United States Geological Survey Professional Paper*, **1066–G**, G1–G26.

—— 1985. Conodonts: Those fascinating little whatzits. *Journal of Paleontology*, 59, 485–494.

—— & BERGSTRÖM, S. M. 1971. Symposium on Conodont biostratigraphy. *Geological Society of America Memoir*, **127**, 1–499.

—— & —— 1974. Provincialism exhibited by Ordovician conodont faunas. *In*: Ross, C. A. (ed.). *Paleogeographic provinces and provinciality. Society of Economic Paleontologists and Mineralogists Special Publication*, **21**, 189–202.

—— & —— 1976. Conodont biostratigraphy of the Middle and Upper Ordovician of the United States Midcontinent. *In*: BASSETT, M. G. (ed.). *The Ordovician System: Proceedings of a Palaeontological Association Symposium, Birmingham, September 1974*, University of Wales Press and National Museum of Wales, Cardiff, 121–151.

—— & —— 1984. Conodont provinces and biofacies of the Late Ordovician. *Geological Society of America Special Paper*, **196**, 69–87.

——, TURCO, E., WARNER, E., Jr., & WILKIE, L. C. 1959. The American Upper Ordovician Standard. 1. Eden conodonts from the Cincinnati region of Ohio and Kentucky. *Journal of Paleontology*, **33**, 1029–1068.

TAYLOR, M. E. 1977. Late Cambrian of Western North America: Trilobite biofacies, environmental significance, and biostratigraphic implications. *In*: KAUFFMAN, E. G. & HAZEL, J. E. (eds.). *Concepts and Methods of Biostratigraphy*. Dowden, Mutchinson, and Ross, Inc., Stroudsburg, 397–425.

TIPNIS, R. S., CHATTERTON, B. D. E. & LUDVIGSEN, R. 1978. Ordovician conodont biostratigraphy of the Southern District of Mackenzie, Canada. *In*: STEICK, C. R. & CHATTERTON, B. D. E. (eds). *Western and Arctic Biostratigraphy*. Geological Association of Canada Special Paper, **18**, 39–91.

VAVRDOVA, M. 1974. Geographical differentiation of Ordovician acritarch assemblages in Europe. *Review of Palaeobotany and Palynology*, **18**, 171–175.

WALLISER, O. H. 1964. Conodonten des Silurs. *Abhandlungen des Hessischen Landesamtes für Bodenforschung*, **41**, 1–106.

WANG ZHI-HAO 1984. Late Cambrian and Early Ordovician conodonts from North and Northeast China with comments on the Cambrian–Ordovician boundary. *In*: *Stratigraphy and Palaeontology of Systemic boundaries in China. Cambrian–Ordovician boundary*, **2**, Anhui Science and Technology Publishing House, 195–258.

WHITTINGTON, H. B. & HUGHES, C. P. 1972. Ordovician geography and faunal provinces deduced from trilobite distribution. *Philosophical Transactions of the Royal Society of London*, **B263**, 235–278.

—— & WILLIAMS, A. 1955. The fauna of the Derfel Limestone of the Arenig district, North Wales. *Philosophical Transactions of the Royal Society of London*, **B238**, 397–427.

——, DEAN, W. T., FORTEY, R. A., RICKARDS, R. B., RUSHTON, A. W. A., & WRIGHT, A. D. 1984. Definition of the Tremadoc Series and the series of the Ordovician System in Britain. *Geological Magazine*, **121**, 17–33.

WILLIAMS, A. 1973. Distribution of brachiopod assemblages in relation to Ordovician paleogeography. *Palaeontological Association London Special Papers in Palaeontology*, **12**, 241–269.

—— 1976. Plate tectonics and biofacies evolution as factors in Ordovician correlation. *In*: BASSETT, M. G. (ed.). *The Ordovician System: Proceedings of a Palaeontological Association Symposium, Birmingham, September 1974*. University of Wales Press and National Museum of Wales, 29–66.

ZENG QINGKUAN, NI SHIZHAO, XU GUANGHONG, ZHOU TIANMEI, WANG XIAO-FENG, LI ZHIHONG, LAI CAIGEN, & XIANG LIWEN 1983. Subdivision and correlation on the Ordovician in the Eastern Yangtze Gorges, China. *Bulletin of the Yichang Institute of Geology and Mineral Resources Chinese Academy of Geological Sciences*, **6**, 21–68.

The relationship of Ordovician graptolite provincialism to palaeogeography

STANLEY C. FINNEY[1] & CHEN XU[2]

[1] Department of Geological Sciences, California State University, Long Beach, CA 90840, USA
[2] Institute of Geology and Palaeontology, Academia Sinica, Nanjing, People's Republic of China

Abstract: Graptolite biogeography is examined for the late Arenig–Llanvirn when provincialism was at its greatest and the Llandeilo–Caradoc when graptolites were cosmopolitan. This examination incorporates new data from China and recently revised palaeogeographic base maps. Water-mass specificity is considered to be the primary factor affecting graptolite biogeography. On the recently revised base maps, the distribution of Pacific and Atlantic provinces in the late Arenig–Llanvirn is in general consistent with the model of water-mass specificity and the hypothesis that climate cooling associated with the onset of continental glaciation lead to the development of provincialism. However, a small clockwise rotation of South America and a slight northward shift in the position of North America is recommended. The climatic-cooling hypothesis includes the proposal that continued cooling confined graptolites to the tropics where the uniform environment precluded provincialism. The distribution of the cosmopolitan fauna of the *Nemagraptus gracilis* Zone in relationship to the Llandeilo–Caradoc palaeogeography is consistent with this hypothesis, except for the positions of the European platform, Britain, Armorica, and the Carpathians in the mid to high southern latitudes. Either the latitudinal positions of these areas must be shifted substantially to the north, or the climatic-cooling hypothesis must be in error.

Between 1960 and 1979, graptolite biogeography received considerable attention (Skevington 1974, 1976; Berry 1979; and references therein), most of which focused on the Early Ordovician when provincialism was at its greatest. This interest was sparked by several factors. Foremost was the fact that it was difficult to correlate zonations between certain continents because of endemism of important index species. In addition, the 1960s and 70s were a time of dramatic advances in understanding the dynamics of the surface environment of the Earth, and significant changes in the Earth's environment during the Ordovician were well documented. These included not only migrating continents, mountain building, and the opening and closing of ocean basins but also the Late Ordovician glaciation and global rises and falls in sea level. As with specialists of other fossil groups, graptolite specialists sought to explain the waxing and waning of provincialism in terms of the Earth's dynamic environment. The attention focused on graptolites resulted in detailed documentation of the available biogeographic and biostratigraphic data, descriptions of the nature and extent of provincialism, and interpretations of the controls on the broad distribution patterns of graptolites.

In this paper, provincialism in the Early Ordovician, in particular the late Arenig–Llanvirn, will again be examined and so too will the breakdown of provincialism in the Llandeilo–early Caradoc. The purpose is not to repeat much of what has already been published. Instead, it is to reevaluate it in terms of newer data now available from China, recently derived and significantly revised palaeogeographic reconstructions, and a better understanding of possible ecological controls on the geographic distribution of graptolites.

Nature and extent of provincialism

The Early Ordovician provincialism that results from the endemism of several distinctive species and genera is expressed in terms of two provinces or regions, the Atlantic typically developed in Britain and Scandinavia and the Pacific typically developed in Australia and North America. Detailed documentation of the history, composition, and reported occurrences of Atlantic and Pacific faunas is available in Berry (1979), Bulman (1971), and Skevington (1969, 1973).

Atlantic faunas in the upper Arenig–Llanvirn are characterized by an abundance of didymograptids with pendent stipes. Although some of these occasionally invaded the Pacific Province, the youngest, *Didymograptus murchisoni*, was widespread in, yet endemic to, the Atlantic Province in the upper Llanvirn, long after pendent didymograptids were extinct in the Pacific Province.

Pacific faunas are much more diverse than Atlantic (e.g. 11 versus 5 genera in the Llanvirn) and are characterized by numerous species of *Isograptus*; the closely related genera *Pseudisograptus*, *Apiograptus*, *Cardiograptus*, and *Oncograptus*; *Paraglossograptus*; and advanced sinograptid genera. Although representatives of some of these Pacific genera have been discovered occasionally in the Atlantic Province, they dominate Pacific faunas, and *Oncograptus* and *Cardiograptus* are entirely endemic to the Pacific Province.

Barriers between the Atlantic and Pacific provinces were not complete. Even at its greatest provincialism was low; for example, Jackson (1969) reports a faunal resemblance of 73–80% for genera of the Atlantic and Pacific faunas in the Arenig to Llanvirn when provincialism was at its maximum. Many genera and species, especially those with monopleural and dipleural scandent rhabdosomes, were pandemic and thus common to both provinces. Species of one province occasionally invaded the other, and mixed province faunas developed in Kazakhstan (Nikitin 1972), Kirgizia (Zima 1974), and Taimyr (Obut & Sobolevskaya 1964).

Immediate post-Llanvirn graptolites are cosmopolitan. The demise of provincialism is reflected in the nearly worldwide recognition of the *Glyptograptus teretiusculus* Zone. Graptolite diversity drops substantially from the Llanvirn (25–30 species) to the lower Llandeilo (less than 10 species). But, by the upper Llandeilo to early Caradoc Zone of *Nemagraptus gracilis*, diversity is again high (more than 30 species), and graptolites are cosmopolitan. The *N. gracilis* Zone has been recognized on every continent except Antarctica and Africa (Finney & Bergstrom 1986).

Controls on the development and subsequent breakdown of upper Arenig–Llanvirn provincialism

Depth stratification (Berry 1962; Erdtmann 1976) and latitudinal variation in surface-water temperature (Bulman 1964; Bouĉek 1972) have each been proposed as primary controls on provincialism in the Early Ordovician. Bulman (1971), concerned with the fact that both pandemic and endemic species occur together, concluded that provincialism resulted from variations in the effect of the physical and biological environment on species with differing degrees of tolerance. Skevington (1974, 1976), however, proposed a temperature control model that addresses Bulman's concern. This model, I believe, explains the broad distribution patterns of graptolites best, although its requires some modification.

Upon plotting localities with Atlantic and Pacific faunas on the Cambrian/Lower Ordovician continental reconstruction of Smith *et al.* (1973), Skevington (1974, text-fig. 1) noted that the

Pacific Province is restricted to localities within 30 degrees of the equator and that the Atlantic Province occurs in southern latitudes higher than 30 degrees. He regarded this as strong evidence that the primary control of Early Ordovician provincialism was latitudinal variation in surface-water temperature with the Pacific and Atlantic provinces corresponding to warm-water tropical and colder-water temperate zones, respectively. The fact that taxonomic diversity increases from the Atlantic to the Pacific province supported Skevington's view.

The development of provincialism in the Early Ordovician and its subsequent demise in the Middle and Late Ordovician is attributed by Skevington to a long interval of climatic cooling that preceeded the Late Ordovician glaciation episode. In the earliest Ordovician, the difference in surface-water temperatures between low and high latitudes was slight. But, by the Arenig, the climate began cooling, surface-water temperatures at high latitudes lowered significantly, the thermal gradient between low and high latitudes steepened, and tropical and temperate zones became distinct. Assuming that surface-water temperature was the primary control on broad distributions of graptolites and that thermal tolerance varied between species, Skevington proposed that those species that were stenothermal with a preference for warm water, such as *Isograptus*, *Oncograptus*, *Cardiograptus*, etc., were restricted to the tropical zone, that those species that were stenothermal with a preference for cold water, such as pendent didymograptids, were restricted to the temperate zone, and that the majority of species, such as monopleural and dipleural biserial graptolites, were eurythermal and thus able to live in both tropical and temperate zones. This variation in environmental tolerances among species allowed for endemic species that characterized the Atlantic and Pacific provinces to co-exist with pandemic species. By the Llandeilo, climatic cooling and the additional lowering of surface-water temperatures restricted graptolite distributions even further so that all species were confined to the tropics; graptolites could no longer survive at high latitudes. In the rather uniform tropical environment, provincialism as it was known in the Early Ordovician disappeared, and a fairly homogeneous fauna persisted until the end of the Ordovician.

When plotted on the palaeogeographic reconstruction of Smith *et al.* (1973), the Atlantic faunas of Britain and Scandinavia are located in the tropics. Skevington favoured a more southerly position for Europe to explain this, instead of the alternative explanation that invokes a cold-water current flowing into the tropical zone. In the reconstructions of Scotese (1986), the Atlantic faunas of Britain and Scandinavia are within the temperate zone, but Atlantic faunas on the North China, South China, and Tibet-W. Yunnan platforms are in the tropics (Figs 1 & 2). Although some modifications are suggested below for Scotese's reconstruction, another control is herein favored to explain the occurrence of cold-water Atlantic faunas in the tropics. This control is

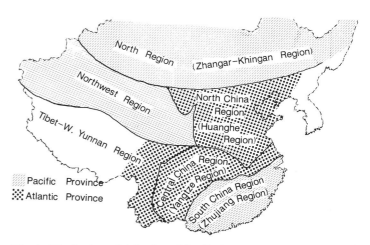

Fig. 1. Distribution of Atlantic and Pacific faunas in China; taken from Mu *et al.* (1980).

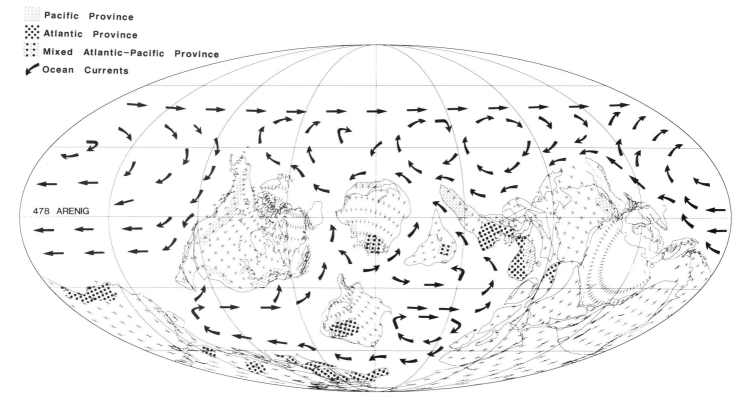

Fig. 2. Upper Arenig–Llanvirn graptolite biogeography plotted on palaeogeographic reconstruction of Scotese (1986) for Arenig.

water-mass specificity, and it can significantly alter the pattern of surface-water temperatures produced by the latitudinal thermal gradient.

The concept of water-mass specificity was first applied to graptolites by Berry (1977; Watkins & Berry 1977) to explain coeval, but dissimilar, faunas in close proximity in the Middle Ordovician of the Northern Appalachians and lateral variation in diversity and density of late Silurian graptolites in Wales. He later (Berry 1979) incorporated water-mass specificity with Skevington's (1974) model of latitudinal variation in surface-water temperature to explain graptolite biogeography from the Early Ordovician to the Devonian, especially for the Ordovician. Subsequently, Finney (1984, 1986) concluded that water-mass specificity was the primary control producing coeval, but dissimilar, faunas locally and regionally in the Middle Ordovician of North America.

The concept of water-mass specificity is appealing because it is well demonstrated for modern plankton (Berry 1977, 1979; and references therein). Bradshaw's (1959) study of planktic foraminifer in the north and equatorial Pacific is especially instructive. These organisms inhabit near-surface waters as graptolites no doubt did. Four faunas are recognized. From north to south, these are the cold-water subarctic fauna, the transition fauna, the warm-water central fauna, and the warm-water equatorial west-central fauna. The distribution of each of these corresponds closely to one or more major oceanic water masses defined by circulation patterns of oceanic currents (Bradshaw 1959, text-figs 4 & 35). These water masses are defined oceanographically by their temperature and salinity, and other subtle physical and chemical factors may also characterize them. The latitudinal thermal gradient greatly influences the temperature of the water masses, but the pattern of oceanic currents greatly modifies the thermal gradient as it is expressed in surface-waters and serves to separate and maintain distinct water masses. For example, the transition fauna is dispersed 20 degrees latitude to the south by the California Current.

The provincialism described by Bradshaw (1959, text-fig. 34) is similar to that of Early Ordovician graptolites. Diversity increases from high to low latitudes. Endemism of cold-water species is low; that of warm-water species is high; a transitional fauna exists; and some species are cosmopolitan. Thus, it is with this concept of biogeographic control, latitudinal variation in surface-water temperature modified by water-mass specificity, that graptolite provincialism is used to examine Scotese's (1986) palaeogeographic reconstructions.

Fortey (1984; Fortey & Cocks 1986) favours a very different control on graptolite biogeography. Because taxa characteristic of the Pacific province are generally found in sediment deposited in extracratonic settings, Fortey postulated that they were restricted to relatively uniform oceanic waters beneath the thermocline and had worldwide distributions. In contrast, Atlantic province taxa were epiplanktonic and restricted to shallow, epicratonic seas, and their distributions in these shallow environments were controlled by water temperature, which varied with latitude. As a result, Atlantic province taxa were isolated on and thus endemic to continental platforms at different latitudes. Although this interpretation is intriguing, the distribution of *Didymograptus murchisoni* is not consistent with it. This distinctively Atlantic province species occurs in deep-water, extracratonic settings (e.g. South Wales (Fortey 1984)), and, no matter which palaeogeographic reconstruction is used, it occurs on continental fragments distributed from tropical to polar regions. In light of this inconsistency, Fortey's interpretation is not considered in this paper.

Graptolite biogeography in China

China was considered a single land mass contiguous with Asia and assigned to the Pacific Province by Berry (1979), Bulman (1971), and Skevington (1973, 1974). The most recent biostratigraphic data on China available to these authors were published in 1963, and most of those data were for the South China region (Fig. 1) from which Pacific faunas were recorded long ago (Hsu 1934). Numerous papers by Chinese graptolite specialists appeared in the late 1970s and 1980s; many of these reported faunas from areas in the interior of China that were only recently explored. Both Pacific and Atlantic faunas were recognized in various parts of China, and it is now realized that modern China is composed of several fragments that were separate platforms in the Ordovician.

Mu *et al.* (1980) described the occurrence of Atlantic and Pacific faunas in six distinct geologic regions of China (Fig. 1), which make up all or part of three plates on Scotese's palaeogeographic reconstructions for the Ordovician (Fig. 2). The South China platform contains two regions, the South China (or Zhujiang) region and the Central China (or Yangtze) region with Pacific and Atlantic faunas, respectively. Likewise, the North China platform contains two regions, the Northwest region and the North China (or Huanghe) region with Atlantic and Pacific faunas, respectively. The Tibet-W. Yunnan region with Atlantic faunas corresponds to the Tibet plate in Scotese's reconstruction; the North (or Zhangar-Khingan) region with Pacific faunas is located on the northern margin of the Siberian plate.

Late Arenig-Llanvirn biogeography

Scotese's reconstruction for the Arenig-Llanvirn (Fig. 2) is substantially different from that of Smith *et al.* (1973) in the high latitudinal position of Britain and the European platform and in the dismemberment of Asia into several plates. Most of the biogeographic data shown are the same as those available to Skevington (1974) except for China, which Skevington assigned wholly to the Pacific Province and situated within the tropical belt.

Consistent with Skevington's results, regions with the Pacific faunas are restricted to a tropical zone between 30 degrees north and south latitudes. Although the European platform, Britain, central Europe, and North Africa with Atlantic faunas are at high latitudes; others areas with Atlantic faunas, i.e. Peru and Bolivia, eastern Tibet, and parts of the South and North China platforms, are adjacent to or within the tropical zone, as too are Taimyr, Kirgizia, and Kazakhstan with mixed Atlantic-Pacific faunas. Thus on Scotese's reconstructions as on the Smith *et al.* reconstruction used by Skevington, there is not a distinct latitudinal separation of Pacific and Atlantic provinces, although there is a general one.

Interpretation of water masses together with palaeogeography offers a better explanation of the biogeography. Generalized oceanic currents are interpreted on the basis of patterns in the modern oceans.

A large clockwise gyre is situated in the Pacific between North America and Australia. Its southern part, the easterly equatorial current straddles the equator, readily connects faunas of western North America with those of Australia and New Zealand, and circulates warm tropical water. No corresponding counter-clockwise gyre is present in the south because the latitudinal position of Gondwanaland prevents the development of West Wind Drift that would help generate it. The occurrence of a Pacific fauna in Argentina is explained by this large Pacific gyre, but not that of the Atlantic fauna in Bolivia and Peru. Perhaps, cold-water currents flowed out of the high southern latitudes along the coast of South America, or perhaps South America was positioned differently than it is on the reconstruction. A clockwise rotation of South America would bring Argentina more into the tropical zone and Bolivia and Peru to a higher latitude.

Part of the warm-water current flowing south along the west coast of North America may have curved easterly around the southwest corner of North America to bring warm-water Pacific faunas to the southern and eastern margins of the continent. The

West Wind Drift in the Southern Hemisphere would have favored this water mass, especially if North America was located slightly more to the north.

Southeasterly currents along the northeast coast of Australia may have carried warm water around Gondwanaland spreading and maintaining Pacific faunas in the Northwest China region on the North China platform and in the North China region on the Siberian platform.

Currents must have been complex in the temperate to polar oceans around the European Platform with clockwise gyres to the east and west. The West Wind Drift at 40 degrees south latitude would serve as a northern barrier for cold-water Atlantic faunas. This current may have been diverted to the northeast along the northwest border of the European platform and the western margin of Gondwanaland, which could explain the mixed province faunas of Taimyr, Kazahkstan, and Kirgizia and the location in the tropics of Atlantic faunas on the North China, South China, and Tibet-W. Yunnan platforms.

Without postulating tectonic translation or palaeogeographic revisions, the only explanation for the close proximity of Atlantic and Pacific faunas on the South China platform are that the inboard position of the South China region within the tropical zone allowed for the establishment of a warm-water environment adjacent to but free of the cold, northeast-flowing current that bathed the Central China region.

These interpretations assume that the polar to equatorial temperature gradient and the positions of winds and ocean currents it generates are the same as today except for modifications due to the shapes and positions of continents. Skevington (1974) considers that this temperature gradient was low in the early Arenig when graptolite faunas were cosmopolitan. A low temperature gradient would produce a broad zone of warm seas and cosmopolitan faunas. A higher temperature gradient than that shown here would constrict the temperate zone, expand the boreal region, but not severely constrict the tropical zone. The contrast between low-latitude, warm-water masses and high-latitude, cold-water masses would intensify, thus increasing the distinctiveness of the provincialism but not markedly changing the biogeography shown here.

Except for South China, the only clear modifications I would make to Scotese's reconstruction on the basis of graptolite biogeography are a clockwise rotation of South America and possibly a northern shift to North America.

Llandeilo–Early Caradoc biogeography

On Scotese's reconstruction (Fig. 3), the *Nemagraptus gracilis* Zone fauna is cosmopolitan and generally restricted to the tropical belt. However, it occurs as far south as 60 degrees latitude on the European platform, Britain, and Armorica and 70 degrees latitude in the Carpathians (Czaplicka 1970, p. 225). These high latitude positions are in conflict with Skevington's distributions plotted on the Smith *et al.* map and with Skevington's basic hypothesis regarding the onset of climatic cooling.

It was the restriction of all localities with *N. gracilis* Zone faunas to within 30 degrees of the equator that led Skevington to conclude that climatic cooling prevented graptolites from surviving at latitudes higher than 30 degrees, that allowed them to persist only within the tropical zone, which because of its uniformity prevented provincialism from developing, and that reflected the onset of dramatic climatic cooling.

The high latitudinal position of *N. gracilis* Zone faunas on Scotese's reconstruction is subject to three different explanations. Climatic cooling may not have begun, in which case the cosmopolitan graptolite fauna and the demise of the upper Arenig–Llanvirn provincialism must be explained by some mechanism other than climatic cooling. The demise of provincialism and the associated decline and subsequent evolutionary increase in graptolite diversity is related to the worldwide Llandeilo regression and subsequent transgression (Fortey 1984). Scotese's reconstruction may be in error with regard to the high latitudinal position of the European platform, Britain, and Armorica, and the Carpathians. Or, if the polar-to-equatorial temperature gradient had increased due to climatic cooling, oceanic circulation

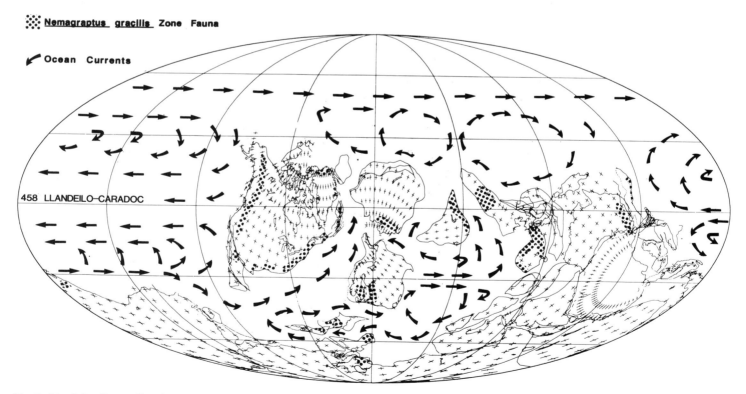

Fig. 3. Llandeilo–Lower Caradoc graptolite biogeography plotted on palaeogeographic reconstruction of Scotese (1986) for Llandeilo–Caradoc.

patterns may have been compressed towards the equator, in which case there may have been space above the north coast of the South American part of Gondwanaland to establish West Wind Drift in the southern-most Pacific. Such a current may have flowed eastwards between North America and Gondwanaland into the oceans surrounding the European Platform and warming the seas permitting a high latitude position for the *N. gracilis* Zone fauna. Of the possible explanations, the last seems unrealistic; the others are reasonable.

Historical versus ecological biogeography

Although, in general, post-Llanvirn Ordovician graptolites were cosmopolitan, graptolite specialists (e.g. Berry 1979; Skevington 1973) have noted that slight provincialism did develop. It usually is local or regional and involves both the endemism of individual species or combinations of species and geographic variation in relative abundances of species composing a fauna. Such provincialism can be explained by water-mass specificity expressed on the local and regional scale; often it represents faunal variation between adjacent onshore and offshore water masses along continental margins (Finney 1984, 1986; Watkins & Berry 1977).

This type of lateral faunal differentiation is an example of ecological biogeography in contrast to historical biogeography. Faunal changes occur over short distances and between environments in close proximity to each other. In contrast, the provincialism of the late Arenig–Llanvirn and its subsequent demise are phenomena of global scale. Their explanations involve not just ecological factors but also phyletic histories, evolving palaeogeography, and major climatic and oceanographic changes.

Conclusions

Because graptolites were planktic, their provincialism was not so much controlled by proximity of continents as it was by the character of the oceans and in particular the near surface waters. The occurrence of a particular faunal province on a continent was really dependent on the character of surface waters that overlied the site at which the graptolitic strata were deposited. That is not to say that palaeogeography, in particular the position of the continents, had no part to play in graptolite biogeography. It did to some degree.

The latitudinal thermal gradient in surface waters was the basic primary control of graptolite provincialism, but it, in turn, was controlled by the global climate of the earth, which was subject to change. The latitudinal thermal gradient of surface waters was, in addition, modified as it is in modern oceans by patterns of oceanic circulation. The influence of palaeogeography on graptolite biogeography lies in the fact that the sizes, shapes, and positions of continents affect circulation patterns as well as determine the latitudinal positions of the continents.

In contrast to benthic organisms, the provincialism of graptolites, even when it is at a maximum, is low. This suggests that graptolite biogeography may not be very helpful in reconstructing palaeogeography. The palaeogeographic changes on Scotese's maps between the times of marked graptolite provincialism in the Arenig–Llanvirn and the cosmopolitan fauna of the late Llandeilo–Caradoc are slight (Figs 2 & 3), again indicating that palaeogeographic changes played a minor role in determining provincialism relative to other factors such as climatic and oceanographic changes.

In both the reconstruction used by Skevington (1974) and that of Scotese (1986), provincialism at its greatest is in general consistent with the latitudinal position of continents. The major departures from this model, such as the Atlantic fauna in the tropics, may reflect circulation patterns of oceanic currents. But, other departures, such as Atlantic faunas in South America, suggest possible revisions to Scotese's reconstruction.

The demise of provincialism by the onset of climatic cooling is consistent with the reconstruction of Smith *et al.* (1973). On Scotese's map, the palaeogeographic changes from the Arenig to the Caradoc are slight. The European platform, Britain, Armorica, and the Carpathians must have migrated northwards much farther than shown, or else one must conclude that the onset of glaciation had not occurred, and instead the decline in provincialism was due to other oceanographic, ecological, or evolutionary factors.

References

BERRY, W. B. N. 1962. Graptolite occurrence and ecology. *Journal of Paleontology*, **36**, 285–293.

—— 1977. Ecology and age of graptolites from graywackes in eastern New York. *Journal of Paleontology*, **51**, 1102–1107.

—— 1979. Graptolite biogeography: a biogeography of some Lower Paleozoic plankton. *In*: GRAY, J. & BOUCOT, A. J. (eds) *Historical Biogeography, Plate Tectonics, and the Changing Environment*. Oregon State University Press, Corvallis, 105–115.

BOUČEK, B. 1972. The paleogeography of Lower Ordovician graptolite faunas: a possible evidence of continental drift. *XXIV International Geological Congress, Section VII*, 266–272.

BRADSHAW, J. S. 1959. Ecology of living planktonic Foraminifera in the North and Equatorial Pacific Ocean. *Cushman Foundation for Foraminiferal Research Contribution*, **10**, 25–64.

BULMAN, O. M. B. 1964. Lower Palaeozoic plankton. *Quarterly Journal of the Geological Society of London*, **119**, 401–418.

—— 1971. Graptolite faunal distribution. *In*: MIDDLEMISS, F. A., RAWSON, P. F. & NEWALL, G. (eds) *Faunal Provinces in Space and Time*. Geological Journal Special Issue, **4**, 47–59.

CZAPLICKA, J. (ed.) 1970. *Geology of Poland, volume 1, Stratigraphy, Part 1, Pre-cambrian and Palaeozoic*. Geological Institute. Publishing House Wydawniclwa Geologiczne, Warsaw.

ERDTMANN, B.-D. 1976. Ecostratigraphy of Ordovician graptoloids. *In*: BASSETT, M. G. (ed.) *The Ordovician System: proceedings of a Palaeontological Association symposium, Birmingham, September 1975*. University of Wales Press and National Museum of Wales, Cardiff, 621–643.

FINNEY, S. C. 1984. Biogeography of Ordovician graptolites in the southern Appalachians. *In*: BRUTON, D. L. (ed.) *Aspects of the Ordovician System*. Palaeontological Contributions from the University of Oslo, **295**, Universitetsforlaget, 167–176.

—— 1986. Graptolite biofacies and correlation of eustatic, subsidence, and tectonic events in the Middle to Upper Ordovician of North America. *Palaios*, **1**, 435–461.

—— & BERGSTROM, S. M. 1986. Biostratigraphy of the Ordovician *Nemagraptus gracilis* Zone. *In*: HUGHES, C. P. & RICKARDS, R. B. (eds) *Palaeoecology and Biostratigraphy of Graptolites*. Geological Society, Special Publication, **20**, 47–60.

FORTEY, R. A. 1984. Global earlier Ordovician transgressions and regressions and their biological implications. *In*: BRUTON, D. L. (ed.) *Aspects of the Ordovician System*. Palaeontological Contributions from the University of Oslo, **295**, Universitetsforlaget, 37–50.

FORTEY, R. A. & COCKS, L. R. M. 1986. Marginal faunal belts and their structural implications, with examples from the Lower Palaeozoic. *Journal of the Geological Society, London*, **143**, 151–160.

HSU, S. C. 1934. The graptolites of the Lower Yangtze Valley. *Monograph of the Natural Research Institute of Geology (Academia Sinica)*, (a), **4**.

JACKSON, D. E. 1969. Ordovician graptolite faunas in lands bordering North Atlantic and Arctic Oceans. *In*: KAY, M. (ed.) *North Atlantic-Geology and Continental Drift*. American Association of Petroleum Geologists Memoir, **12**, 504–511.

MU EN-ZHI, LI JI-JIN, GE MEI-YU, CHEN XU, NI YU-NAN & LIN YAO-KUN. 1980. Ordovician graptolite sequence and biogeographic regions in China. *Scientific Papers on Geology for International Exchange, Stratigraphy and Palaeontology*. Publishing House of Geology, Beijing, 35–42 (in Chinese with English summary).

NIKITIN, I. F. 1972. *Ordovician of Kazahkstan. Part I. Stratigraphy*. Alma-Alta, Publishing House "Nauka" Kazahkstan, USSR (in Russian).

OBUT, A. M. & SOBOLEVSKAYA, R. F. 1964. Ordovician graptolites of

Taimyr. *Akademy Nauk SSSR, Siberia Otdeleniyeh, Institute of Geology i Geofix, Nauchno-issledovatelski Institute of Arktic Geology, GGK, USSR* (in Russian).

SCOTESE, C. R. 1986. Phanerozoic reconstructions: A new look at the assembly of Asia. *University of Texas Institute for Geophysics Technical Report*, **66**.

SKEVINGTON, D. 1969. Graptolite faunal provinces in Ordovician of Northwest Europe. *In*: KAY, M. (ed.) *North Atlantic-Geology and Continental Drift*. American Association of Petroleum Geologists Memoir, **12**, 557–562.

—— 1973. Ordovician Graptolites. *In*: HALLAM, A. (ed.) *Atlas of Palaeobiogeography*. Elsevier Scientific Publishing Company, London, 27–35.

—— 1974. Controls influencing the composition and distribution of Ordovician graptolite faunal provinces. *In*: RICKARDS, R. B., JACKSON, D. E. & HUGHES, C. P. (eds) *Graptolite Studies in Honour of O. M. B. Bulman*. Special Papers in Palaeontology, **13**, 59–73.

—— 1976. A discussion of the factors responsible for the provincialism displayed by graptolite faunas during the early Ordovician. *In*: KALJO, D. & KOREN, T. (eds) *Graptolites and Stratigraphy*. Academy of Sciences of Estonian SSR, Institute of Geology, 180–200.

SMITH, A. G., BRIDEN, J. C. & DREWRY, G. E. 1973. Phanerozoic world maps. *In*: HUGHES, N. F. (ed.) *Organisms and continents through time*. Special Papers in Palaeontology, **12**, 1–42.

WATKINS, R. & BERRY, W. B. N. 1977. Ecology of a Late Silurian fauna of graptolites and associated organisms. *Lethaia*, **10**, 267–286.

ZIMA, M. B. 1974. Ordovician graptolite complexes of northern Kirghizia. *In*: OBUT, A. M. (ed.) *Graptolites of the USSR*. Novosibirsk, Publishing House Nauka Siberian Branch, 36–49 (in Russian).

Graptolite biogeography: implications for palaeogeography and palaeoceanography

WILLIAM B. N. BERRY & PAT WILDE

Marine Sciences Group, 3 Earth Sciences Building, University of California Berkeley, CA 94720, USA

Abstract: Two faunal regions, a cool-water, Atlantic and a tropical-water, Pacific, may be distinguished among Tremadoc into Ashgill planktic graptolite faunas. The zenith of graptolite provincialism was during the Arenig–Llanvirn when Laurentia, Australia–New Zealand, North China, South China, Siberia, Argentine Precordillera, and parts of Kazachkstan were provinces within the Pacific Region. Coeval, Southern Hemisphere, Atlantic Region provinces were: England–Wales, Baltoscania, Bolivia–Peru–Northern Argentina, western Europe, North Africa, and possibly part of modern Tien Shan. South China Arenig–Llanvirn faunas include many incursions of Atlantic Region taxa, probably reflective of current circulation changes linked to development of seasonal monsoons. Mid-Ordovician plate motion included northward movement of Baltoscania into the tropics. That motion resulted in Baltoscania becoming a province in the Pacific Region in the Late Ordovician. Late Ordovician glaciation led to restriction of graptolites to the tropics (Pacific Region) during the latest Ordovician. Deglaciation was followed by re-establishment of a tropical, Pacific Region fauna and a cool-water, Atlantic Region fauna during the Llandovery and Wenlock. Graptolite faunas appear to have been only tropical from the Ludlow to their extinction in the Pragian (Early Devonian). The graptolite regions and provinces are essentially consistent with plate positions suggested by palaeomagnetic and lithofacies data.

Simpson *et al.* (1957) pointed out that animals and plants are distributed across the face of the Earth in certain discrete patterns and that these patterns have both an ecological and an historical component. As any traveller or reader of a global atlas may discern, deserts with essentially similar environmental conditions exist today in Australia, Mongolia, northern Africa, and southern California, among other places. Despite relatively comparable environmental conditions in each of these deserts, animals and plants living in each have little, if any, phyletic relationship to those living in the other deserts. Certain morphological similarities do exist, however, among the plants and animals in each desert. The reasons for the morphological similarities are fundamentally ecological. That is, desert environments are such that only plants and animals with certain morphological and physiological features can survive. The reasons why animals and plants in one desert lack close phyletic relationships with those in other deserts are historical. The historical reasons reflect the fact that each plate of the Earth's surface accumulates a set of organisms in environments on and over it shortly after it begins to form. The organisms come to a new plate from old, already existing plates. Thus, the relationships of the organisms on any plate may have different histories at the outset. Then, as each plate moves, in time, it will pass from one environmental regime to others. For example, a plate may pass from a position near the equator to a position well south or north of the equator. Accordingly, the organisms that stocked the plate's environments initially will change through time. That change, in part, reflects changes in environments resulting from plate motion. From the geological perspective, rocks stacked on a plate bear the record of that plate's history. The rock record includes the polarity of iron particles that bear the remanent magnetism imprinted on them by latitude, the many sedimentological features reflective of environments, and the fossils that also reflect past environmental conditions. Organisms on each plate differ because each plate had a unique set of creatures that came to it and each plate has its own unique history. All the features that reveal a plate's history can and should be used to fit together the history of the Earth's crust, time interval by time interval.

Geographers of the modern floras and faunas tend to recognize an heirarchical order among historical biogeographic patterns (see Simpson *et al.* 1957; George 1962). Relatively broad areas with essentially similar animals and plants are faunal regions. Somewhat smaller areas or areas inhabited by faunas and floras with less strongly marked dissimilarities than those of a region are provinces. For a number of reasons, including such fundamentals as basic identifications of genera and species and a broad knowledge of the world's literature regarding a specific set of plants or animals, recognition of faunal provinces and regions among fossils is difficult. Some degree of common understanding of the basic taxonomy of a group of organisms has to be achieved before provinces and regions based on distribution patterns among fossil taxa can be established. Once the patterns have been established, however, then the reasons for them may be examined in terms of plate positions, time interval by time interval, and oceanic currents, time interval by time interval. Indeed, contributions toward recognition of former plate positions is a fundamental reason to undertake any analysis of the historical biogeographic distribution of organisms.

Graptolite habitats

To enhance understanding of graptolite historical biogeographic patterns, probable graptolite habitats or preferred environments may be noted. Berry *et al.* (1987) pointed out that graptolites appear to be the ancient analogues of certain zooplankton, primarily copepods and euphausids, that are linked closely for their basic nutrition to oxygen-poor, nitrogen compound rich waters. These waters occur about the oxygen minimum zone in the eastern tropical Pacific, the Indian Ocean, the central California upwelling system, and a small portion of the west African upwelling system (see Berry *et al.* 1987). Mullins *et al.* (1985, p. 491) studied the oxygen minimum zone waters off central California, noting that 'the edges or boundaries of the oxygen minimum zone are 'hot spots' of increased biogeochemical activity.' Holligan *et al.* (1984) drew attention to blooms of dinoflagellates in similar waters off both the Peru and west African coasts. Anderson (1982) and Anderson *et al.* (1982) described the chemical conditions at the top of the oxygen minimum zone in the eastern tropical Pacific, noting an interval or zone in which nitrate was reduced to nitrite. Citing those studies, Mullins *et al.* (1985) suggested that bacteria in the oxygen minimum zone were responsible for biologically reducing nitrate to nitrite, and as well, some of the nitrite to molecular nitrogen. Accordingly, Mullins *et al.* (1985) suggested that the bacteria responsible for denitrifying nitrate to nitrite and nitrite to molecular nitrogen constituted a rich nutrient resource. High levels of nitrogen compound availability probably result in the dinoflagellate blooms noted by Holligan *et al.* (1984).

Brinton (1980) reviewed euphausid distribution patterns in the oxygen minimum zone waters in the eastern tropical Pacific, pointing out certain species that are endemic to these waters. Ebeling (1967) cited a distinctive fish fauna that inhabits the oxygen poor but nitrogen compound rich waters of the eastern tropical Pacific. Longhurst (1967) pointed out that populations of copepods and euphausids migrate vertically daily many tens of meters to spend daytime in waters with an oxygen content as low as $0.25 \, ml \, l^{-1}$. Both Brinton (1980) and Longhurst (1967) noted that copepod and euphausid species are depth-distributed in a form of gradient from the core of the oxygen poor water upward to waters with greater oxygen content but lesser nitrogen compound and bacterial content in addition to those species that migrate vertically daily in and out of the oxygen-poor waters. The modern copepods and euphausids described by Brinton (1980) and Longhurst (1967) in and near the oxygen-minimum zone of the eastern tropical Pacific are considered to live today in oceanic environments closely similar to those in which the graptolites lived in the early part of the Palaeozoic (Berry et al. 1987).

Today, the environmental conditions to which graptolites were linked during the Early Palaeozoic are found almost entirely from 30 degrees South to 30 degrees North latitude. That is, they occur in warm waters. As well, these environments occur in waters in which organic productivity is so great that nearly all available oxygen is consumed as organisms die and fall down through the water column. Indeed, bacterial denitrification of nitrate to nitrite will provide some oxygen for decomposition after the oxygen wind-mixed into the surface waters has been used.

Oxygen-minimum zone waters in modern oceans pass downward into increasingly oxygen-laden waters. Such waters form at the poles, sink, and flow to low latitudes. Water in modern deep oceans is relatively oxygen-rich. That water is advected upward to generate a lower limit to oxygen minimum zone water.

Using modern oceanic structure as an analogue, the most richly fossiliferous graptolite-bearing shales and limestones should be reflective of great enough organic productivity to result in formation of oxygen-poor, nitrogen-compound rich waters. The broadest expanse of such waters today is in the tropics. Such waters may form, however, in non-tropical latitudes within highly organically productive upwelling systems and over certain shelves. If organic productivity was great enough to create conditions in which oxygen-poor, nitrogen-compound rich waters came within a few meters of the surface, then shelf seas, even shallow shelf seas, could have been sites in which highly fossiliferous graptolite-bearing strata formed. Furthermore, eustatic sea level rises could have led to transgression of oxygen-poor, nitrogen-compound rich waters over shelves and shelf margins. Regressions could have resulted in draining these waters from shelves. Fortey (1984) pointed out that graptolite faunas were richest at times of transgression across shelves. Regressions from shelves are marked by extinctions among graptolites (Fortey 1984).

Each plate bearing shelf seas and adjacent open ocean environments during the life span of graptolites, that is, during the Tremadoc–Pragian, had the potential for being a homeland for graptolites if oxygen-poor, nitrogen-compound rich waters could form. In general, plates in which such waters formed would have been within the tropics. Plates outside the tropics would have had relatively more meagre graptolite faunas, except in conditions such as coastal upwelling in which organic production was high. Patterns in oceanic currents and direction of current flow as well as distance between plates would condition the degree of faunal provincialism. Plates within the tropics, for example, could have had closely similar faunas if they were within the same current system and were relatively close. Relatively distant plates within the tropics, especially those in different current systems, would have had less highly similar faunas. A plate outside the tropics most likely would have had graptolites with little similarity to coeval faunas on plates within the tropics.

As Berry (1979) described, the richest graptolite faunas are found on the margins of platforms or shelves that were sites of thick sequences of carbonate rocks. Berry (1979) and Ziegler et al. (1984) pointed out that thick shelf sea carbonate successions form primarily in warm tropical conditions. In general, thick

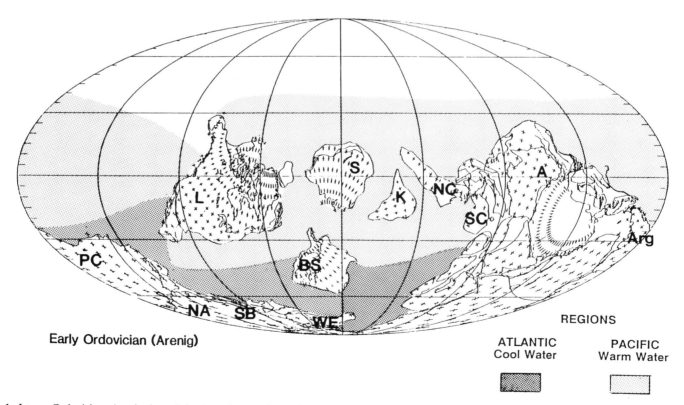

Fig. 1. Lower Ordovician, Arenig time. Atlantic region provinces: SB, England and Wales; BS Baltoscania; PC, Peru; NA, North Africa; WE, Western Europe. Pacific region provinces: A, Australia & New Zealand; L, Laurentia; S, Siberia; K, Kazakhstan; NC, North China; SC, South China; Arg, Western Argentina.

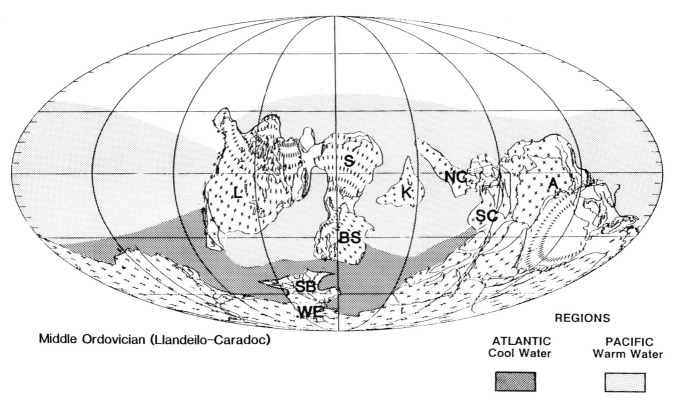

Fig. 2. Middle Ordovician, Llandeilo–Caradoc time. Atlantic region provinces: SB, England and Wales; WE, Western Europe. Pacific Province: A, Australia & New Zealand; L, Laurentia; S, Siberia; K, Kazahkstan; NC, North China; SC, South China, Arg, Western Argentina; BS, Balto-Scania.

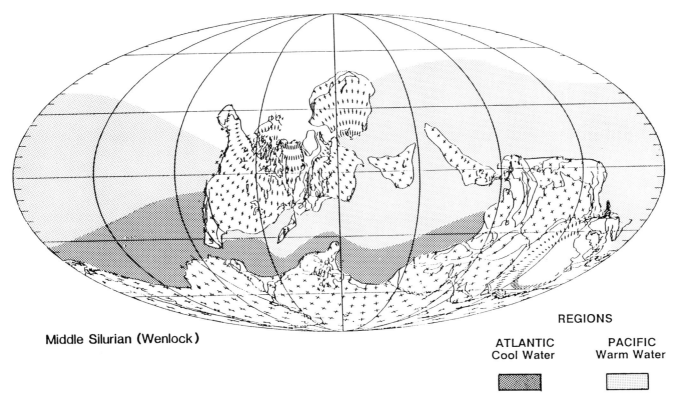

Fig. 3. Middle Silurian–Wenlock Regions.

sequences of carbonates formed in shallow shelf sea environments are indicative of a latitudinal position between about 30 degrees North and 30 degrees South. Association of the most species-diverse and numerically abundant graptolite faunas in rock sequences that are found rimming carbonate platform successions further indicates that the environments preferred by most graptolites are those in the tropics. Berry (1979, figs 1, 2) cited major carbonate platform successions during the Ordovician in modern North America, Asia and part of European Russia, and Australia. Boucot et al. (1968, fig. 1) indicated that, during the Silurian, major carbonate platform successions formed in areas now in North America, European Russia and areas in Asiatic Russia that were not trench or trough areas, and Australian shelf sea or platform areas.

Graptolite historical biogeography

The historical biogeography of graptolites, especially that of Ordovician graptolites, has been reviewed many times (Berry 1959, 1960a, 1967, 1972, 1973, 1979; Boucek 1972, 1973; Bulman 1971; Cooper 1979; Erdtmann 1972; Finney 1984; Finney & Bergstrom 1986; Jackson 1964, 1969, 1974; Koren' 1979; Lenz 1977; Lenz & Jackson 1986; Melchin in press; Mu 1963; Skevington 1968, 1969, 1973, 1974, 1976; Teller 1969). In his reviews of graptolite provinciality, Skevington (1974, 1976) suggested that latitudinal differences in ocean surface water temperatures were a primary influence on graptolite provinces. Koren' (1979) pointed out that the Late Silurian–Early Devonian graptolites occur in deposits that formed within palaeolatitudes of about 30–40 degrees North and 30 degrees South. The youngest graptolites were, in Koren's (1979) view, all within subtropical to tropical ocean waters, and water temperature was the primary environmental control on distribution of the youngest graptolites. Graptolite historical biogeographic distribution seemingly reflects ocean surface water temperatures to some degree. The oceanic surface waters are governed in their position by the geographic position of the plates over which they are situated.

Tremadoc

The earliest planktic graptolites, members of the genera *Anisograptus*, *Adelograptus*, *Clonograptus*, *Rhabdinopora*, and *Staurograptus*, are found spread relatively widely. They occur in Australia and New Zealand, China, North and South America, Asiatic Russia, and many areas in Europe. The systematic position of many Tremadoc taxa needs review and careful analysis. Until such a review and analysis has been carried out, precise identifications of many dictyonemid species and subspecies are uncertain.

Despite the uncertainties concerning Tremadoc graptolite identifications, the available evidence suggests that graptolite provincialism commenced in the Tremadoc. As Jackson (1974) pointed out in his review of Tremadoc graptolites, two faunal regions, an Atlantic and a Pacific, may be recognized among Tremadoc graptolites. Faunas from North America (Berry 1960b; Bulman 1950; Jackson 1974), Australia (Cooper 1979), Argentina (Harrington & Leanza 1957), Taimyr (Obut & Sobolevskaya 1964), and Kazakhstan (Nikitin 1972) comprise the Pacific faunal region. Most of these Tremadoc graptolites occur on or on the margins of carbonate successions that formed in shallow marine platform environments. Apparently endemic taxa are recorded from each of the areas, suggesting that each area or platform was separated from the others to some degree.

Tremadoc graptolites from Baltoscania (Bulman 1954; Tjernvik 1956; and Ulst 1976), Wales (Rushton 1982), North Africa and the St John's, New Brunswick area comprise the Atlantic region fauna. Tremadoc faunas from China (Wang Xiaofeng & Erdtmann, 1987) appear to be Pacific region in overall affinity, but a number of endemic taxa are present.

Staurograptus, *Psigraptus*, certain species of *Anisograptus*, and *Triograptus* typify Pacific region faunas. The Matane Shale graptolites described by Bulman (1950) comprise the richest Tremadoc graptolite fauna recorded to date.

Atlantic region Tremadoc graptolites are poorer in species than coeval Pacific region faunas. Certain clonograptids (*C. tenellus*, for example), *Bryograptus* species, and *Dictyonema desmograptoides* seemingly imply Atlantic region affinities.

Arenig–Llanvirn

Graptolite faunal provincialism and regionalism reached its peak during the Arenig–Llanvirn. Provinces during this time interval have been discussed amply by Berry (1967, 1979), Boucek (1972, 1973), Lenz (1972), and Skevington (1973, 1974, 1976). Both Pacific and Atlantic regions were developed highly in the Arenig–Llanvirn. Pacific region faunas include *Cardiograptus*, *Oncograptus*, pseudisograptids, members of the *Isograptus victoriae* group, *Skiagraptus*, *Apiograptus*, *Paraglossograptus*, and pendeograptids of the *P. fruticosus* group, among many other taxa. Atlantic region faunas are typified by robust didymograptids of the *D. murchisoni* group. *Corymbograptus*, *Gymnograptus*, certain schizograptids, and aulograptids. Boucek (1973) described many of the taxa that characterize Atlantic region faunas.

Atlantic region. Berry (1979) suggested that at least three provinces existed within the Arenig–Llanvirn Atlantic faunal region. The three provinces are: Britain except Scotland and western Ireland, Baltoscania except western Norway (the Bogo Shale area which was in the Pacific region (Schmidt 1987)), and continental Europe. North African Arenig–Llanvirn graptolites (Destombes 1960, 1970; Legrand 1966) probably were within the continental European province.

Bulman (1931) described relatively species-rich Atlantic region Arenig–Llanvirn graptolite faunas from Peru and Chile in South America. Harrington & Leanza (1957) and Turner (1960) recorded Atlantic region Arenig–Llanvirn graptolites from northern Argentina. South American Atlantic region faunas appear to be relatively similar to coeval Baltoscania faunas. The Baltoscanian and South American Arenig–Llanvirn faunas seem to have been in the same province. If so, they may have been close to the boundary with the Pacific region. The degree of similarity suggests that they were within the same current system and may have been at about the same latitude.

Arenig–Llanvirn graptolites from Tien Shan and adjacent southern Kazakhstan have been recorded from five different structural zones (Zima 1976). Atlantic region graptolites occur in some structural belts and Pacific region taxa are cited in others (Nikitin 1972; Zima 1976). The two faunal regions are brought into contact structurally in the mountains of Tien Shan and adjacent Kazakhstan, but they were separated in the Ordovician. The rocks comprising the structural belts of Tien Shan and adjacent southern Kazakhstan may have been close to the Atlantic region–Pacific region border in the Arenig–Llanvirn (Berry 1979).

Pacific region. Berry (1979) pointed out that Pacific region faunas had been described from Australia, North America, western Argentina, Taimyr, and China. Cooper (1979) illustrated and described Pacific region faunas from New Zealand closely similar to those in Australia. Williams et al. (1987) and Williams & Stevens (1987) expanded knowledge of the stratigraphic distribution of the Arenig–Llanvirn Pacific region faunas in Newfoundland. Sheng Shen-Fu (1980) compiled Ordovician graptolite biostratigraphic data for China by region, and Mu En-zhi et al. (1979) discussed and illustrated Arenig–Llanvirn graptolites from southwest China. Zima's (1976) biostratigraphic studies in southern Kazakhstan have clarified understanding of the affinities of Arenig–Llanvirn graptolites found in Kazakhstan. Clearly, Kazakhstan lay in the Pacific region. Cooper & Fortey

(1982) recorded Pacific region faunas from Spitsbergen, and Stone & Strachan (1981) described Pacific region isograptids from Scotland.

Reasoning from the species apparently endemic to an area and from the stratigraphic ranges of genera and species in the several areas in the Pacific region, provinces apparently existed within the region. Australia and New Zealand comprise one province. North America was another province. Faunas from the Bogo Shale in western Norway, western Ireland, Scotland, and Spitsbergen seem to ally these areas to the North American province. Taimyr, Kazakhstan, and south China each were separate provinces within the Pacific region. Faunas from western Argentina are not extensive enough to ascertain if it was a separate province or if it was allied with the North American province. Degree of difference in species present in each of the Pacific region provinces suggests some longitudinal separation of the Australia–New Zealand province, the North American province, the South China province, the Taimyr province, and the Kazakhstan province from each other. Spitsbergen probably was separate from, but in the same current system as, the North American province.

Arenig–Llanvirn faunas from South China merit special attention. Interlayered in the dominantly Pacific region faunas are those (such as *Didymograptus murchisoni*) which clearly indicate Atlantic region affinity. In addition, the South China Arenig–Llanvirn faunas include many apparently endemic taxa. Mu En-zhi *et al*. (1979) discussed these faunas and their stratigraphic occurrences in several stratigraphic sections. These faunas are very rich in species, more so than faunas from any other province. The pattern of graptolite occurrences in South China suggests that the area was situated in a position where Atlantic region waters swept into the Pacific region from time to time. Species and numerical richness of the faunas suggests that the area was close to a major upwelling system which could have been analogous to the present-day system off Peru. Intermingling of faunas from two faunal regions could result from seasonal reversals in oceanic surface current circulation as a consequence of monsoonal conditions. Such current reversals take place today in the Indian Ocean as a result of the seasonal monsoonal system (Wilde *et al*. in press).

Llandeilo

The marked faunal regionalism and provincialism of the Arenig–Llanvirn was reduced markedly during a major sea level change and regression that took place during the Llandeilo. Fortey (1984) drew attention to the regression at this time. Boucek (1973) pointed out that many graptolites became extinct at this time. Sharp reductions took place among dichograptids (that is, didymograptids, tetragraptids, isograptids, phyllograptids, and taxa with similar thecae). Although extinctions among biserial scandent forms did occur as well, reductions among biserial scandent taxa were not so numerous as among dichograptids. Berry *et al*. (1987, p. 110–111) suggested that the biserial scandent colony organization permitted vertical colony migration. Vertical migration for feeding may have been advantageous at a time when oxygen-poor, nitrogen compound rich graptolite habitat waters receded from shelves and became restricted to positions over the slopes and in the open ocean (Berry *et al*. 1987). Dichograptids may have lacked close enough coordination among zooids to permit significant vertical migration.

Extinctions among graptolites during the Llandeilo reduced faunas in both Atlantic and Pacific regions. Despite the many extinctions among graptolites, however, an Atlantic and Pacific region may be distinguished (Skevington 1976). The degree of difference in faunas from the two regions is not as marked as it had been during the Arenig–Llanvirn, yet significant differences exist. For example, dicellograptids and dicranograptids appear in Llandeilo strata in the Atlantic region (Berry 1964). Nemagraptids appear in the British province within the Atlantic region as well during the Llandeilo. The number of endemic biserial scandent taxa in Llandeilo age strata in Pacific region faunas suggests that the provinces recognized in the Arenig–Llanvirn interval probably retained their identity into the post-Llanvirn Ordovician.

Caradoc–Ashgill

The widespread occurrence of *Nemagraptus gracilis* may be used to recognize the early Caradoc in graptolite-bearing strata. Finney & Bergstrom (1986) discussed the distribution and correlation of *Nemagraptus gracilis* zone faunas, noting the many new graptolite lineages that develop during the time of the zone. Fortey (1984) drew attention to a worldwide sea level rise at the time of the *N. gracilis* zone. That sea level rise resulted in transgressions across most cratonal areas and formation of shelf seas with extensive oxygen-poor, nitrogen compound rich waters on their outer margins. These shelf sea environments became sites for marked early Caradoc radiations among such new graptolite lineages as the dicellograptids, the dicranograptids, the leptograptids, the nemagraptids, the orthograptids, the climacograptids, and the glyptograptids. As Berry (1979, p. 11) pointed out, certain of these new lineages (dicranograptids and azygograptids, for example) appear to have originated in the Atlantic region and spread subsequently into the Pacific region. The spread of taxa that originated in the Atlantic region into the Pacific could have been facilitated by northward movement of Baltoscania and Wales during the Llandeilo and Early Caradoc. That plate motion would have permitted members of the new lineages that appeared in Llandeilo strata in the Atlantic region to come into contact with Pacific region ocean surface water currents. Once that happened, then those currents would have carried members of the newly-evolving lineages to new habitats in the newly-invaded region. In time, colonization of new habitats and expansion of populations in them would take place. The widespread and remarkable taxonomic uniformity of the *N. gracilis* zone fauna seen in the stratigraphic record at many Pacific region localities indicates that this sequence of developments did take place. Plate motion coupled with sea level rise seemingly carried newly-evolving lineages into newly-open habitats in which environmentally optimum conditions for graptolites existed. The massive radiation in new lineages described in the stratigraphic record followed.

Skevington (1976) stated that provincialism among graptolites continued through the Late Ordovician. He (Skevington, 1976, p. 196) stated that the faunal 'differences exist at the lowest taxonomic level — that of the species — and the provinces are correspondingly less obvious' than they were in the Arenig–Llanvirn. Skevington (1976, p. 196) went on to say: 'in part, at least, the restricted diversification in rhabdosomal form, in comparison with the Early Ordovician, is one factor contributing to the reduced expression of Late Ordovician provincialism.' Another factor in seeming reduced provincialism of Late Ordovician graptolite faunas, Skevington (1976, p. 196) noted, was 'Late Ordovician proto-Atlantic contraction, achieved by the relative movement of Europe towards North America.' Skevington (1976, p. 196–197) said as well that 'the lowering of surface water temperatures in mid and high latitudes, consequent upon the Late Ordovician glacial phase, appears to have imposed a restriction on the overall distribution of graptolite faunas in comparison with the Early Ordovician.'

Skevington (1976, p. 196–197) pointed out that late Caradoc and Ashgill age graptolites have not been cited from South America, North Africa, and southern Europe. The Ordovician graptolite succession in Wales ends in the mid-Caradoc. British Lake District faunas continue to exist through the Ashgill. As Skevington's (1976) discussion indicates, graptolites had become extinct in all of the Atlantic region, except, possibly, part of Britain, by about mid-Caradoc. Because of equatorward plate motion of that plate bearing what is modern northern Europe during the Llandeilo and onwards, Baltoscania had become a faunal province within the Pacific region by the early part of the Caradoc. Thus, the Atlantic region lost its faunas, both as a consequence of plate motion carrying a plate and its faunas out of a region, and as a result of destruction of graptolite habitats when cold and glacial climates developed over the South Pole. Skevington (1976, p. 197) concluded his analysis of Late Ordovician graptolite provincialism as follows: "The foregoing evidence strongly suggests that in the Late Ordovician all graptolite faunas, with rare exceptions, were confined to the tropical zone, where the general uniformity (measured in terms of surface water temperatures) was impressed upon the composition of graptolite faunas. It is envisaged that the progressively steepening thermal gradients imparted a concomitant restriction upon the areal distribution of graptolites, such that, by the Late Ordovician, the limits of temperature tolerance were located within a few tens of degrees on either side of the palaeoequator. This reduced geographical spread of graptolite faunas increased the possibility of interchange and hence favoured the establishment of more uniform compositional characteristics." As Skevington (1976) described, mostly Pacific region faunas existed after about mid-Caradoc. Although graptolites became essentially tropicopolitan by the Late Caradoc, the provinces recognized among Arenig–Llanvirn faunas continued to exist until graptolites underwent a massive set of extinctions in the late Ashgill.

As that plate bearing what is today northern and central continental Europe moved northward into the Pacific region during the Caradoc and Ashgill, the faunas living in environments on that plate remained provincially distinct from other faunas in the Pacific region. The faunas do reflect, however, alliances with other Pacific region faunas. Motion of that plate led to gradual closing of the former expanse of the Iapetus Ocean. That gradual closing apparently led to development of an eastern North American subprovince. Finney (1986) recorded its fauna from the Viola Spring Limestone in Oklahoma and noted that the fauna had been found in many localities in the eastern part of the United States. Late Ordovician faunas in eastern North America differ from those found in modern western North America (Finney 1986).

Late Ordovician glacio-eustatic sea level changes, greatly steepened thermal gradients in surface waters of the Southern Hemisphere oceans, and generation of oxygen-laden deep ocean waters about ice sheets formed near the South Pole led to marked reduction of the oxygen-poor, nitrogen compound rich environments graptolites apparently preferred (Berry et al. 1987). Consequently, graptolites almost became extinct. The survivors, tiny climacograptids and other diplograptids, apparently persisted because their nutritional needs were less than those of larger colonies or their tolerance for oxygen was greater than that of the graptolites that became extinct. The only part of the world in which graptolites were reduced in numbers for a relatively short interval in the latest Ordovician was South China. There, remarkably rich faunas persisted longer into the Late Ordovician than anywhere else, and the faunas there rebounded from near-extinction more quickly than elsewhere in the world. As noted earlier, this area seems to have lain near a major upwelling system that maintained organic productivity at a high level. Furthermore, it was within the tropics of the time. The British Lake District seems to have been another site at which graptolites rebounded relatively rapidly after near-extinction. There, oceanographic conditions were similar to those in South China (Wilde et al. in press).

Llandovery

The base of the Llandovery and commencement of the Silurian is marked essentially by eustatic sea level rise following deglaciation. As sea level rose, those environments in which graptolites found optimum living conditions developed anew and spread across cratons as part of the shelf sea water. Graptolites of the *Parakidograptus acuminatus* zone at the base of the Llandovery are found widely, although only a few taxa may be found in any single area. Accordingly, recognition of provincialism, if it was present, is difficult. Legrand (1986) reviewed the affinities of the latest Ordovician–earliest Silurian graptolites in North Africa and suggested that they comprised a distinct provincial fauna. The taxa in that fauna probably were derived from survivors of the near-extinction that could tolerate cold, oxygen-laden environments that must have existed in the North African area during the Late Ordovician. Certain species, such as *P. acuminatus*, have been recorded from many areas, however, each area in which early Llandovery graptolites are found has some endemic taxa. That endemism suggests that each of the major graptolitic successions was separate from the others. Significant Llandovery graptolite successions have been found in North America, Britain, many sites in continental Europe, North Africa, China, Australia, Kazakhstan, Taimyr, and Greenland. Aspects of these successions have been described in Berry (1973), Hughes & Rickards (1986), Kaljo & Koren' (1976), Mu En-zhi et al. (1986), Obut (1974), Obut et al. (1985), Obut & Sobolevskaya (1966), and Rickards et al. (1977). Comparisons of the taxa between areas must be at the level of species and subspecies, thus, as Skevington (1976) pointed out for the Late Ordovician, Silurian provincialism is not recognized easily.

Melchin (in press) assessed Llandovery graptolite provincialism. His analysis indicates that, in general, provincialism is consistent with the plate positions suggested in Scotese (1986). Melchin (in press) pointed out as well that the late Llandovery distribution of *Cyrtograptus sakmaricus*, cyrtograptids of the *C. lapworthi* group, and diverse retiolitids in North America and the plates that comprise modern Siberia appears to define a faunal province. That provincial fauna differs from coeval latest Llandovery faunas in Britain, continental Europe, and North Africa. This provincialism may reflect a position in the tropics and north of the equator for the *C. sakmaricus–C. lapworthi* group fauna.

Wenlock–Ludlow–Pridoli–Early Devonian

Commencing in the Wenlock and continuing on through the latter part of the Silurian, lands began to form in certain areas that were shelf seas during the Llandovery (Boucot et al. 1968). Wenlock and Ludlow graptolite faunas thus become somewhat more restricted than they were in the Llandovery. Nonetheless, the basically low-level provincialism recognized among Llandovery faunas continues in the latter part of the Silurian.

During the middle and late Wenlock, graptolites from Portugal, Spain, Sardinia, the Carnic Alps, and the Pyrenees appear to have comprised a unique fauna characterized by taxa with unusually large rhabdosomes. Romariz (1982) gave the name Mediterranean Province to the area in which these unusually robust graptolites had been found. The species richness of the fauna as well as the robustness of many of the colonies implies development near an upwelling system. Potentially, such a system resulted from plate motions in the early part of the Wenlock.

Graptolites almost became extinct in the latest Wenlock (Rickards et al. 1977). That near-extinction terminated the unique Mediterranean Province fauna and led to development of an essentially homogeneous Ludlow fauna.

Koren' (1979) reviewed distributions of Late Silurian (Pridoli) and Early Devonian graptolites, noting that graptolites of this age have been recorded from North America, North Africa, continental Europe, Tien Shan, southeast Asia, China, and Australia. She (Koren', 1979) suggested that the graptolite faunas of this age span were confined to the tropics. By the Late Silurian, North Africa appears to have moved northward to a position close to about 30 degrees south, based on the graptolite faunas found there. Koren' (1979) pointed out that most of the Late Silurian–Early Devonian graptolite occurrences are in rock suites that formed on shelf margins in the tropics. Graptolites living in such areas would have been within equatorial current systems, thus the faunas would tend to be relatively similar. By the Late Silurian–Early Devonian, graptolites were so few in number and so limited in the areas that they inhabited that their distribution patterns are not as helpful in suggesting plate positions as they had been at earlier times.

Conclusions

Graptolite provincialism and regionalism is most easily recognized in the Arenig–Llanvirn. At that time, two regions, an Atlantic and Pacific, may be distinguished. Provinces within these regions reflect plate positions because the provincial faunas developed on the outer parts of and about shelves born on individual plates. The plates are recognized by their terrestrial and marine shelf rock suites.

Provincialism commenced in the Tremadoc, reached its zenith in the Arenig–Llanvirn, and returned to a less pronounced condition in post-Llanvirn Ordovician. Whereas the differences between regions involve different sets of genera as well as species living in the tropical Pacific region waters and the coeval cooler, non-tropical Atlantic region waters, the provincial faunal differences for other parts of the life span of the graptolites involve different sets of species. Life habits of the graptolites apparently resulted in many post-Llanvirn graptolites developing some form of biserial scandent colony. The majority of the Siluro-Devonian graptolite colonies have some form of uniserial scandent colony organization. Because so many of the Siluro-Devonian graptolites are variations of the same colony form, close study of graptolite associations is required to recognize provincial faunas among graptolites of this time interval. The patterns that exist seem to be consistent with plate positions suggested on the base maps provided for the symposium (Scotese 1986).

Plate positions influence ocean surface currents. Wilde *et al.* (in press) described Late Cambrian into Llandovery ocean surface current circulation. That circulation is consistent with and, to some extent, provides an explanation for, the graptolite faunal regions and provinces described herein. The faunal regions and provinces are, in general, consistent with the plate positions suggested on the palaeogeographic base maps provided for the symposium.

The Pacific region was within the tropics. Atlantic region faunas were south of the tropics. Northward drift of the Baltoscania plate during the late Llanvirn–Llandeilo resulted in faunas living in environments over that plate changing affinity from those of the Atlantic region to those of a Pacific region province. Another consequence of that plate motion was repopulation of the Pacific region after the late Llanvirn mass mortality by some taxa whose ancestry was among Baltoscanian faunas when they were a part of the Atlantic region. Northward drift of North Africa during the Ludlow resulted in faunas in that area becoming a component of the Late Silurian–Early Devonian cosmopolitan graptolite fauna. Southern Europe appears to have entered the tropics in the Ludlow as well, based on the graptolite faunas.

Because graptolites seemingly were linked closely to waters over the outer parts of shelves developed on individual plates, drift that resulted in changes in the physical-chemical conditions in these waters had an impact on graptolite development through time. Flowering among graptolites may be linked to eustatic sea level rise and mass mortalities may be linked to sea level lowering. Plate positions resulted in sites of upwelling waters in which graptolites found optimal living conditions, such as in South China for much of the Ordovician and early part of the Silurian. Plate motions resulted in Late Ordovician glaciation that impacted graptolites by leading to destruction of their preferred habitats. Thus, an underlying agent in graptolite biogeography and life history appears to be change in plate positions through time.

References

ANDERSON, J. J. 1982. The nitrite-oxygen interface at the tope of the oxygen-minimum zone in the eastern tropical North Pacific. *Deep-Sea Research*, **29**, 1193–1201.

——, OKUBO, A., ROBBINS, A. S. & RICHARDS, F. A. 1982. A model for nitrite and nitrate distributions in oceanic oxygen-minimum zones. *Deep-Sea Research*, **29**, 1113–1140.

BERRY, W. B. N. 1959. Distribution of Ordovician graptolites. *In*: SEARS, MARY, (ed.) *International Oceanography Congress Preprints*. American Association for the Advancement of Science, 273–274.

—— 1960a. Correlation of Ordovician graptolite-bearing sequences. Internatl. Geol. Congr. 21st. Norden 1960. Rept. Pt. 7, 97–108.

—— 1960b. Graptolite faunas of the Marathon region, west Texas. *Texas University Publication* **6005**.

—— 1964. The Middle Ordovician of the Oslo region, Norway. No. 16. Graptolites of the Ogygiocaris Series. *Norsk Geologisk Tidsskritt*, **44**, 61–169.

—— 1967. Comments on correlation of the North American and British Lower Ordovician. *Geological Society of America Bulletin*, **78**, 419–428.

—— 1972. Early Ordovician bathyurid province lithofacies, biofacies and correlations — their relationship to a proto-Atlantic Ocean. *Lethaia*, **5**, 69–83.

—— 1973. Silurian–Early Devonian graptolites. *In*: HALLAM, A. (ed.) *Atlas of Palaeobiogeography*. Elsevier, Amsterdam, 81–87.

—— 1979. Graptolite biogeography: A biogeography of some Lower Paleozoic plankton. *In*: GRAY, JANE & BOUCOT, A. J., (eds) *Historical biogeography, plate tectonics, and the changing environment*. Oregon State University Press, Corvallis, Oregon 105–115.

—— WILDE, PAT & QUINBY-HUNT, M. S. 1987. The oceanic non-sulfidic oxygen minimum zone: a habitat for graptolites?. *Bulletin of the Geological Society of Denmark*, **35**, 103–114.

BOUCEK, B. 1972. The paleogeography of Lower Ordovician graptolite faunas: A possible evidence for continental drift. *24th International Geological Congress Montreal 1972. Proceedings Section 7*, 266–272.

—— 1973. Lower Ordovician graptolites of Bohemia. *Czechoslovak Academy of Sciences*, Prague.

BOUCOT, A. J., BERRY, W. B. N. & JOHNSON, J. G. 1968. The crust of the earth from a Lower Paleozoic point of view. *In*: PHINNEY, R. A. (ed.) *The History of the Earth's Crust: A symposium*. Princeton University Press, 208–228.

BRINTON, E. 1980. Distribution of euphausiids in the eastern tropical Pacific. *Progress in Oceanography*, **8**, 125–189.

BULMAN, O. M. B. 1931. South American graptolites with special reference to the Nordenskold Collection. *Arkiv for Zoologi* **22A**, 1–111.

—— 1950. Graptolites from the *Dictyonema* Shales of Quebec. *Quarterly Journal of the Geological Society of London*, **106**, 63–99.

—— 1954. The graptolite fauna of the *Dictyonema* Shale of the Oslo Region. *Norsk Geologisk Tiddskrift*, **33**, 1–40.

—— 1971. Graptolite faunal distribution. *In*: MIDDLEMISS, F. A., RAWSON, P. F. & NEWALL, G. (eds) *Faunal provinces in space and time*. Seel House Press, Liverpool, 47–60.

COOPER, R. A. 1979. Sequence and correlation of Tremadoc graptolite assemblages. *Alcheringa*, **3**, 7–19.

—— & FORTEY, R. A. 1982. The Ordovician graptolites of Spitsbergen. *Bulletin of the British Museum of Natural History (Geology)*, **36**, 1–171.

DESTOMBES, J. 1960. Stratigraphie de l'Ordovicien de la partie occidentale du Jbel Zuni, Anti-Atlas occidental (Maroc). *Societe Geologique de France Bulletin series*, **7**, **7**, 747–751.

—— 1970. Cambrien moyen et Ordovicien. *In*: *Colloque internationale sur les correlation du Precambrien, Agadir-Rabat, Mai 1970. Livret-guide de l'excursion Anti-Atlas occidentale et central*. Notes et Memoir de Service Geologique Maroc, **229**, 161–170.

EBELING, A. W. 1967. Zoogeography of tropical deep sea animals. *Studies in Tropical Oceanography*, **5**, 593–613.

ERDTMANN, B-D. 1972. Ordovician graptolite provincialism: Evidence for continental drift? (abs) *24th International Congress, Montreal 1972. Proceedings Section 7*, 274.

FINNEY, S. C. 1984. Biogeography of Ordovician graptolites in the Southern Appalachians. *In*: BRUTON, D. L. (ed.) *Aspects of the Ordovician System*. Palaeontological Contributions, University of

Oslo **295**, 167–176.

—— 1986. Graptolite biofacies and correlation of eustatic, subsidence, and tectonic events in the Middle to Upper Ordovician of North America. *Palaios*, **1**, 435–461.

—— & BERGSTROM, S. M. 1986. Biostratigraphy of the Ordovician *Nemagraptus gracilis* zone. *In*: HUGHES, C. P. & RICKARDS, R. B. (eds). *Palaeoecology and biostratigraphy of graptolites*. Geological Society, Special Publication, **20**, 47–59.

FORTEY, R. A. 1984. Global earlier Ordovician transgressions and regressions and their biological implications. *In*: BRUTON, D. L. (ed.) Aspects of the Ordovician System. *Palaeontological Contributions, University of Oslo* **295**, 37–50.

GEORGE, W. 1962. *Animal geography*. London. Heinemann.

HARRINGTON, H. J. & LEANZA, A. F. 1957. Ordovician trilobites of Argentina. *Kansas University Department of Geology Special Publication*, **1**.

HOLLIGAN, M. R., WILLIAMS, P. J. LeB., PURDUE, D. & HARRIS, R. P. 1984. Photosynthesis, respiration and nitrogen supply of plankton populations in stratified frontal and tidally mixed shelf waters. *Progress in Marine Ecology*, **17**, 201–213.

HUGHES, C. P. & RICKARDS, R. B. (eds) 1986. Palaeoecology and biostratigraphy of graptolites. *Geological Society Special Publication*, **20**.

JACKSON, D. E. 1964. Observations on the sequence and correlation of Lower and Middle Ordovician graptolite faunas of North America. *Geological Society of American Bulletin*, **75**, 523–534.

—— 1969. Ordovician graptolite faunas in lands bordering North Atlantic and Arctic Oceans. *American Association of Petroleum Geologists, Memoir*, **12**, 504–512.

—— 1974. Tremadoc graptolites from Yukon Territory, Canada. *In*: RICKARDS, R. B., JACKSON, D. E. & HUGHES, C. P. (eds). *Graptolite studies in honour of O. M. B. Bulman*. Palaeontological Association Special Papers in Palaeontology, **13**, 35–58.

KALJO, D. & KOREN', T. 1976. *Graptolites and stratigraphy*. Academy of Sciences of Estonian SSR. Institut of Geology, Tallinn.

KOREN', T. 1979. Late monograptid faunas and the problem of graptolite extinction. *Acta Palaeontologica Polonica*, **24**, 79–106.

LEGRAND, P. 1966. Precisions biostratigraphiques sur l'Ordovicien inferieur et le Silurien des chaines d'Ougarta (Sahara algerien). *Societe Geologique de France, Compte Rendu Sommaire des Seances*, **7**, 243–244.

LEGRAND, P. 1986. The lower Silurian graptolites of Oued in Djerane: A study of populations at the Ordovician-Silurian boundary. *In*: HUGHES, C. P. & RICKARDS, R. B. (eds). *Palaeoecology and biostratigraphy of graptolites*. Geological Society Special Publication, **20**, 145–153.

LENZ, A. C. 1972. Ordovician to Devonian history of southern Yukon and adjacent District of MacKenzie. *Bulletin of Canadian Petroleum Geology*, **20**, 321–361.

—— 1977. Some Pacific Faunal Province graptolites from the Ordovician of northwern Yukon, Canada. *Canadian Journal of Earth Science*, **14**, 1946–1952.

—— & JACKSON, D. E. 1986. Arenig and Llanvirn graptolite biostratigraphy, Canadian Cordillera. *In*: HUGHES, C. P. & RICKARDS, R. B. (eds). *Palaeoecology and biostratigraphy of graptolites*. Geological Society, Special Publication, **20**, 27–45.

LONGHURST, A. R. 1967. Vertical distribution of zooplankton in relation to the eastern Pacific oxygen minimum. *Deep-Sea Research*, **14**, 51–63.

MELCHIN, M. J. (in press). Llandovery graptolite biostratigraphy and paleobiogeography, Cape Phillips Formation, Canadian Arctic Islands. *Canadian Journal of Earth Science*.

MU EN-ZHI. 1963. Research in graptolite faunas of Chilianshan. *Scientia Sinica*, **12**, 347–371.

——, BOUCOT, A. J., CHEN XU, & RONG JIA-YU. 1986. Correlation of the Silurian rocks of China. *Geological Society of America Special Paper*, **202**.

——, GE MEIYU, CHEN XU, NI YUNAN & LIN YAOKUN. 1979. Lower Ordovician graptolites of southwest China. *Palaeontologi Sinica New Series B*, **13**.

MULLINS, H. T., THOMPSON, J. B., McDOUGALL, K. & VERCOUTERE, T. L. 1985. Oxygen minimum zone edge effects: evidence from the central California upwelling system. *Geology*, **13**, 491–494.

NIKITIN, I. F. 1972. *Ordovician of Kazakhstan. Part 1. Stratigraphy*. Alma-Alta Publishing House "Nauka" Kazakhstan, USSR. (in Russian).

OBUT, A. M. (ed.). 1974. *Graptolites of the USSR*. Publishing House "Nauka" Siberia Branch, Novosibirsk. (in Russian).

—— & SOBOLEVSKAYA, R. F. 1964. *Ordovician graptolites of Taimyr*. Sibiriskoe Otdelevie Institut Geologii i Geofiziki Akademia Nauk SSSR. (in Russian).

—— & —— 1966. *Lower Silurian graptolites of Kazakhstan*. Sibiriskoe Otdelevie Institut Geologii i Geofiziki Akademia Nauk SSSR. (in Russian).

——, —— & BONDAREV, V. I. 1965. *Silurian graptolites of Taimyr*. Sibiriskoe Otdelevie Institut Geologii i Geofiziki Akademia Nauk SSSR. (in Russian).

RICKARDS, R. B., HUTT, J. E. & BERRY, W. B. N. 1977. Evolution of the Silurian and Devonian graptoloids. *Bulletin of the British Museum of Natural History (Geology)*, **28**, 1–120.

ROMARIZ, C. 1962. Graptoloides des formacoes ftaniticas do Silurico portugues. *Boletim da Sociedade Geologica de Portugal*, **14**, 1–135.

RUSHTON, A. W. 1982. The biostratigraphy and correlation of the Merioneth-Tremadoc Series boundary in North Wales. *In*: DEAN, W. T. & BASSETT, M. G. (eds). *The Cambrian–Ordovician Boundary*. University of Wales Press, 41–59.

SCHMIDT, O. 1987. Lower Ordovician graptolite fauna of the Bogo Shale (West Norway), and its palaeogeoraphical relationships. *Geological Society of Denmark Bulletin*, **35**, 209–215.

SCOTESE, C. R. 1986. Phanerozoic reconstructions: A new look at the assembly of Asia. *University of Texas Institute for Geophysics Technical Report*, **66**.

SHENG SHEN-FU. 1980. The Ordovician System in China. International Union of Geological Sciences Publication, **1**.

SIMPSON, G. G., PITTENDRIGH, C. S. & TIFFANY, L. H. 1957. *Life: An introduction to biology*. Harcourt Brace. New York.

SKEVINGTON, D. 1968. British and North America Lower Ordovician correlation: Discussion. *Geological Society of America Bulletin*, **79**, 1259–1264.

—— 1969. Graptolite faunal provinces in Ordovician of northwest Europe. *In*: KAY, M. (ed.) *North Atlantic — geology and continental drift*. American Association of Petroleum Geologists Memoir, **12**, 557–562.

—— 1973. Ordovician graptolites. *In*: HALLAM, A. (ed.) *Atlas of Palaeobiogeography*. Amsterdam. Elsevier, 27–35.

—— 1974. Controls influencing the composition and distribution of Ordovician graptolite faunal provinces. *In*: RICKARDS, R. B., JACKSON, D. E. & HUGHES, C. P. (eds) *Graptolite studies in honour of O. M. B. Bulman*. Special Papers in Palaeontology, **13**, 59–73.

—— 1976. A discussion of the factors responsible for the provincialism displayed by graptolite faunas during the Early Ordovician. *In*: KALJO, D. & KOREN, T. (eds) *Graptolites and Stratigraphy*. Academy of Sciences of Estonian SSR. Institut of Geology. Tallinn, 180–200.

STONE, P. & STRACHAN, I. 1981. A fossiliferous borehole section within the Ballentrae ophiolite. *Nature*, **293**, 455–457.

TELLER, L. 1969. The Silurian biostratigraphy of Poland based on graptolites. *Acta Geologica Polonica*, **19**, 393–501.

TJERNVIK, T. 1956. On the early Ordovician of Sweden: Stratigraphy and fauna. *Uppsala University Geological Institute Bulletin*, **36**, 107–284.

TURNER, J. C. M. 1960. Faunas graptoliticas de America del Sur. *Asociacion Geologica Argentina Revista*, **14**, 5–180.

ULST, R. 1976. Stratigraphical significance of the Late Tremadocian and the Arenigian graptolites of the Middle East Baltic Area. *In*: KALJO, D. & KOREN, T. (eds) *Graptolites and Stratigraphy*. Academy of Sciences of Estonian SSR. Institut of Geology. Tallinn, 214–221.

WANG XIAOFENG & ERDTMANN, B. D. 1987. Zonation and correlation of the earliest Ordovician graptolites from Hunjiang, Jilin province, China. *Bulletin of the Geological Society of Denmark*, **35**, 245–257.

WILDE, P., BERRY, W. B. N. & QUINBY-HUNT, M. S. in press. *Oceanography in the Ordovician*. Symposium Volume of the Vth International Symposium on the Ordovician System.

WILLIAMS, S. H. & STEVENS, R. K. 1987. Summary account of the Lower Ordovician (Arenig) graptolite biostratigraphy of the Cow Head Group, western Newfoundland. *Bulletin of the Geological Society of Denmark*, **35**, 259–270.

—, BOYCE, W. D. & JAMES, N. P. 1987. Graptolites from the Lower-Middle Ordovician St. George and Table Head groups, western Newfoundland, and their correlation with trilobite, brachiopod, and conodont zones. *Canadian Journal of Earth Sciences*, **24**, 456–470.

ZIEGLER, A. M., HULVER, M. L., LOTTES, A. L. & SCHMACHTENBERG, W. F. 1984. Uniformitarianism and palaeoclimates: Inferences from the distribution of carbonate rocks. *In*: BRENCHLEY, P. J. (ed.) *Fossils and climate*. John Wiley & Sons, Chichester, 3–25.

ZIMA, M. 1976. Graptolite assemblages of the Arenigian and Llanvirnian boundary beds in North Tien Shan. *In*: KALJO, D. & KOREN', T. (eds) *Graptolites and stratigraphy*. Academy of Sciences of Estonian SSR. Institut of Geology, Tallinn, 44–55.

Graptoloid biogeography: recent progress, future hopes

BARRIE RICKARDS[1], SUSAN RIGBY[1] & JONATHAN H. HARRIS[2]

[1] *Department of Earth Sciences, Downing Street, Cambridge CB2 3EQ, UK*
[2] *6 The Roystons, East Preston, W. Sussex, BN16 2TR, UK.*

Abstract: Recent progress in the plotting of biogeographical distributions is shown to be partly due to improved base maps, some of the notable anomalies detected by earlier workers being explained and easily accommodated in the latest maps. Progress is also in part due to increasingly precise systematics, which results in more accurate biostratigraphical correlation: examples include the '*Didymograptus*' *bifidus* controversy and its role in the definition of the Pacific and Atlantic provinces of the Arenig–Llanvirn. Analogy with modern planktonic regimes enables recognition of marker species in specific water masses; and in the distinction of neritic from open ocean plankton. Analysis at species level and lower of Wenlock graptolite assemblages has resulted in the recognition of the Rheic and Mediterranean subprovinces within a Pacific province or realm. Hydrodynamic modelling suggests directions for future research which may provide independent support for biogeographical studies.

Graptoloid plankton originated in the lowest Ordovician (Tremadoc) from early dendroid plankton. This latter included such well known forms as *Dictyonema flabelliforme* Eichwald. Skevington (1973) points to provincialism of the Tremadoc dendroids, the above species and related forms being largely confined to an Atlantic Province and other planktonic dendroids and anisograptids to a second, Pacific Province. These correspond roughly to benthic provinces, but are much broader bands of faunal distribution. Later in the Tremadoc faunas assumed a more cosmopolitan aspect: even forms such as *Psigraptus*, which Skevington (1973) necessarily regarded as a local distribution, are now known to have had a much wider occurrence (Rickards & Stait 1984; Wang & Erdtmann 1987). The duration in time of the dendroid element of the graptolite plankton is unknown, although forms with proximal vane structures are known from the Devonian (Koren' pers. comm.). Benthonic dendroids have yielded no biogeographical data: their record (though spectacular at times, and in some places) is sporadic, and the resultant systematic studies would give an undoubted monographic bias to any attempted distribution plots.

Thus the main work on graptolite biogeography has been achieved with graptoloids, which range from the Tremadoc to the latest Lower Devonian. Graptoloids are often abundant and well preserved, widespread, and evolved rapidly enough from the earliest large dichograptids to give a valuable biostratigraphic framework to biogeographical studies.

Ordovician biogeography

Pacific and Atlantic Provinces

After the earliest Tremadoc provincialism referred to above, the Pacific and Atlantic Provinces became re-established in the middle Arenig and survived through much of the Llanvirn, but by this time their recognition is based upon graptoloids not upon dendroids. It is perhaps a point not sufficiently emphasized in the past that the early provincialism depended upon the rapid evolution and diversification of the earliest macrozooplankton (dendroids), which then achieved cosmopolitan distribution; and that the second wave of provincialism followed exactly the same pattern of explosive evolution and subsequent (post-Llanvirn) cosmopolitan distribution. Such patterns hint at palaeoecological processes rather than tectonic events, although the closure of the Iapetus Ocean has been invoked by several authors to explain the onset of cosmopolitan faunas (e.g. Dewey *et al.* 1970; McKerrow & Cocks 1976).

Early recognition of the Pacific and Atlantic Provinces was beset by difficulties, at least to some extent caused by the base maps on which the data was plotted. Thus Bulman (1964) and Boucek (1972) used Irving's (1964) data, in which the South Pole was located in the South Atlantic of present day geography. However, the same plots upon Smith *et al.* (1973) base maps resulted in various 'anomalies' disappearing. It should be said that the 'anomalies' consist of occurrences of faunas outside the latitudes expected of them. Such 'anomalies', and their solution, need to be viewed most carefully. In the first place, water masses defined by, say, temperature, can occur in a general regime characterized by a different temperature. Secondly, as maps based upon palaeomagnetic data are constantly being improved it follows logically that none of them is 'correct'.

Berry (1960) proposed that distributions were controlled not simply by latitudinal variation in surface temperature, but also by a combination of ocean current dispersal and the effect on these of the distribution of the continents. Looked at in this way 'anomalies' are to be expected, and some of the distributions outlined in this volume by Berry and by Finney may have their explanation contained in such interpretations. Williams (1969) suggested palaeocurrent patterns as controls for graptolite dispersal which in some detail supported Berry's (1960) suggestions.

The definitive paper advocating latitudinal surface temperature gradients as effecting primary control upon the graptoloid plankton, was that of Skevington (1974), where he also summarized other possible mechanisms.

A final point which needs addressing in this section concerns the cause of steepening temperature gradients in post-Llanvirn time. We support Skevington's thesis that the late Ordovician glaciation would necessarily have had an earlier impact in higher latitudes, and that this might well cause provincialism within the plankton as early as the Caradoc. But how can this mechanism of glacial cooling be used to explain the earlier (Tremadoc) phase provincialism of dendroids and its subsequent more cosmopolitan distribution? Was there a much earlier phase of glaciation effecting the same patterns?

Provinces and Subprovinces

It is becoming increasingly clear that the Atlantic and Pacific Provinces can be subdivided into smaller units, and that post-Llanvirn times (including the Silurian; see below) were characterized by endemic faunas. The main features of the Pacific–Atlantic faunas were: (a) considerable diversity and speciation in the (warmer) Pacific Province; (b) endemism at generic level in the Pacific Province (*Brachiograptus*, *Cardiograptus*, *Paraglossograptus*, *Pseudobryograptus*, *Pseudotrigonograptus*, *Allograptus*, *Sinograptus*, *Holmograptus* etc.); (c) Lower diversity in the Atlantic Province, the recognition of which depended upon species occurrence and the absence of the above genera. Certain regions were considered at times to have mixed faunas and to be in a geographically intermediate locale (Skevington,

1974) such as Taimyr, Kazakhstan, Kirgizia, China, South America, eastern North America and western Australia.

Skevington himself draws attention to some late Ordovician provincialism at the species level, though he concluded that the status of late Ordovician provinces is exceedingly slight. In a sense this conclusion results from the contrast with the generic differentiation of earlier provinces: it may not deny the differences which exist. In recent work on the Bendigonian, undoubted Pacific Province, Australian faunas, Rickards & Chapman (in press) compare the faunas of that part of the Bendigonian sequence that they can readily correlate with the Spitsbergen graptolite faunas described by Cooper & Fortey (1982). Of the 32 species or subspecies recorded from Spitsbergen, and 31 from the Bendigonian, only 17 are common to both regions. Whilst this gives an effective and reasonably precise correlation, it cannot mask the obvious endemism. Whether Ekman (1953) is correct in demoting species as identifiers of provinces, it is beyond question that there are considerable faunal differences within the main planktonic provinces so far identified, namely the Pacific and Atlantic. This seems less surprising in light of modern plankton studies which show that in any assemblage the presence of a large number of species will be controlled by local factors, not found throughout a province (see below). Future work will no doubt conveniently subdivide these provinces, and some of the issues referred to above (current circulation, latitudinal temperature gradients, influences of land masses) will be put to the test. Whether the Pacific and Atlantic faunas should continue to be referred to as provinces is open to some question: their areal extent can hardly be much greater without being global! We are inclined to regard them as Pacific and Atlantic Realms, capable of future subdivision into provinces and subprovinces.

Biostratigraphy

Implicit in all the foregoing is that accurate correlation with graptolites is possible. The role of detailed systematic studies and a resultant precise biostratigraphy is often underplayed in global reconstructions. Yet they are vital not only to the interpretation above, but to the continued refinement of it. This was strikingly exemplified in the debate during the 1960's and 1970's over correlation of the British and North American Arenig and Llanvirn: before that dispute was solved, any biogeographic interpretation was necessarily tainted with a strong correlation bias of one form or another. Today systematic revision of Ordovician graptoloids is going on apace and will certainly result in much greater refinement in correlation, and hence in identification of provincial differences. These arguments are strongly supported in Erdtmann's (1976) appraisal of the ecostratigraphy of graptolites.

Silurian biogeography

Using a recent series of publications (Ziegler *et al.* 1981; Scotese *et al.* 1985, 1986; Parrish 1982) we have plotted Wenlock graptolites against palaeocurrent predictions, upwelling zones, and continental distributions. This was preceded by a rigorous systematic analysis of the Wenlock faunas, admittedly involving a series of arbitrary decisions concerning tectonically deformed species in the Mediterranean region. The plot is shown in Fig. 1 where we have transferred the data to the Scotesian maps provided for this volume. Although all the Wenlock graptolite occurrences have not yet been analysed (Fig. 1G) there seem to be three distinct patterns identified by species and subspecies occurrences. Most of the faunas appear to fall into the Pacific province of tropical plankton, the only 'anomaly' being in Khazakstania. However, there is a distinct subprovince in the Baltic region, which we prefer to call the Rheic Subprovince. It is quite possible that this subprovince is actually a reflection of adaptation to a neritic environment. Thirdly, the Mediterranean Subprovince appears to be a high latitude occurrence. It is in this last fauna that we have had to discount several seemingly endemic species and subspecies which are, in fact, tectonically deformed monograptids well known in the rest of the world.

Of the 44 species recorded from the areas so far analysed, 23

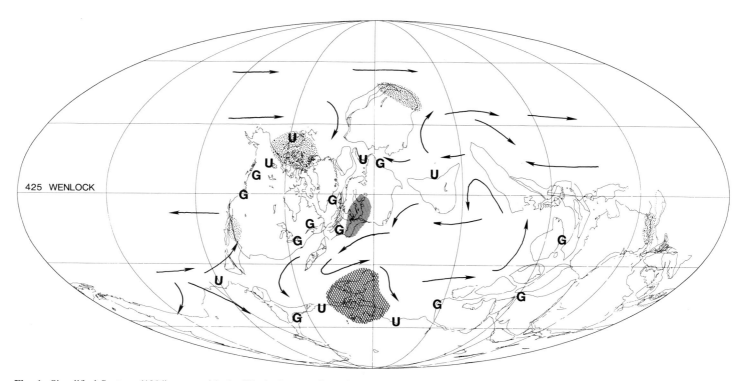

Fig. 1. Simplified Scotese (1986) maps with the Wenlock graptolite subprovinces marked (Rheic, dark stipple; Pacific, light stipple; Mediteranean, diamonds), and unresearched Wenlock graptolite faunas indicated by the letter G. Other parameters after Ziegler *et al.* (1981) and Parrish (1982).

are either restricted or very restricted. In part this may reflect collection failure. For example the *centrifigus* fauna was unknown in the Builth region until recently (Harris, herein). Species diversity increases from the shelf (Rheic) to the offshore (Pacific) regions. Species reported only from the shelf include *Monograptus testis, Retiolites lejskoviensis, R. textor, Pseudoplectograptus praemacilentus* and *Spinograptus spinosus. Cyrtograptus ramosus* has only been found in slope–shelf-edge settings. At least sixteen species occur only in basinal (offshore) records.

The Rheic Subprovince is localised around the northern edge of the Rheic Ocean and within Tornquist's Sea (Poland, Bohemia, Romania, Baltic and Scandinavia). Species found only in the Rheician area are *M. inflexis, M. praecedens, M. solitarius, M. speciosus, M. validus, M. testis bartoszycensis, Barrandeograptus bornholmensis, Gothograptus pseudospinosus* and *M. discoformis*.

The Mediterranean Subprovince occurs in North Africa, Spain, Portugal, S. France, the Pyrenees, and Sardinia and has been recognised previously (see Gueirard *et al.* 1970 for references). Many of the giant species are quite clearly tectonically deformed and have been discarded in our plots. The following may be valid: *M. m. multiferous, M. m. strigosus, M. uncinatus tariccoi, Pristiograptus s. sardous, P. meneghini giganteus*. The subprovince is best developed in the ellesae Zone (Waterlot 1961). Evidence for cold water comes from the associated *Clarkeia* brachiopod faunas.

It is likely that similar analyses will show provincialism in the Ludlow Series. Further, some distributions already suggest neritic versus open ocean occurrences, for example the records of *Pristiograptus tumescens* and *Saetograptus incipiens*. The former is typical of, and abundant in, shelf deposits (e.g. Ludlow) where *S. incipiens* is relatively uncommon except at certain levels. *P. tumescens* is rare in basinal deposits of the same age whilst *S. incipiens* is very abundant. The slope region of Long Mountain displays an overlap of the two (Palmer, pers. comm.). Within the Llandovery basinal environments some interesting distributions have already been detected in preliminary work by the authors. For example, there can be no doubt of the exact correlation of the *argenteus/leptotheca* Zone from Wales to the Lake district: in the Lake District *M. argenteus* is common and *Pribylograptus leptotheca* less so, the ratio being quite dramatically reversed in Wales. The potential for identification on water mass specificity will be considerable as work progresses.

Controls of provincialism

Present day planktonic environments and analogies

Perhaps the next stage in improving the understanding of graptolite distribution should be to investigate the distribution of modern zooplankton. Controls on this distribution are complicated and diverse. The main limiting factors include temperature, salinity, density, turbidity, ocean chemistry, productivity, and the other living components of the environment. These controls are themselves affected by depth, latitude and geography, resulting in the creation of water masses with a unique set of conditions. The water masses themselves can be moved by current action, retaining their individual characteristics for thousands of kilometres from their point of origin (Fig. 2).

In any given assemblage of zooplankton, different controls

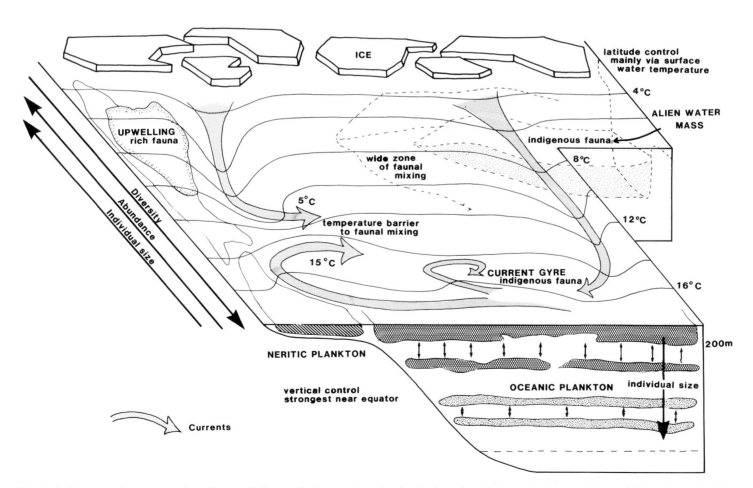

Fig. 2. A diagrammatic representation of some of the practical controls on the distribution of modern zooplankton, each control being the result of a combination of limiting factors such as temperature, density, turbidity and water chemistry.

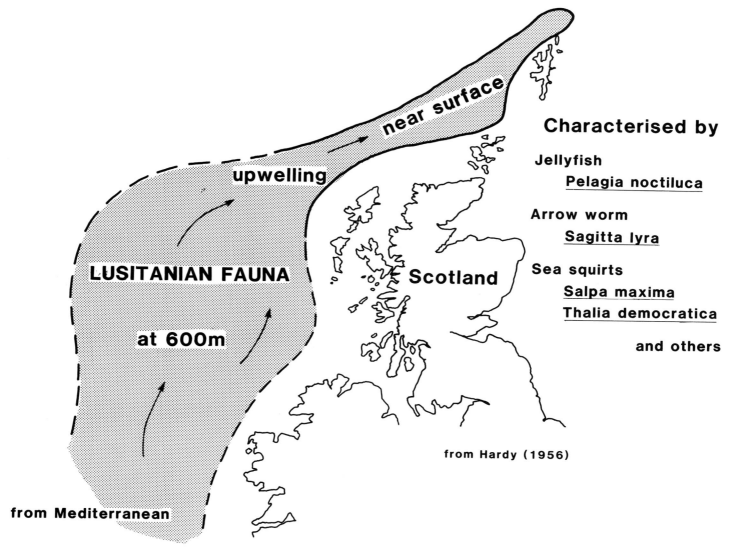

Fig. 3. Occurrence of the Lusitania fauna; explanation in text; after Hardy 1956.

will affect different elements of the fauna. Species can be identified whose distribution is limited by the set of conditions found in a discrete depth or latitudinal range, or water mass. These marker species are the useful biogeographic indicators, occurring in assemblages whose other elements appear to be random, because subject to a different range of controls.

Two examples of the uses and limitations of modern marker species are relevant to the analysis of graptolite distributions. The first, known as the Lusitanian fauna, defines a discrete water mass found to the west of the British Isles (Hardy 1956; Fig. 3). This water mass originates in the Mediterranean and exits at depth into the Atlantic. It travels northwards through the Bay of Biscay, picking up a distinctive fauna which includes *Pelagia noctilucca* (jellyfish), *Sagitta lyra* (arrow worm), and *Salpa maxima* (sea squirt). It then moves around the western edge of Ireland and on around northern Scotland, with its northernmost limit varying from year to year. During this part of its journey, its extent can be determined to a few metres, by the presence or absence of the marker species which have been carried from the Mediterranean or picked up in Biscay. Importantly, the marker species show that the water remains at around 600 m for most of its journey, but rises towards the surface west of the Orkneys. The marker species are unaffected by this change although the assemblage in which they occur changes as depth controlled species appear and disappear.

Apart from suggesting the value of defining marker species amongst the graptolites, two important points emerge from this study. The first is that the Lusitanian marker species are specific to water mass and not to depth; they would be useless for studying depth control. The second is that the fauna can live quite happily in much higher latitudes than the ones in which it originated although it is sometimes unable to breed. The presence or absence of these species would appear anomalous on a plot of geographic distribution, and conversely it might be expected that some anomalies of graptolite distribution will be explained by water mass movements. We feel that Berry (1960) was correct in giving due weight to this factor.

The second example of a modern marker species is from the Southeast Shoal region of the Grand Bank of Newfoundland (Anderson & Gardner 1986). Here, a species of comb jelly (*Pleurobrachia pileus*) lives in shallow water in the neritic zone, where water depth is generally less than 200 m. As the water deepens at the edge of the Bank, it disappears and a species of crustacean (*Calanus finmarchicus*) which also lives at shallow depths, but above deep water, is found instead (Fig. 4). The division between oceanic and neritic plankton is a profound one, with the neritic elements being subject to strong local controls from the input of rivers, local currents and topography. The oceanic plankton is generally much more widespread in its distribution and controls by latitude and depth show up much more

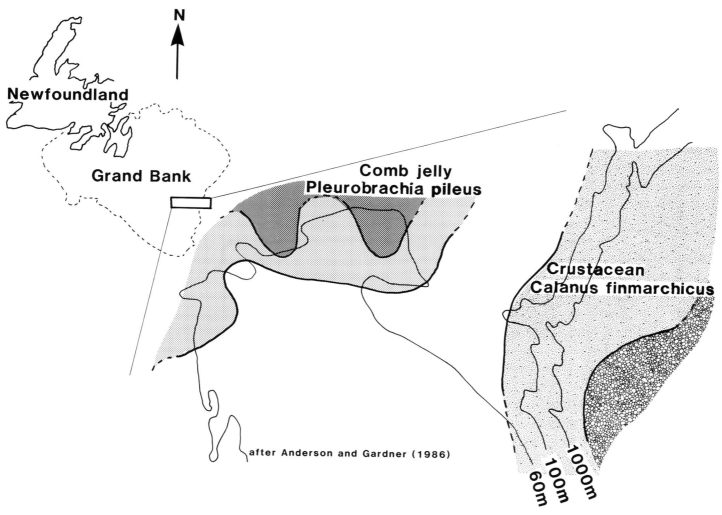

Fig. 4. Occurrence of *Pleurobrachia pileus* (comb jelly) and *Calanus finmarchicus* (crustacean), marker species living at shallow depths respectively over shallow and deep water. After Anderson & Gardner 1986.

clearly. Mixing of the two inevitably occurs, and as always there are many species which are unaffected by the change of water depth. As in all controls of modern zooplankton, even an important division such as this is defined only by a limited number of marker species.

The division of zooplankton into neritic and oceanic elements has been recognised to some extent in graptolite distributions. Berry & Wilde (1978) dealt with upwelling zones which occur along the continental slope, with relation to graptolites, but did not distinguish between neritic and open ocean elements as such. This distinction was made in Finney's (1984) model. It is of interest that the Berry & Boucot (1972) depth zonation rationale might be operable with respect to the open ocean plankton and the continental slope, but it seems less likely to apply to the neritic environment (Fig. 2). There is also a recognisable anomaly concerning Rickards' (1975, *in* Bassett *et al.*) preservation argument, namely that if a real separation of the neritic and open ocean plankton occurred then this model founders for the same reasons as those postulated above for the Berry & Boucot depth zonation idea.

The extent to which modern analogues are relevant to graptolite distributions must be viewed in the light of several differences in the recordings made from each source. Graptolites compose only one element of the Palaeozoic plankton, and their species are defined on different grounds from the species identified in modern plankton. The two may not represent similar levels of relationship, and differences at species level in the modern day might be better represented by sub-species or generic level differences in graptolites. Taphonomic and preservational constraints must also be considered. However, we consider that cautious use of analogies with the modern zooplankton may improve the understanding of graptolite distributions.

Depth zonation

Apart from latitude, the main control advocated for the graptolites has been depth. The idea was introduced into graptolite studies by Ross (1961) and refined by Berry (1962). Erdtmann (1976) related depth zonation ideas directly to major facies associated with the Pacific and Atlantic Provinces: he regarded Atlantic Province graptolite assemblages as being typical of black shales, and Pacific assemblages of carbonate shelf sequences. This is a very approximate generalization and many exceptions to both can be found.

In the more refined depth zonation of Berry & Boucot (1972) for the Silurian, the principle expounded was that upon death the rhabdosomes would drift sideways or downwards, or both, eventually foundering at the depth at which they had lived, or lower. Thus the graptolites occurring in shallower, shelf deposits were those which lived near the surface in the ocean's plankton. It was argued by Rickards (in Bassett *et al.* 1975) that preservational features were difficult to unravel from immediate post-mortem history, and that the forms most common in the offshore, black shale, environments, and those species with most robust rhabdosomes, were those which occurred in the inshore environments. Such occurrences could as easily be explained by chance preservational factors; that is, the chance of a common, robust colony being preserved was actually far greater than that of a rare, delicate species. Calculations to allow for such bias have not been attempted. A much more likely distribution pattern which might, in part at least, result from depth stratification, is one which records presence or absence from a locality: one of exclusiveness. In fact the more convincing of Erdtmann's (1976) distributions are of this kind; but he demotes the role of provincialism itself in both his text and his tables. A question which arises from Erdtmann's work is this: if bathymetry controls the distribution of graptoloids rather than provincialism of Skevington (1974) type, why did major provincialism end after the Llanvirn? Why didn't the supposed bathymetric pattern of the Arenig–Llanvirn period continue? The element of Erdtmann's study which we find most compelling in his assertion that a cool water mass would in places have continued equatorwards at depth. Such a water mass might carry an Atlantic fauna deeper than its normal realm, where it might occur at any depth.

The most striking and authoritative accounts of depth stratification have been argued by Cocks & Fortey (1982), largely with respect to benthos, and by Finney (1984). The latter noted that various post-Llanvirn species were restricted to the inshore shelf regions or to the oceanic regimes. He further postulated depth distributions within the two regimes (1984, fig. 4) as one possible scenario, but preferred the obvious alternative (1984, fig. 5) in which lateral differentiation occurred, characteristic species being confined to particular water masses, such as neritic and open ocean.

Future constraints on analysis of graptolite biogeographical distributions

Analogy with the modern situation leads us to expect a relatively small number of graptolite marker species which define provinces and realms, always mixed with a larger number of endemic forms whose controls are more local or less well understood and the cosmopolitan species which allow accurate biostratigraphical resolution. The difficulty lies in understanding the large scale controls on these realms and provinces, and it is here that a high degree of ambiguity remains.

However, certain constraints can increasingly be applied to graptolite distributions. Thus palaeomagnetic work, such as results in the Scotese reconstructions of this volume, is gradually refining the plots of graptolite data. Global facies analysis (e.g. Boucot *et al.* 1968; Erdtmann 1976; Leggett *et al.* 1981) facilitates recognition and understanding of patterns of graptolite distribution. Further, some of the geochemical studies now being attempted (Wilde & Berry 1982; Berry & Wilde 1978) may result in a more precise recognition of patterns of major biological events in the graptolitic shales. When these constraints are employed, together with the more ambiguous requirements of latitudinal temperature variation, current gyres, alien water masses and neritic–oceanic environments, patterns begin to emerge in graptolite distributions.

Hydrodynamic modelling

There is at least one further approach complimentary to graptolite biogeographic studies which is as yet in its infancy, namely that of hydrodynamic modelling. We have noted that many rhabdosomal shapes occur repeatedly through the Lower Palaeozoic, although of quite different evolutionary origin, and differing in morphological detail (e.g. *Nemagraptus gracilis* and *Sinodiversograptus*). We consider that this represents repeated adaptation to niches which themselves are not necessarily long lived. Identification of such niches, although difficult, will clearly have great impact on interpretations of biogeographic distributions, depth zonation and other parameters affecting planktonic occurrences.

Conclusions

Progress in the past has been hindered not only by the quality of the base maps but by the lack of systematic palaeontological revision of crucial groups of graptolites. Improvements recently in both areas have indicated that the Ordovician provinces are capable of subdivision; whilst the Silurian species distributions suggest that a Pacific (tropical) province has at least two subprovinces, namely the Rheic and Mediterranean. The Rheic Subprovince is probably related to a neritic environment and the Mediterranean Subprovince to a (relatively) cooler high latitude region. Future studies of palaeobiogeographical distribution will be greatly enhanced by comparison with the mechanisms and constraints operating within modern planktonic environments; and by deduction of graptoloid niches and microniches.

References

ANDERSON, J. T. & GARDNER, G. A. 1986. Plankton communities and physical oceanography observed on the Southeast Shoal region, Grand Bank of Newfoundland. *Journal of Plankton Research*, **8**, 1111–1135.

BASSETT, M. G., COCKS, L. R. M., HOLLAND, C. H., RICKARDS, R. B. & WARREN, P. T. 1975. The type Wenlock Series. *Report of the Institute of Geological Sciences 75/13*, i–vi. 1–19.

BERRY, W. B. N. 1960. Correlation of Ordovician graptolite-bearing sequences. *Proceedings of the 21st International Geological Congress, Copenhagen*, **7**, 97–108.

—— 1962. Graptolite occurrence and ecology. *Journal of Palaeontology*, **36**, 185–193.

—— & BOUCOT, A. J. 1972. Silurian graptolite depth zonation. *Proceedings of the 24th International Congress, Montreal*, **7**, 59–65.

—— & WILDE, P. 1978. Progressive ventilation of the oceans — an explanation for the distribution of the Lower Palaeozoic Black Shales. *American Journal of Science*, **278**, 257–275.

BOUCEK, B. 1972. The palaeogeography of Lower Ordovician graptolite faunas: a possible evidence of continental drift. *Proceedings of the 24th International Geological Congress, Montreal*, **7**, 266–272.

BOUCOT, A. J., BERRY, W. B. N. & JOHNSON, J. G. 1968. The crust of the Earth from a Lower Palaeozoic point of view. *In*: PHINNY, R. A. (ed.) *The History of the Earth's Crust*. Princeton University Press, 208–228.

BULMAN, O. M. B. 1964. Lower Palaeozoic plankton. *Quarterly Journal of the Geological Society of London*, **120**, 455–476.

COCKS, L. R. M. & FORTEY, R. A. 1982. Faunal evidence for oceanic separation in the Palaeozoic of Britain. *Journal of the Geological Society, London*, **139**, 465–478.

COOPER, R. A. & FORTEY, R. A. 1982. The Ordovician graptolites of Spitzbergen. *Bulletin of the British Museum (Natural History) Geology*, **36**, 157–302.

DEWEY, J. F., RICKARDS, R. B. & SKEVINGTON, D. 1970. New light on the age of Dalradian deformation and metamorphism in western Ireland. *Norsk Geologisk Tidsskrift*, **50**, 19–44.

EKMAN, S. 1953. *Zoogeography of the Sea*. Sidgwick and Jackson.

ERDTMANN, B.-D. 1976. Ecostratigraphy of Ordovician Graptoloids. *In*: BASSETT, M. G. (ed.) *The Ordovician System: proceedings of a Palaeontological Association symposium, Birmingham, September, 1974*. University of Wales Press and National Museum of Wales, Cardiff, 621–643.

FINNEY, S. C. 1984. Biogeography of Ordovician graptolites in the southern Appalachians. *In*: BRUTON, D. L. (ed.) *Aspects of the Ordovician System*. Palaeontological Contributions of the University of Oslo, **295**, 167–176.

GUEIRARD, S., WATERLOT, G., GHERZI, A. & SAMAT, M. 1970. Sur l'age Llandoverien supérieur a tarannonen inférieur des schistes a Graptolites du Fenouillet, massif des Maures (Var.). *Bulletin Societie Geologique de France*, **12**, 195–9.

HARDY, A. C. 1956. *The Open Sea* (its natural history: the world of Plankton). Collins.

IRVING, E. 1964. *Paleomagnetism*. John Wiley and Sons, New York.
LEGGETT, J. K., COCKS, L. R. M., MCKERROW, W. S. & RICKARDS, R. B. 1981. Periodicity in the early Palaeozoic marine realm. *Journal of the Geological Society, London*, **138**, 167–176.
MCKERROW, W. S. & COCKS, L. R. M. 1976. Progressive faunal migration across the Iapetus Ocean. *Nature*, **263**, 304–305.
PARRISH, J. T. 1982. Upwelling and petroleum source beds, with reference to the Palaeozoic. *Bulletin of the American Association of Petroleum Geologists*, **66**, 750–774.
RICKARDS, R. B. & CHAPMAN, A. in press. Bendigonian Graptolites (Hemichordata) of Victoria. *Memoirs of the National Museum of Victoria*.
—— & STAIT, B. 1984. *Psigraptus*, its classification, evolution and zooid. *Alcheringa*, **8**, 101–111.
ROSS, R. 1961. Distribution of Ordovician graptolites in eugeosynclinal facies in western North America and its palaeogeographic implications. *Bulletin of the American Association of Petroleum Geologists*, **45**, 330–341.
SCOTESE, C. R. 1986. Phanerozoic reconstructions: A new look at the assembly of Asia. *University of Texas Institute for Geophysics Technical Report*, **66**.
——, VAN DER VOO, R. & BASSETT, S. F. 1985. Silurian and Devonian base maps. *In: Evolution and Environment in the Late Silurian and early Devonian*. Philosophical Transactions of the Royal Society, London, **B309**, 57–77.
SKEVINGTON, D. 1973. Ordovician graptolites. *In*: HALLAM, A. (ed.) *Atlas of Palaeobiogeography*. Elsevier Scientific Publishing Company, Amsterdam, 27–35.
—— 1974. Controls influencing the composition and distribution of Ordovician Graptolite faunal Provinces. *In*: RICKARDS, R. B. *et al.* (eds) *Graptolite Studies in honour of O. M. B. Bulman*. Special Papers in Palaeontology, **13**, 59–74.
SMITH, A. G., BRIDEN, J. C. & DREWRY, G. E. 1973. Phanerozoic world maps. *In*: HUGHES, N. F. (ed.) *Organisms and Continents through time*. Special Papers in Palaeontology, **12**, 1–42.
WANG, X. & ERDTMANN, B.-D. 1987. Zonation and correlation of the earliest Ordovician graptolites from Hunjiang, Jilin Province, China. *Bulletin of the Geological Society of Denmark*, **35**, 245–257.
WATERLOT, G. 1961. Contribution a l'etude de la serie stratigraphic gotlandienne et ante-gotlandienne de la vallee du rio Esera (Province de Huesca, Espagne). *Annales Societe geologique du Nord*, **81**, 73–79.
WILDE, P. & BERRY, W. B. N. 1982. Progressive Ventilation of the Oceans — Potential for Return to Anoxic Conditions in the Post-Paleozoic. *In*: SCHLANGER, S. O. & CITA, M. B. (eds) *Nature and Origin of Cretaceous Carbon-rich Facies*. Academic Press, 209–224.
WILLIAMS, A. 1969. Ordovician faunal provinces. *In*: WOOD, A. (ed.) *The Pre-Cambrian and Lower Palaeozoic Rocks of Wales*. University of Wales Press, 117–150.
ZIEGLER, A. & PARRISH, J. T. 1981 Palaeoclimates and climatology — the stability of global atmospheric circulation patterns. *Geological Society of America Abstracts*, **13**, 587.

Cambro–Devonian biogeography of nautiloid cephalopods

REX E. CRICK

University of Texas at Arlington, Arlington, Texas 76019–0049, USA

Abstract: The biogeography of Cambro–Devonian nautiloid cephalopods is documented from their origin during Late Cambrian through to the end of the Devonian. The biogeography developed as the consequence of two related sets of events with different magnitudes. Events controlling the biogeography of first order magnitude were geotectonic, either Gondwana glaciation due to plate movement or contraction in size of oceanic ridges related to decreased rates of sea floor spreading, manifested as first and second order eustatic fall in sea level. These first order events separate nautiloid biogeography into four episodes: (1) Late Cambrian; (2) Ordovician; (3) Silurian through Early Devonian; and (4) Middle through Late Devonian. Each began with an expansion in terms of generic diversity and ended with a crisis of some proportion. Only the end of the Late Cambrian episode cannot be explained by geotectonic events.

Events of the second order are both biologic and geotectonic and involve the interaction between the dispersal characteristics of nautiloid cephalopods and plate movements. Due to lack of a planktonic larva and depth restrictions as a function of shell design, the ability of nautiloids to disperse and colonize shallow shelf seas separated by expanses of ocean or depths exceeding shell design limits was minimal. When the width of oceans or depths decreased through some form of geotectonics, nautiloids rapidly radiated and colonized regions as they became available. Thus second order biogeographic episodes tend to evolve from a collection of disjunct patterns with high faunal similarity among faunas of particular landmasses, but with little similarity among separate landmasses, to rapid development of high similarities both among and within faunas of converging landmasses once the physical barrier of distance and water depth is eliminated.

The purpose of this paper is two-fold: (1) To present the taxonomic, stratigraphic and spatial data for the 811 nautiloid genera known to have lived sometime during the interval of Late Cambrian (Franconian) through Late Devonian (Famennian) in the form of biogeographies at successive age intervals; (2) To compare these biogeographies with the palaeogeographic reconstructions to determine if changes observed in biogeographic patterns through time can be related to, or interpreted as, the result of physical factors controlled by plate movements.

General factors or barriers known to affect the dispersal potential of marine invertebrates were used to interpret the biogeographic patterns of fossil nautiloids. These are: distance separating shallow shelf seas of landmasses, depth of water greater than the implosion limits of the cephalopod phragmocone, palaeolatitude as a general indication of water temperature and salinity, and taxonomic diversity as an indication of palaeolatitude. The emergence or submergence of land barriers and the destruction of shallow shelf seas as a consequence of plate movements are also considered. In the context of the amount of geologic time considered here, such a factor may be illustrative of disjunct patterns after one or more landmasses have joined.

Factors of nautiloid distribution

Cephalopods are often credited with characteristics of distribution that the fossil record simply does not support (Crick 1980; Chamberlain *et al.* 1981). Early reports of 'drifted' shells of *Nautilus* occurring at considerable distances from presumed habitats (e.g. Teichert 1970; Toriyama *et al.* 1965) resulted in the lingering impression that such reports documented the actual post-mortem distribution of shells from the 'typical' region of habitation, the island groups in the Indo-Pacific. Because all cephalopods have the same basic shell design, it is common practice to categorize cephalopods as having had similar patterns of post-mortem dispersal. This clearly is not the case. A body of empirical and experimental evidence exists which documents that shells of *Nautilus* are rarely removed from the region of the parent population by natural circumstance (Saunders & Spinosa 1979; Chamberlain *et al.* 1981). The assumption that all nautiloid cephalopods were equal in terms of post-mortem buoyancy is a gross over-simplification. The majority of Palaeozoic nautiloids designed phragmocones to be counterweighted against positive buoyancy and would have rapidly filled with water and sank after death of the animal (Crick 1988). A detailed analysis of the distributional characteristics of Arenig nautiloids revealed a strong positive relationship between the type of nautiloid shell architecture and the entombing sedimentary environment (Crick 1980). Such a relationship would have been impossible to maintain if such shells were affected by post-mortem distribution. The remainder of the shells of Palaeozoic nautiloids, some coiled as in the tarphycerids, and others without mineralized deposits as in the ellesmerocerids, would have lost positive buoyancy and settled to the bottom because of water influx into the chambered portion of the phragmocone under ambient hydrostatic pressure (Chamberlain *et al.* 1981). Thus the spatial distribution of extinct nautiloids must be considered as representative of their distribution prior to death.

The form of reproduction in fossil nautiloid cephalopods and a life mode as scavengers placed a unique set of controls on dispersal. As such, their distribution was more sensitive to water depth and distance than many other marine invertebrates. The ontogeny of *Nautilus* does not allow for a planktonic larval stage capable of dispersal by oceanic currents, nor is adult *Nautilus* capable of traversing expanses of ocean where water depths exceed 800 m (Saunders & Spinosa 1979). *Nautilus* hatches as a juvenile form of an adult from an egg anchored by the parent in a sheltered, submarine location, thus the newly hatched *Nautilus* is in the same geographic area as its parents. As a vagrant, *Nautilus* is also faced with post-hatching constraints to dispersal in the form of water depths greater than 800 m (Ward *et al.* 1980). Water depth sufficient to implode the shell and lack of planktonic larvae are biogeographic barriers to *Nautilus* and may alone account for its present restriction to regions of the Indo-Pacific.

Evidence that fossil cephalopods were limited by the same biogeographic barriers as *Nautilus* is found preserved in the phragmocone of fossil cephalopods. Complete and well-preserved phragmocones exhibit pre- and post-hatching stages, indicating the absence of a planktonic larval stage (Landman *et al.* 1983). Structural studies of nautiloid phragmocones show that phragmocones of extinct nautiloids would have imploded at even shallower depths than *Nautilus* (Westermann 1973, 1985), and that, unless feeding strategies were vastly different from *Nautilus*, water depths between 300 and 500 m were biogeographic barriers. The distributions of fossil nautiloids illustrated here, and those of Flower (1976) and Crick (1980), indicate that nautiloids were not truly part of the nekton capable of oceanic dispersal, but were members of the shallow-shelf vagrant benthos and were thus capable of dispersal only along continuous or contiguous shelves or over shallow stretches of open ocean. For these reasons, simple distance and the depth of water separating shallow shelf seas was capable of restricting the dispersal of nautiloid cephalopods until such time as the physical environment removed these barriers.

Data, methods, and procedures

Data

The basic framework for the data used here was Moore (1964), but the details of geography and stratigraphy missing from *The Treatise Part K* were added and verified by reference to original works, and assignments were modified when necessary. Additions were made to these data since publication of *Part K* (Moore 1964), using the major works of Balashov (1960, 1968), Barskov (1972), Chen and Teichert (1983a), Crick (1980, 1981, 1988), Ruzhentsev *et al.* (1962), Teichert *et al.* (1979), Zhuravleva (1972, 1974) and supplemented by all stratigraphic and taxonomic data contained in individual systematic works since 1960. A portion of the information contained in these data is summarized in Tables 1 & 2. The number of genera recorded for each age is the total number of genera present and includes those carried over from a preceding age and those that originated within the age. Endemics are those genera which were restricted to one of the landmasses of Figs 1 & 5. The relationship between endemics and total abundance is expected, that is, as abundance declines endemism increases.

The palaeogeographic maps illustrated in Figs 1 & 5 are those of C. R. Scotese, Shell Development Company, Houston. The Arenig map locates all of the general regions used here to depict nautiloid faunas. The composition of these faunas is a composite of localities within the identified regions. The region or fauna of Gondwana passed from Gondwana to Baltica during the Ashgill as a consequence of fragmentation of Gondwana.

Table 1. *Number of genera present on landmasses of Figs 1 and 5 by geological age.* n, number of genera present in age; e, number of endemic genera; %, percentage of endemic genera; LAU, Laurentia; BAL, Baltica; KAZ, Kazakhstania; SIB; Siberia; NC, North China; SC, South China; GON, Gondwana.

	n	e	%	LAU	BAL	KAZ	SIB	NC	SC	GON
Franconian	1	1	100	0	0	0	0	1	0	0
Trempealeauan	40	36	90	3	0	3	1	32	5	0
Tremadoc	55	36	65	37	1	5	14	20	3	3
Arenig	178	120	67	125	37	7	17	34	21	43
Llanvirn	131	76	58	61	34	2	24	50	31	34
Llandeilo	112	70	63	58	40	4	17	26	29	21
Caradoc	135	87	64	93	39	6	18	14	25	29
Ashgill	129	87	67	95	56	5	10	8	9	14
Llandovery	73	52	71	37	34	8	2	4	11	8
Wenlock	165	106	64	88	82	9	2	8	35	27
Ludlow	85	51	60	31	43	11	3	4	14	29
Gedinnian	26	23	88	12	11	3	1	0	0	1
Siegenian	22	19	86	6	8	2	1	0	0	10
Emsian	55	45	82	22	22	9	2	0	0	15
Eifelian	82	62	76	42	46	4	2	0	3	10
Givetian	57	41	72	38	22	2	1	0	1	10
Frasnian	49	40	82	13	32	4	3	0	0	7
Famennian	71	58	82	11	24	40	0	0	2	9

Table 2. *Number of nautiloid genera by order by geological age for Late Cambrian through Late Devonian.* Numbers include genera held over from preceding age and those that originated during the age. Age abbreviations: FRN, Franconian; TRP, Trempealeauan; TRM, Tremadoc; ARN, Arenig; LLV, Llanvirn; LLD, Llandeilo; CAR, Caradoc; ASH, Ashgill; LDV, Llandovery; WNL, Wenlock; LDL, Ludlow; GDI, Gedinnian; SIG, Siegenian; EMS, Emsian; COU, Couvinian; GIV, Givetian; FRS, Frasnian; FAM, Famennian

	FRN	TRP	TRM	ARN	LLV	LLD	CAR	ASH	LDV	WNL	LDL	GDI	SIG	EMS	COU	GIV	FRS	FAM
Endocerida	0	0	15	66	43	24	24	12	3	3	1	0	0	0	0	0	0	0
Intejocerida	0	0	1	3	4	0	0	0	0	0	0	0	0	0	0	0	0	0
Yanheceratida	0	4	0	0	0	0	0	0	0	0	0	0	0	0	0	0	0	0
Actinocerida	0	0	0	9	18	14	17	17	10	10	3	2	1	1	1	0	0	0
Protactinocerida	0	5	0	0	0	0	0	0	0	0	0	0	0	0	0	0	0	0
Ellesmerocerida	1	31	36	56	14	7	6	3	2	1	1	0	0	0	0	0	0	0
Orthocerida	0	0	2	13	21	26	27	26	30	56	43	9	10	8	17	11	11	9
Ascocerida	0	0	0	0	2	4	3	5	2	4	5	0	0	0	0	0	0	0
Oncocerida	0	0	0	0	4	13	29	32	14	47	14	7	7	14	32	8	3	17
Discocerida	0	0	0	1	7	9	18	14	8	27	14	7	2	18	11	19	32	42
Tarphycerida	0	0	1	30	15	11	12	7	1	3	0	0	0	0	0	0	0	0
Barrandeocerida	0	0	0	0	3	4	10	13	3	14	4	1	1	3	3	6	1	0
Nautilia	0	0	0	0	0	0	0	0	0	0	0	1	1	11	17	13	2	1
Totals	1	40	55	178	131	112	135	129	73	165	85	26	22	55	82	57	49	71

Methods

The data were separated into subsets by geologic age and a measure of similarity among faunas was computed using the probabilistic index of similarity (*IS*) of (Raup & Crick 1979). This index is computed as: IS = the probability that k_{exp} will be less than or equal to k_{obs} where k_{exp} is the number taxa expected in common and k_{obs} is the number of taxa observed in common. The index offers the opportunity to rigorously test for significant differences between observed faunal relationships and relationships expected by chance. Other advantages of the index are that all taxa may be used regardless of their distributional characteristics, that all faunas may be used regardless of size, and that faunal similarities of the type usually explained as being the result of sampling error can be tested against a null hypothesis of randomness. If the null hypothesis is accepted, no statistical significance can be assigned these associations. If, however, the null hypothesis is rejected, the probability that the similarity of the faunas being tested is attributable to chance is less than some previously determined level of confidence (95% was used). In practice, the position at which the observed number of shared taxa falls within the distribution of the expected number of shared taxa provides information about the degree to which the observed association rejected or approached rejection of the simulation mode, and this information is used to produce a measure of similarity between faunal pairs (Raup & Crick 1979). Each entry in a similarity matrix represents the relationship between the total population, the size and composition of the faunas, and the number of taxa shared between these faunas. Similarity values range between 1.0 and 0.0 and correspond to the probability that the expected number of taxa in common between two faunas will be less than or equal to the observed number in common. Values equal to or greater than 0.95 indicate associations where the observed number in common is equal to or greater than the expected number in common. Because the observed number in common is significantly large, such associations are significantly similar, which is to say that one or more biogeographic factors functioned to make the faunas more similar than expected. Similarly, values equal to or less than 0.05 indicate associations where the observed number in common is equal to or less than the expected number in common, and because the observed number is significantly small, the faunas are significantly dissimilar for the same reasons given for significantly similar faunas. The program is particularly sensitive to sample size and the frequency characteristics of each taxon. For this reason, faunal pairs of equal size and with the same number of taxa in common, but not the same taxa, may exhibit different *IS* values due to differences in the frequency characteristics of each of the shared taxa. Details and examples of the use of this index are given in Raup & Crick (1979) and Crick (1980).

Although the data can be entered into a host of multivariate techniques designed to reduce the dimensionality of the data to meaningful proportions, the pattern of significantly similar and significantly dissimilar faunal pairs are presented here in the form of matrices (Figs 2, 3, 4, 6 & 7). Faunal pairs which did not receive a plus (+) or minus (−) symbol have less than significant faunal similarities or dissimilarities corresponding to probabilities between 0.95 and 0.05, but are important in determining overall patterns of similarity or dissimilarity.

Nautiloid biogeography

Franconian

The biogeography of Franconian nautiloids is simple and unique in the sense that it begins with the first appearance of nautiloid cephalopods in the region of modern China corresponding to the North China landmass of Fig. 1A. The documentation for the first appearance of the Cephalopoda in China is detailed in several papers by Chen & Teichert (1983*a*, *b*) and Chen *et al.* (1983) and need not be discussed here. It is sufficient to relate that Franconian strata contain no more than three or possibly four species of the ellesmerocerid genus *Plectronoceras* reported from areas which in the Late Cambrian comprised the most equatorial portion of North China (Fig. 1A). All occurrences are within the *Ptychaspis–Tsinania* Zone.

Trempealeauan

In the span of roughly 5 million years from the first appearance of nautiloids, the generic diversity of the group increased from 1 genus to 40 genera (Table 1). The largest generic increase occurred in North China and the first occurrence of nautiloids in the shelf seas of Laurentia, Kazakhstania, Siberia, and South China (Fig. 2A). The regions of the southwest US (U4), Kazakhstania (KZ) and South China (SC) were composed of endemic genera. North China (NC) shared three genera with Inner Mongolia (IM) and one genus with Manchuria (MC). The faunal connection between North China and Inner Mongolia was significantly similar (> 95%) (Fig. 2A). While the faunal similarity among North China and Manchuria was high but not significant (< 95%), the absence of other faunal connections and the fact that *Plectronoceras*, the genus in common, originated in North China suggests that North China, Manchuria and Inner Mongolia should be considered part of the same cephalopod province. The faunal relationship among South China and the other regions or localities was significantly dissimilar (< 5%). The coexistence of two faunas of the size of North China and South China in the proximity suggested by Fig. 1A without shared taxa indicates the existence of some form of major barrier to interchange.

Tremadoc

Cephalopods are notably absent from sediments deposited in the latest Cambrian corresponding to the Fengshan in China and the latest Trempealeauan in North America. Two genera, *Ectenolites* (KZ) and *Clarkoceras* (SC), survived from pre-Fengshan faunas and reappeared in Tremadoc faunas. Consequently, nautiloid faunas of the Tremadoc were compositionally much different from those of the Trempealeauan. The Proactinocerida and Yanhecerida did not survive to the Tremadoc, the Plectronoceratidae were reduced to a few genera, and the Endocerida, Tarphycerida, Intejocerida and Orthocerida appeared (Table 2). Without these considerations, Table 1 is misleading in that it shows only a slight increase in generic diversity between the Trempealeauan and the Tremadoc, but actually reflects the appearance of four new orders and 53 new genera. The oldest ellesmerocerids had the highest diversity and their endemicity was correspondingly low. Endocerids, which largely replaced ellesmerocerids in later faunas, expanded rapidly with uncharacteristically low endemicity. Cephalopods occurred in most of the shallow, low-latitude seas of Laurentia, Kazakhstania, Siberia and made their first appearance on the northern Russian platform (R1) of Baltica and in the eastern Australia (A1) and South America (SA) faunas of Gondwana (Figs 1B & 2B). This expansion, radiation and low endemicity was controlled, at least in part, by a eustatic rise of sea level (Vail *et al.* 1978; Harland *et al.* 1982).

The more ancestral ellesmerocerids dominated the faunas of North China and Laurentia, while the newly evolved endocerids occur on all landmasses. With exception of one endocerid, the connection between Laurentia and North China was all ellesmerocerid, while the connection between Laurentia and Siberia was mostly endocerid except for two ellesmerocerids. Endocerids were also responsible for faunal connections between Kazakhstania, Siberia, Baltica, and Northern Gondwana with assistance of a few ellesmerocerids. One ellesmerocerid has been reported from eastern Australia (A1).

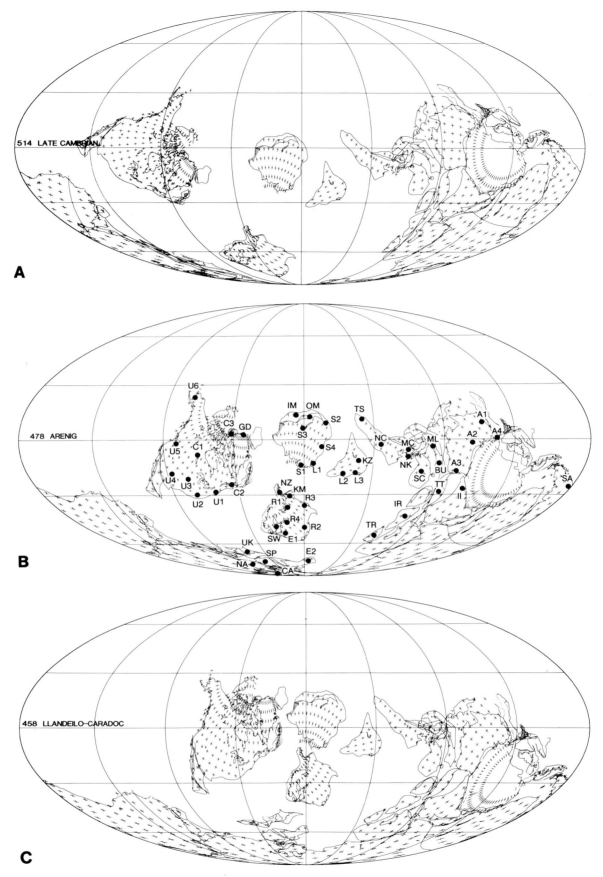

Fig. 1. Cambrian and Ordovician palaeogeography. (A) Late Cambrian (Franconian & Trempealeauan); (B) Early Ordovician (Arenig) with location of faunas; (C) Middle Ordovician (Llandeilo & Caradoc). Identification of faunas: U1, northeastern US; U2, southeastern US; U3, central US; U4, southwestern US; U5, northwestern US; U6, Alaska; C1, central Canada; C2, Maritime Canada; C3, Arctic Canada; GD, Greenland; R1, northern Russian platform; R2, southern Russian platform; R3, eastern Russian platform; R4, western Russian platform; KM, Komi; NZ, Novaya Zemyla; L1, northern Urals (eastern flank); L2, central Urals (eastern flank); L3, southern Urals (eastern flank); KZ, Kazakhstania; S1, northern Siberia; S2, southern Siberia; S3, eastern Siberia; S4, western Siberia (exclusive of Urals); OM, outer Mongolia; IM, inner Mongolia; MC, Manchuria; TS, Tien Shan; NK, north and south Korea; NC, north China; BU, Burma; ML, Malaya; II, India; A1, eastern Australia; A2, central Australia; A3, western Australia; A4, Tasmania; IR, Iran; TR, Turkey; E1, central Europe; E2, Amorica; SW, Sweden; UK, southern Britain; SP, Iberia; NA, north Africa; CA, central Africa; SA, southern South America;

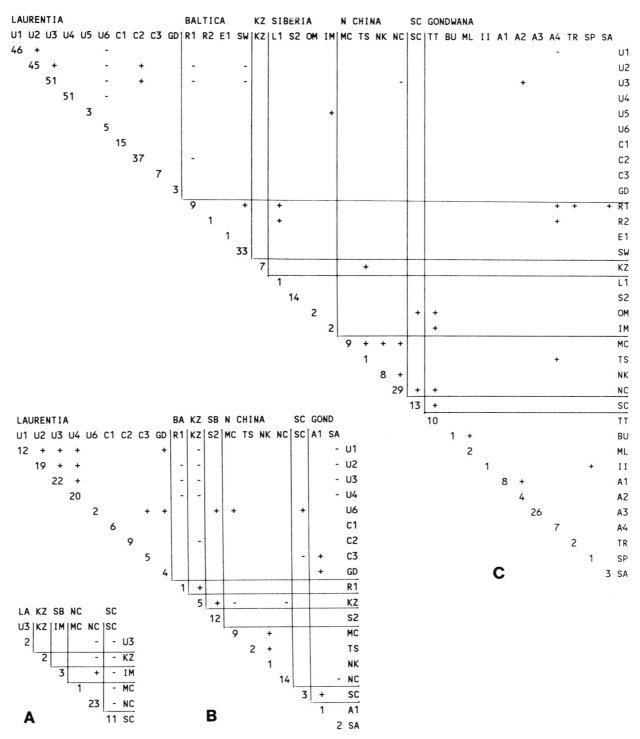

Fig. 2. Matrices of significantly similar and dissimilar faunal pairs for Trempealeauan (**A**), Tremadoc (**B**) and Arenig (**C**). Plus (+) symbol indicates significantly similar pair and minus (−) symbol indicates significantly dissimilar pair. Fauna diversity given on diagonal. Landmass abbreviations: LA, Laurentia; KZ, Kazakhstania; SB, Siberia; NC, North China; SC, South China. Fauna abbreviations are explained in caption to Fig. 1.

Fig. 2B illustrates the pattern of significantly similar and dissimilar faunal pairs. Although diversity is low, the pattern is one that would be expected. Generally, faunas within landmasses are more similar than those among different landmasses. Where this is not the case, important biogeographic information is indicated by significant dissimilarity or similarity among faunas. Most of the faunas of Baltica and Kazakhstania and the South America (SA) fauna of Gondwana are significantly dissimilar with Laurentia indicating faunal isolation. The significantly similar faunas of Siberia (southern Siberia (S2)) and Laurentia (Alaska (U6)) is less difficult to explain than the significant similarity between the eastern Australia (A1; fauna of Gondwana and Alaska (U6), and the Manchuria (MC) fauna of North China and Alaska (U6) (Fig. 2B). The significant similarity between Kazakhstania, Baltica and Siberia and the significant dissimilarity between Kazakhstania and two regions of North China suggests that a biogeographic barrier was operable between North China and Kazakhstania at this time.

Thus as Tremadoc cephalopods began to diversify, most of the radiation was within landmasses as 90% of the taxa were confined by biogeographic barriers to new regions or those occupied during the Trempealeauan. Older forms were able to colonize new regions or to develop faunal connections with existing faunas of other landmasses. The faunas with the highest diversity and those where one or two genera were able to establish connections over considerable geographic difference were located in regions known to have been equatorial to subtropical. These distributions compliment the palaeogeography of Fig. 1.

Arenig

It was during the Arenig that nautiloid generic diversity peaked at 178 genera (Table 1) with the result that nautiloid faunas were much more diverse (Fig. 2C) and all groups more abundant than during the Tremadoc (Table 2). The level of endemicity remained near the level of the Tremadoc.

The oldest group, the Ellesmerocerida, were replaced in numerical superiority by the rapidly diversifying endocerids in all faunas except Gondwana (eastern Australia (A1)) where ellesmerocerids were expanding after their appearance during the Tremadoc. The Actinocerida appeared with a distribution and low endemicity like the Endocerida of the Tremadoc. The orthocerids and tarphycerids diversified and radiated to become biogeographically important. Endocerids, actinocerids, and tarphycerids were the least endemic and thus more important in a biogeographic sense than ellesmerocerids or orthocerids both of which are largely endemic to Laurentia. The number of regions containing nautiloids increased as did the diversity of most faunas. The most notable additions and increases occurred in Siberia, Baltica and Gondwana.

In 1980, Crick (1980) reported 186 nautiloid genera for the Arenig. The difference of 8 genera is the result of the redefinition of lowermost Arenig strata as uppermost Tremadoc, and of uppermost Arenig strata as lowermost Llanvirn. Other aspects of the data have changed as a result of the addition of new genera, especially from North and South China, the redefinition of regions, from changes in taxonomic assignments and from increases and decreases in stratigraphic ranges. The palaeogeography (Fig. 1B) has changed from an earlier Scotese Arenig reconstruction in that North and South China were moved from moderately high northern latitudes to southern low latitudes.

The rapid expansion and radiation of many Arenig orders had a normalizing affect on nautiloid biogeography. Significantly dissimilar faunas were few and the number of significantly similar faunas were reduced within landmasses but increased among landmasses, a pattern which coincides with greater proximity of several landmasses and the radiation of the endocerids. The latter may have been dependent to some degree on the greater proximity of landmasses. Within Laurentia, Alaska (U6) is significantly dissimilar with four regions, a pattern that may reflect the early stages of the first order eustatic fall in sea level before the end of the Arenig (Vail *et al.* 1978; Harland *et al.* 1982). Laurentia remained faunistically distinct from Baltica, and shows only significant similarity with Siberia (Inner Mongolia (IM)) and Gondwana (central Australia (A2)). Faunas of North China were all dissimilar with Laurentia and the major North China fauna (NC) was significantly dissimilar with the central US (U3) fauna (Fig. 2C). Baltica became less similar with Kazakhstania and significantly similar with northwestern Siberia (L1) which works well with the palaeogeography of Fig. 1B. The significant similarity of the northern and southern faunas of the Russian platform (R1 & R2) with Tasmania (A4) and southern South America (SA) would have been difficult but not impossible to achieve. The map projection distorts the distance between Tasmania (A4) and southern South America (SA). The signficantly similar relationship between the northern Russian platform (R1) and Turkey (TR) on Gondwana is in keeping with the reconstruction. The development of significantly similar associations among Siberia, South China, North China and a portion of Gondwana (Tarim-Tibet (TT), Tasmania (A4)) suggests that conditions were correct for exchange between the equatorial regions and regions of higher southern latitude. Such similarities are in keeping with the continued movement of Gondwana into lower southern latitudes.

Llanvirn

The Llanvirn marked the last major ordinal expansion within nautiloid cephalopods and included the Ascocerida, a biogeographically unimportant group, the Barrandeocerida of questionable importance, and the Oncocerida of later importance to Silurian and Devonian faunas (Table 2). Following the eustatic rise in sea level, endemicity was low (58%) and although losses occurred in Laurentia and Gondwana faunas, all landmasses except Kazakhstania contained reasonably diverse faunas (Table 1, Fig. 3A).

The general trend of Llanvirn biogeography was that of the isolation of Laurentian faunas with a growing similarity among Siberia, North China, South China and the proximal faunas of Gondwana (Tarim−Tibet (TT) & Burma (BU)). Laurentian faunas show significant similarity among many Laurentia faunas, and with exception of the significant similarities with two regions, Kazakhstania (KZ) and Tien Shan (TS) in North China, Laurentia was at a peak of isolation (Fig. 3A).

Llandeilo

The overall diversity of the Llandeilo fauna decreased but much of this loss occurred in North China (Fig. 3B) and endemism increased slightly. The genera lost were largely from within the older groups of endocerids and ellesmerocerids. This loss was offset somewhat by expansion within the orthocerids and oncocerids (Table 2).

In general, the Llandeilo biogeography documents an increase in significantly similar faunas among most landmasses and a corresponding decrease in significantly dissimilar faunas. This was presumably the result of continued eustatic rise in sea level combined with increased proximity of most landmasses. The history of the groups involved shows that it is the older more established genera which were responsible for increased faunal connections. The most notable exceptions to this trend are the significants dissimilarities among the major Baltic fauna of Sweden (SW) and several Laurentia faunas (southeastern US (U2), northwestern US (U5) & central Canada (C1)). The expected number of genera in common among Sweden and the Laurentia faunas is greater than the observed number in common suggesting the existence of a controlling factor on dispersal. Siberia, Baltica, North China and South China continue strong faunal relationships with the Tarim−Tibet (TT), Burma (BU), and Malaya (ML) faunas of Gondwana. The North China faunas are significantly similar with South China and these two landmasses were essentially one biogeographic region during this time.

Caradoc

The number of Caradoc nautiloid genera increased to Llanvirn levels while endemism remained essentially unchanged (Table 1). Expansion was mostly within the oncocerids and discosorids while the other orders remained essentially unchanged (Table 2). Endemism was low among all major groups with exception of discosorids, and endocerids, actinocerids, orthocerids and oncocerids dominated Caradoc faunas. Changes in the size and composition of nautiloid faunas and the high percentage of endemism within the discosorids resulted in an increase in provincialism signified by a decrease in significantly similar faunal pairs (Fig. 4A). The general biogeography for this time shows that the same positive or negative aspects developed during the Llandeilo were carried forward into the Caradoc. A few of the localities within the landmasses changed and there is less similarity within most landmasses.

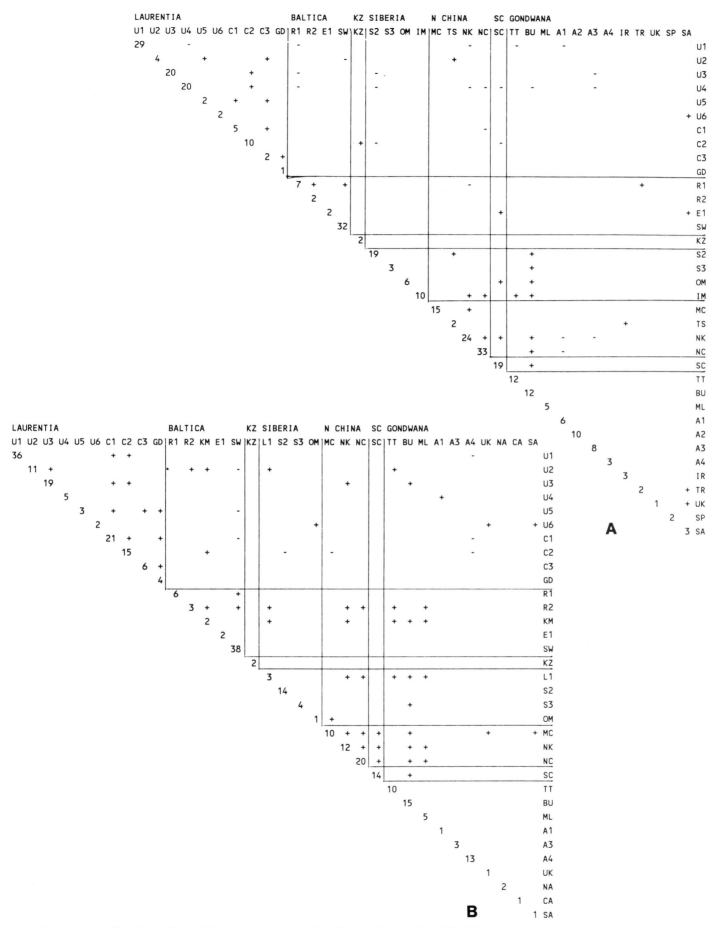

Fig. 3. Matrices of significantly similar and dissimilar faunal pairs for Llanvirn (**A**) and Llandeilo (**B**). See Fig. 2 for an explanation of symbolism and abbreviations.

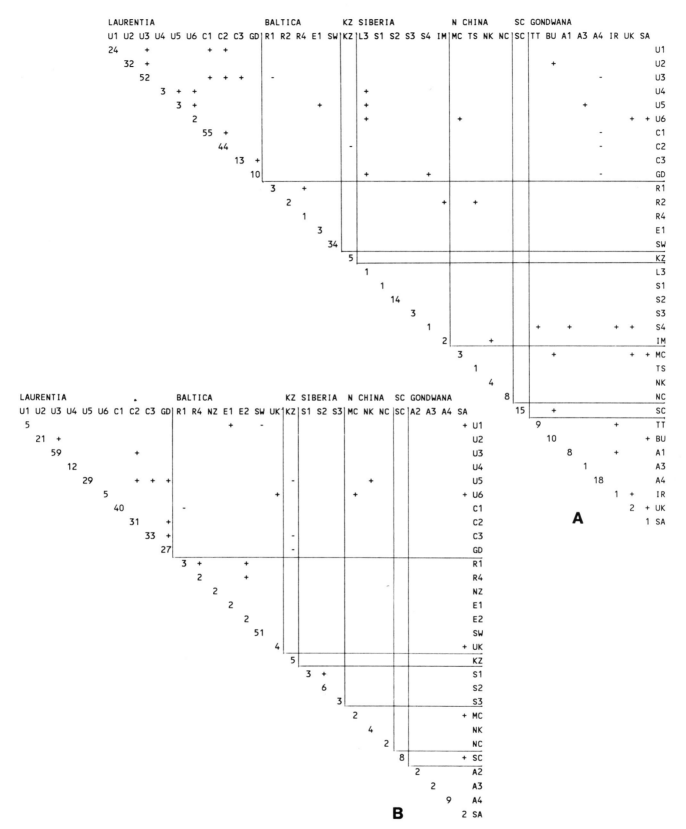

Fig. 4. Matrices of significantly similar and dissimilar faunal pairs for Caradoc (**A**) and Ashgill (**B**). See Fig. 2 for an explanation of symbolism and abbreviations.

Ashgill

The overall diversity of Ashgill faunas remained essentially unchanged from that of the Llandeilo (Table 1). If the late Ordovician glaciation and resulting second order drop in eustatic sea level (Harland *et al.* 1982; Jaanusson 1979, 1984) had an effect on nautiloid diversity, it was mixed among the various groups (Table 2). Endocerids and tarphycerids lost approximately 50% of their genera while the remaining orders increased or decreased slightly in genera. Endemism increased slightly over the Caradoc level. The number of genera per landmass remained essentially unchanged for Laurentia, increased for Baltica, including the addition of the fragment of Gondwana recognized as a portion of the British Isles (UK) (Figs 4B & 5A). The remaining faunas declined in numbers, with North China, Siberia and Gondwana affected the most. The increase in Baltica faunas and the decrease in North China, Siberia and the Australian and South America faunas of Gondwana are consistent with the movement of Baltica into low latitudes and North China, Siberia and a portion of Gondwana into higher latitudes (Fig. 6A). The provinciality of Kazakhstania increased, particularly with respect to Laurentia.

Llandovery

All nautiloid orders experienced a decline in diversity at the close of the Ordovician and Llandovery nautiloid faunas were the least diverse and most endemic since the Tremadoc (Tables 1 & 2). The presumed cause of the decline was the Late Ordovician glaciation and associated second order fall in sea level. The endocerids, ellesmerocerids and tarphycerids, three of the most important components of Ordovician faunas, survived to the earliest Silurian as only a few genera. All remaining orders declined in number of genera except the orthocerids (Table 2). The biogeography of Llandovery nautiloids largely reflects the distribution of orthocerids and associated actinocerids and oncocerids.

As might be expected, the biogeography of Llandovery nautiloids was one of higher than normal provincialism. The growing similarity between Laurentian and Siberian faunas compliments the proximity of Siberia and Laurentia. Although Laurentia and Baltica were essentially the same landmass at this time (cf Figs 7A & 7B), the historical aspect and endemic nature of their faunas plus the developing barriers to migration combined to maintain separate identities into the Devonian.

Wenlock

Nautiloid faunas recorded an impressive expansion and a general decrease in endemism during this period of widespread carbonate shelf seas, particularly in Laurentia and Baltica (Table 1). Expansion affected all orders, but while holdovers from the Ordovician, endocerids, ellesmerocerids, tarphycerids, actinocerids, and ascocerids remained constant in diversity, orthocerids, oncocerids, discosorids and barrandeocerids doubled or tripled in diversity (Table 2). As a result, Wenlock nautiloid biogeography largely reflects the distributional characteristics of the orthocerids, oncocerids and discosorids.

Wenlock nautiloid biogeography continued much the same as that of Llandovery but with higher than normal numbers of shared taxa and the wide distributions of many taxa reduced the number of significant faunas and eliminated significantly dissimilar faunas. Laurentia was much more similar with Baltica than at any time, although only one faunal pair is significantly similar, southern Britain (UK) and northeastern US (U1) (Fig. 6B). The Siberian fauna of the western Russian platform (R4) was significantly similar with Kazakhstania (KZ) and supports the suggested movement of Kazakhstania into closer proximity with Baltica. The western Russian platform (R4) fauna was also significantly similar with the Amorica (E2) fauna of Gondwana as it should have been to reflect the movement together of Gondwana, Laurentia and Baltica. There was sufficient faunal exchange between Sweden (SW) of Baltica and western Australia (A3) of Gondwana to develop a significant similarity among the faunas. This may mark the continued fragmentation between Tarim–Tibet (TT) and western Australia (A3) (Fig. 5A).

Ludlow

In general, nautiloid diversity during the Ludlow decreased to near Llandovery levels but with low endemism (Table 1). The groups which recorded the greatest expansions during the Wenlock, orthocerids, oncocerids, discosorids and barrandeocerids, also experienced the greatest reductions in diversity, a trend which continued through to the Siegenian (Table 2). Endocerids, ellesmerocerids and ascocerids made their last appearance and the actinocerids and barrandeocerids were reduced to a few genera. Losses to the orthocerids and oncocerids were mostly endemic genera and as a consequence levels of endemism for these two groups fell to their lowest levels. The result was a biogeography which was largely dependent on the distribution of the orthocerids and oncocerids and associated discosorids.

Wenlock biogeography continued the trends established during the Llandovery and Wenlock but with fewer faunas. Most faunas were reduced to pre-Wenlock proportions except for that of Gondwana (Amorica (E2), North Africa (NA)) which continued to increase in diversity, a response consistent with its continued movement into lower latitudes. Besides the normal significant similarity among faunal pairs within landmasses, Sweden (SW) rather than southern Britain (UK) was significantly similar with northeastern US (U1), and central Europe (E1) became significantly dissimilar with the northeastern US, indicating perhaps the continued northward rotation of Baltica relative to Laurentia. Although Gondwana and Laurentia continued to move toward one another (Figs 5A & B), the faunas of Laurentia and that of North Africa (NA) were dissimilar to significantly dissimilar (northeastern US (U1) and Maritime Canada (C2)) suggesting that distance and water depth were active biogeographic barriers to strong faunal exchange between Laurentia and this portion of Gondwana. Maritime Canada (C2) became significantly similar with western Australia (A3) of Gondwana, and Sweden (SW) continued its significant similarity with western Australia (A3). The presumed reasons are those given above for the Wenlock. The significant similarity among the faunas of the western Russian platform (R4) and Amorica (E2) continued from the Wenlock as the plates converged. Kazakhstania remains similar to Baltica but the western Russian platform (R4) fauna became significantly similar with that of southern Siberia (S2). Ludlow nautiloid biogeography fits well with the reconstructions of Figs 5A & B.

Gedinnian

Near the end of the Ludlow and early in the Gedinnian, nautiloids and most other marine groups experienced survival crises (Teichert 1986), events which almost certainly reflect the successive second and first order drops in eustatic sea level which occurred at this time (Vail *et al.* 1978; Harland *et al.* 1982). These changes resulted in a loss of 70% of the nautiloid genera and, while diversity would eventually rise to almost Ludlow levels during the Eifelian, nautiloid cephalopods would not regain their prominence in marine faunas (Table 1; Fig. 7A). Endemism was the highest since the Trempealeauan, many faunas were eliminated and all surviving faunas were reduced in diversity. Five of eight nautiloid orders were either eliminated or reduced to one or two genera (Table 2).

Although few faunal pairs are significantly dissimilar, Gedinnian biogeography consists in general of three provinces. Orthocerids dominated Laurentia and the Amorica (E2) fauna of Gondwana, oncocerids were dominant in Baltica and the discosorids are most common in Kazakhstania and Siberia.

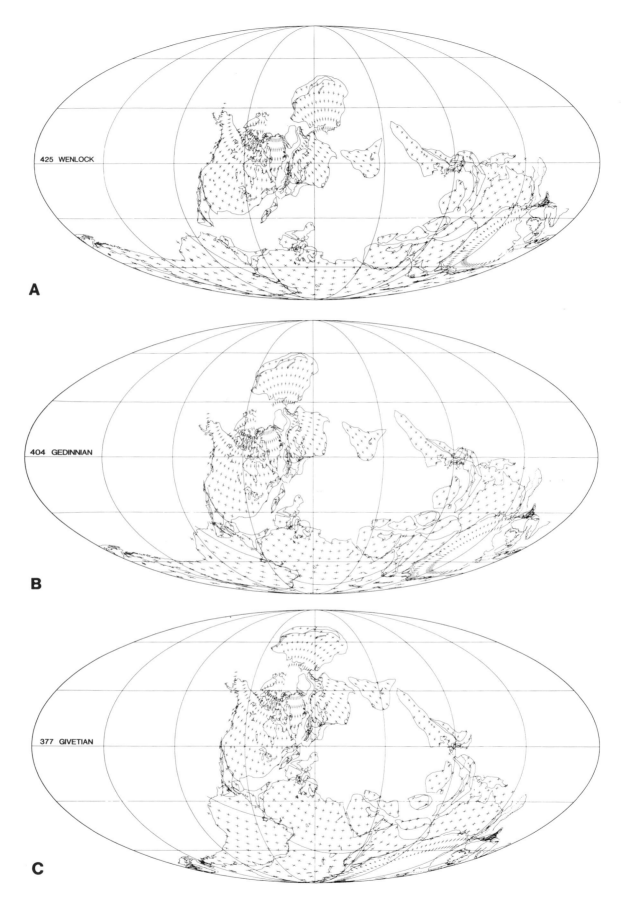

Fig. 5. Silurian and Devonian palaeogeography. (**A**) Middle Silurian (Wenlock); (**B**) Early Devonian (Gedinnian); (**C**) Middle Devonian (Givetian).

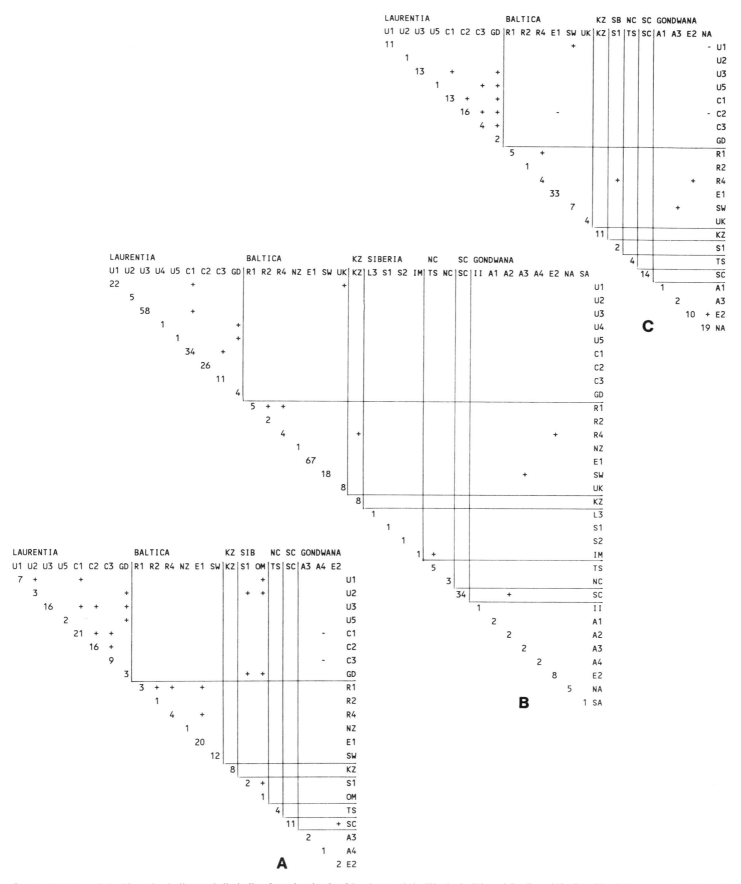

Fig. 6. Matrices of significantly similar and dissimilar faunal pairs for Llandovery (**A**), Wenlock (**B**) and Ludlow (**C**). See Fig. 2 for an explanation of symbolism and abbreviations.

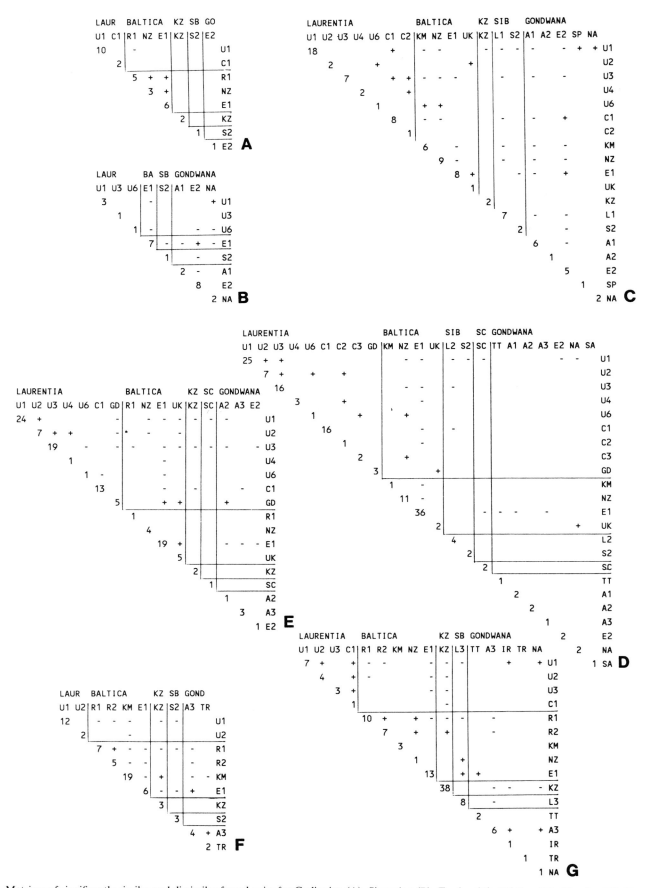

Fig. 7. Matrices of significantly similar and dissimilar faunal pairs for Gedinnian (**A**), Siegenian (**B**), Emsian (**C**), Eifelian (**D**), Givetian (**E**), Frasnian (**F**) and Famennian (**G**). See Fig. 2 for an explanation of symbolism and abbreviations.

Laurentia was significantly dissimilar with the fauna of the northern Russian platform (R1) (Fig. 7A) emphasizing the movement of Baltica into higher latitudes and away from the lower portion of Laurentia (Fig. 5B). Within Baltica the faunal associations are significantly similar as it maintains a characteristic oncocerid fauna. The regions in which nautiloids completely disappeared at the close of the Silurian were, for the most part, inland areas and their loss corresponds very well with the lowest stand of sea level since the Early Cambrian (Harland *et al.* 1982).

Siegenian

Faunal diversity decreased slightly in the Siegenian to reach its lowest point of 22 genera since the Franconian (Table 1). The reduction was confined to the Discosorida except for one actinocerid genus and resulted in a general level of endemism slightly higher than the Gedinnian (Table 2). The Nautilida made their first appearance, but were unimportant to Siegenian biogeography. Faunas of Laurentia and Baltica were again reduced in diversity and Gondwana faunas increased in number and diversity (Fig. 7B).

The number of significantly dissimilar faunas increased (Fig. 7B) and emphasizes the decrease in diversity and resulting endemism. The pattern of similar and dissimilar faunal pairs in Fig. 7B does, however, make biogeographic sense relative to the paleogeography of Fig. 5B. Baltica remains dissimilar with two of the three Laurentia faunas. The northeastern US (U1) fauna was significantly similar with North Africa (NA) as Gondwana continues to move together with Laurentia, and the fauna of Alaska (U6) was significantly dissimilar with both Amorica (E2) and North Africa (NA). The closure of the seaway between Laurentia and Gondwana in the Late Silurian or Early Devonian served as a barrier to migration for taxa along the western portion of Laurentia and the northern regions of Gondwana. The one remaining fauna in Baltica, central Europe (E1), was significantly dissimilar with Siberia and the Gondwana faunas of eastern Australia (A1) and North Africa (NA). Baltica is, however, significantly similar with the Amorica (E2) fauna as Gondwana converges with Baltica.

Emsian

The Emsian marked the beginning of a minor period of expansion for nautiloids which was to last through the Eifelian. The expansion was almost exclusively within the Nautilida, Discosorida, and Oncocerida in order of largest increase, but discosorids were the most abundant due to holdover genera from the Siegenian (Table 2). Endemism decreased slightly with a corresponding increase in the number of regions containing nautiloids.

The pattern of significantly similar and significantly dissimilar faunas (Fig. 7C) shows the effects of increased isolation of regions as Pangaea continued to assemble. This is particularly well illustrated by the relationship of the northeastern US (U1) fauna of Laurentia with those of Baltica, Kazakhstania, Siberia, and Gondwana. The significant dissimilarity between northeastern US (U1), central US (U3) and central Canada (C1) faunas of Laurentia and Komi (KM) and Novaya Zemyla (NZ) faunas of Baltica works well with the palaeogeography of Fig. 5B where Komi and Novaya Zemyla were isolated from most of Laurentia by distance and plate configuration. The significant similarity among Alaska (U6) and Komi and Novaya Zemyla is reasonable given proximity and latitude. The significant similarity among southeastern US (U2) and southern Britain (UK) is also expected. The significantly dissimilar relationships between Kazakhstania and Siberia and northeastern US (U1) is expected by distance of separation and latitude. The same holds for the significant dissimilarity between northeastern US (U1) and eastern Australia (A1). These factors do not account for the significant dissimilarity between faunas of the northeastern US and central US and Amorica (E2) of Gondwana with reasonable proximity but no taxa in common. Amorica (E2) is also significantly dissimilar with faunas of Komi and Novaya Zemyla of Baltica (expected) and significantly similar with central Europe (E1), also expected. The significant dissimilarity of Amorica (E2) with northwestern Siberia (L1) and southern Siberia (S2) and central Australia (A2) is not surprising given the palaeogeography. The significantly similar relationship between the faunas of northeastern US (U1) and those of Iberia (SP) and North Africa (NA) emphasizes again the increased proximity of these regions. The remaining pattern of significantly similar and dissimilar faunal pairs make biogeographic sense in the context of the palaeogeography.

Eifelian

Nautiloids continued to expand during the Eifelian (Table 1) and increases were most pronounced within oncocerids, orthocerids and nautilids (Table 2). Endemism continued to decline with that of orthocerids returning to Wenlock proportions. Diversity increase was greatest among Laurentia and Baltica faunas (Fig. 7D; Table 1) while other faunas remained static or declined in diversity. South China recovered with a small oncocerid and nautilid fauna. The geographic range of discosorids was restricted to Laurentia and Baltica, consequently the oncocerids, orthocerids and nautilids were the important biogeographic groups.

The pattern of significantly similar and dissimilar faunas of Fig. 7D shows far fewer dissimilar faunas than Fig. 7C for the Emsian. This can be attributed to the increased diversity and decreased endemicity of Eifel faunas resulting in increased similarity among most regions. The significant dissimilarity between northeastern US (U1) and North Africa (NA) suggest that the expected number of taxa in common was too small and reflects in part the small size of the North Africa fauna. New material which has just become available and has not been studied in detail (H. Mutvei, Riksmuseet, Stockholm) indicates that the North Africa fauna is much more diverse, but it is not known if the new material will provide a closer link with Laurentia. If the palaeogeography of Fig. 5C is accurate, then there was little to separate the two regions during Eifelian time.

Givetian

Nautiloids experienced another reduction in genera (Table 1) as a result of a 75% decrease in the number of oncocerid genera followed by more modest reductions of 35% and 25% within orthocerids and nautilida respectively (Table 2). The decreases were offset somewhat by a 43% increase in discosorid genera and a 50% increase in barrandeocerids. The level of endemism dropped slightly and can be attributed for the most part to dispersal of discosorids. The single largest change in faunas was the loss of oncocerids from Baltica and the beginning of the dominance of discosorids in Baltic faunas. The fauna size of Kazakhstania and Siberia continued to decline in diversity although orthocerids were replaced by nautilids in the Siberia. The number of Gondwana faunas were reduced.

The most obvious difference between Eifelian and Givetian nautiloid biogeography is the development of significantly dissimilar faunas within Laurentia. Alaska (U6) and Greenland (GD) are dissimilar with most other Laurentia faunas, and must reflect the isolation of these two faunas from the remainder of Laurentia. Greenland remains significantly similar with southern Britain (UK) and becomes significantly similar with central Europe (E1). The remainder of the biogeography is similar to that of the Eifelian with a few changes in significant pairs.

Frasnian

The decline in the number of nautiloid genera continued into the Frasnian and endemism increased (Table 1). This decline is attributed to a second order fall in eustatic sea level which occurred at or near the Givetian–Frasnian boundary (Harland et al. 1982). The result is the continued reduction of oncocerids, the near disappearance of barrandeocerids and an 85% loss of nautilid genera (Table 2). The orthocerids remained static and discosorids appear to have been little affected as they continued to diversify. The near extinction of oncocerids eliminated the South China fauna and left Baltica without oncocerids since the Llandeilo. The Laurentia faunas were reduced to Gedinnian proportions with reductions in all orders and the elimination of nautilids. Kazakhstania and Siberia recovered slightly and Baltica increased due to expansion of discosorids.

The percentage of significantly dissimilar faunas increased sharply. A few new significant associations developed as Pangaea continued to assemble and a portion of Gondwana continued to fragment (Fig. 5C & 7F). Kazakhstania becomes significantly similar for the first time with the Komi (KM) fauna of Baltica, western Australia (A3) of Gondwana becomes significantly similar with central Europe (E1) of Baltica and with Turkey (TR) of Gondwana. The latter associations are reasonable and presumably reflect the development of open marine conditions in the region of western Australia as the Tibet–Tarim block continues to move northward.

Famennian

The total number of nautiloid genera increased by 45% while the percentage of endemics remained the same (Table 1). The increase in diversity occurred within the oncocerids (more than 5 fold) and the discosorids (32%). Orthocerids declined slightly, nautilids were reduced to one genus and barrandeocerids disappeared (Table 2). The most unusual aspect of Famennian diversity is the order of magnitude increase in Kazakhstania fauna (Fig. 7C). This is most likely the result of three factors: intensity of collecting, taxonomic splitting (Teichert et al. 1979), and a real increase in diversity. While the fauna may not have been as diverse as suggested in Fig. 7G, cephalopods of this age were more common in sediments preserved on this landmass than at previous times (Zhuravleva 1972, 1974).

Nautiloid biogeography for the Famennian differs from that of the Frasnian by having a higher percentage of significantly similar faunas and a lower percentage of significantly similar faunas. The biogeography is more like that of the Emsian, with faunas on particular landmasses exhibiting greater similarity within landmasses than between landmasses, that is, a high level of provincialism. Laurentia continued its general pattern of similarities and dissimilarities among other landmasses. The fauna of central Europe (E1) continued to be dissimilar with the other Baltica faunas, but became significantly similar with the faunas of the southern Urals (L3) of Kazakhstania and of Tibet–Tarim (TT) of Gondwana. Within Gondwana, western Australia (A3), North Africa (NA) and Iran (IR) became significantly similar.

Conclusions

The biogeography reported here is far from perfect, but it does show that the distributions of Palaeozoic nautiloid cephalopods were particularly sensitive to distance or water depth separating landmasses. This is particularly clear at times when faunas on different landmasses are known to have been in close proximity but separated by an expanse of ocean which, through either simple distance or water depths exceeding the implosion limit of the phragmocone, prevented the mixing of all or a portion of the faunas. In cases where the taxonomy and biostratigraphy are not in question, there seems no other answer. Because of this sensitivity to distance and water depth combined with the ability of rapid radiation under the proper circumstances, the biogeography of nautiloids appears to proceed rapidly from disjunct distributions to distributions with high levels of similarity. Taking these factors into consideration, the biogeography of Cambro–Devonian nautiloids is reasonably consistent with the palaeogeography used here in comparison. In most cases of disagreement, differences are slight and much larger faunas or better stratigraphic control are required before corrections are proposed.

The time and space relationships among Cambro–Devonian nautiloid genera provide insight into several factors of their behaviour. First, their biogeography can be separated into four periods: (1) pre-Tremadoc; (2) Ordovician; (3) Silurian through Gedinnian or Siegenian; and (4) Emsian through Frasnian–Famennian. The pre-Tremadoc period is unique in that it is marked by a good deal of experimentation among groups and little expansion. The near extinction of Trempealeauan nautiloids left little in the way of history for Tremadoc and later faunas. The next three periods share a pattern which begins with the early expansion of groups which survived some crisis, followed by expansion into most shallow shelf areas positioned in equatorial and subtropical latitudes, and ending with reasonably abrupt declines in generic diversity and in some cases strongly provincial faunas.

The second factor is that nautiloids appear to have been more susceptible to fluctuations in sea level than commonly believed. As vagrant scavengers, this susceptibility presumably reflects the waning of food resources. Support for this observation comes from the crises which serve as an end to one period and the beginning to the next correlate with first or second order falls in eustatic sea level. In some cases, sea level fall can be attributed to Gondwana glaciation and for other was perhaps the result of contraction in size of oceanic ridges related to decreased rates of sea floor spreading (Pitman 1978). In either case the ultimate control was the repositioning of continents as a consequence of plate movement. There is no recorded eustatic fall in sea level during Trempealeauan time, and as yet there is no explanation for the 'outage' of nautiloids which occurred at this time.

The third factor is that the majority of the faunas of each of the four biogeographic periods were controlled by genera of only two or three orders. In each case, the taxa of the dominating orders have characteristics which appear to have placed them at a hydrodynamic advantage over groups which survived from a previous crisis. That is they were better able to compete for resources and able to disperse in relatively short periods of time.

I wish to thank V. Jannusson and H. Mutvei of the Swedish Museum of Natural History for their assistance with Baltic and Gondwana faunas. I also wish to thank the reviewers for their constructive comments and helpful suggestions, and S. McKerrow and C. Scotese for organizing the Symposium on Palaeozoic Palaeogeography and Biogeography convened at Oxford University, August, 1988.

References

BALASHOV, Z. G. 1960. New Ordovician nautiloids of the USSR. *Novye vidy drevnikh rasenii i bespozvonochnykh*, **2**, 123–136, [in Russian].
—— 1968. *Ordovician Endoceratoidea of the USSR*. Leningradskogo Universiteta, Leningrad, [in Russian].
BARSKOV, I. S. 1972. *Late Ordovician and Silurian cephalopod molluscs of Kazahkstan and Middle Asia*. Akademiya Nauk SSSR, Moskva. [in Russian].
CHAMBERLAIN, J. A., Jr., WARD, P. D. & WEAVER, J. S. 1981. Post-mortem ascent of *Nautilus* shells: implications for cephalopod paleobiogeography. *Paleobiology*, **7**, 494–509.
CHEN JUN-YUAN & TEICHERT, C. 1983a, Cambrian Cephalopoda of China. *Palaeontographica*, **181**(A), 1–102.
—— & —— 1983b. Cambrian cephalopods. *Geology*, **11**, 647–650.
——, ——, ZHOU, ZHI-YI, LIN YAO-KUN, WANG ZHI-HAO & XI JUN-HAO 1983. Faunal sequence across the Cambrian–Ordovician boundary in northern China and its international correlation. *Geologica et Palaeontologica*, **17**, 1–15.
CRICK, R. E. 1980. Integration of paleobiogeography and paleogeography: evidence from Arenigian nautiloid biogeography. *Journal of Paleon-*

tology, **54**, 1218–1236.

—— 1981. Diversity and evolutionary rates of Cambro–Ordovician nautiloids. *Paleobiology*, **7**, 216–229.

—— 1988. Buoyancy regulation and macroevolution in nautiloid cephalopods. *Senckenbergiana Lethaea*, **69**, 13–42.

FLOWER, R. H. 1976. Ordovician cephalopod faunas and their role in correlation. *In*: BASSETT, M. G. (ed.) 1976. *The Ordovician System: proceedings of a Palaeontological Association symposium, Birmingham, Sept. 1974*. University of Wales Press and Nature Museum of Wales, Cardiff, 523–552.

HARLAND, W. B., COX, A. V., LLEWELLYN, P. G., PICKTON, C. A. G., SMITH, A. G. & WALTERS, R. 1982. *A Geologic Time Scale*. Cambridge University Press.

JANNUSSON, V. 1979. Ordovician. *In*: ROBINSON, R. A. & TEICHERT, C. (eds) *Treatise on Invertebrate Paleontology, Part A, Introduction*. Geologic Society of America and University of Kansas Press, Lawrence, Kansas, 136–166.

JAANUSSON, V. 1984. What is so special about the Ordovician? *In*: BRUTON, D. L. (ed.) *Aspects of the Ordovician System*. Palaeontological Contributions for the University of Oslo, No. 295, Universitetsforlaget, 1–3.

MOORE, R. C. (ed.) 1964. *Treatise on Invertebrate Paleontology, Part K, Mollusca 3*. Geologic Society of America and University of Kansas Press, Lawrence, Kansas.

LANDMAN, N. H., RYE, D. M. & SHELTON, K. L. 1983. Early ontogeny of *Eutrephoceras* compared to Recent *Nautilus* and Mesozoic ammonites: evidence from shell morphology and light stable isotopes. *Paleobiology*, **9**, 269–279.

PITMAN, W. C., III. 1978. Relationship between eustacy and stratigraphic sequences of passive margins. *Geological Society of America Bulletin*, **89**, 1389–1403.

RAUP, D. M. & CRICK, R. E. 1979. Measurement of faunal similarity in paleontology. *Journal of Paleontology*, **53**, 1213–1227.

RUZHENTSEV, V. E., ZHURAVLEVA, F. A., BALASHOV, Z. G., BOGOSLOVSKIY, B. I. & LIBROVICH, L. S. 1962. Mollusks, Cephalopods I, Nautiloids, Endoceratoids, Actinoceratoids, Bactritoids, Ammonoids. *Osnovy Paleontologii*, YU. A. ORLOV (ed.), 1–438. Moskva. [in Russian].

SAUNDERS, W. B. & SPINOSA, C. 1979. *Nautilus* movement and distribution in Palau, Western Caroline Islands. *Science*, **204**, 1199–1201.

TEICHERT, C. 1970. Drifted *Nautilus* shells in the Bay of Bengal. *Journal of Paleontology*, **44**, 1129–1130.

TEICHERT, C. 1986. Times of crisis in the evolution of the Cephalopoda. *Palaontologische Zeitschrift*, **60**, 227–243.

TEICHERT, C., GLENISTER, B. F. & CRICK, R. E. 1979. Biostratigraphy of Devonian nautiloid cephalopods. *In: The Devonian System*. Palaeontological Association of London, Special Paper, **23**, 259–262.

TORIYAMA, R., SATA, T., HOMADA, T. & KOMALARJUN, P. 1965. Nautilus pompilius drifts on the west coast of Thailand. *Japanese Journal of Geology and Geography*, **36**, 149–161.

VAIL, P. R., MITCHUM, R. M., Jr. & THOMPSON, S., III. 1978. Seismic stratigraphy and global changes of sea level, Part 4: Global cycles of relative changes of sea level. *In*: PAYTON C. (ed.) *Stratigraphic interpretation of seismic data*. American Association of Petroleum Geologists Memoir, **26**, 83–97.

WARD, P. D., GREENWALD, L & ROUGERIE, F. 1980. Shell implosion depth for living *Nautilus macromphalus* and shell strength of extinct cephalopods. *Lethaia*, **13**, 182.

WESTERMANN, G. E. G. 1973. Strength of concave septa and depth limits of fossil cephalopods. *Lethaia*, **6**, 383–403.

WESTERMANN, G. E. G. 1985. Post-mortem descent with septal implosion in Silurian nautiloids. *Palaontologishe Zeitschrift*, **59**, 79–97.

ZHURAVLEVA, F. A. 1972. Devonian nautiloids, Order Discosoridia. *Trudy Paleontologicheskogo instituta A.N. SSSR*, **134**, 1–311, [in Russian].

ZHURAVLEVA, F. A. 1974. Devonian nautiloids, Orders Oncocertida, Tarphyceratida, Nautilida. *Trudy Paleontologicheskogo instituta A.N. SSSR*, **134**, 1–3111, [in Russian].

Early–Middle Palaeozoic biogeography of Asian terranes derived from Gondwana

CLIVE BURRETT[1], JOHN LONG[1] & BRYAN STAIT[2]

[1] *Geology Department, University of Tasmania. P.O. Box 252C, Hobart, Tasmania, Australia, 7001*
[2] *Parliamentary Library, Parliament House, Canberra, ACT, Australia*

Abstract: Contiguity of the Shan-Thai Terrane and NW Australia is suggested for Cambro–Ordovician times by the close faunal affinities seen in Late Cambrian trilobites, Ordovician molluscs, stromatoporoids, brachiopods and conodonts. Taxa such as *Spanodonta* and *Georgina* are found only on these two blocks whilst others have a Shan-Thai North China, Australian distribution. This, with a re-evaluation of early Palaeozoic palaeomagnetism, places Shan-Thai against NW Australia, N China against N Australia, S China against the western Himalayan–Iran region with Indo-China and Tarim lying between S China and Shan-Thai. A palaeomagnetically required anticlockwise rotation of this greater Gondwana from the Early Cambrian to the Middle Ordovician satisfactorily accounts for the changing biogeographic patterns, in particular the differences between North and South China during the Ordovician. Recent studies on microvertebrates and conodonts suggest that Shan-Thai was still very close to Australia in the Middle Devonian, as seen by similar turiniform thelodont species from western Yunnan, northern Thailand, and South Australia, as well as the polygnathid *P. labiosus* lineage, species of which have recently been found in Thailand. Continental fish faunas were highly endemic in South China from Silurian through to Early and Middle Devonian, indicating prolonged isolation of this terrane and separation of South China from Gondwana probably in the Silurian. In the Late Devonian, shared biotic assemblages between North China and South China (endemic Zhongning/Wuting plant flora; endemic antiarch placoderm and polybranchiaspid agnathan fishes); and between these Chinese terranes and Australia (sinolepid antiarchs, earlier appearances of certain antiarchs) indicate close continental proximity of these three major regions. New palaeomagnetic data for Ningxia, together with biogeographic data suggests that this region may have constituted a separate 'Hexizoulang Terrane'.

Although suggested by Argand (1927) it was not until the early 1970's that the concept of Asia as a continental collage was explicitly outlined (Burrett 1973, 1974; Dickinson 1973). It was clear following Hamilton (1970) that Siberia was a separate block and that Kazakhstan was separate and collided with Siberia along the Chingiz suture in the mid-Carboniferous. Southeast Asia and Tarim were recognised as separate blocks. Ordovician faunal differences between North and South China emphasized by Kobayashi (1971), led Burrett (1973, 1974) to suggest that those two blocks behaved independently until collision along the Qinling suture during the Triassic Indosinian orogeny. Whilst palaeomagnetic data from China supports that hypothesis (McElhinny *et al.* 1981; McElhinny 1985; Lin *et al.* 1985*a*, 1985*b*), further biogeographic analysis of Chinese data, with the discovery of intermediate faunas, appears to confound it (Lu *et al.* 1974). Analyses of Cambrian trilobite distributions failed to find provincial dichotomy along the Qinling Mountains for stages of the Cambrian (Jell 1974; Burrett & Richardson 1980). Using new geological and palaeontological data from the Shan-Thai Terrane, Burrett & Stait (1985, 1986) showed that Shan-Thai was almost certainly adjacent to the Australian sector of Gondwana in the early Palaeozoic. Their data also indicated strong connections to North China with minimal affinities with South China.

The Late Palaeozoic–Early Mesozoic position and evolution of the peri-Gondwana Asian terranes has been recently reviewed by Audley-Charles (1988), Metcalfe (1988) and Sengor *et al.* (1988).

A greater Gondwana

The case for a much enlarged Early Palaeozoic Gondwana rests on several arguments.

(1) The *shallow water* faunas of the Shan-Thai Terrane have remarkably close affinities to those of the Australian platform. These similarities are particularly strong in the nautiloid faunas (Stait & Burrett 1982, 1984*b*; Stait *et al.* 1987) where there is a Simpson Index of 0.95 and where genera such as *Georgina* are found only in Australia and Shan-Thai (Burrett & Stait 1985). Because these heavily weighted Ordovician nautiloids are unlikely to have drifted very far *post mortem* it seems probable that these regions were in close proximity (Stait & Burrett 1987). This proximity is also suggested by the distribution of the Ordovician brachiopod *Spanodonta* (Fig. 1) which is found only in Western Australia and Shan-Thai (Laurie & Burrett in prep). The polyplacophoran *Chelodes whitehousei* is also found only in Australia and Shan-Thai (Stait & Burrett 1984*a*).

(2) Several Shan-Thai taxa have affinities with both Australia and North China e.g. the stromatoporoids *Labechia variabilis* and *Rosenella woyuensis* (Webby *et al.* 1984) whilst others such as the nautiloids *Kogenoceras nanpiaoense* and *Wutinoceras robustum* are found in North China. The distinctive Tasmanian conodont *Tasmanognathus* (Fig. 1) is found over most of North China (including Korea) but has not been found on Shan-Thai probably because of a lack of appropriate age (Blackriveran) shallow-water strata on that terrane (An *et al.* 1983; Burrett 1979). The Late Cambrian trilobite fauna of Shan-Thai includes genera such as *Pagodia*, *Tsinania*, *Haniwa* (Fig. 2) and *Quadraticephalus* that are found in North China, South China, Australia, southern Europe and Iran (Shergold *et al.* 1988). The Early Ordovician Gondwana snail *Peelerophon* (Fig. 1) is very common in Shan-Thai (Jell *et al.* 1984).

(3) The Lower Cambrian sequence of rocks and faunas in the east Yunnan part of the South China Terrane is so similar to that in the Salt Range of Pakistan that proximity is indicated (Chang 1981, Table 1). For instance, *Yuehsiensziella* and *Chitidilla* (Fig. 2) are found in similar lithologies and at the same stratigraphic level. Chang (1981) provides several other examples of similarities between South China and the Indian– Iranian sector of Gondwana during the early Palaeozoic.

(4) The Early Cambrian palaeomagnetic data of Lin *et al.* (1985*a*, *b*) gives palaeolatitudes of 37 ± 10 for North China and 10 ± 8 for South China (Fig. 3). These data appear to be reliable but post-Cambrian overprinting has not been definitely ruled out. No reliable palaeomagnetic data are available from the major Chinese terranes for the Late Cambrian–Devonian interval, except for one Early Ordovician determination from North China that indicates near equatorial palaeolatitudes (Fig. 4).

(5) Any arrangement of Asian blocks against Gondwana should preferably have the deep-water facies of each block (and their embedded bordering island arcs) on the outboard side of the proposed continental aggregation.

(6) Any reconstruction has to explain why at certain times there is a very strong faunal difference between North and South China whilst at others the faunas are similar. For instance, in the

From McKerrow, W. S. & Scotese, C. R. (eds), 1990, *Palaeozoic Palaeogeography and Biogeography*, Geological Society Memoir No. 12, pp. 163–174.

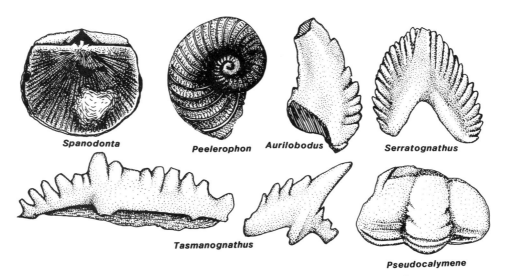

Fig. 1. Ordovician fossils with restricted biogeographic ranges useful in linking Australia and some Asian terranes.

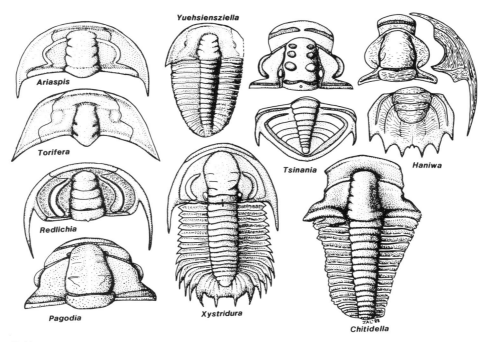

Fig. 2. Some Cambrian trilobites common to Australian and Asian terranes.

Cambrian they are clearly closely related (Jell 1974; Burrett & Richardson 1980) whereas, in the Ordovician, they are almost totally dissimilar. The faunal differences are not only obvious in the trilobites and nautiloids but extend to the conodonts. North China supports a North American Mid-continent-*Tasmanognathus* conodont fauna whereas South China contains a North Atlantic type conodont fauna. The distinctive conodont *Serratognathus* (Fig. 1) is one of the few taxa in common and is also found on Shan-Thai and in the Canning Basin of Western Australia.

The reconstruction

Following from these six major arguments and incorporating the quantitative biogeographic syntheses of Burrett & Richardson (1980) and Jell (1974) for the Cambrian and Burrett (1973) and Whittington & Hughes (1972, 1973) for the Ordovician, we have produced the Early Palaeozoic reconstructions given in Burrett & Stait (1987) and Figs 3 and 4. Thus following point 1, Shan-Thai is placed next to northwest Australia. From point 2, North China is placed in proximity to both Shan-Thai and Australia in the orientation and palaeolatitude suggested by points 4 and 5. South China is placed next to the Himalayan–Iranian sector at a palaeolatitude of 10 ± 8 (point 4) and with the deeper-water facies of the eastern Yangtze Platform outboard and the shallow water facies and land masses of the west being placed inboard (point 5). The positions of the Tarim and Indo-China Terranes are not so well constrained but their overall faunal affinities fit with the tentative placements shown in the figures in Burrett & Stait (1987).

The best Early Palaeozoic palaeomagnetic data for the major Gondwana blocks comes from Australia (Embleton 1979a, b, 1984; Embleton & Giddings 1974; Goleby 1980; Kirshvink 1978a, b; Klootwijk 1980; Luck 1972; McElhinny & Embleton 1974; McElhinny & Luck 1970) and when these are plotted for time

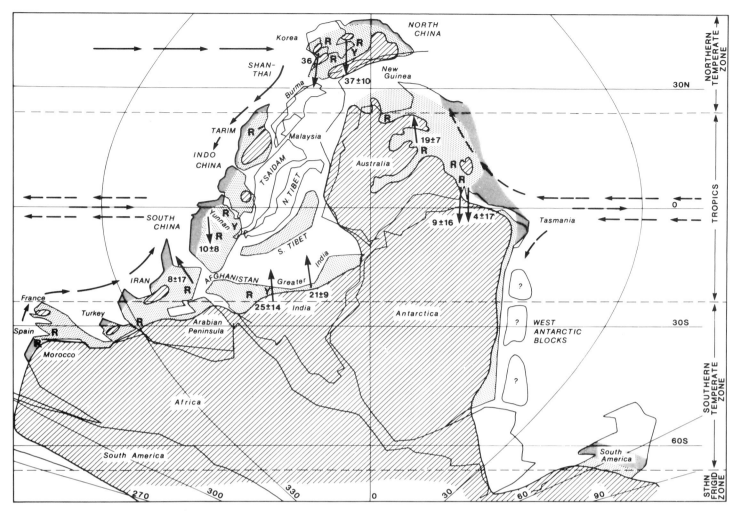

Fig. 3. Late Early Cambrian reconstruction (Upper Olenellus zone). Arrows with numbers are palaeomagnetic declinations, numbers are palaeolatitudes with errors. Dashed arrows are warm ocean currents, solid arrows are cold currents. Sources of palaeomagnetic data: China, Lin et al. (1985a, b) Australia, Embleton (1972a, b) Embleton & Giddings (1974), Kirshvink (1978a, b), Klootwijk (1980), Luck (1972), McElhinny & Embleton (1974), McElhinny & Luck (1970). Australian Cambrian paleogeography mainly from Cook (1982). Asian palaeography from numerous sources. R, *Redlichia*. Y, *Yuehsienziella*. The longitude values are arbitrary in all the figures.

segments of the Cambrian and Ordovician an anticlockwise movement of Australia, and hence of Gondwana, is indicated from the Early Cambrian to the Early Ordovician (Figs 3 & 4). If the Asian terranes are included in Gondwana then these too would have rotated. A major consequence of this rotation would have been to bring the South China Terrane into progressively higher (and hence colder) southerly palaeolatitudes. This explains the faunal similarities during the Early Cambrian (all the terranes were tropical–subtropical) and the differences between South China (cool–cold) and North China (tropical) and the faunal similarities between North China, Australia and Shan-Thai (all tropical) during the Early–Middle Ordovician. Deeper water and hence colder faunas around these tropical terranes would be expected (Cook & Taylor 1975; Burrett et al. 1983, 1984) to have predominantly North Atlantic–South China conodonts and South Chinese–Baltic type trilobites. These are exactly the relationships found in the recently discovered Caradoc-Ashgill deeper-water red limestones in southern Thailand. These are nodular limestones that lack a sessile benthos (Wongwanich et al. in press). However, trilobites are common and include typical deep water genera such as *Nileus*, *Geratrinodus*, *Arthrorhachis*, *Segmentagnostus*, *Cyclopyge*, *Microparia* and *Elongantinileus*. Many of these belong to Pagoda Limestone (South China Terrane) species. In addition there is *Paraphillipsinella globosa*, *Sphaerexochus* cf *fibrisulcatus*, *Dindymenella* cf *sulcata* (all Pagoda Limestone species), the Kazakhstan genus and species *Ovalocephalus kelleri* and the Baltic *Panderia*, *Parvigena* and *Remopleurella burmeisteri* (R. Fortey pers. comm. 1988). The occurrence of South Chinese conodonts in the Canning Basin of Western Australia (such as *Aurilobodus*, Fig. 4) and some trilobites (Legg 1976) may also be accounted for on this model by having the impingement of cold currents into the deeper parts of the Canning Basin during the Early Ordovician (Fig. 4).

Devonian palaeogeography: new data from the South China, North China and Shan-Thai Terranes

We shall now examine the biogeographic data from the South China, North China and Shan-Thai Terranes used in our palaeogeographic reconstructions for the Devonian. In the last decade much new data on middle Palaeozoic vertebrate distributions has come out of China and Australia, with some very recent data relevant to Shan-Thai. Summaries of relevant invertebrate and plant distributions are included in the following discussion. Our concept of the Shan-Thai Terrane is shown in Fig. 5, also illustrating the key localities for Devonian fossil assemblages as discussed in the text.

We have used the latest Devonian palaeomagnetic data from

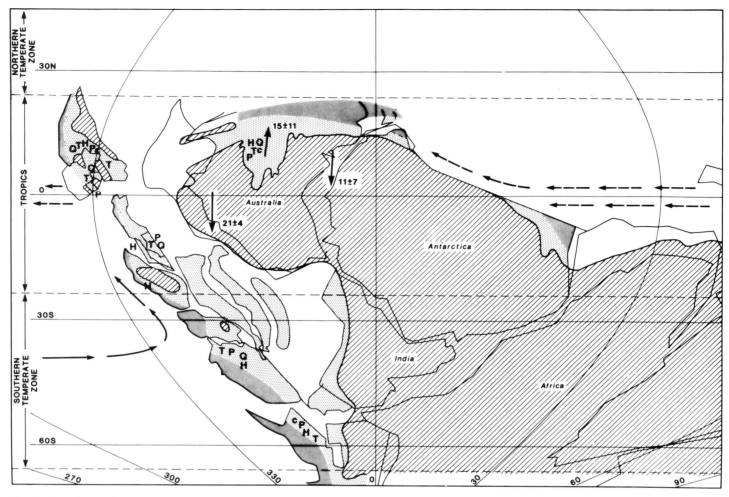

Fig. 4. Early Ordovician (Ibexian–Early Whiterockian) reconstruction. For key and sources of data see Fig. 3. A, *Asaphopsoides*; C, *Chosenia* N, *Neseuretus*; P, *Peelerophon*.

Australia to make the reconstructions shown in Fig. 6. The data of Schmidt *et al.* (1987) for the Early Devonian indicates low (about 44°) southerly palaeolatitudes and an orientation for Australia similar to that of today whereas the Late Devonian data of Hurley & Van der Voo (1987) from the reef complexes of the Canning Basin, Western Australia, indicate a palaeolatitude of about 15° S and an orientation as shown in Fig. 7. Similar low palaeolatitudes are found in the Late Devonian of eastern Australia (Li 1987; Schmidt *et al.* 1987). These data suggest a major rotation of Gondwana in the Middle Devonian. These Late Devonian Australian data bring India into moderate southerly latitudes (confirmed by the data of Chen & Lu 1985, who record palaeolatitudes of 33° S from Southern Tibet), Tarim into the tropics (confirmed by Bai *et al.* 1985 who record 6 ± 7°), and Iran into high palaeolatitudes (confirmed by Wensink *et al.* 1978 who record 50 ± 4°). These and other recent palaeomagnetic data (for instance the paper of Miller & Kent 1988, that places North America at low southerly palaeolatitudes during the Early Devonian) lead to reconstructions that differ from those of Heckel & Witzke (1979) and Scotese *et al.* (1985) which were based on other considerations (palaeoclimatic indicators) and earlier palaeomagnetic results.

South China Terrane

Vertebrates. The highly endemic nature of the South China Devonian vertebrate faunas has been known since the mid 1960s (Liu 1965) although only recently has the extent of this endemism been evaluated. Young (1981) proposed a separate South China vertebrate province based on the unique occurrences of galeaspid agnathans and yunnanolepidoid antiarchs in the Lower Devonian of the Yangzte paraplatform. These highly characteristic forms, some of which are shown in Fig. 7, form the basis for biostratigraphic and biogeographic subdivision of the Early Devonian in China (Pan 1981, Pan & Dineley 1988). A summary faunal list for the fauna from the Xitun Member, Cuifengshan Formation (Upper Lochkovian-Pragian), eastern Yunnan, based on Wang Nianzhong (1984) indicates that of the 18 genera present the only non-endemic is the thelodontid *Turinia*. Other Early Devonian faunas from South China show that of a total 45 genera of fishes present the only non-endemics are the petalichthyid *Lunaspis* (Liu 1981), also known from the Hunsruckscheifer of Germany (Gross 1961) and Australia (Young 1985), and genera represented from microvertebrate remains (*Turinia*, *Nostolepis* and ?*Ohiolepis*: Wang Nianzhong 1984). The latter elements have a more or less cosmopolitan distribution although there are taxonomic problems with defining *Nostolepis* and *Ohiolepis* remains (e.g. Denison 1979). Most of these are placed in endemic families or higher taxonomic levels unique to South China (Order Eugaleaspida or Galeaspidiformes; Order Yunnanolepiformes).

The Eugaleaspida are known from the Lower Silurian to the Upper Devonian in China, although most come from the Lower Devonian of South China (Pan 1984). A possible occurrence of a polybranchiaspid in the Upper Devonian of North China (Ningxia, Zhongning Fm) was noted by Pan (1984) and described by Pan *et al.* (1987). Eugaleaspids are also known from the

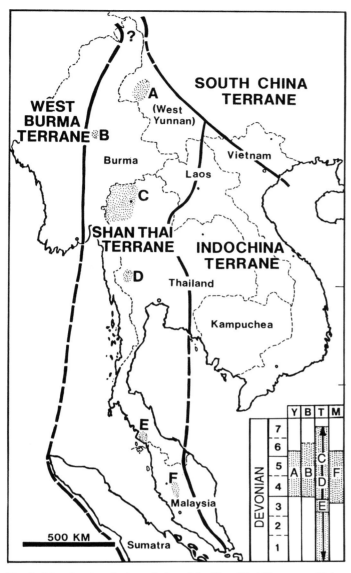

Fig. 5. Map showing the boundaries of the Shan-Thai Terrane, with key Devonian fossil localities shaded.

The dramatic nature of the degree of endemism seen in the early Devonian fish fauna of South China is seen by comparison with the modern Australian mammal fauna. The Australian mammal fauna is dominated by many unique families of marsupials, these are all contained within the two orders Polyprotodonta (carnivorous marsupials; not all of these are Australian, only two suborders, Dasyuromorphia, (Peramelomorphia) and the Diprotodonta (herbivorous marsupials) and these orders fall in the subclass Metatheria (Marsupialia), (Strahan 1983). The Devonian fish fauna of South China contains at least 8 endemic families of fishes and two endemic orders, Polybranchiaspidiformes, Yunnanolepiformes. Considering the incompleteness of the fossil record and the potential for many new discoveries from China, it is clear that this area was an isolated centre for radiation and diversity of many early fish groups comparable with Australia's long isolation after breaking up from Antarctica. The highly endemic nature of the South China fish fauna is a useful biogeographic tool as its breakdown in the Late Devonian gives indications of which crustal blocks were in close proximity, or became open to new migration routes.

Invertebrates and Plants. Many of the marine invertebrate groups of the South China Terrane have been described. Wang *et al.* (1984) found a contrast at the generic level between the Lower and Middle Devonian brachiopods of North Vietnam and South China (South China Terrane) and recognized a distinct South China zoogeographic region. Talent *et al.* (1986) agree and state that the levels of endemism for rugose corals and trilobites are also very high for this region. Zhang Renjie & Pojeta (in Pojeta 1986) monographed the bivalve faunas from the Early and Middle Devonian of Guanxi, and at least half of the many new taxa described are endemic. The high levels of endemism in invertebrates persisted as late as the Famennian — out of 33 taxa of proetid trilobites described from Guanxi, 23 taxa are endemic (Yuan 1988). In general, most of the invertebrates so far described from the South China block are highly endemic as reflected at specific or generic levels. Fossil floras have been described from both microfossils (e.g. Hou 1982) and macrofossils (Cai & Schweitzer, 1983; Li *et al.* 1984), and, again, several endemic species are reported.

It is the vertebrate assemblage that most clearly defines the high degree of endemism in the South China Terrane (Young 1981). The vertebrates show endemism at high taxonomic levels (orders, families, whereas invertebrate and floral assemblages also show strong endemism, but at generic and specific levels (e.g. Hong Fei Hou 1990; Pedder & Oliver 1990).

Luomengshan Mountains which border the South China Terrane. Affinities to the South China fauna are apparent, although the exact palaeogeographic affinities of the Lungmenshan–Qinling region (Yang *et al.* 1981; Hou & Wang 1985) are uncertain.

During the Middle Devonian other non-endemic elements appear in the South China ichthyofauna. The arthrodire *Holonema* (Wudin, Wang Junquing, 1984) and the antiarch *Bothriolepis* (Chang 1965) occur in the Eifelian, and the antiarch *Microbrachius* (Yunnan, Pan 1986) appears in the Givetian. The dinichthyid *Dunkleosteus* is also known from the upper Middle Devonian (Wang Junquing 1982). Despite these, the assemblages remain highly endemic, with high diversity of bothriolepidoid and sinolepidoid antiarchs (Pan 1981), as well as endemic arthrodires (Liu & Wang 1981, Wang & Wang 1983, 1984), petalichthyids and agnathans (Pan 1981, 1984; Wang Junquing 1984). By the Late Devonian the South China fish faunas contain several cosmopolitan genera (e.g. *Bothriolepis*, *Remigolepis* and *Asterolepis*: Pan 1981) although *Remigolepis* (Fig. 8) appears earlier in China (early Frasnian) and Australia (late Frasnian, early Famennian; Young 1974) than for the rest of the world (latest Famennian, Long 1983). Sinolepids (Fig. 8), known throughout South China in the Late Devonian, also occur at this time in North China and eastern Australia.

North China Terrane

Vertebrates. The only vertebrate faunas known are from Ningxia and Gansu Provinces. The Ningxia faunas are Middle and Late Devonian age, although best known is the Late Devonian Zhongning Formation fauna. This formation contains endemic species of the antiarchs *Bothriolepis*, *Remigolepis* and *Sinolepis*, and a polybranchiaspid agnathan (Pan *et al.* 1987). The fauna from near Yumen, Gansu Province, includes arthrodires and antiarchs but details are not yet published (Young 1987a). Immediate affinities of the Zhongning fauna are with the South China block (polybranchiaspids, *Sinolepis*), indicating two possibilities. Firstly that near the end of the Middle Devonian a barrier was removed between North and South China enabling faunal interchange. Alternatively, the terrane containing Ningxia and Gansu Provinces, termed the Hexizoulang Terrane (Li *et al.* 1985) may have originally been attached to the South China Terrane and rifted away sometime after the Devonian. Close faunal and floral affinities between Ningxia and South China, together with the structural/tectonic setting of the Ningxia–Gansu region (being surrounded by major fault systems) would support this view (Yang *et*

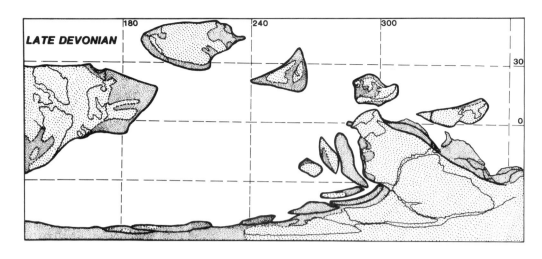
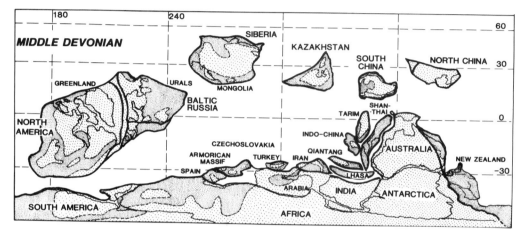
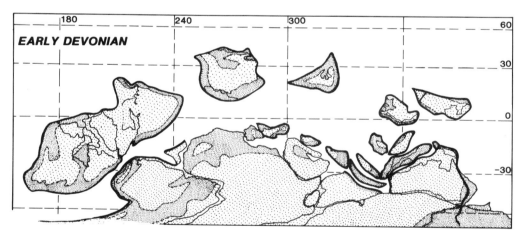

Fig. 6. Palaeogeographic maps for the Devonian, based on new data from vertebrate distributions, and encompassing all recent palaeomagnetic and palaeolithologic data. Land-soft stipple, shallow epicontinental seas-heavy stipple. Major terrane boundaries in heavy lines. Late Devonian position of Hexizoulang (Ningxia) after palaeomagnetic data of Pan *et al.* (1987).

al. 1981). In our reconstructions we have placed Hexizoulang conservatively still as a part of the North China Terrane, but use new palaeomagnetic data from Pan *et al.* (1987) to place this terrane close to eastern Australia. Sinolepid antiarchs (Fig. 8) are known from the Hunter Siltstone near Grenfell, New South Wales, in the uppermost Late Devonian (with *Remigolepis*, Fig. 8, as in both North and South China). This is the only known occurrence of the group outside of North and South China, and is used here, along with other biotic elements in common, (Fig. 9) to place these regions in relatively close positions (Fig. 6).

Invertebrates and plants. The diverse plant flora from the Zhongning Fm, southern Ningxia (Sze & Lee 1945, Xiu-hu *et al.* 1986) includes several forms of endemic genera and species known otherwise only from the Wutung Series of South China. These include *Eolepidodendron wusihense*, *Hamatophyllum verticallatum*, and *Sphenopteridium taihuenensis*. Invertebrates described from the Devonian of North China include conulariids (Zhou 1985), tabulate corals (Li 1983), trilobites (Xiang 1981) and brachiopods (Wang *et al.* 1984). However, of these, only the tabulate corals were found to have biogeographic significance. Yao-Xi (1983)

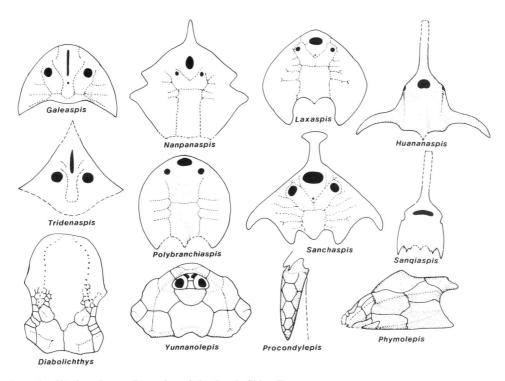

Fig. 7. Endemic fishes from the Silurian–Lower Devonian of the South China Terrane.

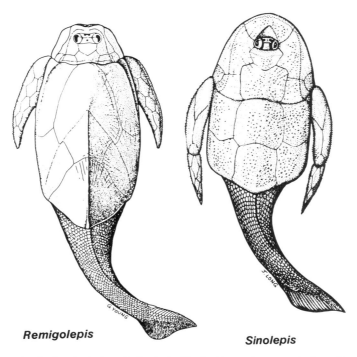

Fig. 8. The Late Devonian freshwater antiarchs *Remigolepis* and *Sinolepis*. Sinolepids are known only from South China, North China and Australia; *Remigolepis* is a late Famennian indicator in Euramerica, but occurs earlier in South China, North China and Australia (Young 1974, 1981; Long 1983).

found that the Silurian and Devonian corals from Gansu Province fell into two biogeographic realms, one assemblage having cosmopolitan genera and one having faunal affinities to the Kazakhstan Terrane (assemblage from Xingan). Until more is known of the vertebrate assemblages from Kazakhstan it is difficult to assess the worth of this information.

Shan-Thai Terrane

Vertebrates. An assemblage of Middle Devonian (Givetian) thelodonts described from Western Yunnan (Shidian) includes *Turinia pagoda*, and two other species, possibly new, of *Turinia* (Wang *et al.* 1986). Other fishes, including arthrodires and anti-

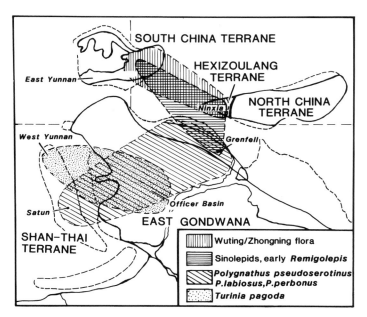

Fig. 9. Biogeographic ranges of certain faunal and floral assemblages. The polygnathid distribution is for the Emsian; *Turinia pagoda*-species group is for Middle Devonian, and the Wuting/Zhongning flora and *Sinolepis/Remigolepis* fauna are both Late Devonian.

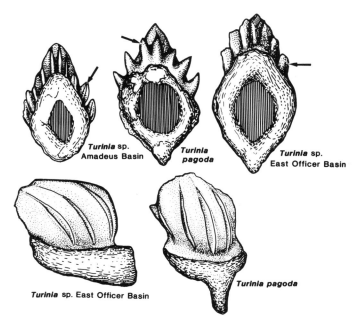

Fig. 10. Scales from the agnathan thelodont *Turinia* from the Amadeus Basin and east Officer Basins, Australia, and from western Yunnan (Shan-Thai Terrane, Fig. 8). These are a closely related species group sharing the development of winglets (indicated by arrows) around the crown circumference (after Wang et al. 1986; Young et al. 1987; Long et al. 1988).

archs are reported to occur by Wang et al. but details are not currently known. From near Maymyo, central Burma, an ichthoduralite was reported (Chhibber 1936) but details are not known. Blieck & Goujet (1978) and Blieck et al. (1984) described an acanthodian scale (*Nostolepis* sp.) and a thelodont scale from near Mae Sariang, northern Thailand. *Nostolepis* is a cosmopolitan genus of uncertain phylogenetic affinities (Denison 1976, 1979). Turner (pers. comm.) believes the Mae Sariang, thelodontid to be a turiniform with resemblances to the west Yunnan species.

Long et al. (1988) have reported *Turinia* cf. *T. pagoda* from the eastern Officer Basin, South Australia. Closer examination of these and other contemporary thelodontids from Australia indicates that only the species from West Yunnan (*T. pagoda*) and forms from the Amadeus Basin (Young et al. 1987) and Officer Basins, Australia, have well-developed winglets projecting from around the circumference of the crown (Fig. 10). Although many of the Australian occurrences of thelodontids have not yet been fully described, the resemblances are striking, and even if different species or subspecies are eventually named for the Australian and West Yunnan specimens under consideration here, it still remains that they are closely related by features not shared with any of the many other known species of *Turinia*. Furthermore, *Turinia* was extinct well before the end of the Early Devonian in Euramerica. Only in Iran (Blieck & Goujet 1978), West Yunnan, Antarctica and Australia do thelodontids (derived forms of *Turinia*, and *Australolepis*) persist into Middle and Late Devonian times. The youngest thelodonts are from the Frasnian Gneudna Formation, Western Australia (Turner & Dring 1981). Therefore, although our knowledge of Shan-Thai Devonian vertebrates is scarce, affinities with East Gondwana are here suggested for the Middle Devonian. Recently we discovered an assemblage of latest Late Devonian microvertebrates from near Mae Sam Lap, northern Thailand. The fauna includes endemic genera and species of chondrichthyans elsewhere known only from South China and Australia (Long & Burrett, in press; Long, in press), and suggests, therefore, that these three terranes were still in close proximity at the end of the Devonian.

Invertebrates. Devonian invertebrates from Shan-Thai are still poorly known. The Middle Devonian brachiopods from Padaukpin, central Burma, were redescribed by Anderson et al. (1969) who suggested that the padaukpinensis fauna had affinities with the South China tonkinensis fauna. Trilobites from Satun Province have been described by Fortey (in press) who suggested affinities with Northern Gondwana countries.

The Emsian was a time of peak endemism for conodont faunas (Klapper & Johnson 1980), with decreasing endemism by the Late Devonian. Emsian conodonts from Satun Province have been described by Long & Burrett (in press), and include a new subspecies of *Polygnathus labiosus* (Fig. 11). The lineage of polygnathids (*P. dehiscens abyssus*, *P. perbonus*, *P. labiosus* and *P. pseudoserotinus*) are known chiefly from Australia and parts of Asia. *P. dehiscens abyssus* is known only from Australia; *P. perbonus* is known from Australia (Philip 1966; Mawson 1987), South China (Wang 1979; Wang & Ziegler 1983) and from central Tadzhikstan, U.S.S.R. (Bardashev 1987); *P. labiosus* and *P. pseudoserotinus* are both known only from Australia (Mawson 1987) and Shan-Thai (Lane et al. 1970; Long & Burrett, in press). Thus the most specialized species of this lineage are also the ones having the most restricted distribution. Fig. 11 shows a sketch cladogram of some of these species with their distributions indicated (data after Mawson 1987, Long & Burrett, in press).

Devonian palaeogeography and biotic dispersal events

Copper (1987) suggested that the rise in late Frasnian sea-levels coupled with increasing global cooling at that time, and the collision between Euramerica with western Gondwana, could have turned warm tropical seas into dry land or cold-water regions. Such an hypothesis was claimed to account for observed biotic dispersal events seen in vertebrate faunas (Young 1981, 1987b) as well as mass extinctions at the Frasnian–Famennian boundary. Young (1987b, p. 300) has questioned the unity of eastern and western Gondwana during the Middle Palaeozoic as different palaeomagnetic data may not agree with palaeoclimatic indicators. Our reconstructions show that the few palaeomagnetic data points for Gondwana (especially Australia) imply a major rotation of the whole supercontinent during the Middle Devonian, assuming

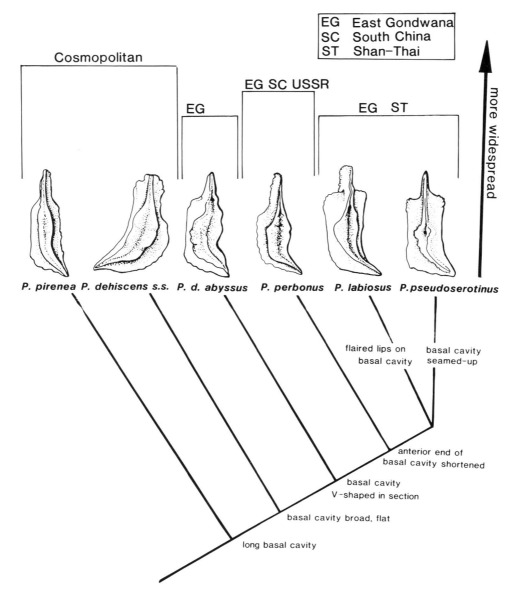

Fig. 11. Shan-Thai, South China and East Gondwana (Australian) polygnathid conodonts with large basal cavities. These species belong to one phylogenetic lineage (morphological data from Mawson 1987) and have restricted distributions during Emsian time.

that the continent was unified. The widespread distribution of certain fresh and brackish-water fishes during the Middle and Late Devonian could have been brought about by two major factors: (1) changing global palaeogeography; (2) increased dispersal capabilities in the fishes themselves; or as is most likely, a combination of both of these factors.

In our reconstructions the distribution of fishes in East Gondwana and Asian terranes is well accounted for throughout the Devonian by close continental positions. The close proximity of the Chinese terranes in the Middle and Late Devonian explains the major chronological anomalies in placoderm distributions such as the early appearances of certain antiarchs (which may have originated in either South China or East Gondwana, Young 1984, 1988). A major difference in our reconstructions compared with those of other workers (e.g. Scotese et al. 1985, Heckel & Witzke 1979) is the more southerly latitude of the northern Gondwana margin (west of Australia to Africa) in the Late Devonian. Such difference may bring into effect temperature barriers to inhibit dispersal rather than enhance faunal interchange. Alternative pathways for fishes through northern sea-routes may explain dispersal between Euramerica and East Gondwana (e.g. Long 1984). We do not see any physical barriers being broken down, and suggest that the major factor responsible for changing fish distribution patterns between East Gondwana–Asia and western Gondwana–Euramerica in the Late Devonian is perhaps the physiological changes in some of the fishes themselves. The most widespread taxa are generally those regarded as the most advanced or phylogenetically derived members of their lineages: the advanced bothriolepids with trifid preorbital recesses, the asterolepidoids *Remigolepis* and *Asterolepis* (Young 1988). Changing global palaeogeography, especially rises in sea-level (Johnson et al. 1985) and global cooling (Fischer 1981) would have effected the routes for dispersal by shallow sea-ways for fishes, and temperature barriers may have caused disjunct populations for some regions. However, if we assume that dispersal was largely achieved by shallow sea pathways, and that larval forms were at the mercy of currents, then there is no problem in explaining widespread dispersal of antiarchs in the Late Devonian. We therefore regard this problem as being a complex one with physiological factors as well as palaeogeographical ones. It is pointed out that even in the Early Devonian, in areas of very high endemism (e.g. South China) some fish taxa are found that have cosmopolitan distributions (*Lunaspis, Nostolepis, Turinia*). As to why these forms could disperse and others living in the same areas could not almost certainly depended on the physiological adaptations and environmental requirements of the species involved.

Following the work of Young (1986) we see that our reconstructions require firm underlying assumptions in order to be of use to future workers. The most crucial of these is that the distribution patterns we have used do indeed reflect real biotic ranges, and are not artefacts of collecting biases or of taxonomic misidentifications (as discussed by Talent *et al.* 1986). Further collecting and description of faunas from the least studied Asian terranes (e.g. Lhasa, Qiantang, Qaidam, Tarim, Indo-China) is required to test the hypotheses presented here and refine palaeogeographic reconstructions.

We thank the following people for helpful discussion of the work: G. Young, J. Shergold and J. Laurie (Bureau of Mineral Resources, Canberra); R. Fortey (British Museum of Natural History); S. Bunopas and T. Wongwanich (Royal Thai Department of Mines). We are indebted to R. Berry for writing the computer program which was used in the reconstructions. This work was funded through grants from the ARC, and the University of Tasmania.

References

AN, T., FANG, Z., WEIDA, X. & ZHANG, Y. 1983. *The conodonts of North China and adjacent regions*. Science Press, Beijing.

ANDERSON, M. M., BOUCOT, A. J. & JOHNSON, J. G. 1969. Eifelian brachiopods from Padaukpin, northern Shan States, Burma. *Bulletin of the British Museum of Natural History (Geology)*, **18**, 107–163.

ARGAND, E. 1927. La tectonique d'Asie. *Colloques Researches 13th Internationale Geologiques Congress 1922, Brussels*, **1**, [for 1924], 1–171.

AUDLEY-CHARLES, M. G. 1988. Evolution of the southern margin of Tethys (North Australia region) from the Early Permian to Late Cretaceous. *In*: AUDLEY-CHARLES, M. & HALLAM, A. (eds) *Gondwana and Tethys*. Geological Society, London, Special Publication, **37**, 79–100.

BARDASHEV, I. A. 1987. Emsian conodonts of the genus *Polygnathus* from central Tadzhikstan. *Paleontological Journal*, **2**, [for 1986], 56–62.

BAI, Y.-H., CHEN, G.-L., SUN, Q., SUN Y.-L., LI Y., DONG, Y. & SUN, D.-J. 1985. Late Palaeozoic polar wander path for the Tarim block and tectonic significance. *Seismology and Geology*, **7**, 71–80.

BLIECK, A. & GOUJET, D. 1978. A propos de nouveau material de Thelodontes (Vertébrés Agnathes) d'Iran et de Thailande: Aperçu sur la repartition geographique et stratigraphique des Agnathes des (regions gondwaniennes) au Paleozoique moyen. *Annals de Societe Geologique du Nord*, **97**, 363–372.

——, ——, JANVIER, P. & LELIEVRE, H. 1984. Microrestes de vertébrés du Siluro-Devonien d'Algerie. de Turquie et de Thailande. *Geobios*, **17**, 851–856.

BURRETT, C. F. 1973. Ordovician biogeography and continental drift. *Palaeogeography Palaeoclimatology, Palaeoecology*, **13**, 161–201.

—— 1974. Plate tectonics and the fusion of Asia. *Earth and Planetary Science Letters*, **21**, 181–189.

—— 1979. *Tasmanognathus* a new Ordovician conodont genus from Tasmania. *Geologica et Palaeontologica*, **13**, 31–38.

—— & RICHARDSON, R. 1980. Trilobite biogeography and Cambrian tectonic models. *Tectonophysics*, **63**, 155–192.

—— & STAIT, B. 1985. South-east Asia as part of an Ordovician Gondwanaland – a palaeogeographic test of a tectonic hypothesis. *Earth and Planetary Science Letters*, **75**, 184–190.

—— & —— 1986. China and southeast Asia as part of the Tethyan margin of Cambro-Ordovician Gondwanaland. *In*: MCKENZIE, K. (ed.) *Shallow Tethys 2*, Balkema, Rotterdam, 65–77.

——, SHARPLES, C., STAIT, B. & LAURIE, J. 1984. A Middle–Upper Ordovician tropical carbonate platform to basin transition, southern Tasmania, Australia. *In*: BRUTON, D. (ed.) *Aspects of the Ordovician System*, Oslo, Universitetsforlaget, 149–157

——, —— STAIT, B., LAURIE, J. 1983. Trilobites and microfossils from the Middle Ordovician of Surprise Bay southern Tasmania, Australia. *Memoirs of the Australasian Association of Palaeontologists*, **1**, 177–193.

CAI, C. & SCHWEITZER, H. J. 1983. Ueber *Zosterophyllum yunnanicum* Hsue aus dem Unterdevon Suedchinas. *Palaeontographica B*, **185**, 1–3.

CHANG, K.-J. 1965. New antiarchs from the Middle Devonian of Yunnan. *Vertebrata Palasiatica*, **9**, 1–14.

CHANG, W. T. 1981. On the northward drift of the Afro-Arabian and Indian plates. *Proceedings of the Royal Society of Victoria*, **92**, 181–185.

CHEN, X.-Y. & LU, L.-Z. 1985. Palaeomagnetism of Palaeozoic formations in Nyalem and Shensha of Tibet. *Acta Geophysica Sinica*, supplement 1 **28**, 211–218.

CHHIBBER, H. L. 1936. *The Geology of Burma*, McMillan & Company, London.

COOK, H. E. & TAYLOR, M. E. 1975. Early Paleozoic continental margin sedimentation, trilobite biofacies and the thermocline, Western United States. *Geology*, **3**, 559–562.

COOK, P. J. 1982. The Cambrian palaeogeography of Australia and opportunities for petroleum exploration. *Journal of the Australian Petroleum Exploration Association*, **22**, 42–64.

COPPER, P. 1987. Frasnian/Famennian mass extinction and cold water oceans. *Geology*, **14**, 835–839.

DENISON, R. H. 1976. Note on the dentigerous jaw bones of Acanthodii. *Neues Jarbuch fur Geologie und Paläontologie, Monatshefte*, **1976**, 395–399.

—— 1979. Acanthodii. *In*: SCHULTZE, H.-P. (ed.) *Handbook of Paleoichthyology*. Fischer-Verlag, Stuttgart, **5**, 1–62.

DICKINSON, W. R. 1973. Reconstruction of past arc-trench systems from petrotectonic assemblages in the island arcs of the Western Pacific. *In*: COLEMAN, P. J. (ed.) *The Western Pacific*. The University of Western Australia Press, Nedlands, 569–601.

EMBLETON, B. J. 1972a. The palaeomagnetism of some Proterozoic–Cambrian sediments from the Amadeus Basin, Central Australia. *Earth and Planetary Science Letters*, **17**, 217–226.

—— 1972b. The palaeomagnetism of some Palaeozoic sediments from Central Australia. *Journal of the Proceedings of the Royal Society of New South Wales*, **105**, 86–93.

—— 1984. Continental palaeomagnetism. *In*: VEEVERS, J. J. (ed.) *Phanerozoic Earth History of Australia*, Clarendon Press, Oxford, 11–16.

—— & GIDDINGS, J. 1974. Late Precambrian and Lower Palaeozoic palaeomagnetic results from South Australia and Western Australia. *Earth and Planetary Science Letters*, **22**, 355–365.

FISCHER, A. G. 1981. Climatic oscillations in the biosphere. *In*: NITECKI, M. (ed.) *Biotic crises in ecological and evolutionary time*, Academic Press, New York, 103–132.

FORTEY, R. A. (in press). A Devonian trilobite fauna from Thailand. *Alcheringa*.

GOLEBY, B. 1980. Early Palaeozoic palaeomagnetism in South East Australia. *In*: MCELHINNY, M. (ed.) *Global Reconstruction and the Geomagnetic Field in the Palaeozoic*, Reidel, Dordrecht, 11–21.

GROSS, W. 1961. *Lunaspis broili* und *Lunaspis heroldi* aus dem Hunsruckschiefer (unterdevon, Rheinland). *Notizblatt Hessisches Landesamt Bodenforsch, Wiesbaden*, **89**, 17–43.

HAMILTON, W. 1970. The Uralides and the motion of the Russian and Siberian Platforms. *Geological Society of America Bulletin*, **81**, 2553–2576.

HECKEL, P. H. & WITZKE, B. J. 1979. Devonian world palaeogeography determined from distribution of carbonates and related lithic palaeoclimatic indicators. *Palaeontological Association Special Paper*, **23**, 99–123.

HONG, FEI HOU & BOUCOT, A. J. 1990. The Balkhash-Mongolia–Okhotsk region of the Old World Realm (Devonian). *In*: MCKERROW, W. S. & SCOTESE, C. R. (eds) *Palaeozoic Palaeogeography and Biogeography*. Geological Society, London, Memoir, **12**, 297–303.

HOU, H.-F. & WANG, S.-T. 1985. Devonian palaeogeography of China. *Acta Palaeontologica Sinica*, **24**, 194–202.

HOU, J.-P. 1982. The spore assemblages of the transition strata between the Devonian System and the Carboniferous System in the Xikuangshan region, central Hunan. *Chinese Academy of Geological Sciences, Bulletin of the Institute of Geology*, **5**, 81–95.

HURLEY, N. F. & VAN DER VOO, R. 1987. Palaeomagnetism of Upper Devonian reefal limestones, Canning Basin, Western Australia. *Geological Society of America Bulletin*, **98**, 123–137.

JELL, P. A. 1974. Faunal provinces and possible planetary reconstruction of the Middle Cambrian. *Journal of Geology*, **82**, 319–355.

——, BURRETT, C. F., STAIT, B. & YOCHELSON, E. 1984. The Early Ordovician bellerophontid *Peelerophon oehlerti* (Bergeron) from Argentina, Australia and Thailand. *Alcheringa*, **8**, 169–176.

JOHNSON, J. G., KLAPPER, G. & SANDBERG, C. A. 1985. Devonian eustatic fluctuations in Euramerica. *Geological Society of America Bulletin*, **96**, 567–587.

KIRSHVINK, J. L. 1978a. The Precambrian–Cambrian boundary problem: palaeomagnetic directions from the Amadeus Basin, Central Australia. *Earth and Planetary Science Letters*, **40**, 91–100.

—— 1978b. The Precambrian–Cambrian boundary problem: magnetostratigraphy of the Amadeus Basin, Central Australia. *Geological Magazine*, **115**, 139–150.

KLAPPER, G. & JOHNSON, J. G. 1980. Endemism and dispersal of Devonian conodonts. *Journal of Paleontology*, **54**, 400–455.

KLOOTWIJK, J. L. 1980. Early Palaeozoic palaeomagnetism in Australia. *Tectonophysics*, **64**, 249–332.

KOBAYASHI, T. 1971. The Cambro-Ordovician faunal provinces and the interprovincial correlation. *Journal of the Faculty of Science, University of Tokyo II*, **18**, 129–299.

LANE, H. R. & ORMISTON, A. R. 1979. Siluro-Devonian biostratigraphy of the Salmontrout River area, east central Alaska. *Geologica et Palaeontologica*, **13**, 39–96.

LAURIE, J. & BURRETT, C. F. in prep. *Spanodonta* (Brachiopoda) from the Ordovician of Malaysia. *Journal of Palaeontology*.

LEGG, D. P. 1976. Ordovician trilobites and graptolites from the Canning Basin, Western Australia. *Geologica et Palaeontologica*, **10**, 1–58.

LI, H., HU, F. & WU, H. 1984. The geological age and flora of the clastic rocks beneath the Huanlong Formation in the Tongling area, Anhui Province. *Bulletin of the Nanjing Institute for Geology and Mineral Resources*, **5**, 152–162.

LI, Y.-P., WILLIAMS, M., ZHU, H., TAN, C. & HE, Z. 1985. Palaeozoic palaeomagnetic results from the Hexizoulang Terrane, China. *Transactions of the American Geophysical Union, Eos*, **66**, 864.

LI, Y. X. 1983. Silurian and Devonian tabulate coral assemblages in the Beishan area, Gansu Province. *Acta Palaeontologica Sinica*, **22**, 71–81.

LI, Z.-X. 1987. New palaeomagnetic results from the Late Palaeozoic rocks of Australia and their tectonic significance. *Proceedings of the Pacrim conference, Gold Coast*, 267–271.

LIN, J. L., FULLER, M. & ZHANG, W. Y. 1985a. Preliminary Phanerozoic polar wander paths for the North and South China blocks. *Nature*, **313** 444–449.

——, ——, & —— 1985b. Palaeogeography of the North and South China blocks during the Cambrian. *Journal of Geodynamics*, **2**, 91–114.

LIU, S.-F. 1965. New Devonian agnathans of Yunnan. *Vertebrata Palasiatica*, **9**, 125–136.

—— 1981. The occurrence of *Lunaspis* in China. *Kexue Tongbao*, **29**, 829–830.

LONG, J. A. 1983. New bothriolepid fishes from the Late Devonian of Victoria, Australia. *Palaeontology*, **26**, 295–320.

—— 1984. The plethora of placoderms — the first vertebrates with jaws? In: ARCHER, M. & CLAYTON, G. (eds) *Vertebrate Zoogeography and Evolution in Australasia*, Hesperian Press, Perth, 285–310.

—— in press. Late Devonian chondrichthyans and other microvertebrate remains from Northern Thailand. *Journal of Vertebrate Paleontology*.

—— & BURRETT, C. F. in press a. Early Devonian conodonts from the Kuan Tung Formation, Thailand: systematics and biogeographic considerations. *Records of the Australian Museum*.

—— & BURRETT, C. F. in press b. Fish from the Upper Devonian of the Shan-Thai Terrane indicate proximity to South China and East Gondwana Terranes. *Geology*.

——, TURNER, S. & YOUNG, G. C. 1988. A Devonian fish fauna from subsurface sediments in the east Officer Basin, South Australia. *Alcheringa*, **12**, 61–78.

LU, Y., CHU, C.-L., CHIEN, Y.-Y., LIN, H., CHOW, T. & YUAN, K. 1974. Bioenvironmental control hypothesis and its application to Cambrian biostratigraphy and palaeozoogeography. *Nanking Institute of Geology, Palaeontology Memoir*, **5**, 27–110.

LUCK, G. R. 1972. Palaeomagnetic results from Palaeozoic sediments of northern Australia. *Geophysics Journal of the Royal Astronomical Society*, **28**, 475–487.

MCELHINNEY, M. W. & EMBLETON, B. J. 1974. Australian palaeomagnetism and the Phanerozoic plate tectonics of eastern Gondwanaland. *Tectonophysics*, **22**, 1–29.

—— & LUCK, G. R. 1970. The palaeomagnetism of the Antrim Plateau Volcanics of Northern Australia. *Geophysics Journal of the Royal Astronomical Society*, **20**, 191–205.

——, EMBLETON, B. J., MA, X. H. & ZHANG, Z. K. 1981. Fragmentation of Asia in the Permian. *Nature*, **293**, 212–216.

MAWSON, R. 1987. Early Devonian conodont faunas from Buchan and Bindi, Victoria, Australia. *Palaeontology*, **3**, 251–297.

METCALFE, I. 1988. Origin and assembly of south-east Asian continental terranes. In: AUDLEY-CHARLES M. & HALLAM A. (eds), *Gondwana and Tethys, Geological Society Special Publication*, **37**, 101–118.

MILLER, J. D. & KENT, D. V. 1988. Paleomagnetism of the Silurian–Devonian Andreas redbeds. Evidence for an Early Devonian supercontinent. *Geology*, **16**, 195–198.

PAN JIANG, 1981. Devonian antiarch biostratigraphy of China. *Geological Magazine*, **118**, 69–75.

—— 1984. The phylogenetic position of the Eugaleaspida in China. *Proceedings of the Linnean Society of New South Wales*, **107**, 309–319.

—— 1986. A new species of *Microbrachius* from the Middle Devonian of Yunnan. *Vertebrata Palasiatica*, **22**, 8–13.

—— & DINELEY, D. 1988. A review of early (Silurian and Devonian) vertebrate biogeography and biostratigraphy of China. *Proceedings of the Royal Society of London*, B, **235**, 29–61.

——, HOU, F., CAO, J.-X., GU, Q.-C., LIU, S.-Y., WANG, J.-Q., GAO, L., & LIU, C. 1987. *Continental Devonian of Ningxia and its biotas*. Geological Publishing House, Beijing, 1–237.

PEDDER, A. E. H. & OLIVER, W. 1990. Devonian coral distributions. In: MCKERROW, W. S. & SCOTESE, C. R. (eds) *Palaeozoic Palaeogeography and Biogeography*. Geological Society, London, Memoir, **12**, 267–275.

PHILIP, G. M. 1966. Lower Devonian conodonts from the Buchan Group, eastern Victoria. *Micropalaeontology*, **12**, 441–460.

POJETA, J. Jr. (ed.) 1986. Devonian rocks and Lower and Middle Devonian pelecypods of Guanxi, China and the Traverse Group of Michigan. *U.S. Geological Survey Professional Paper*, **1394**, 1–108.

SCHMIDT, P. W., EMBLETON, B. J. & PALMER, H. C. 1987. Pre- and post-folding magnetizations for the early Devonian Snowy River Volcanics and Buchan Caves Limestones, Victoria. *Geophysics Journal of the Royal Astronomical Society*, **91**, 155–170.

SCOTESE, C. R., VAN DER VOO, R. & BARRETT, S. F. 1985. Silurian and Devonian base maps. *Philosophical Transactions of the Royal Society of London*, B, **309**, 27–77.

SENGOR, A. M. C., ALTINER, D., CIN, A., USTAOMER & HSU, K. J. 1988. Origin and assembly of the Tethyside orogenic collage at the expense of Gondwanaland. In: AUDLEY-CHARLE, M. & HALLAM, A. (eds) *Gondwana and Tethys*. Geological Society, London, Special Publication. **37**,

SHERGOLD, J., BURRETT, C. F., AKERMAN, T. & STAIT, B. 1988. Late Cambrian trilobites from the Tarutao Formation, Thailand. *New Mexico Institute of Mining and Metallurgy Memoir*.

STAIT, B. & BURRETT, C. F. 1982. *Wutinoceras* (Nautiloidea) from the Setul Limestone (Ordovician) of Malaysia. *Alcheringa*, **6**, 193–196.

—— & —— 1984a. *Chelodes whitehousei* from Tarutao Island, Southern Thailand. *Alcheringa*, **8**, 112.

—— & —— 1984b. Ordovician nautiloids from central and southern Thailand. *Geological Magazine*, **121**, 115–124.

——, WYATT, D. & BURRETT, C. F. 1987. Ordovician nautiloid faunas of Langkawi Islands, Malaysia and Tarutao Island, Thailand. *Neues Jarbuch für Geologie und Paläontologie, Abhandlungen*, **174**, 373–391.

STRAHAN, R. (ed.) 1983. *The Complete Book of Australian Mammals*.

Angus & Robertson, Sydney 1–530.
SZE, H. C. & LEE, H. H. 1945. Palaeozoic plants from Ningshia. *Bulletin of the Geological Society of China*, **25**, 227–260.
TALENT, J. A., GRAITSNOVA, R. T. & YOLKIN, E. A. 1986. Prototethys: Fact or phantom? Palaeobiogeography in relation to the crustal mosaic for the Asia-Australia hemisphere in Devonian–Early Carboniferous times. *In*: MCKENZIE, K. (ed.) *Shallow Tethys 2*, Balkema, Rotterdam, 87–111.
TURNER, S. & DRING, R. 1981. Late Devonian thelodonts (Agnatha) from the Gneudna Formation, Carnarvon Basin, Western Australia. *Alcheringa*, **5**, 39–48.
WANG, J.-Q. 1982. New remains of Dinichthyidae. *Vertebrata Palasiatica*, **20**, 181–186.
—— 1984. Geological and palaeogegraphical distribution of Devonian fishes in China. *Vertebrata Palasiatica*, **22**, 219–229.
—— & WANG, N.-Z. 1983. A new genus of Coccosteidae. *Vertebrata Palasiatica*, **21**, 1–8.
—— & —— 1984. New material of Arthrodira from the Wuding region, Yunnan. *Vertebrata Palasiatica*, **22**, 1–7.
WANG, N.-Z. 1984. Thelodont, acanthodian, and chondrichthyan fossils from the Lower Devonian of Southwest China. *Proceedings of the Linnean Society of New South Wales*, **107**, 419–441.
WANG, C. 1979. Some conodonts from Sipai Formation in Xiangzhou of Guangsu. *Acta Palaeontological Sininca*, **18**, 395–408.
—— & ZIEGLER, W. 1983. Devonian conodont biostratigraphy of Guangxi, South China, and the correlation with Europe. *Geologica et Palaeontologica*, **17**, 75–107.
WANG, Y., BOUCOT, A. J., RONG, J.-Y. & YANG, X.-C. 1984. Silurian and Devonian biogeography of China. *Geological Society of America Bulletin*, **95**, 265–279.
WANG, S.-T., DONG, Z.-Z. & TURNER, S. 1986. Discovery of Middle Devonian Turiniidae (Thelodonti: Agnatha) from western Yunnan, China. *Alcheringa*, **10**, 315–325.
WEBBY, B. D., WYATT, D. & BURRETT, C. F. 1984. Ordovician stromatoporoids from the Langkawi Islands of Malaysia. *Alcheringa*, **9**, 159–166.
WENSINK, H., ZIDJERWALD, J. D. A. & VAREKAMP, J. C. 1978. Palaeomagnetism and ore mineralogy of some basalts of the Geirud Formation of Late Devonian–Early Carboniferous age from the southern Alborz, Iran. *Earth and Planetary Science Letters*, **41**, 441–450.
WHITTINGTON, H. B. & HUGHES, C. P. 1972. Ordovician geography and faunal provinces deduced from trilobite distribution. *Philosophical Transactions of the Royal Society of London*, B, **263**, 235–278.

—— & —— 1973. Ordovician trilobite distribution and geography. *In*: N. F. HUGHES (ed.) *Organisms and Continents through Time*, Special Papers in Palacontology, **12**, 241–270.
WONGWANICH, T., BURRETT, C. F., TSANTHIEN, W. & CHAODUMRONG, P. (in press). Lower to Middle Palaeozoic stratigraphy of mainland Satun Province, Southern Thailand. *Journal of South East Asian Earth Sciences*.
XIANG, L. W. 1981. Some Late Devonian trilobites of China. *Geological Society of America, Special Paper*, **187**, 183–192.
YANG, S.-P., PAN, K. & HOU, H.-F. 1981. The Devonian System in China. *Geological Magazine*, **118**, 113–224.
YOUNG, G. C. 1974. Stratigraphic occurrence of some placoderm fishes in the Middle and Late Devonian. *Newsletters in Stratigraphy*, **3**, 243–261.
—— 1981. Biogeography of Devonian vertebrates. *Alcheringa*, **5**, 225–245.
—— 1984. Comments on the phylogeny and biogeography of antiarchs (Devonian placoderm fishes), and the use of fossils in biogeography. *Proceedings of the Linnean Society of New South Wales*, **107**, 443–473.
—— 1985. Further petalichthyid remains (placoderm fishes, Early Devonian) from the Taemas-Wee Jasper region, New South Wales. *Bureau of Mineral Resources Journal of Australian Geology and Geophysics*, **9**, 121–131.
—— 1986. Cladistic methods in Paleozoic continental reconstruction. *Journal of Geology*, **94**, 523–537.
—— 1987a. Devonian vertebrates of Gondwana. *American Geophysical Union, Geophysics Monograph*, **41**, 41–50.
—— 1987b. Devonian palaeontological data and the Armorica problem. *Palaeogeography, Palaeoclimatology, Palaeoecology*, **60**, 283–304.
—— 1988. Antiarchs (placoderm fishes) from the Devonian Aztec Siltstone, Southern Victoria Land, Antarctica. *Palaeontographica*, A, **202**, 1–125.
——, TURNER, S., OWEN, M., NICOLL, R. S., LAURIE, J. R. & GORTER, J. D. 1987. A new Devonian fish fauna, and revision of post-Ordovician stratigraphy in the Ross River Syncline, Amadeus Basin, central Australia. *Bureau of Mineral Resources Journal of Australian Geology and Geophysics*, **10**, 233–242.
YUAN, J. 1988. Proetiden aus dem Jungeren Oberdevon von Sud-China. *Palaeontographica*, A, **210**, 1–102.
ZHAO, X.-H., WU, X.-Y. & GU, Q.-C. 1986. Late Devonian flora from Southern Ningxia. *Acta Palaeontologica Sinica*, **25**, 544–560.
ZHU, Z.-K. 1985. New materials of Devonian and Permian conulariids from North China. *Acta Palaeontologica Sinica*, **24**, 528–538.

The biogeographic affinities of East Asian corals

LIAO WEI-HUA

Nanjing Institute of Geology and Palaeontology, Academia Sinica, Chi-Ming-Ssu, Nanjing, Peoples' Republic of China

Abstract: In East Asia the oldest tabulate and rugose corals appeared in the Early and Middle Ordovician. All rugose corals and most tabulate corals appear to have become extinct at the end of the Permian.

The Ordovician corals of North China were most closely related to the Americo-Siberian region, but those of South China occupied an independent province. The Silurian corals of Junggar, Hinggan, Mongolia, Altai and Tuva are genera characteristic of the Uralian−Cordilleran region. The South China fauna had a close affinity to that of the East Australia in the Early Silurian, but was more akin to that of the Urals and Central Asia in the Middle and Late Silurian. During the Early and Middle Devonian, there were 5 biogeographic provinces in East Asia, all belonging to the Old World Realm: (1) Arctic province; (2) Junggar−Hinggan province; (3) Uralo−Tian Shan province; (4) Palaeotethyan province and (5) South Chin province.

In East Asia, two distinctive zoogeographic provinces are fairly clearly defined: a southern province, with occurrence of *Kueichouphyllum* and a northern province with occurrences of *Gangamophyllum* during the Early Carboniferous. In addition, during the Early Permian, a Uralo-Arctic province was dominated by Durhaminidae and a Tethyan province by Waagenophyllidae.

Ordovician

In China, the oldest tabulate corals appeared in the late Early Ordovician (late Arenig) and rugose corals in the Middle Ordovician. Based on stratigraphic distribution, the Chinese Ordovician corals may be subdivided (from oldest to youngest) into the following 5 assemblage zones (Lin 1984):

(1) *Rhabdotetradium−Lichenaria* Assemblage zone (Middle Ordovician, Llanvirn);
(2) *Yohophyllum−Ningnanophyllum* Assemblage zone (Middle Ordovician, middle Caradoc);
(3) *Agetolites−Grewingkia* Assemblage zone (Upper Ordovician);
(4) *Calapoecia−Favistella* Assemblage zone (Upper Ordovician, early Ashgill);
(5) *Borelasma−Sinkiangolasma* Assemblage zone (Upper Ordovician, late Ashgill).

It has been considered that only two faunal regions existed in the Ordovician: the Americo-Siberian region and Eurasiatic region. The Middle and Late Ordovician corals of North China were most closely related to the Americo-Siberian region, but those of South China should occupy an independent province (Fig. 1). For example, the Llanvirn *Rhabdotetradium−Lichenaria* fauna and early Ashgill *Calapoecia−Favistella* fauna of North China are very similar to that in the North American−Siberian region. In contrast, the Middle Ordovician *Yohophyllum−Ningnanophyllum* fauna and Upper Ordovician *Agetolites−Grewingkia* fauna of South China occupied an independent biogeographic region, while the late Ashgill *Borelasma−Sinkiangolasma* fauna is similar to that in the Baltic region.

During the Late Ordovician, the coral faunas of Shaanxi, Gansu, Ningxia and Qinghai were related to those in Siberia (Table 1). But the coral fauna of Junggar, in addition to its Siberian affinities, also shows relationships to western North America (similarity coefficient is 49%) and Central Asia (similarity coefficient is 38%); the coral fauna of Zhejiang−Jiangxi (eastern South China) is similar to that in Europe (similarity coefficient is 42%), Siberia and eastern North America (similarity coefficient is 37%), as well as in Australia (similarity coefficient is 35%); and the coral faunas of Sichuan and Guizhou are related to the fauna of eastern North America (similarity coefficient is 43%).

Silurian

The Silurian biogeographic distribution can be divided into two major provinces: (1) The Uralian−Cordilleran province and (2) North Atlantic province.

It was first thought that during the Llandovery, there had been an essentially cosmopolitan North Silurian Realm fauna followed by a later Wenlock, Ludlow and Pridoli division into two regions. Subsequently, some authors (Y. Wang *et al.* 1984) have considered that the Uralian−Cordilleran region was co-existing with the North Atlantic region since the beginning of the Silurian, and that the level of endemism within the Uralian−Cordilleran region decreased from late Llandovery in the Yangtze epicontinental sea (where endemic genera account for 35% of the total; 19 out of 55 genera), showing a high level of similarity with that of East Australia (Table 2). The corals in the north of the Qilian Mountains, and in the central Tianshan Mountains, (Fig. 2) include some Siberian forms, while the corals in the Qomolangma region probably occupied the northern margin of ancient Gondwanaland. Owing to more extensive transgressions in the late Llandovery and Wenlock, the Yangtze fauna has less endemism (25%), and bears a close relationship with faunas of the Urals and Central Asia. Many Central Asian, Uralian and northwestern European forms occurred in West Qingling during the Late Silurian. In strong contrast to the Early and Middle Silurian situation, the west Qinling corals shows a remarkably low similarity to East Australia (Wang Hongzhen *et al.* 1984).

Devonian

It is now generally recognized that there were three first-level marine, biogeographic divisions of the Devonian world (Oliver 1976).

(1) The Old World Realm is the most extensive realm and corals are abundant. It can be subdivided into many provinces.
(2) The Eastern Americas Realm (or Appalachian Realm) was small but very distinct.
(3) The Malvinokaffric Realm is characterized by few corals and can be interpreted as a cold-water area.

The coral faunas of South China, North China and the other parts of East Asia belong to the Old World Realm. Provinciality increased in late Gedinnian to Siegenian time, and reached a maximum during the early Emsian, then decreased through the Middle Devonian to a relatively cosmopolitan coral fauna during the Frasnian. At the end of Frasnian time, a number of important elements became extinct (Boucot *et al.* 1969).

The study of the faunas of China in the Early Devonian permits the discrimination of 4 distinct biogeographic provinces (Liao *et al.* 1988) (Fig. 3).

(1) The Junggar−Hinggan province is characterized by absence of the typical Givetian brachiopod *Stringocephalus* and the presence of small solitary non-dissepimented corals (Table 3).

From McKerrow, W. S. & Scotese, C. R. (eds), 1990, *Palaeozoic Palaeogeography and Biogeography*, Geological Society Memoir No. 12, pp. 175−179.

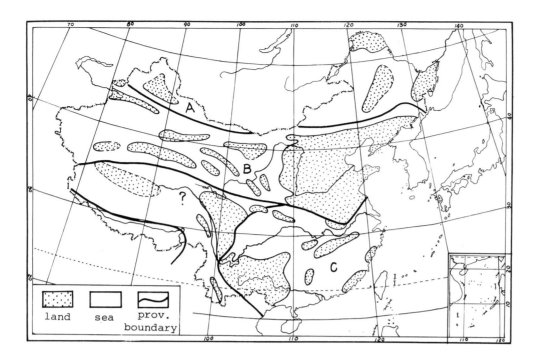

Fig. 1. Ordovician coral biogeographic provinces of China (modified after Wang 1985). A, Junggar Province; B, NW China; C, S China.

Table 1. *Similarity coefficients of Late Ordovician rugose coral faunas in China with various parts of the world*

	West Europe	Central Asia	Siberia	Eastern N America	Western N America	East Australia
Sichuan–Guizhou	(6)34	(5)30	(5)36	(6)43	(3)36	(2)19
Zhejiang–Jiangxi	(6)42	(4)28	(4)37	(4)37	(2)30	(3)35
Tarim–Tianshan	(3)28	(2)19	(2)25	(2)25	(2)40	(1)16
Burhan Buda*	(3)28	(4)38	(2)25	(4)49	(1)19	(2)23
Junggar	(6)37	(5)38	(5)49	(4)38	(5)49	(2)25
Shaanxi–Gansu–Ningxia	(2)18	(2)19	(4)49	(3)37	(1)20	(2)33

(Number in brackets are common genera, after Wang Hongzhen 1985)

* Mts Burhan Buda is in the middle part of the Qinghai province. The common genera in Mts Burhan Buda and eastern N America are *Brachylasma*, *Favistina*, *Streptelasma* & *Tryplasma*.

Table 2. *Comparison of Silurian rugose corals of South China with other parts of the world (upper, number of genera; middle, number of genera in common; lower, similarity coefficient)*

	Siberia	Central Asia	Urals–Tianshan	NW Europe	Western N America	Eastern N America	East Australia
Wenlock and late Llandovery	40	X	24 13 40	28 14 61	29 10 32	X	17 10 36
Early and middle Llandovery	55	X	26 18 49	21 16 47	X	X	25 21 58

(Modified from Wang Hongzhen *et al.* 1984)

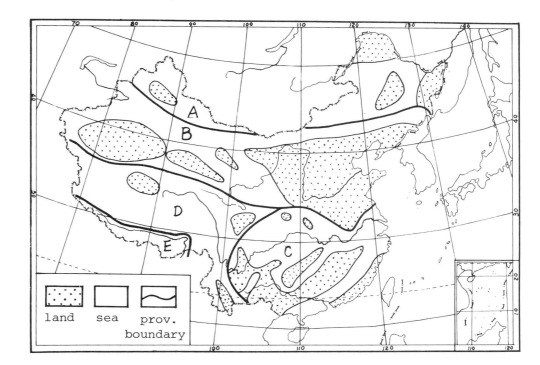

Fig. 2. Silurian coral biogeographic provinces of China (modified after Li 1987). A, Junggar–Hinggan Province; B, Tianshan–Qilian Province; C, Yangtze Province; D, N Tibet Province; E, Qomolangma Province.

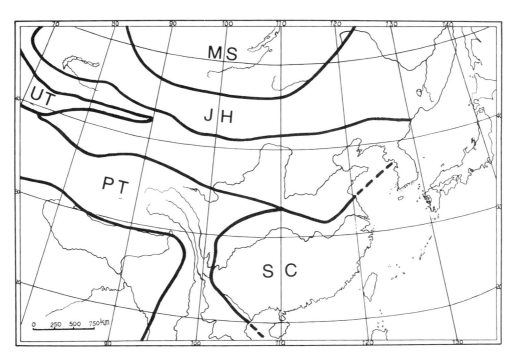

Fig. 3. Devonian coral biogeographic provinces of E Asia. MS, Mongolo–Sayan Province; JH, Junggar–Hinggan Province; UT, Uralo-Tien Shan Province; SC, S China Province; PT, Palaeotethyan Province.

Table 3. *Comparison of total number of genera and endemic forms of Lower Devonian rugose corals in China*

	S China	Tibet–W Yunnan	Junggar–Hinggan	S Tianshan
Total genera	42	29	72	X
Endemic genera	12	0	4	X
Percent endemic	29	0	6	X

This coral fauna is characterized by the presence of European and North American forms.

(2) The Tianshan province is characterized in the Early Devonian by extensive development of *Chlamydophyllum*, *Alaiphyllum* and others. It is probably related to the Urals.

(3) The Palaeotethyan province including Tibet and western Yunnan, yields a lot of Australian elements in the Early Devonian, such as *Martinophyllum*, *Embolophyllum*, *Loyolophyllum*, *Lyrielasma* and *Carlinastraea*.

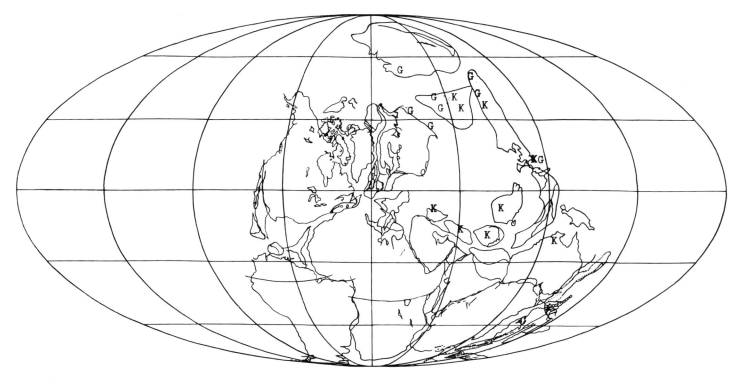

Fig. 4. Early Carboniferous (Visean) coral distributions. K, *Kueichouphyllum*; G, *Gangamophyllum*.

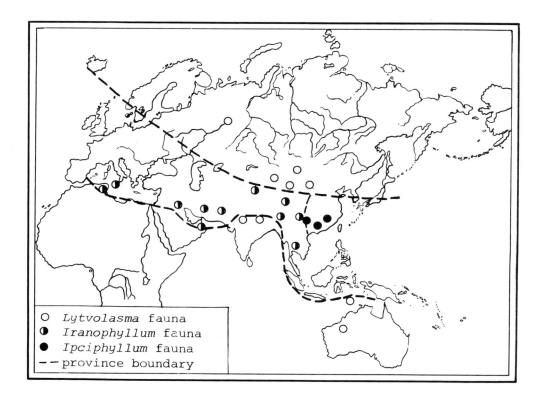

Fig. 5. Early Permian coral biogeographic provinces of Asia after Wu *et al.* 1982).

(4) The South China province was an area of high Lower Devonian endemism, especially in the Siegenian and the lower Emsian. However, from the late Emsian on, there appeared many West European and western North American, as well as some Australian genera, for instance *Trapezophyllum, Acanthophyllum, Dohmophyllum, Psydracophyllum, Breviseptophyllum, Utaratuia, Sociophyllum, Dendrostella*, and others.

Carboniferous

Two distinctive zoogeographical realms are apparent for the Carboniferous: America and Eurasia. In the latter realm, two provinces are fairly clearly defined: a southern province with *Kueichouphyllum* and a northern province with *Gangamophyllum* during the Early Carboniferous.

There are many small ceratoid non-dissepimented corals in Junggar, northern Xinjiang, associated with ammonoid faunas during the Tournaisian. Lower Carboniferous corals in Qinghai, Gansu, Ningxia and the Tianshan Mountains of NW China are characterized by many European and Central Asian elements, as well as some North American genera, such as *Caninia, Lophophyllum, Siphonophyllia, Carruthersella, Cyathoclisia, Humboldtia, Gangamophyllum, Dorlodotia, Palaeosmilia, Siphonodendron, Caninophyllum* and others (Fig. 4).

The Early Carboniferous corals of the South China epicontinental sea can be designated as the *Pseudouralinia–Kueichouphyllum* fauna, which is characterized by endemic genera of so-called 'South China type' (Luo 1984). In the Tethyan province, including southern Qinghai, northern Tibet, western Yunnan and northern Sichuan, typical genera of South China are often associated with the typical genera of the Tianshan–Qilian Mountains. This may be called the mixed fauna.

Permian

During the Early Permian, there are also two provinces in East Asia (Hill 1981): (1) Uralo-Arctic province with Durhaminidae dominant; (2) Tethyan province with Waagenophyllidae dominant.

The north and south parts of Asia belong to the cold-water *Lytvolasma* fauna, while the middle part of Asia is the warm-water Tethyan province, termed the *Iranophyllum* fauna, and its eastern extension in the South China province, namely the *Ipciphyllum* fauna (Fig. 5).

All rugose corals and nearly all tabulate corals appear to have become extinct at the top of the Permian Period (Changhsing Formation) in South China.

References

BOUCOT, A. J., JOHNSON, J. G. & TALENT, J. A. 1969. Early Devonian brachiopod zoogeography. *The Geological Society of America Special Paper*, **119**, 1–86.

HILL, D. 1981. Treatise on Invertebrate Palaeontology (F), Coelenterata Supplement 1, Rugosa and Tabulata. TEICHERT, C. (ed.) Vol. 1–2. Geological Society of America & University of Kansas Press.

LI, Z. 1987: On the Silurian Biogeography of China. *Collected Papers of Lithofacies and Palaeogeography*, **3**, Geological Publishing House. Beijing, China, 125–138.

LIAO, W & RUAN, Y. 1988. Devonian of East Asia. *In*: MCMILLAN, N. J. (ed.) *Devonian of the World—Proceedings of the Second International Symposium on the Devonian System*. Vol. 1. Canadian Society of Petroleum Geologists, Calgary, Canada. 597–606.

LIN, B. 1984. New Developments in coral biostratigraphy of the Ordovician of China. *Palaeontographica Americana*, **54**, 444–447.

LUO, J. 1984. Early Carboniferous rugose coral assemblages and palaeobiogeography of China. *Palaeontographica Americana*, **54**, 427–432.

OLIVER, W. A. Jr. 1976. Biogeography of Devonian rugose corals. *Journal of Paleontology*, **50**, 365–373.

WANG, H. 1985. Systematics and Palaeobiogeography of the Middle and Late Ordovician rugose corals of China. *Earth Science, Journal of Wuhan College of Geology*, **10**, 19–34.

——, LI, Z. & WANG, Z. 1984. Silurian and Early Devonian rugose coral biogeography of China. *Palaeontographica Americana*, **54**, 423–426.

WANG, Y., BOUCOT, A. J., RONG, J. & YANG, X. 1984. Silurian and Devonian biogeography of China. *Geological Society of America Bulletin*, **95**, 265–279.

WU, W., LIAO, W. & ZHAO, J. 1982. Palaeozoic rugose corals from Tibet. *In*: *The Series of the Scientific Expedition to the Qinghai-Tibet Plateau*, Palaeontology of Tibet, **4**, Science Press, Beijing, 144–176.

Brachiopod zoogeography across the Ordovician–Silurian extinction event

P. M. SHEEHAN & P. J. COOROUGH

Department of Geology, Milwaukee Public Museum, 800 West West Wells St., Milwaukee, WI 53233 USA

Abstract: The extinction event at the close of the Ordovician, one of the largest of the Phanerozoic, had a profound effect on the zoogeographic distribution of brachiopods. Strong endemism in the Ashgill was particularly apparent in the epicontinental seas covering the primary lithospheric plates. Areas in the open ocean and around the margins of the continents had more widely distributed faunas. Glacio-eustatic decline in sea level and climatic deterioration caused by the developing North African glaciation have been widely invoked as the cause of the extinction event. The Hirnantian Stage roughly corresponds to the glacial maximum, and Hirnantian faunas were somewhat more cosmopolitan than were early–middle Ashgill faunas. At the end of the Hirnantian, sea level rise and climatic amelioration accompanying the end of the glacial maximum apparently caused another wave of extinctions. Early Silurian brachiopods were far more cosmopolitan, especially in the epicontinental seas, than they had been in the Late Ordovician. There is a strong correlation between wide geographic distribution of genera and survival of the extinction event. Provincial patterns are consistent with the symposium maps.

Zoogeographic patterns provide important constraints on plate tectonic reconstructions. However, many factors other than geographic distance influence faunal similarities. We examine here the effects of a major extinction event on global biogeography. The widely recognized cosmopolitan nature of the Silurian is found to have resulted from the extinction event, rather than changing geographic positions of plates. In general, faunas following major extinction events are cosmopolitan, and they provide little assistance in determining the positions of plates.

The Ordovician–Silurian extinction event was one of the five largest in the Phanerozoic (Raup & Sepkoski 1982). Most extinction events result in cosmopolitan faunas (Jablonski 1986), and the terminal Ordovician event is no exception (Boucot 1975; Sheehan 1975; Sheehan & Coorough 1986). However, quantitative data documenting this phenomenon are lacking.

The palaeozoogeography of brachiopods from both the Late Ordovician and the Early Silurian has been examined by numerous previous authors (e.g. Williams 1969, 1973; Boucot & Johnson 1973; Boucot 1975, 1979; Spjeldnaes 1961, 1967, 1981; Wang *et al.* 1984; Jaanusson 1973, 1979; Rozman 1977). Historically, brachiopod workers have focused on either Ordovician or Silurian faunas, and as a result few studies have applied comparable techniques to faunas of both Ordovician and Silurian age. In this investigation we apply simple statistical treatments to faunal lists assembled from more than 200 sources. Ten distinct geographic regions have sufficient data for analysis. The regions correspond to continental or oceanic plates or to the margins of continental plates (Fig. 1). Faunal lists were compiled for three time intervals: the early–middle Ashgill (prior to the Hirnantian), the Hirnantian (Upper Ashgill), and the early–middle Llandovery. The lists have been deposited with the British Library at Boston Spa, W Yorkshire, UK, as Supplementary Publication No. SUP18059 (14 pp.) and are also available from the Geological Society Library.

In line with the intent of this symposium we attempt to follow the zoogeographic relationships of brachiopods from distinct geographic regions through time on the symposium maps. We do not attempt to recognize formal zoogeographic provinces.

Epicontinental seas harbour relatively endemic faunas compared with more cosmopolitan faunas at the margins of the continents and in shallow water settings adjacent to islands in open oceans. Most of the zoogeographic studies cited above recognized well-differentiated zoogeographic provinces in the Ordovician. Silurian faunas on the other hand have been found to be among the most cosmopolitan known, although Wang *et al.* (1984) found that China was an unusually provincial region during the Silurian. Few authors have considered the causes of the changes in provinciality at the systemic boundary, but the most common theme has invoked continental drift (e.g. McKerrow & Cocks 1976; Cocks *et al.* 1979), although the effects of the extinction have also been suggested (Lesperance & Sheehan 1976b).

During the Hirnantian, epicontinental seas were drained by glacioeustatic sea-level decline (Sheehan 1973, 1975, 1988; Berry & Boucot 1973; Lenz 1976; Brenchley & Newall 1980, 1984; Brenchley 1984). During this regression and associated with a distinct interval of extinction, the *Hirnantia* Fauna became widespread (Rong 1983; Brenchley & Newall 1984; Brenchley 1984; Sheehan 1979; Lesperance & Sheehan 1976a). At the end of the Ordovician another wave of extinctions, associated with a rapid rise in sea level and presumed climatic amelioration (Brenchley 1984), produced a cosmopolitan fauna in the Early Silurian.

Methods

Ashgill through middle Llandovery faunal lists were compiled from the literature. More than 200 faunal lists cited in Coorough (1986) were expanded by an additional 30 faunal lists (Supplementary Publication). The faunal lists were assigned to lithospheric plates based on maps furnished for this symposium. Faunas from the margins of the Laurentian plate, from the open ocean islands in the region of Great Britain, Ireland, and Belgium, and Southern Europe which was marginal to North Africa were recorded separately. The ten regions (Fig. 1A) with their abbreviations in parentheses are: (1) Laurentia (L); (2) the margins of Laurentia (including south-central Alaska, northern Greenland, Gaspe, Maine, and the Klamath Mountains of California) (Lm); (3) Baltica (B); (4) the British Isles and Belgium (En); (5) southern Europe (Eu); (6) North Africa (Af); (7) Kazakhstan (K); (8) Siberia (S); (9) Kolyma (Kol); and (10) South China (C).

The only significant departure from the maps adopted for this symposium is that Northern Ireland and southern Scotland are included in the same region (En) as the remainder of the British Isles and Belgium. Although the region to the north of the Iapetus suture is commonly believed to have been associated with the margin of North America, the Late Ordovician brachiopods have a much stronger affinity with the British Isles. Of 62 early–middle Ashgill genera recorded in the Irish–Scottish region north of the suture, 18 genera have been found in the British Isles south of the suture but not the margin of Laurentia, whereas only one genus has been previously recorded in the margin of Laurentia but not elsewhere in the British Isles. Of the remaining genera, 23 have been found both south of the suture and in the margin of Laurentia, and 20 genera have not been recorded in either of the areas. The existence of the Iapetus Ocean is not in question, but it is possible that northern Ireland and the Girvan area may not have been near the margin of Laurentia in the Late Ordovician.

Early–middle Llandovery brachiopods were very uncommon

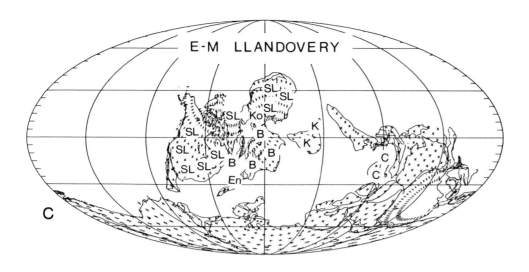

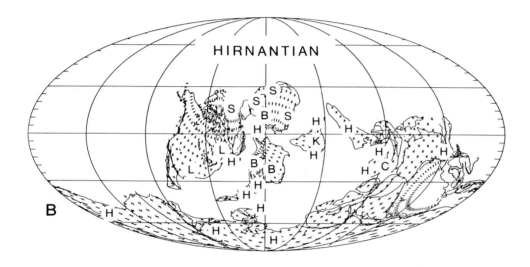

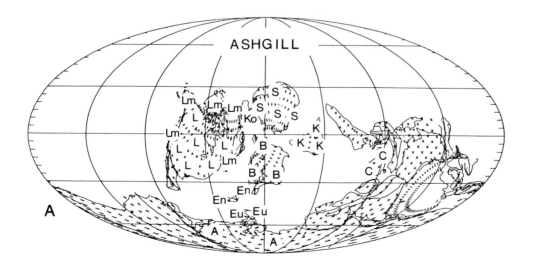

Fig. 1. Brachiopod Zoogeography. (**A**) Brachiopod zoogeography during the early–middle Ashgill. (**B**) Brachiopod zoogeography during the Hirnantian. *Hirnantia* fauna occurrences are designated by the letter H. (**C**) Brachiopod zoogeography during the early–middle Llandovery.

in North Africa and Southern Europe (Berry & Boucot 1967), and as a result only eight regions could be recognized in the Silurian. Very limited data is available for a number of other important plates (e.g. Australia, South America and North China).

Lists of genera and their regional distribution have been compiled for the early–middle Ashgill, Hirnantian and early–middle Llandovery (Supplementary Publication). The regional faunas are then compared using the Provinciality Index (PI) of Johnson (1971) where

$$PI = \frac{C}{2E}$$

when C = the number of genera in common between two regions and E equals the number of genera endemic to the region with fewer genera. The Provinciality Index provides a straightforward method of comparison of regions with samples of differing sizes. A PI = 1.0 occurs when 50% of all genera are cosmopolitan if the samples are of the same size (or normalized to twice the smaller fauna if the sample sizes differ), and a PI = 0.5 occurs when the number of cosmopolitan genera equals the number of endemic genera in the smaller fauna. Although this index is not ideal, it does provide a method to compare samples of differing sizes by normalizing to the smaller sample.

Numerous difficulties confront any zoogeographic evaluation. In this study some of the most important problems include inaccurate correlation within or between regions, incomplete description of faunas from many regions, inconsistent identifications of brachiopods by different workers (lumpers or splitters), incomplete coverage of the literature, the availability of more outcrop for study of early–middle Ashgill strata than either Hirnantian or early–middle Llandovery, and the likelihood that early–middle Ashgill brachiopods have been more intensively studied than those from the other intervals.

The recent set of correlation charts for Ordovician strata published by the IGCP and Silurian correlation charts edited by Berry & Boucot (GSA Special Papers series) were used for correlation where available. Three relatively long time intervals have been chosen to compensate for the inaccuracy of correlation between distinct zoogeographic regions. Only robust trends in the data base should be trusted, because small differences are most likely due to the inherent inaccuracies of the data base.

Stratigraphic distribution of brachiopod genera

The early–middle Ashgill has 221 genera, the Hirnantian 124 genera, and the early–middle Llandovery 133 genera (Supplementary Publication). Although the decline in genera was caused primarily by the extinction event, some of the differences in diversity are probably due to factors outlined previously. Early–middle Ashgill strata have more exposure than Hirnantian and early–middle Llandovery strata, and this bias tends to inflate the early–middle Ashgill figure. The early–middle Ashgill and early–middle Llandovery time intervals were probably longer than the Hirnantian, and this tends to undervalue the Hirnantian number. Finally, since the early–middle Llandovery and the Hirnantian are substantial intervals of geological time, some of the genera recorded must have evolved or gone extinct during these time intervals, making the figures overestimates of standing generic diversity at the beginning of the intervals (following the two primary pulses of extinction).

Marginal and open ocean v. epicontinental regions

Johnson (1974) proposed the term 'perched faunas' for zoogeographic realms that evolved in epicontinental seas. In the early–middle Ashgill the perched faunas included those from the five continental plates examined in this study. Glacio-eustatic re-

Table 1. *Provinciality Indexes for regions during the early–middle Ashgill.*

Lm	0.48								
B	0.48	0.59							
En	0.66	1.47	2.36						
Eu	0.14	0.33	0.47	1.03					
Af	0.00	0.06	0.40	1.75	1.00				
K	0.41	0.42	0.59	0.68	0.23	0.00			
S	0.44	0.48	0.37	0.68	0.28	0.06	0.73		
Kol	0.27	0.44	0.56	0.56	0.15	0.06	0.92	2.33	
C	0.16	0.37	0.42	1.00	0.47	0.14	0.60	0.42	0.56
	L	Lm	B	En	Eu	Af	K	S	Kol

gression and associated climatic deterioration at the end of the Ordovician would have had the greatest effect in epicontinental seas (Sheehan 1988). The early–middle Ashgill endemic genera which became extinct were concentrated in these epicontinental seas (Supplementary Publication).

Open-ocean environments and areas marginal to the continental plates had more cosmopolitan taxa than did those in epicontinental seas. Four of these regions were studied in the Ashgill: British–Belgian (En), Laurentian Margin (Lm), Kazakhstan (K) and Southern Europe (Eu). Region En, although commonly allied with Baltica (B), is actually in an open ocean position (Fig. 1A). Region Lm was marginal to Laurentia (L), and region Eu was marginal to the North African part of Gondwanaland (Af). Kazakhstan is composed of several island arcs. These marginal faunas (Table 1) have PIs with a mean of 0.65 ($n = 30$) in the Ashgill compared to all other Ashgill faunas have PIs with a mean of 0.44 ($n = 15$). Thus the marginal and open ocean regions were distinctly more cosmopolitan than were the epicontinental seas.

Similar open ocean cosmopolitanism compared to epicontinental sea endemism has been found in the Ordovician for conodonts (Bergstrom 1973; Sweet & Bergstrom 1974, 1984), chitinozoans (Laufeld 1979), bryozoans (Anstey 1985), trilobites (Ross 1975; Fortey 1975, especially fig. 5) and in the Devonian for brachiopods (Savage *et al.* 1979), conodonts (Klapper & Johnson 1980) and corals (Oliver 1976). The presence of cosmopolitan faunas in open oceans and around the margins of continents was a common zoogeographic pattern in the Paleozoic and is a fundamental corollary to the concept of perched faunas.

Marginal, as opposed to epicontinental sea environments, were not differentiated for the faunas from Siberia, South China and parts of Baltica. As a result, faunas from these regions probably are a mixture of cosmopolitan marginal faunas and endemic epicontinental sea faunas. When faunal distributions are better understood from these regions, provincialism of these epicontinental seas may be determined to have been even stronger than recorded here. Visual scanning of the fauna lists and geographic relationships indicate that the Altay–Sayan and Kolyma areas may have been marginal to Siberia and Novaya Zemleya marginal to Baltica.

Ashgill biogeography

Brachiopods from the primary continental plates (B, L, Af, S, C) have PIs indicating substantial endemism (Table 1). Kolyma and Siberia have very similar faunas (PI = 2.33) and the PIs compared with other regions track each other closely (Table 1). Kolyma could simply be part of Siberia or it could be a marginal fauna, although the close tracking of Kolyma and Siberia PIs with other regions does not follow the pattern for other marginal regions. On the other hand, since marginal faunas are not recognized for Siberia, the tracking of Kolyma may be due to combining marginal with epicontinental sea faunas in the list of Siberian genera.

The Lm and En regions have similar faunas (PI = 1.47), but the PIs of the two regions track other regions quite differently. For example, En correlates very strongly with B (PI = 2.36) and Af (PI = 1.75), while Lm has only a modest correlation with B (PI = 0.59) and a very weak correlation with Af (PI = 0.06).

A relatively strong temperature gradient from carbonate dominated tropical and subtropical regions to clastic dominated southern high-latitude regions has been widely recognized in the Ashgill (Spjeldnaes 1961, 1967; Havlicek 1976). Southern high-latitude deposition has been reviewed by Robardet & Dore (1988). North Africa, which was within 30 degrees of the south pole, had very low diversity (9 genera) which may be due to the high latitude position. However, there are relatively few Af samples, and they occur in a small geographic area (See Schopf et al. 1978). The greatest generic diversity is in En which was near 45° S latitude. In addition, four zoogeographically distinct epicontinental sea provinces were within 30° of the equator (Fig. 1A). Brachiopod zoogeography is consistent with the symposium map (Fig. 1A) with the exception of northern Ireland and southern Scotland as discussed above.

Hirnantian biogeography

The Hirnantian Stage corresponds with world-wide glacio-eustatic sea level decline (Sheehan 1988). Epicontinental seas retreated to the margins of the plates, as is reflected by the distribution of localities on Fig. 1B. The ten regions maintained patterns of endemism that were quite similar to the early–middle Ashgill (Table 2, Fig. 1B). Af remained endemic compared with other epicontinental sea plates. Relatively strong endemism was maintained in S, K and L, although all three became much more similar to B than they had been earlier in the Ashgill. Coincident with the glacio-eustatic regression, the *Hirnantia* Fauna became widely distributed (Rong 1984a, b; Brenchley & Cullen 1984) Typical examples of this unique fauna have been found in Af, Eu, En, B, K, Lm, C, Kol, Argentina (Bennedetto 1986), North China (Su 1980), and Australia (Laurie 1982). The *Hirnantia* Fauna is concentrated in the southern hemisphere, and southward facing margins of the continents in the tropics, such as Gaspé and Maine in North America, Kolyma adjacent to Siberia and western Kazakhstan (Fig. 1B). Exceptions do occur, for example eastern Kazakhstan. Apparently contemporaneous occurrences of virgianid faunas are concentrated north of the primary occurrences of the *Hirnantia* fauna in Siberia, North Greenland, and Kazakhstan, and less commonly in Sweden, Norway and Estonia. Brachiopod zoogeography is consistent with the symposium map (Fig. 1B).

Early–middle Llandovery biogeography

The glacio-eustatic transgression just before the end of the Hirnantian was accompanied by a final pulse of brachiopod extinctions. As the transgression proceeded, deposition resumed in the epicontinental seas. A completely new zoogeographic regime was present in the Silurian. The platform regions had much more

Table 2. *Provinciality Indexes for regions during the Hirnantian.*

Lm	0.39								
B	1.10	1.28							
En	1.10	0.50	0.70						
Eu	0.07	0.39	0.89	0.89					
Af	0.08	0.20	0.38	0.38	0.38				
K	0.17	0.47	0.82	0.41	0.33	0.14			
S	0.17	0.50	1.75	1.00	0.40	0.08	0.50		
Kol	0.39	0.77	1.25	0.67	0.39	0.14	0.58	0.63	
C	0.30	0.37	1.50	0.75	1.29	0.50	0.35	0.40	0.32
	L	Lm	B	En	Eu	Af	K	S	Kol

Table 3. *Provinciality indexes for regions during the early and middle Llandovery.*

Lm	0.50						
B	1.36	2.25					
En	0.68	0.77	1.05				
K	0.50	0.23	1.07	0.42			
S	2.10	1.00	0.95	0.47	1.33		
Kol	0.32	0.50	1.00	0.32	0.19	1.30	
C	0.43	0.47	0.76	0.43	0.35	0.58	0.63
	L	Lm	B	En	K	S	Kol

	L	Lm	B	En	K	S	Kol	Average
Ashgill	0.16	0.37	0.42	1.00	0.60	0.42	0.56	0.50
Early–middle Llandovery	0.43	0.47	0.76	0.43	0.35	0.58	0.63	0.52

Fig. 2. Endemicity of South China during the Ashgill and Llandovery. Note that unlike other regions, in the Early Silurian endemicity of the South China region remains similar to that found in the Ashgill.

Table 4. *Provinciality Indexes for early–middle Llandovery Provinces, assuming Laurentia and Siberia are one province and the Laurentian Margin and Baltica are another province.*

B-Lm	0.89				
En	0.42	1.11			
K	2.25	1.07	0.42		
Kol	1.30	1.30	0.32	0.19	
C	0.52	0.93	0.43	0.35	0.63
	L-S	B-Lm	En	K	Kol

cosmopolitan faunas than previously (Table 3). S and L have a PI of 2.10 and B and Lm have a PI of 2.25. Although distinct zoogeographic regions are shown on Fig. 1C, these regions carry faunas that are not nearly as distinct as they had been in the Ordovician.

Geographic distance between plates may have exerted a primary control on brachiopod distribution. Only C maintained a fauna that is distinct from the other regions, and had PIs reflecting endemism similar to that in the Ordovician (Fig. 2). China was set well away from the other areas with data (Fig. 1C). Geographically, B was quite close to L, S and K and has PIs of 1.36, 0.95 and 1.07 respectively. L and K, which are separated by B, have the least in common with a PI of 0.50.

Table 4 was constructed by calculating the PIs for the combined regions. L and S are combined into a single province based on a PI of 2.10 (Table 3 and Fig. 1C). A distinct Lm province disappeared and regions previously part of the Lm province became allied with the closest plate (compare Figs 1B and 1C). By the Early Silurian, Baltica lay very close to the northeastern margin of North America. The region of Gaspe and Maine, which previously had been a marginal region to Laurentia, had a fauna with Baltic affinities. Similarly, north Greenland, situated between Laurentia and Siberia, became part of the combined S-L Province.

Extinction and survival of genera

Of 211 genera recorded in the early–middle Ashgill, 92 (44%) did not survive into the Hirnantian. A substantial portion of these genera probably died during the early–middle Ashgill rather than at the base of the Hirnantian. In addition to the 92 genera that have not been found later than the early–middle Ashgill, 49 early–middle Ashgill genera (23%) were recorded in the Hirnantian but not the Silurian. Thus 67% of the early–middle

Ashgill genera were not recorded in the Silurian. These data show the maximum magnitude of generic extinctions during the environmental perturbations at the end of the Ordovician.

In the Early Silurian there are 15 Lazarus genera that were found in the early–middle Ashgill but not in the Hirnantian. These 15 genera indicate the quality of the data base, because although 124 genera were recorded in the Hirnantian, at least 15 more genera must have existed during this interval but were not recorded.

Of the 124 genera in the Hirnantian, 103 were recorded previously in the Ashgill. Thus only 21 new genera (17%) appeared during the Hirnantian. Of the 124 genera 63 (51%) were not recorded from the Silurian. It must be remembered that the Hirnantian fauna had already been culled by the extinctions at the boundary between the middle Ashgill and the Hirnantian. Again, it is uncertain what proportions of these genera died at the end of the Hirnantian as opposed to during the Stage. Of the 21 genera first recorded in the Hirnantian, 14 (66%) did not survive into the Silurian.

Of the 130 genera in the Early Silurian, 77 were recorded previously in the Ordovician. Thus 53 genera (41%) appeared in the Early Silurian. Although it is tempting to compare the generic origination rates for the Early Silurian (41%) and the Hirnantian (17%) it must be remembered that the Early Silurian could have been a longer time interval than was the Hirnantian. In the future the authors will attempt to subdivide the Early Silurian into early and middle Llandovery, and patterns revealed in this study might be more instructive. If Hirnantian origination rates were exceptionally low, both Early Silurian intervals should have comparatively high origination rates.

Ordovician v. Silurian provincialism

Although as Wang *et al.* (1984) noted, some provinciality certainly existed in the Early Silurian, provinciality was not nearly as strong as it had been in the Ordovician. Regions that previously had an endemic fauna became part of the same province, resulting in fewer provinces in the Silurian. Other regions that remained distinct in the Silurian were decidedly less endemic than they had been in the Ordovician (with the exception of China).

Figure 3 compares the provinciality between all regions in the three time intervals under consideration. Note that because southern Europe and northern Africa have no Early Silurian shelly faunas there are fewer regions in the Silurian. Decidedly more regions had a PI below 0.5 in the Ordovician than did post-extinction faunas. The average PI between the regions changed from 0.58 in both the early–middle Ashgill and the Hirnantian to 0.78 in the Early Silurian.

Even more striking changes are apparent when faunas are compared from the regions of epicontinental seas, where the effect of the extinction of perched faunas is apparent. During the early–middle Ashgill, distinct provinces existed on each of the primary continental plates that were studied, but subsequent faunas were progressively less endemic. PIs comparing Baltica, Laurentia, South China and Siberia with each other range from 0.16 to 0.48 (mean = 0.38) for the early–middle Ashgill, from 0.17 to 1.75 (mean = 0.87) for the Hirnantian, and from 0.43 to 2.10 (mean = 1.03) for the early–middle Llandovery. Comparing B, L, C & S to all regions reveals a similar trend, with average PIs for the four regions of 0.56 ($n = 30$) for the early–middle Ashgill, 0.65 ($n = 30$) for the Hirnantian and 0.89 ($n = 22$) for the early–middle Llandovery. The progressive increase in cosmopolitanism in epicontinental sea faunas reflects the tendency for endemic genera to suffer greater extinction than widely distributed genera during the Late Ordovician. The preferential survival of cosmopolitan taxa is examined in detail below. The pattern of extinction in the four regions supports the finding that the terminal Ordovician extinction was concentrated in the perched faunas that lived in epicontinental seas.

PI =	<0.5	0.5–1.0	>1.0
Early–middle Ashgill	27	13	5
Hirnantian	24	14	7
Early–middle Llandovery	10	11	7

Fig. 3 Comparisons of PI values between regions for each time interval. During both the early–middle Ashgill and the Hirnantian most PI values were less than 0.5, while in the early–middle Llandovery most PI values were greater than 0.5. Therefore, Ordovician faunas were more provincial than Silurian faunas.

Although biases exist in the data-base, most should not affect the basic trends reported here. For example, the small areal distribution of Hirnantian shelly faunas is compensated for by nature of the Provinciality Index, which normalizes to the smaller sized fauna. The long time intervals studied might allow new genera to evolve in isolated areas, but this would bias the data against the conclusion that post-extinction faunas were more cosmopolitan than the early–middle Ashgill fauna. This is especially true for the Early Silurian data which do not allow a census of genera living immediately after the extinctions.

Preferential survival of widely distributed genera

A clear correlation exists between the distribution of genera among the regions and the survival of genera during the extinction interval. The number of regions from which each genus was recorded was determined for each time interval from the fauna lists (Supplementary Publication). Figure 4 shows the distribution of early–middle Ashgill genera by number of regions in which they were found. Also shown are the number of genera that survived into the Hirnantian and the number that survived into the Llandovery. Similarly, Fig. 5 shows the distribution of Hirnantian genera by number of regions in which they were found. Also shown are the number of genera that survived into the Llandovery.

The diagrams show a strong correlation between wide geographic distribution and survival of the extinction event. For example, more than half of the early–middle Ashgill genera recorded in 6 or more regions survived into the Silurian while only 14% of the genera recorded in only one region survived into the Silurian. The finding that widely distributed genera are more likely to survive an extinction event than are narrowly distributed genera is not unexpected. Jablonski (1986) found a similar pattern at the Cretaceous–Tertiary boundary and reviewed similar patterns at other extinction events.

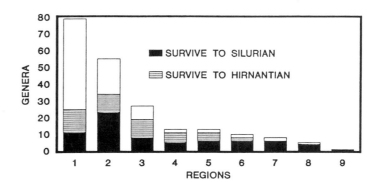

Fig. 4. Histogram showing distribution of early–middle Ashgill genera by the number of regions in which they were recorded. The black region at the bottom of each column records the number of genera that survived into the Silurian. The region with horizontal lines in each column records the number of genera identified in the Hirnantian but not the Silurian, and the unshaded region records the number of genera not found above the Ashgill. Note that there is strong preferential survival of widely distributed species.

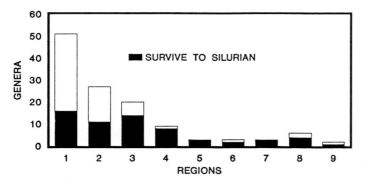

Fig. 5. Histogram showing distribution of Hirnantian genera by the number of regions in which they were recorded. The black region at the bottom of each column records the number of genera that survived into the Silurian. Note strong preferential survival of widely distributed species, but some widely distributed species did not survive into the Silurian.

One remarkable departure from this pattern is found among some members of the *Hirnantia* Fauna. Elements of the *Hirnantia* Fauna first appear in high southern latitudes (Havlicek 1970, 1971; Havlicek & Massa 1973). The fauna expanded dramatically during the Hirnantian Stage during which time the fauna was very widely distributed (Fig. 1B). At the beginning of the Silurian this fauna disappeared. Figure 5 reveals that there were four genera that had very wide distribution (6 or more regions), but which nonetheless became extinct at the end of the Ordovician. The four genera, *Eostropheodonta* (8 regions), *Leptaenopoma* (6 regions), *Paramalomena* (8 regions) and *Plectothyrella* (9 regions), were all members of the *Hirnantia* Fauna. It is tempting to suggest that in this case wide geographic distribution did not afford protection from extinction because the environment to which the *Hirnantia* Fauna was adapted disappeared. The rapid transgression in the Early Silurian was caused by glacio-eustatic sea level rise which heralded the end of the African glacial episode. If, as several authors have suggested, the *Hirnantia* Fauna was adapted to cool waters, this environment may have suddenly disappeared as global climate ameliorated and high latitude shelves were inundated by the transgression.

Discussion

Provinciality results from the evolution of organisms in relative isolation. The Early Silurian fauna was composed of widely distributed genera as an after effect of the extinction. The time frame for the evolution of new zoogeographic entities appears to have been on the order of tens of millions of years. Boucot & Johnson (1973), Boucot (1975) and Rozman (1986) noted that the Silurian fauna began to develop some evidence of endemism in the Wenlock, but it was not until the Early Devonian that strong endemism characterized the world brachiopod biota. Thus, much of the Silurian (nearly 30 million years) was needed for the zoogeographic patterns to recover from the Late Ordovician extinction event.

Recovery of community patterns after the extinction required an order of magnitude less time than did recovery of zoogeographic patterns. For about 3 to 5 million years after the extinction, Early Silurian communities were unstable and rapidly changing in composition prior to becoming stable in the late Llandovery (Sheehan 1975, 1982). Intervals of similar length were needed for recovery of communities following other extinction events (Sheehan 1985). The differing lengths of time required for communities and zoogeographic patterns to recover from extinction events may reflect basic differences of integration between communities and provinces.

On the other hand, the destruction of highly endemic early-middle Ashgill provincial patterns occurred in the geologically brief interval of the Hirnantian Stage. The rapidity of change and the coincidence of the biogeographic changes with one of the major extinction events of the Phanerozoic does not support suggestions that the change to cosmopolitanism in the Silurian was due to continental drift bringing zoogeographically distinct regions together. When biogeography is used in studies of plate distributions, care must be taken following intervals of extinction because geographic proximity may not be needed to explain faunal similarities.

References

ANSTEY, R. L. 1985. Bryozoan provinces and patterns of generic evolution and extinction in the Late Ordovician of North America. *Lethaia*, **19**, 33–51.

BENNEDETTO, J. L. 1986. The first typical *Hirnantia* fauna from South America (San Juan Province, Argentine Precordillera). *In*: RACHEBOEUF, P. R. & EMIG, C. C. (eds) *Les Brachiopodes fossiles et actuels*. Biostratigraphie du Paleozoique, **4**, 439–447.

BERGSTROM, S. M. 1973. Ordovician conodonts *In*: HALLAM, A. (ed.) *Atlas of Palaeobiogeography*. Elsevier, Amsterdam, 47–58.

BERRY, W. B. N. & BOUCOT, A. J. 1967. Pelecypod–graptolite association in the Old World Silurian. *Geological Society of America Bulletin*, **78**, 1515–1522.

—— & —— 1973. Glacio-eustatic control of Late Ordovician–Early Silurian platform sedimentation and faunal changes. *Geological Society of America Bulletin*, **84**, 275–284.

BOUCOT, A. J. 1975, *Evolution and extinction rate controls*. Elsevier, Amsterdam, 427p.

—— 1979. Silurian. *In*: BERGGREN, W. A. *et al*. (eds) *Treatise on Invertebrate Paleontology, Part A: Introduction*. University of Kansas Press, Lawrence, A167–A182.

—— & Johnson, J. G. 1973. Silurian brachiopods. *In*: HALLAM, A. (ed.) *Atlas of Palaeobiogeography*. Elsevier, Amsterdam, 59–65.

BRENCHLEY, P. J. 1984. Late Ordovician extinctions and their relationship to the Gondwana glaciation. *In*: BRENCHLEY, P. J. (ed.) *Fossils and Climate*. John Wiley & Sons, Chichester, 291–327.

—— & CULLEN, B. 1984. The environmental distribution of associations belonging to the *Hirnantia* fauna — evidence from North Wales and Norway, *In*: BRUTON, D. L. (ed.) *Aspects of the Ordovician System*. Palaeontological Contributions from the University of Oslo, **295**, 113–126.

—— & NEWALL, G. 1980. A facies analysis of Upper Ordovician regressive sequences in the Oslo region, Norway — A record of glacio-eustatic changes. *Palaeogeography, Palaeoclimatology, Palaeoecology*, **31**, 1–38.

—— & —— 1984. Late Ordovician environmental changes and their effect on faunas. *In*: BRUTON, D. L. (ed.) *Aspects of the Ordovician System*. Palaeontological Contributions from the University of Oslo, **295**, 65–79.

COCKS, L. R. M., MCKERROW, W. S. & LEGGETT, J. K. 1979. *Silurian Palaeogeography of the margins of the Iapetus Ocean in the British Isles*. International Geological Correlation Program, The Caledonides in the USA. Blacksburg, Virginia, 49–55.

COOROUGH, P. J. 1986. *Brachiopod provinciality in the Late Ordovician–Early Silurian*. MSc Thesis. University of Wisconsin, Milwaukee.

FORTEY, R. A. 1975. Early Ordovician trilobite communities. *Fossils and Strata*, **4**, 339–360.

HAVLICEK, V. 1970. Heterorthidae (Brachiopoda) in the Mediterranean Province. *Sbornik Geol. Ved. Paleont.*, **12**, 7–39.

—— 1971. Brachiopodes de l'Ordovicien du Maroc. *Notes et Memoires, Service Geologique du Maroc*, **230**, 1–135.

—— 1976. Evolution of Ordovician brachiopod communities in the Mediterranean Province. *In*: BASSETT, M. G. (ed.) *Aspects of the Ordovician System*. University of Wales Press, Cardiff, 349–358.

—— & Massa, D. 1973. Brachiopodes de l'Ordovicien supérieur de Libye occidentale implications stratigraphiques regionales. *Geobios*, **6**, 267–290.

JAANUSSON, V. 1973. Ordovician articulate brachiopods. *In*: HALLAM, A. (ed.) *Atlas of palaeobiogeography*. Elsevier, Amsterdam, 20–25.

—— 1979. Ordovician. *In*: ROBISON, R. A. & TEICHERT, C. (eds) *Treatise* on *Invertebrate Paleontology, Part A. Introduction*. Lawrence, University of Kansas Press, A136–A166.

JABLONSKI, D. 1986. Background and mass extinctions: The alternation of macroevolutionary regimes. *Science*, **231**, 129–133.

JOHNSON, J. G. 1971. A quantitative approach to faunal province analysis. *American Journal of Science*, **270**, 257–280.

—— 1974. Extinction of perched faunas. *Geology*, **2**, 479–482.

KLAPPER, G. & JOHNSON, J. G. 1980. Endemism and dispersal of Devonian conodonts. *Journal of Paleontology*, **54**, 400–455.

LAUFELD, S. 1979. Biogeography of Ordovician, Silurian, and Devonian chitinozoans. *In*: GRAY, J. & BOUCOT, A. J. (eds) *Historical biogeography, plate tectonics and the changing environment*. Corvallis, Oregon State University Press, 75–90.

LAURIE, J. R. 1982. *The taxonomy and biostratigraphy of the Ordovician and Early Silurian articulate brachiopods of Tasmania*. PhD Thesis, University of Tasmania.

LENZ, A. C. 1976. Late Ordovician–Early Silurian glaciation and the Ordovician–Silurian boundary in the northern Canadian Cordillera. *Geology*, **4**, 313–317.

LESPERANCE, P. J. & SHEEHAN, P. M. 1976a. Brachiopods from the Hirnantian Stage (Ordovician–Silurian) at Perce, Quebec, Palaeontology, 19, 719–731.

—— & —— 1976b. The Silurian — An interval with cosmopolitan faunas in the Appalachian–Caledonian Orogen. *Geological Society of America. Abstracts with Programs*, **8**, 218.

MCKERROW, W. S. & COCKS, L. R. M. 1976. Progressive faunal migration across the Iapetus Ocean. *Nature*, **263**, 304–306.

OLIVER, W. A. 1976. Presidential Address: Biogeography of Devonian rugose corals. *Journal of Paleontology*, **50**, 365–373.

RAUP, D. M. & SEPKOSKI, J. J. 1982. Mass extinctions in the marine fossil record. *Science*, **215**, 1501–1503.

ROBARDET, M. & DORE, F. 1988. The Late Ordovician diamictic formations from southwestern Europe: North-Gondwana glaciomarine deposits. *Palaeogeography, Palaeoclimatology, Palaeoecology*, **66**, 19–31.

RONG JIA-YU 1984a. Distribution of the *Hirnantia* fauna and its meaning, *In*: Bruton, D. L. (ed.) *Aspects of the Ordovician System*. Palaeontological Contribution, University of Oslo, **295**, 101–112.

—— 1984b. Ecostratigraphic evidence of the Upper Ordovician regressive sequences and the effect of glaciation (In Chinese with English abstract). *Journal of Stratigraphy*, **1984**, 3, 19–29.

ROSS, R. J. 1975. Early Paleozoic trilobites, sedimentary facies, lithospheric plates, and ocean current. *Fossils and Strata*, **4**, 307–329.

ROZMAN, KH. S. 1977. Biostratigrafiya i zoogeografiya Verkhnego Ordovika Severnoy Azii i Severnoy Ameriki (po Brakhiopodam), (Biostratigraphy and Zoogeography of Upper Ordovician of North Asia and North America). *Akademy Nauk USSR. Order of the Red Banner of Labour Geological Institute, Transactions*, **305**, 1–171.

—— 1986. The Early Silurian brachiopods of the genus *Tuvaella* from Mongolia. *Paleontological Journal*, **2**, 24–34.

SAVAGE, N. M., PERRY, D. G. & BOUCOT, A. J. 1979. A quantitative analysis of Lower Devonian brachiopod distribution. *In*: GRAY, J. & BOUCOT, A. J. (eds) *Historical Biogeography, Plate Tectonics and the Changing Environment*. Oregon State University Press, Corvallis, 169–200.

SCHOPF, T. J. M., FISHER, J. B. & SMITH, C. A. F. 1978. Is the marine latitudinal diversity gradient merely another example of the species area curve?, *In*: BATTAGLIA, B. & BEARDMORE, J. A. (eds) *Marine Organisms: Genetics, Ecology and Evolution*. Plenum Press, New York, 365–386.

SHEEHAN, P. M. 1973. The relation of Late Ordovician glaciation to the Ordovician-Silurian changeover in North America brachiopod faunas. *Lethaia*, **6**, 147–154.

—— 1975. Brachiopod synecology in a time of crisis (Late Ordovician–Early Silurian). *Paleobiology*, **1**, 205–212.

—— 1979. Swedish Late Ordovician marine benthic assemblages and their bearing on brachiopod zoogeography. *In*: GRAY, J. & BOUCOT, A. J. (eds) *Historical Biogeography, Plate Tectonics, and the Changing Environment*. Oregon State University Press, Corvallis, 61–73.

—— 1982. Brachiopod Macroevolution at the Ordovician–Silurian boundary, *In*: MAMET, B. & COPELAND, M. J. (eds) *Third North American Paleontological Convention, Proceedings*, **2**, 477–481.

—— 1985. Reefs are not so different — They follow the evolutionary pattern of level-bottom communities. *Geology*, **13**, 46–49.

—— 1988. Late Ordovician events and the terminal Ordovician extinction. *In*: WOBERG, D. L. (ed.). *Rousseau Flower Volume*. New Mexico Bureau of Mines and Mineral Resources Memoir, **44**, 31–49.

—— & COOROUGH, P. J. 1986. Brachiopod zoogeography at the terminal Ordovician extinction. *Fourth North American Paleontological Convention, Abstracts with Programs*, A42.

SPJELDNAES, N. 1961. Ordovician climatic zones. *Norsk geologiske Tidsskrift*, **41**, 45–77.

—— 1967. The palaeogeography of the Tethys region during the Ordovician. *In*: ADAMS, C. G. & AGER, D. V. (eds) *Aspects of Tethyan biogeography*. Systematics Association Publication, **7**, 45–57.

—— 1981. Lower Palaeozoic Palaeoclimatology, *In*: HOLLAND, C. H. (ed.) *Lower Palaeozoic of the Middle East, Eastern and Southern Africa, and Antarctica*. John Wiley & Sons, Chichester, 199–256.

SU, Y. 1980. *Palaeontological Atlas of northeast China (in Chinese). 1: Paleozoic*, 261–301.

SWEET, W. C. & BERGSTROM, S. M. 1974. Provincialism exhibited by Ordovician conodont faunas, *In*: Ross, C. A. (ed.) *Paleogeographic Provinces and Provinciality*. Society of Economic Paleontologists and Mineralogists, Special Publication, **21**, 189–202.

SWEET, W. C. & BERGSTROM, S. M. 1984. Conodont provinces and biofacies of the Late Ordovician. *In*: CLARK, D. L. (ed.) *Conodont Biofacies and Provincialism*. Geological Society of America Special Paper, **196**, 69–87.

WANG YU, BOUCOT, A. J., RONG, JIA-YU & YANG XUE-CHANG. 1984. Silurian and Devonian biogeography of China. *Geological Society of America Bulletin*, **95**, 265–279.

WILLIAMS, A. 1969. Ordovician faunal provinces with reference to brachiopod distribution. *In*: WOOD, A. (ed.) *The Pre-Cambrian and Lower Palaeozoic rocks of Wales*. University of Wales Press, Cardiff, 117–150.

—— 1973. Distribution of brachiopod assemblages in relation to Ordovician palaeogeography. *Special Papers in Palaeontology*, **12**, 241–269.

Silurian–Devonian Biogeography

Silurian biogeography

A. J. BOUCOT
Department of Zoology, Oregon State University, Corvallis, Oregon 97331−2914, USA

Abstract: Silurian biogeography is characterized by a globally moderate climatic gradient, on which is superimposed a low latitude, warm region, moderate level of longitudinal provincialism. The cool to cold region Malvinokaffric Realm is bounded to the north by the warm water North Silurian Realm. The North Silurian Realm is divided into North Atlantic and Uralian−Cordilleran Regions. The North Atlantic Region, in turn, is divided into North American and European Provinces, whereas a Mongolo−Okhotsk Subprovince is recognized within the Uralian−Cordilleran Region. There is no biogeographic evidence favouring shallow water regions in the mid to high latitudes of the Northern Hemisphere. The nonmarine biogeography of the Silurian is largely unknown.

Biogeographically the Silurian is an average period, average in the sense that it displays a moderately high global climatic gradient latitudinally, and a moderately high longitudinal level of provincialism at lower latitudes. It is sandwiched between the later Ordovician, which is characterized by a far higher climatic gradient featuring widespread Southern Hemisphere glaciation and significantly a higher level of lower latitude provincialism, and the Early Devonian, with its similar level of climatic gradient on which is superimposed a much higher level of lower latitude provincialism than even that present in the later Ordovician.

Silurian biogeography is important for the geologist because of the constraints it imposes on the reliability of fossil-based correlations. The moderate levels of provincialism result in lower correlation reliability than one would obtain in a more cosmopolitan time interval such as the Late Devonian or Early Triassic.

Our knowledge of Silurian biogeography is based almost entirely on evidence derived from marine fossils, mainly level bottom and pelagic taxa, with very little from nonmarine forms. This situation will probably not change very much in the future, except possibly for higher land plant spore data supplemented by that derived from vertebrates, because of the overall scarcity of freshwater and terrestrial organisms.

Knowledge of Silurian marine biogeography has accumulated largely in the past two decades. Earlier there was not much consistent interest in the topic, although earlier workers were aware of the fact that the faunas in one part of the world differed significantly from those in other parts. These regional differences were commonly dismissed as due to 'biofacies' distinctions, without much concern whether environmental influences, reproductive isolation, or both were part of the cause.

An understanding of Silurian biogeography has suffered from the lack of agreement on Silurian palaeogeography. Reasons for dismissing the Silurian palaeogeographies based exclusively or largely on the evidence of remanent magnetism (Meyerhoff 1970 provides cogent reasons about why such data has not proved reliable in the Silurian) were previously discussed (Boucot & Gray 1983; Boucot 1985 a,b). In this paper I arbitrarily employ a pangaeic palaeogeography (Fig. 1) not because I think it is 'the' palaeogeography, but because the alternatives published by others raise more problems than they solve (Fig. 2).

Biogeographic units

Silurian biogeographic conditions persisted, with some modification, from the Ordovician. The Silurian, as a whole, was dominated by moderate levels of latitudinal provincialism and a weaker level of longitudinal provincialism (Boucot & Johnson 1973; Boucot 1975; Wang Yu *et al.* 1984; Boucot 1985*a*; this paper). Levels of longitudinal provincialism decreased markedly from those of the later Ordovician, and increased markedly during the subsequent Early Devonian. The dominant elements are a Southern Hemisphere Malvinokaffric cool or cold water Realm bounded by a carbonate rich North Silurian Realm (Fig. 1).

The Silurian begins with the disappearance of evidence favouring widespread continental glaciation centered in Africa and South America, as far west as the central Andean region. The glaciation was responsible for a high level of regression which is followed in the Llandovery by progressive transgression that culminates globally in the Ludlow. The progressive nature of the transgression may be interpreted as due to slow isostatic rebound of the African and South American shield regions that bore a load of ice during the Ashgill−earlier Llandovery, or alternatively, continental regions lost to the geologic record in much of central Africa that might still have been heavily ice covered and undergoing slow deglaciation during much of the Llandovery. However, this last possibility would involve no transport of glacial marine debris elsewhere, i.e. completely landlocked ice masses, as compared with the early part of the Llandovery and the upper Ashgill. Of course, both of these alternatives might have been intimately involved to have produced a complex final product.

Malvinokaffric Realm

The Malvinokaffric Realm during the Silurian is a cool or cold water, high southern latitude unit. The Malvinokaffric Realm of the Ordovician is strictly equivalent biogeographically and environmentally to the Malvinokaffric Realm of the Silurian, although the major terminal extinction event at the end of the Ordovician completely wiped out the fauna, to be replaced by a Silurian item with somewhat uncertain taxonomic origins in terms of the pre-existing Malvinokaffric Realm. The Malvinokaffric Realm of the earlier Devonian is also strictly equivalent environmentally, and in a large part geographically, to that of the Silurian, although differing completely in taxonomic terms because of the almost total extinction of the Silurian megafauna.

The Malvinokaffric Realm during the Silurian, as during the Ordovician and Devonian, lacks evidence of limestone, carbonate-based reefs, laterites, evaporites, bauxites and redbeds. This situation is consistent with a cool to cold climatic regime, and contrasts with the carbonate rich Silurian record known elsewhere. The Malvinokaffric Realm fauna is characterized (Boucot 1975, Table V, p. 299, brachiopods) by a low total number of superfamilies, as well as genera and species. Many major groups such as the conodonts (the conodont animal was probably a warm water creature), corals, stromatoporoids, calcareous algae and bryozoans appear to be almost totally absent as in the Ordovician and Devonian of the Realm, as is also the case for virtually all pelmatozoan taxa. Individual level bottom communities contain far fewer species than are found in contemporary communities in the carbonate rich realm.

North Silurian Realm

The North Silurian Realm (Boucot 1975) includes the carbonate facies equivalents to all of the extra-Malvinokaffric Realm

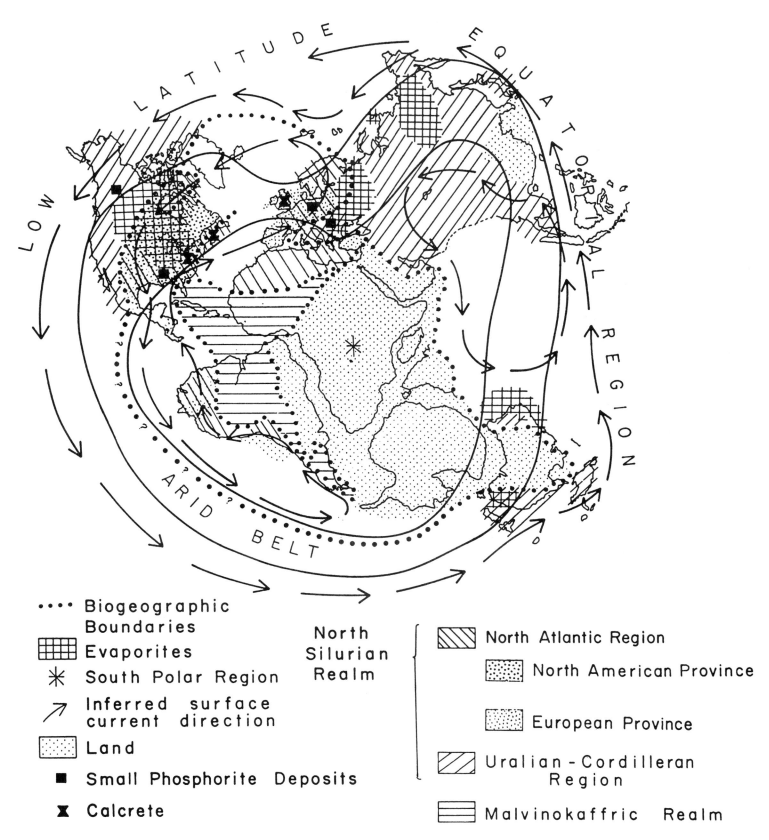

Fig. 1. Upper Silurian marine level-bottom biogeography, plotted on a pangaeic base (after Boucot & Gray 1979). Note that the Bosphorus region is included within the European Province of the North Atlantic Region because of evidence provided by such items as the presence of *Dayia* in Pridoli age beds, as well as of true *Stricklandia* in the Lower Silurian (this is a major change from the reconstruction for this region provided by Boucot & Gray (1979)).

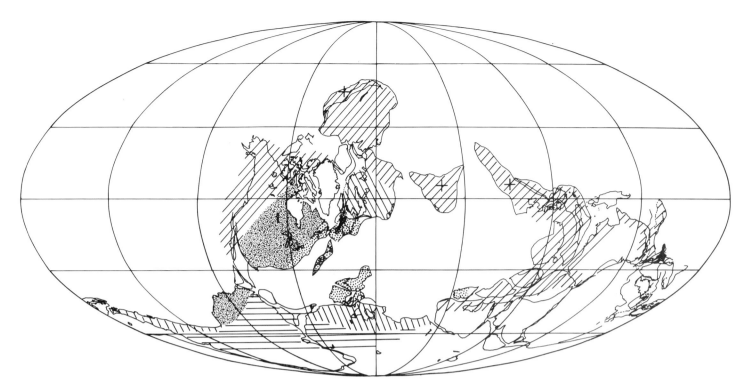

Fig. 2. Upper Silurian marine level-bottom biogeography, plotted on the Symposium base map for the Ludlow. Note the physical difficulty of devising any reasonable surface current circulation pattern capable of keeping the following items (Mongolia, 'Kazakhstania', and Heilungjiang, indicated with a 'plus', which yield *Tuvaella*, and the Tasman genus *Maoristrophia* in Mongolia in the Silurian) in reproductive communication: (1) Mongolia, at highest latitudes on the north side of Siberia, with 'Kazakhstania' and Heilunjiang (the Mongolo–Okhotsk Subprovince; note that this problem is greatly exacerbated during the Lower Devonian; see Hou & Boucot (1990) for levels of provincialism here, and the problem of reproductive communication with the Tasman and Eastern Americas units. Note too, that this 'problem' begins in the Cambrian and only ends in the earlier Middle Devonian, the Eifelian.). (2) The Istanbul Region European Province faunas with those of the main body of that Province. (3) The problem of selecting barriers to reproductive communication adequate to maintain the integrity of the North Atlantic Region, which finds itself in the midst of the Uralian–Cordilleran Region. (4) Finding a mechanism for keeping the European Province reproductively isolated from the North American Province. (5) Note too the anomaly of having equatorial Late Silurian evaporites in North America and Europe, as well as in Siberia, whereas those in Australia appear reasonable. Shading as for Fig. 1.

Ordovician biogeographic units. The North Silurian Realm is subdivided into North Atlantic and Uralian–Cordilleran Regions (Boucot 1975; Wang Yu *et al.* 1984). The North Atlantic Region, beginning in the later Llandovery, is divided into North American and European Provinces that are described and defined below. The Uralian–Cordilleran Region includes a Mongolo–Okhotsk Subprovince, and a second, province level unit for everything else. In terms of taxonomic continuity from the Ordovician it is clear that the Ordovician North American, Baltic and Siberian unit faunas were most heavily extinguished from the shallower parts of the platform, say Benthic Assemblages 1 through 3, whereas the survivors are derived chiefly from the Benthic Assemblage 4–5 regions. Jannusson (1979) noted this fact when commenting that his Hiberno–Salairian biota seemed to supply most of the survivors, although I would regard his unit as a major ecologic rather than as a biogeographic unit, i.e. it includes deeper water faunas from many major later Ordovician biogeographic units beginning in about the Llandeilo (Pratt Ferry Formation faunas). In any event, it appears that Silurian biogeographic units, other than the division into Malvinokaffric and extra-Malvinokaffric, are unlike those of the Ordovician, i.e. there appears to have been a completely different pattern of post-Ordovician, extra-Malvinokaffric Realm biogeographic units. Whereas, the change from the Silurian into the Devonian is completely transitional, with no major biogeographic changes of any kind within the marine environment.

The carbonate rich North Silurian Realm, has geographic limits extending from the western portion of the Andean region (the southernmost locality know to date is in northwesternmost Argentina; Isaacson *et al.* 1976) north to the Merida Andes of Venezuela (actually to an unknown point somewhere between the Merida Andes Silurian south of Lake Maracaibo and the Malvinokaffric Realm Silurian on the north side of the Amazon on the Rio Trombetas). From northern South America the boundary extends northerly almost to Florida (Laufeld 1979; some Florida Silurian chitinozoans are Malvinokaffric Realm types; Pojeta *et al.* 1976; some Florida Late Silurian bivalves are European Province types; this suggests that Florida was an area with some European Province benthos mixed with some Malvinokaffric Realm plankton) and the adjacent Appalachians (the coastal Silurian, in the Northern Appalachians, from Boston to Nova Scotia belongs to the North Silurian Realm, as does that in northeastern Florida), and then easterly to include part of North Africa in the Malvinokaffric Realm. Brittany during the latest Silurian has some beds with the Malvinokaffric Realm genus *Clarkeia*, which may be regarded as either a brief, disjunct occurrence or alternatively as a brief extension of the Realm. The Prague area, the Carnic Alps and the Sea of Marmora region near Istanbul yield rich North Silurian Realm faunas. Most of North Africa and the Mediterranean region lack significant shelly faunas except for occasional representatives of Bohemian types orthoceroid limestones and associated bivalve faunas as in Sardinia. Malvinokaffric Realm occurrences are also present in the Cape Mountains of South Africa (Gray *et al.* 1986; i.e. beds no older than the *persculptus* Zone) in the Table Mountain Sandstone above Ordovician higher land plant spore, trilobite chitinozoan and conodont bearing beds of the Soom Shale Member. No fossiliferous Silurian rocks are known in Antarctica (except for some contentious carbonate blocks from moraines) or the Falkland Islands, where one might predict Malvinokaffric Realm Silurian by comparison with the overlying typical non-carbonate Malvinokaffric Lower Devonian. Neither are fossiliferous, marine Silurian rocks known south of Buenos Aires in the Sierra de la Ventana where they might be expected. It is entirely possible that the area from the southern Sahara to South Africa, to the Sierra de la Ventana plus Falkland Islands region, and Antarctica, might have been nonmarine parts of Malvinokaffric Realm affinities during the Silurian, as there are unfossiliferous beds beneath the fossiliferous Lower Devonian in some places that could be of Silurian age despite their lack of good marine fossils (nearshore marine Llandovery on the eastern edge of the Paraná Basin in Brazil; Gray et al. 1985; Boucot et al. 1986; and on the western side in Paraguay. The simplest way to describe the distribution of North Silurian Realm elsewhere is to state that the known Silurian of Australia−New Zealand, North America, and all of Eurasia (except for the small part of Europe cited above) belongs to it. The Silurian of the Near East, particularly in the area of the Bosphorus and in Jordan, is also of North Silurian Realm type, although there is still some uncertainty about that of Arabia and the adjacent Zagros (Llandovery only known in both).

The North Silurian Realm is divided into North Atlantic and Uralian−Cordilleran Regions (Boucot 1975), with a Mongolo-Okhotsk Subprovince of the Uralian−Cordilleran Region being present in east−central Asia (Wang Yu et al. 1984). The Uralian−Cordilleran Region has not been recognized in western North America for the Llandovery (rich post-Llandovery, Silurian, Uralian−Cordilleran Region faunas are present), from the Yukon to southern California, but this probably reflects the almost total absence to date of any diverse Llandovery shelly faunas in central Nevada together with adjacent parts of California and Idaho, that would make recognition possible, but is present in North Greenland. The North Atlantic Region is divided into North American and European Provinces.

Significant carbonate reefs incorporating framework building organisms are unrecognized in the Llandovery, but appear in the Late Wenlock in many places within the North Silurian Realm. The taxa present in these reefs appear to have had local, level bottom antecedents. These reef taxa also participate to varying degrees in the Region level endemism shown by the level bottom biota. The reef endemism is notable among the well studied pentameroid brachiopods (Boucot & Johnson 1979) for which the Midcontinent genera and those of the Urals-Tien Shan-Altai are very distinct from each other, i.e. this amounts to North Atlantic Region and Uralian−Cordilleran Region endemism shown by reef related pentameroids. Such carbonate reefs are abundant in both the late Wenlock and Ludlow, but become relatively uncommon during the Pridoli.

North American and European Provinces of the North Atlantic Region. Witzke et al. (1979) made it clear that during the Late Silurian there were major biogeographical differences between the attached echinoderms of North America and Europe, including differences at the family level, as well as at generic and specific levels. Copeland & Berdan (1977) made the same distinction for Beyrichiacean ostracode provincialism for Silurian and Early Devonian time. Blieck (pers. comm.) also points out how well this provincial concept fits the vertebrates, i.e. the *Thelodus parvidens* fauna originally described from Europe, but now well known from coastal Acadia (White Rock Formation, Bear River, Nova Scotia; Bouyx & Goujet 1985; Arisaig Nova Scotia; Boucot et al. 1974,; Long Reach Formation, southern New Brunswick; Turner 1973). I suggested (Boucot 1985a) that these relations be formalized in terms of North American and European Provinces of the North Atlantic Region of the North Silurian Realm. An older generation of North American palaeontologists (Williams 1912, is a good example) commented on the 'European' and 'Trans-Atlantic' affinities of these coastal faunas. But, their faunal analyses are flawed in many regards, probably being influenced in no small part by the siliciclastic lithofacies that is similar in many regards to that present in the contemporary Anglo-Welsh Basin, as contracted to the North American Platform carbonate rocks, whereas the majority of the shelly fossils are shared between the two regions. It is now clear, using brachiopods, plus a few other shelly invertebrates, that coastal Acadia (southern and eastern New England, coastal New Brunswick, and Nova Scotia) belong to the European Province. The subsurface Late Silurian of northeastern Florida (Pojeta et al. 1976) is assigned to the European Province on the basis of the Cone well bivalves, although recognizing the closeness of the Malvinokaffric Realm in terms of Laufeld's (1979) chitinozoans. Some of the European provincial brachiopods include *Dayia*, *Plectotreta*, *Pembrostrophia*, *Shaleria* (*Janiomya*), *Shaleria* (*Shaleria*), *Quadrifarius*, and *Protochonetes*, as well as beyrichiacean ostracodes (Copeland & Berdan 1977), and some of the North American brachiopods include *Eccentricosta*, *Onychotreta*, *Strophonella* (*Strophonella*), *Costistrophonella*, *Plicostricklandia*, *Supertrilobus*, *Stenopentamerus*, *Apopentamerus Callipentamerus*, plus *Pentamerus* and *Pentameroides* in the Wenlockian, and *Coelospira* (it has a single European occurrence in Gotland). Note that the level of provincial endemism shown by the brachiopods is considerably lower than that shown by the stalked echinoderms or by the ostracodes. For the Lower Silurian our resolution is considerably lower with the brachiopods (although better with Copeland & Berdan's ostracodes!): I can think here only of the American endemic *Microcardinalia*, but the Late Llandovery stalked echinoderms, according to Witzke et al. (1979) are just as endemic as earlier (pre-late Llandovery echinoderm data are inadequate for making a determination). In many ways the North American Province of the Late Silurian is the precursor both faunally and geographically of the subsequent Appohimchi Subprovince of the Eastern Americas Realm, whereas the European Province of the Late Silurian is the faunal and geographic precursor of the Rhenish−Bohemian Region of the Old world Realm: note the discordant changes in biogeographic ranks here.

It is of some interest that these two provincial level subdivisions of a major Silurian Region should be the precursors of two units occupying similar positions within two distinctive realms in the Devonian. This situation raises the puzzling question about how best to explain the Old World Realm of the Devonian, i.e. whether is was derived from a mixture of the Silurian European Province of the North Atlantic Region plus the Silurian Uralian−Cordilleran Region, or whether it is better viewed as derived only from the Uralian−Cordilleran Region.

Ecologic−evolutionary units

The pre-C_3, Late Llandovery part of the Silurian belongs to Ecologic−Evolutionary Unit V, and the remainder to the older part of VI (Boucot 1983). Ecologic−Evolutionary Unit V is essentially the relict fauna remaining after the terminal Ordovician extinction event (below the zone of *persculptus* if one assigns it to the Ordovician rather than to the Silurian), whereas Ecologic−Evolutionary Unit VI begins with the major dispersal event of Uralian−Cordilleran Region taxa into the North Atlantic Region in about C_2-C_3 time. As far as can be determined the biogeography of both units is relatively similar. The only exception might be that in the Late Silurian, although not certainly in the Late Llandovery part of VI, one can now recognize European and North American Provinces within the North Atlantic Region, whereas this has not yet been proved in Unit V. This difference may, however, be merely a sampling artifact. In other words, the Ecologic−Evolutionary Unit boundaries in the Silurian have no biogeographic unit expression, although they do represent a major dispersal event (Wang Yu et al. 1984 provide data additional to that in Boucot 1975).

Climatic gradients

Widespread evidence for glaciation is lacking in the Silurian, in contrast to the Ashgill. However, the Malvinokaffric Realm rocks and fossils provide good evidence for a moderately high global climatic gradient throughout the Period. The virtual absence of evaporites in the Llandovery, as contrasted with their widespread occurrence in the Late Silurian, may be taken as evidence for a marked broadening of an arid belt from the Early into the Late Silurian.

Orogeny

The Silurian Period is virtually anorogenic. There are some minor bits and pieces of unrest evident in the Tasman Region, and in the Late Silurian (Salinic Disturbance) of parts of the Northern Appalachians, but these add up to little more than local faulting or folding in very restricted areas. It is hard to see any correlation between Silurian biogeography and orogeny.

Levels of provincialism

In the Silurian the brachiopods, trilobites and tetracorals show levels of endemism and provincialism corresponding to what has been discussed above at the region level: the regions were primarily based on the brachiopods. Both conodonts and graptolites appear to be very cosmopolitan in both the Upper and Lower Silurian of the North Silurian Realm, although a few groups of graptolites (see Berry, in Boucot 1975,) have distribution patterns in the Upper Silurian that parallel those of the shelly groups. Stridsberg (pers. comm.) finds no significant provincialism between the North American Platform nautiloids and those of the Baltic region or Bohemia. Recall also that conodonts are unknown from the Malvinokaffric Realm, and that climacograptids occur in both realms during the Llandovery, but persist through the end of the Silurian in the Malvinokaffric Realm. The Silurian ostracodes and pelmatozoans, however, appear to be more provincial than the groups cited above (Lundin 1971; Polonova 1971; Copeland 1977a, b, for ostracodes: Witzke et al. 1979, for pelmatozoans). With these two groups province level groups can be recognized in the North Atlantic Region, for which there is some minimal support from other more cosmopolitan groups such as the brachiopods.

Although there is widespread evidence for a Silurian higher land plant flora (Gray 1985, 1988) there is not enough evidence with which to define phytogeographic units, although McGregor (1984) discusses some aspects of spore provincialism that might, when more data is available, support some type of provincial scheme.

Causation

We do not yet have any evidence concerning the factor or factors responsible for isolating some elements of the Uralian–Cordilleran Region from the North Atlantic Region, nor for their subregional units. The North Atlantic Region has a lengthy common boundary with the Malvinokaffric Realm, whereas the Uralian–Cordilleran Region does not, so that a temperature gradation away from that cooler Realm can be appealed to, but the evidence is still weaker than one would like. Evaporite bodies signifying hypersaline conditions and land barriers cannot be appealed to either (evaporites are virtually absent in the Early Silurian).

Inspection of Fig. 1 makes it clear that the Late Silurian evaporites, and a few calcretes, may be interpreted to have occurred in a low, but not equatorial, latitude belt reflecting conditions of aridity-low rainfall.

In general the brachiopod fauna of the pre-C_3 part of the Upper Llandovery consists of relict Ordovician genera and subgenera derived from more offshore, B.A. 4–5, portions of the fauna. During the early part of the Upper Llandovery, near the C_2–C_3 boundary, there is sudden dispersal of previously endemic genera representing a number of families from the Uralian–Cordilleran Region into the North Atlantic Region (Wang Yu et al. 1984). This invasion marks the Ecologic–Evolutionary Unit V–VI boundary.

Conclusions

The Silurian is a Period characterized overall by a relatively uniform level of moderate provincialism combined with a moderately high global climatic gradient. It is sandwiched between the later Ordovician and earlier Devonian, both being times of considerably higher levels of provincialism, and the later Ordovician one of the limited number of Phanerozoic intervals characterized by higher latitude continental glaciation that affected the sealevel regions. There is no evidence in the Silurian for moderate or high northern latitude landmasses or marine biogeographic units, i.e. there may have been only a Northern Hemisphere oceanic realm. We have no good understanding of the isolating factor(s) responsible for the existence of the North Atlantic and Uralian–Cordilleran Regions, nor for their subdivisions. The existence of the Malvinokaffric Realm, however, is easily explained in terms of the global climatic gradient of the time.

It must be emphasized that until the Silurian biota has been subjected to far more biogeographic analysis it would be premature to conclude that the present state of our understanding is anything more than very preliminary.

Appendix

At present, our knowledge of Silurian marine biogeography is based on summaries of the brachiopods (Boucot 1975), graptolites (Berry 1979; Jaeger 1976), corals (Kaljo & Klaamann 1973; Leleshus 1971; Pickett 1975; Hill & Jell 1969; Wang & He 1981), attached echinoderms (Witzke et al. 1979; Ausich 1978), ostracodes (Lundin 1971; Polonova 1971; Copeland 1977a, b; Copeland & Berdan 1977; Stone & Berdan 1984), conodonts (Jeppson 1974); chitinozoans (Laufeld 1979), trilobites (Chlupac 1975; Schrank 1977; Chatterton & Campbell 1980) and agglutinated foraminifera (Poyarkov 1977). Several other groups have been partially treated, such as the gastropods (Forney et al. 1980), rostroconchs (Pojeta 1979), bivalves (Pojeta et al. 1976) and carpoids (Derstler 1979). A number of potentially useful groups such as the nautiloids and stony bryozoans remain largely unstudied. Careful consideration of the acritarchs in a systematic, well documented manner would undoubtedly be of great value. Some attention has been given to the vertebrates, particularly the thelodonts (Turner & Tarling 1982), but much more data is needed here.

References

Ausich, W. I. 1978. *Pisocrinus* from California, Nevada, Utah and Gaspe Peninsula. *Journal of Paleontology*, **52**, 487–491.

Berry, W. B. N. 1979. Graptolite biogeography: A biogeography of some Lower Paleozoic plankton. *In*: Gray, J. & Boucot, A. J. (eds) *Historical Biogeography, Plate Tectonics, and the Changing Environment*. Oregon State University Press, Corvallis, 105–115.

Boucot, A. J. 1975. *Evolution and Extinction Rate Controls*. Elsevier, Amsterdam.

—— 1983. Does evolution take place in an ecological vacuum? II. *Journal of Paleontology*, **57**, 1–30.

—— 1985a. Late Silurian–Early Devonian biogeography, provincialism, evolution and extinction. *Philosophical Transactions of the Royal Society, London, B*, **309**, 323–339.

—— 1985b The relevance of biogeography to palaeogeographical reconstruction. *Philosophical Transactions of the Royal Society, London, B*, **309**, 79–80.

——, Dewey, J. F., Dineley, D. L., Fletcher, R., Fyson, W. K., Griffin, J. G., Hickox, C. F., McKerrow, W. S. & Ziegler, A. M.

1974. Geology of the Arisaig Area, Antigonish County, Nova Scotia. *Geological Society of America Special Paper*, **139**.

—— & GRAY, J. 1979 Epilogue: A Paleozoic Pangaea?. *In*: GRAY, J. & BOUCOT, A. J. (eds) *Historical Biogeography, Plate Tectonics, and the Changing Environment*. Oregon State University Press, Corvallis, 465–482.

—— & GRAY, J. 1983. A Paleozoic Pangaea. *Science*, **222**, 571–581.

—— & JOHNSON, J. G. 1973. Silurian Brachiopods *In*: HALLAM, A. (ed.) *Atlas of Palaeobiogeography*. Elsevier, Amsterdam, 59–65.

—— & JOHNSON, J. G. 1979. Pentamerinae (Silurian Brachiopoda) *Palaeontographica*, Abteilung A, Stuttgart, **163**, 87–129.

——, ROHR, D. M., GRAY, J., de FARIA, A. & COLBATH, G. K. 1986. *Plectonotus* and *Plectonotoides*, new subgenus of *Plectonotus* (Bellerophontacea; Gastropoda) and their biogeographic significance. *Neues Jahrbuch fur Geologie und Palaontologie*, Abhandlungen, Stuttgart, **173**, 167–180.

BOUYX, E., GOUJET, D. 1985. Decouverte de Vertebres dans le Silurien superieur de la zone de Meguma (Nouvelle Ecosse, Canada) implications paleogeographiques. *Compte Rendus de l'Academie de Science*, Paris, 301, Serie II, numero 10, 711–714.

CHATTERTON, B. D. E. CAMPBELL, K. S. W. 1980. Silurian trilobites from near Canberra and some related forms from the Yass Basin: *Palaeontographica*, Abteilung A, Stuttgart, **167**, 77–119.

CHLUPAC, I. 1975. The distribution of phacopid trilobites in space and time. *Fossils and Strata*, Marburg, **4**, 399–408.

COPELAND, M. J. 1977a. Early Paleozoic Ostracoda of eastern Canada. *In*: SWAIN, F. (ed.) *Stratigraphic Micropaleontology of Atlantic Basins and Borderlands*, Elsevier, Amsterdam, 1–17.

—— 1977b. Early Paleozoic Ostracoda from Southwest District of Mackenzie and Yukon Territory. *Geological Survey of Canada Bulletin*, **275**.

—— & BERDAN, J. 1977. Silurian and Early Devonian Beyrichiacean ostracode provincialism in northeastern North America. *Geological Survey of Canada Paper, Report of Activities, Part B*, **77**, 5–24.

DERSTLER, K. 1979. Biogeography of the stylophoran carpoids (Echinodermata). *In*: GRAY, J. & BOUCOT, A. J. (eds) *Historical Biogeography, Plate Tectonics, and the Changing Environment*, Oregon State University Press, Corvallis, 91–104.

FORNEY, G., BOUCOT, A. J. & ROHR, D. M. 1980. Silurian and Lower Devonian zoogeography of selected molluscan genera. *In*: GRAY, J., BOUCOT, A. J. & BERRY, W. B. N. (eds) *Communities of the Past*, Hutchinson Ross Publishing Company, Stroudsburg, 119–164.

GRAY, J. 1985. The microfossil record of early land plants: advances in understanding of early terrestrialization, 1970–1984. *Philosophical Transactions of the Royal Society, London*, series B, **309**, 167–195.

—— 1988. Land plant spores and the Ordovician–Silurian boundary, *In*: COCKS, L. R. M. & RICKARDS, R. B. (eds) *A Global analysis of the Ordovician-Silurian boundary*, Bulletin of the British Museum (Natural History). Geology Series, **43**, 351–358.

——, COLBATH, G. K., DE FARIA, A., BOUCOT, A. J. & ROHR, D. M. 1985. Silurian-age fossils from the Paleozoic Parana Basin, southern Brazil. *Geology*, **13**, 521–525.

——, THERON, J. N. & BOUCOT, A. J. 1986. Age of the Cedarberg Formation, South Africa and early land plant evolution. *Geological Magazine*, **123**, 445–454.

HILL, D. & JELL, J. S. 1969. On the rugose coral genera *Rhizophyllum* Lindstrom, *Platyphyllum* Lindstrom and *Calceola* Lamarck. *Neues Jahrbuch fur Geologie und Palaontologie*, Monatsheft, **9**, 534–551.

HOU HONG FEI & BOUCOT, A. J. 1990. The Balkash–Mongolia–Okhotsk region of the Old World Realm. *In*: MCKERROW, W. S. R. & SCOTERE, C. R. (eds) *Palaeozoic Palaeogeography and Biogeography*. Geological Society, London, Memoir, **12**, 297–303.

ISAACSON, P. E., ANTELO, B. & BOUCOT, A. J. 1976. Implications of a Llandovery (Early Silurian) brachiopod fauna from Salta Province, Argentina. *Journal of Paleontology*, **50**, 1103–1112.

JAANUSSON, V. 1979. Ordovician. *In*: ROBISON, R. A. & TEICHERT, C. (eds) *Treatise on Invertebrate Paleontology*, Part A, Introduction, Geological Society of America, Incorporated, and University of Kansas Press, 136–166.

JAEGER, H. 1976. Das Silur und Unterdevon vom thuringischen Typ in Sardinien und seine regionalgeologische Bedeutung. *Nova Acta Leopoldina*, Neuefolge, Nummer. 224, 45, Franz-Kossmat-Symposium, 263–299.

JEPPSON, L. 1974. Aspects of late Silurian conodonts. *Fossils and Strata*, Marburg, **6**.

KALJO, D. L. & KLAAMANN, E. 1973. Ordovician and Silurian corals. *In*: HALLAM, A. (ed.) *Atlas of Palaeobiogeography*, Elsevier, Amsterdam, 37–46.

LAUFELD, S. 1979. Biogeography of Ordovician, Silurian and Devonian chitinozoans. *In*: GRAY, J. & BOUCOT, A. J. (eds) *Historical Biogeography, Plate Tectonics, and the Changing Environment*, Oregon State University Press, Corvallis, 75–90.

LELESHUS, V. L. 1971. Paleozoogeography in Ordovician, Silurian, and Early Devonian on basis of tabulatomorph corals and the boundaries of the Silurian System. *International Geology Review*, **13**, 427–434.

LUNDIN, R. F. 1971. Possible paleoecological significance of Silurian and Early Devonian Ostracode Faunas from the Mid-continental and Northeastern North America. *In*: OERTLI, H. (ed.), *Bulletin du centre de la Recherche*, Pau-SNPA, supplement, 5, Paleoecologie des Ostracodes, Pau (1970), 853–868.

MCGREGOR, D. C. 1984. Late Silurian and Devonian spores from Bolivia. *Academia Nacional de Ciencias* Cordoba, Argentina, **69**.

MEYERHOFF, A. A. 1970. Continental drift: Implications of paleomagnetic studies, meteorology, physical oceanography, and climatology, *Journal of Geology*, **78**, 1–51.

PICKETT, J. 1975. Continental reconstruction and the distribution of coral faunas during the Silurian. *Journal and Proceedings of the Royal Society of New South Wales*, **108**, 147–156.

POJETA, J., JR. 1979. Geographic distribution of Cambrian and Ordovician Rostroconch Mollusks. *In*: GRAY, J. & BOUCOT, A. J. (eds) *Historical Biogeography, Plate Tectonics, and the Changing Environment*, Oregon State University Press, Corvallis, 27–36.

——, KRIZ, J. & BERDAN, J. M. 1976. Silurian–Devonian Pelecypods and Paleozoic Stratigraphy of Subsurface Rocks in Florida and Georgia and Related Silurian Pelecypods from Bolivia and Turkey. *United States Geological Survey Professional Paper*, **879**.

POLONOVA, E. N. 1971. Biogeographical types of Early Devonian ostracodes. *Bulletin de la centre du Recherche*, Pau-SNPA, 5 supplement, 843–852.

POYARKOV, B. V. 1977. Biogeografiya foraminifer devona. Akademiya Nauk S. S. S. R., Vipusk 347, Sibirskoe Otdelenie., 8–29.

SCHRANK, E. 1977. Zur Palaobiogeographie silurischer Trilobiten. *Neues Jahrbuch fur Geologie und Palaontologie*, Abhandlungen, **155**, 108–136.

STONE, S. M. & BERDAN, J. M. 1984. Some Late Silurian (Pridolian) ostracodes from the Roberts Mountains, central Nevada. *Journal of Paleontology*, **58**, 977–1009.

TURNER, S. 1973. Siluro–Devonian thelodonts of the Welsh Borderland. *Journal of the Geological Society of London*, **129**, 557–584.

——, TARLING, D. H. 1982. Thelodont and other Agnathan distributions as tests of Lower Palaeozoic continental reconstructions. *Palaeogeography, Palaeoclimatology, Palaeoecology*, **39**, 295–311.

WANG HONG-ZHEN & HE, XIN-YI. 1981. Silurian rugosan coral assemblages and paleobiogeography of China. *Geological Society of America Special Paper*, **187**, 55–63.

WANG YU, BOUCOT, A. J., RONG JIA-YU & YANG XUE-CHANG. 1984. Silurian and Devonian biogeography of China. *Bulletin of the Geological Society of America*, **95**, 265–279.

WILLIAMS, H. S. 1912. Correlation of the Paleozoic faunas of the Eastport Quadrangle, Maine. *Bulletin of the Geological Society of America*, **23**, 349–356.

WITZKE, B. J., FREST, T. J. & STRIMPLE, H. L. 1979. Biogeography of the Silurian–Lower Devonian echinoderms. *In*: GRAY, J. & BOUCOT, A. J. (eds) *Historical Biogeography, Plate Tectonics, and the Changing Environment*, Oregon State University Press, Corvallis, 117–129.

Distributions and extinctions of Silurian Bryozoa

M. E. TUCKEY

Geraghty & Miller, Suite 145, 50 W. Big Beaver Road, Troy, Michigan USA 48084

Abstract: Gradient and cluster analyses of bryozoan genera, from a worldwide bryozoan data base has revealed biogeographic patterns for the Silurian. In the Llandovery, three provinces can be distinguished: (1) a north American–Siberian Province; (2) a Baltic Province; and (3) a Mongolian Province, which includes localities on the Siberian and South China plates.
During the Wenlock, the North American–Siberian and Baltic Provinces merged into a single province as the Iapetus Ocean continued to close, with the Mongolian Province still present. A single cosmopolitan province developed during the Ludlow, and little biogeographic differentiation can be observed among Ludlow and Pridoli faunas. Formation of the Mongolian Province may be related to latitudinal climatic zonation, as the northern portion of the Siberian plate had moved into temperate latitudes in the Early Silurian. Loss of the north temperate Mongolian Province in the Ludlow, suggests that the Siberian plate may have moved southward in the Late Silurian.
Increases in extinctions of bryozoan species and genera coincided with loss of provinciality in the Silurian. Merging of the Baltic and North American–Siberian Provinces in the Wenlock coincided with large increases in extinctions in North America and Baltica. Wenlock extinctions may be related to climatic change caused by continental convergence, or competition between faunas. Loss of the Mongolian Province in the Ludlow also coincided with increases in extinctions, which may be related to southward movement of the Siberian plate into a warmer climatic zone.

The Silurian has been characterized as a period of low provinciality for marine invertebrates, with little biogeographic differentiation being observed in faunas from major continents. Biogeographic analyses have been conducted for brachiopods (Boucot & Johnson 1973; Cocks & McKerrow 1973), graptolites (Berry 1973, 1979), corals (Kaljo & Klaaman 1973) and echinoderms (Witzke *et al.* 1979). General reviews of Silurian biogeography have been provided by Ziegler *et al.* (1977, 1981), Boucot (1979) and Spjeldnaes (1981). Astrova (1965) described biogeographic patterns of Silurian bryozoans in the Soviet Union; however, no worldwide biogeographic studies have been done on Silurian bryozoans.

In this study, bryozoan biogeographic distributions are analysed for each stage of the Silurian, using quantitative techniques and data drawn from a global bryozoan data base. Extinctions of bryozoan genera are also analysed in relation to changes in provinciality observed in the Silurian.

Methods

The multivariate statistical techniques of gradient analysis and cluster analysis were used to quantitatively assess faunal similarities between localities. Gradient analytical methods of reciprocal averaging, polar ordination and detrended correspondence analysis have been used extensively in community ecology and paleoecology to quantitatively define the distributions of taxa along environmental gradients. In these methods, axes of variation are defined by two poles (the samples most distant from one another) and the remaining samples are ordinated along the axes with respect to their similarity to the poles.

Reciprocal averaging has been used by Cisne & Rabe (1978) in Ordovician paleoecological studies. Anstey *et al.* (1987*a*) used reciprocal averaging and polar ordination in studies of Late Ordovician biogeography and paleobathymetry. Polar ordination was also used by Raymond (1987) to define Devonian phytogeographic provinces. Detrended correspondence analysis (hereafter called DCA) was used by Anstey *et al.* (1987*b*) in a study of Late Ordovician palaeocommunities. DCA has been labelled as an improvement over reciprocal averaging, as it eliminates the distorting effects of axis compression and arching (Gauch 1982).

In this study, DCA was used for distinguishing biogeographic units. The input data matrix for DCA was composed of the number of species per genus present at each locality (Appendix 1). DCA was run with a separate data matrix for each series of the Silurian. Localities of less than five genera were not included in the analysis. Biogeographic patterns were distinguishable on plots of locality scores for DCA axes one v. two.

Cluster analysis was used as a backup technique to assess faunal similarities. Williams (1973) and Raymond (1987) have previously used cluster analysis for biogeographic analyses. In this study, clustering was done with the average linkage between group method. The input data set consisted of a matrix of Simpson's indices of faunal similarity. In some cases cluster analysis results differed from those of DCA. Faunal associations between locations indicated by cluster analysis are shown on DCA plots as dotted lines linking localities across provincial boundaries.

Age determinations of faunas from North America and the British Isles have been taken from the stratigraphic correlation charts of Berry & Boucot (1970, 1974). Manten (1971) was used as a reference for age determinations of the Silurian formations of Gotland. Ages of faunas from the Soviet Union and Eastern Europe were accepted as listed in the literature.

Llandovery

Three bryozoan faunal provinces can be distinguished in the Llandovery: a Baltic Province, a North American–Siberian Province and a Mongolian Province (Fig. 1). Evidence of closing of the Iapetus Ocean is indicated by the inclusion of Anticosti Island, on the northeastern coast of North America, in the Baltic Province. Llandoverian formations of Anticosti Island and the Baltic island of Gotland have these genera in common: *Asperopora, Ceramopora, Corynotrypa, Cuneatopora, Cyphotrypa, Fenestella, Glauconomella, Hallopora, Nematopora, Phaenopora, Ptilodictya, Semicoscinium, Thamniscus* and *Eridotrypa*. Sheehan (1975) found that North American and Baltic brachiopod provinces merged in the Llandovery when Baltic genera invaded the North American continent following the Late Ordovician extinctions. Eight genera from the Ashgill of Baltica, *Asperopora, Clathropora, Cheilotrypa, Eridotrypella, Fistulipora, Hennigopora, Rhinopora* and *Thamniscus* newly appear on the North American continent during the Llandovery. Three of these genera, *Asperopora, Cheilotrypa* and *Thamniscus* newly appear at Anticosti Island, giving the fauna a Baltic aspect. Other localities in the Baltic Province include Estonia, Norway and Podolia.

The North American–Siberian Province includes localities from the Podkammenaya–Tunguska and Viluya River Valleys of the Siberian Platform, and the midcontinent regions of North America. Late Ordovician bryozoan faunas of North America

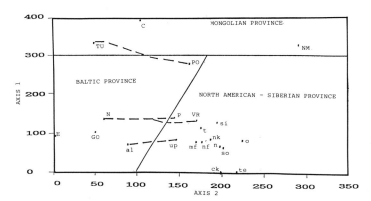

Fig. 1. Llandovery DCA axes 1 v. 2. Symbols: al, Anticosti Island; C, central China; ck, central Kentucky; E, Estonia; Go, Gotland; mf, Meaford; N, Norway; n, New York; nf, Ontario-Niagara Falls region; nk, northern Kentucky; NM, northwestern Mongolia; o, Oklahoma; P, Podkammenaya Tunguska River; PO, Podolia; si, southern Indiana; so, southern Ohio; t, Toronto; te, Tennessee; TU, Tuva; up, Michigan Upper Peninsula; VR, Viluya River. Dotted lines between localities across provincial boundaries indicate additional faunal similarities detected by cluster analysis.

were partitioned into several distinct biomes (Anstey 1986); however, Llandovery faunas of North America were homogeneous. Siberian localities also have faunal similarities with Baltica, as indicated in the cluster analysis results (dotted lines in Fig. 1). The Podkammenaya−Tunguska River Valley locality shares five of its seven genera with Norway; however, it also shares 5 genera with Meaford, Ontario. This reflects the cosmopolitanism of many genera in the Llandovery. Appearing in the Llandovery is a third faunal province, the Mongolian Province, which contains faunas from Tuva, northwestern Mongolia and central China. Podolia was linked with this province by the cluster analysis. Tuva and northwestern Mongolia were situated on the northern portion of the Siberian plate, while central China rested on the South China plate. Faunal provinces of the Llandovery are plotted on the Silurian palaeocontinental reconstruction (Fig. 2). Silurian brachiopods show a similar provincialism in this region as Boucot & Johnson (1973) described a provincial *Tuvaella* Community fauna from the late Llandovery−Wenlock of southeast Kazakhstan, Tuva, the Altai Mountains, Mongolia and Manchuria. Ziegler *et al.* (1977) suggested that Silurian provinciality was caused by climatic zonation, with the Mongolian region situated in the north temperate realm. A comparison of palaeocontinental reconstructions from the Ashgill and Llandovery (Scotese 1986) reveals that the Siberian continent moved northward during this time interval and provinciality may have developed as the northern portion of the Siberian plate moved into north temperate realms in the late Llandovery.

The fauna from central China is a low diversity fauna of five genera from the late Llandovery Cuijiago and Lojoping Formations of northern Sichuan and southern Shaanxi provinces. Because of its low diversity, biogeographic conclusions are tentative. However, its affinities with the Mongolian Province in the Llandovery, and also in the Wenlock may indicate that the South China plate was also in a north temperate latitude at this time. Scotese (1986) positioned South China near the equator, in accordance with Early Cambrian and Permian paleomagnetic data. South China's faunal similarity with Mongolia in the Llandovery and Wenlock suggests that it may have drifted northward in the Ordovician−Silurian and returned to an equatorial latitude by the Permian.

Podolia (western Ukraine) is regarded as belonging to the Baltic province, although it shares two genera in common with central China, *Fistulipora* and *Hennigopora*. Podolia was located on the southern portion of the Baltic plate at this time, and the faunal affinities between Podolia and the Mongolian province can perhaps be explained by the similar late Llandovery ages of their faunas rather than by geographic proximity.

Wenlock

During the Wenlock, the Baltic and North American−Siberian Provinces merged. All Baltic and North American localities group as a single cluster (Fig. 3). Also included in the Baltic−North American−Siberian Province is a fauna from the Wenlock of Kazakhstan. Kazakhstan is pictured as a separate continent located in the tropical climatic zone east of Baltica and North America (Fig. 4). Baltic and North American localities share a number of common genera in the Wenlock, among them: *Asperopora, Ceramopora, Corynotrypa, Fenestella, Fistulipora, Hallopora, Monotrypa, Ptilodictya* and *Sagenella*. A somewhat unusual fauna was described from northwestern Illinois by Grubbs (1939). This fauna occurred in the Niagaran reefs of the Racine Dolomite, of

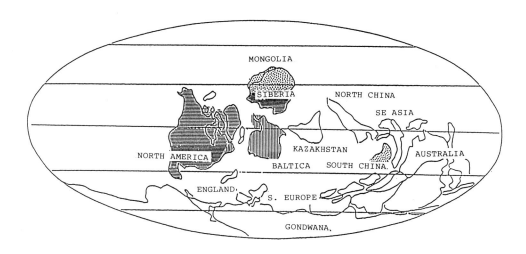

Fig. 2. Llandovery Faunal Provinces. Palaeogeographic reconstruction after Scotese (1986).

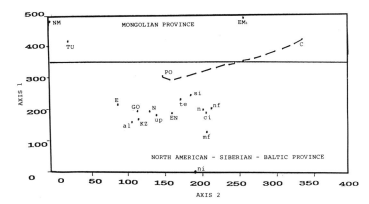

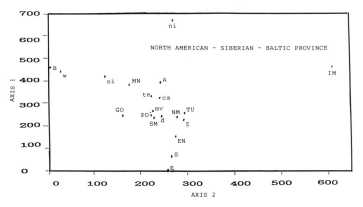

Fig. 3. Wenlock DCA axes 1 v. 2. Symbols: al, Anticosti Island; C, central China; ci, central Indiana; E, Estonia; EM, eastern Mongolia; EN, England; GO, Gotland; KZ, Kazakhstan; mf, Meaford; N, Norway; n, western New York; nf, Ontario-Niagara Falls area; ni, northwestern Illinois; NM, northwestern Mongolia; PO, Podolia; si, southern Indiana; te, Tennessee; TU, Tuva; up, Michigan upper Peninsula. Dotted lines connecting localities across provincial boundaries indicate additional faunal similarities detected by cluster analysis.

Fig. 5. Ludlow DCA axes 1 v. 2. Symbols: A, Australia; B, Bohemia; ca, Canadian Arctic; d, Dolgiy Island; E, Estonia; EN, England; GO, Gotland; i, northern Indiana; IM, Inner Mongolia; MN, Montagne Noire; mv, Moldavia; ni, northwestern Illinois; NM, northwestern Mongolia; PO, Podolia; S, Sweden; SM, southern Mongolia; te, Tennessee; TU, Tuva, w, Wisconsin; Z, Novaya Zemyla-Vaygach-Pay Khoy.

Wenlock–Ludlow age, and included endemic genera such as *Pholidopora* and *Arthrostylus*, which explain its isolated position on the DCA plot (Fig. 3).

Also reappearing in the Wenlock is the Mongolian faunal province from the northern Siberian plate. The province is composed of faunas from the Wenlock of northwestern Mongolia, Tuva, eastern Mongolia and central China. Podolia was again linked with the Mongolian Province in the cluster analysis. The complete merging of the North American and Baltic Provinces in the Wenlock slightly preceded closing of the Iapetus Ocean, as Late Silurian folding in Scotland and Norway suggests that the Northern Iapetus had closed by Ludlow or Pridoli time (Cocks & McKerrow 1973).

Ludlow

The Ludlow was a time of cosmopolitanism among the Bryozoa. The Mongolian Province of Llandovery–Wenlock time disappeared, as faunas from Mongolia and Tuva now show high faunal similarities with European and American faunas (Fig. 5).

Distinctive faunas again occur in the Niagaran reefs of northwestern Illinois, and also in the Quganhebu and Xibiehu Formations of Inner Mongolia. Ludlow faunal gradients are controlled by the presence of distinctive faunas at single localities rather than by provinciality. The fauna from Inner Mongolia was located on the North China plate, and contains the genera *Anaphragma*, *Eridotrypa*, *Homotrypa*, *Paralioclema* and *Stictopora*. Although Llandovery–Wenlock faunas from the South China plate had faunal similarities with the Mongolia–Tuva region, this fauna from the North China plate is distinctive in nature.

The cosmopolitan Baltic–North American–Siberian Province consists of faunas from southern and northwestern Mongolia, Tuva, Gotland, Novaya Zemyla–Vaygach–Pay Khoy, Podolia, Sweden, Estonia, Moldavia, England, Tennessee, Arctic Canada, Australia, Wisconsin, northern Indiana, northwestern Illinois, Bohemia and Montagne Noire (Fig. 6). These localities show a high degree of similarity to one another and contain common Late Silurian genera such as *Fistulipora*, *Fenestella*, *Hallopora* and *Monotrypa*.

The disappearance of the Mongolian Province in the Ludlow suggests that the Siberian plate may have moved southward into

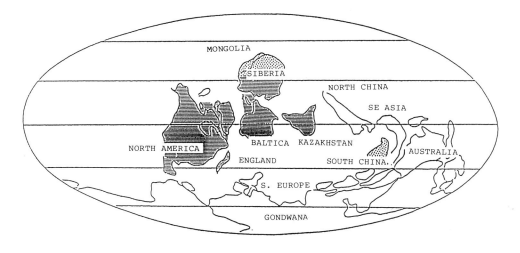

Fig. 4. Wenlock Faunal Provinces. Paleogeographic reconstruction after Scotese (1986).

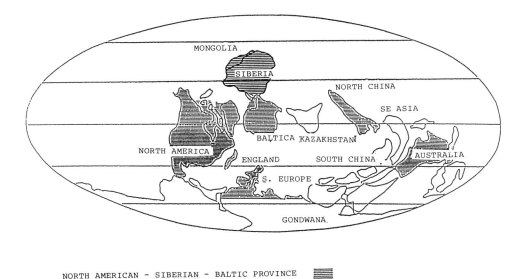

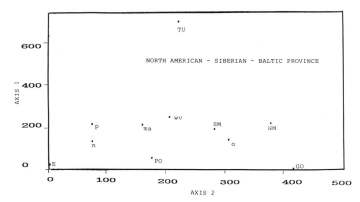

Fig. 6. Ludlow Faunal Provinces. Palaeogeographic reconstruction after Scotese (1986).

a warmer climatic zone in the Late Silurian. This explanation is consistent with Devonian biogeographic studies of corals (Pedder & Oliver 1988) and miospores (Streel *et al.* 1988) which postulate a Devonian location for Siberia at a more southerly latitude than indicated on the continental reconstructions of Scotese (1986).

Pridoli

The cosmopolitanism of the Ludlow continued into the Pridoli. There is little biogeographic differentiation into provinces among faunas from Estonia, Podolia, Gotland, northwestern Mongolia, southern Mongolia, Pennsylvania, Maryland, West Virgina, New York, Oklahoma and Tuva (Fig. 7). The fauna from the Taugantelyski Formation of Tuva shows a high degree of dissimilarity with other faunas. It is a low diversity fauna of five genera: *Amplexopora*, *Eridotrypella*, *Eridotrypa*, *Heterotrypa* and *Stigmatella*. This fauna has a distinctly Ordovician aspect to it as most of these genera were abundant in the Caradoc and Ashgill. However, Ludlow faunas from Tuva also contain these 'Ordovician' genera along with more typical Silurian genera such as *Fistulipora*, *Hallopora* and *Lioclema*.

Another highly endemic fauna is found in the reef community of the Hamra Formation in Gotland. Along with *Fenestella* and *Fistulipora* are found the endemic genera *Saffordotaxis*, *Flabellotrypa* and *Sagenella*. These faunas from Tuva and Gotland are interpreted to be communities within the cosmopolitan Baltic–North American–Siberian Province. Pridoli faunal provinces are shown in Fig. 8.

Silurian extinctions

The Silurian has not been recognized as a major period of extinction, however Middle to Late Silurian generic extinctions have been reported for corals (Kaljo & Klaaman 1973), brachiopods (Boucot 1975), nautiloids, ostracodes and crinoids (Sepkoski 1986). Raup & Boyajian (1988) have also recognized generic extinctions for foraminifera, corals, bryozoans, brachiopods, bivalves, gastropods, cephalopods, echinoderms and arthropods.

Silurian extinctions of bryozoan genera parallel changes in provinciality. Numbers of Llandovery extinctions were low in North America, Baltica and Mongolia (Fig. 9). During the Wenlock, a large increase in generic extinctions occurred in North America and Baltica coincident with the merging of the North American–Siberian and Baltic Provinces. Wenlock extinctions in Mongolia remained low. An increase in extinctions occurred in Mongolia in the Ludlow when the Mongolian Province ceased to exist and all continents were part of the same cosmopolitan province. Ludlow extinctions in North America and Baltica declined from their high values in the Wenlock. All three regions had increases in extinctions in the Pridoli, with North America and Mongolia having their greatest numbers of extinctions during this time.

Extinctions caused by continental convergence were postulated by Valentine & Moores (1970, 1972) as a method for explaining the Permo-Triassic extinctions. Continental convergence is associated with climatic change towards increased seasonality. Convergence is also associated with marine regressions, which contribute to climatic instability and may also destroy epeiric sea habitats. Aside from climatic effects, provinciality itself is a factor in species diversity. Creation of dispersal barriers through either climatic or geographic effects allows additional opportunities for speciation, while elimination of dispersal barriers should lower speciation rates. Boucot (1975) observed that extinction rates of Siluro–Devonian brachiopods increased during times of decreasing provincialism and suggested that competition may have been a factor.

Generic extinctions may be associated with either high species extinction rates or low speciation rates. Estimated rates of speciation and species extinction reflect the effects of continental convergence between North America and Baltica. Merging of

Fig. 7. Pridoli DCA axes 1 v. 2. Symbols: E, Estonia; GO, Gotland; ma, Maryland; n, New York; NM, northwestern Mongolia; o, Oklahoma; p, Pennsylvania; PO, Podolia; SM, southern Mongolia; TU, Tuva; wv, West Virginia.

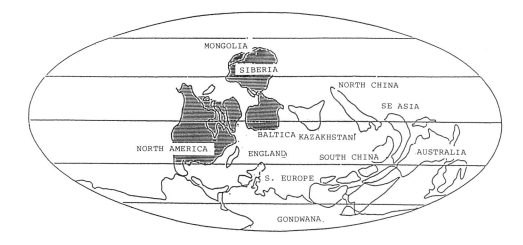

Fig. 8. Pridoli Faunal Provinces. Palaeogeographic reconstruction after Scotese (1986).

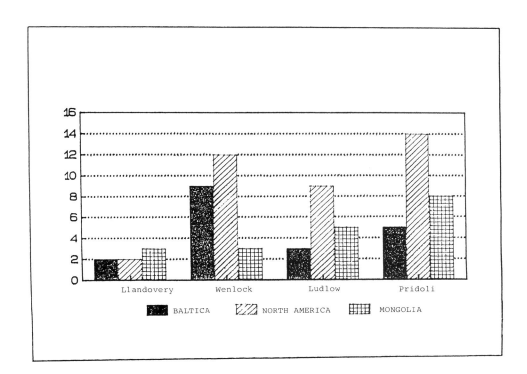

Fig. 9. Extinctions of Silurian bryozoan genera in North America, Baltica and Mongolia.

the two provinces in the Wenlock brought about a large increase in species extinction rate for the combined continents of North America and Baltica (Table 1). These extinctions may be due to the above mentioned climatic effects of convergence, and/or to competition between faunas. Species extinction rates fell for the two continents in the Ludlow, however Ludlow and Pridoli speciation rates became significantly lower. Low Late Silurian speciation rates show the effects of elimination of dispersal barriers on faunas and may also be related to a Late Silurian eustatic regression which began in the late Ludlow (McKerrow 1979). Kaljo & Klaaman (1973) attributed Late Silurian coral extinctions to this regression and it is the most likely cause of increases in extinctions of Pridoli bryozoan genera on all continents.

Ludlow extinctions in Mongolia were coincident with a large

Table 1. *Speciation and species extinction rates, estimated from first and last appearances of Silurian bryozoan species, for Mongolia and the combined continents of North America and Baltica. The time scale employed uses estimated durations of 10 Ma for the Llandovery, 7 Ma for the Wenlock and Ludlow, and 6 Ma for the Pridoli*

	Speciations/Ma		Extinctions/Ma	
	Mongolia	America–Baltica	Mongolia	America–Baltica
Llandovery	4.8	15.8	3.0	8.9
Wenlock	2.7	13.9	3.9	24.8
Ludlow	7.3	7.0	11.4	12.3
Pridoli	1.0	8.0	—	—

increase in species extinction rate and a small rise in speciation rate. The high incidence of Ludlow extinctions and the disappearance of the Mongolian Province suggests the possibility that this north temperate zone fauna was eliminated by climatic change, possibly due to southward motion of the Siberian plate. Boucot & Johnson (1973), however, recognized the existence of the *Tuvaella* brachiopod fauna in this region during the Late Silurian.

References

ANSTEY, R. L. 1986. Bryozoan provinces and patterns of generic evolution and extinction in the Late Ordovician of North America. *Lethaia*, **19**, 33–51.

——, RABBIO, S. F. & TUCKEY, M. E. 1987a. Bryozoan bathymetric gradients within a Late Ordovician epeiric sea. *Paleoceanography*, **2**, 165–176.

——, ——, & —— 1987b. Major community gradients and biome patterning on the Late Ordovician epeiric sea in North America. *Geological Society of America Abstracts*, **19**, 218.

ASTROVA, G. G. 1965. Morphologiya, istoriya razvitiya i sistema Ordovikskiy i Siluriyskiy Mshanok. *Akademiya Nauk SSSR Trudy Paleontologicheskogo Institut*, **106**.

BERRY, W. B. N. 1973. Silurian–Early Devonian graptolites. *In*: HALLAM, A. (ed.) *Atlas of Palaeobiogeography*. Elsevier, Amsterdam, 81–88.

—— 1979. Graptolite biogeography: A biogeography of some Lower Paleozoic plankton. *In*: GRAY, J. & BOUCOT, A. J. (eds) *Historical Biogeography, Plate Tectonics and the Changing Environment*. Proceedings of the 37th Annual Biology Colloquium, Oregon State University Press, Corvallis, 105–116.

—— & BOUCOT, A. J. 1970. Correlation of the North American Silurian rocks. *Geological Society of America Special Paper*, **102**.

—— & ——. 1974. Correlation of the Silurian rocks of the British Isles. *Geological Society of America Special Paper*, **154**.

BOUCOT, A. J. 1975. *Evolution and Extinction Rate Controls*, Elsevier, Amsterdam.

—— 1979. Silurian. *In*: ROBISON, R. A. (ed.) *Treatise on Invertebrate Paleontology Part A*. Geological Society of America and University of Kansas Press, A167–182.

—— & JOHNSON, J. G. 1973. Silurian brachiopods. *In*: HALLAM, A. (ed.) *Atlas of Palaeobiogeography*. Elsevier, Amsterdam, 59–66.

CISNE, J. L. & RABE, B. D. 1978. Coenocorrelation: Gradient analysis of fossil communities and its applications in stratigraphy. *Lethaia*, **11**, 341–64.

COCKS, L. R. M. & MCKERROW, W. S. 1973. Brachiopod distributions and faunal provinces in the Silurian and Lower Devonian. *In*: HUGHES, N. F. (ed.) *Organisms and Continents through Time*. Special Papers in Palaeontology, **12**, 291–304.

GAUCH, H. G. 1982. *Multivariate Analysis in Community Ecology*. Cambridge University Press, Cambridge.

GRUBBS, D. M. 1939. Fauna of the Niagaran nodules of the Chicago area. *Journal of Paleontology*, **13**, 543–60.

KALJO, D. & KLAAMAN, E. 1973. Ordovician and Silurian corals. *In*: HALLAM, A. (ed.) *Atlas of Palaeobiogeography*. Elsevier, Amsterdam, 37–46.

MANTEN, A. A. 1971. *Silurian Reefs of Gotland*. Elsevier, New York.

MCKERROW, W. S. 1979. Ordovician and Silurian changes in sea level. *Journal of the Geological Society of London*, **136**, 137–145.

PEDDER, A. E. H. & OLIVER, W. A. 1988. Devonian coral distributions as a test of paleogeographic models. *Palaeozoic Biogeography and Palaeogeography Symposium Abstracts*, 57.

RAUP, D. M. & BOYAJIAN, G. E. 1988. Patterns of generic extinction in the fossil record. *Paleobiology*, **14**, 109–125.

RAYMOND, A. 1987. Paleogeographic distribution of Early Devonian plant traits. *Palaios*, **2**, 113–132.

SCOTESE, C. R. 1986. Phanerozoic reconstructions: A new look at the assembly of Asia. *University of Texas Institute for Geophysics Technical Report*, **66**.

SEPKOSKI, J. Jr. 1986. Phanerozoic overview of mass extinction. *In*: RAUP, D. M. & JABLONSKI, D. (eds) *Patterns and Processes in the History of Life*. Springer-Verlag, Berlin, 277–295.

SHEEHAN, P. M. 1975. Brachiopod synecology in a time of crisis (Late Ordovician–Early Silurian). *Paleobiology*, **1**, 205–217.

SPJELDNAES, N. 1981. Lower Paleozoic paleoclimatology. *In*: HOLLAND, C. H. (ed.) *Lower Paleozoic of the Middle East, Eastern and Southern Africa and Antarctica*. John Wiley and Sons, New York, 199–256.

STREEL, M., FAIRON-DEMARET, M. & LOBOZIAK, S. 1988. Givetian to Famennian phytogeography of Euramerica and western Gondwana based on miospore distribution. *Palaeozoic Biogeography and Palaeogeography Symposium Abstracts*, 65.

VALENTINE, J. W. & MOORES, E. M. 1970. Plate-tectonic regulation of faunal diversity and sea level: a model. *Nature*, **228**, 657–659.

—— & —— 1972. Global tectonics and the fossil record. *Journal of Geology*, **80**, 167–184.

WILLIAMS, A. 1973. Distribution of brachiopod assemblages in relation to Ordovician palaeogeography. *In*: HUGHES, N. F. (ed.) *Organisms and Continents through Time*. Special Papers in Palaeontology, **12**, 241–269.

WITZKE, B. J., FREST, T. J. & STRIMPLE, H. L. 1979. Biogeography of the Silurian-Lower Devonian echinoderms. *In*: GRAY, J. & BOUCOT, A. J. (eds) *Historical Biogeography, Plate Tectonics and the Changing Environment*. Proceedings of the 37th Annual Biology Colloquium, Oregon State University Press, Corvallis, 117–129.

ZIEGLER, A. M., HANSEN, K. S., JOHNSON, M. E., KELLY, M. A., SCOTESE, C. R., & VAN DER VOO, R. 1977. Silurian continental distributions, paleogeography, climatology and biogeography. *Tectonophysics*, **40**, 13–51.

——, BAMBACH, R. K., PARRISH, J. T., BARRETT, S. F., GIERLOWSKI, E. H., PARKER, W. C., RAYMOND, A. & SEPKOSKI, J. J. Jr. 1981. Paleozoic biogeography and climatology. *In*: NIKLAS, K. J. (ed.) *Paleobotany, Paleoecology and Evolution*. Praeger Publishing, New York, 231–267.

Appendix 1. *Silurian Bryozoan Data Base. Numbers of species per genus, recorded at each locality.*

	Llandovery																					Wenlock															
	Al	Ch	Es	Go	SI	CK	NK	Me	Mi	NM	No	NY	SO	Ok	On	Po	PT	Te	To	Tu	VR	Al	Ch	En	Es	Go	Il	CI	SI	Ka	Me	Mi	EM	NM	NY	No	On	Po
Acanthoclema				1				1																	1						1				1	1		
Acanthotrypina				1																					1						1				1	1		
Anplexopora										1				1		2						1										1						1
Anaphragma																																						
Anisotrypa																																						
Anomalotoechus																																1						
Archaeofenestella			1																			3	1	2														
Arthrostylus																										1												
Asperopora	1		3	4			1	1							1							4			4	3	2		1					6		3	1	
Aspidopora			1	1						2																												
Astroviella																																						
Atactoporella							1																									1	1					
Atactotoechus									1																													1
Batostoma							1									2																1						
Bythopora			1	1																					1				1					2	2			
Calanotrypa																						1																
Callocladia																																						
Calloporella																											1											
Ceramopora	1			1	3						1	1									1	1			1		1	3						2	1	3	1	
Ceranoporella			1					1																	1	1								2		2		
Chasmatopora		1		1	1	1	1	1			1	1		1			1	1		2									1	1								
Cheilotrypa	1			1																									1	1					1		1	
Clathropora			2	2	1					1		1		1															1						2		2	
Clonopora																																						
Constellaria													3																									
Corynotrypa	2		1	1	1						1											3			1			1	1					2		1		
Crepipora			1																									1										
Cuneatopora	1		1	1																		3	1															
Cyclotraypa			2																																			
Cyphotrypa	1	2	1	1													3								1	1			1									
Dianesopora			1							1																			3	1				2		1		
Diploclena	1																								1	1								2		1		
Diplostenopora																																						
Diplotrypa		1		1							1						1												1					2	3	2		
Discotrypa																						1																
Discotrypina																1																						
Duncanoclema																																						
Ensipora	1		1			1		2		1		1			1			5				1	1			1					1	2		1		1		
Eostenopora				1																			1	1		1									1			
Eridocampylus																																						
Eridotrypa	1		1	1				1									2					3	2	1		1	1				1	1		4		3		
Eridotrypella			1						1													1				1								1		1		
Favositella				1																		3	1															
Fenestella	3		1				2		1	2			1				1		1	2		4		2	1	4	1	3	2	1				2	1	2		
Fistulicanta																			1																			
Fistulipora		1	1	1			1			1	1			1								3			4	3	1	1						4		3	1	
Fistuliporella																										1												
Fistuliranus		1																										1										
Flabellotrypa																																						
Glauconomella	1		1	1						1												1			1				1									
Graptodictya															3																							
Hallopora	2		5	3	1	2	3	1		1			1	1	3	1		2	5			1	2		2		3	2		1	1			2	1	3	1	
Haplotrypa			1												1											1												1
Hederella																																						
Helopora	4			1			1	1			1	1			1		1	1	1	1												1			1			
Henieridotrypa																																						
Hemiphragma																																						
Henitrypa			1	1							1	1																										
Hennigopora		2	1	2										1		1					1					2								1		1	1	
Hernoidea																																						
Heterotrypa			1			2																				1					2							
Homotrypa				1	1					1					1	1																						
Idiotrypa			1																															1		1		
Isotrypa																									1													
Leptotrypa																							1															
Leptotrypella																																		1				

			Ludlow																		Pridoli										
	Te	Tu	Au	Bo	CA	DI	En	Go	Il	NI	IM	Mv	NM	SM	MN	NZ	Po	Sv	Te	Tu	W	Es	Go	Ma	NM	SM	NY	Ok	Pe	Po	Tu	WV
Acanthoclema	1																															
Acanthotrypina																																
Amplexopora		2															1			2											2	
Anaphragna						1			1																							
Anisotrypa																														1		
Anomalotoechus																		1		1												
Archaeofenestella																																
Arthrostylus							1																									
Asperopora						3												2														
Aspidopora																																
Astroviella																		1		2		1										
Atactoporella		1																														
Atactotoechus													4													2						
Batostoma		2					1											1							1							1
Bythopora					1	2			1																1							
Calanotrypa						1																										
Callocladia						1																	1									
Calloporella																																
Ceramopora	1			1				1					2	1			2	1		1				1	1		1				1	
Ceramoperella							1																									
Chasmatopora		1									1																					
Cheilotrypa			1			1				2					1		1														4	1
Clathropora																																
Clonopora							1																									
Constellaria		2											1												1							
Corynotrypa	1						2																									
Crepipora																												1				
Cuneatopora																																
Cyclotrypa				1																												
Cyphotrypa		2											2	2									1	1	2	1						2
Dianesopora	2																															
Diploclema																																
Diplostenopora																									1		1					1
Diplotrypa		1				1														1												
Discotrypa							1									1															1	
Discotrypina		1																		1												
Duncanoclema																									1							
Ensipora																																
Eostenopora													1			1				1					1	1				1		
Eridocampylus																				1												
Eridotrypa				1			2	2		1	1						2		1						1						5	1
Eridotrypella					1						1					1							1			4			1		1	1
Favositella							1					4					1								1	4			2			1
Fenestella	1		1	7	1			1	1	4				1	1		1		1		1	2	1	2			1	2	3	1		1
Fistulicanta																				2												
Fistulipora	3	1	1	1	1	1	3	3		2		3	8	1	1	2	3	1	2	5		1			5	1		3		1		1
Fistuliporella																							3									3
Fistuliranus													1	1	1							1										
Flabellotrypa																						1										
Glauconomella		1	1														1	1										1				
Graptodictya																																
Hallopora	1	2	1				1	1		1		3	1				1		1	2					2	1						
Haplotrypa																																
Hederella				2																							1	1				
Helopora																																
Henieridotrypa													1		1										1							
Hemiphragna		1																	1													
Henitrypa			1																													
Hennigopora		1													1		1	1											1	1		
Hernoidea			1																									1				
Heterotrypa		4	1			2					1						4														3	
Honotrypa								2											1													
Idiotrypa																																
Isotrypa																																
Leptotrypa												1											1			1			1	1		1
Leptotrypella					1	2					1											1			1				1			1

	AI	Ch	Es	Go	SI	CK	NK	Me	Mi	NM	No	NY	SO	Ok	On	Po	PT	Te	To	Tu	VR	AI	Ch	En	Es	Go	Il	CI	SI	Ka	Me	Mi	EM	NM	NY	No	On	Po
Lichenalia										1														1									1					
Lioclena		2	8			1						9	3		1	2					9			1	6	3			2									3
Lioclemella							1																															
Loculipora																																						
Mediapora																	3															1						
Meekopora																																						
Mesotrypa																																						
Metadictya																																						
Mitoclena																																			1			
Mongoloclema																																						
Monotrypa			1	3								1												3	1			1	3						3	1	2	
Monotrypella																								2		1												
Monticulipora									1						3									1												1		
Moyerella			3										1			1																						
Nekhorosheviella																					2																	
Nenatopora	1		1	1				1			1				1										1	2		1								1		1
Neotrematopora																																			1			
Nicholsonella	1																	1																		1		
Oandeullina		1																																				
Orbignyella		2																3																		1	1	
Orbignyopora																												1								1	1	
Orthopora																																						
Pachydictya	1				1	1		1			2			1			1	1									1	1										
Paleschara																												2										
Paralioclema																																						
Pesnastylus																																						
Petalotrypa																																						
Phaenophragna									3																										3			
Phaenopora	1		1	4	2	2	2	1	2	1	2	4	2	4	1	5		4	5	7					2					1		2	1		1	1
Phaenoporella																	1																					
Pholidopora																								1														
Phyllodictya																																						
Phylloporina			1									1			1			1			1	1					1					1	1
Pinnatopora																																						
Polypora															1					3									1	2								
Polyporella															1																							
Prasopora													2						1																			
Profistulipora																																						
Pseudohornera															1			1	1	2															2	1	
Ptilodictya	3		3	3	2		1	1		1	1		3		1	1		1	1			3	3	4	1		2								2		1
Ptiloporella				1																						1									1	1	
Reptaria																																						
Reteporina			1																			1	1					1										
Rhinopora			1	1	1	1				2	1	1	1		1	1													1							1	1
Rhombopora																							1															
Saffordotaxis			1																						1													
Sagenella			1	1								1											1		1		1	1								1	1
Sceptropora													1																									
Semicoscinium	1	2	1	1			1			1			1					1				1	2		1	1										1	1
Semiopora																																						
Spatiopora																						1	1	1												1	1
Sphragiopora			1																																			
Stictopora		1						1					1		2	1						1	1													1		
Stictoporella	1	1																				2	1	1	1													
Stictotrypa																								2												1	1
Stigmatella			1				1																					1				1	1					
Stomatopora	1			1																									1									
Streblotrypa																																						
Taeniodictya																																					1
Thanniscus	1		1																						2	1										1	1
Trachytoechus																																						
Trematopora		2		2													1	1			1				4	2					2					3	3
Trenatoporina																																		1				
Trigonodictya	2		1		5	2	2	2	1		1	5	1	2			1	2			2	1			1	1			1	1							1	1
Unitrypa																					1																	
Wolinella			1																									1										

Abbreviations: AI, Anticosti Island; Au, Australia; Bo, Bohemia; CA, Canadian Arctic; Ch, South China; DI, Dolgiy Island; En, England; Es, Estonia; Go, Gotland; Il, Illinois; CI, central Indiana; NI, northern Indiana; Si, southern Indiana; Ka, Kazakhstan; CK, central Kentucky; NK, northern Kentucky; Ma, Maryland; Me, Meaford Ontario; Mi, Michigan; EM, eastern Mongolia; IM, Inner Mongolia; NM, northwest Mongolia; SM, southern Mongolia; Mv Moldavia; MN, Montagne Noire; NY, New York; No, Norway; NZ, Novaya Zemlya; SO, southern Ohio; Ok, Oklahoma; On, Ontario–Niagara Falls; Pe, Pennsylvania; Po, Podolia; PT, Podkamenaya Tunguska River-Siberia; Sw, Sweden; Te, Tennessee; To, Toronto Ontario; Tu, Tuva; VR, Viluya River-Siberia; WV, West Virginia; W, Wisconsin.

	Te	Tu	Au	Bo	CA	DI	En	Go	Il	NI	IM	Mv	NM	SM	MN	NZ	Po	Sv	Te	Tu	W	Es	Go	Ma	NM	SM	NY	Ok	Pe	Po	Tu	WV
Lichenalia							1		1		1														1				1		1	
Lioclema						3				1				7		1				1				1	3							1
Lioclemella			2					1											1	1										1	1	
Loculipora			1		1														1	1										2	1	
Mediapora																																
Meekopora				1															1										1		1	
Mesotrypa		1	1																					1								
Metadictya		5	2																					2								
Mitoclema																			1										1			
Mongoloclema																																
Monotrypa	1	3		1	1		2			2				5				2	3					2								
Monotrypella																																
Monticulipora																																
Moyerella																																
Mekhorosheviella																																
Menatopora							1																									
Neotrematopora																																
Nicholsonella																																
Oandeullina																																
Orbignyella							2																									
Orbignyopora	1																							1		1						
Orthopora												1								2							1	1			1	
Pachydictya																																
Paleschara																																
Paralioclena										1		1								1												
Pesnastylus		1																														
Petalotrypa																																1
Phaenophragna																																
Phaenopora		5																														
Phaenoporella																																
Pholidopora								1																								
Phyllodictya																									1							
Phylloporina																																
Pinnatopora								1																								
Polypora			2					2									1			1												
Polyporella																																
Prasopora						1																										
Profistulipora																2																
Pseudohornera	1		1						1	1																						
Ptilodictya				2	2												1			2	1								1			
Ptiloporella																																
Reptaria			1																					1								
Reteporina							1																									
Rhinopora																																
Rhombopora						2																										
Saffordotaxis															1					1												
Sagenella	1					1	1												1	1												
Sceptropora																																
Semicosciniun		2						1							1			2												2		1
Semiopora																											1					
Spatiopora	1					1	1		1														1									
Sphragiopora																																
Stictopora							1	1																								
Stictoporella		1		1																												
Stictotrypa							1																									
Stigmatella		1								1	2							2			1									1		
Stomatopora	1					1																										
Streblotrypa																	1															
Taeniodictya																																
Thamniscus																								1								
Trachytoechus												1																				
Trematopora	1	2																									1				1	
Trematoporina										3									1	1												
Trigonodictya							1												1										1			
Unitrypa																																
Wolinella																																

Palaeobiogeography of Middle Palaeozoic organic-walled phytoplankton

G. KENT COLBATH
Department of Earth Sciences, Cerritos Community College, Norwalk CA 90650, USA

Abstract: Potential advantages to the use of Palaeozoic organic-walled phytoplankton over other fossil groups are their presence in large numbers in small samples (particularly cores) and their partial independence from lithofacies. Reasons why this potential has not been fully realized are discussed, and the history of study is briefly reviewed. The 'latitude parallel' distribution model proposed for Silurian phytoplankton does not hold up when plotted on a more recent plate reconstruction, although some important insights are forthcoming. Ordovician microfloras appear broadly consistent with proposed plate models, and the diachronous distribution of the genus *Frankea* can be explained by combining the plate model with a restricted, high southern latitude distribution. Ten Frasnian microfloras are compared in detail using G. G. Simpson's Index and cluster analysis. Microfloras from Brazil, Ghana and Algeria cluster together, distinct from outlying areas, while the Carnarvon Basin microflora in Australia is considered relatively endemic. The Frasnian phytoplankton data support a plate reconstruction in which a large seaway separated Africa and South America from North America and Europe.

Organic-walled phytoplankton are widespread in fine-grained marine Palaeozoic sedimentary rocks which have not been subjected to extremes of oxidation or heating. Specimens range in size from 5μm to a few milimetres, and may occur in concentrations of up to 100 000 individuals per gram of rock. Large numbers are thus readily recovered from relatively small core samples.

Some forms are considered the phycomota of single-celled prasinophycean green algae (Tappan 1980; Colbath 1983), and can be unequivocally considered phytoplankton. Other forms ('acritarchs') do not have established affinities with living taxa. Many exhibit excystment or dehiscence structures suggesting an origin as resting or reproductive stages (Le Hérissé 1984), and exhibit morphologies consistent with a planktonic habitat. Maximum diversity of these forms is observed in open shelf settings (Dorning 1981; Colbath, unpublished observations), suggesting that most species originated within the plankton, rather than as disseminules of benthic plants. Changes in relative species abundances are observed in transects perpendicular to the palaeoshoreline, but are not tightly correlated with lithofacies.

Palaeozoic organic-walled phytoplankton thus possess several intrinsic advantages over other types of fossils for the study of palaeobiogeography: (1) Large, living dinoflagellates are only capable of swimming at speeds of 1–2 metres per hour, largely for purposes of vertical migration (Taylor 1980 p. 39). Smaller flagellates are much slower. In contrast to adult fish and some cephalopods, phytoplankton are incapable of active dispersal against even relatively weak oceanic currents, and thus their distribution reflects physical processes. (2) In contrast to benthic invertebrates with planktotrophic larvae, phytoplankton assemblages are not tightly correlated with lithofacies. (3) Large numbers of specimens can be recovered from subsurface samples.

As noted by Kennett *et al.* (1985 p. 198), 'Modern planktonic microfossil groups represent sensitive tracers of surface and near-surface water masses'. Water mass distribution, in turn, is a function of climatic gradients and plate distribution. Assuming that comparison to modern forms is valid, Palaeozoic phytoplankton biogeography should provide some information about the palaeolatitude of ancient continents, and reflect the presence of longitudinal barriers to oceanic circulation.

Perusal of available literature reveals that in spite of these advantages, phytoplankton studies have played a relatively minor role in the study of Palaeozoic biogeography. Difficulties with using phytoplankton for the study of Palaeozoic biogeography can be attributed to four sources: (1) there is little concensus on phytoplankton systematics at the species level; (2) the biostratigraphy of many published microfloras is poorly documented; (3) sampling is spotty, both geographically and stratigraphically; (4) data on modern preservable phytoplankton are not generally presented in a way which facilitates comparison with Palaeozoic assemblages.

History of study

The pioneering work concerning Palaeozoic phytoplankton biogeography was published by Cramer and colleagues in the late 1960s–early 1970s (see reviews by Cramer 1971; Cramer & Díez 1972). Cramer recognized several 'palynofacies' for the Silurian which he regarded as distributed in bands parallel to palaeolatitude. Cramer's data were plotted on the 'Bullard' plate tectonic fit, which is substantially different from more recent models of Early Palaeozoic plate configurations. Full details of Cramer's analysis have never been published, and no explicit criteria were established for distinguishing local environmental effects from the effect of latitudinal change.

In spite of these reservations, most subsequent authors have either cited Cramer's work uncritically, or ignored questions of biogeography. However, Tappan (1980 p. 200–201) noted that criteria for 'palynofacies' recognition are ambiguous, and that some of the boundaries between units correlate with distance from shoreline.

Relatively little new information directly concerned with biogeography has been published subsequent to that of Cramer. Vavrdová (1974) noted that Ordovician phytoplankton described from northern Europe are distinct from those from southern Europe, and suggested that two provinces can be recognized on that basis. Nautiyal (1977) applied Cramer's model to published data for the Devonian, and also used the 'Bullard' fit. He recognized four biofacies which he related to presumed palaeolatitude. Brito (1978) plotted the distribution of the genus *Maranhites* on a reconstruction of the Gondwana continents, and briefly discussed its biostratigraphic significance.

Silurian phytoplankton biogeography

In arguing that Silurian palynofacies reflected palaeolatitude, Cramer (1971 p. 33) acknowledged that oceanic circulation patterns can produce significant spatial deviations in the latitudinal distribution of species (the Gulfstream provides an example in which a sharp biogeographic boundary is not parallel to latitude). He considered the presence of such strong currents unlikely over the broad epeiric sea indicated by the 'Bullard' fit (see also Cramer & Díez 1972 p. 118). Cramer's latitude parallel model is thus partially dependent on the plate tectonic configuration adopted. Although there is little doubt that phytoplankton distribution should at least partially reflect palaeolatitude, the latitude parallel model should be viewed with some scepticism.

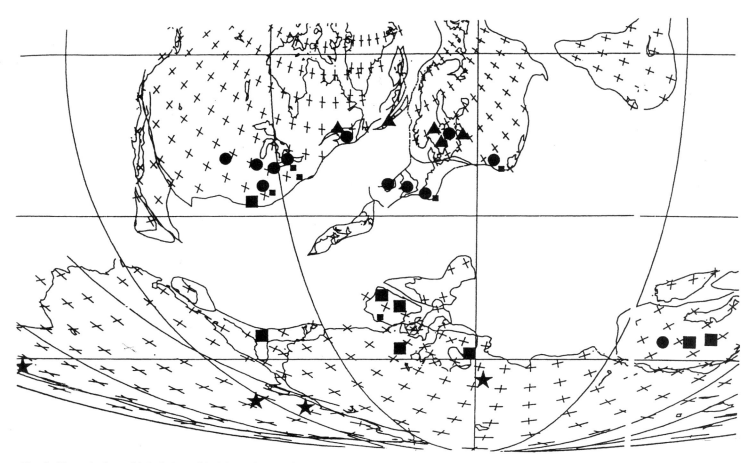

Fig. 1. Phytoplankton 'biofacies' modified from Cramer (1971, text-figure 4A) plotted on Llandovery base map of Scotese (1986). Some similar points consolidated for purposes of clarity. Stars, 'impoverished' *Neoveryhachium carminae* biofacies (*sensu* Cramer 1971 p. 22); large squares, *Neoveryhachium carminae* biofacies; small squares, *Neoveryhachium carminae* biofacies with elements of *Domasia–Deunffia* biofacies; circles, *Domasia–Deunffia* biofacies; triangles, Baltic biofacies. Note that on this reconstruction, the boundary between the *Neoveryhachium carminae* and *Domasia–Deunffia* biofacies is not parallel to palaeolatitude in North America.

Because Cramer's papers on Silurian phytoplankton biogeography are widely cited, it is instructive to plot his data on a more recent plate reconstruction. In Fig. 1, data adapted from Cramer (1971, fig. 4A) are plotted (with modification) on the map for the Llandovery supplied here (Scotese 1986).

Although the latitude parallel model is not strongly supported by the new plot of Cramer's data, in some ways the new plot is superior to the old. Cramer (1971 p. 26) was among the first to note that assemblages from the Florida subsurface have much stonger affinities with those from Spain or North Africa than with Georgia and Alabama. This association is readily explained if Florida is considered part of Gondwana rather than North America during the Silurian (Fig. 1).

A major deviation from the latitude parallel model is the presence of assemblages attributed to the *Neoveryhachium carminae* biofacies in the Appalachian region of the United States. These assemblages appear to be distributed parallel to the margin of the depositional basin rather than to palaeolatitude, and are not in close proximity to other assemblages within the biofacies. Unless this plate reconstruction is grossly in error, these assemblages appear to reflect local environmental conditions rather than biogeography.

Cramer's biogeographic model thus contains some useful insight, but cannot be accepted uncritically. Silurian data clearly demonstrate the importance of establishing explicit criteria (independent of plate tectonic models) for recognizing biogeographic patterns.

Ordovician phytoplankton biogeography

Vavrdová (1974) recognized two provinces for Ordovician phytoplankton in Europe. The Baltic (or Boreal) Province included the northern Soviet Union, Sweden, Poland, northern Germany, and 'probably' part of the British Isles. The Mediterannean Province included Belgium, France, Spain, northern Africa, southern Germany, central Bohemia, and Bulgaria. This geographic differentiation was poorly expressed in the Tremadoc, but became well-developed by the Arenig.

The northward drift of the Baltic plate throughout the Ordovician indicated by the maps for this symposium (Scotese 1986) provides a ready explanation for the increasing distinction between the northern Soviet Union, Sweden and Poland compared to the southern European countries. Northern Germany and Belgium are problematical, however, because both appear on the same plate in the reconstruction.

Here I have plotted the distribution of the distinctive genus *Frankea* for the Arenig–Llanvirn (Fig. 2) and Caradoc (Fig. 3). As noted elsewhere (Colbath 1986), *Frankea* is restricted to the Arenig–Llanvirn in southern and central Europe, but is present in rocks of undoubted Caradoc age in North Africa. The genus is unknown in Baltic and North American assemblages (excluding Florida).

This distribution pattern can be explained rather elegantly if *Frankea* was restricted to high latitudes (> approximately 60° S). According to the reconstruction, England, Belgium and northern Germany began to drift northward in the early part of the

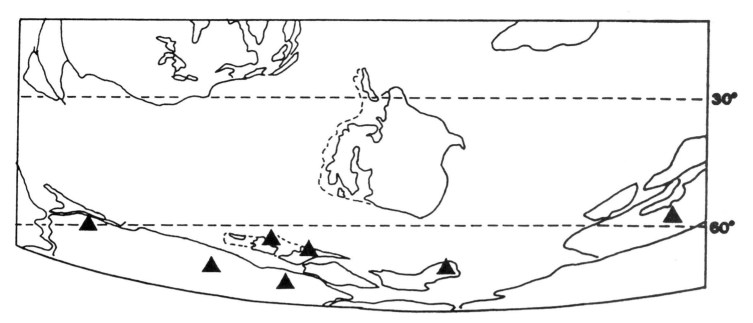

Fig. 2. Distribution of the phytoplankton genus *Frankea* in the Arenig–Llanvirn plotted on the reconstruction of Scotese (1986). Triangles, known records as listed by Colbath (1986), with the addition of records from Morrocco (Elaouad-Debbaj 1984) and Bohemia (Vavrdova 1986). The Bohemian and German records are combined as a single triangle. The Saudia Arabian record is listed only as Ordovician, and thus is poorly constrained in age. Note that *Frankea* is restricted to rocks located at approximately 60° S latitude or higher.

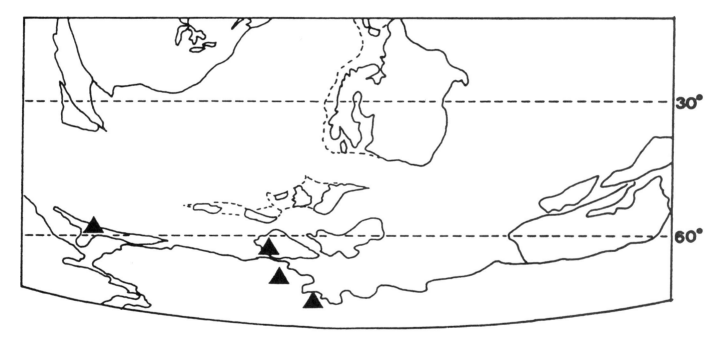

Fig. 3. Distribution of the phytoplankton genus *Frankea* in the Caradoc plotted on the Llandeilo–Caradoc reconstruction in Scotese (1986). Triangles, records listed by Colbath (1986) with the addition of a Late Ordovician record from southern Spain (Cramer 1971). Note that *Frankea* is absent from England, Belgium, Germany and Bohemia, all of which drifted north from their positions in the Arenig–Llanvirn based on the reconstruction.

Ordovician, reaching a position of approximately 50° S by the Llandeilo–Caradoc. This apparently placed them outside of the geographic range of *Frankea*, which was still present in North Africa (and possibly Spain and Florida).

Frasnian phytoplankton biogeography

The discussions above are rather general, and do not provide firm bases for considering the relationship between phytoplankton distribution and plate tectonics. I have chosen the Frasnian for more systematic study because: (1) there are a number of extensively described and illustrated microfloras published for the stage; (2) published plate reconstructions for the Frasnian–Famennian are in conflict due to problems with Carboniferous overprinting on many palaeomagnetic records (reviewed by Piper 1987 p. 258). Analysis of Frasnian phytoplankton may thus have significance for plate reconstructions.

Methods

Ten Frasnian microfloras used in the present study are listed in Table 1. A paper documenting additional phytoplankton from the Givetian–Frasnian section in northern France was received too late for inclusion (Le Hérissé & Deunff 1988), but does not appear to contradict conclusions derived from the present data base.

Where possible, plates and descriptions were examined carefully to ensure uniformity of systematic treatment. Species lists are considered much less reliable records of the microflora due to major differences in taxonomic concepts among phytoplankton workers. For example, Amirie (1984) recognized 15 species in the genus *Micrhystridium* from the German Frasnian. Based on my experience with these notoriously variable forms, I've reduced the number of species to two. Failure to resolve such contrasting taxonomic concepts would seriously distort subsequent analysis of the species lists. Microfloras which are not fully described and illustrated thus provide inherently less reliable data than those so described and illustrated.

Similarities among the ten microfloras were computed using G. G. Simpson's index (Table 2). Simpson's index considers only presence or absence of species, but does compensate for disparities in total numbers of species. Standardized relative abundance data are insufficiently available to permit quantitative analysis but such data are considered in a secondary, qualitative role.

Similarity values were clustered using weighted average linkage pair-group analysis (MVSP statistical package, public domain software by W. L. Kovach). Weighted pair-group averaging introduces less distortion into the lower part of the dendrogram than does unweighted pair-group averaging (Gauch 1982), which is appropriate for present purposes. Clusters were established by inspection of the dendrogram.

Results

The cluster dendrogram for the ten selected microfloras (Table 1) is presented in Fig. 4. Ghana, Brazil and Algeria form one relatively tight cluster, while the remaining localities (exclusive of the Carnarvon Basin) form a second, somewhat looser cluster. The Carnarvon Basin microflora is relatively endemic, exhibiting a consistently lower similarity with assemblages from other continents than does the Canning Basin microflora. Of particular interest, the Canning microflora exhibits a higher similarity with Iowa than with the Carnarvon microflora.

Microfloras from Ghana, Brazil and Algeria are not as exhaustively described as are those from the other areas, suggesting some possible bias in the analysis. Examination of the species lists, however, indicates that the distribution of several distinctive taxa supports the clustering as illustrated. Species of *Umbellasphaeridium* are restricted to Ghana, Brazil and Algeria in the Frasnian. *Cymatiosphaera perimembrana* Staplin, 1961, '*Villosacapsula*' *ceratioides* (Stockmans & Willière, 1962) and species of *Hapsidopalla* are only recorded outside of these areas. *Stellinium micropolygonale* (Stockmans & Willière, 1960) and *Daillydium pentaster* (Staplin, 1961) are known from Algeria and outlying areas, but not from Ghana or Brazil. *Maranhites brasiliensis*

Table 2. Similarity matrix for Frasnian floras*

	CN	CR	GH	IW	AB	FR	BG	WG	AG†
CR	31	–	–	–	–	–	–	–	–
GH	28	17	–	–	–	–	–	–	–
IW	36	27	17	–	–	–	–	–	–
AB	34	26	17	31	–	–	–	–	–
FR	20	15	28	22	26	–	–	–	–
BG	28	25	17	36	32	39	–	–	–
WG	18	11	11	22	31	18	36	–	–
AG	26	22	33	22	13	26	17	13	–
BZ	15	10	35	10	10	15	10	15	45

* Abbreviations as in Table 1.
† Similarities are percentages based on G. G. Simpson's Index, where the number of species in common between two samples is divided by the smaller of the two sample sizes, then multiplied by 100%.

Table 1. Frasnian phytoplankton floras used for biogeographic analysis

	Location	Species*	Description of Flora	References
CN	Canning Basin, Western Australia	74	Fully described and illustrated, shallow boreholes in reef complex	Colbath (in press)
CR	Carnarvon Basin, Western Australia	58	Fully described and illustrated, borehole material	Playford & Dring (1981) Playford (1981)
GH	Ghana	18	Partially illustrated with species list, outcrop material	Bär & Riegel (1974)
IW	Iowa, USA	41	Fully described and illustrated, outcrop material	Wicander & Playford (1985)
AB	Alberta, Canada	35	Partially described and illustrated, boreholes in reef complex	Staplin (1961) Turner (1986)
FR	Boulonnais, France	46	Partially described and illustrated with species list, outcrop material	Deunff (1981)
BG	Belgium	28	Fully described and illustrated, outcrop material	Stockmans & Willière (1962) Martin (1982, 1985)
WG	Rheinland, West Germany	45	Fully described and illustrated, outcrop material	Amirie (1984)
AG	Algeria	23	Partially illustrated with range chart, borehole zones L4 and L5	Jardiné et al. (1974)
BZ	Brazil	20	Partially (?) described and illustrated, borehole material pooled from two separate basins	Brito (1965a, b, 1971)

* Number of species in flora after synonymy.

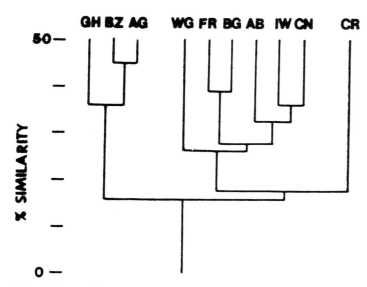

Fig. 4. Cluster diagram, Frasnian phytoplankton floras. Weighted average pair-group clustering, abbreviations as in Table 1.

Brito, 1965 is a dominant part of the microflora in Brazil, Ghana and Algeria (Brito 1978), is rare in France (Deunff 1981; Le Hérissé & Deunff 1988), and is less than 1% of any sample in the Canning Basin (Colbath in press). It appears that there is some meaning to the clustering, and that the pattern is not a mere artifact of methodology.

The ten microfloras are all from open-shelf, cratonic deposits (a shallow-water microflora from Libya was deliberately excluded), and the presence or absence of associated reefs does not strongly correlate with calculated similarity coefficients. The clustering is thus considered to reflect biogeographic rather than local ecological differences among the assemblages.

Two alternative plate reconstructions are available for the Frasnian–Famennian based on differing interpretations of African palaeomagnetic data. The first reconstruction (reconstruction 'A') is the one provided for this symposium (Scotese 1986), and shows North Africa and South America juxtaposed with Europe and North America. In the second reconstruction ('B') a wide seaway is considered to separate these areas (Scotese 1984). As a consequence of treating the Gondwana continents as a single unit, Australia occupies a much higher latitude in configuration A than in B.

The alternate plate reconstructions are illustrated in Figs 5 & 6. The Ghana–Brazil–Algeria cluster is represented by triangles, the Carnarvon Basin as a square, and the remaining localities as circles.

Discussion

In attempting to apply Cramer's latitude parallel model to Devonian phytoplankton, Nautiyal (1977) pooled information from microfloras ranging in age from Lower to Upper Devonian. Frasnian data provide little support for Nautiyal's interpretation. Nautiyal (1977, fig. 1) considered North Africa to belong in the same biofacies (II) as central North America, whereas in the present analysis Algeria clusters with Ghana and Brazil. The

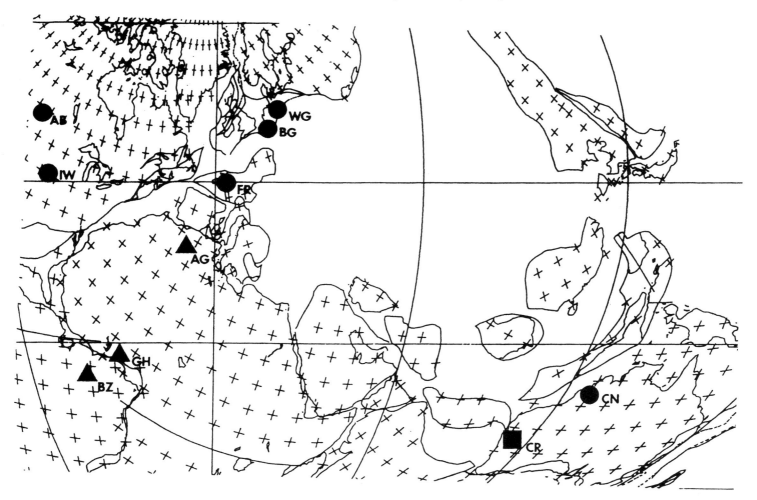

Fig. 5. Frasnian phytoplankton floras plotted on Famennian plate reconstruction A as provided for this symposium (Scotese 1986). Abbreviations as in Table 1, symbols based on cluster analysis as explained in the text.

Biogeography of Silurian Stromatoporoids

HELDUR NESTOR

Institute of Geology, Academy of Sciences of the Estonian SSR, Estonia Avenue 7, Tallinn 200101, USSR

Abstract: Stromatoporoids were widespread in the Ordovician, Silurian and Devonian shallow seas with carbonate types of sedimentation. Their occurrences in reefs and association with bahamitic type carbonate sediments, sometimes replaced by evaporites, allow them to be considered as stenothermal warm-water organisms which lived in the tropical to subtropical climatic belt between 40° N and S. The rapid eustatic deepening after the latest Ordovician glaciation gave rise to ecological stress in the shallow-water stromatoporoids and suppressed their adaptive radiation. Therefore a taxonomically monotonous cosmopolitan fauna of stromatoporoids formed at the beginning of the Llandovery. Few endemic genera existed during the late Llandovery and Wenlock. In Ludlow time a low degree of provincialism appeared when differences developed between the European and Asiatic faunas; these vanished again in the Pridoli. The data on stromatoporoid distribution are in good agreement with the current palaeogeographical reconstructions.

Stromatoporoids were widespread in the Ordovician, Silurian and Devonian shallow seas with carbonate types of sedimentation. They colonized ocean-facing continental shelves, epicontinental seas and fringing areas of island arcs. They played a significant role in the coral–stromatoporoid–algal reef and bank communities as frame-builders and deposit-binders (Heckel 1974; Copper 1974; Webby 1984; Nestor 1986). The range of habitats occupied by Silurian stromatoporoids and the main ecological constraints controlling their distribution have been discussed previously (Nestor 1977, 1984).

The occurrence of stromatoporoids in reefs together with corals, and their frequent association with carbonates of bahamitic type, sometimes laterally replaced or vertically succeeded by evaporites, show that they lived in warm shallow seas of the tropical to subtropical climatic belt (Webby 1980; Nestor 1984). Analysing the distribution pattern of Ordovician stromatoporoids, Webby (1980) reached the clear conclusion that their occurrences were limited to warm waters between 30°N and S of the equator. Heckel & Witzke (1979) also treated stromatoporoids as useful indicators of warm, tropical to subtropical climate and supposed that the distribution of Devonian stromatoporoids was restricted to the low–middle latitudes between 35°N and 40°S. More or less the same limits of latitudinal distribution may be expected for the Silurian stromatoporoids.

The unevenness of the data on the taxonomy and distribution of stromatoporoids makes it difficult to be positive about the existence of biogeographical provinces of the Silurian stromatoporoid fauna. The present state of knowledge does not enable one to trace the geographical distribution of vicarious species or subspecies used as the main tool in the present-day biogeography. At the generic level, provincialism of the Silurian stromatoporoids is rather weakly expressed. True endemic genera or higher taxa are poorly known. Supposed 'endemics' often turn out to be exotic forms occurring in small numbers, or taxa not generally accepted. Palaeobiogeographical peculiarities of stromatoporoid faunas from different regions are expressed mostly by the presence of the so-called temporary endemics, i.e. the genera, making their first appearance in one region and spreading afterwards into other areas. The use of such forms for establishing provincialism and migration routes is also limited as global stratigraphic correlations can be carried out mostly at the series level which is too long a time interval compared to the migration rate of the marine organisms.

These difficulties apply not only to the biogeographical analysis of stromatoporoids, but to the palaeontological material as a whole. Their affect is stronger the worse the state of knowledge of the material analysed. Silurian stromatoporoids have only been studied thoroughly in a few regions. Monographical studies deal with their taxonomy and distribution in the Great Lakes and Hudson Bay areas (Parks 1907, 1908, 1909), North Appalachians (Parks 1933; Stearn & Hubert 1966), New York State (Stock 1979), Baffin Island (Petryk 1967), Estonia (Riabinin 1951; Nestor 1964, 1966), Gotland (Mori 1968, 1970), Podolia (Riabinin 1953; Bolshakova 1973), the Urals (Bogoyavlenskaya 1973); the Siberian Platform (Nestor 1976), the Altai Mountains (Khalfina 1960) and Japan (Sugiyama 1940). In a series of papers Lesovaya (1962, 1971, 1972; Lesovaya & Zakharova 1970) has described stromatoporoids from Central Asia, Dong with co-authors (Dong 1984; Dong & Wang 1984; Dong & Yang 1978; Yang & Dong 1980) from different parts of China and Birkhead (1976, 1978) from New South Wales. In the present review only data on the distribution have been used which are contained in papers accompanied by taxonomic descriptions and figures. In several cases the original names of the genera have been revised, especially with respect to the old papers in which the definitions of the genera were considerably wider than nowadays. Abundant data by Yavorsky and Riabinin have been omitted as they were unsatisfactorily linked to stratigraphic sequences.

A preliminary survey on the biostratigraphy of the Ordovician and Silurian stromatoporoids together with corals was published nearly twenty years ago (Kaljo *et al.* 1970). Meanwhile a lot of new data have been added, therefore the above-mentioned survey has become obsolete in many aspects.

In the present biogeographical analysis use was made of the recent Palaeozoic reconstructions by Scotese (1986).

Distribution

Investigators of different groups of fossils have emphasized the relatively high provincialism of the Ordovician faunas and cosmopolitan nature of the Silurian ones. Webby (1980) has demonstrated that during the Caradoc and early Ashgill a rather clear differentiation of the stromatoporoid faunas existed, but towards the end of the Ordovician a noticeable impoverishment and unification of faunas took place which led to the vanishing of biogeographical provinces. At that time the labechiids were in decline and clathrodictyids were emerging to become the dominant stromatoporoid group. Such a tendency is well observed in the Ashgill sequences of Estonia and Anticosti Island (East Canada) where the transition from the Ordovician to the Silurian is well characterized by stromatoporoids. In Estonia, the species of the well-known Ordovician labechiid genera *Cystostroma* and *Stromatocerium* are still present in the regional Pirgu stage (middle Ashgill), while in the overlying Porkuni stage (upper Ashgill) they are replaced by clathrodictyid-dominated association including *Clathrodictyon*, *Ecclimadictyon* and *Pachystylostroma*. Therefore the Porkuni fauna of stromatoporoids has Llandovery rather than Ordovician affinities (Nestor 1964).

Analogous alterations have been observed in the Ashgill stromatoporoid succession of Anticosti Island (Webby 1980). A

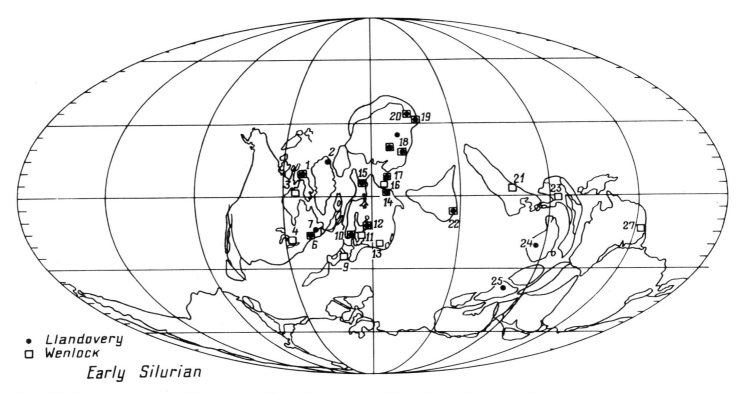

Fig. 1. Distribution of stromatoporoids in the early Silurian (Llandovery and Wenlock) plotted on the modified base map by Scotese (1986). Mollweide projection. Numbers of regions correspond to those in the Tables 1 and 2.

cylindrical labechiid *Aulacera* is the most common form until the top of member 5 of the Ellis Bay Formation. *Clathrodictyon* appears first in the overlying bioherms of member 6 of the Ellis Bay Formation. Consequently both in the Estonian and Anticosti Ashgillian stromatoporoid successions a distinct change in faunal composition from a labechiid-dominated assemblage to clathrodictyid-dominated association has been established. In both sequences it coincides with the substitution of more argillaceous rocks by biohermal and sparitic limestones which is likely to be related to a sudden glacio-eustatic shallowing in Hirnantian time.

Llandovery

The main regions from where the Llandovery stromatoporoids have been described are shown in Fig. 1. Their distribution nowhere exceeds the latitudes 40° N and S. The distribution of the genera in these regions is given in Table 1. The early and middle Llandovery fauna of stromatoporoids as well as other groups was taxonomically rather monotonous and consisted mainly of the representatives of *Clathrodictyon* and *Ecclimadictyon* which had attained their world-wide distribution by the end of the Ordovician. Probably the rapid eustatic deepening after the end Ordovician. glaciation gave rise to ecological stress in the stromatoporoids as stenothermal shallow-water organisms and suppressed their adaptive radiation. The lower and middle Llandovery stromatoporoids have been described from very few regions: Anticosti Island, Estonia, Baffin Island of Arctic Canada, southern Tian Shan, Siberian Platform, islands of Novaya and Severnaya Zemlya, USSR. The fauna of the stromatoporoids of those areas is strikingly conspicuous with respect to its exceptional uniformity: mostly consisting of *Clathrodictyon boreale*, *Cl. kudriavzevi* and *Ecclimadictyon microvesiculosum* groups with single representatives of *Labechia*, *Pachystylostroma*, *Intexodictyon* or *Plectostroma* (see Table 1). The last genus deserves attention as the earliest representative of the order Actinostromatida.

The stromatoporoids recorded from the Llandovery come mostly from the upper Llandovery. The most widespread is the species community associated with the *Pentamerus oblongus* beds. The typical species *Clathrodictyon variolare*, *Ecclimadictyon fastigiatum* etc., can be easily identified, at least from Anticosti Island to Severnaya Zemlya. It shows that the cosmopolitan nature of the Llandovery faunas, emphasized by many authors, is also valid with respect to stromatoporoids. Dong & Yang Jingzhi (1978) pointed to the surprising similarity of the Llandovery faunas of Guizhou province of South China and Estonia, which possess many species and most of the genera is common.

In the late Llandovery, the representatives of several new phylogenetic stocks of stromatoporoids appeared, e.g. the earliest member of the order Stromatoporida: genus *Stromatopora* (Norway, ? Arctic Canada); the first clathrodictyids with regular laminae: *Simplexodictyon* (Norway, western slope of the Urals); clathrodictyids with additional long pillars: *Oslodictyon* (Anticosti, Norway, Estonia) and '*Gerronostroma*' (Arctic Canada). It looks as if, in the late Llandovery, Norway became a centre of stromatoporoid evolution. Actually there may be other reasons for such a phenomenon. With the progressive transgression, the latest Llandovery environments became too deep for stromatoporoids in many regions. In Norway, Ringerike, from where the above-mentioned ancestral forms of new phylogenetic stocks come (Mori 1978), suitable facies for stromatoporoids reach up to the Llandovery–Wenlock boundary, allowing stromatoporoids to survive in a stratigraphic interval badly represented in other sequences. The only real endemics in the Llandovery deposits may be the forms described by Petryk (1967) from the Baillarge Formation of Baffin Island (Arctic Canada) as *Stromatopora baillargense* and *Gerronostroma juvense*, which probably represent a new genus.

Wenlock

The distribution of the Wenlock stromatoporoids is given in Fig. 1 and in Table 2. As in the Llandovery, most of the occur-

Table 1. *Distribution of the Llandovery stromatoporoid genera*

	Arctic Canada (1)	N Greenland (2)	N Appalachians (6)	Anticosti Island (7)	Norway (10)	Estonia (12)	W Urals (14)	Novaya Zemlya (15)	Severnaya Zemlya (17)	Siberian Platform (18)	Altai (19)	Tuva (20)	Tian Shan (22)	Yangzi Platform (24)	Iran (25)
<u>Clathrodictyon</u>	+	+		+	+	+	+	+	+	+	+	+	+	+	
Clavidictyon									+					+	
Cystocerium											?			+	
<u>Ecclimadictyon</u>	+	+	+	+	+	+		+	+			+	+	+	+
Forolinia				?	+	+		+						+	
'Gerronostroma'	+														
Intexodictyon	?			?		+	+							+	
<u>Labechia</u>				+		+	+		+	+	+			+	
Ludictyon														?	
Oslodictyon				+	+	+									
Pachystylostroma		?				+								+	
Plectostroma						+	+						+		
Rosenella						+								+	
Simplexodictyon					+		+								
Stelodictyon						?								?	
Stromatopora	?				+										

The names of the cosmopolitan genera are underlined. Numbers refer to the location on the map (Fig. 1).

rences of the Wenlock stromatoporoids are related to low latitudes within the limits of 30° N and S. Fig. 1 shows the regions of South Siberia (Altai, Tuva) up to 40–45° N. Such a latitudinal position disagrees with the fact that reefs of Wenlock age have been reported from the Tuva region. They contain rich coral and stromatoporoid faunas that suggests locations at lower latitudes.

In consideration of both the diversity (30 genera) and geographic extent, the Silurian stromatoporoid fauna reached its acme of evolution in Wenlock time when all the main phylogenetic stocks of the Palaeozoic stromatoporoids finally appeared. Besides *Clathrodictyon* and *Ecclimadictyon*, wide distribution was also achieved by *Simplexodictyon*, *Stelodictyon* and *Stromatopora*. Very typical were *Actinodictyon*, *Yabeodictyon*, *Neobeatricea* which radiated from the ancestral form *Ecclimadictyon*. Fine-reticulated actinostromatids *Densastroma*, *Araneosustroma*, *Pseudolabechia* and their probable descendant *Vikingia* made their first appearance.

The distribution of several clearly distinct genera was confined to certain closely grouped regions. For example, *Vikingia* and *Araneosustroma* have been recorded from the East-European Platform, the islands of Novaya and Severnaya Zemlya and the Siberian Platform. *Clavidictyon* (although mostly questionable identifications) has been reported from Japan, the Siberian Platform, Severnaya Zemlya, and New South Wales. The distribution of *Oslodictyon* was restricted to North America and Europe as earlier. In North America the first genuine representative of *Plexodictyon* appeared, a very characteristic genus of Ludlow and Pridoli faunas.

In spite of some peculiarities, it is impossible to distinguish palaeogeographical provinces in the Wenlock stromatoporoid fauna. In New South Wales (Birkhead 1976), all the species are in common with North America, Baltic or the Urals. Although the identification of some species is somewhat doubtful, the close similarity of these remote faunas is apparent.

The most endemic Wenlock fauna is in the eastern slope of the Urals. It contains representatives of a very peculiar genus *Gerronodictyon*, and the species content of other genera is rather different from other areas. This suggests that the eastern Urals, which represented an island arc setting, were distinct from the North European plate during the Silurian. The stromatoporoid fauna of the Kawauti Series of Kitakami Mountainland of Japan (Sugiyama 1940) is rather peculiar. The exact dating of the fauna is not clear and it is only surmised that it approximately corresponds to the Wenlock–Ludlow interval.

Ludlow

The Ludlow stromatoporoids were almost as widely distributed and taxonomically diverse as the Wenlock ones (Fig. 2, Table 3), though on the North American and Siberian Platforms epicontinental seas regressed considerably and finds of stromatoporoids become very rare.

The occurrences plotted on the basemap (Fig. 2) show a large latitudinal dispersion of the Ludlow stromatoporoid fauna in comparison with the earlier epochs of the Silurian. The Altai region, placed at the margin of the Siberian plate, lies almost at the latitude of 50° N; this seems rather problematic considering the diverse coral-stromatoporoid fauna present. Of the stromatoporoid localities placed within the limits of the Gondwana supercontinent special attention is paid to those of Iran (Flugel 1969) and Turkey (Wiessermel 1939). According to the reconstruction used, these areas were located close to 40° S, i.e. at the supposed critical latitude for the stromatoporoid distribution. A typical species of *Lophiostroma* has been described and figured from the Bosporus area, but in the Late Silurian, all the other known localities of this genus were confined to the East European Platform (Estonia, Gotland, Podolia). Nevertheless, it is still questionable whether the presence of *Lophiostroma* in the Bosporus area points to the closer connections of this region with the East European Platform, because some species included in this primitive genus have been described even from the Ordovician of North China and the Siberian Platform.

In the Ludlow, *Densastroma*, *Plexodictyon* and *Parallelostroma* became cosmopolitan genera of stromatoporoids. The order Clathrodictyida, although still the most abundant subdivision, began to give up its dominant position. For example, the previously

Table 2. *Distribution of the Wenlock stromatoporoid genera*

	1 Arctic Canada	3 Hudson Bay area	4 Great Lakes area	6 N Appalachians	9 Welsh Borderland	10 Norway	11 Gotland	12 Estonia	13 Podolia	14 W Urals	15 Novaya Zemlya	16 E Urals	17 Severnaya Zemlya	18 Siberian Platform	19 Altai	20 Tuva	21 Inner Mongolia	22 Tian Shan	23 Japan	27 New South Wales
Actinodictyon		+					+			+	+		+		+					
Actinostromella							?										?			
Araneosustroma								+			+		+							
Clathrodictyon	+	+	+	+	+		+	+	+					+			+		+	
Clavidictyon													?	?					+	?
Cystocerium													+	+						
Densastroma			+	+	+		+	+							+		+		?	
Ecclimadictyon			+	+	+	+	+	+	+	+	+	+	+	+	+			+		+
Gerronodictyon												+								
'Gerronostroma'	+														?				?	+
Intexodictyon														+					?	
Labechia					+		+			+	+		+		+					+
Neobeatricea											+	+	+	+						
Oslodictyon		+					+													
Pachystylostroma							+						+	+						
Parallelostroma A	+			+			+		+		+		+	+	+		+			
Plectostroma		+	?		+	+	+							+				?		+
Plexodictyon		?	?																	
Plumatalinia							?													
Pseudolabechia							?		+			+	?							
Rosenella		?					+													
Rosenellinella											?					+				
Simplexodictyon			+	+	+		+	+		+	+	+	+	+	+		+	+	+	
'Solidostroma'											+									
Stelodictyon		+	+	+	+	+	+		+		+	+	?	+						+
Stromatopora	+	+	+	+	+	+	+	+			+		+	+	+	+	+	+	?	?
Syringostromella							+		?					+			+	+	+	
Vikingia							+	+	+		+			+						
Yabeodictyon	+		+				+	+					+	+	+					
Yavorskiina			+										+	+	+					

The names of the cosmopolitan genera are underlined. Numbers refer to the location on the map (Fig. 1).

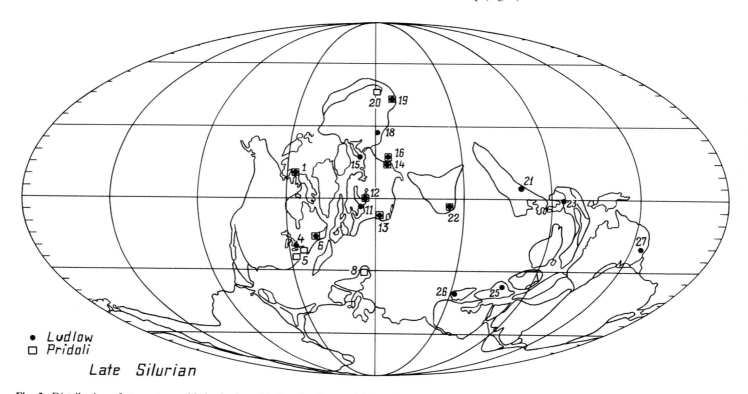

Fig. 2. Distribution of stromatoporoids in the late Silurian (Ludlow and Pridoli). Numbers of regions correspond to those in the Tables 3 and 4.

Table 3. *Distribution of the Ludlow stromatoporoid genera*

	1 Arctic Canada	4 Great Lakes Area	6 N Appalachians	11 Gotland	12 Estonia	13 Podolia	14 W Urals	15 Novaya Zemlya	16 E Urals	18 Siberian Platform	19 Altai	21 Inner Mongolia	22 Tian Shan	23 Japan	25 Iran	26 Turkey	27 New South Wales
Actinostromella				+	+	+											
Amnestostroma									+								
Araneosustroma					+				+			?					
Clathrodictyella	?						+		+				+				
Clathrodictyon			+	+	+	+	+		+		+	?	+			+	+
Densastroma			+	+	+	+	+		+		+	+	+				+
Ecclimadictyon		+	+	+					+	+							
Gerronodictyon									+								
Gerronostroma								+						?			
Hexastylostroma													+				
Intexodictyon																	+
Labechia		+															
Lophiostroma				+	+	+											
'Parallelopora'				+	+												
Parallelostroma A+B			+	+	+	+	+	+	+	+		+					+
'Paramphipora'				?									+				?
Plectostroma	+	?	+	+	+				+								
Plexodictyon	+		+	+	+		+	+	+			+	+				+
Pseudolabechia				+	+		+										
Rosenella																	+
Schistodictyon												?	+				+
Simplexodictyon	+			+	+	+			+	+			+				+
Stelodictyon		+							+								
Stromatopora		?	+	+	+	+					+	+					
Syringostromella		?		+	+				+	+		+					
Trigonostroma									+								
Yabeodictyon	+			+			+						+				

The names of the cosmopolitan genera are underlined. Numbers refer to the location on the map (Fig. 2).

widespread genus *Ecclimadictyon* was replaced by the descendants *Plexodictyon* and *Schistodictyon*. Representatives of the order Labechiida became extremely rare. At the same time the role of actinostromids, especially that of the fine-reticulated forms *Actinostromella*, *Araneosustroma*, and *Densastroma* increased. The variety and frequency of the representatives of the order Stromatoporida increased also, particularly those of *Parallelostroma*.

In general, the cosmopolitan nature of the stromatoporoid fauna continued during the Ludlow, but the restricted distribution of some specific forms showed a certain tendency to provincialism. In Europe (Estonia, Gotland, Podolia) such genera were clearly *Lophiostroma*, some fine-reticulated actinostromatids (*Actinostromella*, '*Parallelopora*', *Araneosustroma*) and also *Pseudolabechia*. On the other hand, fine cylindrical amphiporids (*Clathrodictyella*, '*Paramphipora*') made a very early appearance in some regions of Asia, first in Tian Shan and the eastern Urals. Besides, the distribution of *Gerronodictyon*, *Trigonostroma* and *Amnestostroma* was restricted to the eastern slope of the Urals, while *Schistodictyon* spread only in Central Asia and New South Wales. Further investigations and comparison with the distribution of other groups will show whether the above-mentioned differences between European and Asian stromatoporoid faunas are sufficient for distinguishing provinces or subprovinces of stromatoporoids or not. Taking into consideration that Birkhead (1976, 1978) has described many species from the Ludlow of New South Wales which also occur in Gotland, Estonia and Middle Asia, it seems more likely that clear provincialism of stromatoporoid faunas did not exist in Ludlow times either.

Pridoli

The Pridoli stromatoporoids are less well known and have a more restricted diversity and distribution than the Ludlow faunas (Fig. 2, Table 4). The generic composition did not change essentially, only representatives of some little–known genera (*Habrostroma*, *Perplexostroma*, *Praeidiostroma*) were added. As before, the cosmopolitan genera were *Densastroma*, *Parallelostroma*, *Plexodictyon* and *Simplexodictyon*, but the representatives of the most typical Silurian genus *Clathrodictyon* were almost completely absent. The Pridoli stromatoporoid fauna has been studied most completely in the sequences of the western slope of the Appalachians, Podolia, the eastern slope of the Urals and Tian Shan. The North American fauna had a comparatively low diversity and consisted mainly of the most widespread genera, having several species in common with Estonia and Podolia. The stromatoporoid fauna in Podolia, the eastern Urals and Tian Shan was richer, with some specific, but little known genera.

The suggestion of a low degree of provincialism, observed in the Ludlow stromatoporoid fauna, disappeared in the Pridoli. The distribution of amphiporids (*Clathrodictyella*, '*Paramphipora*', *Stellopora*) earlier restricted to Asia, now expanded also to Europe, more exactly, to Podolia. On the other hand, previous

Table 4. *Distribution of the Pridoli stromatoporoid genera*

	1 Arctic Canada	5 New York, Virginia	8 Bohemia	12 Estonia	13 Podolia	14 W Urals	16 E Urals	19 Altai	20 Tuva	22 Tian Shan
Actinostromella				+						?
Amnestostroma							+			
Araneosustroma								+		
Clathrodictyella				?	+	+	+			+
Densastroma		+		+	+		+	+	+	
Ecclimadictyon							+			
Gerronostroma						+	+		+	
Habrostroma		?								?
Lophiostroma				+	+					
Pachystylostroma				+						
Parallelostroma B		+		+	+	+	+			+
'Paramphipora'					+					
Perplexostroma					+					
Plectostroma		?		+	+					+
Plexodictyon	+	+			+	+	+	+	+	+
Praeidiostroma							+			
Schistodictyon										+
Simplexodictyon		?	+		+		+		+	+
Stelodictyon			?					+		
Stellopora					+		+			
Stromatopora		+	+							?
Syringostromella					+		+		+	?
Vikingia					+					
Yabeodictyon	+									+

The names of the cosmopolitan genera are underlined. Numbers refer to the location on the map (Fig. 2).

'European genera', *Actinostromella* and *Araneosustroma* spread during the Pridoli to Central Asia.

Provincialism and continental reconstructions

The relatively cosmopolitan nature of many Silurian faunas is a generally accepted opinion (Hill 1959; Boucot & Johnson 1973; Pickett 1975). The only clearly distinct faunal province was the Malvinokaffric Province with the *Clarkeia* fauna. It embraced marginal seas of Gondwanaland at high latitudes in South America and South Africa (Boucot & Johnson 1973). The stromatoporoids were entirely absent here.

In the Northern Hemisphere there is no definite counterpart of the *Clarkeia* community, as a cold-water Arctic to temperate zone fauna is unknown, though the *Tuvaella* fauna in Central Asia (Southeast Kazakhstan, Altai, Tuva, Mongolia) (Boucot & Johnson 1973) may turn out to have been in northern temperate zone position. However, the occurrences and taxonomic content of stromatoporoids do not support such an opinion. The Late Silurian basemap needs some revision, as South Siberia and Mongolia have been placed at more northerly latitudes than those in which stromatoporoids usually occurred.

Different opinions have been expressed about the biogeographic subdivision of the so-called Silurian Cosmopolitan Province, which, apart from the Malvinokaffric Province, included all the main areas of Silurian sedimentation in North America, North and East Europe, Siberia, Central Asia, China, Japan, Australia and the Near East. Boucot and Johnson (1973) stated that beginning with late Wenlock time, it is possible to divide the Silurian Cosmopolitan Province into two sub-provinces: the North Atlantic and the Uralian–Cordilleran, which extended on through the end of the period and formed a precursor of the much higher provincialism of the Devonian. Roughly the same two sub-provinces (or provinces) can be defined on the basis of the coral faunas. Pickett (1975) distinguished on the one hand a western and northern province, including North America, Europe and the northern USSR, and on the other a southern and eastern province including Australia, Japan, Central and South-East Asia and India. Kaljo & Klaamann (1973) have defined two provinces for the coral faunas of the Late Silurian calling them the European and the Asiatic. Contrary to Boucot & Johnson, they claimed that by the end of the Silurian the provincial differences of the coral faunas decreased.

The analyses of stromatoporoid distribution also showed that the strongest evidence of provincialism appeared in Ludlow time as is most clearly revealed by differences between the European and the Asiatic faunas. In the Pridoli, these differences vanished, but it may be due to poorer knowledge and more restricted occurrence of the Pridoli stromatoporoids.

In general, the low degree of differentiation of Silurian faunas and its gradual progression during the period is in good accordance with the present palaeogeographical reconstructions. In the Silurian, the vast southern landmass of Gondwana only reached into the equatorial zone in its Australian sector. The remaining, much smaller, land areas were situated in low and middle latitudes, but nowhere did they form a substantial north–south trending land barrier to east–west migration of tropical marine faunas. It was only in the Devonian that the collision of North America and Gondwana formed such a geographical barrier which prevented equatorial migration of faunas and caused their clear differentiation into provinces.

I am very grateful to C. R. Scotese who supplied me with the Palaeozoic reconstructions as basemaps for plotting the data on stromatoporoid distribution. I thank also L. Lippert who drew the diagrams, and A. Noor for her help in preparing the manuscript.

References

BIRKHEAD, P. K. 1976. Silurian stromatoporoids from Cheesmans Creek, with a survey of some stromatoporoids from the Hume Limestone Member, Yass, New South Wales. *Records of Geological Survey of New South Wales*, **17**(2), 87–122.

—— 1978. Some stromatoporoids from the Bowspring Limestone Member (Ludlovian) and Elmside Formation (Gedinnian), Yass Area, New South Wales. *Records of Geological Survey of New South Wales*, **18**(2), 155–168.

BOGOYAVLENSKAYA, O. V. 1973. Siluriiskie stromatoporoidei Urala (Silurian stromatoporoids from the Urals). Nauka, Moskva [in Russian].

BOLSHAKOVA, L. N. 1973. Stromatoporodei silura i nizhnego devona Podelii. (Silurian and Lower Devonian stromatoporoids of Podolia). Academiya nauk SSSR, Trudy Paleontologicheskogo Institute, 141, Nauka, Moskva [in Russian].

BOUCOT, A. J. & JOHNSON, J. G. 1973. Silurian brachiopods. In: HALLAM A. (ed.) *Atlas of Palaeobiogeography*. Elsevier, Amsterdam, London, New York, 59–65.

COPPER, P. 1974. Structure and development of Early Paleozoic reefs. *Proceedings of the Second International Coral Reef Symposium, Brisbane*, **1**, 365–386.

DONG, DE-YUAN. 1984. Silurian and Lower Devonian stromatoporoids from Darhan Mumingnan Joint Banner, Inner Mongolia. In: LI, WEN-GUA, RONG, JIA-YU & DONG, DE-YUAN (eds). *Silurian and Devonian rocks and faunas of the Bateaobao area in Darhan Mumingan Joint Banner, Inner Mongolia*. The Peoples' Publishing House of Inner Mongolia, 57–77 [in Chinese with English summary].

DONG, DE-YUAN & WANG BAO-YU. 1984. Palaeozoic stromatoporoids from Xinjiang and their stratigraphical significance. *Bulletin Nanjing Institute of Geology and Palaeontology, Academia Sinica*, **7**, 237–

286 [in Chinese with English summary].

Dong, De-yuan & Yang, Jing-zhi. 1978. Lower Silurian stromatoporoids from Northeastern Guizhou. *Acta Palaeontologica Sinica*, **17**(4) 421–436 [in Chinese with English summary].

Flugel, E. 1969. Stromatoporen aus dem Silur des ostlichen Iran. *Neues Jahrbuch fur Geologie und Palaeontologie, Monatshefte*, **4**, 209–219.

Heckel, P. H. 1974. Carbonate buildups in the geologic record: a review. *In*: Laporte, L. F. (ed.) *Reefs in time and space*. Selected examples from the Recent and Ancient. Special Publication of the Society of Economical Paleontologists and Mineralogists, Tulsa, **18**, 90–154.

—— & Witzke, B. J. 1979. Devonian world palaeogeography determined from distribution of carbonates and related lithic palaeoclimatic indicators. *In*: House, M. R., Scrutton, C. T. & Bassett, M. G. (eds.) *The Devonian System*. Special Papers in Palaeontology, **23**, 99–123.

Hill, D. 1959. Distribution and sequence of Silurian coral faunas. *Journal and Proceedings of the Royal Society of New South Wales*, **92**, 151–173.

Kaljo, D. & Klaamann, E. 1973. Ordovician and Silurian corals. *In*: Hallam, A. (ed.) *Atlas of Palaeobiogeography*. Elsevier, Amsterdam, 37–45.

——, —— & Nestor, H. 1970. Palaeobiogeographic review of Ordovician and Silurian corals and stromatoporoids. In: Zakonomernosti rasprostraneniya palezoiskih korallov SSSR (Trudy II Vsesoyuznogo symposium po izucheniyu iskopayemyh korallov SSSR, 3). Nauka, Moskva 6–15 [in Russian].

Khalfina, V. K. 1960. Stromatoporoidea. Siluriiskaya Sistema. *In*: Khalfin, L. L. (ed.) *Biostratigrafiya paleozoya Sayano-Altaiskoi Gornoi oblasti, 2, Srednii palaeozoi*. Trudy Sibirskogo, nauchno issledovatelskogo instituta geologii, geofiziki i mineralnogo syrya, **20**, 44–56.

Lesovaya, A. I. 1962. Stromatoporoidei ludlova severnogo sklona Turkestanskogo khrebta. (Ludlow stromatoporoids of the northern slope of Turkestan range). *In: Stratigrafiya i paleontologiya Uzbekistana i sopredelnykh raionov, 1*. Akademiya nauk Uz. SSR, Taschkent, 128–146 [in Russian].

—— 1971. Stromatoporoidei pogranichnykh sloev silura i devona Zheravshanskogo khrebta. (Stromatoporoids of the Silurian and Devonian boundary beds of Zheravshan range). *In: Rugosy i stromatoporoidei paleozoya SSSR Trudy Ii Vsesoyuznogo symposiuma po izucheniyu iskopaemykh korallov SSSR, 2*, Nauka, Moskva, 112–125 [in Russian].

—— 1972. Novye siluriiskie i nizhnedevonskie stromatoporeidei yuzhnogo Tian-Shanya. (New Silurian and Early Devonian stromatoporoids of the southern Tian Shan). *In: Novye dannye po faune paleozoya i mezozoya Uzbekistana*. Akademiya nauk Uz. SSR, Izdatelstvo FAN, Tashkent, 46–52 [in Russian].

—— & Zakharova, V. M. 1970. Novye stromatoporoidei is verhnego silura Turkestanskogo khrebta. (New stromatoporoids from the upper Silurian of the Turkestan range). *Paleontologicheskii zhurnal*, Moskva, **2**, 47–51 [in Russian].

Mori, K. 1968. Stromatoporoids of Gotland. Part 1. Acta Universitatis Stockholmiensis, *Stockholm Contributions in Geology*, **19**, Stockholm, Almqvist & Wiksell.

—— 1970. Stromatoporoids from the Silurian of Gotland. Part 2. Acta Universitatis Stockholmiensis, *Stockholm Contributions in Geology*, **22**, Stockholm, Almqvist & Wiksell.

—— 1978. Stromatoporoids from the Silurian of the Oslo Region, Norway. *Norsk Geologisk Tidsskrift*, **58**, 121–144.

Nestor, H. 1964. *Stromatoporoidea ordovika i llandoveri Estonii*. (Ordovician and Llandoverian stromatoporoids of Estonia). Tallinn [in Russian with English summary].

—— 1966. *Stromatoporoidea wenloka i ludlova Estonii*. (Wenlockian and Ludlovian Stromatoporoidea of Estonia). Tallinn, Valgus [in Russian with English summary].

—— 1976. *Rannepaleozoiskie stromatoporoidei basseina reki Moiero* (sever Sibirskoi platformy). Early Palaeozoic stromatoporoids from the Moiero River, north of the Siberian Platform. Tallinn, Valgus [in Russian with English summary].

—— 1977. On the ecogenesis of the Palaeozoic stromatoporoids. *In: Second Symposium international sur les coraux et recifs coralliens fossiles*. Memoires du Bureau de Recherches Geologiques et Minieres, **89**, 249–254.

—— 1984. Autecology of stromatoporoids in Silurian cratonic seas. *Special Papers in Palaeontology*, **32**, 265–280.

—— 1986. Role of stromatoporoids in the building of Early Palaeozoic reefs. *In*: Sokolov, B. S. (ed.) *Fanerozoiskie rify i korally SSSR* (Irudy V Vsesoyuznogo sympoziuma po korallam i rifam, Dushanbe, 1983). Nauka, Moskva, 202–208 [in Russian].

Parks, W. A. 1907. Stromatoporoids of the Guelph Formation in Ontario. *University of Toronto Studies, Geological Series*, **4**.

—— 1908. Niagara stromatoporoids. *University of Toronto Studies, Geological Series*, **5**.

—— 1909. Silurian stromatoporoids of America (exclusive of Niagara and Guelph). *University of Toronto Studies, Geological Series*, **6**.

—— 1933. New species of stromatoporoids, sponges and corals from the Silurian strata of Baie des Chaleurs. *University of Toronto Studies, Geological Series*, **33**.

Petryk, A. A. 1967. Some Silurian stromatoporoids from Northwestern Baffin Island, District of Franklin. *Geological Survey of Canada Paper*, 67–7.

Pickett, J. 1975. Continental reconstructions and the distribution of coral faunas during the Silurian. *Journal and Proceedings, Royal Society of New South Wales*, **108**, 147–156.

Riabinin, V. N. 1951. Stromatoporoidei Estonskoi SSR. (Silur i verkhi ordovika). (Stromatoporoids of Estonian SSR, Silurian and uppermost Ordovician). Trudy vsesoyuznogo neftyanogo nauchno – issledovatelskogo geologo – razvedochnogo instituta (VNIGRI), *Novaya Seria*, **43**, Leningrad-Moskva [in Russian].

—— 1953. Siluriiskie stromatoporoidei Podolii. (Silurian stromatoporoids of Podolia). Trudy vsesoyuznogo neftyanogoh nauchno – issledovatelskogo geologo – razvedochnogo instituta (VNIGRI), *Novaya Seria*, **67**, Leningrad-Moskva [in Russian].

Scotese, C. R. 1986. Phanerozoic reconstructions: a new look at the assembly of Asia. *University of Texas Institute for Geophysics, Technical Report*, **66**.

Stearn, C. W. & Hubert, C. 1966. Silurian stromatoporoids of the Matapedia–Temiscouata area, Quebec. *Canadian Journal of Earth Sciences*, 3, **31**, 31–48.

Stock, C. W. 1979. Upper Silurian (Pridoli) Stromatoporoidea of New York. *Bulletin of American Paleontology*, 76, 308, 289–389.

Sugiyama, T. 1940. Stratigraphical and paleontological studies of the Gotlandian deposits of the Kitakami Mountainland. *Scientific Reports Tohoku Imperial University, Sendai, Japan* (2), Geology, **21**(2), 81–146.

Webby, B. D. 1980. Biogeography of Ordovician stromatoporoidea. *Palaeogeography, Palaeoclimatology, Palaeoecology*, **32**, 1–19.

—— 1984. Ordovician reefs and climate: a review. *In*: Bruton, D. L. (ed.) *Aspects of the Ordovician System*. Palaeontological Contributions from the University of Oslo, 295, Universitetsforlaget, 89–100.

Weissermel, W. 1939. Neue Beitrage zur Kenntnis der Geologie, Palaeontologie und Petrographie der Umgegend von Konstantinopel. 3. Obersilurische und devonische Korallen, Stromatoporiden und Trepostome von der Prinzeninsel Antirovitha und aus Bithynien. *Abhandlungen der preussischen geologischen Landesanstalt, Neue Folge*, 190.

Yang, Jing-zhi & Dong, De-yuen. 1980. Discussion on the Early Silurian strata in southwestern Hubei and northeastern Guizhou in light of fossil stromatoporoids. *Acta Paleontologica Sinica*, **19**(9), 393–404 [in Chinese with English summary].

The Silurian and Early Devonian biogeography of ostracodes in North America

JEAN M. BERDAN

Department of Paleobiology, E-308 National Museum of Natural History, Smithsonian Institution, Washington, DC 20560, USA

Abstract: Silurian and Early Devonian ostracode associations in North America represent at least three ecotypes, a leperditicopid association, a large beyrichiacean association, a mixed association, and possibly a fourth, spinose podocopid association, or Thuringian ecotype. Comparison of the large beyrichiacean association and mixed association ecotypes indicates the presence of three informal ostracode provinces, the Appohimchi, Baltic-British and Cordilleran, which remained relatively constant in geographical position throughout the Silurian and Early Devonian. Plotting the provinces on palaeogeographic maps suggests that temperature was not an important factor in delimiting the provinces, and land barriers, or possibly deep-water troughs as barriers, were the cause of provincial development. Benthic ostracodes differed from other benthic invertebrates such as brachiopods and corals in developing provincialism in the late Llandovery and having it start to decline in the Pragian through the Emsian.

Environments

Most Silurian and Early Devonian ostracodes appear to have been benthic, although a few widely distributed forms such as the entomozoids are generally thought to have been pelagic. The distribution of the benthic forms was controlled by environmental factors such as salinity, temperature, type of substrate and depth of water, as is the case with other benthic invertebrates such as brachiopods and corals. Of these, only the type of substrate is readily determinable in the geological record; the other factors are largely inferred. The effects of water depth and substrate are a function of the energy level, that is, whether the rocks containing ostracodes were deposited above or below wave base.

About 1970, students of Palaeozoic ostracodes became interested in relating the occurrence of ostracode assemblages to the possible palaeoenvironment of the rocks in which they were found, and several papers were published in a volume on the *Paleoecology of Ostracodes* (Oertli, ed., 1971). Most of these were concerned with post-Early Devonian faunas, but two (Lundin 1971; Polenova 1971) discussed Early Devonian and Silurian–Devonian ostracode assemblages. Becker (1975) proposed three ostracode ecotypes, discussed further by Gooday & Becker (1979) and Becker (1982), for the Middle and Late Devonian of Europe. Becker's ecotypes are the Eifeler ecotype, developed in a shallow-water, high energy environment, the Thuringian ecotype in lower energy, deeper water environment, and the Entomozoid ecotype, indicative of deeper water than any of the other assemblages. In North America, Warshauer & Smosna (1977) recognized three ostracode 'communities' in the Late Silurian (Pridoli) Tonoloway Limestone at Pinto, Maryland, a '*Leperditia* community', a '*Welleria–Dizygopleura* community' and a '*Zygobeyrichia–Halliella* community'. The first of these has also been discussed by Berdan (1969), Copeland (1971, 1976), Copeland & Berdan (1977) and Berdan (1984). This association recurs throughout the Silurian and Early Devonian in facies which on lithological grounds indicate very shallow water tidal flat, lagoonal or deltaic environments and brackish or hypersaline waters (Berdan 1984). Warshauer & Smosna (1977) considered that their '*Welleria–Dizygopleura* community' represented a normal marine environment which was subtidal and relatively shallow, extending perhaps to the level of wave base, and their '*Zygobeyrichia–Halliella* community' as a deeper subtidal environment at or near the level of wave base. Both of these 'communities' probably fall within the Eifeler ecotype of Becker or the 'beyrichiid type' of ostracode fauna discussed by Polenova (1971), as there is no physical or faunal evidence that any part of the Tonoloway was deposited in the deeper water of the Thuringian ecotype.

Recently, Wang (1988) has proposed five ostracode associations for the Palaeozoic, a leperditiid association, a palaeocopid association, a smooth-podocopid association, a spinose-podocopid association and an entomozoacean association. Of these, the leperditiid association corresponds to that discussed by previous authors, the palaeocopid association appears to represent the Eifeler ecotype of Becker (1975) and the 'beyrichiid type' of Polenova (1971) and the spinose-podocopid type is comparable to the Thuringian ecotype of Becker (1975). Neither the smooth-podocopid association nor the entomozoacean association have yet been recognized in the Silurian and Early Devonian of North America.

In North America, a leperditicopid association is well developed throughout the Silurian and Early Devonian. A large beyrichiacean association, indicative of normal marine waters above or slightly below wave base, is also present. Seaward of this, and presumably below wave base, a diverse ostracode assemblage is found in fine-grained sediments. This association includes palaeocopes such as bolliids and amphissitids, metacopes such as thlipsurids and pachydomellids, podocopes such as beecherellids, as well as spinose forms such as aechminids and tricorninids more typical of the Thuringian ecotype of Becker (1975). It is more comparable to the 'healdiid–aparchitid type' of fauna of Polenova (1971) and is apparently a mixture of Becker's Eifeler and Thuringian ecotypes. In eastern North America, this association is represented by the ostracodes of the Lower Devonian Kalkberg Limestone of New York, the Silurian Newsom Shale and Brownsport Formation of Tennessee and the Silurian Henryhouse and Devonian Haragan Formations of Oklahoma (Lundin 1971). It is not yet certain that the true Thuringian ecotype of Becker, in which the ostracodes are commonly silicified, is present in eastern North America, although an undescribed silicified fauna from the Lower Devonian Port Ewen Formation of New York might qualify. The mixed association is widespread and fairly common.

In summary, there are four ostracode associations related to increasing water depth in North America, the leperditicope assemblage, the large beyrichiacean assemblage ('beyrichiid type' of Polenova, 1971, Eifeler ecotype of Becker 1975; palaeocopid association of Wang 1988), the mixed association ('healdiid–aparchitid type' of Polenova, 1971), and possibly a tricorninid–aechminid association (Thuringian ecotype of Becker, 1975; spinose-podocopid type of Wang, 1988). These can be recognized in widely separated geographic areas and recur throughout the Silurian and Early Devonian when the appropriate environmental conditions were present.

Provinces

For purposes of correlation and zonation by ostracodes, it is necessary to compare ostracodes from similar associations; when this is done, this group of fossils is useful in intrabasinal and interbasinal correlation. Recognition of the environmentally controlled associations makes possible the comparison of similar ecotypes from eastern and western North America with those of Europe. Although many of the ostracode genera appear to be widespread and cosmopolitan, there are striking differences in certain groups, and it appears that there are three distinct ostracode provinces in North America. For the purposes of this report, a faunal province is considered to be an area which contains a distinctive group of genera essentially restricted to that area and which do not appear to be controlled by lithological facies or environmental factors. No attempt has been made to quantify the number of endemic and cosmopolitan genera, as discussed by Boucot (1975, p. 266–267) because the faunas of many formations known to contain ostracodes have yet to be described. However, enough is known about the distribution of ostracode faunas to allow the recognition of informal ostracode provinces.

The three provinces in North America are the Appohimchi, Baltic–British and Cordilleran. These provinces are based on comparison of large beyrichiacean associations and mixed associations; the leperditicope association has not been compared because of difficulties in discriminating between taxa from the literature, and the tricorninid–aechminid association is too little known as yet. The Appohimchi ostracode province is essentially the same as the Appohimchi Subprovince based on brachiopods of Boucot (1975) and the Appohimchi Province of Oliver (1977) based on corals, all in the Eastern Americas Realm. The Baltic–British ostracode province is in the Old World Realm of Boucot (1975) but as used here does not exactly correspond to other subdivisions of this Realm. The Cordilleran ostracode province approximates the Uralian–Cordilleran Region of Boucot (1975), but shows more endemism. From late Llandovery through at least Lochkovian time, the Appohimchi and the Baltic–British ostracode provinces can be directly compared, as both represent the large beyrichiacean association. The Cordilleran province is less directly comparable, because according to Murphy (1977), the ostracode-bearing limestones in the lower part of the sequence were deposited as allochthonous debris fans or turbidites that slid into a deep, subtidal basin from shallow-water environments, and the fauna is consequently mixed. However, Copeland (1977a) described ostracodes from platform limestones in northwestern Canada, and ostracodes from allochthonous deposits in Nevada include taxa swept in from shallow-water ecosystems.

North American ostracode provinces have previously been discussed by Copeland & Berdan (1977) and Berdan (1983). They have been plotted on the base maps of the atlas of Phanerozoic Reconstructions by Scotese (1986) for the Llandovery, Wenlock and Ludlow Series and the Lochkovian and Emsian Stages. The geographic extent of the provinces is based largely on Copeland (1977a, 1989) for the Cordilleran province and on Abushik (1971) and Siveter (1978) for the Baltic–British province. The extent of the Appohimchi province is based on numerous sources in the literature and on personal knowledge. It should be noted that use of the older literature has required a certain amount of interpretation, as many of the taxa described are no longer assigned to their original genera. For example, none of the species assigned to the genus *Kloedenia* by Ulrich & Bassler (1923) are now considered to belong in that genus, and the same is true of *Octonaria* and *Thlipsura*, all of these in the Appohimchi province. Ages of the formations from which ostracodes have been described are based on Berry & Boucot (1970) for the Silurian and on Boucot *et al.* (1969) for the Early Devonian.

Llandovery Series

In most parts of the world (Fig. 1), only the upper part of this series contains ostracode faunas (Abushik 1971, 1988, pers. comm.; Copeland 1977b). Apparently, after the faunal crisis at

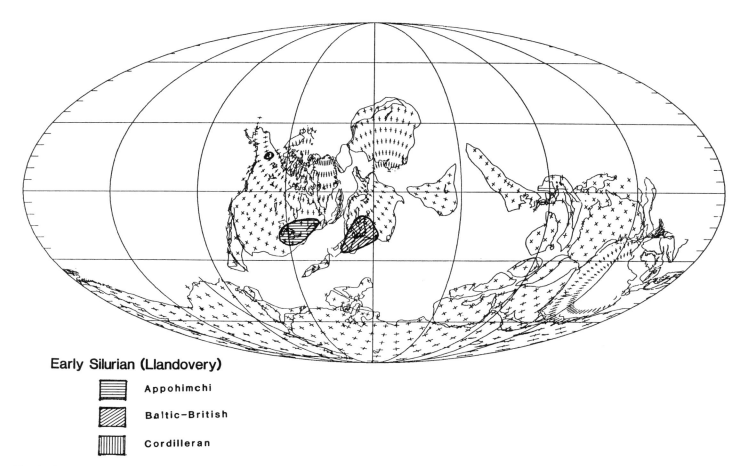

Fig. 1. Ostracode distributions in the Llandovery of North America and Europe.

the end of the Ordovician the ostracodes recovered slowly, and many new groups, notably the beyrichiaceans, appeared in the faunas. By late Llandovery time, the Appohimchi ostracode province had developed a diversity of zygobolbid genera such as *Zygobursa, Zygobolba, Zygosella, Zygobolbina, Bonnemaia*, and *Mastigobolbina*, to such an extent that Ulrich & Bassler (1923) proposed nine faunal zones based on species of this group. In contrast, Martinsson (1962) described only two genera from the Baltic–British province, neither of which appear to be closely related to the Appohimchi forms. The corresponding large beyrichiaceans of the Baltic–British province were species of *Beyrichia* s.s., of which only one species from Anticosti, Quebec, Canada has been found in the Appohimchi province (Copeland 1982). The Cordilleran province was similar to the Baltic–British province in the presence of *Beyrichia*, but also included the first representatives of an endemic group of eurychilinid ostracodes such as *Yukonibolbina* (Copeland 1989), which have not as yet been recorded from the Llandovery of either of the other provinces. The extent of the Cordilleran province in Llandovery time is based on the work of Copeland (1989) and may be too small, because of lack of knowledge of ostracode faunas of this age in the western United States. The Baltic–British ostracode province is apparently absent in North America, although several genera such as *Apatobolbina, Craspedobolbina* and *Noviportia*, as well as *Beyrichia*, occur in both the upper Llandovery of Anticosti Island and the Island of Gotland in the Baltic (Copeland 1982; Schallreuter & Siveter 1985). Possibly the northern part of the Appohimchi province had some connection with the Baltic–British province and was less endemic than the southern part in late Llandovery time.

Wenlock Series

Figure 2 shows the extent of the Appohimchi province in eastern North America, the Baltic–British province in Great Britain, the Baltic area and Podolia, and the Cordilleran province in western North America. The Coastal Volcanic Belt of Gates (1969) and Berry & Boucot (1970), or the Fundy Belt of Copeland (1977b) extends from Massachusetts, USA to Nova Scotia, Canada, and has been included in the Baltic–British province on rather tenuous evidence. One locality in the Eastport quadrangle, Maine, has yielded Wenlock ostracodes, but these are from a xenolith in volcanic tuff (Gates pers. comm., 1974) and the exact horizon from which they came is not known. Unfortunately, these are the only ostracodes which were described in the Eastport Folio (Bastin & Williams 1914). As in the late Llandovery, the Appohimchi province has zygobolbids, specifically the genus *Drepanellina*, but other prominent elements in the fauna are aechminids (*Paraechmina* spp.), kloedenellids (*Dizygopleura* and *Kloedenella*) and the metacope thlipsurids (*Thlipsuroides*). The Baltic–British province contains large beyrichiaceans such as *Beyrichia, Bingeria, Sleia and Strepula*, (Martinsson 1962; Siveter 1978) and also thlipsurids (*Thlipsura, Octonaria*) and primitiopsids (*Primitiopsis, Venzavella*) as shown by Siveter (1978) and Sethi (1979). None of these genera occur in the Appohimchi province, although the beyrichiaceans *Apatobolbina* and *Craspedobolbina* are common to both. The area shown as the Cordilleran province is based on the work of Copeland (1977a) and, as with the Llandovery, may be too small, due to lack of knowledge of ostracode-bearing Wenlock formations farther south in North America. The Cordilleran province contains Baltic–British

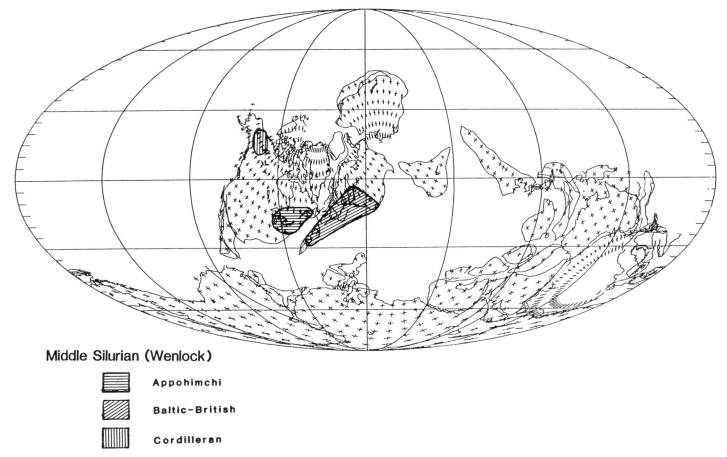

Fig. 2. Ostracode distributions in the Wenlock of North America and Europe.

beyrichiacean genera such as *Beyrichia*, *Cornikloedenina* and *Welleriella*, as described by Copeland (1977*a*). Other genera also appear to have Baltic–British affinities.

Ludlow Series

By Ludlow time, the zygobolbids had disappeared from the Appohimchi ostracode province and had been replaced by the beyrichiaceans *Zygobeyrichia*, *Pintopsis*, *Velibeyrichia* and *Huntonella* (Fig. 3). Metacopid ostracodes are represented by the thlipsurids *Thlipsorothella*, *Thlipsuroides* and *Thlipsuropsis*, and among the podocopes are the pachydomellids *Pachydomella*, *Phanassymetria* and *Tubulibairdia*. As shown by Lundin (1971), this province comprises at least two ecotypes, one in the mid-continent region dominated by thlipsuraceans and healdiaceans, and one in northeastern North America dominated by beyrichiaceans and kloedenellaceans. The Baltic–British province again was characterized by species of *Beyrichia*, as well as *Calcaribeyrichia*, *Neobeyrichia*, *Sleia*, *Hemsiella*, *Lophoctenella*, *Macrypsilon* and *Strepula*. The thlipsurids present are *Thlipsura* s.s. and *Octonaria* s.s., and primitiopsids, which have not been reported from the Appohimchi province, are represented by *Primitiopsis* and *Clavofabella*. Martinsson (*in* Berry & Boucot 1970) noted that ostracodes from the Edmunds Formation in the Coastal Volcanic Belt of Maine suggest an upper Ludlow correlation. The Cordilleran province has Baltic–British genera such as *Beyrichia* and *Calcaribeyrichia*, as well as *Cornikloedenina*, *Welleriella* and the primitiopsids *Scipionis* and *Undulirete*, but also has the Appohimchi genus *Pintopsis*. In addition, two endemic genera, *Yukonibolbina* and *Berdanopsis*, are present. Most of the information about Ludlow ostracodes in the Cordilleran province is based on Copeland (1977*a*). The southern extension of the Cordilleran province in Ludlow time is highly speculative; a few scrappy collections exist which suggest that Ludlow ostracodes occur in the western United States, but there is no definitive proof that these ostracodes are really Ludlow.

Pridoli Series

During Pridoli time, ostracodes are varied and abundant in all three provinces. The Appohimchi province extended from the Eastern Townships of Quebec, Canada (and possibly farther north in Canada) and west central Maine (Boucot 1969) south to north central Alabama (Berdan *et al.* 1986). West of the Appalachians beds of this age are missing due to an unconformity. This province is characterized by beyrichiaceans such as *Welleria*, *Welleriopsis*, *Kloedeniopsis*, *Zygobeyrichia*, *Bolbiprimitia* and *Dibolbina*, and the thlipsurids *Thlipsurella* and *Thlipsuropsis*, as well as kloedenellids such as *Dizygopleura* and *Kloedenella*. Primitiopsids such as *Limbinaria*, *Leiocyamus* and *Venzavella* are also present. In contrast, the Baltic–British province includes the beyrichiaceans *Calcaribeyrichia*, *Gannibeyrichia*, *Hemsiella*, *Kloedenia*, *Londinia*, *Lophoctenella*, *Macrypsilon*, *Neobeyrichia*, *Nodibeyrichia* and *Sleia*, as well as *Thlipsura* s.s. Martinsson (1967; *in* Berry & Boucot 1970; 1977) recognized two successive faunas in this province, the lower one with species of *Frostiella* and *Londinia*, and the upper one with *Nodibeyrichia* and *Kloedenia*. As noted by Martinsson, many of the same genera are present at Arisaig, Nova Scotia and Eastport, Maine. In particular, *Calcaribeyrichia*, *Hemsiella*, *Lophoctenella*, *Londinia*, *Macrypsilon* and *Sleia* occur in the Leighton Shale and *Nodibeyrichia* is in the overlying Hersey Shale in Maine. At

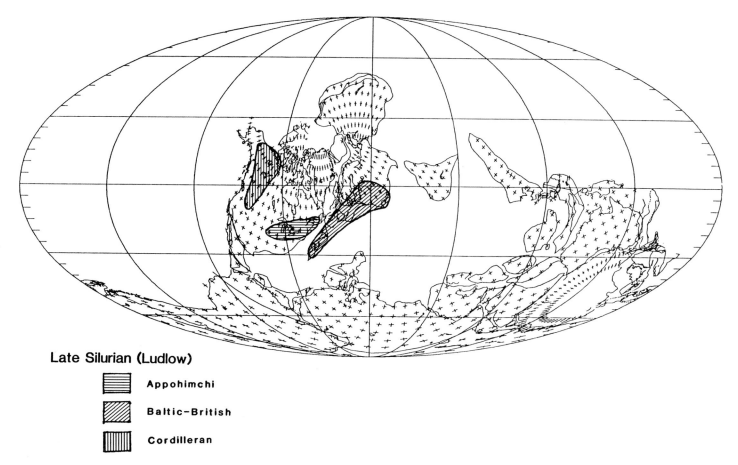

Fig. 3. Ostracode distributions in the Ludlow of North America and Europe.

Arisaig, the lower part of the Stonehouse Formation contains *Hemsiella*, *Neobeyrichia*, *Londinia* and *Kloedenia*, and the upper part in the shore section contains *Hemsiella*, *Kloedenia*, *Macrypsilon*, *Neobeyrichia*, *Nodibeyrichia* and *Sleia* (Copeland 1960, 1964). Neither *Neobeyrichia* nor *Kloedenia* have as yet been found in the Eastport area. However, it is of stratigraphic interest that *Londinia* consistently appears lower in the section than *Nodibeyrichia* in widely separated areas. The Cordilleran province shares the genera *Beyrichia*, *Calcaribeyrichia* and *Gannibeyrichia* with the Baltic−British province, but also has the endemic genera *Alaskabolbina*, *Berdanopsis*, *Nevadabolbina* and *Yukonibolbina* (Copeland 1977a; Stone & Berdan 1984). The Pridoli marks the first appearance of the genus *Treposella*, which recurs throughout the Cordilleran province through the Emsian, but is not known in eastern North America before the Middle Devonian.

Lower Devonian

Lochkovian Stage

In the lowest Lower Devonian the ostracode provinces are still clearly defined and in approximately the same areas as in the Upper Silurian (Fig. 4), although the increasing development of the Old Red Sandstone Continent restricted the northern extent of the Baltic−British province, which at this time was best developed in Podolia. The Appohimchi province has large beyrichiaceans such as *Kloedeniopsis*, *Lophokloedenia* and *Welleriopsis*, as well as a number of endemic thlipsurid genera such as *Craterellina*, *Neocraterellina*, *Neothlipsura*, *Rothella*, *Thlipsorothella*, *Thlipsurella* and *Thlipsuropsis*. In the Baltic−British province, on the other hand, the beyrichiaceans are represented by *Cornikloedenina* and the metacopes by *Leptoprimitia*, neither of which occur in either the Appohimchi or Cordilleran provinces. The Eastport of the Coastal Volcanic Belt of Maine contains an undescribed quadrilobate beyrichiacean genus which is allied to *Cornikloedenina*, but is even more like *Carinokloedenia* from the Pragian of Podolia and the Lochkovian of northern France (Groos-Uffenorde 1986). The Eastport is certainly Lower Devonian, but it is difficult to determine the precise horizon in the Lower Devonian because the fauna is new. The Cordilleran province continues to have *Beyrichia*, and also contains endemic genera such as *Yukonibeyrichia*, *Yukonibolbina*, *Eurekabolbina*, *Dolichoscaphoides*, *Nevadabolbina* and *Pseudomyomphalus* (Berdan 1986). Some genera with Appohimchi affinities, such as *Chironiptrum*, *Kirkbyella*, *Prosumia* and *Saccarchites* are also present in the southern part of the province.

Pragian Stage

The ostracode provinces are still more or less distinct, but there are more genera in common and their areal extent differs somewhat from that of the Lochkovian. The Appohimchi province reaches from the Gaspé Peninsula in Canada (Copeland & Lesperance 1980) down the Appalachians to western Maryland and northeastern West Virginia in the United States. The westward extension of this ostracode province is not certain, as no ostracodes have been described from formations of Pragian age west of the Appalachians. By Pragian time, most of the formations from which ostracodes are known, such as the Shriver Chert of Pennsylvania (Swartz 1936) and the Port Ewen of New

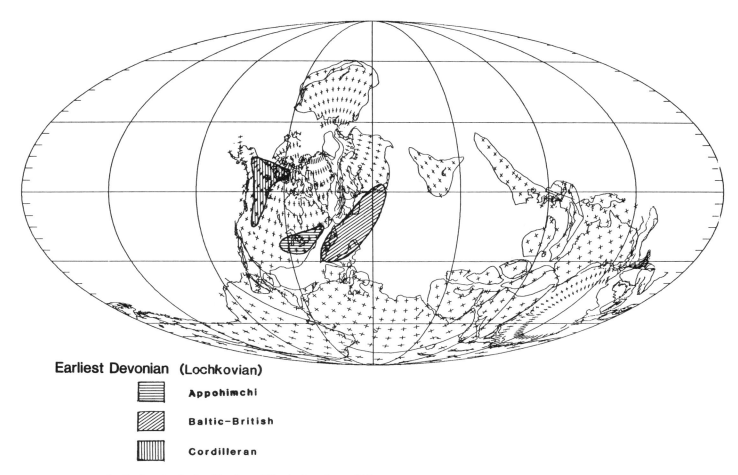

Fig. 4. Ostracode distributions in the Lochkovian of North America and Europe.

York (Berdan 1981) had shifted to an intermediate or even Thuringian ecotype, presumably indicating somewhat deeper water, and the faunas are thus less directly comparable to those of the Baltic–British province. With the exception of *Zygobeyrichia* from the Chapman Sandstone of Maine, (Ulrich 1916), large beyrichiaceans are lacking, and the small beyrichiacean *Huntonella* is associated with metacopes such as *Craterellina*, *Neothlipsura*, *Strepulites* and *Thlipsurella*. Several species of the genus *Bollia* are also present. The 'Baltic–British' province became restricted to the north due to the continued development of the Old Red Sandstone Continent, and is not represented in North America. It is also absent in Great Britain and the Baltic area and is best developed in Podolia, where large beyrichiaceans such as *Carinokloedenia*, *Welleriella* and *Zygobeyrichia* are present. Pebbles from an Emsian conglomerate which have been dated as Pragian by nowakiids (Groos-Uffenorde & Jahnke 1973) also contain *Welleriella* associated with *Leptoprimitia* and *Jenningsina*, Weyant (1965, 1966) has also described Pragian ostracodes from Normandy. The occurrence of *Zygobeyrichia*, an Appohimchi genus, in the 'Baltic–British' province suggests that some connection had been established between the two provinces by this time, although the metacope *Leptoprimitia* is still endemic to the 'Baltic–British' province. The Cordilleran province continues to be characterized by large species of *Beyrichia*, as well as the endemic genera *Alaskabolbina* and *Yukonibeyrichia* and the metacopes *Neocraterellina* and *Neothlipsura*. Hollinacean genera include *Abditoloculina*, *Adelphobolbina*, *Hollina*, *Hollinella*, *Parabolbina* and *Tetrasacculus*. There are several taxa in common between the Appohimchi province and the Cordilleran province; among them are *Thlipsurella*, *Parahealdia?*, *Neothlipsura*, *Parabolbina* and *Ulrichia*, which indicates a decrease in provincialism (Berdan 1983).

Emsian Stage

By Emsian time the provinces are even less clearly delimited (Fig. 5). Most of the Appohimchi province ostracodes belong to the intermediate or even the Thuringian ecotype, with the exception of a small fauna from the Rickard Hill Member of the Schoharie Formation in Albany County, New York. This assemblage includes large beyrichiaceans such as *Parabingeria* and *Schohariella*, and on physical grounds indicates a high energy environment (Berdan 1971). The intermediate or mixed ecotype is noted for several species of *Bollia*, the hollinacean genera *Ctenoloculina*, *Hollinella*, *Parabolbina* and *Tetrasacculus*, and the thlipsurid metacopes *Neothlipsura*, *Stibus*, *Strepulites*, *Thlipsurella* and *Thlipsurina*, as well as *Favulella* and *Ranapeltis* (Bassler 1941; Swain 1953; Berdan 1981). The 'Baltic–British' province is not present in North America in Emsian rocks, and, as with the Pragian, is put in quotation marks because it is also no longer present in either Great Britain or the Baltic area. However, it is present in the Ardenno-Rhenish Massif (Groos & Jahnke 1970; Groos–Uffenorde & Jahnke 1973; Becker & Bless 1974), southern France (Feist & Groos-Uffenorde 1979), Spain (Becker & Sanchez de Posada 1977) and Algeria (Le Fèvre 1971). The large beyrichiaceans *Carinokloedenia*, *Zygobeyrichia* and *Welleriella* are present in the Eifeler ecotype of the Ardenno-Rhenish Massif, but these are mostly lacking in southern areas, where hollinaceans such as *Ctenoloculina*, *Parabolbina* and *Semibolbina* and metacopes such as *Polyzygia*, *Leptoprimitia*, *Loquitzella* and *Zeuschnerina* are conspicuous, as well as many tricorninids. The northern part of the Cordilleran province continues to have large beyrichiaceans such as *Beyrichia* (*Beyrichia*) and *Beyrichia* (*Scabribeyrichia*), as well as many hollinacean genera such as *Abditoloculina*, *Abortivelum*, *Adelphobolbina*,

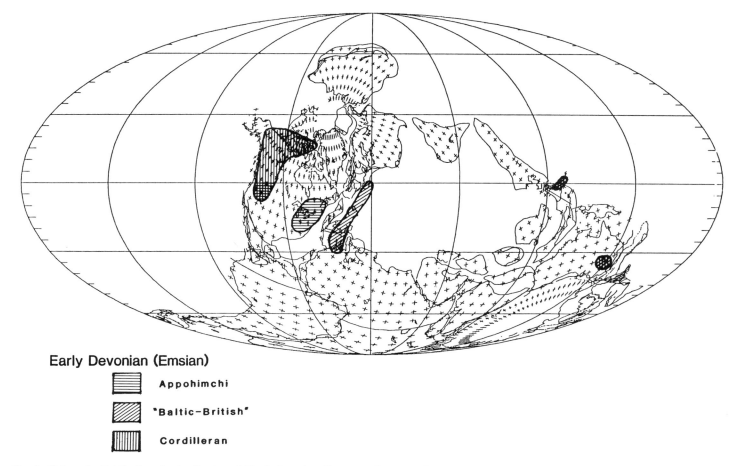

Fig. 5. Ostracode distributions in the Emsian of North America, Europe and eastern Gondwana.

Falsipollex, Flaccivelum, Hollina, Hollinella, Infractivelum, Parabolbina, Arctobolbina, Glyptobolbina and *Prosthenantrum* (Berdan & Copeland 1973; Weyant 1975) and the endemic genera *Alaskabolbina* and *Yukonibeyrichia* are also present in this area. However, in the southern part of the province, in Nevada, large beyrichiaceans are represented only by *Arikloedenia (Crassikloedenia)*, although hollinaceans such as *Hollina, Hollinella, Falsipollex, Parabolbina, Tetrasacculus, Abditoloculina, Flaccivelum?* and *Subligaculum* are common (Berdan 1977; Kennedy 1977). Metacopes include *Strepulites* and, in the lower part, *Neothlipsura*. The endemic genera are not present and, as in the Pragian, the fauna contains many Appohimchi forms. However, some of these, such as *Abditoloculina, Hollina* and *Hanaites*, do not occur in the Appohimchi province until the Middle Devonian (Berdan 1983). This suggests that, as indicated for the corals by Oliver & Pedder (1979), there was a marine connection around the southern end of the Transcontinental Arch which permitted an exchange of ostracode taxa. The late appearance of some of the hollinacean genera in the Appohimchi province suggests that the hollinaceans may have evolved in the Cordilleran province and migrated eastward at the beginning of Middle Devonian time. Emsian ostracodes of either a mixed ecotype or Thuringian ecotype have been described from New South Wales, Australia (Reynolds 1978) and central Japan (Kuwano 1987); these faunas have some taxa in common with the 'Baltic–British' province, but some also occur in the Cordilleran province.

Summary and conclusions

Although some Ordovician ostracodes are cosmopolitan (Schallreuter & Siveter 1985), comparison of similar ecotypes between western and eastern North America and western Europe indicates the presence of three ostracode provinces in the Silurian and Early Devonian. These provinces were distinct from the upper Llandovery through the Lochkovian, but show a greater mixture of taxa in Pragian and Emsian time.

Definition of the provinces has been based mainly on two groups of ostracodes, the palaeocope beyrichiaceans and the metacope thlipsurids and ropolonellids; if it had been possible to include the leperditicope association, the extent of the provinces in North America as shown on figures 1 through 5 would probably be quite different. It is not yet certain that the leperditicopes were provincial, but the restricted environment in which they lived may have acted as a barrier to the other ostracodes which lived in a more normal marine environment, and may have promoted provincialism in the other groups. The beyrichiaceans were most provincial during the Silurian, although a few genera were cosmopolitan. For example, *Garniella* was described from the Wenlock and Ludlow of the Baltic–British province (Martinsson 1962) but occurs in the Pridoli of the Appohimchi province (Berdan 1972) and the Cordilleran province (Stone & Berdan 1984). *Craspedobolbina*, occurring from the upper Llandovery through the lower Ludlow of the Baltic–British province, is also found in the upper Llandovery of the Appohimchi province (Martinsson 1968; Copeland 1982) and the Pridoli of the Cordilleran province (Stone & Berdan 1984).

Tricorninids, *Berounella*, beecherellids, bairdiids and amphissitids occur in all three provinces in the intermediate or in the Thuringian ecotype in the Lower Devonian. Collections from the Thuringian ecotype are commonly silicified, which facilitates identification of the spinose forms. Other cosmopolitan groups are the aechminids and pachydomellids, especially *Tubulibairdia*. On the other hand, the genus *Bollia*, originally described from the Wenlock of England and reported from the Lochkovian of northern France (Groos–Uffenorde 1986), is represented by many species throughout the Lower Devonian of the Appohimchi province, although only one small species is present in the Cordilleran province, in the allochthonous Lower Devonian Slaven Chert. However, even the cosmopolitan genera tend to show provincialism at the species level.

During Pragian and Emsian time, provincialism began to be less noticeable. An influx of Appohimchi genera such as *Ulrichia, Placentella, Parahealdia* and *Neothlipsura* took place in the southern end of the Cordilleran province in Nevada (Berdan 1983); none of these are present in the northern part of the province in Alaska and Yukon Territory (Berdan & Copeland 1973; Copeland 1977a). Although the differences between the northern and southern parts of the Cordilleran province suggest migration of the Appohimchi genera around the southern end of the Transcontinental Arch, the failure of these taxa to spread to the northern part of the province may have been due to latitudinal differences.

In general, the informal ostracode provinces discussed above agree geographically with the more formal units proposed for brachiopods (Boucot 1974, 1975) and corals (Oliver 1977; Oliver & Pedder 1979). However, unlike the brachiopods (Boucot 1974), the ostracodes show distinct provinciality through the late Llandovery and Wenlock as well as in the Ludlow and Pridoli. Also, both the brachiopods (Boucot 1974) and the corals (Oliver & Pedder 1979) are considered to be highly provincial during the Emsian. This is not the case with the ostracodes of the Cordilleran province, as with the exception of the northern part, there appear to be as many or more taxa in common with the Appohimchi province as there were in the Pragian.

As may be seen from the maps (Figs 1–5) of the Llandovery through Emsian, the ostracode provinces generally lie between palaeolatitudes 30°S and 30°N and, except for the Appohimchi province, cross the palaeoequator from the Ludlow through the Emsian. Furthermore, although the provinces expand and contract during this period of time, they remain in essentially the same positions with respect to modern geography. Boucot (1974) suggested several possible causes of provincialism, among which were highly differentiated climatic regimes (basically latitudinal temperature variations), presence of ocean currents with water masses possessing different properties than water on either side (also basically temperature differences) and presence of land barriers. The presence of a deep water trough as a barrier has also been suggested (Copeland 1977b; Copeland & Berdan 1977).

The position of the ostracode provinces with respect to the palaeoequator suggests that latitudinal temperature differences were not the cause of ostracode provincialism, with the possible exception of the differences between the northern and southern parts of the Cordilleran province during the Emsian. It seems unlikely that strong currents with marked temperature differences would have developed in the epicontinental seas of North America during the Silurian and Early Devonian. During most of the time under consideration, the Transcontinental Arch served as a land barrier between the Cordilleran and Appohimchi provinces. In the Silurian, the barrier between the Appohimchi province and the Baltic–British province may have been a deep-water trough which the benthic ostracodes could not cross (Copeland 1977b), but during the Early Devonian, Heckel & Witzke (1979) indicate the Acadian Mountains as a land barrier between these two provinces.

In summary, it seems most likely that the barriers separating the provinces were either land or deep water. Obviously, further study of the distribution of Silurian and Early Devonian ostracode faunas is necessary, including more systematic description of faunas, especially with regard to the possible southward extension of the Cordilleran province in Llandovery and Wenlock time.

References

ABUSHIK, A. F. 1971. Ostrakod'i opornogo razreza Silura-Nizhnego Devona Podolii. *In*: IVANOVA, V. A. (ed.) *Ostrakod'i iz Oporn'ikh*

Razrezov Evropeiiskoii Chasti SSSR. Izdatel'stvo "Nauka", Moskva, 7–133.

BASSLER, R. S. 1941. Ostracoda from the Devonian (Onondaga) chert of west Tennessee. *Washington Academy of Science Journal*, **31**, 21–27.

BASTIN, E. S. & WILLIAMS, H. S. 1914. *Eastport Folio, Maine*. United States Geological Survey Folio 192.

BECKER, G. 1975. Palaöokologische Analyse der Ostracoden-Faunen. *In*: BANDEL, K. & BECKER, G. 1975. Ostracoden aus paläozoischen pelagischen Kalken der Karnischen Alpen (Silurium bis Unterkarbon). *Senckenbergiana lethaea*, **56**, 1, 1–83 (58–61).

—— 1982. Ostracoden-Entwicklung im Kantabrischen Variszikum (Nordspanien). *In: Subsidenz-Entwicklung im Kantabrischen Variszikum und an passiven Kontinentalrändern der Kreide; Teil l Variszikum*. Neues Jahrbuch für Geologie und Paläontologie, Abhandlungen, **163**, 2, 153–163.

—— & BLESS, M. J. M. 1974. Ostracode stratigraphy of the Ardenno-Rhenish Devonian and Dinantian. *In*: BOUCKAERT, J. & STREEL, M., (eds) *International Symposium on Belgian Micropaleontological Limits, from Emsian to Visean, Namur, Sept. 1st to 10th, 1974*. Publication No. 1. Geological Survey of Belgium.

—— & SANCHEZ DE POSADA, L. C. 1977. Ostracoda aus der Moniello-Formation Asturiens (Devon: N-Spanien). *Palaeontographica, Abt. A*, **158**, 115–203.

BERDAN, J. M. 1969. Possible paleoecologic significance of leperditiid ostracodes (abstract). *Geological Society of America Special Paper*, **121**, 337.

—— 1971. Some ostracodes from the Schoharie Formation (Lower Devonian) of New York. *In*: DUTRO, J. T., JR. (ed.) *Paleozoic perspectives: A paleontological tribute to G. Arthur Cooper*. Smithsonian Contribution to Paleobiology, **3**, 161–174.

—— 1972. *Brachiopoda and Ostracoda of the Cobleskill Limestone (Upper Silurian) of central New York*. United States Geological Survey Professional Paper 730.

—— 1977. Early Devonian ostracode assemblages from Nevada. *In*: MURPHY, M. A., BERRY, W. B. N. & SANDBERG, C. A. (eds) *Western North America: Devonian*. University of California Riverside Campus Museum Contribution, **4**, 55–64.

—— 1981. Ostracode biostratigraphy of the Lower and Middle Devonian of New York. *In*: OLIVER, W. A., JR. & KLAPPER, G. (eds) *Devonian biostratigraphy of New York, Part 1*. International Union of Geological Sciences Subcommission on Devonian Stratigraphy, Washington, D.C., 83–96.

—— 1983. Biostratigraphy of Upper Silurian and Lower Devonian ostracodes in the United States. *In*: MADDOCKS, R. F. (ed.) *Applications of Ostracoda*. Department of Geosciences, University of Houston-University Park, Houston, Texas, 313–337.

—— 1984. *Leperditicopid ostracodes from Ordovician rocks of Kentucky and nearby States and characteristic features of the Order Leperditicopida*. United States Geological Survey Professional Paper 1066-J.

—— 1986. New ostracode genera from the Lower Devonian McMonnigal Limestone of central Nevada. *Journal of Paleontology*, **60**, 361–378.

—— & COPELAND, M. J. 1973. *Ostracodes from Lower Devonian formations in Alaska and Yukon Territory*. United States Geological Survey Professional Paper 825.

——, BOUCOT, A. J. & FERRILL, B. A. 1986. The first fossiliferous Pridolian beds from the southern Appalachians in northern Alabama, and the age of the uppermost Red Mountain Formation. *Journal of Paleontology*, **60**, 180–185.

BERRY, W. B. N. & BOUCOT, A. J. 1970. *Correlation of the North American Silurian rocks*. Geological Society of America Special Paper 102.

BOUCOT, A. J. 1969. Geology of the Moose River and Roach River Synclinoria, northwestern Maine. *Maine Geological Survey Bulletin*, **21**.

—— 1974. Silurian and Devonian biogeography. *In*: Ross, C. A. (ed.) Paleogeographic provinces and provinciality. Society of Economic Paleontologists and Mineralogists Special Publication, **21**, 165–176.

—— 1975. *Evolution and extinction rate controls*. Elsevier Scientific Publishing Company, Amsterdam, Oxford, New York.

——, JOHNSON, J. G. & TALENT, J. A. 1969. *Early Devonian brachiopod zoogeography*. Geological Society of America Special Paper 110.

COPELAND, M. J. 1960. Ostracoda from the Upper Silurian Stonehouse Formation, Arisaig, Nova Scotia, Canada. *Palaeontology*, **3**, 93–103.

—— 1964. Stratigraphic distribution of Upper Silurian Ostracoda, Stonehouse Formation, Nova Scotia. *Geological Survey of Canada Bulletin*, **117**, 1–13.

—— 1971. Biostratigraphy of some Early Middle Silurian Ostracoda, eastern Canada. *Geological Survey of Canada Bulletin*, **200**, Part I, 1–18.

—— 1976. Leperditicopid ostracodes as biostratigraphic indices: *Canada Geological Survey Paper*, **76–1B**, 83–88.

—— 1977a. Early Paleozoic Ostracoda from southwestern District of MacKenzie and Yukon Territory. *Geological Survey of Canada Bulletin*, **275**.

—— 1977b. Early Paleozoic Ostracoda of eastern Canada. *In*: SWAIN, F. M. (ed.) *Stratigraphic micropaleontology of Atlantic Basin and Borderlands*. Elsevier Scientific Publishing Company, Amsterdam, 1–17.

—— 1982. An occurrence of the Silurian ostracode Beyrichia (Beyrichia) from Anticosti Island, Quebec. *Geological Survey of Canada Paper*, **82–1B**, 223–224.

—— 1989. Silicified Late Ordovician–Early Silurian ostracodes from the Avalanche Lake area, southwestern District of MacKenzie. *Geological Survey of Canada Bulletin*, **341**, (in press).

—— & BERDAN, J. M. 1977. Silurian and Early Devonian beyrichiacean ostracode provincialism in northeastern North America. *Geological Survey of Canada Paper*, **77–1B**, 15–24.

—— & LESPERANCE. 1980. The occurrence of Ostracoda with "southern" Appalachian affinities in the Lower Devonian Shiphead Formation, Forillon Peninsula, Gaspe, Quebec. *Geological Survey of Canada Paper*, **80–1B**, 255–258.

FEIST, R. & GROSS-UFFENORDE, J. 1979. Die "Calcaires à polypiers siliceux" und ihre Ostracoden-Faunen (Oberes Unter-Devon; Montagne Noire, S-Frankreich). *Senckenbergiana lethaea*, **60** (1/3), 83–187.

GATES, O. 1969. Lower Silurian–Lower Devonian volcanic rocks of New England coast and southern New Brunswick. *American Association of Petroleum Geologists Memoir*, **12**, 484–503.

GOODAY, A. J. & BECKER, G. 1979. Ostracods in Devonian biostratigraphy. *The Palaeontological Association Special Papers in Palaeontology*, **23**, 193–197.

GROOS, H. & JAHNKE, H. 1970. Bemerkungen zu unterdevonischen Beyrichien (Ostracoda) aus dem Rheinischen Schiefergebirge und dem Harz. *Göttinger Arbeiten zur Geologie und Paläontologie*, **5**, 37–48.

GROOS-UFFENORDE, H. 1986. Ostracodes. *In*: RACHEBOEUF, P. R. (ed.) *Le Groupe de Lievin, Pridoli-Lochkovien de l'Artois (N. France)*. Biostratigraphie du Paleozoique 3, Université de Bretagne Occidentale, 175–184.

—— & JAHNKE, H. 1973. Die Fauna der Kalkgerölle aus dem unterdevonischen Konglomerat bei Marburg. *Hessisches Landesamt für Bodenforschung, Abhandlungen; Notizblatt*, **101**, 80–98.

HECKEL, P. H. & WITZKE, B. J. 1979. Devonian world palaeogeography determined from distribution of carbonates and related lithic palaeoclimatic indicators. *In: The Devonian System*. The Palaeontological Association, Special Papers in Palaeontology, **23**, 99–123.

KENNEDY, P. J. 1977. The Lower Devonian ostracode sequence at Table Mountain, near Eureka, Nevada. *In*: MURPHY, M. A., BERRY, W. B. N. & SANDBERG, C. A. (eds) *Western North America: Devonian*. University of California Riverside Campus Museum Contribution, **4**, 80–88.

KUWANO, Y. 1987. Early Devonian conodonts and ostracodes from central Japan. *Bulletin, National Science Museum, Tokyo, Ser. C*, **13**, 77–105.

LE FÈVRE, J. 1971. Paleoecological observations on Devonian ostracodes from the Ougarta Hills (Algeria). *Bulletin du Centre de Recherches Pau-SNPA*, supplément au volume 5, 817–841.

LUNDIN, R. F. 1971. Possible paleoecological significance of Silurian and Early Devonian ostracode faunas from midcontinental and northeastern North America. *Bulletin de Centre de Recherches Pau-SNPA*, supplément au volume 5, 853–868.

MARTINSSON, A. 1962. Ostracodes of the family Beyrichiidae from the Silurian of Gotland. *Bulletin of the Geological Institutions of the*

University of Uppsala, **41**.

—— 1967. The succession and correlation of ostracode faunas in the Silurian of Gotland. *Geologiska Föreningens i Stockholm Förhandlingar*, **89**, 350–386.

—— 1968. The Appalachian species of the Silurian ostracode genus Craspedobolbina. *Geologiska Föreningens i Stockholm Förhandlingar*, **90**, 302–308.

—— 1977. Palaeocope ostracodes. *In*: MARTINSSON, A. (ed.) *The Silurian–Devonian boundary*. International Union of Geological Sciences, Series A, **5**, 327–332.

MURPHY, M. A. 1977. Nevada. *In*: MARTINSSON, A. (ed.) *The Silurian–Devonian boundary*. International Union of Geological Sciences, Series A, **5**, 264–271.

OERTLI, H. J. (ed.) 1971. *Colloque sur la Paléoécologie des Ostracodes*. Bulletin du Centre de Recherches Pau-SNPA, supplément au volume 5.

OLIVER, W. A., JR. 1977. Biogeography of Late Silurian and Devonian rugose corals: *Palaeogeography, Palaeoclimatology, Palaeoecology*, **22**, 85–135.

—— & PEDDER, A. E. H. 1979. Biogeography of Late Silurian and Devonian rugose corals in North America. *In*: GRAY, J. & BOUCOT, A. J. (eds) *Historical biogeography, plate tectonics, and the changing environment*. Oregon State University Press, Corvallis, Oregon, 131–145.

POLENOVA, E. N. 1971. Biogeographical types of Early Devonian ostracodes. *Bulletin du Centre de Recherches Pau SNPA, supplément au volume 5*, 843–852.

REYNOLDS, L. 1978. The taxonomy and palaeoecology of ostracodes from the Devonian *Receptaculites* Limestone, Taemas, New South Wales, Australia. *Palaeontographica, Abt. A*, **162**, 144–203.

SCHALLREUTER, R. E. L. & SIVETER, D. J. 1985. Ostracodes across the Iapetus ocean. *Palaeontology*, **28**, 577–598.

SCOTESE, C. R. 1986. *Phanerozoic reconstructions: a new look at the Assembly of Asia*. University of Texas Institute for Geophysics Technical Report 66.

SETHI, D. K. 1979. Palaeocope and eridostracan ostracodes. *In*: JAANUSSON, V., LAUFELD, S. & SKOGLUND, R. (eds) *Lower Wenlock faunal and floral dynamics — Vattenfallet section, Gotland*. Geologiska Undersökning, Serie C, Nr. 762, Avhandlingar och Uppsatser, Arsbok, **73**, 142–166.

SIVETER, D. J. 1978. The Silurian. *In*: BATE, R. & ROBINSON, E. (eds) *A stratigraphical index of British Ostracoda*. Geological Journal Special Issue 8, Seel House Press, Liverpool, 57–100.

STONE, S. M. & BERDAN, J. M. 1984. Some Late Silurian (Pridolian) ostracodes from the Roberts Mountains, central Nevada. *Journal of Paleontology*, **58**, 977–1009.

SWAIN, F. M. 1953. Ostracoda from the Camden chert, western Tennessee. *Journal of Paleontology*, **27**, 257–284.

SWARTZ, F. 1936. Revision of the Primitiidae and Beyrichiidae, with new Ostracoda from the Lower Devonian of Pennsylvania. *Journal of Paleontology*, **10**, 541–586.

ULRICH, E. O. 1916. Ostracoda. *In*: WILLIAMS, H. S. & BREGER, C. L. The fauna of the Chapman sandstone of Maine, including descriptions of some related species from the Moose River sandstone. *United States Geological Survey Professional Paper*, **89**, 289–293.

ULRICH, E. O. & BASSLER, R. S. 1923. Systematic paleontology of Silurian deposits; Ostracoda. *Maryland Geological Survey, Silurian*, 500–704.

WANG, S. 1988. Late Paleozoic ostracode associations from South China and their paleoecological significances. *Acta Palaeontologica Sinica*, **27**, 91–102 (98–102 in English).

WARSHAUER S. M. & SMOSNA R. 1977. Paleoecologic controls of the ostracode communities in the Tonoloway Limestone (Silurian; Pridoli) of the central Appalachians. *In*: LOFFLER, H. & DANIELOPOL, D. (eds) *Aspects of ecology and zoogeography of recent and fossil Ostracoda*. Dr W. Junk b. v. Publishers, The Hague, 475–485.

WEYANT, M. 1965. Beyrichiidae (Ostracodes) du Dévonien inférieur de la Normandie. *Bulletin de la Société Linnéenne de Normandie*, 10 Serie, **6**, 76–92.

—— 1966. Représentants de quelques familles d'Ostracodes du Dévonien inférieur de la Normandie (Leperditiidae, Bolliidae, Arcyzonidae, Bassleratiidae, Kloedenellidae, Thlipsuridae, incertae familiae). *Bulletin de la Société Linnéenne de Normandie*, 10 Serie, **7**, 117–138.

—— 1975. Ostracodes Devoniens du Sud-Ouest de l'île Ellesmere (Archipel arctique canadien). *Geobios*, **8**, fasc. 6, 361–408.

Constraints on Silurian and Early Devonian phytogeographic analysis based on megafossils

DIANNE EDWARDS

Department of Geology, University of Wales College of Cardiff PO Box 914, Cardiff, CF1 3YE, UK

Abstract: The detection of phytogeographic patterns in Silurian and Early Devonian land vegetation is hampered by the dearth of occurrences of megafossils, the absence of consistently precise correlation, unreliability of identification, and the lack of sufficient detailed sampling and information on sediments to permit evaluation of taphonomic influences on composition of assemblages. Interpretation is further complicated by the exceptional evolutionary position of the plants themselves and the lack of extant representatives. There is little information on whole plants, their life histories, their ecological and climatic tolerances. The composition of Ludlow, Pridoli and Lochkovian/Gedinnian assemblages is analysed. Those from Kazakhstan and Siberia are particularly enigmatic. The assemblages are plotted on the appropriate continental reconstructions. Most are clustered on the southeast margin of Laurussia and very few occur at high latitudes. In Ludlow time, *Baragwanathia* dominated assemblages in Australia in contrast with the rhyniophytoid assemblages of Laurussia. New information from Kazakhstan and northwest China adds to the data base in the Pridoli.

Differences in Gedinnian assemblages from South Wales, the Welsh Borderland and Scotland, where sediments have been intensively studied and correlation is based on fish faunas and, more usefully, on palynomorphs, are related to taphonomy, to local vegetation distribution pattern, and to evolutionary changes. World-wide analysis of occurrences shows four distinct assemblages, based on Laurussia, Kazakhstan, Australia and Siberia, the latter being a rare example of high latitude vegetation.

My remit for the symposium was to consider land vegetation and continental reconstructions in the Silurian and Early Devonian. The most numerous and diverse assemblages are Pragian and Emsian. They have recently been subjected to multivariate analyses by Raymond *et al.* (1985) and Raymond (1987), who attempted to detect floristic patterns, and to give explanations for them. Although I have some reservations on their handling of the data base, theirs are up-to-date analyses and supercede earlier and more traditional efforts (e.g. Edwards 1973; Petrosyan 1968). In addition, as Ann Raymond and I are currently reassessing Pragian and Emsian assemblages, I here propose to concentrate on the Silurian and Lochkovian, a period critical for the early evolution of vascular plants. As Raymond *et al.* emphasized, pre-Pragian assemblages are not amenable to complex statistical analyses. Problems relate to quantity and quality of data, and to the unique evolutionary position of the plants involved. The Silurian and Early Devonian saw the origin, spread and diversification of vascular plants. Information is derived, for the most part, from fragmentary allochthonous fossils and we are almost completely ignorant of the nature of whole plants, of their life histories, habitats and climatic tolerances. In addition the very simplicity of their axial organization is such that convergence must have been common, a further complication for identification and hence the detection of floristic patterns.

Quality controls on analysis of data

Most of the limitations of the available data are self-evident, but I discuss them here to emphasize the problems, often peculiar to the Silurian and Early Devonian, and to account for the lack of progress in phytogeographic analysis of this time interval.

Stratigraphic accuracy and precision

At a time of rapid change in plant evolution, accuracy in both stratigraphic position and age determination is clearly desirable. 'Chrono-umbrellas' such as 'Late Silurian' and 'Lower Devonian' are no longer adequate. Ideally, too, there should be the opportunity for sampling throughout a well-dated succession in a restricted geographic area. It is perhaps surprising that good correlation based on independent faunal evidence is achieved in the Silurian where all megafossils occur in marine sediments (Richardson & Edwards 1989), some associated with graptolites (Table 1; Obrhel 1962; Garratt 1978; Edwards *et al.* 1983). In the Devonian, assemblages are found in marine and continental facies, the latter traditionally correlated by means of fossil fish (e.g. White 1961) which can be scarce in certain critical successions such as the Lower Old Red Sandstone of the Brecon Beacons area, South Wales (Loeffler & Thomas 1980). The development of a spore-based zonation for the Old Red Sandstone in Europe and North America by McGregor, Richardson, Streel and co-workers also allows correlation between continental and marine strata, and in the Gedinnian, further subdivision (e.g. Richardson *et al.* 1982; Richardson & McGregor 1986; Steemans 1982; Streel *et al.* 1987). I use the rejected Gedinnian (Ziegler & Klapper 1985) rather than Lochkovian in this context, because fine resolution between the Rhenish and Bohemian facies has not yet been achieved using spores.

I confess to some disquiet when the age of sediments with plant megafossils is obtained from the products of the plants (i.e. spores) themselves, particularly where such localities are geographically widely separated. However, spore-based zonations such as Richardson & McGregor's (1986) are developed from assemblages in a number of facies and can be tested against an independent faunal biostratigraphy. They seem to indicate that migration cannot be detected in the time resolution available, which is consistent with our information on the spread of recent pteridophytes. In addition, the spores for which the parent plants are known form only a small percentage of the dispersed spore assemblages (Allen 1980; Gensel 1980).

In the past, in the absence of faunas, the plants themselves have been used to date the enclosing sediments. Thus, for example, *Zosterophyllum* has been considered indicative of a Gedinnian age, although different species of this genus have been recorded throughout the Lower Devonian. A slightly different approach was adopted by Rayner (1988). He chose to prefer a basal Devonian age for a plant assemblage in the Witpoort Sandstone (lower part of the Witteberg Group) in the Cape Fold Belt, South Africa, based on the levels of organization exhibited by the plants, even though estimates of age based on spores, vertebrates and invertebrates in other sediments in the area range from the Middle Devonian to the Early Carboniferous (see references in Anderson & Anderson 1985).

The most critical appraisal of age determination is necessary

Table 1. *Silurian localities with plant megafossils (see also Fig. 1)*

Age	Authors	Geographic area	Composition	Basis for age
Llandovery (?Telychian)	Schopf et al. (1966)	Maine, USA	*Eohostimella*	invertebrates
Homerian	Edwards et al. (1983)	Tipperary, Ireland	*Cooksonia*	graptolites
1. Gorstian	Edwards et al. (1979)	Powys, Wales	*Cooksonia*	graptolites
1. Ludfordian	Edwards & Rogerson (1979)	Powys, Wales	*Cooksonia, Steganotheca*	invertebrates acritarchs
2. ?late Ludlow	Tims & Chambers (1984)	Victoria, Australia	*Baragwanathia, Salopella, Hedeia*	graptolites
3. Ludlow/?Ludfordian	Edwards (work in progress)	North Greenland	*Salopella*	graptolites
4. Pridoli (?ultimus)	Obrhel (1962)	Bohemia, Czechoslovakia	*Cooksonia*	graptolites
5. Pridoli (?ultimus)	Lang 1937; Edwards & Fanning (MS)	Hereford, England	*Cooksonia, Salopella*	invertebrates and spores
5. Pridoli (early)	Edwards (1979)	Dyfed, Wales	*Cooksonia, Tortilicaulis, Psilophytites*	spores
5. Pridoli (early)	Edwards & Rogerson (1979)	Dyfed, Wales	*Cooksonia, Steganotheca*	spores
7. Pridoli (bouceki)	N. Petrosyan (pers. comm.)	Kazakhstan, USSR	*Cooksonia, Zosterophyllum*	graptolites
6. Pridoli (late)	Ishchenko (1975)	Podolia, USSR	*Cooksonia, Eorhynia (Salopella), ?Zosterophyllum, Lyopodolica*	invertebrates
7. Pridoli (late)	Senkevich (1975)	Balkhash area, Kazakhstan	*Cooksonella* sp. *?Baragwanathia* sp. *?Taeniocrada* sp. *Jugumella burubaensis*	graptolites
8. Pridoli (late)	Dou, Cai & Edwards (work in progress)	Junggar Basin Xinjiang, China	*Cooksonella* sp. *?Lycopodolica*	graptolites
9. Pridoli (?late)	Banks (1973)	New York State, USA	*Cooksonia*	conodonts
10. Pridoli	Daber (1971)	Libya	*Cooksonia*	graptolites
11. Indet. Silurian/?Wenlock	Klitzsch et al. (1973)	Libya	*?Cooksonia, ?Steganotheca, ?Dawsonites, Protolepidodendron, Archaeosigillaria, Protosigillaria, Precyclostigma,* cf. *Palaeostigma*	field relations

where the composition of assemblages is at variance with their presumed stratigraphic position. The most debated example is the Lower *Baragwanathia* assemblage in Victoria, Australia, recently dated as upper Ludlow on the presence of monograptids in beds alternating with the plants (Garratt 1978; Garratt & Rickards 1984; Garratt et al. 1984). Still of equivocal age is the Libyan assemblage in the Acacus Formation in the Mourzouk Basin, where fragmentary, sterile presumed rhyniophytoids and *Dawsonites* occur with lycophytes typical of Middle and Upper Devonian sediments elsewhere (Klitzsch et al. 1973). From field relationships, the upper part of the Acacus Formation has been considered upper Llandovery to Wenlock (Boucot & Gray 1982) and certainly earlier than Lower Siegenian (Klitzsch et al. 1973). Spore assemblages elsewhere in the Basin range from Llandovery to Ludlow (Douglas & Lejal-Nicol 1981), but there is no direct dating of the sediments enclosing the plants. Finally, a new genus, *Pinnatiramosus*, with highly and regularly branched axes and tracheids exhibiting convincing pitting has been described from supposedly Wenlock strata in Guizhen Province, China (Gang 1985). In this region the Silurian is immediately overlain by the Permian, and there is the possibility that these complex plants belong to the younger strata (Holland, pers. comm.).

Recognition of facies bias

Silurian assemblages occur in marine sediments which range from open sea (e.g. in the *Monograptus ultimus* zone in Bohemia, Obrhel 1962) to marginal (e.g. near the base of the Pridoli and just above the local equivalent of the Ludlow Bone Bed at Perton Lane, Hereford (Lang 1937; work in progress in Cardiff)). Differences in the composition of these approximately coeval assemblages, probably result from these facies differences, with the English one showing greater diversity. Meyen (1987) was of the opinion that these early 'higher plants' (rhyniophytoids) were aquatic to semi-aquatic rather than terrestrial because herbaceous land plants 'have never been encountered incorporated in sediments as megafossils' (p. 292). I believe that they were terrestrial plants and when buried in clastic sediments persisted to become fossils because they possessed peripheral thick-walled support tissues, adaptations to remain upright in times of water stress

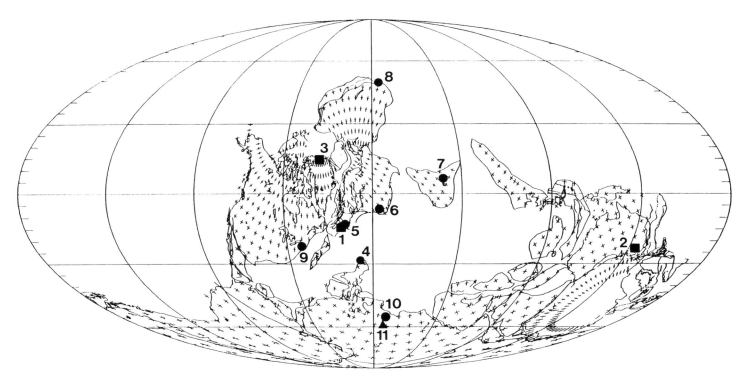

Fig. 1. Occurrences of Ludlow (■) and Pridoli (●) assemblages. ▲, uncertain age.

(Edwards 1980; Edwards *et al.* 1986). Furthermore, as a corollary, I have suggested that plants such as *Rhynia gwynne-vaughanii* that depended on turgor for their rigidity would be under-represented, if present at all, in allochthonous sediments.

Problems associated with sorting of the transported plant material in Gedinnian Old Red Sandstone sediments occurring in a relatively small geographic area are discussed later. Unfortunately extensive assemblages in the marine Rhenish facies of Belgium and Germany are rare (e.g. Schweitzer 1983; Steemans & Gerrienne 1984), but as in the succeeding early Pragian assemblages there appear to be no major differences in composition. However when Pragian and Emsian assemblages from various parts of Laurussia (Old Red Sandstone Continent) were compared Ziegler *et al.* (1982) commented on the low generic diversity in assemblages collected from the north-west side where vegetation may have been under the influence of inferred cold eastern boundary currents. An alternative explanation in two cases (Alaska: Churkin *et al.* 1969 and Arctic Canada: Hueber 1971) is that impoverishment results from transport; the two assemblages being associated with graptolites. Similarly the Lochkovian plant assemblage in the Bohemian graptolitic facies contains only one kind of rhyniophytoid, a marked contrast to the diversity seen in several coeval localities in Britain (Table 2).

Accuracy of identification and affinity

Examples of identification problems range from the very obvious, such as the identification of conodont bedding-plane assemblages in the Lower Silurian of South Africa as vascular plants or their algal ancestors (Theron & Kovacs Endrody 1986) through the more problematic, such as the presumed algal affinity of *Powysia bassettii* (Edwards 1977) to the downright misleading, where *bona fide* names are applied to very fragmentary fossils. Examples of generic names based on reproductive structures used in such 'optimistic' identifications are *Cooksonia*, *Dawsonites* and *Psilophyton*. In extreme cases the problems are compounded when the plants are then used to date the sediments (e.g. Lemoigne 1967: 'Lower' Devonian of western Sahara). Where there is doubt over identification taxa should be prefixed with a question mark or 'cf'. Our studies on rhyniophytoids from southern Britain (Edwards & Fanning 1985 and work in progress) make us particularly wary of fragments where little more than a solitary sporangium is preserved. Thus, for example, isolated reniform sporangia with short subtending axes, could belong to *Cooksonia caledonica*, to *Renalia* or even to *Zosterophyllum*. Vegetative remains to be treated with caution in the perusal of lists of taxa include *Taeniocrada*, which I believe should be used as a form genus for an axis with a narrow central strand (Edwards & Edwards 1986), *Psilophytites*, *Zosterophyllum* (for K- and H-branching), *Drepanophycus* and even *Baragwanathia*. The latter was used for a compression fossil of a small 'leafy' shoot (*c.* 2–3 mm in diameter) from the Villavicencio Formation in Mendoza Province, Argentina (Cuerda *et al.* 1987). Regardless of the accuracy of the attribution, the authors seem unaware of the Silurian *Baragwanathia*, and use its presence to confirm a Lower Devonian age for the sediments.

Finally there are the problems of analysis of less accessible assemblages, where photographic illustrations are often inadequate and treatment of taxa may well suffer from what Raymond (1987) tactfully calls 'regional taxonomic bias'. Such comments are particularly pertinent to the extensive assemblages from the Soviet Union and in the context of this paper, to the megafossils described by Senkevich (1975) from the Pridoli/Gedinnian of Kazakhstan (see Tables 1 and 2).

Comments on the composition of Silurian assemblages

The list in Table 1 is reproduced with minor additions from Richardson & Edwards (1989), and the localities are plotted on Fig. 1.

The Ludlow map also shows Pridoli localities. The most important additions since a recent review (Richardson & Edwards 1989) are the assemblages in Kazakhstan and Xinjiang Province, northwest China, which were previously thought to be Gedinnian but following a reassessment of the associated graptolites are considered to be of Pridoli age. The Chinese assemblage, dis-

Table 2. *Gedinnian localities with plant megafossils (see also Fig. 2)*

Age	Authors	Geographic area	Composition	Facies	Basis for age
1. *uniformis* zone	Obrhel (1968)	Bohemia, Czechoslovakia	*Cooksonia downtonensis* (≡ *C. hemisphaerica*)	Marine (deep)	graptolites
2. *micrornatus-newportensis* (lower-middle)	Lang (1927) Edwards (1975)	Forfar, Scotland	*Zosterophyllum myretonianum Cooksonia caledonica*	ORS (internal facies)	spores, fish
2. *micrornatus-newportensis* (lower-middle)	Edwards (1972)	Arbilot	*Z. fertile*	ORS (internal facies)	spores, fish
3. *micrornatus-newportensis* (lower)	Edwards & Fanning (1985)	Targrove, Shropshire	*C. hemisphaerica, C. pertoni, C. cambrensis, C. caledonica, Salopella* (2 new sp) + various other unnamed rhyniophytoids	ORS fluviatile (distal)	spores, fish
3. *micrornatus-newportensis* (lower)	Edwards & Fanning unpublished	Brown Clee Hill, Shropshire	As for Targrove plus *Tortilicaulis, Sporogonites*	ORS fluviatile (distal)	spores
4. *micrornatus-newportensis* (upper)	Leclercq (1942)	Nonceveux, Belgium	*Z. fertile*	Marine (Rhenish)	spores, fish
3. ?	Edwards & Kenrick unpublished	Cwm Mill, Gwent, Wales	*Z. fertile, Cooksonia* sp.	ORS fluviatile (distal)	?
5. upper Gedinnian	Schweitzer (1983).	Rhineland, Germany	*Drepanophycus spinaeformis Taeniocrada* sp.? *Zosterophyllum rhenanum*	Marine littoral	field reln.
3. upper Gedinnian	Edwards & Richardson (1974) and unpublished	Newton Dingle + environs.	*Z.? fertile, Salopella allenii*	ORS fluviatile (medial)	spores
3. *breconensis-zavallatus*	Edwards & Kenrick (unpublished)	Allt Ddu, Brecon Beacons, Powys, Wales	*Salopella allenii, Salopella* sp., *C.* cf *caledonica Cooksonia* sp. cf *Psilophyton princeps* var *ornatum* (= zosterophyll) *Gosslingia breconensis Z. fertile, Z.* sp.	ORS fluviatile (medial)	spores
4. *breconensis-zavallatus*	Steemans & Gerrienne (1984)	Gileppe, la Vesdre, Belgium	*Gosslingia breconensis* & other abundant remains (work in progress)	Marine (Rhenish)	spores
6. ? basal Gedinnian	Senkevich (1975)	Balkhash, Kazakhstan	*Cooksonella sphaerica, Taeniocrade pilosa, Tastaephyton bulakus, Mointina quadripartita, Jujumella burubaensis, J. jugata, Balchaschella tenera*	Marine	faunas/ graptolites
7. Gedinnian indet.	Stepanov (1975)	Kuzbass, Siberia	*Zosterophyllum, C. pertoni, Stolophyton acyclicus, Juliphyton glazkini, Uksunaiphyton ananievi, Pseudosajania pimula, Salairia bicostata*	Marine	?
8. Gedinnian	Li & Tsai (1978)	E. Yunnan, south-west China	*Zosterophyllum* sp.	?	?
9. ? Lochkovian undet.	J. Tims (pers. comm.)	Tyers, Victoria, Australia	*Baragwanathia longifolia, Zosterophyllum* n.sp., *Baragwanathia* n.sp.	Marine	faunas (corals)
10. Gedinnian indet.	Høeg (1942)	Spitsbergen	Sterile remains only: *Hostinella, Taeniocrada Zosterophyllum*	ORS (internal)	spores
6. upper Gedinnian — basal Siegenian	Senkevich (1975)	Balkhash	*C. crassiparietilis, Zosterophyllum* sp. *Tastaephyton bulakus, Jugumella burubaensis, J. jungata*	Marine	?
11. ?Gedinnian	Janvier *et al.* (1987)	Viet Nam	*Cooksonia*	ORS	plants & fish

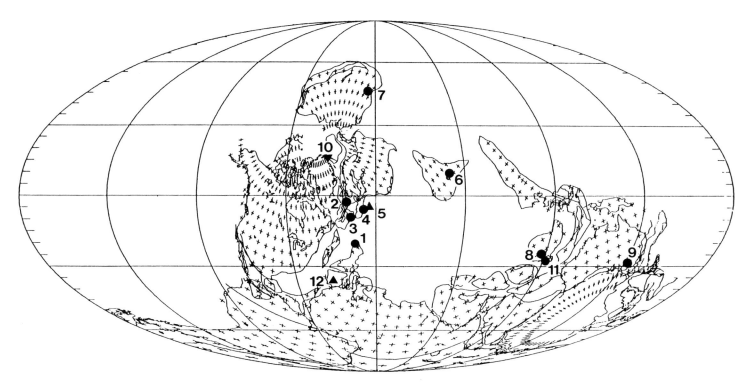

Fig. 2. Occurrences of Gedinnian (Lochkovian) assemblages. (▲, sterile remains).

covered by Dou & Sun (1983) and currently being studied by Cai and the present author, contains *Junggaria* Dou, a rhyniophytoid in which sporangia with broad, sharply truncated borders, sometimes dissected, terminate dichotomously or pseudomonopodially branching, hairy axes. This plant is probably identical to *Cooksonella* originally described from the Balkhash region of Kazakhstan (Senkevich 1975, pers. comm.; Nikitin & Bandaletov 1986). Also found in the Xinjiang assemblage was a small unbranched axial fragment covered with distally directed needle-like appendages. Where these are absent their attachment sites are marked by small raised areas. No regular arrangement of these appendages could be detected. The specimens named ?*Baragwanathia* in the Tokrau horizon assemblage from the adjacent Balkhash region of Kazakhstan may be the same plants as the Chinese examples. Also similar are the 'leafy' fragments named *Lycopodolica* by Ishchenko (1975) found in the Skala horizon in Podolia and those called *Baragwanathia* from Argentina (Cuerda *et al.* 1987). Preservation in all cases is insufficient to allow assignment with confidence even to a higher taxonomic rank (e.g. Lycophyta, Algae).

Research in Cardiff by Fanning (1987) has demonstrated greater generic diversity at the famous Downtonian Perton Lane exposure than that recognized by Lang (1937). All the new taxa are best described as rhyniophytoid.

Table 3. *Stratigraphic positions of Gedinnian localities with plant megafossils in Britain*

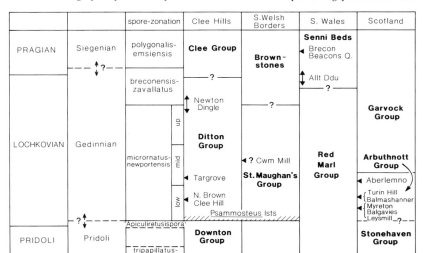

(Based on Richardson *et al.* 1984, and unpublished palynological data from Dr J. B. Richardson).

Comments on geographic distribution of Silurian megafossils

I have nothing to add on the Wenlock *Cooksonia* assemblage in Ireland (Edwards *et al.* 1983). Ludlow and Pridoli occurrences are plotted on the same map, because the latter is usually interpreted as of short duration (*c.* 2 Ma) and although the Ludlow is longer (<7 Ma), the plant-bearing localities of that age are mostly in the upper part of the series. Considering continental positions, the period marks the final stages in the closure of Iapetus, and thus Laurussia may be considered as a single land mass more or less straddling the equator. The rhyniophytoid dominated Ludlow assemblages from Greenland and Britain suggest a uniform vegetation, at least in coastal areas. The composition of the two Australian assemblages differs from the remainder in that *Baragwanathia* occurs with rhyniophytoids. Whether or not this difference is sufficient to support a hypothesis that there were two distinct floristic provinces in Ludlow times remains to be seen.

In the Pridoli most occurrences are on the southeast margin of Laurussia with the greatest numbers in Britain. The palaeogeographic positions of the Xinjiang and Balkhash assemblages are more problematic. I have plotted the latter on the Kazakhstan plate: its distinctive composition adds weight to Petrosyan's suggestion that the later Devonian vegetation of Kazakhstan belonged to a distinct floral province. Raymond (1987) also recognized a major phytogeographic unit based on Kazakhstan in the Lower Devonian, but linked it with north Gondwana. The latter can be discounted as it was based on a single assemblage (*viz.* Lemoigne: western Sahara, Lower Devonian) which was misplotted and provides an excellent example of 'optimistic' identification based on very fragmentary remains.

If *Cooksonella* and *Junggaria* are indeed congeneric then the Kazakhstan assemblage may be linked with that from northwest Xinjiang which I have plotted on the Siberian plate (Chen Xu pers. comm.). Although these localities are in proximity today, the palaeocontinental reconstructions result in a high latitude position for the Junggar plants. While this may be taken as evidence for a similar flora, it is perhaps more probable that at least the Altay−Sayan region of the Siberian plate is not in its correct palaeoposition on the Scotese reconstruction. Such uncertainly is all the more regrettable in that, apart from Libyan assemblages, all other occurrences are at low latitudes. Thus the Kazakhstan occurrences are at approximately the same palaeolatitudes as the Laurussian, and perhaps shared a tropical, possibly seasonal climate. Even the Australian localities at slightly higher latitude would have been warm. The proximity of the various land masses suggests that there would have been no major barriers to spore dispersal and thus, provided that suitable habitats were available, a uniform vegetation might be anticipated (Barratt 1985; Streel *et al.* 1990). The demonstration of distinct floras at such an early stage in land colonization by higher plants could then be taken as evidence for a number of centres of origin, but more records of both micro- and megafossils are clearly needed from Gorstian and older sediments.

The best-dated Silurian locality at a higher latitude is Daber's report of *Cooksonia* from Libya (Daber 1971). The identification of the plant is less convincing, as is the dating of the other Libyan assemblages. We have no direct information on the climatic tolerances of these early land plants. Being free-sporing and homosporous, the independent sexual gametophyte would have lived in damp habitats, and required a film of water for fertilization. The nature of the gametophytes remains elusive (but see Remy 1982; Schweitzer 1983). I have already suggested that thick-walled peripheral tissues were an adaptation for survival from water stress (Edwards 1979*a*). Raymond (1987) postulated that clustering of sporangia, e.g. into strobili, in later Lower Devonian plants gave protection from dessication. Undoubtedly clustering of sporangia during development would have so protected them, but their arrangements in strobili or trusses is so widespread that they were probably evolved to increase spore production per major axis. While, apart from extreme cases, availability of water is probably of major importance in affecting the local distribution of plants, sensitivity to extremes of temperature is the most important single factor in influencing their world-wide distribution. Research has concentrated on seed and particularly crop plants. Information on morphological and anatomical indicators of climate is similarly orientated, and comparisons of fossils possessing anatomically simple axial organization with living leaf-dominated vegetation are unrewarding. However, 'pre-lycophytes' such as *Baragwanathia* and to a lesser extent *Drepanophycus spinaeformis* superficially resemble the extant lycophyte *Lycopodium s.l.*, a genus showing remarkably little morphological variation, but ranging from high latitudes to the tropics, although gametophytes are subterranean in the former. *D. spinaeformis* is very widespread in the Lower Devonian, while certain herbaceous lycophytes are cosmopolitan later in the Devonian (Edwards & Benedetto 1985).

The major hindrance to global phytogeographic analyses is the absence of microfossil and megafossil evidence from high latitudes in Silurian and later Devonian sediments (Edwards 1989). A fascinating exception relates to the distribution of permanent (obligate) spore tetrads in the Ordovician and Early Silurian of Gondwana (e.g. Gray 1985, 1988; Gray *et al.* 1982). Gray has postulated that the spore producers colonized land surfaces and were physiologically similar to bryophytes, a hypothesis that ties in with the tolerance of some members of that group to exceedingly low temperature.

Comments on the composition of Gedinnian assemblages

Data are summarized in Table 2. The most intriguing are the most inaccessible, viz. the Kazakhstan Balkhash assemblage (Senkevich 1975) and that from the Kuzbass (Stepanov 1975). The Siberian fossils are well illustrated, except for an unconvincing *Cooksonia pertoni*, and are quite unlike any from elsewhere. Their affinities are obscure. The *Unocatoella verticillata* − *Zosterophyllum* sp. assemblage from the lowest part of the Cufengshan Formation in east Yunnan, southwest China, is said to be Gedinnian (Lee & Tsai 1978). The fossil identified as *Zosterophyllum* sp. is a fragment of a spike with few complete or attached sporangia. Better documentation of both fossils and age determination is required. The plant assemblage in typical Old Red Sandstone sediments in adjacent northern Viet Nam (Janvier *et al.* 1987) contains axes with isotomous, anisotomous and H-branching and one with some evidence for spine bases. Isolated sporangia range from rounded to oval to reinform in shape with the latter possessing a border. One was called *Cooksonia* sp., although its large size and lack of any axial system cast doubt on this generic assignation. The Lower Devonian, possibly Gedinnian, age of the sediments was based on these plants and two placoderm fragments. In Laurussia, Spitsbergen and German assemblages are sterile, while the two symbols in the British Gedinnian represents more occurrences than the rest of the world put together. The recent report of *Sciadophyton steinmanni* from the Gedinnian of Badajoz, Spain (number 12 on Fig. 2) by C. Alvarez-Ramis at a meeting in Lille in 1988 demonstrates the presence of vascular plants in north Gondwana (see also Obrhel 1968). The genus is thought to be the gametophyte of *Zosterophyllum rhenanum* and *Stockmansella langii* (Schweitzer 1983). Rayner's South African record is omitted, because there is no compelling evidence for its Early Devonian age (Rayner 1988, and above).

Gedinnian assemblages in Britain (Fig. 3)

The best known Early Devonian assemblages in Britain occur in Scottish Dittonian localities and are characterized by *Zosterophyllum myretonianum* (Lang 1927; Edwards 1975) the zonal

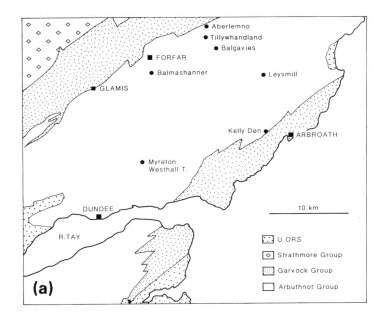

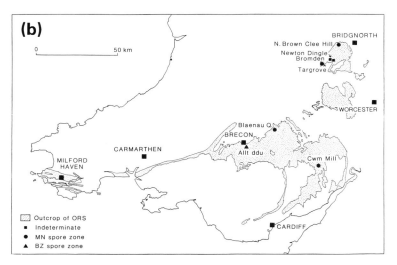

Fig. 3. Localities in the Gedinnian of: (a) the Angus region (outcrop based on Armstrong & Paterson 1970); (b) the Welsh Borderland and South Wales.

index fossil of Banks' Zone II (Banks 1980). The species is found in at least six localities in the Dundee Formation (Arbuthnott Group) in the Dundee–Forfar region of Angus. The plants, sometimes associated with fish and arthropods, are preserved in siltstones and flaggy sandstones interbedded with medium to coarse-grained, cross-bedded sandstones. Correlation is based on faunas (Westoll 1977) and palynomorphs (Richardson *et al.* 1984). Most of the plant-localities occur in the lower part of the *micrornatus-newportensis* spore biozone, but Aberlemno, where *Z. myretonianum* exists with *Cooksonia caledonica* (Edwards 1970) and spiny axes, is in the middle part. The approximately coeval Kelly Den locality contains *Z.* cf. *fertile* (Edwards 1972). The fossils themselves are usually abundant with mats of tangled drifted sterile axes sometimes covering bedding planes: individual axes may be tens of centimetres long.

In contrast, *Z. myretonianum* has not been unequivocally demonstrated in Welsh Borderland Dittonian localities, although *Z. fertile* has been recorded (e.g. Cwm Mill). Recent intensive reconnaissance in this area and in South Wales has produced at least ten new localities with abundant sterile axial remains as well as the fertile records listed in Table 2. Correlation is based on vertebrates (Ball & Dineley 1961; White 1961) and palynomorphs (Richardson *et al.* 1984). Assemblages in finer grained lithologies are dominated by fragmentary rhyniophytoids. That at Targrove, near Ludlow, occurs at the top of a fining-upwards cycle and contains a number of *Cooksonia* and *Salopella* species, oval terminal sporangia and bifurcating sporangia (Edwards & Fanning 1985). A number of grey siltstone horizons recently discovered in stream sections on the north side of Brown Clee Hill, Shropshire, appear to the naked eye unfossiliferous, but on dissolving the matrix abundant, but minute, well-preserved sporangia and fragments of axes are recovered. Taxa present are much the same as at Targrove, but details of sporangium wall, stomata and *in situ* spores are preserved (Edwards *et al.* 1986). Some *Salopella* and *Cooksonia* sporangia terminate dichotomously branching systems, but their overall length is less than a centimetre. *Zosterophyllum* has not been recorded except perhaps as isolated sporangia which are reniform, possess a border and contain retusoid spores. In some cases in situ spores provide evidence for greater diversity than that apparent from morphology alone (e.g. at least two different kinds of spores have been isolated from sporangia assigned to *Salopella* on shape alone, (but not from the same sporangium), and morphologically similar bifurcating sporangia contain spores similar to those in *Salopella* or *Tortilicaulis*).

Differences between the Scottish and Welsh Borderland Zone megafossil assemblages may be facies controlled, but it is also possible that they reflect local differences in composition of vegetation. Evidence for this comes from Richardson's research on spores. He noted that although the overall composition of spore assemblages in the two regions is sufficiently similar to permit confident correlation, a notable difference is the abundance of species of *Aneurospora* and *Streelispora* in the Welsh Borderland (Richardson *et al.* 1984). These taxa have been isolated from sporangia of *Cooksonia pertoni* and possibly certain salopellas at Targrove and Brown Clee Hill localities (Fanning *et al.* 1988). Thus it is possible that these plants were less abundant in Scotland, where they have not been recorded as megafossils, or that they grew in low-lying areas closer to the sea and thus their spores are under-represented in sediments thought to be deposited in lakes and rivers in intermontane basins (Simon & Bluck 1982)

In slightly younger horizons in the Welsh Borderland, coarser sediments contain longer and wider axes with *Salopella allenii* and occasional spikes of *Zosterophyllum*. Typical are the Newton Dingle assemblage (Edwards & Richardson 1974) and those from two new localities near Bromden. John Richardson has isolated spores from the upper *micrornatus–newportensis* spore biozone from a few metres below the Newton Dingle assemblage and believes the latter to be in the late Gedinnian. Field relationships suggest a similar age for the Bromden plants. We have yet failed to find any useful plant megafossils in fine-grained sediments at similar stratigraphic levels in the immediate area.

A new locality (Allt Ddu) at the base of the Senni Beds some 3 km NNE of Pen-y-Fan in the Brecon Beacons is placed in the *breconensis-zavallatus* spore biozone (Hassan 1982; Richardson pers. comm.) and provides striking evidence for the earliest diversification of zosterophylls. Abundant spiny axes of plants called cf *Psilophyton princeps* by Croft & Lang (1942) as well as rhyniophytoids such as *Cooksonia* and *Salopella* are present (Fig. 4). A similar assemblage with some additional taxa such as *Drepanophycus spinaeformis* occurs in basal Siegenian sediments in the Brecon Beacons Quarry, a few kilometres to the southwest (Richardson pers. comm.). Unfortunately the sediments immediately underlying the plant beds at Allt Ddu are unfossiliferous red marls and so it is impossible to determine if this earliest record coincides with the earliest radiation. The Welsh Borderland Newton Dingle assemblage is probably slightly older (Richardson pers. comm.) and is generically impoverished. This difference is unlikely to be facies-related as the coarser fluvial sediments

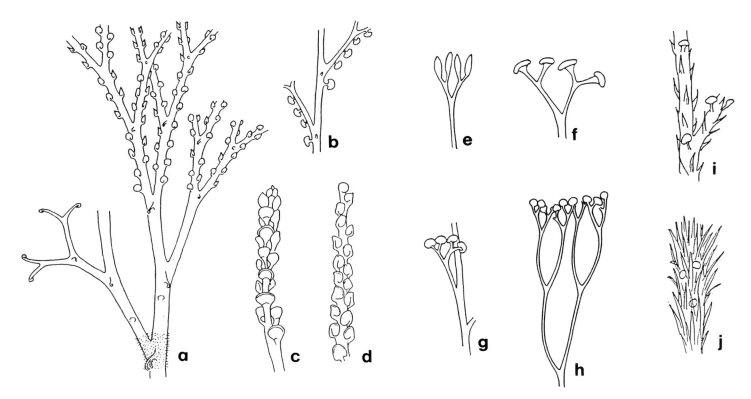

Fig. 4. Sketches of representative Gedinnian plants. (**a**) cf. *Psilophyton princeps* (Croft & Lang 1942); (**b**) *Gosslingia breconensis*; (**c**) *Zosterophyllum myretonianum*; (**d**) *Z. fertile*; (**e**) *Salopella* sp (**f**) *Cooksonia pertoni*; (**g**) *Renalia* sp; (**h**) *C. caledonica*; (**i**) *Drepanophycus spinaeformis*; (**j**) *Baragwanathia longifolia*. (a) and (b) occur only in the *breconensis−zavellatus* spore zone.

present towards the top of the Ditton Group (Allen 1975) suggest similar depositional environments in the two areas. It could also reflect the relatively poor exposure in the Newton Dingle stream section, but it is noteworthy that there are neither spiny axes nor branching systems with sub-axillary branches (axillary tubercles) in the abundant sterile axes found at a number of localities in the Welsh Borderland.

Thus in Britain evidence points to a major ?evolutionary floral change near the top of the Gedinnian and is reinforced by records of *Drepanophycus spinaeformis* in the upper Gedinnian of Germany (Schweitzer 1983) and an extensive assemblage including *Gosslingia breconensis* in the *breconensis-zavallatus* spore biozone at Gileppe, Belgium (Steemans & Gerrienne 1984).

Comments on geographic distribution of Gedinnian megafossils

Differences in the composition of British Gedinnian assemblages thus may result from taphonomic effects, from local variation in distribution of vegetation or from a major evolutionary event. The small amount of progress so far achieved in recognizing and disentangling these factors has been made possible because of the relatively large number of localities in a restricted geographical area. Records in other parts of the world are not so detailed nor so precisely dated and, in the light of experiences in Britain, the problems of interpretation are horrendous.

Based on the palaeogeographic reconstructions of Scotese, the relative positions of the continents in the Early Devonian (Fig. 1) are little changed from the Silurian (Fig. 2) except that a sinistral movement of Laurussia and Siberia results in the European localities being closer to the equator and the Siberian at higher latitudes, and hence in a cooler climate. This could account for the unique composition of the Siberian assemblages. The land masses were still close together and there would have been no major oceanic barrier to northwards spread from the Russian platform in the east. However, inferred high ground to the west of the presumed coastal habitats of the Kuzbass plants would have formed an insuperable barrier with western parts of northern Laurussia. The oldest and very fragmentary records on the South China plate are at approximately the same palaeolatitude as the probable Lochkovian assemblage in Victoria, Australia. They share the cosmopolitan *Zosterophyllum*, but there is no evidence for the endemics that characterize later extensive south China assemblages, nor of *Baragwanathia* itself. Apart from *Cooksonella*, the Kazakhstan assemblage contains only endemics; my comments on the Silurian assemblages from the same area apply equally well here.

Thus on the very limited data available there appear to be four regions (Laurussia, Siberia, Kazakhstan, Australia) yielding assemblages of distinctive composition. Whether or not these should be translated into floral provinces awaits the discovery of considerably more and well-dated assemblages.

Apologia

The intention of this account is to emphasize the pitfalls when handling Silurian and Early Devonian megafossils, but is not meant to be critically destructive. Research in Britain, in particular, has demonstrated that it is possible to improve the database even in an area which has an extensive research history. Other sources of optimism are the detailed descriptions of new taxa from China, the discovery of assemblages in South America and the Palaeozoic initiative in Antarctica by T. N. Taylor and colleagues at the Byrd Polar Research Centre, Ohio.

I thank U. Fanning and P. Kenrick for access to unpublished research data, but am particularly grateful to J. B. Richardson whose generosity in providing unpublished information on the age of the plant assemblages in South Wales and the Welsh Borderland made possible my analysis. This biostratigraphical project in southern Britain was supported by NERC Research Grant, and I thank C. Cleal and M. Rowlands and P. Tarrant for fieldwork assistance.

References

ALLEN, J. R. L. 1975. The Devonian rocks of Wales and the Welsh Borderland. *In*: OWEN, T. R. (ed.) *The Upper Palaeozoic and post-Palaeozoic rocks of Wales*. University of Wales Press, Cardiff, 47–84.

ALLEN, K. C. 1980. A review of *in situ* late Silurian and Devonian spores. *Review of Palaeobotany and Palynology*, **29**, 253–270.

ARMSTRONG, M. & PATERSON, I. B. 1970. *The Lower Old Red Sandstone of the Strathmore region*. Report No. 72/12. Institute of Geological Sciences.

ANDERSON, J. M. & ANDERSON, H. M. 1985. *Palaeoflora of Southern Africa. Prodromus of South Africa megafloras: Devonian to Lower Cretaceous*. Balkema, Rotterdam.

BALL, H. W. & DINELEY, D. L. 1961. The Old Red Sandstone of Brown Clee Hill and the adjacent area. 1 Stratigraphy. *Bulletin of the British Museum (Natural History) Geology*, **5**, No. 7, 175–242.

BANKS, H. P. 1973. Occurrence of *Cooksonia*, the oldest vascular land plant macrofossil, in the Upper Silurian of New York State. *Journal of the Indian Botanical Society*, **50A**, 227–235.

BANKS, H. B. 1980. Floral assemblages in the Siluro–Devonian. *In*: DILCHER, D. L. & TAYLOR, T. N. (eds). *Biostratigraphy of fossil plants*. Dowden, Hutchinson & Ross, Stroudsberg, Pa. 1–24.

BARRETT, S. F. 1985. Early Devonian continental positions and climate; a framework for palaeophytogeography. *In*: TIFFNEY, B. H. (ed.). *Geological Factors and the evolution of plants*. Yale University Press, New Haven, 93–127.

BOUCOT, A. J. & GRAY, J. 1982. Geologic correlates of early land plant evolution. *Third North American Paleontological Convention, Proceedings*, **1**, 61–66.

CHURKIN, M., EBERLEIN, G. D., HUEBER, F. M. & MAMAY, S. H. 1969. Lower Devonian land plants from graptolitic shale in south-eastern Alaska. *Palaeontology*, **12**, 559–73.

CROFT, W. N. & LANG, W. H. 1942. The Lower Devonian flora of the Senni Beds of Monmouthshire and Breconshire. *Philosophical Transactions of the Royal Society of London, Series B*, **231**, 131–163.

CUERDA, A., CINGOLANI, C., ARRONDO, O., MOREL, E. & GANUZA, D. 1987. Primer registro de plantas vasculares en la Formacion Villavicencio, Precordillera de Mendoza, Argentina. *IV Congresso Latinoamericano de Paleontologia, Bolivia (1987)*, **1**, 179–183.

DABER, R. 1971. Cooksonia: one of the most ancient psilophytes – widely distributed, but rare. *Botanique (Nagpur)*, **2**, 35–39.

DOU YAWEI & SUN ZHE HUA. 1983. Devonian plants of Xinjiang. *In*: Palaeontological Atlas of northwestern China. Regional Geological Survey Team & Institute of Geological Science of the Geological Bureau, Xinjiang. Geological Survey, Department of Petrology Bureau, Xinjiang, 561–594 (in Chinese).

DOUGLAS, J. G. & LEJAL-NICOL, A. 1981. Sur les premières flores vasculaires terrestres datée du Silurian: Uhe comparaison entre la "Flore a *Baragwanathia*" d'Australie et la "Flore a Psilophytes et Lycophytes" d'Afrique du Nord. *Compte Rendu Académie Sciences, Paris*, **292**, 685–7.

EDWARDS, D. 1970. Fertile Rhyniophytina from the Lower Devonian of Britain. *Palaeontology*, **13**, 451–461.

—— 1972. A *Zosterophyllum* fructification from the Lower Old Red Sandstone of Scotland. *Review of Palaeobotany and Palynology*, **14**, 77–83.

—— 1973. Devonian floras. *In*: HALLAM, A. (ed.). *Atlas of palaeobiogeography*, 105–115. Elsevier, Amsterdam.

—— 1975. Some observations on the fertile parts of *Zosterophyllum myretonianum* Penhallow from the Lower Old Red Sandstone of Scotland. *Transactions of the Royal Society of Edinburgh*, **69**, 251–265.

—— 1977. A new non-calcified alga from the Upper Silurian of mid-Wales. *Palaeontology*, **20**, 823–832.

—— 1979a. The early history of vascular plants based on late Silurian and early Devonian floras of the British Isles. *In*: LEAKE, B. E., HARRIS, A. L. & HOLLAND, C. H. (eds). *The Caledonides of the British Isles – reviewed*. Geological Society, London, Special Publication, **8**, 405–410.

—— 1979b. A late Silurian flora from the Lower Old Red Sandstone of southwest Dyfed. *Palaeontology*, **22**, 23–52.

—— 1980. Early land floras. *In*: PANCHEN, A. L. (ed.). *The terrestrial environment and the origin of land vertebrates*. Systematics Association. **15**, Academic Press, London, 55–85.

—— 1989. Devonian paleobotany: problems, progress and potential. *In*: TAYLOR, T. N. & TAYLOR, E (eds). *Antarctic paleobiology and its role in the reconstruction of Gondwana*. Springer, Berlin. (In press).

—— & BENEDETTO, J. L. 1985. Two new species of herbaceous lycopods from the Devonian of Venezuela with comments on their taphonomy. *Palaeontology*, **28**, 599–618.

—— & EDWARDS, D. S. 1986. A reconsideration of the Rhyniophytina Banks. *In*: SPICER, R. A. & THOMAS, B. A. (eds). *Systematic and taxonomic approaches in Palaeobotany*. The Systematics Association, Clarendon Press, Oxford; **31**, 199–220.

—— & FANNING, U. 1985. Evolution and environment in the late Silurian–early Devonian: the rise of the pteridophytes. *Philosophical Transactions of the Royal Society of London, series B*, **309**, 147–165.

—— & RICHARDSON, J. B. 1974. Lower Devonian (Dittonian) plants from the Welsh Borderland. *Palaeontology*, **17**, 311–324.

—— & ROGERSON, E. C. W. 1979. New records of fertile Rhyniophytina from the late Silurian of Wales. *Geological Magazine*, **116**, 93–98.

——, BASSETT, M. G. & ROGERSON, E. C. W. 1979. The earliest vascular land plants: continuing the search for proof. *Lethaia*, **12**, 313–324.

——, FANNING, U. & RICHARDSON, J. B. 1986. Stomata and sterome in early land plant. *Evolutionary Trends in Plants*, **2**, 13–24.

——, FEEHAN, J. & SMITH, D. G. 1983. A late Wenlock flora from Co. Tipperary, Ireland. *Botanical Journal of the Linnean Society*, **86**, 19–36.

FANNING, U. 1987. *Late Silurian–early Devonian plant assemblages in the Welsh Borderland*. PhD Thesis, University of Wales.

——, RICHARDSON, J. B. & EDWARDS, D. 1988. Cryptic evolution in an early land plant. *evolutionary trends in plants*, **2**, 13–24.

GARRATT, M. J. 1978. New evidence for a Silurian (Ludlow) age for the earliest *Baragwanathia* flora. *Alcheringa*, **2**, 217–224.

—— & RICKARDS, R. B. 1984. Graptolite biostratigraphy of early land plants from Victoria, Australia. *Proceedings of the Yorkshire Geological Society*, **44**, 377–384.

——, TIMS, J. D., RICKARDS, R. B., CHAMBERS, T. C. & DOUGLAS, J. G. 1984. The appearance of *Baragwanathia* (Lycophytina) in the Silurian. *Botanical Journal of the Linnean Society*, **89**, 355–358.

GENG BAO-YIN. 1985. Anatomy and morphology of *Pinnatiranosus*, a new plant from the Middle Silurian (Wenlockian) of China. *Acta Botanica Sinica*, **28**, 667–670 (in Chinese).

GENSEL, P. G. 1980. Devonian *in situ* spores: a survey and discussion. *Review of Palaeobotany and Palynology*, **30**, 101–132.

GRAY, J. 1985. The microfossil record of early land plants: advances in understanding of early terrestrialization, 1970–1984. *Philosophical Transactions of the Royal Society of London, series B*, **309**, 167–195.

—— 1988. Land plant spores and the Ordovician–Silurian boundary. *Bulletin of the British Museum (Natural History) Geology*, **43**, 351–358.

——, MASSA, D. & BOUCOT, A. J. 1982. Caradocian land plant microfossils from Libya. *Geology*, **10**, 197–201.

HASSAN, A. M. 1982. *Palynology, stratigraphy and Provenance of the Lower Old Red Sandstone of the Brecon Beacons (Powys) and the Black Mountains, (Gwent and Powys) South Wales*. PhD thesis. King's College, University of London.

HØEG, O. A. 1942. The Downtonian and Devonian flora of Spitsbergen. *Norges Svalbard-og Ishavs-Undersøkelser*, **83**, 1–228.

HUEBER, F. M. 1971. Early Devonian land plants from Bathurst Island, District of Franklin. *Geological Survey of Canada, Paper* **71–28**, 1–1.

ISHCHENKO, T. A. 1975. *The late Silurian flora of Podolia*. Institute of Geological Science, Academy of Science of the Ukrainian SSR., Kiev, 1–80.

JANVIER, P., BLIECK, A., GERRIENNE, P. & THANH TONG-DZUY. 1987. Faune et flore de la Formation de Sika (Dévonien inférieur) dans la presqu'île de Dô Son (Viêt Nam). *Bulletin du Muséum d' Histoire Naturelle. Paris* 4ᵉ ser, **9**, Section C, 291–301.

KLITSCH, E., LEJAL-NICOL & MASSA, D. 1973. Le Siluro-Dévonien a Psilophytes et Lycophytes du bassin de Mourzouk (Libye). *Comptes rendus hebdomadaires des Sceances de L'Académie-des Sciences,*

Paris, **277**, D. 2465–2467.

LANG, W. H. 1927. Contributions to the study of the Old Red Sandstone Flora of Scotland. VI. On *Zosterophyllum myretonianum*, Penh., and some other plant remains from the Carmyllie Beds of the Lower Old Red Sandstone. *Transactions of the Royal Society of Edinburgh*, **55**, 443–455.

—— 1937. On the plant remains from the Downtonian of England and Wales. *Philosophical Transactions of the Royal Society of London, series B*, **227**, 245–291.

LECLERCQ, S. 1942. Quelques plantes fossiles recueillies dans le Devonian inférieur des environs de Nonceveux (Bordure orientale du bassin de Dinant). *Société géologique de Belgique*, **65**, B148–B211.

LEE, H-H. & TSAI, C-Y. 1978. III. *Devonian floras of China*. Papers for the International Symposium on the Devonian System 1978, Nanjing, 1–14.

LEMOIGNE, Y. 1967. Reconnaissance paléobotanique dans le Sahara occidental (Région de Tindouf et Gara-Djebilet). *Annales de la Société géologique du Nord*, **87**, 31–38.

LI XING-XUE & TSAI CHONG-YANG. 1978. A type-section of Lower Devonian strata in southwest China with brief notes on the succession and correlation of its plant assemblages. *Acta Geologica Sinica*, **1**, 1–14.

LOEFFLER, E. J. & THOMAS, R. G. 1980. A new pteraspidid ostracoderm from the Devonian Senni Beds Formation of South Wales and its stratigraphic significance. *Palaeontology*, **23**, 287–296.

MEYEN, S. V. 1987. *Fundamentals in Palaeobotany*. Chapman & Hall, London.

NIKITIN, I. F. & BANDALETOV, S. M. 1986. *The Tokrau horizon of the Upper Silurian Series*: Balkhash Segment. Alma-Ata Nauka (in Russian).

OBRHEL, J. 1962. Die Flora der Pridoli-Schichten (Budnany-Stufe) des mittelböhmischen Silurs. *Geologie*, **11**, 83–97.

—— 1968. Die Silur und Devonflora des Barrandiums. *Paläontologische Abhandlungen B*, **2**, 635–793.

PETROSYAN, N. M. 1968. Stratigraphic importance of the Devonian flora of the USSR. *In*: OSWALD, D. H. (ed.). *International Symposium on the Devonian System, Calgary, 1967*. Alberta Society of Petroleum Geologists, Calgary, 579–586.

RAYMOND, A. 1987. Paleogeographic distribution of early Devonian plant traits. *Palaios*, **2**, 113–132.

RAYMOND, A., PARKER, W. C. & BARRETT, S. F. 1985. Early Devonian phytogeography. *In*: TIFFNEY, B. H. (ed.). *Geological factors and the evolution of plants*. Yale University Press, New Haven and London, 129–167.

RAYNER, R. J. 1988. Early land plants from South Africa. *Botanical Journal of the Linnean Society*, **97**, 229–237.

REMY, W. 1982. Lower Devonian gametophytes: relation to the phylogeny of land plants. *Science, Washington*, **215**, 1625–1627.

RICHARDSON, J. B. & EDWARDS, D. 1989. Sporomorphs and plant megafossils. *In*: HOLLAND, C. H. & BASSETT, M. G. (eds). *A global standard for the Silurian System*, 216–226.

—— & MCGREGOR, D. C. 1986. Silurian and Devonian spore zones of the Old Red Sandstone continent and adjacent regions. *Geological Survey of Canada, Bulletin*, **364**, 1–79.

——, FORD, J. H. & PARKER, F. 1984. Miospores, correlation and age of some Scottish Lower Old Red Sandstone sediments from the Strathmore region (Fife and Angus). *Journal of Micropalaeontology*, **3**, 109–124.

——, STREEL, M., HASSAN, A. & STEEMANS, PH. 1982. A new spore assemblage to correlate between the Breconian (British Isles) and the Gedinnian (Belgium). *Annales de la Société Géologique de Belgiques*, **105**, 135–143.

SCHOPF, J. M., MENCHER, E., BOUCOT, A. J. & ANDREWS, H. N. 1966. Erect plants in the early Silurian of Maine. *US Geological Survey. Professional Paper*, **550-D**, D69–D75.

SCHWEITZER, H-J. 1983. Die Unterdevonflora des Rheinlandes. *Palaeontographica*, **B189**, 1–138.

SENKEVICH, M. A. 1975. New Devonian psilophytes from Kazakhstan. *Esheg Vses Paleontol Obschestva*, **21**, 288–298 (in Russian).

SIMON, J. B. & BLUCK, B. J. 1982. Palaeodrainage of the southern margin of the Caledonian mountain chain in the northern British Isles. *Transactions of the Royal Society of Edinburgh: Earth Sciences*, **73**, 11–15.

STEEMANS, PH. 1982. Gedinnian and Siegenian spore stratigraphy in Belgium. *Courier Forschungsinstitut Senckenberg*, **55**, 165–180.

—— & GERRIENNE, PH. 1984. La micro- et macroflore du Gedinnian de la Vesdre, Belgique. *Annales de la Société Géologique de Belgique*, **107**, 51–71.

STEPANOV, S. A. 1975. Phytostratigraphy of the key sections of the Devonian of the marginal parts of the Kuznetsk. *Transactions of the Silurian Institute of Geology, Geophysics and Mineral Resources*, **211**, 1–150.

STREEL, M., FAIRON-DEMARET, M. & LOBOZIAK, S. 1990. Givetian to Frasnian phytogeography of Euramerica and western Gondwana based on miospore distribution. *In*: MCKERROW, W. S. & SCOTESE, C. R. (eds) *Palaeozoic Palaeogeography and Biogeography*. Geological Society, London, Memoir, **12**, 291–296.

——, HIGGS, K., LOBOZIAK, S., RIEGEL, W. & STEEMANS, PH. 1987. Spore stratigraphy and correlation with faunas and floras in the type marine Devonian of the Ardenne–Rhenish Regions. *Review of Palaeobotany and Palynology*, **50**, 211–219.

THERON, J. N. & KOVACS ENDRODY, E. 1986. Preliminary note and description of the earliest known vascular plant, or an ancestor of vascular plants, in the flora of the Lower Silurian Cedarberg Formation, Table Mountain Group, South Africa. *South African Journal of Science*, **82**, 102–106.

TIMS, J. D. & CHAMBERS, T. C. 1984. Rhyniophytina and Trimerophytina from the early land floras of Victoria, Australia. *Palaeontology*, **27**, 265–279.

WESTOLL, T. S. 1977. Northern Britain. *In*: HOUSE, M. R. (ed.). *A correlation of the Devonian rocks in the British Isles*. Geological Society of London, Special Report, **7**, 66–93.

WHITE, E. I. 1961. The Old Red Sandstone of Brown Clee Hill and the adjacent area. II. Palaeontology. *Bulletin of the British Museum (Natural History) Geology*, **5**, No. 7, 243–310.

ZIEGLER, A. M., BAMBACH, R. K., PARRISH, J. T., BARRETT, S. F., GIERLOWSKI, E. H., PARKER, W. C., RAYMOND, A. & SEPKOSKI, J. J. 1982. Palaeozoic biogeography and climatology. *In*: NIKLAS, K. J. (ed.). *Palaeobotany, paleoecology, and evolution*, **2**, Praeger, New York, 231–266.

ZIEGLER, W. & KLAPPER, G. 1985. Stages of the Devonian System. *Episodes*, **8**, 104–109.

Devonian vertebrate distribution patterns and cladistic analysis of palaeogeographic hypotheses

GAVIN C. YOUNG

Division of Continental Geology, Bureau of Mineral Resources, PO Box 378, Canberra, ACT, Australia 2601

Abstract: Early Devonian vertebrate faunal provinces are clearly defined for four regions: Euramerica, Siberia, China, and East Gondwana. The presence of osteostracans in southwest Siberia (Tuva, Minusa Basins) suggests either proximity of the Siberian block to Euramerica, or palaeogeographic separation of the Tuva region. The Knoydart fauna of Nova Scotia demonstrates that the Avalon Terrane was connected to Euramerica by Gedinnian time. Widespread antarctilamnid sharks in Gondwana suggest a distinctive Gondwana vertebrate fauna, isolated by marine barriers from Euramerica in the Early–Middle Devonian. Late Devonian patterns indicate faunal communication between Gondwana and Euramerica by Frasnian time, and between China and East Gondwana in the late Famennian. The Late Devonian base maps require anomalously wide latitudinal distributions for some taxa. Displacement of Turkey along the northern margin of Gondwana provides an intermediate occurrence of phyllolepid placoderms between disjunct distributions in Euramerica and East Gondwana, but the fossil data do not necessarily corroborate geological evidence for displacement. Biogeographic data generally must be interpreted in the context of palaeogeographic hypotheses, and lack of integration with geological and geophysical data sets has been a major problem. Hierarchical analysis using cladistic techniques has the potential for integrating biological, geological, and geophysical data, as illustrated in a cladistic analysis of the Williams and Hatcher model of Appalachian terranes. As an adjunct to map representation, an area cladogram enables a historical sequence of palaeogeographic events to be represented on a single diagram, together with crucial supporting evidence; it presents an analysis rather than synthesis of empirical data, and the hypothesis is more exposed to falsification.

This paper has three main aims: first, to summarize briefly the taxonomic data base for Devonian vertebrates as currently known; secondly, to use this data base to outline distribution patterns for Devonian vertebrates with respect to the global palaeogeographic reconstructions of Scotese (1986); and finally to consider the theoretical approach used in construction of Palaeozoic palaeogeographic maps, and the way biogeographic data may be used as tests.

Today freshwater fishes make up about 41% of the modern fauna of bony fishes, even though they occupy less than 0.01% of the world's water (Horn 1972). It is not possible to arrive at comparable figures for the Devonian, but somewhat more than half of known forms probably occurred in coastal sections of rivers and lakes, and are preserved as fossils in fluviatile red bed deposits of Old Red Sandstone type; that is, largely external continental facies rocks with marine intercalations (Gray 1988). As well as strictly freshwater forms, such environments no doubt also contained many euryhaline 'peripheral division' fishes, but an assessment of which taxa belonged to which ecological grouping can only be made through an analysis of distribution patterns in relation to phylogeny. For this reason the taxonomic data base is of primary importance in understanding the significance of Devonian fish distributions.

Taxonomic data base

By the beginning of the Devonian period there were in existence four major groups of agnathans or jawless vertebrates, and four major groups of gnathostomes or jawed fishes (Fig. 1). Based on recent summaries of known vertebrate fossils (Blieck 1984; Campbell & Barwick 1987; Carroll 1988; Denison 1978, 1979; Janvier 1985b; Young unpublished), the following approximate numbers of named Palaeozoic genera may be given for the major Devonian groups: osteostracans, 35; galeaspids, 30; anaspids, 10; thelodontids, 16; heterostracans, 110; acanthodians, 60; placoderms, 250. Osteichthyans, chondrichthyans, and tetrapods, with about 85, 22, and 4 known Devonian genera respectively, attained a much greater diversity in post-Devonian times. Amongst osteichthyans (bony fishes) the major Devonian groups are osteolepiform crossopterygians (32 genera) and lungfishes (24 genera). Many new taxa have been added to all groups in recent years. For example, placoderms (armoured fishes) are the most diverse single Devonian group, and the above assessment represents an increase of over one third the number of genera listed in an authoritative summary of the group presented only 10 years ago (Denison 1978). Probably a third as many placoderm genera again are held in scientific collections, and await description. Many such genera are monotypic, or contain only a few species, but some are very diverse (e.g. *Bothriolepis*, with over 70 named species). Recent taxonomic contributions include both major revisions of Euramerican faunas (e.g. Blieck 1984; Goujet 1984; Janvier 1985a), and descriptions of faunas from new areas (for example: Middle East, Blieck et al. 1980; Far East, Thanh & Janvier 1987; China, Wang 1984, Pan & Dineley 1988; Australia, Long & Turner 1984; Antarctica, Young 1988, 1989; southern Africa, Chaloner et al. 1980; South America, Goujet et al. 1984, Gagnier et al. 1988, Janvier & Suarez-Riglos 1986). Special mention must be made of the vertebrate faunas from the Devonian of China, where over 70 genera (90% endemic) are now recorded, some 57% of which have entered the literature since 1975. It is clear that the data base is currently undergoing rapid expansion, and for the first time may now be considered of global extent (Fig. 2). Previously there has been a taxonomic bias towards the biogeographically restricted Devonian vertebrate faunas of the Euramerican region (Young 1981); these have been used to reach spurious conclusions about certain global phenomena (e.g. regarding extinction events; McGhee 1982). No doubt the current data base is still very incomplete, but there is now some knowledge of Devonian vertebrate faunas from most regions thought to be discrete crustal blocks during the Palaeozoic on geological evidence, permitting some preliminary biogeographic input to global palaeogeographic reconstructions.

Palaeogeographic base maps

The four Devonian maps of Scotese (1986) show Euramerica in close proximity to Gondwana, with Kazakhstan and Siberia as isolated blocks, and various Middle Eastern and Asian terranes associated with the northern margin of Gondwana. During the Devonian a clockwise rotation is depicted, such that both Australia and north America lie largely in 0–30° south palaeolatitude in the Gedinnian, but by Givetian time Australia is mainly south of 30° south, and north America straddles the equator. In the Famennian, Australia is shown extending to 60° south, and most

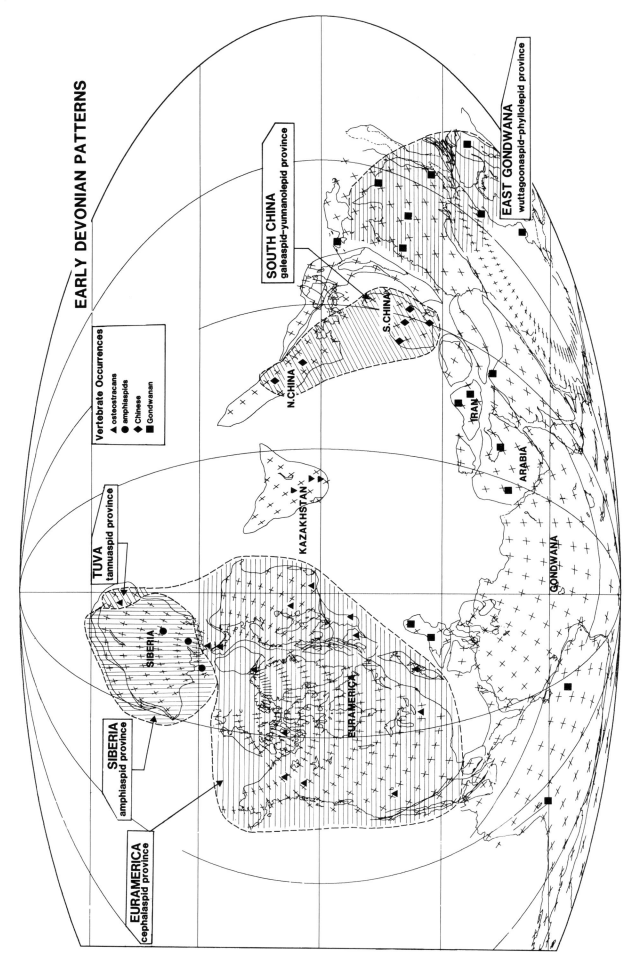

Fig. 3. Summary of Early Devonian vertebrate distribution patterns, using the Gedinnian (Lochkovian) reconstruction of Scotese (1986) as a base map.

all other members of this well known and widespread group except for *Dianolepis* from China. From near Tashkent in the Late Devonian (Frasnian) comes an endemic species of *Bothriolepis* (*B. turanica*).

North China. North China is placed just in the northern hemisphere northwest of Australia. Of the two vertebrate faunas from here the western locality (Yumen, Gansu Province) contains antiarchs and arthrodires (Wang 1984). The antiarch is tentatively identified as *Hunanolepis*, a possible close relative of *Stegolepis* from Kazakhstan. From Ningxia a Middle Devonian quasipetalichthyid and Late Devonian galeaspid and sinolepid have been described (Pan *et al.* 1987), which indicate connection with South China at least by Middle Devonian time. The Ningxia region is part of the 'Hexizoulang Terrane' (Li *et al.* 1985), which may have had a separate history to the rest of the North China block. A possible connection with central Asia is also indicated by the quasi-petalichthyid features of undescribed placoderms from Altai Sayan (Tuva).

South China. South China is adjacent to Gondwana in a position northwest of Australia, and a connection to North China is indicated. In the Gedinnian–Emsian South China has the same palaeolatitude as eastern Australia, yet their vertebrate faunas are very different. The South China fauna contains over 45 genera, most of which are endemic, and belong to endemic higher taxa (Eugaleaspida, Yunnanolepiformes) of subordinal or higher rank. This highly diverse fauna contrasts with the scarcity of pre-Emsian Devonian vertebrates from Australia, where the oldest faunas contain only acanthodians and thelodonts. Apart from the occurrences in North China just mentioned, the South Chinese fauna otherwise occurs only in northeast Vietnam, north of the Red River suture. Thanh & Janvier (1987) have established for the first time from described material the presence here of both galeaspids and yunnanolepids. Hsu *et al.* (1988) have recently proposed that the Huanan Mountains were formed by continent collision, and that there was no 'South China Platform'. Devonian vertebrate localities southeast of their postulated suture include those of south Jiangxi Province and north Guangdong Province (e.g. Wang 1984). Pan (1981) reported the antiarch *Hunanolepis* from the latter, and Pan & Dineley (1988, fig. 7) show a distinctive endemic fauna including this genus distributed widely in southeast and central China, suggesting that significant separation across the postulated suture was unlikely in the Middle Devonian.

Burrett, Long & Stait (1990) compare the high endemism of South China with the modern marsupial fauna of Australia. However the freshwater fish fauna of modern Australia is depauperate, with only 18 families compared to the much more diverse faunas of Africa (32 families) and South America (46 families; Berra 1981). Using this analogy, isolation is a necessary but not a sufficient condition for the development of a highly endemic freshwater fish fauna, and an equatorial position with adequate rainfall to support extensive rivers and lakes is another prerequisite (clearly tectonism is a potential complicating factor here). By analogy with the Amazon basin of today, I therefore suggest a more northerly position for South China (just as proposed on other grounds by Hou & Wang 1985, fig. 1), straddling the equator, and in the equatorial doldrums rain belt, to account for its highly diverse Devonian fish fauna. The position of Australia in the subtropical dry belt provides an explanation for its impoverished pre-Emsian Devonian fish fauna, as it partly does today for the low diversity of the modern Australian freshwater fish fauna compared to that of southeast Asia.

The affinities of the South China fauna change through the Devonian. Strong isolation is indicated for the Early Devonian, but a possible sister group relationship between the Chinese galeaspids and the osteostracans of Euramerica (Young in press) would indicate some earlier (?Silurian) connection. The antiarchs *Microbrachius* and *Byssacanthus* also occur on the Baltic plate, suggesting a diminished marine barrier in the earlier Middle Devonian, whilst the Givetian bothriolepid antiarchs (?non-marine) show affinities with eastern Australia and Victoria Land, Antarctica (Young 1988). In the Famennian one endemic Chinese group of antiarch placoderms (sinolepids) also occurs in eastern Australia.

Gondwana. On the maps Gondwana lies with its northern margin mainly just south of 30°S, with Australia forming a northern projection to the equator in the Gedinnian. The East Gondwana province of Young (1981) was identified only in Australia and possibly Antarctica. On the evidence of the biostratigraphic distribution of thelodonts and other groups (Young 1988), it seems that the endemism of the East Gondwana region persisted into the Givetian–Frasnian (Young 1989). Emsian marine faunas of eastern Australia show some affinity in the placoderms with South China.

Because of its size the existence of a single distinctive Gondwana vertebrate fauna is not established (the characteristic placoderm *Wuttagoonaspis* is only known from central and southeastern Australia), but many new vertebrate localities are now known from Gondwana (summarized in Young 1987a). A significant recent addition to the data base concerns the fish faunas from the Devonian of South America. In earlier literature *Machaeracanthus* spines were recorded but are not yet confirmed from the Parnaiba Basin of Brazil (not the Parana basin as stated in Young 1987b). This genus also occurs in the Ellsworth Mountains and Ohio Range of Antarctica, and may represent part of a Malvinokaffric element in the vertebrate fauna. On the other hand *Machaeracanthus* is also known from southern Europe, and the placoderm *Bolivosteus* described by Goujet *et al.* (1985) from Bolivia has its closest known relative in *Gemuendina* from the German Rhineland. Gagnier *et al.* (1988) have described the Seripona fauna from Bolivia, which contains *Turinia* scales similar to those from Antarctica, a new species of the shark *Antarctilamna*, and various acanthodian remains, suggesting affinities with East Gondwana, and possibly south China. The occurrence of *Antarctilamna* is significant, being previously recorded from eastern Australia and Antarctica, and Iran. A poorly known Early–Middle Devonian fish fauna from the Jauf region of Saudi Arabia (Boucot 1984) includes similar remains (Forey, pers. comm.). This suggests continuity in the vertebrate faunas across Gondwana, and on present evidence the Bolivian faunas are readily distinguished from those of Euramerica. But a marine barrier between north and south America presents a problem for the invertebrate distributions of this region (Fig. 4). Some form of (intermittent?) barrier is required to explain the occurrence of endemic species in the Nevadan subprovince of the Eastern Americas Realm, when the two other subprovinces of this realm (Appohimchi and Amazon–Colombian) share many common species (Boucot 1975). One or more vicariance events might be related to the appearance of this barrier (arrows in Fig. 4). Land barriers to the east (dashed lines, Fig. 4) to separate the Rhenish–Bohemian province of the Old World Realm seem unlikely in view of the apparent north–south distinctiveness of the vertebrate faunas just discussed, but latitudinal separation is a possible explanation considering the much more southerly position recently suggested for Laurentia on palaeomagnetic data (Miller & Kent 1988). To the south (dotted line, Fig. 4), the affinities between the Appohimchi and Malvinokaffric faunas (e.g. Boucot 1975) are consistent with this being a climatically controlled rather than a geographic barrier (Copper 1977).

Late Devonian

On the Late Devonian map (Fig. 5) the distribution of four widespread placoderm taxa is shown in relation to the psammosteid heterostracans of Euramerica: *Bothriolepis*, the most widely distributed Devonian fish, *Remigolepis*, and phyllolepids and

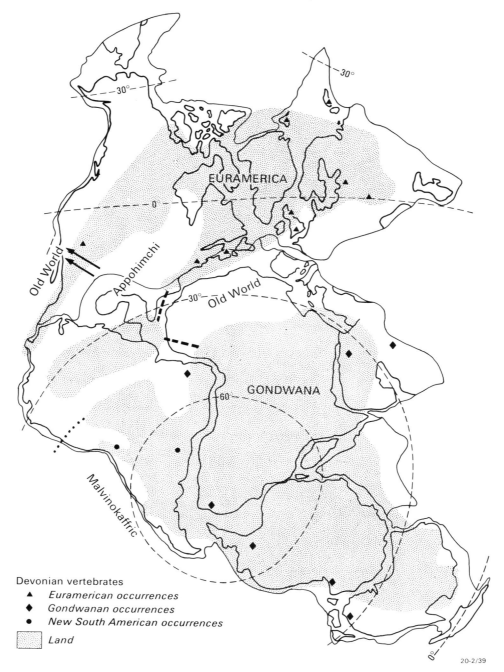

Fig. 4. Palaeogeographic relationship between Euramerica and Gondwana in the Early and Middle Devonian. Selected vertebrate localities shown. Palaeogeography and invertebrate faunal provinces generalised after Boucot (1975), Copper (1977), and Oliver (1977). For details of Devonian vertebrate occurrences see Fig. 2, and Young (1987a, fig. 1; 1987b, fig. 2). Faunal barriers (dashed and dotted lines) are discussed in the text.

sinolepids. More restricted distributions of phyllolepids and bothriolepids in the earlier part of their stratigraphic range are indicated; these may be inferred to indicate barriers preventing faunal interchange. Phyllolepids may have originated in East Gondwana, with their appearance in Euramerican sequences at or near the Frasnian–Famennian boundary signifying a persistent terrestrial connection being established with Gondwana for the first time (Young 1981, 1984, 1987b). Independent arguments related to cold-water oceanic circulation and the Frasnian–Famennian extinction event support this palaeogeographic conclusion (Copper 1986). Although 'fish could swim' (Hurley & Van der Voo 1987, p. 144), it is germane to point out that the group in question here (phyllolepid placoderms) are readily recognised, yet were apparently absent from Late Devonian fish faunas of Siberia, and North and South China, even though other groups preserved in similar deposits were able to cross intervening barriers. For this reason phyllolepids have been interpreted as a strictly non-marine group (Young 1981).

The Taymyr Suture forms the northern margin of the Siberian block, and the presence of a Late Devonian *Psammosteus* on Severnaya Zemlya to the north of the suture (the Kara block) indicates Baltican affinity, and is entirely consistent with the Early Devonian evidence from here (see above). However the Taymyr region is not separated from the Siberian block on the maps. From the Tuva and Minusa basins of the Altai–Sayan in southwestern Siberia Late Devonian fish are recorded from

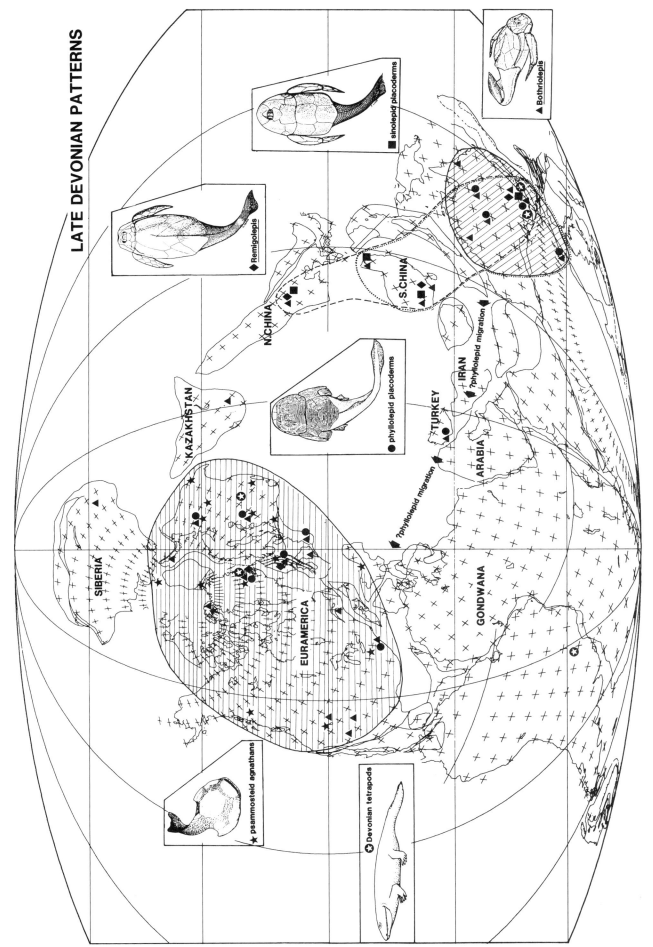

Fig. 5. Summary of Givetian–Famennian vertebrate distribution patterns, using the Givetian reconstruction of Scotese (1986) as a base map.

the Frasnian ('*Bothriolepis cellulosa*', *B. siberica*, *Dipterus*, *Megistolepis*, etc.) and Famennian (*B. siberica*, *Thaumatolepis*?). Noteworthy is the absence of *Phyllolepis* from the latter fauna.

With Siberia in its reconstructed position these occurrences of *Bothriolepis* lie at a palaeolatitude of 60°N, comparable with the 60–70°S palaeolatitude for the Aztec Siltstone fauna of southern Victoria Land, Antarctica (Fig. 5). This wide distribution for what might be considered tropical fishes calls for explanation. In particular, the Aztec fauna of southern Victoria Land is a diverse assemblage containing over 30 taxa. Various palaeoenvironmental indicators for the Aztec Siltstone strongly suggest a hot, seasonally wet and dry climate, with extended periods of aridity (McPherson 1980). A similar problem is presented by the Famennian vertebrate faunas of East Greenland and eastern Australia, which have at least six genera in common, but on the Famennian reconstruction of Scotese (1986) the former is about 20°N, and the latter about 55°S (cf. Heckel & Witzke 1979, where both lie between 15–30°S). Wider latitudinal distribution of Late Devonian fishes could be interpreted as a manifestation of global warming, but some authors (e.g. Copper 1977, 1986) have suggested global cooling in the Late Devonian. By comparison, of 157 families of freshwater fishes in the modern fauna (Berra 1981), only 8 (5%) have distributions more than 50° from the equator in both hemispheres.

To summarize for Late Devonian vertebrates, the southerly position of Australia–Antarctica and northerly position of Siberia are problematic, suggesting that the palaeomagnetic evidence for these positions is seriously in error (as stated a decade ago by Heckel & Witzke 1979). The close resemblance between the faunas of Euramerica and East Gondwana calls for explanation, particularly because of the striking differences in their Early Devonian vertebrate faunas. The maps (Figs 3 & 5) show no obvious palaeogeographic change to explain this difference in biogeographic pattern. The disjunct distribution of phyllolepids (in Euramerica and East Gondwana) is alleviated by their discovery in Turkey (Janvier 1983), which on the maps is midway between the two areas. This is the best evidence possible, from a single new discovery, for faunal continuity along the northern margin of Gondwana, a region poorly sampled for Devonian vertebrates. However it could also be argued that this new phyllolepid discovery represents evidence that Turkey formed part of Euramerica. This exemplifies a common dilemma in interpreting biogeographic data as tests of palaeogeographic reconstructions, discussed in more detail below.

To conclude this section I consider some recent palaeomagnetic data which indicate considerably altered Devonian positions for the various blocks. Miller & Kent (1988) show Laurentia extending to almost 45° south in the Early Devonian, then drifting north to give a mid-east coast palaeolatitude of about 15° in the Late Devonian. The Early Devonian position corresponds closely with that inferred from palaeoclimatic data by Heckel & Witzke (1979, fig. 3A), but these authors envisaged little change in position through the Middle and Late Devonian. Miller & Kent (1988) suggest an Early Devonian supercontinent configuration, with North America against northwestern South America, essentially as proposed by McKerrow & Ziegler (1972). Eastern Americas Realm brachiopods in Colombia and Venezuela (Boucot 1975) require Early Devonian proximity, but the distinctive vertebrate faunas discussed above indicate a persistent 'Theic Ocean' between Euramerica (Baltica, Laurentia, Avalonia) and Gondwana at that time, which can be accommodated by longitudinal separation.

For the Late Devonian, the new southern position for Euramerica means that the position of Gondwana based on the Canning Basin (Australia) pole of Hurley & Van der Voo (1987) no longer implies a wide oceanic separation of Gondwana and Euramerica. This Gondwana pole, and an alternative palaeomagnetic interpretation of Klootwijk & Giddings (1988), both retain Australia in low southern palaeolatitudes, as required by palaeoclimatic and faunal data, in contrast to the much more southerly position on the reconstruction of Scotese (1986). Van der Voo (1988, p. 321) commented that because of cosmopolitan faunas there were no constraints on the amount of latitudinal separation between North Africa and Euramerica for the Late Devonian, but Young (1987b, p. 292) noted close faunal similarities between the Famennian fishes of Morocco and eastern North America, which is inconsistent with the more complex model proposed by Van der Voo (1988, fig. 8), involving collision of Laurentia and Gondwana (including Avalonia, Armorica) near the Siluro–Devonian boundary, followed by separation of Gondwana from Laurentia (+Avalonia and Armorica) during the Devonian. The good faunal evidence from freshwater fishes of initial terrestrial connections occurring at or near the Frasnian–Famennian boundary (Young 1981, 1984, 1987b), together with a body of other evidence from various invertebrate groups (summarized in Young 1987b), and palaeoclimatic and oceanic circulation interpretations (Copper 1986) suggesting a Late Devonian ocean closure, are in conflict with palaeomagnetic data of an opening ocean (Van der Voo 1988, p. 322). One could postulate westward longitudinal motion of Gondwana during the Devonian (Hargraves *et al.* 1987) which resulted in fortuitous collision with the trailing edge of a northwardly drifting Euramerica, to permit exchange of continental faunas and floras, but this contorted explanation leaves unexplained the cause of the Acadian orogeny, the increasing faunal affinities between North America and Africa through the Early–Middle Devonian, the occurrence of the same genera of fishes in the Famennian of Ohio and Morocco, etc. My judgement therefore is that the palaeomagnetic data are still seriously in error.

Biogeographic tests of Palaeogeographic hypotheses

The general integration of biological and non-biological data sets as they apply to Palaeozoic palaeogeography remains an unresolved problem. Since the advent of plate tectonics, qualitative biological (i.e. biogeographic) data have been viewed by some as incapable of providing decisive evidence of past continental positions. Jardine & McKenzie (1972) stated that 'it is ... no longer profitable for biologists to speculate about the past arrangement of land masses'. In contrast, geophysical (especially palaeomagnetic) data, which provide quantitative rather than qualitative information, have been regarded as essential in providing a basic palaeogeographic framework for biogeographic interpretation. Scotese (1984) expressed the view that 'paleomagnetism is the key to our eventual understanding of Paleozoic plate interactions'. Attempts to make these data sets more comparable have involved quantification of biogeographic data, mainly through use of various coefficients measuring faunal similarity between areas, but no real integration has ensued. Van der Voo (1988) has recently commented on the lack of cross-referencing between the three major subdisciplines (palaeomagnetism, palaeoclimatology, biogeography) which contribute to Palaeozoic palaeogeographic reconstructions.

For the modern biota the main concern in biogeography has been to explain distributions of animals and plants in terms of geological theories, but in recent years it has been argued that modern distributions, like modern geography, can be completely known, and therefore analysed as a biological pattern in their own right, without reference to geography or to geological hypotheses. Overlooked is the fact that modern geography is the result of historical geological processes, just as biogeographic data are the result of an historical process (i.e. evolution in space). Any interpretation of fossil distributions must take note of geological ideas, concerning for example the past distribution of continents and oceans, because palaeogeography cannot be regarded as completely known. Palaeobiogeographic data have thus been used to assess geological hypotheses, and better integration of biological and geological data is clearly desirable.

It has become almost a convention that a single geological

hypothesis (or in the present case one set of maps) is provided as a framework to be tested by biogeographic data. From the biological viewpoint such hypotheses are generally too rudimentary to explain the very detailed historical processes indicated by biogeographic data. But there are also many different and opposing ideas in geology, and if presented as detailed competing palaeogeographic hypotheses, then the biogeographic data could be used effectively to test and choose between them.

It is my view that the highly complex and well organised historical set of biological data has not been fully exploited in palaeogeographic reconstruction, and the same may apply to a whole range of qualitative geological data. A possible reason is confusion about the distinction between experimental and historical investigations (e.g. Van der Voo 1988). In the experimental sciences a hypothesis may be subjected to a single crucial test, but historical investigations have no single data set which provides unique access to the truth. Van der Voo (1988, p. 312) has claimed two advantages of palaeomagnetic interpretations over techniques of palaeoclimatology and biogeography: they are more quantitative, and results can be repeatedly tested by carefully designed experiments, leading to unambiguous acceptance or rejection of the data. But such tests are at the intra-discipline experimental level, and do not differ in kind from repeated tests (perhaps involving new field investigations, stratigraphic or age-dating studies, more extensive faunal or geochemical analyses, etc.) that a particular evaporitic sequence in a particular area indicates a latitudinal position of about 10–30° at a particular time, or that particular fossil faunas were isolated by barriers at a particular time.

To derive an historical conclusion concerning palaeogeography from such evidence all three methods (palaeomagnetism, palaeoclimatology, biogeography) employ the same principle: the criterion of concordance with a general pattern. For palaeomagnetism the underlying assumption of a geocentric co-axial dipole is justified by general agreement in pattern between palaeomagnetic and palaeoclimatic data, but there is no way a single palaeomagnetic datum which diverges widely from a general pattern can be unambiguously accepted. If there is a difference between the three types of evidence, it is perhaps that at the experimental level palaeoclimatic or biogeographic data, because they are qualitative, initially involve a much lower level of ambiguity than palaeomagnetic data. A good example is the Msissi Norite, repeatedly cited in the literature from 1974–1986 as the most reliable Devonian palaeomagnetic pole for Gondwana, but now rejected, whereas the palaeoclimatic indicators with respect to the Devonian position of North America carry the same message now as they did when Heckel & Witzke (1979) predicted a discrepancy of up to 30° in the palaeomagnetic data. In biogeography, 'objections to palaeogeographic reconstructions may always be overcome by *ad hoc* hypotheses about barriers or seaways' (Van der Voo 1988), but *ad hoc* hypotheses are also available to explain anomalous palaeoclimatic or palaeomagnetic data, in the latter case the much over-used conclusion that a divergent palaeomagnetic result indicates palaeogeographic separation, regardless of strong contrary evidence from other fields.

In historical investigations the most powerful principle at our disposal is similar to Whewell's (1840) idea of 'consilience'; we accept with greater confidence a hypothesis formulated to explain patterns inferred from one data set, if it also accounts for patterns inferred from a completely unrelated set of data. The hypothesis has been corroborated, perhaps 'tested', but neither data set on its own gives a unique, or even a small number of solutions to the problem at hand. To use palaeomagnetic data as a primary constraint in constructing palaeogeographic maps, subject to conclusive tests from other data, is hardly satisfactory when the palaeomagnetic data themselves may be subject to very different interpretations (e.g. Livermore *et al.* 1985, fig. 2). To suggest that such maps should remain immune from criticism until all possible longitudinal permutations of continental positions have been considered (Livermore *et al.* 1985) is an *ad hoc* protection from falsification. As with palaeoclimatic and palaeomagnetic evidence, interpretation of Palaeozoic biogeographic data depends on the inferred palaeogeographic setting, and the biogeographic evidence must be assessed in the context of other geological or geophysical evidence. In the example of Turkey mentioned in the preceding section, the biogeographic evidence must be weighed against geological or geophysical evidence for its displacement along the northern margin of Gondwana, but this evidence is not provided on the reconstructions. There is a similar problem with the osteostracans from the Early Devonian of the Tuva region discussed above.

Problems of map representation

It is therefore appropriate to consider the use and deficiencies of map representation in the context of testing palaeogeographic hypotheses. Maps display the spatial relationships of various entities, and geological, topographic, or vegetation maps showing distribution of outcrop, vegetation types, etc., are all familiar. The features displayed change with time, but attempts to incorporate a historical element on a map tend to confuse rather than edify; an out-of-date road map may have some historical interest, but for the motorist going from A to B it is a confusing document; to be useful the road map must show the current road network.

With palaeogeographic maps the difficulty of representing history is partly overcome by producing maps in series, to summarize palaeogeographic change through time. But this presents the problem of whether each map should represent a single time plane, or an average over a time interval; the former may suffer from lack of information in certain areas, and the latter may juxtapose features that were chronologically separate.

The data of complex hypotheses may be represented in various ways, which fall into two general categories. First are those that synthesize, by bringing together all available data in a coherent statement emphasizing or assuming the reliability and consistency of the evidence, the success of the investigation, and the strength of the hypothesis. This is of course a legitimate and useful approach, and currently seems the predominant one in historical earth science. However the same data may also be represented in an analytical way, which emphasizes instead the conflicts and inconsistencies in the evidence, the inadequacies of the data, and the vulnerability of the hypothesis. Amongst the various means of data representation relevant to palaeogeography I judge standard palaeogeographic map reconstructions, and accretionary history diagrams of the type used by Williams & Hatcher (1983, fig. 2), to fall within the first category, and apparent polar wander path (APWP) representation of palaeomagnetic data to fall within the second. Emulating palaeomagnetism by APWP representation of palaeoclimatic data (Scotese 1986) permits the testing of these two data sets against each other. Biogeographic data can also provide evidence for palaeolatitude, through recognition of cold and warm assemblages, diversity gradients, etc., and this type of evidence can potentially be analysed using APWP in the same way. However the main strength of biogeographic data is in providing a different type of evidence, which does not translate into a simple indicator of palaeolatitude, but concerns instead connections or barriers between regions, as reflected in reproductive isolating mechanisms producing endemic taxa in different areas (faunal provinces). Although sometimes thought to be a special attribute of biogeographic data, this is not the case; the past distributions of land and sea, palaeocirculation patterns, and some of the major criteria for determining accretionary history in terrane analysis, such as overlap assemblages and sedimentological linkage, are affected by or primarily concerned with connections between regions. Moreover, wide palaeolatitudinal separation indicated by palaeomagnetic data in itself may preclude connections between areas.

Data concerning only palaeolatitude can thus be integrated

and analysed using APWP representation, but integration of these with biogeographic and other evidence concerned with connections or barriers between areas has traditionally been achieved by plotting all the data on a palaeogeographic reconstruction which as far as possible does not violate any of the available data. The problem is, as noted above, that this is a synthetic rather than analytical representation, which largely renders the data immune from test.

Maps are important and probably essential in synthesizing pictorially the evidence of palaeogeography of an area, but as a means of portraying a complex historical hypothesis they are deficient in their inadequate representation of historical change, in synthesizing data which may not be time equivalent, and in the difficulty of displaying all the empirical data on which the reconstruction is based.

Cladistic techniques in palaeogeographic analysis

An alternative which overcomes these difficulties is provided by cladistics. This method was applied to some problems of Palaeozoic palaeogeography by Young (1986), in an attempt to reduce the complexity of ideas in the literature by representing competing hypotheses on branching diagrams, or cladograms. These summarized fragmentation or fusion histories for certain regions for specific periods of the geological past. By comparison of branching patterns, conflicting viewpoints on hypothesized collision sequences were made explicit. The example used involved only four areas (Laurentia, Baltica, Armorica, Gondwana), but it was noted that the complexity of the problem would be greatly increased if microplates or suspect terranes were taken into account, or if all other hypothesized palaeocontinents were incorporated in a global cladogram.

The evidence on which the various hypotheses were based was discussed in a general way, but not incorporated on the cladograms. Yet it is in the organisation of data supporting competing hypotheses that cladistics can make a most important contribution to historical analysis in geology. The resolution of conflicting data has become increasingly difficult with the wider use of large computer-manipulated data sets. Cladistics is concerned with hierarchical analysis, and may be applied to any problem or data set which can be viewed as having hierarchical organisation. Cladistic analysis embodies the hypothesis that data relevant to the problem can be organised hierarchically (Brady 1983). Cladistic representation of evolutionary history (as a dichotomously branching hierarchical system) is one of many different approaches to historical biotic change, each of which has led to advances in different directions, or highlighted new and different classes of problems. In palaeogeography, an approach emphasizing the history of fusion or fragmentation of discrete regions, rather than their spatial positions (on a map), imparts an obvious hierarchical structure to hypotheses (Young 1986). Given appropriate geological definitions of fragmentation or fusion, all observations relevant to fragmentation or fusion history, whether geological, geophysical or biological, can be organised hierarchically for the areas under consideration.

For any area assumed to have a separate history, the potential number of geological observations, and the subset of observations relevant to fragmentation or fusion history, is infinite. A cladistic approach predicts that all data in this infinite subset can be organised hierarchically to conform to a particular cladogram, which represents the unique fusion or fragmentation history of the areas concerned. Specific observations which do not conform, and suggest a different cladogram, provide an empirical test of the hypothesis. Within cladistics elaborate techniques have been developed for assessing competing hypotheses using the criterion of parsimony, and for handling complexities of conflicting data which far exceed those currently exposed to criticism and refutation in the literature of palaeogeographic reconstruction (Fig. 6). Congruence amongst data sets organised hierarchically may be assessed using component analysis (Nelson & Platnick 1981), and in complex examples it is not obvious by casual inspection that different branching patterns are consistent with a single hypothesis (Fig. 6).

It has been suggested that such branching diagrams based on geological information are not true cladograms because they are not derived from explicit analysis of geographical or geological characters in the way that synapomorphies are analysed in systematics (Rosen 1988, p. 449). The special nature of synapomorphies (the shared derived attributes by which sister—group relationships of organisms are identified) is often erroneously attributed to high morphological complexity, but the only real criterion for validity is constancy of distribution within the hierarchical system of a synapomorphy scheme (Patterson 1982). This attribute applies in exactly the same way to geological or geophysical data supporting an area cladogram. This is best illustrated by analysing a hypothesis from the geological literature.

My example is the terrane model for the Appalachian foldbelt presented by Williams & Hatcher (1982, 1983). Three main orogenic phases are recognised in the Appalachians: an Early Palaeozoic Taconic, a Middle Palaeozoic Acadian, and a Late Palaeozoic Alleghenian orogeny. Williams & Hatcher's model involved many allochthonous terranes, some of which make up the Acadia landmass of published reconstructions.

This model is a complex palaeogeographic hypothesis which could profoundly affect any biogeographic interpretation of fossil occurrences not only in this region, but globally. In fact, there are many Devonian vertebrate localities in the Appalachians, but their Devonian palaeogeographic settings are very uncertain. For example Perroud et al. (1984, figs 2 & 3) showed the Piedmont Terrane adjacent to Gondwana from Ordovician to Devonian, and on palaeomagnetic evidence suggested that this terrane may not have docked until Permian time. Various authors (e.g. Scotese 1984; Van der Voo 1988) have suggested that the Avalon and Meguma Terranes may also have originated in Gondwana, and did not join Laurentia until the Middle Devonian.

Williams & Hatcher presented a diagrammatic summary of accretionary history according to their model (1983, fig. 2), using criteria for determining the time of accretion formulated by Coney and others in their work in the western cordillera of North America. These are: (1) overlap assemblages (the age of strata covering a terrane boundary); (2) sedimentologic linkage (the depositional age of detritus shed from one terrane onto another); and (3) similar postdocking intrusive (a), metamorphic (b), structural (c), palaeontological (d), or palaeomagnetic (e) histories.

Within these classes of evidence are specific empirical observations regarding the accretion of each terrane. The following are extracted from the summary of Williams & Hatcher (1983), and include what they regard as the defining features of their

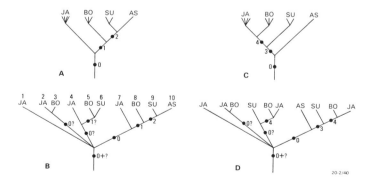

Fig. 6. Four cladograms for 10 hypothetical species distributed in four areas (Java, Borneo, Sumatra, Asia) with an analysis of their components. The congruent pairs (A, B) and (C, D) imply two different patterns of area relationship (redrawn from Nelson & Platnick 1981, figs 6.16, 6.17).

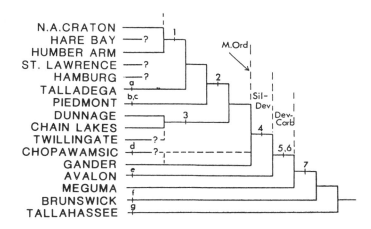

Fig. 7. Cladogram summarising accretion history for the Appalachian foldbelt, according to the terrane model of Williams & Hatcher (1983). Features which define terranes (a–g), and empirical evidence for docking history (1–7) are listed and discussed in the text.

terranes (letters on the geological cladogram of Fig. 7), and the observations on which their collision sequence is based (numbers in Fig. 7). Although the sense of the branching diagram in Fig. 7 is reversed (i.e. converging rather than diverging through time as is the case with a biological cladogram), the empirical observations (1–7) are entirely comparable to shared derived characters which define the branching points of a biological cladogram. These may be listed as follows (page numbers are from Williams & Hatcher 1983; numbers and letters in square brackets refer to the three evidence categories listed above).

(1) (p. 35). Middle Ordovician Long Point Group overlaps Humber Arm terrane and miogeocline. [1]
(2) (p. 39). Deformed Piedmont rocks overlapped by Silurian and Devonian of Gaspe Synclinorium. [1]
(3) (p. 40). Olistostrome of Dunnage terrane containing rocks (St Daniel Formation) resembling and probably derived from Chain Lakes terrane, and overlain by Middle Ordovician shales of the Beuceville Formation. [1,2]
(4) Middle Ordovician–Silurian Aroostook-Metapedia Group overlaps North American craton, Dunnage, and Gander terranes from Quebec to New Brunswick and Maine. Also Middle Ordovician Davidsville Group in Newfoundland overlaps Dunnage and Gander. [1]
(5) (p. 43). Devonian plutons cutting Avalon–Gander boundary in Newfoundland, and similar Devonian plutonic history in New Brunswick and Maine. [3a]
(6) Silurian and Devonian conglomerates containing metamorphic and plutonic rocks derived from the Gander Terrane. [2]
(7) (p. 44). 'Similar' Carboniferous overlap assemblage [1]

Similarly the defining characters of each terrane are represented on the cladogram as letters.

(a) (p. 37). Talladega terrane defined by 'stratigraphic contrasts with nearby miogeocline'.
(b,c) (p. 38). Piedmont terrane suspect because of (b) contrasting stratigraphy, and (c) presence of ultramafics.
(d) (p. 41). Chopawamsic. Mid–Late Ordovician brachiopod fauna which has 'diverse elements compared to faunas of the North American miogeocline'.
(e) (p. 42). Avalon. Distinctive Atlantic realm Cambrian trilobite faunas and 'similar rocks of same age in Wales, Brittany, Spain, and Morocco'.
(f) Lower frequency, much more symmetrical long wavelength magnetic and gravity anomalies of Brunswick Terrane.
(g) High amplitude, high frequency magnetic anomalies. Fossils in Early Palaeozoic rocks (drill-holes) which are 'akin to African forms'.

The problems of data organisation displayed by constructing the cladogram of Fig. 7 are strikingly similar to those encountered in phylogenetic analysis. As in that case, the branches for which there is no evidence are of interest (I have represented as best I can the Williams & Hatcher hypothesis even if evidence is lacking). The first four terranes are Taconic allochthons sitting above the miogeocline, all thought to have been emplaced at about the same time (Middle Ordovician). However specific evidence is only presented for the Humber Arm terrane (1), an overlap assemblage of the Long Point Group with the miogeocline. The Talladega Terrane is defined as a suspect terrane because of stratigraphic contrasts with the nearby miogeocline, but early linkage is said to be indicated by correlations between older metaclastics, and marbles and quartzites, with basal clastics (Ocoee Series) and less deformed equivalents of the miogeocline. As expressed here the same evidence has been used to indicate both differences and similarities. This is contradictory, and it is not possible on the evidence presented to formulate a defining character for this branching point (the Ordovician linkage discussed here is omitted from Fig. 7).

An assumption of the Williams & Hatcher model is that accretionary history progressed from inside outward, with more outboard terranes accreted after inboard. The Dunnage and associated terranes are inboard of the Gander Terrane, but there is no evidence presented which excludes the Dunnage and Gander coming together first, to form a composite terrane. Character 4 (Fig. 7) is an overlap assemblage of both terranes on the North American craton, so is consistent with either possibility. On the evidence presented rather than the authors' intentions this point on the cladogram should have been represented as a trichotomy. The Piedmont terrane was mentioned above as a possible fragment of Gondwana in the Devonian, which on palaeomagnetic evidence did not arrive in its present position until the Permian. On the cladogram however the Piedmont Terrane is shown as accreting in the Ordovician. Williams & Hatcher (1983) discuss this point and refer to 'unassailable' evidence of Middle Ordovician–Silurian overlap between the Dunnage and Gander terranes and the miogeocline. These terranes are outboard of the Piedmont, thus implying that the Piedmont terrane 'represents rocks at the deformed precarious edge of eastern North America, rather than a far-travelled alien terrane'. Again the quantitative data of palaeomagnetism are in conflict with qualitative geologic evidence. Nevertheless, even if consistent amongst themselves in relation to a proposed APWP for the North American craton, the immediate test of such palaeomagnetic data lies not with additional palaeomagnetic investigation, but with the field interpretation of the conflicting geological evidence. These data represent competing hypotheses about the history of this terrane, and the conflict can only be resolved by referring back to the empirical data on which each was based. The poor formulation of palaeogeographic hypotheses in the current literature means that many such inconsistencies are no doubt hidden among a confusing mass of data.

The cladogram of Fig. 7 summarizes one particular model of terrane accretion for the Appalachians, by presenting clearly and succinctly the crucial evidence relating to the historical sequence of accretionary events, whether it be geological, geophysical, or biological. No doubt many other geological data are relevant to this history, and could be added to the cladogram. They may be congruent with this pattern, or incongruent, in which case they would imply a competing hypothesis and a different historical sequence. In this way the hypothesis can be tested by new data, or by reanalysis of existing data to resolve apparent conflicts, using standard techniques of cladistic analysis.

I conclude by summarizing three major attributes possessed by an area or terrane cladogram such as represented in Fig. 7, but lacking in traditional map presentations of a complex historical hypothesis:

(1) A cladogram can represent a historical sequence on a single diagram.

(2) It emphasizes the analysis of data, rather than the synthesis of a very large dataset in an uncritical way, i.e. the approach is eclectic and analytical, not synthetic.

(3) The empirical data on which the hypothesis is based are clearly displayed, so the crucial evidence supporting the hypothesis may be restudied to resolve conflicts in the data, and the hypothesis is thereby highly exposed to refutation.

I thank the symposium organisers and sponsors for financial support which enabled me to attend the symposium. C. Burrett and J. Long provided an unpublished manuscript, and P. Forey and A. Blieck gave some unpublished information. I benefited from discussions at the symposium, particularly with A. Boucot, C., Burrett, D. Kent, M. Streel, R. Van der Voo and B. Witzke. Reviewers' comments, and W. S. McKerrow's editorial advice, assisted in preparation of the final manuscript. Published with the permission of the Director, Bureau of Mineral Resources, Canberra.

References

AFANASSIEVA, O. & JANVIER, P. 1985. *Tannuaspis, Tuvaspis*, and *Ilemoraspis*, endemic osteostracan genera from the Silurian and Devonian of Tuva and Khakassia (USSR). *Geobios*, **18**, 493–506.

BERRA, T. M. 1981. *An Atlas of Distribution of the Freshwater Fish Families of the World*. University of Nebraska Press.

BLIECK, A. 1984. *Les heterostraces pteraspidiformes, agnathes du Silurien–Devonien du continent nord-atlantique et des blocs avoisinants: revision systematique, phylogenie, biostratigraphie*. Cahiers de Paleontologie, Section Vertebres. Paris: Editions du Centre National de la Recherche Scientifique.

——, GOLSHANI, F., GOUJET, D., HAMDI, A., JANVIER, P., MARK-KURIK, E. & MARTIN, M. 1980. A new vertebrate locality in the Eifelian of the Khush-Yeilagh Formation, eastern Alborz, Iran. *Palaeovertebrata*, **9-V**, 133–154.

BOUCOT, A. J. 1975. *Evolution and extinction rate controls*. Elsevier, Amsterdam.

—— 1984. Old World Realm (Rhenish–Bohemian region), shallow-water, Early Devonian brachiopods from the Jauf Formation of Saudi Arabia. *Journal of Paleontology*, **58**, 1196–1202.

——, JOHNSON, J. G. & TALENT, J. A. 1969. Early Devonian brachiopod zoogeography. *Geological Society of America Special Paper*, **199**, 1–113.

—— & GRAY, J. 1979. *Historical Biogeography, Plate Tectonics, and the Changing Environment*. Oregon State University Press.

BRADY, R. H. 1983. Parsimony, hierarchy, and biological implications. *In*: PLATNICK, N. I. & FUNK, V. A. (eds) *Advances in Cladistics, Volume 2*. Proceedings of the 2nd meeting of the Willi Hennig Society. Columbia University Press, New York, 49–60.

BURRETT, C., LONG, J. & STAIT, B. 1990. Early–Middle Palaeozoic biogeography of Asian terranes derived from Gondwana. *In*: MCKERROW, W. S. & SCOTESE, C. R. (eds) *Palaeozoic Palaeogeography and Biogeography*. Geological Society, London, Memoir, **12**, 163–174.

CAMPBELL, K. S. W. & BARWICK, R. E. 1987. Palaeozoic lungfishes — a review. *Journal of Morphology, Supplement*, 1 [for 1986], 93–131.

CARROLL, R. L. 1988. *Vertebrate Paleontology and Evolution*. Freeman & Co., New York.

CHALONER, W. G., FOREY, P. L., GARDINER, B. G., HILL, A. J. & YOUNG, V. T. 1980. Devonian fish and plants from the Bokkeveld Series of South Africa. *Annals of the South African Museum*, **81**, 127–157.

COCKS, L. R. M. & FORTEY, R. A. 1982. Faunal evidence for oceanic separations in the Palaeozoic of Britain. *Journal of the Geological Society of London*, **139**, 465–478.

COPPER, P. 1977. Paleolatitudes in the Devonian of Brazil and the Frasnian–Famennian mass extinction. *Palaeogeography, Palaeoclimatology, Palaeoecology*, **21**, 165–207.

—— 1986. Frasnian–Famennian mass extinction and cold-water oceans. *Geology*, **14**, 835–839.

DENISON, R. H. 1955. Early Devonian vertebrates from the Knoydart Formation of Nova Scotia. *Fieldiana, Zoology*, **37**, 449–464.

—— 1978. *Placodermi Handbook of Paleoichthyology, volume 2* Schultze, H.-P. (ed.) Gustav Fischer Verlag, Stuttgart, New York.

—— 1979. *Acanthodii. Handbook of Paleoichthyology, Volume 5*. SCHULTZE, H.-P. (ed.) Gustav Fischer Verlag, Stuttgart, New York.

DINELEY, D. L. 1967. The Lower Devonian Knoydart faunas. *Journal of the Linnean Society (Zoology)*, **47**, 15–29.

FRIMAN, L. & JANVIER, P. 1986. The Osteostraci (vertebrate, Agnatha) from the Lower Devonian of the Rhenish Slate Mountains, with special reference to their anatomy and phylogenetic position. *Neues Jahrbuch fur Geologie und Palaontologie, Abhandlungen*, **173**, 99–116.

GAGNIER, P.-Y., TURNER, S., FRIMAN, L., SUAREZ-RIGLOS, M. & JANVIER, P. 1988. The Devonian vertebrate and mollusc fauna from Seripona (Dept. of Chuquisaca, Bolivia). *Neues Jahrbuch fur Geologie und Palaontologie, Abhandlungen*, **176**, 269–297.

GOUJET, D. 1984. *Les poissons placoderms du Spitsberg. Arthrodires Dolichothoraci de la Formation de Wood Bay (Devonien Inferieur)* Cahiers de Paleontologie, Section Vertebres. Paris: Editions du Centre National de la Recherche Scientifique.

——, JANVIER, P. & SUAREZ-RIGLOS, M. 1984. Devonian vertebrates from South America, *Nature*, 312, 311.

——, JANVIER, P. & SAUREZ-RIGLOS, M. 1985. Un nouveau rhenanide (Vertebrata, Placodermi) de la Formation de Belen (Devonien moyen), Bolivie. *Annales de Paleontologie*, **71**, 35–53.

GRAY, J. 1988. Evolution of the freshwater ecosystem: the fossil record. *Palaeogeography, Palaeoclimatology, Palaeoecology*, **62**, 1–214.

HALSTEAD, L. B. 1985. The vertebrate invasion of fresh water. *Philosophical Transactions of the Royal Society of London*, **B309**, 243–258.

HARGRAVES, R. B., DAWSON, E. M. & VAN HOUTEN, F. B. 1987. Paleomagnetism and age of mid-Palaeozoic ring complexes in Niger, West Africa, and tectonic implications. *Royal Astronomical Society Geophysical Journal*, **90**, 704–729.

HECKEL, P. H. & WITZKE, B. J. 1979. Devonian world palaeogeography determined from distribution of carbonates and related lithic palaeoclimatic indicators. *Special Paper of the Palaeontological Association*, **23**, 99–123.

HORN, M. H. 1972. The amount of space available for marine and freshwater fishes. *Fisheries Bulletin*, **70**, 1295–1297.

HOU, H. F. & WANG, S. T. 1985. Devonian palaeogeography of China. *Acta Palaeontologica Sinica*, **24**, 186–197.

HSU, K. J., SUN, S., LI, J., CHEN, H., PEN, H. & SENGOR, A. M. C. 1988. Mesozoic overthrust tectonics in south China. *Geology*, **16**, 418–421.

HURLEY, N. F. & VAN DER VOO, R. 1987. Paleomagnetism of Upper Devonian reefal limestones, Canning Basin, Western Australia. *Geological Society of America Bulletin*, **98**, 138–146.

JANVIER, P. 1983. Los vertebres devoniens de la Nappe Superieure d'Antalya (Taurus Lycien occidental, Turquie). *Geologie Mediterraneenne*, **10**, 1–13.

—— 1985a. *Les Cephalaspides du Spitsberg*. Cahiers de Paleontologie, Section Vertebres. Paris: Editions du Centre National de la Recherche Scientifique.

—— 1985b. Environmental framework of the diversification of the Osteostraci during the Silurian and Devonian. *Philosophical Transactions of the Royal Society of London*, **B309**, 259–272.

—— & SUAREZ-RIGLOS, M. 1986. The Silurian and Devonian vertebrates of Bolivia. *Bulletin de l'Institut Francais d'Etudes Andines*, **15**, 73–114.

JARDINE, N. & MCKENZIE, D. 1972. Continental drift and the dispersal and evolution of organisms. *Nature*, **235**, 20–24.

KLOOTWIJK, C. & GIDDINGS, J. 1988. An alternative APWP for the Middle to Late Palaeozoic of Australia — implications for terrane movements in the Tasman Fold Belt. *Ninth Australian Geological Convention, Abstracts*, **21**, 219–220.

LI, Y., MCWILLIAMS, M., ZHU, H., TAN, C. & HE, Z. 1985. Palaeozoic paleomagnetic results from the Hexizoulang Terrane, China. *Transactions of the American Geophysical Union EOS*, **66**, 864.

LIVERMORE, R. A., SMITH, A. G., & BRIDEN, J. C. 1985. Palaeomagnetic constraints on the distribution of continents in the Late Silurian and Early Devonian. *Philosophical Transactions of the Royal Society of London*, **B309**, 29–56.

LONG, J. A. & TURNER, S. 1984. A checklist and bibliography of Australian fossil fishes. *In*: ARCHER, M. & CLAYTON, G. (eds) *Vertebrate Zoo-

geography and Evolution in Australasia. Hesperian Press, Perth, 235–254.

MALINOVSKAYA, S. P. 1973. *Stegolepis* (antiarch, Placodermi), a new Middle Devonian genus from central Kazakhstan. *Paleontological Journal*, **7**, 189–199.

—— 1977. The systematic position of the antiarchs from central Kazakhstan. *In*: MENNER, V. V. (ed.) *Essays on phylogeny and systematics of fossil agnathans and fishes*: 29–35. Nauka, Moscow [in Russian].

MARK-KURIK, E. 1973. *Kimaspis*, a new palaeacanthaspid from the Early Devonian of Central Asia. *Eesti NSV Teaduste Akadeemia Toimetised*, **22(4)**, 322–30.

—— 1974. Discovery of new Devonian fish localities in the Soviet Arctic. *Eesti NSV Teaduste Akadeemia Toimetised*, **23**, 332–335.

—— & NOVITSKAYA, L. 1977. The Early Devonian fish fauna on Novaya Zemlya. *Eesti NSV Teaduste Akadeemia Toimetised*, **26**, 143–149.

McGHEE, G. R. 1982. The Frasnian–Famennian extinction event: a preliminary analysis of Appalachian marine ecosystems. *Geological Society of America, Special Paper*, **190**, 491–500.

McKERROW, W. S. & ZIEGLER, A. M. 1972. Palaeozoic oceans. *Nature*, **240**, 92–94.

McPHERSON, J. G. 1980. Genesis of variegated redbeds in the fluvial Aztec Siltstone (Late Devonian), southern Victoria Land, Antarctica. *Sedimentary Geology*, **27**, 119–142.

MILLER, J. D. & KENT, D. V. 1988. Paleomagnetism of the Silurian–Devonian Andreas redbeds: evidence for an Early Devonian supercontinent? *Geology*, **16**, 195–198.

NELSON, G. & PLATNICK, N. 1981. *Systematics and biogeography: cladistics and vicariance*. Columbia University Press, New York.

NOVITSKAYA, L. I. 1971. *Les amphiaspides (Heterostraci) du Devonien de la Siberie*. Cahiers de Paleontologie, Section Vertebres. Paris: Editions du Centre National de la Recherche Scientifique.

OLIVER, W. A. 1977. Biogeography of Late Silurian and Devonian rugose corals. *Palaeogeography, Palaeoclimatology, Palaeoecology*, **22**, 85–135.

PAN, J. 1981. Devonian antiarch biostratigraphy of China. *Geological Magazine*, **118**, 69–75.

——, HUO, F., CAO, J., GU, Q., LIU, S., WANG, J., GAO, L. & LIU, C. 1987. *Continental Devonian system of Ningxia and its Biotas*. Geological Publishing House, Beijing.

—— & DINELEY, D. L. 1988. A review of early (Silurian and Devonian) vertebrate biogeography and biostratigraphy of China. *Proceedings of the Royal Society of London*, **B235**, 29–61.

PATTERSON, C. 1982. Morphological characters and homology. *In*: JOYSEY, K. A. & FRIDAY, A. E. (eds) *Problems of phylogenetic reconstruction*. Systematics Association Special Volume **21**, 21–74.

PERROUD, VAN DER VOO, R. & BONHOMMET, N. 1984. Palaeozoic evolution of the Armorica plate on the basis of paleomagnetic data. *Geology*, **12**, 579–584.

ROSEN, B. R. 1988. From fossils to earth history: applied historical biogeography. *In*: MYERS, A. A. & GILLER, P. S. (eds) *Analytical Biogeography*. Chapman and Hall, London, 437–481.

SCOTESE, C. R. 1984. An introduction to this volume: Palaeozoic paleomagnetism and the assembly of Pangea. *In*: VAN DER VOO, R., SCOTESE, C. & BONHOMMET, N. (eds) *Plate Reconstruction from Palaeozoic Paleomagnetism*. American Geophysical Union, Geodynamics series, **12**, 1–10.

—— 1986. Phanerozoic reconstructions: a new look at the assembly of Asia. *University of Texas Institute for Geophysics Technical Report*, **66**, 1–54.

TALENT, J. A., GRATSIANOVA, R. T., SHISHKINA, G. R. & YOLKIN, E. A. 1987. Devonian faunas in relation to crustal blocks: Kazakhstan, Mongolia, northern China. *Courier Forschunginstitut Senckenberg*, **92**, 225–233.

—— & YOLKIN, E. A. 1987. Transgression–regression patterns for the Devonian of Australia and southern West Siberia. *Courier Forschunginstitut Senckenberg*, **92**, 235–249.

THANH, T. D. & JANVIER, P. 1987. Les vertebres Devoniens du Vietnam. *Annales de Paleontologie*, **73**, 165–194.

VAN DER VOO, R. 1988. Palaeozoic paleogeography of North America, Gondwana, and displaced terranes: comparisons of paleomagnetism with paleoclimatology and biogeographical patterns. *Geological Society of America Bulletin*, **100**, 311–324.

WANG, J. Q. 1984. Geological and paleogeographic distribution of Devonian fishes in China. *Vertebrata Palasiatica*, **22**, 219–229 [in Chinese with English summary].

WHEWELL, W. 1840. *The Philosophy of the Inductive Sciences*. Parker, London.

WILLIAMS, H. & HATCHER, R. D. 1982. Suspect terranes and accretionary history of the Appalachian orogen. *Geology*, **10**, 530–536.

—— & —— 1983. Appalachian suspect terranes. *Geological Society of America, Memoir*, **158**, 33–53.

YOUNG, G. C. 1981. Biogeography of Devonian vertebrates. *Alcheringa*, **5**, 225–243.

—— 1984. Comments on the phylogeny and biogeography of antiarchs (Devonian placoderm fishes), and the use of fossils in biogeography. *Proceedings of the Linnean Society of New South Wales*, **107**, 443–473.

—— 1986. Cladistic methods in Palaeozoic continental reconstruction. *Journal of Geology*, **94**, 523–537.

—— 1987a. Devonian vertebrates of Gondwana. *In*: McKENZIE, G. D. (ed.) *Gondwana Six. Stratigraphy, Sedimentology, and Paleontology*. American Geophysical Union, Geophysical Monograph, **41**, 41–50.

—— 1987b. Devonian palaeontological data and the Armorica problem. *Palaeogeography, Palaeoclimatology, Palaeoecology*, **60**, 283–304.

—— 1988. Antiarchs (placoderm fishes) from the Devonian Aztec Siltstone, southern Victoria Land, Antarctica. *Palaeontographica*, **A202**, 1–125.

—— 1989. The Aztec fish fauna of southern Victoria Land – evolutionary and biogeographic significance. *In*: CRAME, J. A. (ed.) *Origin and Evolution of the Antarctic Biota*. Geological Society of London Special Publication, **47**, 43–62.

—— in press. The first armoured agnathan vertebrates from the Devonian of Australia. *In*: CHANG, M. M. (ed.) *Problems in Early Vertebrate Evolution*. Proceedings of 1987 Beijing Symposium.

Biogeography of the Devonian stromatoporoids

C. W. STOCK

Department of Geology, University of Alabama, Tuscaloosa, Alabama 35487−0338, USA

Abstract: Stromatoporoids were a common component of shallow carbonate environments of North America, Eurasia, and Australia during the Devonian. They were least abundant during the Early Devonian. After that time abundance increased, and remained high steadily through the Frasnian. At the Frasnian−Famennian boundary the number of stromatoporoids was greatly diminished, but they did not become extinct until the end of the Devonian (at the end of the Strunian). The geographic extent of stromatoporoids expanded and contracted concurrently with increases and decreases in total population size. Provincialism at the genus level prevailed during the Early Devonian, with stromatoporoids inhabiting the Old World and Eastern Americas Realms; none are known from the Eastern Americas during the Siegenian. For the remainder of the Devonian stromatoporoids were cosmopolitan at the genus level. The abundance of stromatoporoids varied directly with eustatic sea level during the Devonian. Variations in depositional conditions apparently controlled the local distribution of genera.

Stromatoporoids, an extinct group of sessile benthic organisms, secreted a calcium carbonate exoskeleton called a coenosteum. Many palaeontologists believe they are a kind of sponge, although some workers believe that Palaeozoic forms represent cnidarians or cyanobacterial stromatolites. The stromatoporoids appeared in the Early Ordovician, and generally increased in number during the remainder of the Ordovician, the Silurian, and the Devonian. They experienced a near-extinction at the close of the Frasnian, and disappeared entirely at, or near the end of, the Devonian. Devonian stromatoporoids occur in large numbers in carbonate rocks of North America, Eurasia, and Australia; however, they are not known from the Antarctic, South American, Indian, and African parts of Gondwanaland, with the exception of Morocco.

Studies of stromatoporoid depositional environments indicate that they lived in shallow, warm, clear, usually well-agitated, normal marine environments. By the Devonian they inhabited level bottoms, banks, and reefs (e.g. Dolphin & Klovan 1970; Nestor 1977); the twig-like *Amphipora* was able to live in restricted lagoons of above-normal salinity (Fischbuch 1962). As such, stromatoporoids are recognized as major indicators of warm temperatures in Devonian seas (e.g. Heckel & Witzke 1979).

As with any palaeontological study that is world-wide in scope, the stromatoporoid data base is not uniform. Locality and stratigraphic information ranges from the general to the specific. The exact collection site may be pinpointed, or only a region, such as 'southern Urals', may be given. The chronostratigraphic occurrence may be given simply as 'Upper Devonian', or it may be as precise as 'lower Frasnian'. The lithostratigraphic unit of occurrence may be given, but the age assigned to that unit can change as the biostratigraphy is refined. Nearly three-quarters of the 97 stromatoporoid genera used in the Devonian have been named since 1950; 44 of those genera were proposed by Soviet workers. Taxonomy varies among workers, and regional biases are apparent. There is disagreement over the validity of genera; for example Stearn (1980) rejected 17 Devonian genera proposed between 1955 and 1979. The identifications of described stromatoporoids were checked where possible. Unfortunately in some cases the written descriptions and illustrations proved insufficient for total verification of genus assignments, but these identifications were accepted where not demonstrably incorrect. The summary work of Flügel & Flügel-Kahler (1968) proved particularly useful in evaluating identifications in older publications.

To bring some order and uniformity to this study, certain criteria were established: (1) the taxonomic levels of genus and order, based on the classification of Stearn (1980), were accepted as most reliable (exceptions to the total acceptance of Stearn's classification were the removal of *Stachyodes* and *Clathrocoilona* from Order Stromatoporellida, and their placement in Order Stromatoporida); (2) only faunas that had been described in some detail, including photomicrographs, were employed, plus a few listings by established stromatoporoid workers; (3) only faunas with chronostratigraphic information down to stage (e.g. Givetian) were employed.

Publications including the systematics of stromatoporoid faunas rarely contain palaeoecological information, and publications dealing with stromatoporoid palaeoecology rarely deal with the identification of the specimens. Even where the co-occurrence of taxa is reported, it is unlikely that their presence in the same bed will be noted. Consequently the use of standard groupings of taxa, such as the community and the assemblage, is impossible in most cases. A new term that can be applied to nearly all published information is needed; the geographic and chonostratigraphic entity employed in the present paper is the 'association'. An association is an artificial grouping that includes all the stromatoporoids that occur in a particular stage in a particular region (e.g. Frasnian of north-central Iowa, Eifelian of the southern Urals). Figures 2−5 illustrate the locations of the associations for each stage. A total of 203 associations were determined, based on information from 190 publications. It must be realized that throughout the world many stromatoporoids await description. In the United States alone, there are undescribed Devonian stromatoporoid faunas in at least 17 states, mostly in the mountainous West.

The only previous plotting of Devonian stromatoporoid distributions on a palaeogeographic base was presented by Flügel (1975), who illustrated the occurrences of Middle Devonian stromatoporoids on a map proposed by House (1973). He also illustrated the distribution of Middle Devonian and Frasnian stromatoporoids on present-day maps.

Stromatoporoid biogeography

Stromatoporoid abundance varied a great deal during the Devonian. This is reflected in the number of associations reported (Fig. 1), and the geographic, primarily latitudinal, range of the group for each age.

Early Devonian

Stromatoporoids were least abundant during the Early Devonian; there were more than twice as many associations determined for the Middle Devonian (85) than for the Early Devonian (41).

Gedinnian Age. No Gedinnian stromatoporoids have been reported from China (Fig. 2). Order Stromatoporida was the most widespread, present in 82% of the 17 associations delineated; *Parallelostroma* and *Stromatopora* were the most prominent gen-

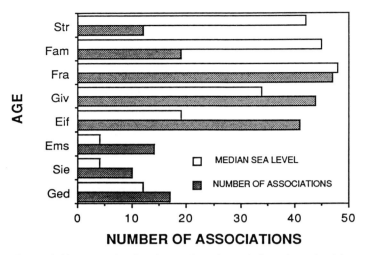

Fig. 1. A histogram showing the number of associations determined for each age of the Devonian, plotted alongside the relative median eustatic sea level for each age according to Johnson *et al.* (1985). Note the concurrent rise and fall of abundance and sea level during the Devonian.

Table 1. *Associations by stage*

Stage	Duration (Ma)	Associations	Normalised (per Ma)
Strunian	4.5	12	2.7
Famennian	5.5	19	3.5
Frasnian	9	47	5.2
Givetian	9	44	4.9
Eifelian	8	41	5.1
Emsian	10	14	1.4
Siegenian	5	10	2.0
Gedinnian	6	17	2.8

Durations based on McKerrow *et al.* (1985), with the Famennian–Strunian proportions based on Haq & Van Eysinga (1987).

era. Orders Actinostromatida, featuring *Atelodictyon*, *Actinostroma*, and *Plectostroma*, and Clathrodictyida, with *Amphipora* and *Anostylostroma* foremost, were less prominent, each occurring in 53% of the Gedinnian associations. The Actinostromatida are not known from North America and western and central Europe, and the Clathrodictyida were rare in North America. Order Stromatoporellida is found in only 29% of the associations, and is also not reported from North America; Family Stictostromatidae, characterized later by *Stictostroma* and *Stromatoporella* had not yet appeared. The genera included above represent a mixture of forms typical of the Silurian and Devonian. For example, *Parallelostroma* and *Plectostroma* were prominent in the Silurian, but *Atelodictyon*, *Actinostroma*, and *Amphipora* are characteristic of the Devonian. Latitudinal limits on stromatoporoid distribution were 55°N and 33°S (Fig. 2).

Siegenian Age. During the Siegenian stromatoporoid abundance reached a Devonian minimum (Fig. 1), with only ten associations identified, none from North America (Fig. 2); however, when normalised for the duration of the Siegenian (5 Ma; Table 1), it has the second lowest abundance. The Stromatoporida were again widespread, as they would be through the Frasnian, occurring in 70% of the associations; again *Parallelostroma* and *Stromatopora* predominated. The Actinostromatida are also found in 70% of the associations, but are not known from China; *Atelodictyon* and *Actinostroma* were most prominent. The Clathrodictyida are present in 60% of the associations; twig-like genera such as *Amphipora* were rare. The Stromatoporellida remained uncommon (30% of associations), and are not known from Europe and the Siberian plate. Latitudinal constraints changed little from the Gedinnian (Fig. 2).

Emsian Age. During the Emsian stromatoporoids retreated from the northern hemisphere, but extended into the southern hemisphere (Fig. 3). Stromatoporoids experienced a small rebound at this time, with 14 associations known, but this increase is due in a large part to the greater length of the Emsian (10 Ma; Table 1), resulting in the lowest Devonian rate of 1.4 associations per Ma. The Clathrodictyida, featuring *Clathrodictyon* and *Anostylostroma*, and Stromatoporida, dominated by *Stromatopora* and *Ferestromatopora*, were most abundant, present respectively in 71% and 64% of the associations. Twig-like clathrodictyids remained rare. The Actinostromatida, with *Actinostroma* most prominent, apparently were absent from North America. The Stromatoporellida, with *Stromatoporella* emerging into a prominent role, have not been reported from China. Both of the previous two orders were uncommon, found in only 29% of associations each.

Middle Devonian

The Middle Devonian represents an epoch of overall maximum abundance of stromatoporoids.

Eifelian Age. Forty-one associations were delineated for the Eifelian, nearly triple that of the Emsian. The Stromatoporida are known from 68% of the associations, and underwent a great diversification of prominent genera: *Stromatopora*, *Syringostroma*, *Parallelopora*, *Stachyodes*, *Clathrocoilona*, and *Habrostroma*. The Clathrodictyida were nearly as widespread, found in 66% of the associations; predominant genera were: *Anostylostroma*, *Clathrodictyon*, *Amphipora*, and *Pseudoactinodictyon*. The Stromatoporellida, featuring *Stromatoporella*, *Trupetostroma*, and *Stictostroma*, and known from 54% of the associations, first became an important component of the world stromatoporoid fauna. Although accounting for only 46% of the Eifelian associations, the Actinostromatida recovered from an Emsian low, and lived in many areas; however, they are known only from one part of North America, in Kentucky and Indiana. *Actinostroma* and *Atelodictyon* were the most common actinostromatid genera. Latitudinal extremes for the Eifelian represent a major colonization of the northern hemisphere (Fig. 3).

Givetian Age. The Givetian is marked by a slight increase in the number of associations with 44, and maximum cosmopolitanism. The Clathrodictyida are known from 89% of the associations; the most prominent genera are *Amphipora* and *Anostylostroma*. The Stromatoporida and Stromatoporellida are each known from 80%. There is still a diversity of common stromatoporid genera: *Stromatopora*, *Stachyodes*, *Parallelopora*, *Clathrocoilona*, *Ferestromatopora*, and *Syringostroma*. The predominant stromatoporellids are: *Trupetostroma*, *Stromatoporella*, *Hermatostroma*, and *Stictostroma*. The Actinostromatida, again featuring *Actinostroma* and *Atelodictyon*, are found in 61% of the associations; they occurred in many areas, but appear to have been absent from eastern and central North America.

Late Devonian

At the beginning of the Late Devonian the trends of the Middle Devonian appear to continue; stromatoporoids were widespread and diverse. However in the latter part of the epoch their numbers greatly decreased, and finally the typical Palaeozoic stromatoporoids became extinct.

Frasnian Age. Although cosmopolitanism decreased for the four orders mentioned above, the Frasnian stromatoporoids displayed

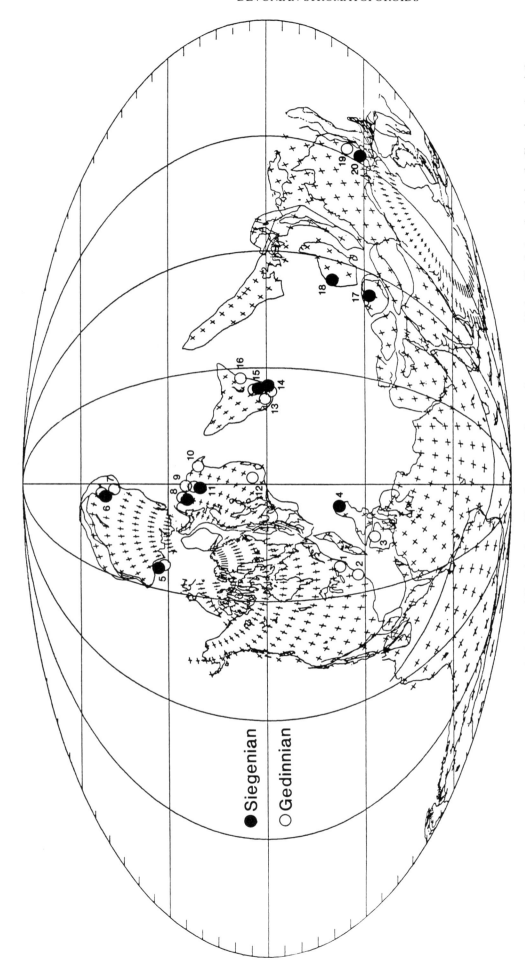

Fig. 2. Distribution of stromatoporoid associations during the Gedinnian and Siegenian Ages. Note the apparent absence of stromatoporoids from North America during the Siegenian. Association localities: (1) New York; (2) Virginia; (3) Spain; (4) Czechoslovakia; (5) Ulachan-Sis Range; (6) Altai, Salair; (7) Kuznetsk Basin; (8) northern Urals; (9) eastern slope Urals; (10) southern Urals; (11) western slope Urals; (12) Podolia; (13) Uzbekistan; (14) Turkestan & Zeravshan Ranges; (15) Central Asia; (16) Tien Shan; (17) Vietnam; (18) Sichuan; (19) New South Wales; (20) Victoria. Locality 5 is on the Kolyma plate. Localities 1 and 2 constitute the Eastern Americas Realm; the rest are Old World Realm. Gedinnian base map of Scotese (1986).

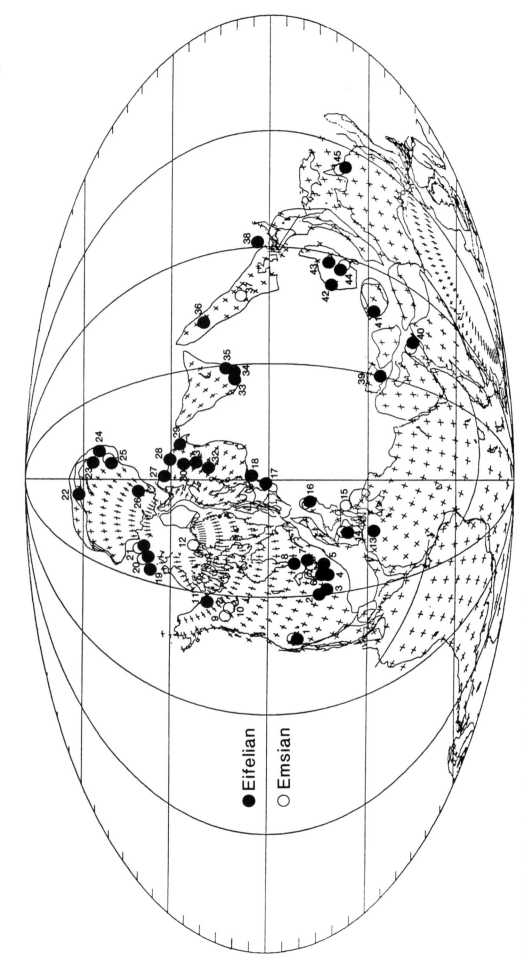

Fig. 3. Distribution of stromatoporoid associations during the Emsian and Eifelian Ages. Note the increase in Eifelian associations over those of the Emsian. Association localities: (1) Nevada; (2) Missouri; (3) Illinois; (4) Indiana; (5) Ohio; (6) Michigan: (7) southern Ontario; (8) northern Ontario; (9) Yukon; (10) (11) Northwest Territories; (12) Ellesmere Island; (13) Morocco; (14) Spain; (15) Carnic Alps; (16) Czechoslovakia; (17) Belgium; (18) Poland; (19) Omulev Mountains, Omolon; (20) Kolyma Basin, Kolyma River; (21) Ulachan-Sis Range; (22) Mongolia; (23) (24) Altai-Sayan, Salair; (25) Kuznetsk Basin; (26) western Siberia; (27) northern Urals; (28) eastern slope Urals; (29) southern Urals; (30) western slope Urals; (31) Russian Platform; (32) Leningrad region; (33) Uzbekistan; (34) Zeravshan Range; (35) Isfar, Kashkadar; (36) Xinjiang; (37)Inner Mongolia; (38) northeastern China; (39) Turkey; (40) Afghanistan; (41) Vietnam; (42) Sichuan; (43) Hunan; (44) Guangxi; (45) northern Queensland. Localities 19–21 are on the Kolyma plate. Localities 6 and 7 constitute the Eastern Americas Realm during the Emsian; the rest are Old World Realm. Emsian base map of Scotese (1986).

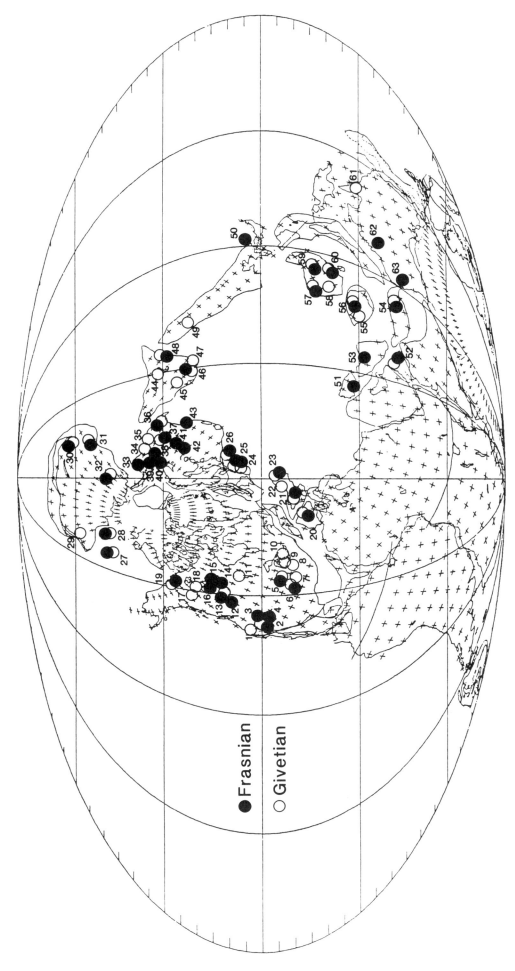

Fig. 4. Distribution of stromatoporoid associations during the Givetian and Frasnian Ages, times of maximum abundance. Association localities: (1) California; (2) (3) Nevada; (4) Utah; (5) Iowa; (6) Missouri; (7) Illinois; (8) Indiana; (9) Michigan; (10) southern Ontario; (11) Manitoba; (12)–(16) Alberta; (17) Yukon; (18) (19) Northwest Territories; (20) Spain; (21) France; (22) southern West Germany; (23) Czechoslovakia; (24) Belgium; (25) northern West Germany; (26) Poland; (27) Omolon; (28) Ulachan-Sis Range; (29) Sette-Daban Range; (30) Altai-Sayan, Salair; (31) Kuznetsk Basin; (32) western Siberian Platform; (33) Yogorsk Pennisula; (34) eastern slope northern Urals; (35) southern Urals; (36) western slope Urals; (37) western slope Urals; (38) Voivo-Vozh; (39) Pechora Basin; (40) Timan; (41) Russian Platform; (42) Leningrad region; (43) lower Volga, Volgograd; (44) Karaganda; (45) Kazakhstan; (46) Uzbekistan; (47) Zeravshan Range; (48) Tien Shan; (49) Xinjiang; (50) northeastern China; (51) Turkey; (52) Afghanistan; (53) Caucusus; (54) Xizang; (55) Yunnan; (56) Vietnam; (57) Sichuan; (58) Guizhou; (59) Hunan; (60) Guangxi; (61) northern Queensland; (62) Canning Basin; (63) Carnarvon Basin. Localities 27 and 28 are on the Kolyma plate. Givetian base map of Scotese (1986).

a maximum latitudinal range of 126° (limits: 76 °N, 50 °S) (Fig. 4), and a maximum number of associations (47). This was also a time of the greatest number of associations per Ma (5.2; Table 1). The Stromatoporida, Stromatoporellida, and Clathrodictyida are each found in 74% of the associations, and the Actinostromatida in 57%; all were widespread. The most prominent genera in each order were as follows: Stromatoporida: *Stromatopora*, *Stachyodes*, *Clathrocoilona*, *Ferestromatopora*; Stromatoporellida: *Trupetostroma*, *Stromatoporella*, *Hermatostroma*, *Stictostroma*; Clathrodictyida: *Amphipora*, *Anostylostroma*, *Hammatostroma*, *Pseudoactinodictyon*; Actinostromatida: *Actinostroma*, *Atelodictyon*. For the first time in the Devonian, Order Labechiida appears in several European associations (11%). This group dominated in the Ordovician, rapidly decreased during the Silurian, and remained as a rare component of the stromatoporoid fauna through most of the Devonian.

Famennian Age. There appear to be two ways of viewing the Famennian. One is to regard it as the terminal age of the Devonian (e.g. McKerrow *et al.* 1985; Haq & Van Eysinga 1987), but another is to hold the Famennian to be next-to-last, with the Strunian directly preceding the Carboniferous (e.g. Stearn *et al.* 1987). Sorauf & Pedder (1986) included the Strunian, also known as Etroeungtian (Haq & Van Eysinga 1987), as latest subdivision of the Famennian. Matters are complicated further by others who regard the Strunian/Etroeungtian as an earliest Carboniferous age (e.g. Bogoyavlenskaya 1982). In that many publications describing stromatoporoids employ a separate, post-Famennian time interval, and given the general uncertainty of the age of the Strunian in regard to the Famennian, the two are used here as separate ages.

The Famennian brought with it two major changes: a drastic decrease in the number of associations to 19, and a reshuffling of dominance among the stromatoporoid orders. Now Order Labechiida became most prominent, with 58% of associations, but it is not known from Australia (Fig. 5). Predominant genera were *Labechia*, *Rosenella*, and *Stylostroma*. Of secondary importance were orders Clathrodictyida and Stromatoporida (*Stromatopora* most common), occurring in 47% of associations each. The Stromatoporellida (32%) and Actinostromatida (26%) were even less abundant, featuring *Trupetostroma* and *Atelodictyon*, respectively. The latter two orders have not been reported from North America. No Famennian stromatoporoids have been described from China.

Strunian Age. The Strunian represents the narrowest latitudinal range of Devonian stromatoporoids at 83° (limits: 62 °N, 21 °S; Fig. 5). The stromatoporoid fauna did not change much from the Famennian, but there were fewer associations (12; back to Early Devonian levels), and the relative abundance of the orders changed. The Clathrodictyida, with *Anostylostroma* most prominent, are found in 67% of the associations, the Actinostromatida, featuring *Actinostroma* and *Atelodictyon*, are in 58% and the Labechiida and Stromatoporellida are in 50% each. The predominant labechiids were *Rosenella*, *Stromatocerium*, *Stylostroma*, *Labechia*, and *Pennastroma*; the predominant stromatoporellid was *Stromatoporella*. The Stromatoporida fell to an all-time Devonian low of 25%; *Stromatopora* was most common. There are no Strunian stromatoporoids known from North America, Australia, and the Kazakhstan plate. The Actinostromatida and Labechiida have not been reported from the Siberian plate; the latter are also not known from western Europe. The Stromatoporellida are not known from eastern Europe.

Discussion

The information outlined above suggests several questions. What controlled the abundance of stromatoporoids at various times during the Devonian, and their extinction at the end of the Strunian? What relation did palaeolatitudinal changes have to changes in abundance? Why are stromatoporoids found in Australia, but not in South America? Why did the stromatoporoids extend farther into the northern hemisphere than into the southern hemisphere? Was there provincialism among the stromatoporoids during the Devonian, or are differences in stromatoporoid faunas due to local environmental conditions?

Abundance

Enough work has been done on Devonian stromatoporoids to provide what is probably an accurate reflection of relative abundance at a particular time (Fig. 1). The Siegenian has the fewest associations with 10. After that time there is a steady increase in the number of associations to the Frasnian with 47, followed by a decrease during the Famennian and Strunian. These trends are affected only slightly when the number of associations per age are normalized (Table 1).

Two factors controlling stromatoporoid distributions seem to have limited their geographic extent: temperature and depth. Stromatoporoids required warmer water temperatures than did many contemporary benthic taxa, e.g. the rugose corals. There are rugose corals known from the South American Devonian (Oliver 1977), but stromatoporoids are lacking. Devonian rugose corals also ranged into deeper, hence cooler, environments than stromatoporoids (e.g. Lecompte 1954). Increases in world temperature would have expanded the tropical and subtropical areas available for stromatoporoid habitation. During the Gedinnian and Siegenian stromatoporoids had an overall latitude range of 88°. This increased progressively to 92° in the Emsian and 112° in the Eifelian. After a drop to 102° in the Givetian, a peak of 126° was reached in the Frasnian. Latitudinal extremes then fell to 107° in the Famennian and 83° in the Strunian. Evidence for global cooling of the Early Devonian is not apparent, but much has been said concerning cooling during the Late Devonian (e.g. Copper 1977; Johnson *et al.* 1985).

When world-wide sea level is high, large areas of continents are flooded, producing an increase in shallow shelf habitats conducive to stromatoporoid colonization. One of the most recent summaries of relative sea level throughout the Devonian is that of Johnson *et al.* (1985). They show a minimum eustatic sea level at both ends of the Pragian (Siegenian), followed by an overall sea level rise, peaking in the late Frasnian. Sea level began to fall during the Famennian and Strunian (included in the Famennian by Johnson *et al.*). The overall shape of the Johnson *et al.* (1985) sea level curve shows trends similar to that derived from the total number of associations per age (Fig. 1). Certainly overall sea level controlled at least the habitable area available to stromatoporoids at a particular latitude.

The factors cited above are also most likely responsible for the stromatoporoids' extinction. Stromatoporoid abundance decreased significantly at the end of the Frasnian, but no orders and few genera became extinct at that time. Of prominent Devonian genera, only *Pseudoactinodictyon*, *Hermatostroma*, and *Ferestromatopora* have not been reported from Famennian or Strunian strata.

Longitudinal and latitudinal asymmetry

The absence of stromatoporoids from the Devonian of western South America, coeval with their presence at similar latitudes in Australia, is most likely due to differences in temperature or regional depositional conditions. During the Devonian the world consisted of two ocean basins (Figs 2–5), a very large basin, the remains of which form today's Pacific Ocean, and a smaller, restricted basin which, with some modification, led to the Tethys Sea. Western South America was on the eastern edge of the large basin, and eastern Australia was on the western edge; both were well into the southern hemisphere. If a typical counter-

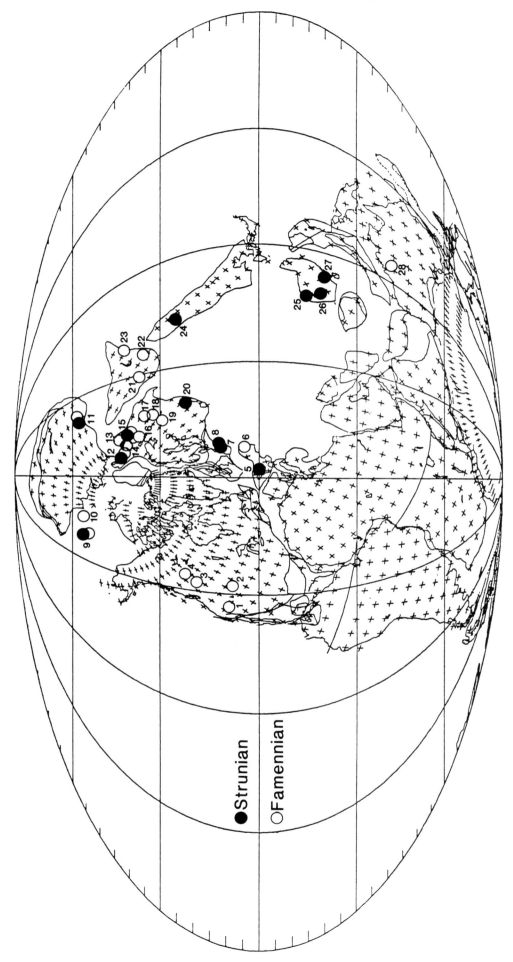

Fig. 5. Distribution of stromatoporoid associations during the Famennian and Strunian. Note the apparent absence of stromatoporoids from China during the Famennian, and from Australia and North America during the Strunian. Association localities: (1) Nevada; (2) Wyoming; (3) (4) Alberta; (5) France; (6) Czechoslovakia; (7) Belgium; (8) West Germany; (9) Omolon; (10) Ulachan-Sis Range; (11) Kuznetsk Basin; (12) Novaya Zemlya; (13) Yogorsk Pennisula; (14) Bolshaya Zelenets & Dolgi Islands; (15) Pechora Basin; (16) Timan; (17) western slope Urals; (18) (19) Russian Platform; (20) Donetz Basin; (21) Kazakhstan; (22) Chatkago-Kuraminsk; (23) Tien Shan; (24) Xinjiang; (25) Sichuan; (26) Guizhou; (27) Guangxi; (28) Canning Basin. Localities 9 and 10 are on the Kolyma plate. Famennian base map of Scotese (1986).

clockwise southern subtropical gyre (Apel 1987) existed in that Devonian ocean, it would have brought cool, polar water up the South American coast, and warm, tropical water down the Australian coast. It is also possible that western South America was tectonically active during the Devonian, resulting in a clastic influx that choked off any stromatoporoid that tried to live there.

The greater extension of stromatoporoids into the northern hemisphere than into the southern hemisphere (Figs 2–5) is difficult to explain. There has been some discussion of a southern polar ice cap on Gondwanaland beginning in the Famennian (Johnson *et al.* 1985), but it is unlikely that one existed at the Devonian north pole. The large ocean basin of that time probably contained a clockwise northern subtropical gyre that carried warm water along southern China and Kazakhstan up to Siberia and Mongolia. This warm water could have prevented the freezing of sea water at the pole, and accounted for the presence of stromatoporoids on the east coast of the Devonian Siberia, but what of the stromatoporoids on the west coast? Today the northern hemisphere contains two-thirds of the earth's land area; consequently there are greater annual temperature extremes in the north (Lutgens & Tarbuck 1986). During the Devonian a reverse situation existed, with Gondwanaland dominating the southern hemisphere. The annual mean low temperature in the Devonian northern hemisphere at any one latitude would have been higher than for that same latitude in the southern hemisphere. Another possibility is that the reconstructions of Scotese (1986) employed here have the Siberian plate too far north.

Provincialism

In order to better understand the geographic distribution of Devonian stromatoporoids, their occurrences will be compared to those already established for brachiopods (Johnson & Boucot 1973; Boucot 1975) and rugose corals (Oliver 1977); both of these schemes agree well. Basically three realms were in existence during the Devonian: (1) Malvinokaffric Realm, including southern South America and southern Africa; stromatoporoids are unknown, (2) Eastern Americas Realm, including eastern North America and northern South America; stromatoporoids have not been reported from South America, and (3) Old World Realm, including western North America, the Canadian Arctic, Eurasia, Australia and northern Africa. Brachiopods displayed provincialism from the Gedinnian through middle Givetian (Johnson & Boucot 1973), and the rugose corals from the Gedinnian through the Givetian (Oliver 1977). Cosmopolitanism began in the lat Givetian for the brachiopods, and the Frasnian for the rugose corals, and extended into the Famennian (and Strunian?).

The distribution of stromatoporoid orders was noted above; the four orders that were prominent throughout the Devonian were cosmopolitan from the Eifelian to the close of the Devonian. Orders Clathrodictyida and Stromatoporellida were found in the Eastern Americas and Old World Realms during the Emsian. Stromatoporoids are not known from the Eastern Americas Realm during the Siegenian. Orders Stromatoporida and Clathrodictyida occurred in both realms during the Gedinnian.

The distributions of 16 genera prominent during the Devonian were assessed: *Actinostroma*, *Atelodictyon*, *Clathrodictyon*, *Pseudoactinodictyon*, *Gerronostroma*, *Anostylostroma*, *Amphipora* (including *Paramphipora*), *Stromatoporella*, *Stictostroma*, *Trupetostroma*, *Hermatostroma*, *Stromatopora*, *Ferestromatopora*, *Clathrocoilona* (including *Synthetostroma*), *Parallelopora*, and *Stachyodes*; *Syringostroma* was not included because many species lacking megapillars have been mistakenly assigned to it (Fagerstrom 1982).

Provincialism was strong during the Early Devonian. Of the 16 genera, only *Anostylostroma* has been reported from both the Eastern Americas and Old World realms, stromatoporoids are not known from the Eastern Americas Realm during the Siegenian, and only four genera occurred in both realms during the Emsian: *Anostylostroma*, *Stromatoporella*, *Stictostroma*, and *Stachyodes*.

Twelve genera were found in both realms during the Eifelian. All genera restricted to one realm occurred in the Old World Realm: *Actinostroma*, *Hermatostroma*, *Ferestromatopora*, and *Clathrocoilona*. Fourteen of the 16 genera were cosmopolitan during the Givetian, having occurred in the Eastern Americas and Old World Realms; *Clathrodictyon* and *Gerronostroma* have not been reported from the Eastern Americas Realm. Therefore the Middle Devonian was a time during which the stromatoporoids experienced a higher degree of cosmopolitanism than did the brachiopods and rugose corals.

Late Devonian cosmopolitanism was in full force for the stromatoporoids, as it was for the brachiopods and rugose corals, during the Frasnian. However, the Famennian and Strunian display a decrease in the number of stromatoporoid associations determined and in their geographic range. Stearn *et al.* (1987) have summarized the geographic distribution of Famennian and Strunian stromatoporoids. Briefly, they suggested four assemblages, and pointed out that labechiid stromatoporoids dominated in many assemblages, but were absent from a few. In addition they postulated that cool water favoured labechiid genera over those in orders more abundant before the Famennian. Another possible explanation involves on-shore, off-shore environmental (e.g. depth) differences. In her summary of the distribution of Famennian and Etroeungtian (Strunian) stromatoporoids, Bogoyavlenskaya (1982) reported that labechiids were found on Novaya Zemlya and on the western slope of the Urals, but were absent from stromatoporoid assemblages collected from the eastern slope of the Urals. According to Hamilton (1970) Novaya Zemlyan and western Uralian strata are miogeosynclinal in origin, whereas those strata on the eastern side have a eugeosynclinical origin. An analogous, although not as dramatic, situation existed in the Early Devonian (Gedinnian) of New York. There the Manlius Formation contains abundant *Habrostroma*, assigned questionably to *Parallelostroma* by Stock (1988). The Coeymans Formation was deposited simultaneously with much of the Manlius, but in a more off-shore, higher-energy environment (Rickard 1975). Stromatoporoids are common to abundant in the Coeymans, but *Habrostroma* is rare. Instead, *Parallelostroma* and *Parallelostroma*-like forms predominate, with *Anostylostroma* in a secondary role.

Another aspect revealed in the distribution of the 16 prominent genera has to do with origin and migration of faunas through time. It is apparent that these genera that existed during the Gedinnian were restricted to a small part of the Old World Realm, particularly what was then the eastern part of the Kazakhstan plate (Uzbekistan, Zeravshan Range, Tien-Shan) and the northeastern part of the Siberian plate (Kuznetsk Basin, Salair). At the same time the Eastern Americas Realm was dominated by minor Devonian genera such as *Habrostroma* and *Parallelostroma*. Five of the genera are not known from the Gedinnian: *Pseudoactinodictyon*, *Stromatoporella*, *Stictostroma*, *Ferestromatopora*, and *Parallelopora*. One must wonder what is contained in the Gedinnian stromatoporoid faunas of western North America.

By the Siegenian, 14 of the 16 genera were present and were concentrated in three areas, eastern Kazakhstan (Zeravshan Range), Victoria (Australia), and Vietnam. Thirteen genera have been reported from the Emsian including four that appeared during the repopulation of the Eastern Americas Realm: *Anostylostroma*, *Stromatoporella*, *Stictostroma*, and *Stachyodes*. It was in the Emsian of the Michigan–Ontario area that Family Stictostromatidae (*Stromatoporella*, *Stictostroma*) first flourished. All 16 genera were present by the Eifelian with the initial appearance of *Pseudoactinodictyon* in Missouri, Ohio, Ontario (Eastern Americas Realm), and Guangxi (China-Old World Realm).

During the Frasnian Age both brachiopods and rugose corals

existed in the Appalachian Basin (Johnson & Boucot 1973; Oliver 1977), however one must look west to Iowa and Missouri for the easternmost Frasnian stromatoporoids. This discrepancy is mostly due to the stromatoporoids' inability to withstand the influx of clastic material generated during the Acadian Orogeny.

Conclusions

(1) Stromatoporoids are a sensitive indicator of warm palaeo-temperatures, and of the absence of significant clastic sedimentation.

(2) The state of the literature on Devonian stromatoporoids is such that detailed biogeographic interpretations are difficult or impossible to accomplish.

(3) At the epoch level, stromatoporoids were least abundant during the Early Devonian, and most abundant during the Middle Devonian.

(4) At the age level, stromatoporoids showed a gradual decrease in abundance from the Gedinnian to the Emsian, when a Devonian low was reached, followed by a sharp increase in the Eifelian, and near stability through the Frasnian; a major extinction at the end of the Frasnian dramatically reduced stromatoporoid abundance, a trend that continued through the Famennian and Strunian when the typical Palaeozoic stromatoporoids became extinct.

(5) Increases in stromatoporoid abundance were accompanied by rises in eustatic sea level as established by Johnson et al. (1985); decreases in abundance occurred when eustatic sea level fell.

(6) Stromatoporoid extinction was probably brought on by a combination of falling sea level and falling temperatures.

(7) Two realms, the Eastern Americas and Old World realms, existed for stromatoporoids during the Early Devonian, although none are known from the Siegenian of the Eastern Americas Realm.

(8) The presence of stromatoporoids on the western edge of the proto-Pacific and their absence from the eastern edge is probably due to the influence of the southern subtropical gyre, but local clastic influx in western South America may have prevented stromatoporoid growth.

(9) The extension of stromatoporoids further into the northern hemisphere in relation to the southern may be due to warmer temperatures in the north, a consequence of most land area being preset in the southern hemisphere, or to a too far northward plotting of the Siberian plate on the palaeogeographic reconstructions of Scotese (1986).

(10) Certain stromatoporoid genera responded differently to different environmental conditions, such as differences in water depth.

References

APEL, J. R. 1987. *Principles of Ocean Physics*. Academic Press, New York.
BOGOYAVLENSKAYA, O. V. 1982. Stromatoporaty pozdnego devona-rannego karbona. *Paleontologicheskii Zhurnal*, 1982(1), 33–38. [Translated in *Paleontological Journal*, **15**, 29–36.]
BOUCOT, A. J. 1975. *Evolution and Extinction Rate Controls*. Elsevier Scientific Publishing Company, Amsterdam.
COPPER, P. 1977. Paleolatitudes in the Devonian of Brazil and the Frasnian–Famennian mass extinction. *Paleogeography, Palaeoclimatology, Palaeoecology*, **21**, 165–207.
DOLPHIN, D. R. & KLOVAN, J. E. 1970. Stratigraphy and palaeoecology of an Upper Devonian carbonate bank, Saskatchewan River Crossing, Alberta. *Bulletin of Canadian Petroleum Geology*, **18**, 289–331.
FAGERSTROM, J. A. 1982. Stromatoporoids of the Detroit River Group and adjacent rocks (Devonian) in the vicinity of the Michigan Basin. *Bulletin of the Geological Survey of Canada*, **339**.
FISCHBUCH, N. R. 1962. Stromatoporoid zones of the Kaybob reef, Alberta, *Journal of the Alberta Society of Petroleum Geologists*, **10**, 62–72.
FLÜGEL, E. 1975. Fossile Hydrozoen — Kenntnisstand und Probleme. *Paläontologische Zeitschrift*, **49**, 369–406.
—— & FLÜGEL-KAHLER, E. 1968. Stromatoporoidea. *Fossilium Catalogus, I: Animalia*, **115**, 116.
HAMILTON, W. 1970. The Uralides and the motion of the Russian and Siberian Platforms. *Geological Society of America Bulletin*, **81**, 2553–2576.
HAQ, B. U. & VAN EYSINGA, F. W. B. 1987. *Geological Time Table*. Elsevier Scientific Publishing Company, Amsterdam.
HECKEL, P. H. & WITZKE, B. J. 1979. Devonian world palaeogeography determined from distribution of carbonates and related lithic palaeoclimatic indicators. *In*: HOUSE, M. R., SCRUTTON, C. T. & BASSETT, M. G. (eds) *The Devonian System*. Special Papers in Palaeontology, **23**, 99–123.
HOUSE, M. R. 1973. An analysis of Devonian goniatite distributions. *In*: HUGHES, N. F. (ed.) *Organisms and Continents through Time*. Special Papers in Palaeontology, **12**, 305–317.
JOHNSON, J. G. & BOUCOT, A. J. 1973. Devonian brachiopods. *In*: HALLAM, A. (ed.) *Atlas of Palaeobiogeography*. Elsevier Scientific Publishing Company, Amsterdam.
——, KLAPPER, G. & SANDBERG, C. A. 1985. Devonian eustatic fluctuations in Euramerica. *Geological Society of America Bulletin*, **96**, 567–587.
LECOMPTE, M. J. 1954. Quelques données rélatives à la genèse et aux charactères écologiques des "récifs" du Frasnien de l'Ardenne. *L'Institut Royal des Sciences Naturelles de Belgique, Victor Van Straelen, Jubilaire*, **1**, 153–94. [Translated in *International Geology Review*, **1**, 1–24.]
LUTGENS, F. K. & TARBUCK, E. J. 1986. *The Atmosphere*, 3rd ed. Prentice-Hall, Englewood Cliffs, New Jersey.
MCKERROW, W. S., LAMBERT, R. ST. J. & COCKS, L. R. M. 1985. The Ordovician, Silurian and Devonian periods. *In*: SNELLING, N. J. (ed.) *The Chronology of the Geological Record*. Geological Society, London, Memoir, **10**, 73–80.
NESTOR, H. A. 1977. On the ecogenesis of the Paleozoic stromatoporoids. *2nd Symposium International sur les Coraux et Récifs Coralliens Fossiles*. Memoires de Bureau de Recherches Géologiques et Minièrs de France, **89**, 249–254.
OLIVER, W. A., JR. 1977. Biogeography of Late Silurian and Devonian rugose corals. *Palaeogeography, Palaeoclimatology, Palaeoecology*, **22**, 85–135.
RICKARD, L. V. 1975. Correlation of the Silurian and Devonian rocks in New York State. *Map and Chart Series of New York State Museum and Science Service*, **24**.
SCOTESE, C. R. 1986. Phanerozoic reconstructions: a new look at the assembly of Asia. *University of Texas Institute for Geophysics Technical Report*, **66**.
SORAUF, J. E. & PEDDER, A. E. H. 1986. Late Devonian rugose corals and the Frasnian–Famennian crisis. *Canadian Journal of Earth Sciences*, **23**, 1265–1287.
STEARN, C. W. 1980. Classification of the Paleozoic stromatoporoids. *Journal of Paleontology*, **54**, 881–902.
——, HALIM-DIHARDJA, M. K. & NISHIDA, D. K. 1987. An oil-producing stromatoporoid patch reef in the Famennian (Devonian) Wabamun Formation, Normandville field, Alberta. *Palaios*, **2**, 560–570.
STOCK, C. W. 1988. Lower Devonian (Gedinnian) Stromatoporoidea of New York: redescription of the type specimens of Girty (1895). *Journal of Paleontology*, **62**, 8–21.

Rugose coral distribution as a test of Devonian palaeogeographic models

A. E. H. PEDDER[1] & W. A. OLIVER, JR.[2]

[1] *Institute of Sedimentary & Petroleum Geology, 3303, 33rd Street NW, Calgary, Alberta, Canada, T2L 2A7*
[2] *United States Geological Survey, E-305, National History Building, Smithsonian Institution, Washington, DC, 20560, USA*

Abstract: A data bank, based on stage by stage distributions of 420 rugose coral genera in 25 regions of the world is analysed, mostly by means of Otsuka coefficients, to test an Emsian reconstruction of the world proposed by Scotese. Devonian rugose corals inhabited a narrower range of facies than some other benthic groups, and even without regard to facies, provide a tool for testing geographic reconstructions. Basin dwelling coral genera typically have longer temporal and broader geographic ranges than corals living in shallower environments, and are less suitable for palaeogeographic studies. They are treated separately in this work.

For the most part, conclusions drawn from the analysis are either consistent with, or positively supportive of, the Scotese reconstruction. However, large but poorly known rugose coral faunas from Mongolia and the Amur Basin are at about 60°N in the reconstruction, and other well known coral faunas, from Altai–Sayan, are at 45°–50°N. In the light of known distributions of both modern corals and Devonian southern hemisphere corals, in all recently proposed palaeogeographic reconstructions, it is questionable that the original latitude of any large northern hemisphere Devonian coral fauna would have exceeded 45°.

Because rugose corals are extinct and have left no known record of their soft tissues, palaeontologists are forced to make certain assumptions. One assumption underlying our current work is that the larval life-styles of Devonian rugose corals were much like those of scleractinian corals and other living benthic groups.

Some scleractinian larvae settle within a few days of spawning (Babcock & Heyward 1986; Heyward *et al.* 1987). But other marine larvae, especially those of tropical species, have the ability to remain in the larval stage for much greater periods of time, the longest being more than 300 days. Long-living larvae, adapted to a planktonic existence and regularly found in the open seas, were said to be teleplanic by Scheltema (1971). Velocities of ocean currents are typically in the order of $1-4$ km h^{-1}, so that even the shorter-living teleplanic larvae are capable of being transported 1000 km or more. The 300 day duration of the longest living teleplanic larvae is more than twice the time theoretically required for a passive object to drift across the Atlantic Ocean, either on one of the main gyres, or in the Equitorial Undercurrent (Scheltema 1968, 1971, 1972).

An examination of the stratigraphic and geographic distribution of rugose coral genera shows that they had different temporal ranges in different regions of the Palaeozoic world. Causes of the provinciality of coral faunas stemming from these discrepant distributions were undoubtedly complex. Factors involved would have included the patterns and speeds of ocean currents, latitudinal temperature gradients in the oceans, distribution of land and sea on the continental crust, distribution of facies favourable to rugose corals, including shallow staging posts in the large oceans, and whether the larval stages of individual species comprising a genus were teleplanic or not.

The corollary to this is that a full understanding of the provinciality of Devonian rugose coral faunas should lead to an understanding of the factors controlling the provinciality, which would be of considerable importance in reconstructing Devonian palaeogeography.

We are currently building a data base on the distribution of Devonian rugose genera. It is not complete, but is sufficiently advanced to be used as a preliminary test of Devonian palaeogeographic models. The magnitudes of provinciality varied during the Palaeozoic. One of the times of maximum provinciality was the middle and late Early Devonian. For this reason, we are using the Scotese Emsian map as the basis of our discussion.

Devonian rugose coral distribution data base

Our data base is a table recording the occurrences of approximately 420 Devonian rugose coral genera, in essentially stage-level increments, in 25 regions of the world, shown as letters A–Y in Fig. 1. The assembly of such a large table is an enormous task. More than 4000 species have been proposed for Devonian rugose corals and we are aware of many undescribed forms. Descriptions of Devonian rugose corals are spread over about 2500 publications written in at least 11 languages. We have not confined ourselves to published data; our records from North Africa, Spain, France, Novaya Zemlya, Eastern Australia and all of North America depend heavily on unpublished data. From the start, we have made every effort to be objective and consistent. Each entry on the table is based on a reappraisal of the generic identity of the relevant material in terms of uniformly applied generic diagnoses. Age determinations are based exclusively on evidence provided by other animal groups, especially conodonts. Data from unreliably dated faunas have been discarded.

With the exception of the *patulus* conodont zone, which we use in the old broad sense for an interval of time straddling the Lower–Middle Devonian boundary, our other units are the standard Devonian stages recommended by the IUGS Subcommission on Devonian Stratigraphy, However, use of the *patulus* Zone, broad sense, means that our Emsian Stage lacks the uppermost conodont zone (*patulus* Zone s.s.) and our Eifelian Stage lacks the lowermost conodont zone (*partitus* Zone). The Famennian Stage is not included in the data base compilation. This is because the well known late Frasnian faunal crisis extinguished almost all corals inhabiting shelf and slope facies (Sorauf & Pedder 1986). The rare Famennian shelf-dwelling coral faunas are largely underscribed, and, for the most part, comprise forerunners of later Carboniferous faunas, rather than survivors of the Frasnian crisis.

In addition to the requirements for firm and consistent generic identifications, and the need for accurate age determinations, compilation of the data base also requires rational geographic groupings of faunal lists in terms of known Devonian plate boundaries. Uncertainties regarding the plate tectonic history of certain regions have posed many problems for our data compilation. Faunas from suspect terranes, such as parts of western North America and eastern Australia, and faunas from possible micro-

From McKerrow, W. S. & Scotese, C. R. (eds), 1990, *Palaeozoic Palaeogeography and Biogeography*, Geological Society Memoir No. 12, pp. 267–275.

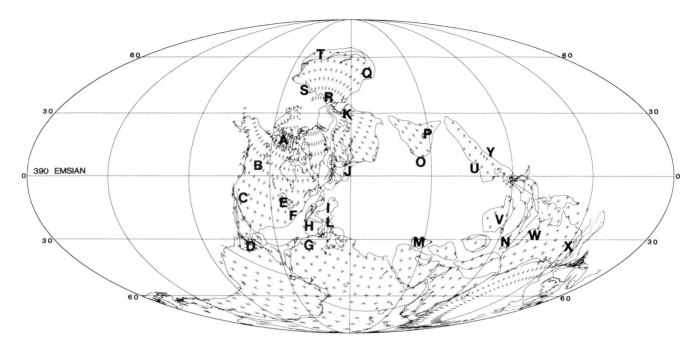

Fig. 1. Coral faunas analyzed are grouped geographically as shown by letters A to Y, here located on the Emsian palaeogeographic reconstruction of Scotese (1986). A, Canadian Arctic Islands; B, Western Canada and neighbouring northeastern Alaska; C, Great Basin, Nevada; D, Venezuela and Columbia; E, Michigan and Illinois Basins; F, Appalachian Belt; G, North Africa, including Algeria, Morocco and Spanish Sahara; H, Spain and the Pyrenees; I, Armorica, Montagne Noire, northern Vosges (Saxothuringian plate), Thuringia and Czechoslovakia; J, Britain, northern France, Belgium, central Germany (Rhenish Massif) and Poland; K, northeastern Europe, western Urals and Novaya Zemlya; L, Carnic Alps and southeastern Austria, separated from I on the basis of Weijermars' (1987) tectonic map of the Mediterranian Collision Zone between the African and Eurasian plates; M, Turkey; N, Northern Zizang (Tibet), northwestern Yunnan and Burma (Shan Plateau); O, Tien Shan, Devonian position discussed in text; P, Lake Balkhash region, Kazakhstan, and Hoboksar region of northern Xinjiang; Q, Altai–Sayan, including the Salair, Kuznetsk Basin, Minusa Depression, Gorniy Altai and Rudniy Altai; R, Taimyr; S, Indigirka River region; T, Mongolia; U, Qinling Mountains region, central China; V, South China, including the Longmenshan Mountains and part of northern Vietnam; W, Western Australia; X, Eastern Australia; Y, Zhusilengharhan area of western Nei Mongol (Inner Mongolia) and the Erdaogou area of Jilin Province, China.

plates such as those from the eastern slopes of the Urals, are excluded from our current computation. However, it is much more difficult to avoid problems of this nature in southeast Asia, because of the complex and imperfectly understood plate histories of the entire region. We expect to have to modify columns in the table covering this region, as plate histories of the region become better known.

Use of the data base

The long term aim is to subject the data to computer cluster analysis using various numerical coefficients. For the moment, our treatment is less rigorous. We have tested the similarity of faunas from different regions (Fig. 1) to see how similar or dissimilar they are to each other. The index used is the Otsuka Coefficient, which has been used often for binary similarity comparisons by biogeographers. It is expressed as:

$$\left(\frac{C}{\sqrt{N_1 N_2}}\right) \times 100$$

where N_1 is the number of taxons present in one region, N_2 is the number of taxons present in the other region, and C is the number of taxons common to both regions (Cheetham & Hazel 1969). In comparing faunas separated by distances as great as those separating the regions shown on Fig. 1, and where the taxon is a Devonian genus of mainly shelf dwelling rugose corals, an Otsuka value of less than 20 indicates low similarity, and a value of more than 30 indicates high similarity.

In a compilation of mondial scope it is impractical to attempt detailed sorting of occurrences on the basis of facies associations, however desirable this might be. This is unfortunate, but is probably less disadvantageous to analyses based on Devonian rugose corals than to analyses based on other predominantly benthic phyla, because Devonian rugose corals were abundant and taxonomically diverse in a narrower range of facies than some other benthic groups.

In terms of Boucot's (1975) Benthic Assemblage classification with depth ratings of 1 (supratidal) to 7 (deep), Devonian Rugosa are virtually restricted to BA3, BA4 and BA5. But we think that we can recognize the relatively deep basin dwelling Devonian rugose corals, that inhabited the BA zone 5, by their distinct morphology and their associations with cephalopods and dacryoconarid tentaculites in death assemblages. Fig. 2 illustrates typical corals of shelf and slope facies (BA3, 4) on the left and basin genera (BA5) on the right. All are drawn to the same scale. *Lythophyllum* is the most widely distributed Lower and Middle Devonian rugose genus, and *Temnophyllum* is a common Middle and Upper Devonian genus. Both of the species shown are typical of their genus and facies in respect to size. The basin dwelling genera depicted are also typical in respect to size, and are clearly very much smaller. Most basin dwelling genera are nondissepimented and some have complicated skeletal structures, such as plates sloping in different directions inside and outside pairs of contratingent septa. Many of them also develop various axial structures, which are rare in other pre-Carboniferous Rugosa. Many basin dwelling corals had long temporal and broad geographic ranges and were generally either immune to the effects that caused the late Frasnian crisis, or were isolated from them, in their deep, presumably dark and relatively cold niches (Sorauf & Pedder 1986).

Pragian rugose coral faunas are scarce in central Europe. The

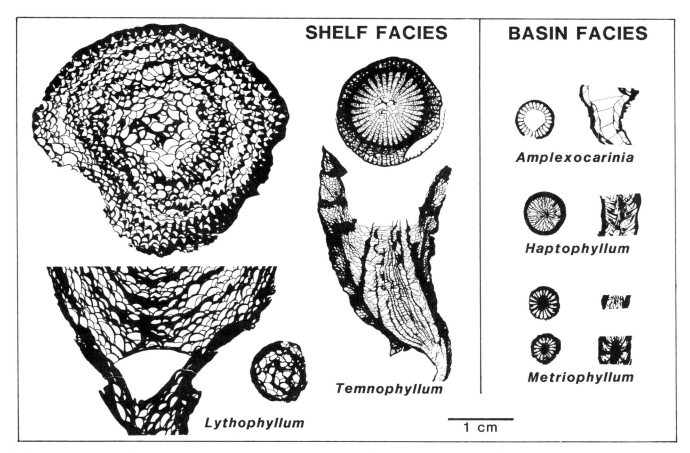

Fig. 2. Two major groups of Devonian coral bearing facies are distinguishable. Devonian rugose coral inhabitants of shallow reef, shelf and slope facies include both colonial and solitary species. Adult solitary forms, such as the species of *Lythophyllum* and *Temnophyllum* shown here, are normally more than 20 mm in diameter. Corals inhabiting deeper basin facies, such as the three genera on the right side of the figure, are exclusively solitary. Most of them are less than 15 mm in diameter and many have distinctive internal morphologies not found in shallow facies corals.

only two that have been adequately described and dated come from the Tentaculiten Knollenkalk of Saxo-Thuringia (Weyer 1984) and the Koneprusy Limestone of Bohemia (Oliver & Galle 1971). Table 1 lists the known genera from these formations. The

Table 1. *Generic composition of the Pragian Tentaculiten Knollenkalk and Koneprusy rugose coral faunas*

Central European Pragian coral faunas	
Tentaculiten Knollenkalk Saxo–Thuringia (basin facies)	Koneprusy Limestone Czechoslovakia (shallow facies)
Neaxon	*Lythophyllum*
Palaeocyathus	*Iowaphyllum*
Sutherlandinia	*Rhizophyllum*
'*Petraia*'	*Chlamydophyllum*
'*Amplexus*'	*Syringaxon?*
	Pseudamplexus
	Acanthophyllum
	Lyrielasma
	Pseudochonophyllum

Both faunal lists are likely to be more or less complete since the faunas have been studied for many years. Despite their similar age and close geographic proximity, in Devonian as well as present time, the two faunas are entirely different (zero Otsuka coefficient). This is because the Tentaculiten Knollenkalk is basin facies, whereas the Koneprusy is a shallower platform and reef facies.

basin facies Tentaculiten Knollenkalk does not share a single genus with the shallow shelf facies of the Koneprusy Limestone. The two faunas inhabited different plates: the Tentaculiten Knollenkalk on the Saxothuringian Plate (Fig. 1J); the Koneprusy on the Moldanubian Plate (Fig. 1I). However, they are not distinct for this reason, because the plates evidently began to collide shortly after the Pragian (Vollbrecht *et al.* 1988). They differ because they lived in entirely different facies, and consideration of these faunas affords a fine example of how important it is to be able to recognize facies in comparing isochronous benthic faunas from restricted geographic areas. Apart from Table 7, which compares the principal Pragian to Eifelian basin facies coral faunas of the world and shows the high degree of similarity between them, all of the other tables consider only the shallower water faunas.

Results obtained from the coral distribution data base

Our results are most conveniently discussed by considerations of individual regions and discussion of how their faunas compare with those of other regions. Before proceeding with this, however, comment should be made on the positioning of the faunas from Mongolia, Inner Mongolia and the Amur Basin (Fig. 1T) on the Scotese (1986) Emsian map. Latitude 35°N is about the northernmost limit of coral communities in the present day Indo-Pacific Province (Tribble & Randall 1986). The large, normal, Lower Devonian coral faunas are not rigorously dated and therefore do not appear on any of our tables, but their very high 60° latitude on the Scotese Emsian map inevitably raises questions about this part of the map.

Appalachian Belt, Eastern North America

We have described and discussed previously, the very high level of endemism in the Early and Middle Devonian coral faunas of eastern North America (Fig. 1E, F), and more recently have discussed the origins, migrations and extinctions of all the North American Devonian rugose faunas (Oliver & Pedder 1979; in press). Table 2 is based on our most recent data, and reinforces what has already been said. The Appalachian faunas (F) are very unlike the Old World Realm faunas of western and Arctic Canada (A, B) and only somewhat more like the Great Basin faunas (C). North African corals (G) are more similar, reflecting the closeness of these areas fairly early in the Devonian. The high level of similarity between the Appalachian (F) and Illinois–Michigan basins (E), reflects the cohesion of the Eastern American Realm (D, E, & F), and is in contrast with the other areas. The low similarities with Europe (H, I, J, K, L) are in accord with the increasing distances of these areas from eastern North America.

The Transcontinental Arch was an effective barrier between eastern and western North America (Oliver & Pedder 1979; in press). The Old Red Continent separated the Appalachian Basin from Africa and Europe (op. cit.) so that such migration as took place was apparently around its southern end, possibly by way of northern South America (D). The decreasing coefficients of similarity between eastern America and areas G, H, J, I, K and L (Table 2), are very compatible with the relative positions of these areas on the Scotese maps (e.g. Fig. 1). North America is an excellent example of a major plate with land barriers providing the explanation for significant differences between contemporary faunas on the same and adjacent plates.

Canadian Arctic Islands

Table 3 compares the Canadian Arctic Islands (Fig. 1A) with regions B, S, R and K. Pragian coefficients are omitted because only three genera of this age are known from the Canadian Arctic Islands. We are not able to comment on the extraordinary lack of similarity between the Eifelian faunas of the Canadian

Table 2. *Otsuka coefficients between shallow facies coral faunas of the Appalachian Belt (Fig. 1F) and surrounding regions according to the Scotese (1986) Emsian reconstruction*

	Appalachian belt (F) Otsuka comparisons (shallow facies)			
	Lochkovian	Pragian	Emsian	Eifelian
Illinois, Michigan Basins (E)	–	–	80	75
Canadian Arctic Islands (A)	10	(0)	0	10
Western Canada, NE Alaska (B)	7	0	4	10
Great Basin USA (C)	17	17	12	27
North Africa (G)	–	37	(13)	41
Spain, Pyrenees (H)	–	–	21	18
Armorica, Montagne Noire, Vosges, Czechoslovakia (I)	(0)	0	7	10
Britain, N France, Belgium, central Germany, Poland (J)	–	–	(0)	14
NE Europe, W Urals, Novaya Zemlya (K)	15	0	4	4
Carnic Alps (L)	–	0	(0)	0

Lack of similarity between eastern (E, F) and western (B, A) regions is due to the barrier effect of the Transcontinental Arch. Connections between Spain and especially North Africa are entirely consistent with the Scotese Emsian reconstruction. A dash, here and in other tables, indicates a lack of known coral faunas. Brackets enclose coefficients where one or both faunas consists of less than five genera.

Table 3. *Otsuka coefficients between the Canadian Arctic Islands (Fig. 1A) and surrounding regions according to the Scotese Emsian reconstruction*

	Canadian Arctic Islands (A) Otsuka comparisons (shallow facies)		
	Lochkovian	Emsian	Eifelian
Western Canada, NE Alaska (B)	37	50	52
Great Basin USA (C)	35	12	39
Appalachian Belt (F)	10	0	10
Indigirka River (S)	(0)	45	31
Taimyr (R)	–	32	0
NE Europe, W Urals, Novaya Zemlya (K)	46	35	36
Britain, N France, Belgium, central Germany, Poland (J)	–	–	28
Jilin NE China (Y)	48	–	–

Similarity between region A and regions B, S, R (except Eifelian) and K is consistent with the Scotese Emsian reconstruction. The lack of similarity between region A and E (not listed) and F is due to the barrier effect of the Transcontinental Arch. The lack of similarity between the Eifelian coral faunas of the Canadian Arctic Islands and those of Taimyr may be an artifact of the small size (6 genera) of the Taimyr fauna as it is presently known. The early Eifelian age (*costatus* Zone) of the known Canadian Arctic coral faunas may also be a factor.

Arctic Islands and Taimyr (R). The Taimyr fauna is small, with only six genera, and the area should be carefully recollected before attempts are made to explain this apparent anomaly. Apart from Taimyr, the Otsuka comparisons on Table 3 are consistent with the Scotese Emsian map. The strong similarities shown between western Canada and the Indigirka River region(s) and Novaya Zemlya–Northern Urals (K) are to be expected. Northern Europe (J) is not similar to the Canadian Arctic Islands, but its Lower Devonian coral faunas are almost non-existent and were separated from the Canadian Arctic Islands by the Old Red Continent.

Altai–Sayan

Large Early Devonian and Eifelian rugose coral faunas from the Salair, Kuznetsk Basin, Minusa Depression, Gorniy Altai and Rudniy Altai regions are united under this heading, and are located in Fig. 1 by the letter Q. The very high northern latitude of 50° or more, shown on the Scotese Emsian reconstruction for the Altai–Sayan region, seems incompatible with distribution of living major scleractinian faunas, which are confined to the 35°N–35°S range. The absence of Devonian rugose coral faunas from southern latitudes of more than 45° in the Scotese Emsian reconstruction is also highly significant in this context.

In addition to the comparisons between Altai–Sayan and Kazakhstan (P) and Tien Shan (O) shown in Tables 4 & 5, we have calculated Otsuka coefficients between Altai–Sayan and northeastern Europe (Russian Platform), western Urals and Novaya Zemlya (Fig. 1K) as: Lochkovian, 48; Pragian, 45; Emsian, 39; Eifelian, 44.

Our coral data suggest that the Altai–Sayan region should be brought south, to be closer to the western Urals and Tien Shan. It would be better if evidence could be found to detach Altai–Sayan from the Devonian Siberian Craton, because the position in the Scotese reconstruction of the present northern margin of the Siberian Craton (R and S on Fig. 1) is satisfactory in respect to coral faunas of the Canadian Arctic Islands.

Table 4. *Otsuka coefficient between the Lake Balkhash region of Kazakhstan (Fig. 1P) and surrounding regions according to the Scotese Emsian reconstruction*

	Balkhash Kazakhstan (P) Otsuka comparisons (shallow facies)			
	Lochkovian	Pragian	Emsian	Eifelian
Tien Shan (O)	28	17	21	21
Altai–Sayan (Q)	23	15	32	38
Taimyr (R)	–	(0)	13	19
NE Europe, W Urals, Novaya Zemlya (K)	20	19	23	30
Britain, N France, Belgium, central Germany, Poland (J)	(0)	–	–	33
Qinling Mountains central China (U)	37	–	10	25
Jilin NE China (Y)	16	–	–	–
South China (V)	–	(0)	21	29

The generally rather low Otsuka coefficients are not easily explained in terms of the Scotese model. The Emsian and Eifelian similarity between Kazakhstan and Altai–Sayan (Fig. 1Q) seems anomalous, but the high latitudes (45°–50°N) of the Altai–Sayan region in the model suggest that it is the positioning of the Q faunas that is more at fault.

Table 5. *Otsuka coefficients between shallow facies coral faunas from southern Tien Shan (Fig. 1O) and other regions*

	Tien Shan (O) Otsuka comparisons (shallow facies)			
	Lochkovian	Pragian	Emsian	Eifelian
Qinling Mountains central China (U)	21	–	21	27
South China (V)	–	(0)	45	42
Jilin NE China (Y)	35	–	–	–
Balkhash Kazakhstan (P)	28	17	21	21
NE Europe, W Urals, Novaya Zemlya (K)	58	42	44	39
Altai–Sayan (Q)	44	50	58	36
Britain, N France, Belgium central Germany, Poland (J)	–	–	–	38
Armorica, Montagne Noire, Vosges, Czechoslovakia (I)	(18)	43	43	29
N Zizang (Tibet), NW Yunnan, Burma (N)	30	48	(17)	18

Similarities between southern Tien Shan and Altai–Sayan (Q) are consistently high, and are cited as one of two reasons for believing that Altai–Sayan, at 45°–50°N, is positioned too far north on the Scotese Emsian reconstruction. The other reason being that no Devonian rugose fauna is known from high latitudes in the Devonian southern hemisphere.

Kazakhstan

Kazakhstan (Fig. 1P) has given so much trouble to Devonian biostratigraphers in the last two decades or so, that we feel especially obliged to compare it with other regions closest to it on the Scotese Emsian map. Correlations used for the comparison given in Table 4 are outlined in Fig. 3, which is a simplified version of Fig. 2 of Talent *et al.* (1987). Authors of the Talent *et al.* paper are respected brachiopod and trilobite workers and

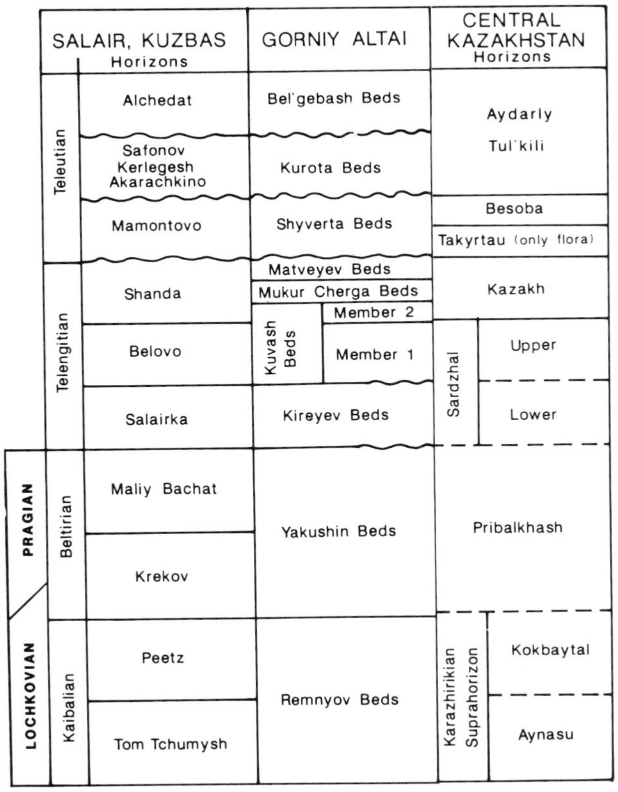

SALAIR, KUZBAS Horizons	GORNIY ALTAI	CENTRAL KAZAKHSTAN Horizons
Teleutian		
Alchedat	Bel'gebash Beds	Aydarly
Safonov Kerlegesh Akarachkino	Kurota Beds	Tul'kili
Mamontovo	Shyverta Beds	Besoba
		Takyrtau (only flora)
Telengitian		
Shanda	Matveyev Beds / Mukur Cherga Beds	Kazakh
	Kuvash Beds Member 2	
Belovo	Member 1	Sardzhal Upper
Salairka	Kireyev Beds	Lower
PRAGIAN Beltirian		
Maliy Bachat	Yakushin Beds	Pribalkhash
Krekov		
LOCHKOVIAN Kaibalian		
Peetz	Remnyov Beds	Karazhirikian Suprahorizon Kokbaytal
Tom Tchumysh		Aynasu

CORRELATION OF KEY DEVONIAN SEQUENCES OF CENTRAL ASIA (after Talent et al., 1987)

Fig. 3 Orthodox correlations of key coral bearing Devonian sequences in central Asia, based heavily on brachiopods and trilobites (after Talent *et al.* 1987). Left and centre columns are Q on Fig. 1; the right column is P on Fig. 1. Low Otsuka coefficients of 23 and 15 for the Lochkovian and Pragian shallow coral faunas of the two regions, suggest that the regions were more distant from each other in Early Devonian time than they are today.

we conclude that correlation of the Aynasu Formation with Lochkovian formations of the Salair and Gorniy Altai (Fig. 1Q) is consistent with trilobite and brachiopod distributions. Fig. 4 explains the relationship between the *Monograptus perneri* graptolite zone recognized by Koren' (1983) in one section, and various rugose coral bearing horizons of the Aynasu Formation. If physical correlations of the Aynasu sections by Russian stratigraphers, the graptolite evidence of Koren', and the opinions of the trilobite and brachiopod workers are all accepted, a Lochkovian age for the Aynasu coral horizons seems probable. But if the coral faunas were considered alone, their age would likely be determined as Silurian.

Fig. 5 relates the Aynasu coral occurrences to a Devonian shoreline map of the Lake Balkhash area published by Nekhoroshev (1977). The Irtysh Suture and the West Jungar Ophiolite Belt which have been added to Nekhoroshev's figure are probable margins of the Kazakhstan Plate. On the basis of Nekhoroshev's palaeogeographic reconstruction, the Aynasu coral localities are situated in deep embayments where 'Silurian' corals were possibly able to survive into Lochkovian time.

Comparisons between coral faunas from around Lake Balkhash and those of other regions are laid out in Table 4. The Lake Balkhash Lower Devonian faunas are similar to contemporaneous faunas in Tien Shan (O), northeastern Europe, Western Urals and Novaya Zemlya (K) and southern China (V), but the only strong connection in Emsian time is with Altai–Sayan (Q). Lake

some similarity between western Canada and eastern Australia. This may be the result of transport of teleplanic larvae on an ocean gyre, which would have been possible on the Scotese map if part of region B extended into the southern Devonian hemisphere. However, several of the genera common to both regions are near cosmopolitan cystiphyllid and ptenophyllid genera. Qualitatively, the most significant single feature of the eastern Australian and southern Chinese Devonian coral faunas is the abundance and early appearance of phacellophyllid genera. This distinctive family did not arrive in western Canada until Late Devonian time.

Basin Facies Faunas.

For the purposes of calculating Otsuka coefficients, *Combophyllum*, *Adradosia*, *Palaeocyathus* and all laccophyllid, metriophyllid and polycoeliid genera are counted as basin dwelling forms. Together, they total 40 genera and represent about 10.5 per cent of the Devonian genera recognized in our data base. Most faunas of this facies include fewer than 10 coral genera; some of the largest are from North Africa (Fig. 1G). Table 7 lists Otsuka coefficients between North Africa and other regions with good Lower Devonian or Eifelian basin coral faunas. One might assume that there would be fewer barriers, other than land and shallow water areas, to the migration of 'deep' water corals than to shallow water genera. In general the comparisons bear this out. Except for the Tien Shan (O; see discussion), there seem to have been relatively easy routes for these corals across or around the central ocean area of the map (Fig. 1). Eastern North America (F) was accessible only by shallow water routes around the Old Red Continent. Because many of the comparatively few basin dwelling genera have wide geographic distribution and long stratigraphic ranges, Otsuka values in Table 7 are high and not necessarily significant. However, the table reinforces the notion, indicated in Table 2, of a Pragian and Emsian connection between the Appalachian Belt, North Africa and Spain. Such a connection is clearly feasible on the Scotese Emsian model.

Conclusions

(1) Distribution of Early Devonian coral faunas on the North American Plate was largely controlled by the barrier effect of the Transcontinental Arch.

(2) With the exception of a small Eifelian fauna from Taimyr, Devonian coral distributions in the far northern regions of the present world are consistent with the Scotese Emsian model.

(3) Rugose coral distribution in the Appalachian Belt, North Africa and Spain are well explained by the Scotese Emsian map.

(4) The Lower Devonian position of southern Tien Shan appears to be close to the western Urals, in some part of the northern interior ocean on the Scotese reconstruction.

(5) Rugose coral distributions generally support the arrangement on the Scotese Emsian map of Australia, Southern China, northern Zizang (Tibet), northwestern Yunnan and the Shan Plateau of Burma.

(6) Poorly known faunas from Mongolia and the Amur Basin (Fig. 1T) appear to be positioned too high, at 60°N, on the Scotese Emsian map.

(7) Well known and well dated faunas from Altai–Sayan (Fig. 1Q) also appear to be positioned too high, at 45–50°N, because no Devonian coral fauna is known from latitudes this high in the southern Devonian hemisphere as it has been reconstructed on the Scotese map.

References

Aristov, V. A. & Chernyshuk, V. P. 1988. Pereotlozhenie konodontov i ego znachenie dlya resheniya nekotorykh voprosov geologii. *Moskovskoe obshchestvo Ispytateley Prirody, Byulleten', Otdel Geologicheskiy*, **63**(6), 40–56.

Babaev, A. M. 1988. Tektonicheskaya granitsa mezhdu Gissaro-Alaem i Tadzhikskoy depressiey. *Moskovskoe obshchestvo Ispytateley Prirody, Byulleten', Otdel Geologicheskiy*, **63**(3), 22–30.

Babcock, R. C. & Heyward, A. J. 1986. Larval development of certain gamete-spawning scleractinian corals. *Corals Reefs*, **5**, 111–116.

Bondarenko, O. V., Stukalina, G. A. & Ushatinskaya, G. T. 1975. Kharakteristika stratigraficheskoy skhemy verkhnego silura i nizhnego devona Tsentral'nogo Kazakhstana. In: Menner, V. V. (ed.), *Kharakteristika fauny pogranichnykh sloev silura i devona Tsentral'nogo Kazakhstana. Materialy po geologii Tsentral'nogo Kazakhstana*, **12**, 5–39.

Boucot, A. J. 1975. Evolution and extinction rate controls. *Development in Palaeontology and Stratigraphy*, **1**. Elsevier Scientific Publishing Company, Amsterdam, Oxford, New York.

Boulin, J. 1988. Hercynian and Eocimmerian events in Afghanistan and adjoining regions. *Tectonophysics*, **148**, 253–278.

Cheetham, A. H. & Hazel, J. E. 1969. Binary (presence–absence) similarity coefficients. *Journal of Paleontology*, **43**, 1130–1136.

Guo Sheng-zhe, 1986. On determination of convergence time between Siberian Plate and Sino-Korean Plate and its biostratigraphic evidence [Chinese with English abstract]. Shenyang Institute of Geology and Mineral Resources, Chinese Academy of Geological Sciences, **14**, 128–136.

Harrington, H. J., Burns, K. L. & Thompson, B. R. 1973. Gambier–Beaconsfield and Gambier–Sorell fracture zones and the movement of plates in the Australia–Antarctica–New Zealand region. *Nature, Physical Science*, **245**, 109–112.

Heyward, A., Yamazato, K., Yeemin, T. & Minei, M. 1987. Sexual reproduction of corals in Okinawa. *Galaxea*, **6**, 331–343.

Koren', T. N. 1983. New late Silurian monograptids from Kazakhstan. *Palaeontology*, **26**(2), 407–434.

Leitch, E. C. & Scheibner, E. 1987. Stratotectonic terranes of the Eastern Australian Tasmanides. In: Leitch, E. C. & Scheibner, E. (eds), *Terrane accretion and orogenic belts*. American Geophysical Union, Geodynamics Series, **19**, 1–19.

Nekhoroshev, V. P. 1977. Devonskie mshanki Kazakhstana. *VSEGEI, Trudy*, **186**, 1–192

Oliver, W. A., Jr. & Galle, A. 1971. Rugose corals from the Upper Koneprusy Limestone (Lower Devonian) in Bohemia. *Sbornik Geologickych Ved, Paleontologie*, **14**, 35–106.

Oliver, W. A., Jr. & Pedder, A. E. H. 1979. Biogeography of Late Silurian and Devonian rugose corals in North America. In: Gray, J. & Boucot, A. J. (eds), *Historical biogeography, plate tectonics, and the changing environment*, Oregon State University Press, Corvallis, 131–145.

Table 7. *Otsuka coefficients between regions with important Early Devonian and Eifelian basin dwelling coral faunas*

	North Africa (G) Otsuka comparisons (basin facies)		
	Pragian	Emsian	Eifelian
Appalachian Belt (F)	61	(38)	25
Spain, Pyrenees (H)	–	60	61
Armorica, Montagne Noire, Vosges, Czechoslovakia (I)	62	53	(0)
Britain, N France, Belgium, central Germany, Poland (J)	(0)	(19)	45
Turkey (M)	–	36	–
Tien Shan (O)	50	40	(35)
Balkhash Kazakhstan (P)	67	48	45
Eastern Australia (X)	33	33	–

The generally long temporal and broad geographic ranges of genera included in the table give abnormally high otsuka values. Most coral faunas from basin facies comprise less than 10 genera; in this table, brackets indicate that one of the faunas considered consists of three or less genera.

—— & —— (in press). Origins, migration and extinctions of Devonian Rugosa in North America. *In*: *Proceedings of the Fifth International Symposium on Fossil Cnidaria*, Brisbane.

SCHELTEMA, R. S. 1968. Dispersal of larvae by equatorial ocean currents and its importance to the zoogeography of shoal-water tropical species. *Nature*, **217**, 1159–1162.

—— 1971. The dispersal of the larvae of shoal-water benthic invertebrates over long distances by ocean currents. *In*: CRISP, D. J. (ed.), *Fourth European Marine Biology Symposium, Bangor*, University Press, Cambridge, 7–28.

—— 1972. Dispersal of larvae as a means of genetic exchange between widely separated populations of shoal-water benthic invertebrate species. *In*: BATTAGLIA, B. (ed.) *5th European Marine Biology Symposium*, Venice, 101–114, Piccini Editore, Padua.

SCOTESE, C. R. 1986. Phanerozoic reconstructions: a new look at the assembly of Asia. *University of Texas Institute for Geophysics, Technical Report*, **66**, 1–54.

SORAUF, J. E. & PEDDER, A. E. H. 1986. Late Devonian rugose corals and the Frasnian–Famennian crisis. *Canadian Journal of Earth Sciences*, **23**, 1265–1287.

TALENT, J. A., GRATSIANOVA, R. T., SHISHKINA, G. R. & YOLKIN, E. A. 1987. Devonian faunas in relation to crustal blocks: Kazakhstan, Mongolia, Northern China. *Courier Forschungsinstitut Senckenberg*, **92**, 225–233.

TRIBBLE, G. W., RANDALL, R. H. 1986. A description of the high-latitude shallow water coral communities of Miyaka-jima, Japan. *Coral Reefs*, **4**, 151–159.

VEEVERS, J. J., JONES, J. G., POWELL, C. M. & TALENT, J. A. 1984. Chapter VI. Synopsis. *In*: VEEVERS, J. J. (ed.) *Phanerozoic earth history of Australia, Oxford Monographs on Geology and Geophysics*, **2**, Clarendon Press, Oxford, 351–364.

VOLLBRECHT, A., WEBER, K., SCHMOLL, J. 1988. Structural model for the Saxothuringian–Moldanubian suture in the Variscan basement of the Oberpfalz (Northeastern Bavaria, FRG) interpreted from geophysical data. *Tectonophysics*, **157**, 123–133.

WEIJERMARS, R. 1987. A Revision of the Eurasian–African plate boundary in the western Mediterranean. *Geologische Rundschau*, **76**, 667–676.

WEYER, D. 1984. Korallen in Palaozoikum von Thuringen. *Hallesches Jahrbuch fur Geowissenschaften*, **9**, 5–33.

WILLIAMS, E. 1978. Tasman Fold Belt System in Tasmania. *Tectonophysics*, **48**, 153–205.

ZHANG, Z. M., LIOU, J. G. & COLEMAN, R. G. 1984. An outline of the plate tectonics of China. *Bulletin of the Geological Society of America*, **95**, 295–312.

Early and Middle Devonian gastropod biogeography

R. B. BLODGETT[1], D. M. ROHR[2] & A.J. BOUCOT[3]

[1] US Geological Survey, Branch of Paleontology & Stratigraphy, National Center/MS 970, Reston, Virginia 22092, USA
[2] Department of Geology, Sul Ross State University, Alpine, Texas 79832, USA
[3] Department of Zoology, Oregon State University, Corvallis, Oregon 97331, USA

Abstract: Early and Middle Devonian gastropods show biogeographic patterns remarkably similar to those of better studied faunal groups of this same interval, notably articulate brachiopods, rugose corals, and trilobites. Three biogeographic realms are recognized: Old World, Eastern Americas, and Malvinokaffric Realms. Gastropod diversity is highest in the Old World Realm and lowest in the Malvinokaffric Realm. The diversity patterns and the high degree of shell ornamentation suggest that the Old World Realm was generally warmer than the Eastern Americas Realm, and that both were considerably warmer than the cool temperate to cold polar waters of the Malvinokaffric Realm. The utility of gastropods for fine-scale delineation of biogeographic units is illustrated for the Eifelian of western North America. At least two subprovincial units (the Alaska–Yukon and Nevada subprovinces) can be recognized. Eifelian gastropods from interior and southeastern Alaska belong to a single unit (the Alaska–Yukon Subprovince), and are most closely related to coeval faunas of northwestern Canada, suggesting little displacement of most of Alaska's so-called 'suspect' terranes. Plotting the data on the Devonian palaeogeographic maps of Scotese results in several suggested emendations: (1) North America should be moved south by 10–20°; (2) Australia is too far south on the Emsian and Givetian reconstructions, it should be in a more palaeotropical position; and (3) Siberia is too far north, it too should also be placed in a palaeotropical position.

The study of Devonian marine biogeography has sparked considerable interest during the past two decades. It has taken on even more significance in its role and implications in supporting various proposed 'mobilist' palaeogeographic reconstructions for this interval. The positioning of former Devonian continental land masses must make sense in terms of their contained fossil fauna and flora, allowing for their development under appropriate climatic conditions (i.e. fossil reef buildups should not be expected in polar regions) as well as allowing for reproductive communication between biogeographically similar marine faunas. Brachiopods have been the subject of the greatest amount of attention (Boucot 1974, 1975; Boucot et al. 1969; Johnson 1970, 1971; Johnson & Boucot 1973; Savage et al. 1979; and Wang Yu et al. 1984), followed secondly by rugose corals (Oliver 1973, 1976, 1977; Oliver & Pedder 1979, 1984), and thirdly by trilobites (Ormiston 1972, 1975; Kobayashi & Hamada 1975; Eldredge & Ormiston 1979). Other Devonian faunal and floral groups have only received very scant attention to date. The general picture emerging from these studies indicates the existence of three first order level biogeographic units which existed for much of the Devonian: the Old World Realm, the Eastern Americas Realm, and the Malvinokaffric Realm. The early Early Devonian (Lochkovian) is characterized by a moderate level of provincialism, followed by an acme of provincialism in middle to late Early Devonian (Pragian–Emsian) time. The Middle Devonian evinces progressively waning provincialism (the Eifelian remaining still strongly provincial, and subsequent Givetian being only moderately provincial), with the Late Devonian being nearly cosmopolitan, at least at the generic level. The cosmopolitan distribution in the Late Devonian may be exaggerated due to the absence or poor representation shallow marine benthic faunas in areas earlier occupied by Malvinokaffric Realm faunas.

Previous studies of Devonian gastropod biogeography

Gastropods have received little study in terms of Devonian global palaebiogeography, and up to the present decade no papers had been published on this subject. Forney et al. (1981) discussed the global distribution of selected molluscan genera in the Silurian and Lower Devonian. Gastropod genera with Devonian representatives utilized in that study were *Boiotremus*, *Euomphalopterus*, *Oriostoma*, and *Poleumita*. The distribution of these genera as well as that of the bivalve genus *Hercynella* are in agreement with biogeographic patterns previously established by brachiopod studies for the Lower Devonian.

Blodgett et al. (1986, 1988) analysed the distribution of Lower Devonian gastropods in the Western Hemisphere and found them to be consistent with previously utilized biogeographic units. Other significant conclusions reached in the latter studies were the recognition that Old World Realm faunas exhibited the greatest taxic diversity, and that these faunas were also much more highly ornamented than the remaining realms. On the basis of diversity patterns, as well as on other biotic and abiotic factors, it was suggested that the Early Devonian equator probably passed through Alaska and the Canadian Arctic Islands, and not obliquely across the North American continent (Nevada to New York) as suggested by earlier palaeomagnetic-based studies. The remaining two realms (Eastern Americas and Malvinokaffric) were also found to yield characteristic gastropod faunas. The former was moderately diverse, and during most of the Early Devonian it was typified by a plexus of spinose platyceratids and other endemic platyceratid genera (*Strophostylus* and *Crossoceras*). The latter realm is most notable for its extreme low diversity (containing only four of the seventeen superfamilies then extant); characteristic taxa included *Plectonotus* (*Plectonotus*) and large species of *Tropidodiscus*. The significantly lower faunal diversity of the latter was suggested to be consistent with its presumed position close to the Lower Devonian South Pole.

In the Eifelian (early Middle Devonian), both the Old World and Eastern Americas Realms were recognized in North America, each characterized by a relatively high degree of endemism (Blodgett et al. 1987). Typical Old World Realm elements include: *Hypomphalocirrus*, *Mastigospira*, *Odontomaria*, *Buechelia*, *Platyceras* (*Praenatica*), *P.* (*Prosigaretus*), *Plagiothyra*, *Strobeus*, and a number of other genera. Typical endemic Eastern Americas Realm taxa include: *Pleuronotus*, *Trochonema* (*Trochonemopsis*), *Elasmonema*, *Isonema*, *Turbonopsis*, *Palaeotochus*, as well as the entire plexus of spinose platyceratids. In addition, two new Eifelian palaeobiogeographic subprovinces were defined for gastropod faunas of the Cordilleran Region of the Old World Realm: the Alaska–Yukon and Nevada Subprovinces. The former includes faunas from west-central Alaska (Nixon Fork terrane) and east-central Alaska (Livengood terrane), while the latter includes faunas presently known only from Nevada. The former is characterized by greater diversity, more highly ornamented shells, and by a rich accompanying calcareous algal flora (absent in Nevada).

From McKerrow, W. S. & Scotese, C. R. (eds), 1990, *Palaeozoic Palaeogeography and Biogeography*, Geological Society Memoir No. 12, pp. 277–284.

The differences were ascribed to the positions of these two subprovinces in equatorial and warm, subtropical regions, respectively. The subprovince assignment of intervening areas (District of Mackenzie and Canadian Arctic Islands) were not discussed due to the poorly known character of their gastropod faunas.

Method of study and suggestions for future work

In this paper we extend our previous assessment of both Lower and Middle Devonian gastropod biogeography to include rocks of the Eastern Hemisphere, with the aim to establish global palaeobiogeographic patterns for this group, time interval by time interval. This study is based in part on analysis of previously published faunas in the older literature. This is complicated by the fact that outside of a few scattered investigators, little serious attention has been paid to Devonian gastropods subsequent to the First World War. This stands in marked contrast to the prominent position gastropods achieved in the studies by the great Nineteenth-Century Devonian investigators such as Goldfuss, D'Archiac and De Verneuil, the Sandberger brothers, Holzapfel and Oehlert. Of course in terms of mere number of Devonian gastropod taxa named, the most outstanding contributor has been Jaroslav Perner, who in his three tomes (1903, 1907 & 1911) proposed the largest number of Devonian gastropod genera and species. Work subsequent to World War One has been restricted to only a small number of investigators, working primarily on American, Czech, German, and Australian faunas. One of the tedious tasks associated with the assessment of the older literature is the utilization of greatly outdated taxic assignments, especially in the literature of the previous century. Reassignment of these faunas must rely on careful examination of line drawings, often of variable quality, and frequently not showing apertural characters which are most necessary in the delineation of gastropod genera and higher taxa. Also invaluable in the study of the European faunas was the help extended by a number of colleagues in allowing the examination of material in various repositories, or in arranging for loan of critical specimens. We are especially thankful to P. R. Racheboeuf, F. Langenstrassen, H. Jahnke, O. Walliser, H. Jaeger E. Schindler, S. Clasen, I. Chlupáč and V. Havlíček.

A serious problem hindering a complete assessment of Devonian gastropod biogeographic patterns is the dearth of knowledge concerning such faunas from large areas of the globe like Siberia, China, and New Zealand. Each of these areas has been the subject of considerable discussion concerning the biogeographic affinities of other Devonian biotic elements (most notably brachiopods, rugose and tabulate corals) found therein. Little can be said about the biogeographic affinities of gastropods from these areas until more basic taxonomic studies have been completed. It is our hope that the results presented here will encourage other workers to be watchful in their future collecting, and to collect and submit Devonian gastropods to qualified specialists for further study.

Devonian gastropod palaeoecology

The palaeoecological setting and community analysis of Devonian gastropods has received little comment in the literature. In our experience we have recognized gastropods in a number of shelf environments of varying depths, though in general, gastropods are most abundant and common in shallow shelf, nearshore settings (i.e. interidal zone, lagoons or reefs). These are all in the shallower part of the photic zone, a constraint which appears to be due to the dominance of Devonian (and Palaeozoic) gastropod faunas by the archaeogastropods. Modern archaeogastropods are overwhelmingly herbivorous, and it is commonly presumed that their Palaeozoic antecedents were also predominantly so. Hence, shallow photic zone depths would be most favoured by such a group since it would support a large algal flora which presumably provided a primary food source. Linsley (1979) considered that all archaeogastropods may not have always been algal grazers, since living pleurotomariaceans live below the photic zone. Nevertheless, we that feel this generalization probably holds true for Palaeozoic forms, due to their overwhelming abundance in very shallow-marine settings, an interpretation based both on sedimentological and biotic factors. The Alaskan Middle Devonian gastropod-dominated communities (often with over 40 gastropod species present at a collection horizon) provide especially compelling evidence for the essential correctness of this supposition. These faunas are nearly always accompanied by a rich and diverse calcareous green algal flora (Blodgett 1987; Blodgett et al. 1987; Poncet & Blodgett 1987) containing both dasycladacean and udoteacean elements. Typical genera encountered include *Coelotrochium* (a dasycladacean), *Lancicula* (a udoteacean), receptaculitids (*Sphaerospongia tessellata*), and charophytes (*Sycidium*). Silurian gastropod faunas are also dominated by archaeogastropods, and Peel (1984) considered that most Silurian gastropods were probably microherivores or browsers on algae or colonial animals.

Early Devonian gastropod biogeography

Early Devonian gastropods evidence biogeographic patterns remarkably similar to those previously established on the basis of articulate brachiopods, rugose corals, and trilobites (Blodgett et al. 1986, 1988; Forney et al. 1981). A detailed biogeographic analysis for Early Devonian gastropods in the Western Hemisphere was presented by Blodgett et al. (1988). Global endemism increased from a moderate level in the Lochkovian, to a very high level in the Pragian–Emsian. As with other faunal groups, three first-order level biogeographic units (realms) can be delineated. Gastropod diversity is highest in the Old World Realm, which is recognized in Eurasia, northern Africa, Australia, and western, Arctic, and northeasternmost North America. This realm, based on its biotic character and associated climatically sensitive lithologic features of its strata (presence of calcareous green algae (Boucot et al. 1988; Poncet & Blodgett 1986), oolites (Boucot et al. 1988), reef buildups, evaporite deposits, etc.) is interpreted to have been in a warm, tropical to subtropical palaeoclimatic setting. This interpretation is supported by the more highly ornamented nature of shells found in the gastropod faunas of this realm (Blodgett & Rohr 1989). Typical Early Devonian taxa restricted to the Old World Realm include: *Paleuphemites*, *Coelocylcus*, *Boiotremus*, *Aspidotheca*, *Kodymites*, *Sinistracirsa*, *Oehlertia*, *Hesperiella*, *Stenoloron*, *Platyceras* (Praenatica), *P.* (*Prosigaretus*), oriostomatids, *Michelia*, *Coelocaulus*, *Scalaetrochus*, *Euomphalopterus*, *Planotrochus*, *Mitchellia*, *Stylonema*, and *Katoptychia*.

The Eastern Americas Realm occupied much of the eastern and mid-continent region of North America, as well as northern South America. On the basis of its biotic and lithologic content, this realm is interpreted to have been in tropical to cool temperate palaeolatitudes. Three palaeoclimatic regimes can be recognized for both the Early and Middle Devonian: Hudson Bay Platform–Michigan Basin (tropical to subtropical with abundant evaporites and oolites; gastropod diversity is also higher here than elsewhere in the realm), Appalachian Basin (subtropical to warm temperate), and the Amazon–Colombian area (moderate to cool temperate with limited or no carbonates). Early Devonian endemic taxa include spinose platyceratids (found in all subprovince of the realm), *Crossoceras* and *Strophostylus*.

Nevadan gastropods confirm the shifting biogeographic boundaries recognized previously in other faunal groups. In Lochkovian time the area was part of the Old World Realm. During Pragian–early Emsian time Nevada contained Appalachian gastropod genera (Nevadan Subprovince of the Eastern Americas Realm). Late in the Emsian (*pinyonensis* Zone) mixing occurred with elements of both realms co-occurring.

In Eifelian time the gastropod faunas are almost wholly again of Old World Realm character (Nevada Subprovince of the Cordilleran Region of the Old World Realm).

The Malvinokaffric Realm of southern and central South America, Falkland Islands, South Africa, and Antarctica contains a strongly depauperate fauna with only four superfamilies represented from the 17 extant in the Early Devonian (18 in the Middle Devonian). This realm appears to have been situated in cool temperate to even colder polar waters of the Devonian Southern Hemisphere, based on its total lack of carbonates, complete absence of biohermal buildups, greatly diminished taxonomic diversity of nearly all invertebrate phyla, as well as the complete absence of warm-water biotic elements such as stromatoporoids, conodonts, gypidulinid and atrypid brachiopods.

Lochkovian

The biogeographic distribution of well documented Lochkovian gastropod faunas (Fig. 1) is in close agreement with that of other studied faunal groups. Although their reported occurrences are somewhat patchy, Old World Realm gastropods are known from western and Arctic North America, as well as from Nova Scotia (see Blodgett et al. 1988 for detailed listing of Lower Devonian gastropods from the Western Hemisphere). They are also present in France (Massif Armoricain), Czechoslovia (Perner 1903, 1907, 1911; Horný 1963), Belgium (Asselberghs 1930), Germany, Podolia (Siemiradzki 1906), the Urals, North Africa (Termier & Termier 1950; Sougy 1964), Kazakhstan (Rohr et al. 1979), and Australia.

Eastern Americas Realm Lochkovian gastropod faunas are known from the Appalachians (Gaspé, Maine, New York, Maryland, Virginia) and the central United State (Tennessee, Missouri); no occurrences have been noted as yet of Lochkovian age marine strata in northern South America (later the site of Eastern Americas Realm faunas). No Lochkovian age gastropods have as yet been recognized from rocks of the Malvinokaffric Realm.

Pragian–Emsian

This interval, including the two uppermost stages of the Lower Devonian, has been lumped together for convenience. Biogeographically distinctive gastropod faunas of this age are reported from a number of places globally (Fig. 2). Old World Realm gastropod faunas are recognized in western (only as far south as British Columbia) and Arctic North America, throughout Europe (France (Oehlert 1877, 1888), Belgium (Maillieux 1932), Germany (Dahmer 1917, 1921, 1926), the Carnic Alps (Frech 1894; Spitz 1907; Jhaveri 1969)), the Urals, North Africa (Termier & Termier 1950; Sougy 1964), Saudi Arabia (Boucot et al. 1988), Australia (Tassell 1976, 1977, 1978, 1980, 1982) and New Zealand.

Eastern Americas Realm gastropod faunas of this interval are recognized from the Appalachians (Gaspé to Alabama), central US (Tennessee, Missouri, Oklahoma, Texas), and northern South America. The most typical gastropod elements of the realm during this interval are the platyceratids, many belonging to a seemingly endemic plexus of spinose species (which may upon further study be delineated into new subgenera). Nevadan gastropods from this interval are overwhelmingly of Eastern Americas Realm affinities, in contrast to their Old World Realm affinities, both before and after.

The taxonomically non-diverse Malvinokaffric gastropod faunas are recognized in central and southern South America, the Falkland Islands, South Africa, Ghana, and Antarctica.

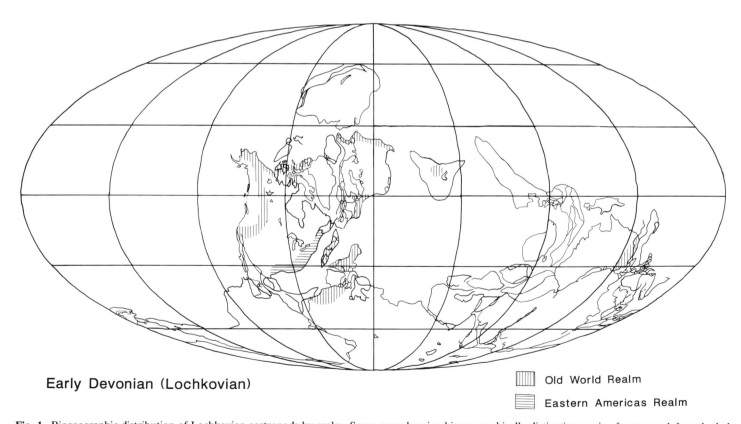

Fig. 1. Biogeographic distribution of Lochkovian gastropods by realm. Some areas bearing biogeographically distinctive marine faunas are left unshaded on this and the following figures due to lack of knowledge about gastropods from these areas. Base map is that for the Gedinnian (approximately equivalent to the Lochkovian) from Scotese (1986).

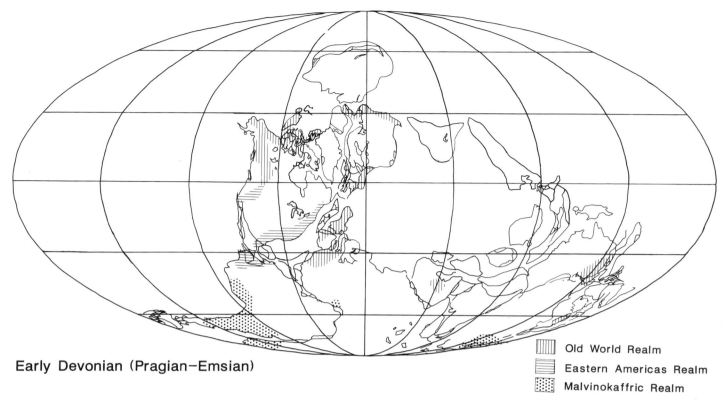

Fig. 2. Biogeographic distribution of Pragian–Emsian gastropods by realm. Some areas bearing marine faunas of this age are left unshaded due to lack of knowledge about gastropods from these areas. The area of the Great Basin of North America is shown as belonging to the Eastern Americas Realm. During the Pragian–early Emsian, gastropods from this area were predominantly of Eastern Americas Realm affinity. Late in the Emsian (*pinyonensis* Zone) mixing occurred with elements of both realms co-occurring. Base map is that for the Emsian from Scotese (1986).

Only a handful of species can be recognized throughout the realm, the most typical elements being *Plectonotus* (*Plectonotus*) and large species of *Tropidodiscus*.

Middle Devonian gastropod biogeography

Middle Devonian gastropods are well known from both the Old World and Eastern Americas Realms, and show strong provincialism in the Eifelian, with distinct but somewhat lessened provincialism still present in the Givetian. The general decline of provincialism from the Eifelian into the Givetian is a feature which has also been noted in other better studied faunal groups of this interval. As in the Early Devonian, taxic diversity is greater in Old World Realm faunas than in that of the corresponding Eastern Americas Realm. This again is due, we feel, to a generally warmer, more tropical positioning of areas occupied by the Old World Realm faunas (Blodgett *et al.* 1988; Boucot *et al.* 1988). Typical Middle Devonian Old World Realm taxa include: *Pedasolia*, *Hypomphalocirrus*, *Mastigospira*, *Odontomaria*, *Platyschisma*, *Buechelia*, *Scalitina*, n. gen. aff. *Scalitina*, *Baylea*, *Catantostoma*, n. gen. aff. *Porcellia*, *Platyceras*, (*Praenatica*), *P.* (*Prosigaretus*), *Oriostoma* (oriostomatids declined rapidly and became extinct near the end of the Eifelian), *Plagiothyra*, n. gen. neritopsinid, nodose *Murchisonia* (*Murchisonia*) (abundant in Givetian reefs), *Astralites*, *Scoliostoma*, *Spanionema*, and *Strobeus*.

Typical Middle Devonian, endemic Eastern Americas Realm gastropod taxa include: *Pleuronotus*, *Trochonema* (*Trochonemopsis*), *Elasmonema*, *Isonema*, *Murchisonia* (*Hormotomina*), *Turbonopsis*, *Palaeotrochus*, and the entire plexus of spinose platyceratids. Again endemism is much more strongly marked in this realm during the Eifelian, than in the subsequent Givetian. Limited faunal communication during the Middle Devonian between the Appalachian area and northwestern Europe (Old World Realm) is suggested by the rare occurrence of such a typical Appalachian taxon as a spinose *Platyceras* (*Platyceras*) in the Eifelian of Germany.

Eifelian

Eifelian gastropods show a considerable degree of endemism (Blodgett 1987; Blodgett *et al.* 1987), and all three realms are recognized as present in this interval (Fig. 3). Old World Realm gastropod faunas are recognized in the Eifelian of western and Arctic North America (Blodgett 1987, 1988; Blodgett & Rohr 1989; Cleland 1911; LaRocque 1949; Linsley 1978; Tolmachoff 1926), Europe (Goldfuss 1844; Spriestersbach 1942; Whidborne 1889–1892), North Africa (Termier & Termier 1950), and the Kuznetsk Basin of Siberia (Butusova 1960). Nevadan Eifelian gastropods are of notable Old World Realm character, in marked contrast to their Eastern Americas Realm affinities in the Pragian-early Emsian, and mixed character in the later Emsian. No Eifelian gastropods are known from Australia, presumably due to the erosion or non-deposition of strata during the Tabberabberan Orogeny.

Eastern Americas Realm gastropods are known from Eifelian strata of the eastern North America (Hall 1879; Kindle 1901; Linsley 1968; Meek 1873; Nettelroth 1889; Rollins *et al.* 1971; Stauffer 1909) and northern South America (Colombia, Venezuela and the Amazon Basin). The area of the Michigan Basin is shown as one of mixed biogeographic realm affinities during the Eifelian (Fig. 3.). In fact, it is predominantly of Eastern Americas Realm affinity for much of the Eifelian, with only a short-lived invasion of strongly Old World Realm forms recognized in late Eifelian gastropods of the Rogers City Limestone of Michigan and Lake Church Formation of Wisconsin (Blodgett

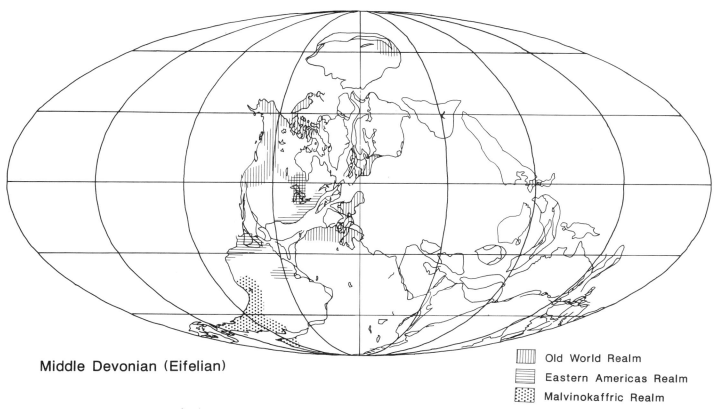

Fig. 3. Biogeographic distribution of Eifelian gastropods by realm. Note area of realm boundary mixing indicated in the Michigan Basin. Base map is that for the Givetian from Scotese (1986).

1988). Re-establishment of Eastern Americas Realm affinities are evidenced by the succeeding, latest Eifelian age gastropods from Michigan. Malvinokaffric Realm Eifelian gastropods are shown as being present in western and southern South America (Fig. 3), though their occurrences are often in strata whose correlation could also be with the Emsian.

An excellent example of the potential utility of gastropods for the discrimination of detailed biogeographic units, providing the sampling base is sufficient, is demonstrated by Eifelian gastropods in western North America. There, extremely large, predominantly silicified faunas collected from across the breadth of Alaska by one of us (RBB) and equivalent material from Nevada (made available by J. G. Johnson) allows the recognition of least two subprovincial biogeographical units (Alaska–Yukon and Nevada Subprovinces) within the Eifelian of the Cordilleran Region of the Old World Realm (Blodgett 1987; Blodgett et al. 1987). In interior and southeastern Alaska Eifelian age gastropod faunas are remarkably similar, and their close affinities suggest they belong to a single biogeographic unit (Alaska–Yukon Subprovince). Eifelian gastropods of the Great Basin represent a distinct unit, which was termed the Nevada Subprovince. The presence of a nearly homogeneous Eifelian gastropod fauna across the breadth of interior and southeastern Alaska, and the alliance of this and other accompanying faunal groups with coeval faunas from northwestern Canada, suggest that much of Alaska was more or less in place relative to North America in Devonian time, and not formed by the accretion of numerous disparate tectonostratigraphic terranes during the Mesozoic, as others have previously suggested (Coney et al. 1980; Jones & Silberling 1979; Jones et al. 1981, 1982). Both palaeobiogeographic and stratigraphic evidence from Devonian strata of Alaska have been previously used to support this position by Blodgett (1983) and Blodgett & Clough (1985), as well as by a number of recently published studies of Early and Middle Palaeozoic faunas and floras in the past few years (Poncet & Blodgett 1987; Potter et al. 1988; Rigby et al. 1988; Rohr & Blodgett 1985). Most of the terrane boundaries in Alaska are nothing more than strike-slip faults, requiring only minor dislocation of terrane blocks northward along such well-known features as the Tintina and Denali Faults, not long-distance trans-oceanic voyages as more fanciful interpretations have suggested.

Givetian

The biogeographic distribution of Givetian gastropods (Fig. 4) by realm is very similar to that shown for the Eifelian. One major difference is the absence of Malvinokaffric Realm faunas. No definitive Givetian localities bearing gastropods are known from areas previously assigned to this realm earlier in the Devonian. Old World Realm faunas are known from western and Arctic America (LaRocque 1949; Linsley 1978; Whiteaves 1892), Europe (D'Archiac & DeVerneuil 1842; Goldfuss 1844; Holzapfel 1895; Kirchner 1915; Sandberger & Sandberger 1850–1856; Whidborne 1889–1892), North Africa (Sougy 1964) south China (Yunnan), and Australia (Heidecker 1959). The European Givetian faunas are highly distinctive and were prominently figured in many monographs, especially in the nineteenth century, from both reefal and lagoonal settings in England and Germany. Perhaps the most distinctive elements among Givetian Old World Realm gastropods are the diverse and rapidly evolving plexus of nodose members of *Murchisonia* (*Murchisonia*). This nodose group appears to be limited wholly to strata of Givetian–Frasnian age. Givetian Old World Realm gastropods are best known from North America in the faunas of the Winnipegosis Formation of Manitoba (Whiteaves 1892; Linsley 1978). It should be noted that the area of the Michigan Basin was one of faunal realm mixing, as in the Eifelian, with strongly provincial faunas of either Old World or Eastern Americas Realm affinity found

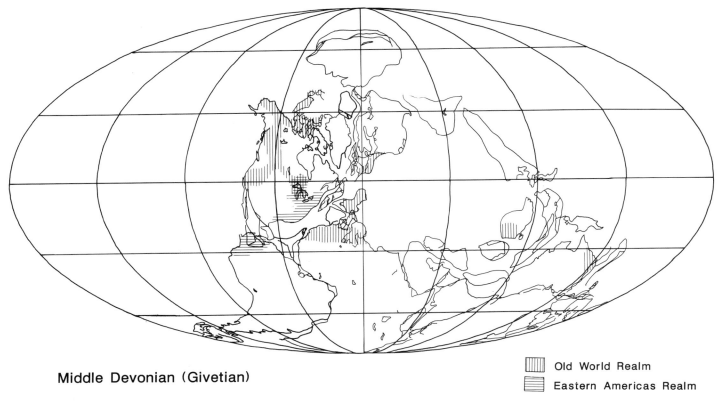

Fig. 4. Biogeographic distribution of Givetian gastropods by realm. Note area of realm boundary mixing indicated in the Michigan Basin. Base map is that for the Givetian from Scotese (1986).

restricted to seemingly separate horizons. The Givetian gastropods of south China (Yunnan) are known from Mansuy (1912), and by themselves are quite remarkable since many of the species are either conspecific or very close to species described from equivalent strata of Germany. Australian Givetian faunas are still poorly known, all illustrated representatives being from Queensland. However, the presence of several prominent, endemic genera (*Burdikinia*, *Austerum*, *Labrocuspis*), which appear to be locally common in the megafauna, suggests that Australia was still characterized by a greater degree of endemism than other, better known parts of the Old World Realm, which now show little differentiation of endemic gastropod genera. Givetian gastropods of the Eastern Americas Realm (Hall 1879; Fraunfelter 1973, 1974) exhibit a lesser degree of endemism than their Eifelian antecedents. Endemism is still evident, however, as witnessed by the common occurrence of spinose platyceratids.

Conclusions

Devonian gastropods show biogeographic patterns similar to those noted in other better studied faunal groups for this time interval, most notably the articulate brachiopods, rugose corals, and trilobites. Taxonomic studies of Devonian gastropods tend to show a very strong geographical bias, with the faunas of some areas being relatively well known during certain intervals (notably central Europe), while those from other areas (i.e. China and Soviet Asia) are barely known at all. These strong regional biases in our knowledge are due to the lack of personnel able to undertake taxonomic studies of what many palaeontologists of late have tended to regard as a relatively 'useless' group for Palaeozoic biostratigraphic studies. We feel this view is in error. Emerging detailed knowledge of Eifelian age gastropod faunas of western North America currently allow delineation of at least two subprovincial level biogeographic units (Blodgett 1987; Blodgett et al. 1987). The remarkable faunal similarities of Eifelian gastropods from interior and southeastern Alaska suggest that they all belong to the same unit (Alaska–Yukon Subprovince). These faunas are also closely allied to coeval faunas of northwestern Canada, indicating little transport of the so-called terrane 'blocks', suggested to comprise most of Alaska. This relatively local derivation, and minor ascribed motion to the terranes is in accord with the emerging pattern from other recently published palaeobiogeographic studies of Early and Middle Palaeozoic faunas of Alaska.

Devonian gastropod diversity patterns, degree of shell ornamentation, as well as a number of other biotic and abiotic factors (discussed above) suggest that in general the Old World Realms faunas were warmer than that of Eastern Americas Realm, and that both were considerably warmer than the Malvinokaffric Realm faunas, the latter which are considered to have been situated in cool temperate to even colder polar waters of the southern hemisphere.

Plotting of the Devonian gastropod biogeographic data (Figs 1–4) on the maps utilized for this symposium (Scotese 1986) result in several suggested emendations: (1) North America has been placed too high, it is suggested that it should be moved south by 10–20° in order to place Alaska and the Canadian nearly astride the palaeo-equator during Early Devonian–Eifelian time in accord with evidence provided above; (2) the position of Australia on the Emsian and Givetian maps is too far south of the palaeo-equator, the character of the fossil fauna and flora suggest it was in a more tropical (palaeo-equatorial) position; and (3) although described Devonian gastropods are few from Siberia, the richly, diverse calcareous green algal flora known from the Kuznetsk Basin indicates that Siberia is placed too far north in these reconstructions, and must have been situated in a palaeotropical belt during the Early and Middle Devonian.

References

ARCHIAC, E. J. A. D' & DE VERNEUIL, E. P. 1842. On the fossils of the older deposits in the Rhenish provinces, preceded by a general survey of the fauna of Palaeozoic rocks, and followed by a tabular list of the organic remains of the Devonian System in Europe. *Transactions of the Geological Society of London*, **6**, 303–410.

ASSELBERGHS, E. 1930. Description des Faunes marines du Gedinnien de l'Ardenne. *Mémoires du Musée Royal d'Histoire Naturelle de Belgique*, **41**.

BLODGETT, R. B. 1983. Paleobiogeographic affinities of Devonian fossils from the Nixon Fork terrane, southwestern Alaska. *In*: STEVENS C. H. (ed.), *Pre-Jurassic rocks in Western North American Suspect Terranes*. Pacific Section, Society of Economic and Petroleum Geologists, 125–130.

—— 1987. *Taxonomy and paleobiogeographic affinities of an Early Middle Devonian (Eifelian) gastropod faunule from the Livengood quadrangle, east-central Alaska*. PhD thesis, Oregon State University, Corvallis.

—— 1988. *Wisconsinella*, a new raphistomatinid gastropod genus from the Middle Devonian of Wisconsin. *Journal of Paleontology*, **62**, 442–444.

—— & CLOUGH, J. G. 1985. The Nixon Fork terrane – part of an *in-situ* peninsular extension of the Palaeozoic North American continent. *Geological Society of America, Abstracts with Programs*, **17**, 342.

—— & ROHR, D. M. 1989. Two new Devonian spine-bearing pleurotomariacean gastropod genera from Alaska. *Journal of Paleontology*, **63**, 47–53.

——, —— & BOUCOT, A. J. 1986. Lower Devonian gastropod biogeography of the Western Hemisphere. *Geological Society of America, Abstracts with Programs*, **18**, 543.

——, —— & —— 1987. Early Middle Devonian (Eifelian) gastropod biogeography of North America. *Geological Society of America, Abstract with Programs*, **19**, 591.

——, —— & —— 1988. Lower Devonian gastropod biogeography of the Western Hemisphere. *In*: MC MILLAN, N. J., EMBRY, A. F. & GLASS, D. J. (eds) *Devonian of the World*. Canadian Society of Petroleum Geologists Memoir, **14**, Vol. 3, 281–294.

BOUCOT, A. J. 1974. Silurian and Devonian biogeography. *In*: ROSS, C. A. (ed.) *Paleogeographic Provinces and Provinciality*. Society of Economic Paleontologists and Mineralogists, Special Publication, **21**, 165–176.

—— 1975. *Evolution and Extinction Rate Controls*. Elsevier, Amsterdam.

——, JOHNSON, J. G. & TALENT, J. A. 1969. Early Devonian Brachiopod Zoogeography. Geological Society of America, Special Paper, **119**.

——, ROHR, D. M. & BLODGETT, R. B. 1988. A marine invertebrate faunule from the Tawil Sandstone (basal Devonian) of Saudi Arabia and its biogeographic-paleogeographic consequences. *In*: WOLBERG, D. L. (ed.), *Contributions to Paleozoic Paleontology and Stratigraphy in honor of Rousseau H. Flower*. New Mexico Bureau of Mines & Mineral Resources Memoir, **44**, 361–372.

BUTUSOVA, I. P. 1960. Nekotorye gastropody mamontovskikh sloev srednego devona kuznetskogo basseina. VSEGEI, *Informatsionnyi Sbornik*, **35**, 81–89.

CLELAND, H. F. 1911. The fossils and stratigraphy of the Middle Devonic of Wisconsin. *Wisconsin Geological and Natural History Survey, Bulletin*, **21**.

CONEY, P. J., JONES, D. L. & MONGER, J. W. H. 1980. Cordilleran suspect terranes. *Nature*, **288**, 329–333.

DAHMER, G. 1917. Studien über die Fauna des Oberharzer Kahlebergsandsteins. I. *Jahrbuch der Königlich Preussischen Geologischen Landesanstalt für 1916*, **37**, 443–526.

—— 1921. Studien über die Fauna des Oberharzer Kahlebergsandsteins. II. *Jahrbuch der Preussischen Geologischen Landesanstalt für 1919*, **40**, 161–306.

—— 1926. Die Fauna der Sphärosideritschiefer der Lahnmulde. Zugleich ein Beitrag zur Kenntnis unterdevonischer Gastropoden. *Jahrbuch der Preussischen Geologischen Landesanstalt für 1925*, **46**, 34–67.

ELDREDGE, N. & ORMISTON, A. R. 1979. Biogeography of Silurian and Devonian trilobites of the Malvinokaffric Realm. *In*: GRAY, J. & BOUCOT, A. J. (eds) *Historical Biogeography, Plate Tectonics, and the Changing Environment*. Oregon State University Press, Corvallis, 147–167.

FORNEY, G. G., BOUCOT, A. J. & ROHR, D. M. 1981. Silurian and Lower Devonian zoogeography of selected molluscan genera. *In*: GRAY, J., BOUCOT, A. J. & BERRY, W. B. N. (eds) *Communities of the Past*. Hutchinson Ross, Stroudsberg, Pennsylvania, 119–164.

FRAUNFELTER, G. H. 1973. Mollusca from the Lingle and St. Laurent Limestones (Middle Devonian) of southern Illinois and southeastern Missouri. *Southern Illinois Studies*, **11**.

—— 1974. Invertebrate megafauna of the Middle Devonian of Missouri. *Southern Illinois Studies*, **13**.

FRECH, F. 1894. Ueber das Devon der Ostalpen. III. (Die Fauna des unterdevonischen Riffkalkes. I). *Zeitschrift der Deutschen geologischen Gesellschaft*, **46**, 446–479.

GOLDFUSS, G. A. 1844. *Petrefacta Germaniae*. Volume 3, Dusseldorf.

HALL, J. 1879. Containing descriptions of the Gasteropoda, Pteropoda, and Cephalopoda of the Upper Helderberg, Hamilton, Portage and Chemung groups. *Natural History of New York, Palaeontology*, **5**, part 2, Albany.

HEIDECKER, E. 1959. Middle Devonian molluscs from the Burdekin Formation of North Queensland. *University of Queensland Papers, Department of Geology*, **V**, no. 2.

HOLZAPFEL, E. 1895. Das Obere Mitteldevon (Schichten mit *Stringocephalus burtini* und *Maeneceras terebratum* im Rheinischen Gebirge. *Abhandlungen der Königlich Preussischen geologischen Landesanstalt*, Neue Folge, Heft 16.

HORNÝ, R. J. 1963. Lower Palaeozoic Bellerophontina (Gastropoda) of Bohemia. *Sborník Geologických Věd, ráda P, svazek* **2**, 57–164.

JHAVERI, R. B. 1969. Unterdevonische Gastropoden aus den Karnischen Alpen. *Palaeontographica Abt. A*, **133**, 146–176.

JOHNSON, J. G. 1970. Taghanic onlap and the end of North America Devonian provinciality. *Geological Society of America Bulletin*, **81**, 2077–2105.

—— 1971. A quantitative approach to faunal province analysis. *American Journal of Science*, **270**, 257–280.

—— & BOUCOT, A. J. 1973. Devonian brachiopods. *In*: HALLAM, A. (ed.) *Atlas of Palaeobiogeography*. Elsevier, New York, 89–96.

JONES, D. L. & SILBERLING, N. L. 1979. *Mesozoic stratigraphy – the key to tectonic analysis of southern and central Alaska*. U.S. Geological Survey Open File Report, 79–1200.

—, COX, A., CONEY, P. & BECK, M. 1982. The growth of western North America. *Scientific American*, **247**, no. 5, 70–84.

——, SILBERLING, N. L., BERG, H. C. & PLAFKER, G. 1981. *Map showing tectonostratigraphic terranes of Alaska, columnar sections, and summary descriptions of terranes*. U.S. Geological Survey Open File Report, 81–792.

KINDLE, E. M. 1901. The Devonian fossils and stratigraphy of Indiana. *Indiana Department of Geology and Natural Resources, Twenty-fifth Annual Report*, 529–758, 773–775.

KIRCHNER, H. S. 1915. Mitteldevonische Gastropoden von Soetenich in der Eifel. *Verhandlungen des Naturhistorischen Vereins der preussischen Rheinlande und Westfalens*, Einundsiebzigster Jahrgang, 1914, Zweite Hälfte, 189–261.

KOBAYASHI, T. & HAMADA, T. 1975. Devonian trilobite provinces. *Japan Academy, Proceedings*, **51**, 447–451.

LAROCQUE, A. 1949. New uncoiled gastropods from the Middle Devonian of Michigan and Manitoba. *University of Michigan, Contributions from the Museum of Paleontology*, **7**, 113–122.

LINSLEY, R. M. 1968. Gastropods of the Middle Devonian Anderdon Limestone. *Bulletins of American Paleontology*, **54**, 333–465.

—— 1978. The Omphalocirridae: a new family of Palaeozoic Gastropoda which exhibits sexual dimorphism. *Memoirs of the National Museum of Victoria*, **39**, 33–54.

—— 1979. Gastropods of the Devonian. *In*: HOUSE, M. R., SCRUTTON, C. T. & BASSETT, M. G. (eds) *The Devonian System*. Palaeontological Association, Special Papers in Palaeontology, **23**, 249–254.

MAILLIEUX, E. 1932. La Faune de l'Assise de Winenne (Emsien Moyen). *Mémoires du Musée Royal d'Histoire Naturelle de Belgique*, **52**.

MANSUY, H. 1912. Étude géologique de Yun-nan Oriental, Part II, Paléontologie. *Mémoires du Service Géologique de l'Indo-Chine*, **I**, Fascicule II.

MEEK, F. B. 1873. Descriptions of invertebrate fossils of the Silurian and

Devonian systems. *Geological Survey of Ohio*, **1**, part 2, 1–243.

NETTELROTH, H. 1889. Kentucky fossil shells; a monograph of the fossil shells of the Silurian and Devonian rocks of Kentucky. *Kentucky Geological Survey*.

OEHLERT, D. P. 1877. Sur les fossiles dévoniens du département de le Mayenne. *Bulletin de la Société géologique de France, 3e série*, **5**, 578–603.

—— 1888. Descriptions de quelques espèces dévoniennes du département de la Mayenne. *Bulletin de la Société d'Etudes Scientifiques d'Angers*, **1887**, 65–120.

OLIVER, W. A., JR. 1973. Devonian coral endemism in eastern North America and its bearing on paleogeography. *In*: HUGHES, N. F. (ed.) *Organisms and Continents through Time*. Palaeontological Association, Special Papers in Palaeontology, **12**, 318–319.

—— 1976. Biogeography of Devonian rugose corals. *Journal of Paleontology*, **50**, 365–373.

—— 1977. Biogeography of Late Silurian and Devonian rugose corals. *Palaeogeography, Palaeoclimatology, Palaeoecology*, **22**, 85–135.

—— & PEDDER, A. E. H. 1979. Rugose corals in Devonian stratigraphic correlation. *In*: HOUSE, M. R., SCRUTTON, C. T. & BASSETT, M. G. (eds) *The Devonian System*. Palaeontological Association, Special Papers in Palaeontology, **23**, 233–248.

—— & —— 1984. Devonian rugose coral biostratigraphy with special reference to the Lower–Middle Devonian boundary. *Geological Survey of Canada Paper*, **84–1A**, 449–452.

ORMISTON, A. R. 1972. Lower and Middle Devonian trilobite zoogeography in northern North America. *24th International Geological Congress*, Section **7**, 594–604.

—— 1975. Siegenian trilobite zoogeography in Arctic North America. *In*: MARTINSSON, A. (ed.) *Evolution and morphology of the Trilobita, Trilobitoidea, and Merostomata*. Fossils and Strata, **4**, 391–398.

PEEL, J. S. 1984. Autoecology of Silurian gastropods and monoplacophorans. *In*: BASSETT, M. G. & LAWSON, J. D. (eds) *Autoecology of Silurian organisms*. Palaeontological Association, Special Papers in Palaeontology, **32**, 165–182.

PERNER, J. 1903. *In*: BARRANDE, J. *Système silurien du centre de la Bohême*, Vol. 4, *Gastéropodes*, Tome 1 (Patellidae et Bellerophontidae), Prague.

—— 1907. *In*: BARRANDE, J. *Système silurien du centre de la Bohême*, Vol. 4, *Gastéropodes*, Tome 2, Prague.

—— 1911. *In*: BARRANDE, J. *Système silurien du centre de la Bohême*, Vol. 4, *Gastéropodes*, Tome 3, Prague.

PONCET, J. & BLODGETT, R. B. 1987. First recognition of the Devonian alga *Lancicula sergaensis* Shuysky in North America (west-central Alaska). *Journal of Paleontology*, **61**, 1269–1273.

POTTER, A. W., BLODGETT, R. B. & ROHR, D. M. 1988. Paleobiogeographic relations and paleogeographic significance of Late Ordovician brachiopods from Alaska. *Geological Society of America, Abstracts with Programs*, **20**, A339.

RIGBY, J. K., POTTER, A. W. & BLODGETT, R. B. 1988. Ordovician sphinctozoan sponges of Alaska and Yukon Territory. *Journal of Paleontology*, **62**, 731–746.

ROHR, D. M. & BLODGETT, R. B. 1985. Upper Ordovician Gastropoda from west-central Alaska. *Journal of Paleontology*, **59**, 667–673.

——, BOUCOT, A. J. & USHATINSKAYA, G. T. 1979. Early Devonian-gastropods from the north Pribalkhash region, central Kazakhstan, USSR. *Journal of Paleontology*, **53**, 981–989.

ROLLINS, H. B., ELDREDGE, N. & SPILLER, J. 1971. Gastropoda and Monoplacophora of the Solsville Member (Middle Devonian, Marcellus Formation) in the Chenango Valley, New York State. *Bulletin of the American Museum of Natural History*, **144**, 129–170.

SANDBERGER, G. & SANDBERGER, F. 1850–1856. *Die Versteinerungen des rheinischen Schichtensystems in Nassau*. Wiesbaden.

SAVAGE, N. M., PERRY, D. M. & BOUCOT, A. J. 1979. A quantitative analysis of Lower Devonian brachiopod distribution. *In*: GRAY, J. & BOUCOT, A. J. (eds) *Historical Biogeography, Plate Tectonics, and the Changing Environment*. Oregon State University Press, Corvallis, 169–200.

SCOTESE, C. R. 1986. Phanerozoic reconstructions: A new look at the Assembly of Asia. *University of Texas Institute for Geophysics Technical Report*, **66**.

SIEMIRADZKI, J. 1906. Die Palaozoischen gebilde Podoliens. *Beiträge zur Paläontologie und Geologie Österreich-Ungarns und des Orients*, **19**, 173–286.

SOUGY, J. 1964. Les formations paléozoiques du Zemmour noir (Mauritanie septentrionale); etude stratigraphique, pétrographique et paléontologique. *Université de Dakar. Annales de la Faculté des Sciences*, **15**.

SPITZ, A. 1907. Die Gastropoden des Karnischen Unterdevon. *Beiträge zur Paläontologie und Geologie Österreich-Ungarns und des Orients*, **20**, 115–190.

SPRIESTERSBACH, J. 1942. Lenneschiefer (Stratigraphie, Fazies und Fauna). *Abhandlungen des Reichsamts für Bodenforschung*, Neue Folge, Heft 203.

STAUFFER, C. R. 1909. *The Middle Devonian of Ohio*. Geological Survey of Ohio, Fourth Series, Bulletin 10.

TASSELL, C. B. 1976. A revision of the gastropod fauna of the Lilydale Limestone (Early Devonian) of Victoria. *Memoirs of the National Museum of Victoria*, **37**, 1–22.

—— 1977. Gastropods from some Early Devonian limestones of the Walhalla Synclinorium, central Victoria. *Memoirs of the National Museum of Victoria*, **38**, 231–245.

—— 1978. Gastropods from the Early Devonian Bell Point Limestone, Cape Liptrap Peninsula, Victoria. *Memoirs of the National Museum of Victoria*, **39**, 19–32.

—— 1980. Further gastropods from the Early Devonian Lilydale Limestone, Victoria. *Records of the Queen Victoria Museum Launceston*, **69**.

—— 1982. Gastropods from the Early Devonian "*Receptaculites*" Limestone, Taemas, New South Wales. *Records of the Queen Victoria Museum Launceston*, **77**.

TERMIER, G. & TERMIER, H. 1950. Invertébrés de l'Ère Primaire, Fascicule III, Mollusques. *Notes et Mémoires*, **78**, Fascicule III.

TOLMACHOFF, I. P. 1926. On the fossil faunas from Per Schei's Series D from Ellesmere Land with exception of brachiopods, corals, and cephalopods. *Report of the second Norwegian Arctic Expedition in the "Fram" 1898–1902*, **38**.

WANG YU, BOUCOT, A. J., RONG JIA-YU & YANG XUE-CHANG. 1984. Silurian and Devonian biogeography of China. *Geological Society of America Bulletin*, **95**, 265–279.

WHIDBORNE, G. F. 1889–1892. *A monograph of the Devonian fauna of the South of England*, Volume 1, *The fauna of the limestone of Lummaton, Wolborough, Chircombe Bridge, and Chudleigh*. Palaeontographical Society, London.

WHITEAVES, J. F. 1892. The fossils of the Devonian rocks of the islands, shores, or immediate vicinity of Lakes Manitoba and Winnipegosis. *Geological Survey of Canada, Contributions to Canadian Palaeontology*, **1**, part 4, 255–359.

Biogeography of Devonian Algae

J. PONCET

URA 1364 du CNRS, Département de Géologie de l'Université de Caen 14032 Caen Cédex, France

Abstract: Most of the Devonian calcareous algae are distributed in the Palaeotethyan Realm on carbonate platforms located around the Palaeotethys Ocean. Possible connections by hypothetical palaeocurrents between the different palaeobiogeographical areas have been reported. Palaeogeographic Devonian reconstructions are tested using the latitudinal constraints due to the ecology of calcareous algae.

Knowledge of Devonian calcareous algae has greatly increased during the last twenty years, thanks to the publication of several monographs. Although some regions have hitherto been little studied or not examined at all, it is possible to start a study of the biogeography of Devonian algae. In this work, taxa, such as the genus *Receptaculites* with poorly established systematic positions, have not been discussed.

To study palaeogeographical distributions, the following palaeogeographical areas are recognized: North America (Laurentia), North Europe (Baltica), South Europe (Armorica), Asia (Central Siberia, Kazakhstania, China), and Gondwana.

Citing full toxonomic references would result in an excessively long list and therefore a selected bibliography is provided at the end of this paper.

Early Devonian (Fig. 1)

North America

Calcareous algae are scanty. The microflora is limited to the genera *Renalcis* (problematical blue-green algae) and *Garwoodia* (Porostromata) described respectively from Alaska and New York. Climate does not appear to be a limiting factor because contemporaneous sediments have been deposited under the control of a warm climate.

North Europe

No algae have been recorded from North Europe.

South Europe

The main assemblages are located in the Armorican Massif. The 'Lancicula flora' (Poncet 1982) composed of the following Udoteaceae: *Lancicula, Litanaia, Abacella, Paradella* and *Uva* is well represented. It is also present in the Carnic Alps and Yugoslavia. In the Armorican Massif the genera *Unella* (Dasycladale) and *Clibeca* (Udoteaceae) are endemic.

Asia

During the Early Devonian algae are most diversified in the Urals and Central Siberia. The *Lancicula* flora is present with Dasycladales (*Anthracoporella, Rhabdoporella, Dasyporella*), Codiaceae and Red Algae. A restricted *Lancicula* flora has been described from Southwestern Asia (Turkey).

Gondwana

South America and Africa are devoid of calcareous algae. Some specimens belonging to the *Lancicula* flora and to the genus *Renalcis* occur in Southeastern Australia (New South Wales).

Remarks

Because the *Lancicula* flora is well represented in Asia, South Europe and Australia, a link clearly appears between these different palaeogeographic areas. Plotted on the palaeogeographic Early Devonian reconstruction of Scotese (1986) the calcareous algal microfloras are essentially located in the western part of the Palaeotethys Ocean. There, the algae lived on carbonate platforms built on passive margins. Curiously the genus *Renalcis* present from the Cambrian to the Upper Ordovician, but absent during the Silurian, reappears in the Early Devonian.

Distribution of algae falls between 65°N and 35°S.

Middle Devonian (Fig. 2)

North America

As in the Early Devonian, the North American algal flora remains scanty. The ubiquitous genus *Sphaerocodium* (Porostromata) has been found in Washington. The genus *Solenopora* (Red Algae) is known from the Canadian Arctic Archipelago. The genus *Lancicula* appears in Alaska. It reached Alaska by way of a circum-equatorial current (Poncet & Blodgett 1987).

North Europe

By Middle Devonian times carbonate sediments were of bahamitic type with coral-stromatoporoid reefs in the Ardennes. Reefs were colonized by a very diversified algal flora (31 taxa have been listed). The genus *Pseudopaleoporella* (Codiaceae) occurs in reefs from Southwestern England (Devonshire).

South Europe

In the Cantabrian Mountains (Spain) the three ubiquitous Porostromata: *Sphaerocodium, Ortonella* and *Girvanella* are associated with coral-stromatoporoid reefs of Eifelian age. The Dasycladacean algae: *Coelotrichium, Scribroporella* and *Zeapora* have been described in Germany and Austria.

Asia

In the Urals and Central Siberia the algal microflora is very impoverished by comparison with that of the Early Devonian. It is represented by only two genera: *Litanaia* and *Bicorium* (Red Algae).

Gondwana

On the North Gondwana platform in the Tafilalet (Morocco) *Sphaerocodium* makes its first appearance in reefal carbonates. In Queensland and Victoria (Australia) the algal microflora is

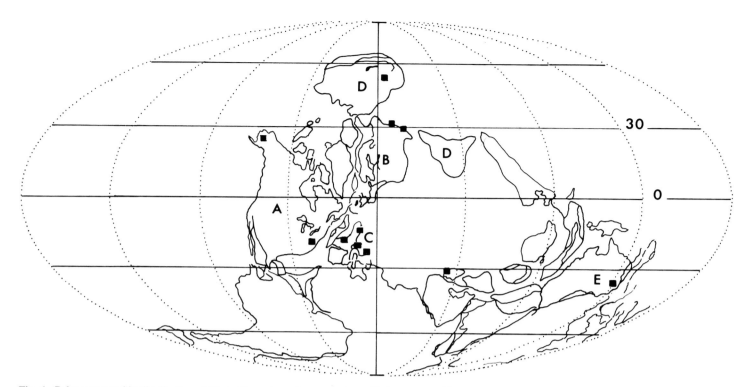

Fig. 1. Palaeogeographic distribution of Early Devonian calcareous algae. (Palaeogeographic map of the Early Devonian after Scotese 1986). A, North America; B, North Europe; C, South Europe; D, Asia; E, Australia.

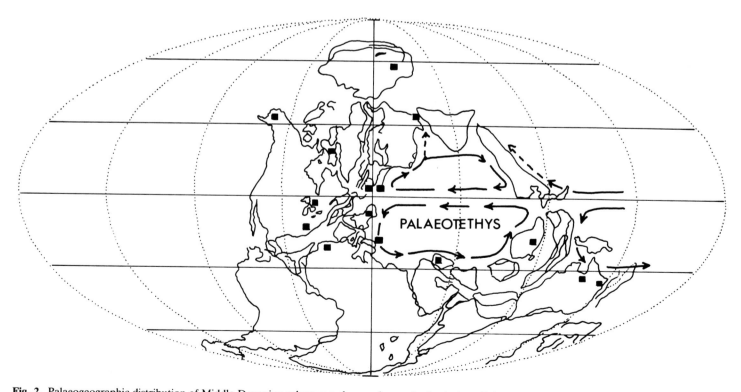

Fig. 2. Palaeogeographic distribution of Middle Devonian calcareous algae and oceanic circulation. (Palaeogeographic map of the Middle Devonian after Scotese 1986).

impoverished. It is only represented by three taxa: *Sphaerocodium*, *Solenopora* and *Lancicula*.

Remarks

The Middle Devonian algal microflora is more diverse than in the Early Devonian. In the western part of the Palaeotethys Ocean calcareous algal microfloras were still abundant on carbonate platforms of North Europe. Carbonate platforms appeared in the Ardennes. Reefs with numerous algal taxa flourished on these platforms.

In Australia, by Middle Devonian times algae migrated from the southeast to the north. This migration suggests a latitudinal shift of Australia.

There is a great affinity between Australian and North American algal microfloras. This denotes a good communication between these two areas.

It should be noted that algae are unreported in most of the Gondwana platforms. Their absence may be due to a climatic zonation, present since the Early Devonian.

It is important to mention the first report of *Issinella* because the Palaeoberesellids are a new lineage which originates from the Issinellids. Palaeoberesellids partly renewed the algal flora and were very prolific during Carboniferous times.

Algae are distributed between 55°N and 35°S.

Late Devonian (Fig. 3)

North America

Renalcis, *Jansaella* (? Chlorophytes) and *Parachaetetes* (Red algae) have been recorded in Alberta and *Courvoisiella* (Chlorophytes) in Western Virginia. *Jansaella* and *Courvoisiella* are endemic. This endemism may result from the Acadian orogenesis and from the Old Red Continent which acted as a barrier between North America and Europe.

North Europe

In North Europe the only important microflora is located in Frasnian reefs in the Ardennes. The genus *Epimastopora* seems to be endemic in these assemblages. *Sphaerocodium* has been recorded in the rocks of the Russian Platform near Moscow.

During the Late Devonian the number of taxa drastically declined in this area: from the Middle Devonian high of 84 to 17 in the Late Devonian.

South Europe

No algae have been recorded from South Europe.

Asia

By comparison with the Middle Devonian microflora, there is an increase in taxa in the Late Devonian microflora.

Calcareous algae are located in the Urals and in the Eastern Siberia where the genus *Exvotarisella* is reported from the Devonian-Carboniferous boundary. The genera *Uraimella* and *Katavella* (Red Algae) are endemic in the Urals. The genus *Epiphyton* (? Red Algae) is known from Tadzhikistan.

Gondwana

Algae are located in Morocco and Australia. The Australian microflora evolves similarly to the Asiatic microflora, and shows an increase in the number of taxa it contains. This microflora is located in Western Australia (Canning basin, Gulf basin). The genus *Tharama* (? Red Algae) seems to be endemic.

Remarks

By Late Devonian times, Western Australia, the Urals and Kazakhstan have a relatively close microfloral connection with 25% of the taxa exhibiting common occurrences. In other respects,

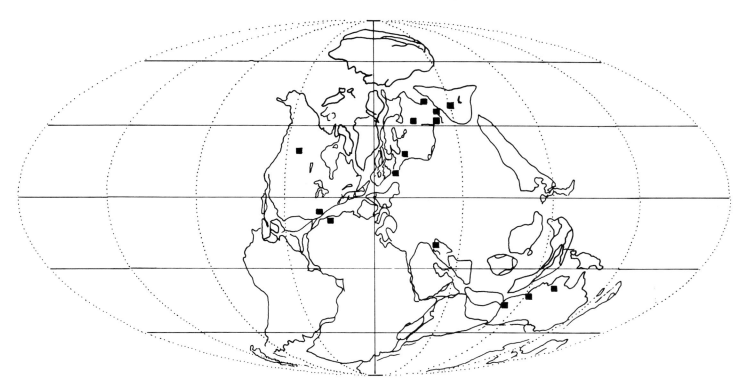

Fig. 3. Palaeogeographic distribution of Late Devonian calcareous algae. (Palaeogeographic map of the Late Devonian after Scotese 1986).

a lesser affinity appears between the Ardennes and the Urals and Kazakhstan areas with 16.5% of the taxa in common. These affinities indicate that a probable current connected these areas.

During Late Devonian, plotted algal microfloras fall between latitudes 65°N and 50°S.

Algal microfloras are reduced in North Europe and disappear in South Europe. The reduced flora in North Europe (Belgium, France) results from a predominant detrital sedimentation (sandstones and shales) on platforms.

In Australia algal microfloras are located on the west edge. Thus, during Devonian times algae migrated from South-East to West. This migration may be viewed upon as a response to displacement of Australia towards higher southern latitudes.

The global algal microflora marks a decrease in taxa: from the Middle Devonian high of 39 to 14 in the Late Devonian. This partial extinction has also been recognized in other micro- and macro-organisms.

The Palaeotethyan Realm

During Devonian times in Scotese's reconstructions most of the calcareous algal floras grew on platforms located around the Palaeotethys Ocean. In this ocean, which straddled the palaeo-equator, superficial warm currents were induced by easterly tropical winds. The disposition of continents impedded the penetration of cold waters into this ocean. The east-west circulation was blocked by North and South Europe and subdivided into two subtropical gyres which were possible agents for the dispersal of calcareous algae in the Palaeotathyan Realm.

The constraint of calcareous algae for palaeogeographic reconstruction

The occurrences of calcareous algae plotted on palaeogeographic global maps related to timespan with a contrasted palaeoclimate (such as the Early and Middle Devonian times) must have a mainly inter-tropical distribution. In this respect, divergences appear according to the different authors' maps. For example we may compare the reconstructions 2 and 4 where the occurrences of Middle Devonian calcareous algae are plotted.

On the Scotese reconstruction (Fig. 2) the calcareous algae are distributed between 55°N and 35°S. On the Heckel and Witzke reconstruction (Fig. 4) the algal floras are distributed between 0° and 35°S. In fact, the latter is more compatible with the ecological constraints of calcareous algae.

From that consideration it results that calcareous algae (especially Udoteaceae and Dasycladaceae), which are good palaeoclimate indicators, give precious latitudinal data and must be taken into account for the palaeogeographic reconstructions.

Appendix

Recorded Devonian calcareous algae (genera)

B-G, Blue-Green Algae; P Porostromata; C, Chlorophytes; R, Rhodophytes; ?, Incertae sedis.

1, USA (New York, Washington, W Virginia, Alaska); 2, Canada (Alberta, Ellesmere Island); 3, England (Devonshire); 4, France (Armorican Massif, Boulonnais, Ardennes); 5, Spain (NW Spain); 6, Morocco (Tafilalet, Aoujgal); 7, Algeria (Kabylie); 8, Belgium (Ardennes); 9, Germany (Eifel, Westphalia); 10, Austria; 11, Italy (Carnic Alps); 12, Yugoslavia (Bosnia); 13, Turkey (N Anatolia; Chio Island); 14, USSR (Urals, Central Asia, Kazakhstan, Tadzhikistan); 15, China (S China); 16, Australia (New South Wales, Victoria, Queensland, W Australia).

Early Devonian genera of calcareous algae

Abacella (C, 14), *Amicus* (C, 14), *Anthracoporella* (C, 13, 14), *Catenophycus* (C, 14), *Clibeca* (C, 4), *Dasyporella* (C, 14), *Eovelebitella* (C, 4, 14), *Garwoodia* (P, 1), *Girvanella* (P, 4, 11, 13), *Hedstroemia* (P, 14, 16), *Lancicula* (C, 4, 11, 12, 13, 14, 16), *Litanaia* (C, 12, 13, 14, 16), *Litopora* (C, 13, 14, 16), *Ortonella* (P, 4, 11, 14), *Parachaetetes* (R, 13), *Paradella* (C, 14), *Poncetella* (C, 4), *Poncetellina* (C, 4), *Pseudochaetetes* (R, 4), *Renalcis* (? B-G, 1, 4, 13, 16), *Rhabdoporella* (C, 14), *Sphaerocodium* (P, 4, 14, 16), *Solenopora* (R, 14), *Thibia* (C, 14), *Unella* (C, 4), *Uva* (C, 13, 14, 16).

Middle Devonian genera of calcareous algae

Aphralisia (? C, 8), *Asphaltina* (?, 8), *Bevocastria* (P, 4, 5, 8), *Bicorium* (R, 14), *Coelotrochium* (C, 9), *Frutexites* (? R, 8), *Garwoodia* (P, 8), *Girvanella* (P, 4, 5, 8, 15), *Givetianella* (C, 8), *Issinella* (C, 8), *Kamaena* (C, 8), *Kamaenella* (C, 8), *Labyrinthocanus* (?, 8), *Lancicula* (C, 8, 14, 16), *Litanaia* (C, 14), *Mitcheldeania* (P, 4, 8), *Nostocites* (?, 8), *Ortonella* (P, 4, 5, 8, 15), *Palaeoberesella* (C, 8), *Palaeomicrocodium* (?, 8), *Paralitanaia* (C, 8), *Poncetellina* (C, 8), *Proninella* (?, 8), *Pseudohedstroemia* (P, 8), *Pseudoissinella* (C, 8), *Pseudopaleoporella* (C, 3, 8), *Rectangulina* (P, 8), *Reisteignella* (C, 8), *Rhabdoporella* (C, 7), *Solenopora* (R, 2, 4, 5, 6, 16), *Sphaerocodium* (P, 1, 4, 5, 6, 16), *Sphaeroporella* (? C, 8), *Tharama* (? R, 8), *Triangulinella* (C, 8), *Vermiporella* (C, 8), *Zeapora* (C, 10).

Late Devonian genera of calcareous algae

Catenophycus (C, 14), *Courvoisiella* (C, 1), *Epimastoporella* (C, 4), *Epiphyton* (? R, 8, 14, 16), *Exvotarisella* (C, 14, 16),

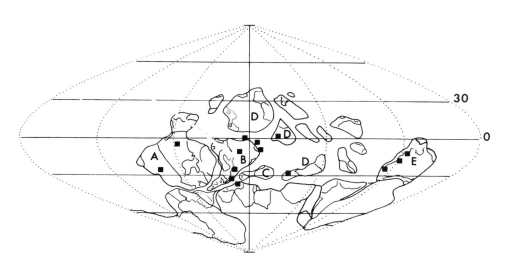

Fig. 4. Palaeogeographic distribution of Middle Devonian calcareous algae. (Palaeogeographic map after Heckel & Witzke 1979). Key as for Fig. 1.

Girvanella (P, 16), *Issinella* (C, 14), *Jansaella* (? C, 1), *Katavella* (R, 14), *Litanaia* (C, 16), *Ortonella* (P, 16), *Palaeomicrocodium* (?, 16), *Parachaetetes* (R, 1, 14), *Poncetellina* (C, 16), *Renalcis* (? B-G, 1, 4, 16), *Rhabdoporella* (C, 4, 14, 16), *Solenopora* (R, 13, 14, 16), *Sphaerocodium* (P, 4, 14, 16), *Uraimella* (R, 14).

References

HECKEL, P. H. & WITZKE, B. J. 1979. Devonian world palaeogeography determined from distribution of carbonates and related lithic palaeoclimatic indicators. *In*: HOUSE, M. R., SCRUTTON, C. T. & BASSETT, M. G. (eds). *The Devonian System*. Special Papers in Palaeontology, **23**, 99–123.

PONCET, J. 1982. L'apport des Udoteaceae (Algues vertes calcaires) dans la paléogéographie mondiale éodévonienne. *Bulletin de la Société géologique de France*, **24**, 5/6, 1087–1090.

—— & BLODGETT, R. B. 1987. First Recognition of the Devonian Alga *Lancicula sergaensis* Shuysky in North America (West–Central Alaska). *Journal of Paleontology*, **61**, 6, 1269–1273.

SCOTESE, C. R. 1986. Phanerozoic Reconstructions: A New Look at the Assembly of Asia. *Univ. Texas Institute for Geophysics Technical Report*, **66**.

Selected Bibliography

BASSOULLET, J. P., BERNIER, P., DELOFFRE, R., GÉNOT, P., PONCET, J. & ROUX, A. 1983. Les Algues Udotéacées du Paléozoïque au Cénozoïque. *Bulletin Centres Recherche Exploration – Production Elf-Aquitaine*, **7**, 2, 449–621.

DELOFFRE, R. 1988. Nouvelle taxonomie des Algues Dasycladales. *Bulletin Centres Recherche Exploration – Production Elf-Aquitaine*, **12**, 1, 165–217.

EMBERGER, J. 1978. Les Algues (Chlorophyceae, Prasinophyceae, Rhodophyceae) du Dévonien. Essai d'un inventaire bibliographique, géographique, stratigraphique. *Bulletin de l'Institut de Géologie du Bassin d'Aquitaine*, Talence, n° spécial.

MAMET, B. & PRÉAT, A. 1982. *Givetianella tsienii*, une Dasycladacée nouvelle du Givétien de la Belgique. *Bulletin Société belge Géologie*, **91**, 4, 209–216.

—— & —— 1985a. Sur quelques algues vertes nouvelles du Givétien de la Belgique. *Revue de Micropaléontologie*, **28**, 1, 67–74.

—— & —— 1985b. Sur la présence de *Palaeomicrocodium* (Algue?, Incertae sedis?) dans le Givétien inférieur de Belgique. *Geobios*, **18**, 3, 389–392.

—— & —— 1986. Algues givétiennes du Bord Sud du Bassin de Dinant et ses régions limitrophes. *Annales de la Société géologique de Belgique*, **109**, 431–454.

—— & ROUX, A. 1974. Algues dévoniennes et carbonifères de la Téthys occidentale. *Revue de Micropaléontologie*, **18**, 134–187.

—— & —— 1983. Algues dévono-carbonifères de l'Australie. *Revue de Micropaléontogie*, **26**, 63–131.

RIDING, R. 1979. Devonian calcareous algae. *In*: HOUSE, M. R., SCRUTTON, C. T. & BASSETT, M. G. (eds). *The Devonian System*. Special Papers in Palaeontology, **23**, 141–144.

YANG ZHENGQIANG 1985. The Fossil Algae and Cryptoalgal Limestone of Middle Devonian in South China and their Palaeoenvironmental Significance. *Bulletin of the Yichang Institute of Geology and Mineral Resources of the Chinese Academy of Geological Sciences*, Beijing, **9**, 88.

Givetian–Frasnian phytogeography of Euramerica and western Gondwana based on miospore distribution

MAURICE STREEL[1], MURIEL FAIRON-DEMARET[1] & STANISLAS LOBOZIAK[2]

[1] *Paléontologie, Université, 7 place du vingt-août, B-4000 Liege, Belgium*
[2] *Paléobotanique, Université des Sciences et Techniques de Lille, URA 1365, F-59655 Villeneuve d'Ascq, France*

Abstract: The Givetian and Frasnian miospore distributions in western Gondwana and southern Euramerica show a rather uniform vegetation prevailing from palaeo-polar to palaeo-tropical regions. Similar climatic conditions are certainly required to explain this but it is concluded from a discussion on the dispersal of homosporous vegetation that no wide oceans separated these regions at the time. Frasnian northern Euramerica vegetation seems different and might correspond to an equatorial belt. Heckel & Witzke's palaeogeographical reconstruction fits much better with the miospore distribution than other maps.

A recent comparison (Loboziak *et al.* 1989) between Middle–Upper Devonian miospore assemblages from Libya and Brazil led to the following palaeoclimatological conclusions: a rather uniform vegetation and therefore uniform climate prevailed at the time from palaeo-subtropical to palaeo-polar regions in western Gondwana: within the area studied, the climate became progressively more uniform from the Givetian to the Frasnian.

These conclusions were reached after studying borehole sequences originating from nearshore marine palaeoenvironments, it is important to note that both sequences were studied by the same palynologists, thus eliminating investigator bias. Loboziak *et al.* (1989) observed that characteristic miospores entered the geological record in the same order of succession in both sequences (Fig. 1) and that the small quantitative differences between these sequences were more obvious in Givetian than in Frasnian assemblages (Fig. 2). These comparisons can now be extended to other regions.

Comparison between western Gondwana and southern Euramerica

Non-palynological dating methods have not been used, but the first occurrences of each characteristic miospore of the Libyan and Brazilian assemblages are the same as those of the western European succession (southern Euramerica in Devonian time), where they form the basis for a miospore zonation which has been related to well documented megafloras and marine faunas (Streel *et al.* 1987). We have applied the same quantitative analysis, as that performed on the Libyan and Brazilian material, to some previously studied samples from the Boulonnais area (Loboziak & Streel 1981) and some samples currently being studied from the Eifel area (Fig. 2).

Note that the percentages computed for each sample are based on the number of specimens of each species, and therefore give information on the relative abundance of different components of the vegetation. Ignoring the morphologically simple miospores, most of the 'elaborate miospores' (Fig. 2) from the Givetian and Frasnian of western Europe are of the same species as in Brazil and Libya. This allows the same miospore zonation to be applied across all these regions.

The conclusion of Loboziak *et al.* (1989) concerning the uniform distribution of vegetation in western Gondwana must now be extended to take account of the information from southern Euramerica: if we accept a palaeo-southern tropical position for this region, as suggested by most of the available palaeogeographical reconstructions, we have to assume that a rather uniform vegetation, and therefore a rather uniform climate, prevailed during Givetian to Frasnian time from palaeo-tropical to palaeo-polar regions.

The similarity of Givetian and Frasnian floras across western Gondwana and southern Euramerica, as suggested by the quantitative analyses of miospores, raises the question of the existence of 'A newly opened ocean ... forming between the Gondwana craton and Laurentia (with Avalonian/Armorican accreted terranes left behind, adjacent to Laurentia) during the Middle and Late Devonian' (Van der Voo 1988, p. 322). Is an ocean more than 2500 km wide (as shown by Van der Voo, 1988, fig. 5) compatible with the exchange of plant dispersal (propagules) devices and thereby the migration of vegetation?

Land plant miospores as evidence in palaeogeographical reconstructions.

Land plants 'break up as fossils into leaves and stems (...) and a multitude of spores, pollen, and seeds which all act as small dispersants moving downstream with clastic sedimentation into all resulting sediments wherever they are deposited' (Hughes 1988). Small spores and pollen (miospores) are normally preserved even in fine-grained clastic marine sediments because they are protected by chemically very resistant envelopes.

Some factors affecting spore numbers and composition in sediments.
Miospores are produced in enormous amounts by land plants and their usual method of dispersal is by wind. However, Raynor *et al.* (1976) state that 'most (modern fern) spores released will settle to the ground or to other vegetation within relatively short distances, (only) a few may be carried long distances, particularly during periods of strong winds and good atmospheric mixing'. Even very light grass pollen grains are almost all deposited by the wind at a very short distance from the vegetation-source (Moseholm *et al.* 1987).

After transportation to the sedimentary basins through flooding and fluvial systems, most of the miospores will be deposited nearshore so that only a small proportion of the miospores produced will be carried off-shore by the water-currents (Muller 1959).

Near-shore deposits will still contain high concentrations of miospores (several thousands to several tens of thousands in each gram of a suitable sediment). Off-shore sediments contain a much lower concentration and Traverse & Ginsburg (1966) have shown that the resulting assemblages are species poor and contain miospores with a restricted range of diameters because of sorting during transport. In the present paper, we have examined only rich assemblages from near-shore marine sediments. We believe that these spores were derived from land plants that grew in the near hinterland, that is in area of a few tens of square kilometres, and that they were transported from this area through the fluvial system. This relatively large catchment area would have contained a variety of different land plant communities and

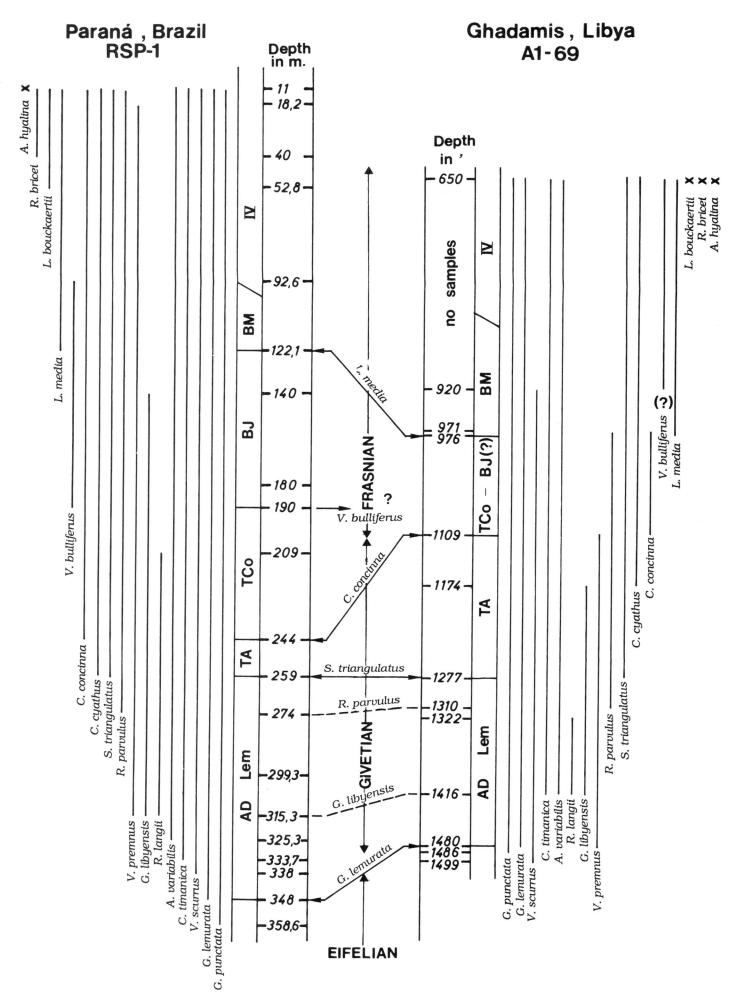

Fig. 1. Comparison between the stratigraphical range of the main characteristic Givetian–Frasnian miospores in two boreholes from Brazil (Loboziak *et al.* 1988) and Libya (Loboziak & Streel 1989).

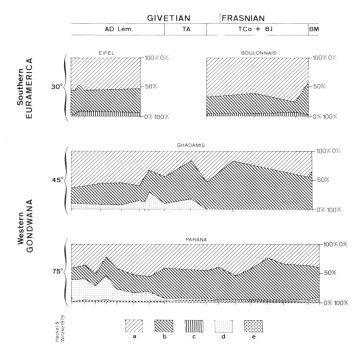

Fig. 2. Quantitative distribution of various miospore classes in different regions of western Gondwana and southern Euramerica. a, morphologically simple (smooth or finely ornamented, one-layered) miospores. b–e, morphologically elaborate miospores; b, occurring in all regions; c, never found in Gondwana; d, only found in Gondwana; e, only found in Parana (Brazil). Stratigraphic scale as on Fig. 1. Note that percentages are of specimens, not of species.

thus, unlike other groups of (hyp)-autochthonous organisms, the resulting spore assemblages are less likely to reflect very small scale environmental variation. Some of these miospores may also have been derived from more distant localities through air or water currents but research on modern sediments demonstrates that they are always greatly outnumbered by the local production.

Migration of homosporous land plants by long distance wind dispersal of spores. In theory, a few spores carried casually over long distances are all that is necessary to allow land plants to migrate to a suitable environment. This applies only to homosporous plants, the isolated spores of which have the potential of producing new sporophytes. Heterosporous and seed plants, arising in Middle and Late Devonian times, evolved a restricted number of large propagules (megaspores and seeds) which made the possibility of long distance dispersal improbable.

Also, the megaspores of heterosporous plants are usually shed before fertilization, so even if carried a long distance, they would have to wait for a microspore of the same species to arrive before a new sporophyte could be produced, another unprobable event. A good example of the limitations this imposes on dispersal can be seen in the geographic distribution of the modern heterosporous *Selaginella* which has been shown by Tryon (1971) to rely entirely on a step by step migration processes resulting in small scale extensions of its geographic range.

References to present-day dispersal processes of homosporous plants are rather scarce and we can discard those related to bryophytes because of the rather small size (usually less than 25 μm) of their spores in comparison to those of most of the supposedly homosporous Middle and Late Devonian plants and modern ferns discussed here (usually 50 to 150 μm).

Nearly all modern ferns are homosporous, with spores that can readily be dispersed in the air over long distances, and a review of various aspects of spore dispersal indicates that these ferns have a great capacity for dispersal by wind (Tryon 1970). This contradicts the view of Copeland (1940) who wrote 'Ten miles of open water is no barrier to their spread, once they are picked up by the wind. A thousand miles, though, seems to be, in general, an impassable barrier'. In the context of this paper, however, we are not so much concerned with chance dispersal over long distances but rather with the probability, given enough time, of the spread of an entire homosporous plant vegetation containing similar components. Studies on isolated islands (Tryon 1970), endemic areas, and geographic speciation in modern tropical American ferns (Tryon 1972) demonstrate that long-distance migration, involving the transport of fern spores over distances of more than 1600 km to regions where the environment is equivalent to that of the source area, normally induces genetic isolation in the new colony due to a lack of re-immigration. Geographic isolation may be even more effective if the new environment is not equivalent to that of the source area. Tryon (1970) presents evidence that distances greater than 800 km are sufficient for isolation of an insular fern flora because of the small size of the newly colonized territory. The chances of re-immigration are of course greater with time but when the isolation distance is large, chance needs more time, and the speciation might be quicker than the re-immigration process. It is instructive to compare the present flora of the 10 million year old island for St Helena with those of the geologically younger islands of Ascension and the Tristan da Cunha group. All these islands are at least 2000 km from the nearest continents. Ascension and the Tristan da Cunha group have a much lower percentage of endemic species (24% and 40% respectively) than has St Helena (52%) (Muir & Baker 1968).

Ten million years is about the same timespan covered by the sampling in the three areas studied in the Givetian and Frasnian. During this timespan, about ten bio-events (the first occurrence of some of the listed species in Fig. 1) arose in the same order of succession in all these areas, indicating similar vegetation patterns. Similar climatic conditions are certainly required to explain this but a wide ocean separating these regions would have prevented it.

Plant macrofossil evidence

The similarity of Givetian and Frasnian floras, as interpreted through the microfossil evidence, across western Gondwana and southern Euramerica appears to contradict the analysis of Devonian floras made by Edwards (1973) which was based on macrofossils; she concluded (page 109) that a distinction between present-day Southern and Northern Hemisphere floras is present in the Middle Devonian. It should be noted however that this conclusion is based on a few inadequately dated South African and Argentinian fossil floras. In any case, being isolated discoveries, they might well correspond to the few percent of miospores restricted to 'the highest palaeolatitudes' (those restricted to the Parana basin, on Fig. 2). On another hand, Edwards & Benedetto (1985, p. 616) provided 'evidence for a uniform vegetation between a part of north Gondwana (Venezuela) and the Old Red Continent in mid- to late-Devonian times, but (their data) in isolation do not provide compelling evidence for global uniformity during that period. The similarities may simply reflect the palaeogeographic proximity of the localities or their occurrence in encompassing more than one palaeocontinent'. There is no detailed analysis using multivariate statistical tool available for the Middle and Upper Devonian which might be compared to the Lower Devonian analysis proposed by Raymond *et al.* (1985).

A reappraisal of the published Givetian and Frasnian plant megafossils made by one of us (M. F-D) does not show any clear subdivision of floras across the regions discussed here. It does not demonstrate any continuity either as information has only been obtained from isolated localities where cosmopolitan taxa, if present are often less abundant than 'endemic' species. We

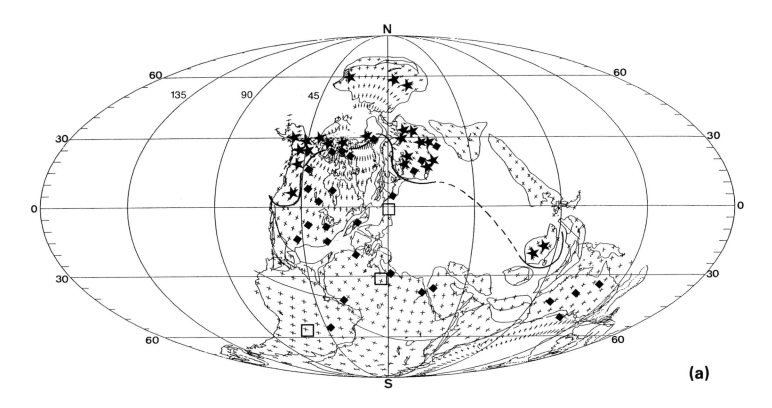

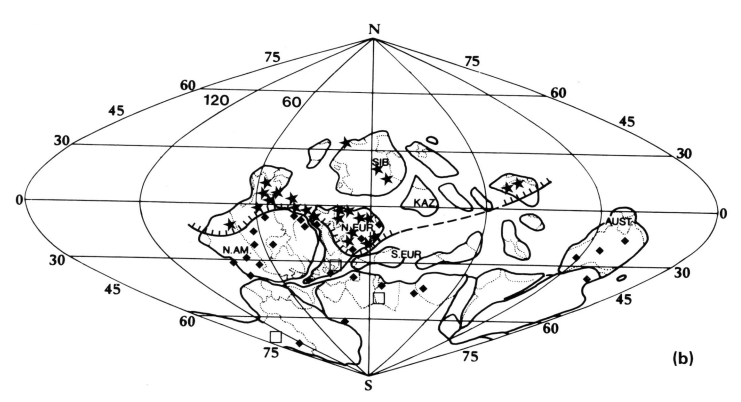

Fig. 3. Palaeophytogeographic reconstructions based on maps by (a) Scotese (1986) and (b) Heckel & Witzke (1979). Black stars, *Archaeoperisaccus* distribution after McGregor (1979) and Streel (1986). Black diamonds, *Geminospora lemurata* distribution after McGregor (1979) and Streel (1986). White square, regions where quantitative data (see Fig. 2) have been evaluated by the authors.

believe that these megafloral data are too sporadic to be compared with the miospore data. As Hughes (1988, p. 547) remarked: 'In practice, megafossil paleobotany can be regarded as a spatter of very small beautifully formed islands set in an almost unlimited ocean of palynomorphs which inevitably represent the vast majority of fossil plant information that is available'.

Comparison between southern and northern Euramerica

We were not able to study a sequence of samples from the Givetian and Frasnian of northern Euramerica. Therefore, comparisons with the above data cannot have the same security, as we have had to rely on descriptions and photographs of miospores made by other authors. Moreover, part of northern Euramerica was studied by Russian palynologists at a time when they were often using their own nomenclature and poor illustrations. Richardson & McGregor (1986, fig. 6) have checked the first occurrence of ten zonally significant miospores in a selection of Givetian and Frasnian spore-bearing strata controlled by conodonts from the western (Canada) and the eastern (USSR) parts of northern Euramerica and from southern (Boulonnais, France) Euramerica. Only four species show north-south correlations. Amongst these four species, *Geminospora lemurata* and *Samarisporites triangulatus* are now considered by us to enter the geological record respectively in the latest Eifelian and the middle Givetian of the Ardenne–Rhenish basins. Despite these difficulties, some observations have been made which indicate that at least the Frasnian miospore assemblages are somewhat different from those described in southern Euramerica and western Gondwana. Loboziak & Streel (1981) for instance found it difficult to correlate conodont dated Frasnian miospore zonations from the USSR and northern France (Boulonnais area). McGregor (1981) remarks that many of the species of the Givetian and the Frasnian of the USSR that are regarded as stratigraphically important by Soviet workers have not been recognized in other regions. He notes that a section on Melville Island in the Canadian arctic provides a palynological bridge between the European USSR and the Boulonnais, as it contains species so far found only in one or the other of these regions. The still unpublished thesis of Braman (1981) contains the most detailed data on the miospores distribution in the middle (?) and late Frasnian (*Polygnathus asymmetricus* to *Palmatolepis gigas* conodont zones of the Imperial Fomation, District of Mackenzie and Yukon) for the western part of northern Euramerica. One of us (M. S.) has checked that out of the 95 species recorded, 26 were species of 'elaborate' morphology not present in southern Euramerica and 20 were of undifferentiated simple morphology. Unfortunately, counts made on numbers of specimens of these species are poorly known except that one species (*Hymenozonotriletes deliquescens*) varies from 19.3 to 26.3% of the total assemblage. *H. deliquescens* and several species of *Archaeoperisaccus* are good characteristic species for this timespan in the northern Euramerica (See also McGregor, 1981, figs 3 and 8).

McGregor (1979) and Streel (1980, 1986) have plotted, on different palaeogeographical maps, the distribution of *Archaeoperisaccus* versus other Frasnian miospore assemblages containing *Geminospora lemurata*. These data are reproduced here (Fig. 3) on two different maps, one proposed for the Givetian by Scotese (1986), the other for the Middle Devonian by Heckel & Witzke (1979). It should be noted that the last reconstruction is based on the supposition that the palaeomagnetic poles of the earth at that time did not coincide with the rotational poles.

Heckel & Witzke's reconstruction allows the location of the *Archaeoperisaccus* flora of northern Euramerica (and South China) in an equatorial belt (between 35° N and 20° S) and indicates that the rest of the floras in the southern hemisphere were situated between 10° and 75° palaeolatitudes. Following the reconstruction by Scotese, the *Archaeoperisaccus* flora of northern Euramerica (and South China) would have been randomly distributed between the 60° northern palaeolatitude and the 30° southern palaeolatitude. This asymmetrical floral distribution would imply a strongly asymmetrical distribution of climatic belts. Moreover, the western European flora studied in the early part of this paper would occupy an equatorial position with the same characteristics as in palaeo-subtropical and palaeo-polar regions. It would be hard to understand why tropical floras north of the palaeo-equator have no counterpart in the south, especially since no major physical barriers (mountain ranges or wide oceans) can be distinguished between the two realms. Thus, we believe that Heckel & Witzke's reconstruction fits much better with the miospore distribution than other maps.

Conclusion

Givetian–Frasnian miospore distributions in western Gondwana and Euramerica suggest a greater contrast in the vegetation between supposedly palaeo-equatorial and palaeo-tropical regions than between palaeo-tropical and palaeo-polar regions. They also favors palaeogeographic models that show a close proximity between Gondwana and Euramerica in the southern hemisphere during this time.

M. J. M. Bless and P. Kenrick are thanked for stimulating suggestions and corrections of language which contributed much to improve the manuscript.

References

BRAMAN, D. R. 1981. *Upper Devonian–Lower Carboniferous miospore biostratigraphy of the Imperial Formation, District of Mackenzie and Yukon*. Thesis, Department of geology and geophysics, Calgary, Alberta.

COPELAND, E. B. 1940. Antarctica as the source of existing ferns. *Proceedings of the Sixth Pacific Sciences Congress*, **4**, 625–627.

EDWARDS, D. 1973. Devonian floras. *In*: HALLAM, A. (ed.) *Atlas of palaeobiogeography* Elsevier, 105–115.

—— & BENEDETTO, J. L. 1985. Two new species of herbaceous lycopods from the Devonian of Venezuela with comments on their taphonomy. *Palaeontology*, **28**, 3, 599–618.

HECKEL, P. H. & WITZKE, B. J. 1979. Devonian world palaeogeography determined from distribution of carbonates and related lithic palaeoclimatic indicators. *In*: HOUSE, M. R., SCRUTTON, C. T. BASSETT, M. G. (eds) *The Devonian System* Palaeontological Association Special paper, **23**, 99–123.

HUGHES, N. F. 1988. Plant fossils in Earth Sciences. *Geological Magazine*, **125**, 547–549.

LOBOZIAK, S. & STREEL, M. 1981. Miospores in Middle-Upper Frasnian to Famennian sediments partly dated by conodonts (Boulonnais, France). *Review of Palaeobotany and Palynology*, **34**, 49–66.

—— & —— 1989. Middle–Upper Devonian miospores from the Ghadamis Basin (Tunisia–Libya): systematics and stratigraphy. *Review of Palaeobotany and Palynology*, **58**, 173–196.

——, —— & BURJACK, M. I. A. 1988. Quelques données nouvelles sur les miospores dévoniennes du Bassin de Paraná (Brésil). *Sciences géologiques*, **40**, 381–391.

——, —— & —— 1989. Déductions paléoclimatiques d'une comparaison entre des assemblages de miospores du Dévonien moyen et supérieur de Libye et du Brésil. *Geobios*, **22**, 247–251.

MCGREGOR, D. C. 1979. Spores in Devonian stratigraphical correlation. *In*: HOUSE, M. R., SCRUTTON, C. T., BASSETT, M. G. (eds) *The Devonian System* Palaeontological Association. Special paper, **23**, 163–184.

—— 1981. Spores and the Middle-Upper Devonian boundary. *Review of Palaeobotany and Palynology*, **34**, 25–47.

MOSEHOLM, L., WEEKE, E. & PETERSEN, B. 1987. Forecast of pollen concentrations of Poaceae (Grasses) in the air by time series analysis. *Pollen et Spores*, **29**, 305–322.

MUIR, M. D. & BAKER, I. 1968. The early Pliocene flora of St. Helena. *Palaeogeography, palaeoclimatology and palaeoecology*, **5**, 251–268.

MULLER, J. 1959. Palynology of recent Orinoco delta and shelf sediments.

Micropaleontology, **5**, 1, 1–32.
RAYMOND, A., PARKER, W. C. & BARRETT, S. F. 1985. Early Devonian phytogeography. *In*: TIFFNEY, B. H. (ed.) *Geological factors and the evolution of plants*. Yale university press, 129–167.
RAYNOR, G. S., OGDEN, E. C. & HAYES, J. V. 1976. Dispersion of fern spores into and within a forest. *Rhodora*, 78, 473–487.
RICHARDSON, J. B. & MCGREGOR, D. C. 1986. Silurian and Devonian spore zones of the Old Red Sandstone Continent and adjacent regions. *Geological Survey of Canada Bulletin*, **364**.
SCOTESE, C. R. 1986. Phanerozoic reconstructions: a new look at the assembly of Asia. *University of Texas Institute for Geophysics Technical Report*, **66**.
STREEL, M. 1980. Evidences palynologiques sur les relations entre le climat et la distribution géographique des flores dévoniennes et dinantiennes. *Mémoires du Muséum national d'histoire naturelle-série B*, **27**, 261–267.
—— 1986. Miospore contribution to the Upper Famennian-Strunian event stratigraphy. *Annales de la Société géologique de Belgique*, **109**, 75–92.
——, HIGGS, K., LOBOZIAK, S., RIEGEL, W. & STEEMANS, P. 1987. Spore stratigraphy and correlation with faunas and floras in the type marine Devonian of the Ardenne-Rhenish regions. *Review of Palaeobotany and Palynology*, **50**, 211–229.
TRAVERSE, A. & GINSBURG, R. N. 1966. Palynology of the surface sediments of Great Bahama Bank, as related to water movement and sedimentation. *Marine Geology*, **4**, 417–459.
TRYON, R. 1970. Development and evolution of fern floras of oceanic islands. *Biotropica*, **2**, 76–84.
—— 1971. The process of evolutionary migration in species of Selaginella. *Brittonia*, **23**, 89–100.
—— 1972. Endemic areas and geographic speciation in tropical American ferns. *Biotropica*, **4**, 121–131.
VAN DER VOO, R. 1988. Paleozoic paleogeography of North America, Gondwana, and intervening displaced terranes: comparisons of paleo-magnetism with paleoclimatology and biogeographical patterns. *Geological Society of America Bulletin*, **100**, 311–324.

The Balkhash–Mongolia–Okhotsk Region of the Old World Realm (Devonian)

HOU HONG-FEI[1] & A. J. BOUCOT[2]

[1] *Institute of Geology, Chinese Academy of Geological Sciences, Baiwanzhuang Road, Beijing 100037, People's Republic of China*
[2] *Department of Zoology, Oregon State University, Corvallis, OR 97331, USA*

Abstract: Reconsideration of the Emsian brachiopods from parts of central Asia extending from the Lake Balkhash Region easterly through Heilungjiang leads us to conclude that they should be separated as a distinct region, termed the Balkhash–Mongolia–Okhotsk Region, of the Old World Realm. This conclusion was arrived at after considering the high level of endemic genera, plus the significant admixture of genera common with the Eastern Americas and Tasman Realms.

It has been long recognized that the earlier Devonian of Central Asia, including here southeastern Kazakhstan, parts of the Altai-Sayan, Mongolia and Inner Mongolia, and Heilungjiang, is characterized by a unique mixture of typical European and Appalachian type brachiopods (see Bublichenko 1958, for a typical statement). We now formally characterize the European component as Old World Realm type and the Appalachian component as Eastern Americas Realm type. For some time it has been recognized that there is an additional, Tasman Region, Old World Realm component. The overall biogeographic assignment and ranking of this globally unique mixture, and a reasonable explanation for its existence is the purpose of this paper.

Chinese Devonian biogeographic data

Devonian biogeography of China, based largely on the distribution of brachiopods, has been discussed by many (Hou *et al.* 1979; Hou 1981; Wang Yu & Rong 1983). Wang Yu *et al.* (1984) analysed the available brachiopod data, published and unpublished, from the Silurian and Devonian in some detail. The recognition of the South China Region of the Old World Realm as a very distinct unit is widely agreed. The Early Devonian biogeography of China has been summarized (Hou & Wang 1985; Hou *et al.* 1985) as consisting of four regions: Junggar–Hinggan, South Tianshan, South China and W. Yunnan–Tibet. The brachiopod record in South Tianshan and W. Yunnan–Tibet is still poor, although they may be assigned to the Uralian and Rhenish–Bohemian Regions respectively of the Old World Realm. There are preliminary investigations and descriptions based on a large quantity of brachiopods from North China, the vast area from Junggar on the west to the Greater and Lesser Khingan Ranges on the east.

In China, as elsewhere in the world, the Emsian transgression reached into most basins, and was the time of highest Devonian provincialism for the level bottom faunas. During the Emsian, the articulate brachiopods are the most widespread, and most abundant fossils currently useful for biogeographic purposes. We will, therefore, discuss Emsian provincialism within China and adjacent areas. For this purpose it is best to consider generic rather than specific data in order to avoid taxonomic confusion.

The problem

The Chinese part of the Central Asian area (Fig. 1) includes the Junggar–Hinggan Region from west to east (both sides of the Junggar Basin, Beishan Mountain in northern Gansu Province), most of Inner Mongolia, and the northern part of Heilungjiang (northeast China). This Chinese area has characters in common with the non-Chinese parts of the Central Asian area in terms of the Emsian brachiopod fauna and in terms of the geologic evolution of the region. Therefore, we will discuss the provincialism of the region as a whole, including Kazakhstan, Mongolia, and parts of the Soviet Far East.

We ask three questions concerning the provincialism of this large region. (1) Is this a single biogeographic unit, or more than one distinct unit? (2) Is this an ecologic unit or a biogeographic unit (or units)? (3) If a biogeographic unit (or units), what are its unique features, and what is its biogeographic rank?

Figure 2 is a correlation chart for the Early Devonian of Central Asia. Note that most areas of early Lower Devonian strata are not well exposed, and that knowledge of early and middle Lower Devonian brachiopods is still relatively incomplete. Available data indicate that the earlier Lower Devonian brachiopods are of Old World Realm aspect, and that Eastern Americas Realm brachiopods appeared in the middle of the epoch (see Boucot *et al.* 1969, for a brief discussion of the brachiopod genera characteristic of the two realms). The later Lower Devonian epoch considered here corresponds largely to the Sardjal and Kazakh Formations, and equivalent formations, of approximate Emsian age. The Emsian brachiopods of this large area are relatively well known.

Discussion

(1) Is this a single biogeographic unit, or more than one distinct unit?

There are a number of discussions of this problem. Some workers (Boucot 1975; Rzhonsnitskaya *et al.* 1978) have recognized Junggar–Balkhash and Mongolo–Okhotsk as separate units at the region level (Boucot kept the Junggar-Balkhash as a terrigenous facies of the Uralian Region). Wang Yu *et al.* (1984) were uncertain whether or not two units should be recognized, and indicated that if two units are involved they would probably be of provincial or subprovincial rank. Talent *et al.* (1987, p. 277) evaluated the brachiopod faunas from Kazakhstan, Mongolia and Northern China and argued that '... they belong to at least different provinces and, probably different zoogeographic regions or even realms.' They concluded further that probably the '... Junggar mini-continent functioned as a 'staging point' for limestone faunal interchange' (p. 278).

Investigators (Hou *et al.* 1979; Yang *et al.* 1981; Zhang 1983) working in Xinjiang agree on the sedimentary and faunal similarity of the Devonian on both sides of the Junggar Basin. The Devonian Armentai Mountain, Baytik Mountain, and Kokserge Mountain northeast of the Kelameli Ophiolite zone of eastern Junggar are easily correlated, unit by unit, not only in lithology, but also in faunal assemblage, with those of western Junggar north to the Darbut Ophiolite Zone. The fauna (including brachiopods, corals and trilobites) in eastern and western Junggar is similar to those of Kazakhstan, Mongolia and the Hinggan Ranges. Reproductive communication between the eastern and western Junggar Basin was easy. Our knowledge of the Pribalkhash region brachiopods is still drawn largely from Kaplun's publications of the 1960s, and those from Xinjiang are only partly described. This situation does limit our biogeographic understanding somewhat.

From MCKERROW, W. S. & SCOTESE, C. R. (eds), 1990, *Palaeozoic Palaeogeography and Biogeography*, Geological Society Memoir No. 12, pp. 297–303.

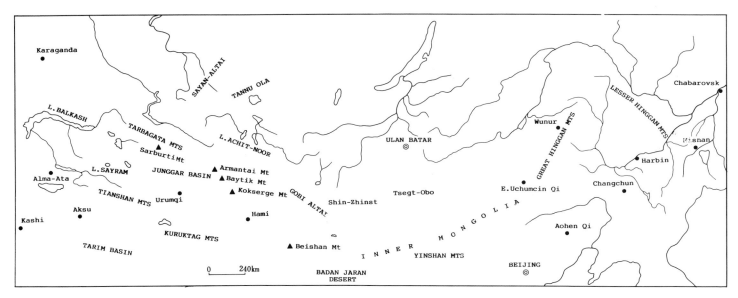

Fig. 1. Localities in the Central Asian region.

Table 1 indicates the high level of faunal similarity shown by the brachiopods of the Junggar–Balkhash and Mongolo–Okhotsk regions, based on recent literature (Kaplun & Modzalevskaya 1978; Nilova 1973; Ushatinskaya & Nilova 1975; Hamada 1971; Su 1976, 1980; Alekseeva et al. 1981; Zhang 1983; Ushatinskaya 1983) and collections deposited in the Institute of Geology CAGS. More than 50 genera are recorded from the late Early Devonian in both regions, of which 17 genera (33%) are in common. These common genera are most abundant in the late Early Devonian, but very few genera are reported only from the middle Early Devonian. Among the large number of genera remaining (Table 2) are cosmopolitans with a wide geographic distribution, endemic ones known from only limited areas, and rare ones only known from one or two localities. Thus, the characteristic genera limited to the Junggar–Balkhash and Mongolo–Okhotsk Regions are very few. *Discomyorthis* and *Chonostrophiella* appear to be limited to the latter region, while *Xingjiangospirifer* and a new spiriferid (genotype *Delthyris minimus* Kaplun) is only recorded in the former region. The existence of only one or two diagnostic genera is an inadequate basis for defining biogeographic regions, provinces, or subprovinces. Therefore, we suggest that the Junggar–Balkhash and Mongolo–Okhotsk regions be combined as a single unit to be known as the Balkhash–Mongolo–Okhotsk Region.

(2) Is this an ecologic unit or a biogeographic unit (or units)?

This is an ever puzzling question, i.e., does a biotically unique region merely represent an ecologically unique part of a larger biogeographic unit, or does it represent a unique biogeographic unit. Before discussing the dilemma further, let us point out that palaeogeographically the western and southern boundaries of the Balkhash–Mongolia–Okhotsk Region were bounded by the Kazakhstan Isthmus, central Tianshan Complex and North China Block during Devonian time, i.e., a linear landmass (Fig. 3). This linear landmass separated the Balkhash–Mongolia–Okhotsk Region marine waters from those to the south in the area of the southern Tianshan (Soviet, Chinese, Mongolian), and potentially limited reproductive communication and direct faunal interchange. Such reproductive isolation may, however, have been the more direct result of generating separate water masses and surface

CENTRAL KAZAKHSTAN (Kaplun et al. 1978)	W. JUNGGAR (Hou et al. 1979; Yang et al. 1989; Zhang et al. 1983)	E. JUNGGAR	NW MONGOLIA (Alekseeva et al.1981)	WEST INNER MONGOLIA (Zhang 1981)	GREAT HINGGAN RANGE		LESSER HINGGAN (Hamada 1971; Xue et al. 1981)	FAR EAST (Kaplun et al. 1978)
					South part (Su et al. 1976)	North part (Wang et al. 1984)		
Kazakh	Mangkelu Fm.	Zuomubasite Fm.	Osugan Fm.	Zhusileng Fm.	Wenduerao-baote Fm.	Beikuang Fm.	Helongmeng Fm.	Imatchin-skaya Fm.
Sardjal			Upper Tulgen Fm.		Aubaoting-hundi Fm.	Wunuer Fm.	Jingshui Fm.	
Pribalkhash	Mangger Fm.	Taherbasite Fm.	Lower Tulgen Fm.		Baruntehua Fm.	Luotuo-shan Fm.	Handaqi Mem. Niqiuhe Fm.	Bolshenever-skaya Fm.
Kakbaital			Tsagansala-gol Fm.		not exposed		Xigulanha Fm.	
Ainasu	Utubulake Fm.	?	Nulshotgol Fm			?		

Fig. 2. Correlation of the Lower Devonian of the Junggar–Balkhash and Mongolo–Okhotsk regions.

Table 1. *The common genera in the Junggar–Balkhash and Mongolo–Okhotsk regions of late Early Devonian age*

	JUNGGAR–BALKHASH REGION			MONGOLO–OKHOTSK REGION			
	Balkhash area	W. Junggar	E. Junggar	Lake Achitnoor, NW Mongolia	E. Uchumucin, S. Great Hinggan	Lesser Hinggan	Far East
Coelospira	Kaplun & Modzalevskaya (1978, p. 107, pl. 6, figs 1–5)	in coll.	Zhang (1983, p. 338, pl. 97, fig. 9)		Su (1976, p. 203, pl. 103, figs 15–22 his *Coelospirella*)	Hamada (1971, p. 72, pl. 24, figs 10–17, his *Bifida*)	
Pacificocoelia	Kaplun (1961, p. 83, pl. 12, figs 1–12). Kaplun & Modzalevskaya (p. 111, pl. 6, figs 11–18)		Zhang (1983, p. 378, pl. 90, fig. 5)		Su (p. 204, pl. 91, figs 2–6)		Kaplun & Modzalevskaya (p. 111, pl. 6, figs 19, 20)
Dalejina	Kaplun & Modzalevskaya (1978, p. 38)			Alekseeva et al. (1981, p. 23, pl. 3 figs 7–10)			
Sinostrophia	Kaplun & Modzalevskaya (p. 86, pl. 4, figs 1–4)	Zhang (1983, p. 272, pl. 86, fig. 15)	Zhang (1983, p. 86, pl. 86, fig. 15)	Alekseeva et al. (1981, pl. 53, pl. 11, figs 1–8)	Su (p. 173, pl. 98, figs 1–3)	Hamada (1971, p. 53, pl. 11–17)	Kaplun & Modzalevskaya (p. 86, pl. 4, figs 5–8)
Maoristrophia	Kaplun (1961, p. 73, pl. 9, figs 1–6), Kaplun & Modzalevskaya (p. 95, pl. 5, figs 1–5)			Alekseeva et al. (1981, p. 64, pl. 15, figs 7–9)			
Rhytistrophia	Kaplun & Modzalevskaya (p. 92, pl. 2, figs 5–6)	in coll.	Zhang (1983, p. 272, pl. 86, fig. 1)	Alekseeva et al. (1981, p. 140, tab. 3)	Su (1976, p. 176, pl. 95, figs 7–10, pl. 96, figs 1–4, 7–8)	Su (1980, p. 281, pl. 123, figs 17–20)	Kaplun & Modzalevskaya (1978, p. 92, pl. 2, figs 7, 8, pl. 3, fig. 1)
Leptaenopyxis	Ushatinskaya & Nilova (1975, p. 101, pl. 25, figs 4, 5)	Zhang (1983, p. 271, pl. 87, fig. 5)	Zhang (p. 271, pl. 87, fig. 4)	Alekseeva et al. (1981, p. 48, pl. 8, fig. 5; pl. 9, figs 1–7; pl. 10, figs 1–6)	Su (1976, p. 170, pl. 94, figs 1–7; pl. 110, figs 5–7)	Hamada (1971, p. 48, pl. 7, figs 3–5)	Kaplun & Modzalevskaya (1978, p. 83, pl. 1, figs 10–13)
Phlidostrophia	Kaplun & Modzalevskaya (p. 89, pl. 5, figs 9–13)					Hamada (1971, p. 57, pl. 18, figs 1–3)	
Leptostrophiella	Kaplun (1961, p. 78, pl. 10, figs 1–3; pl. 9, fig. 10) Kaplun & Modzalevskaya (p. 93, pl. 3, figs 2, 3)		Zhang (1983, p. 272, pl. 86, figs 2, 4, 3, 11)	Alekseeva et al. (1981, p. 55, pl. 12, figs 1–14; pl. 13, figs 1–10)		Hamada (1971, p. 48, pl. 9, figs 1–6; pl. 10, figs 1–4)	Kaplun & Modzalevskaya (1978, p. 93, pl. 3, fig. 4)
Wilsoniella	Ushatinskaya & Nilova, (1975, p. 111, pl. 28, fig. 2)		in coll.	Alekseeva et al. (1981, p. 82, pl. 17, figs 1–3; pl. 18, 1–4)	Su (1976, p. 191, pl. 112, figs 1–6)	Hamada (1971, p. 64, pl. 21, figs 5–7; pl. 22, figs 1–5; pl. 23, figs 1–3)	
Tridensilis		Zhang (1983, p. 320, pl. 97, figs 7, 8)	Hou (1960, p. 57, pl. 1, figs 8, 9)		Su (1976, p. 193, pl. 102, figs 1–4; pl. 112, figs 7–12)	Su (1980, p. 257, pl. 129, figs 1–5)	

(Table 1. Continued)

	JUNGGAR–BALKHASH REGION			MONGOLO–OKHOTSK REGION		
Reeftonia	Zhang (1983, p. 263, pl. 88, figs 4, 5)	in coll.	Alekseeva *et al.* (1981, p. 14, pl. 1, figs 7–16; pl. 2, figs 1–7)	Su (1976, p. 166, pl. 93, figs 6–14)	Hamada (1971, p. 43, pl. 5, figs 1–13; pl. 6, figs 1–8)	
Megakozlowskiella	Zhang (1983, p. 363, pl. 91, figs 3, 4)	Zhang (1983, p. 363, pl. 91, figs 1, 2)		Su (1976, p. 219, pl. 109 figs 1–5)		Kaplun & Modzalevskaya (1978, p. 117, pl. 7, figs 14–16)
Fallaxispirifer	Kaplun & Modzalevskaya (1978, p. 113, pl. 6, figs 21–24, their *E. togatus*)	Zhang (1983, p. 368, pl. 97, fig. 1)		Su (1976, p. 226, pl. 108, figs 1–5)	Su (1980, p. 316, pl. 133, figs 7–13)	
"*Paraspirifer*"	Zhang (1983, p. 361, pl. 89, figs 1, 2)	Zhang (1983, p. 361, pl. 89, figs 4, 5; pl. 90, fig. 4)		Su (1976, p. 220, pl. 107, figs 1–6)	Hamada (1971, p. 78, pl. 27, figs 2–7; pl. 28, fig. 2)	Kaplun & Modzalevskaya (1978, p. 115, pl. 7, fig. 21)
"*Fimbrispirifer*"	Zhang (1983, p. 139, pl. 97. fig. 5)	in coll.			Hamada (1971, p. 50, pl. 28, fig. 1)	
"*Ivanothyris*" (*grandis*-type)	Kaplun & Modzalevskaya (1978, p. 116, pl. 7, figs 12, 13)		Alekseeva *et al.* (1981, p. 92, pl. 19, figs 1–6; pl. 25, figs 7, 8)			
Xinjingospirifer	Kaplun (1961, p. 101, pl. 17, figs 5–9)	in coll.	Zhang (1983, p. 362, pl. 90, figs 1–3)		Su (1980, p. 316, pl. 131, figs 8–13)	

water currents than the existence of the landmass itself, i.e., land masses by themselves need not be a barrier to reproductive communication unless they are responsible for generating separate water masses. However, whether or not a barrier, ocean or continent existed somewhere between the Altai-Sayan to the north and the region is still under discussion. The situation just described is very different from that of the Rhenish–Bohemian Region where a strong case has been made for local ecologic controls only.

Next, consider that the carbonate facies from within the Balkhash–Mongolia–Okhotsk Region occurs along the Gobi Altai (Shin–Zhinst and Tsegt–Obo areas). The Emsian Tchulun bed is characterized by tabulate corals, bryozoans, a few stromatoporoids, and brachiopods. Alekseeva *et al.* (1981) recorded the brachiopods *Reeftonia* sp., *Leptagonia goldfussiana* (Barr.), *Glossoleptaena arguta* Grats., *Uncinulus* ex gr. *parallelopipedus* (Bronn), *Atrypa krekovskensis* Rzhon., *Carinatina arimaspis* (Eich.), *Megakozlowskiella perlamellosa* (Hall), *Mongolospira turgensis* Aleks., *Schizophoria* sp., *Latonotoechia* cf. *latona* (Barr.), *Spinatrypa spinosaeformis* Khod. This small fauna, derived from carbonate rocks, is similar to that from the region's terrigenous rocks, with the addition of *Carinatina*. The carbonate zone extends eastward to the Wunuer, middle part of the Greater Hinggan Range, where the brachiopods of the Emsian Beikung and Wunuer Formations are common in the region (see Wang Yu *et al.* 1984, p. 273). Regrettably, these fossils have not been described. Their identifications require confirmation. So far as we know no characteristic Uralian elements, such as *Karpinskia*, have been described in the carbonate facies of the Balkhash–Mongolo–Okhotsk Region. In fact, there are also two different biofacies in the Uralian Region, the brachiopod representative of terrigenous facies being a *Lissatrypa* fauna, quite different with those of the region.

Based on the above, together with the unique admixture of Eastern Americas and Tasman genera, we conclude that the region is distinct biogeographically from the Uralian Region, and that its unique properties are not ecologic.

(3) If a biogeographic unit (or units), what are its unique features, and what is its biogeographic rank?

Many brachiopod workers mentioned that the characteristic of the brachiopod fauna in the region is one of mixing, that is, the presence of a number of genera restricted elsewhere to the Eastern Americas Realm plus a few genera from the Tasman Region. Boucot (Boucot 1975, p. 304; Wang *et al.* 1984, p. 277) listed the following Eastern Americas Realm taxa: *Pacificocoelia*, *Meristella*, *Leptostrophia* s.s., *Coelospirella*, *Megakozlowskiella*, *Orthostrophia*, *Chonostrophiella*, *Rhytostrophia*, *Discomyorthis* and probably *Costistrophonella*, and the Tasman genera *Maoristrophia* and *Reeftonia*. These facts substantiate reproductive communication into the Balkhash–Mongolia–Okhotsk Region from the Eastern Americas Realm and the Tasman Region.

A large number of endemic genera make up the other characteristic feature of the region. Some spiriferids mistakenly assigned to *Ivanothyris* (*grandis* type), *Delthyris* (*minimus* type), *Fimbrispirifer* (aff. *divaricatus*), *Paraspirifer* (*gigantea* type) and part of *Megakozlowskiella* are new. Other described endemic genera include *Mongolella* Alek., *Wilsoniella* Khalf, *Sinostrophia* Hamada, *Fallaxispirifer* Su, *Tridensilis* Su, *Trilobostrophia*

Table 2. *Late Lower Devonian brachiopod genera additional to those in Table 1*

Tastaria (Kaplun & Modzalevskaya 1978, p. 95, pl. 3, figs 10–12; FE; Su 1976, p. 174, pl. 94, figs 8–11; SH)
Discomyorthis (Kaplun & Modzalevskaya 1978, p. 81, pl. 1, figs 4–8; FE; Alekseeva *et al.* 1981, p. 25, pl. 2, fig 18; pl. 3, figs 1–6; pl. 24, figs 10, 12; M; Hamada 1971, p. 41, pl. 3–4; LH)
Rotundostrophia (Kaplun & Modzalevskaya 1978, p. 98, pl. 3, fig. 5; FE)
Mesodouvillina (Ushatinskaya 1983, p. 44, pl. 5, figs 1–16; B)
Chonostrophia (Kaplun & Modzalevskaya 1978, p. 101, pl. 5, figs 15–15; B)
Chonostrophiella (Hamada 1971, p. 61, pl. 20, figs 3–13; LH)
Notoparmella (Kaplun & Modzalevskaya 1978, p. 112, pl. 7, figs 7–10; FE)
Xingjiangospirifer (Zhang 1983, p. 362, pl. 90, figs 1–3; EJ, WJ; Kaplun & Modzalevskaya 1978, p. 116, pl. 7, fig. 11; B)
Struveina (Kaplun & Modzalevskaya 1978, pl. 6, figs 25–30; B)
Isorthis Alekseeva *et al.* p. 11, pl. 1, figs 1–6; M)
Leptagonia (Alekseeva *et al.* 1981, p. 45, pl. 8, figs 6–10; M)
Schizophoria (Hamada 1971, p. 37, pl. 1, fig. 2; LH; Su 1976, p. 164, pl. 93, figs 11–14; SH; Alekseeva *et al.* 1981, p. 35, pl. 5, figs 5–9, 11, 12; pl. 6, figs 1–10; pl. 7, figs 1–11; pl. 8, figs 1–4; M)
Areostrophia (Alekseeva *et al.* 1981, p. 67, pl. 15, figs 3–6; pl. 16, figs 1–10; M)
Platyorthis (Alekseeva *et al.* 1981, p. 27, pl. 3, figs 11–18; pl. 4, figs 1–10; pl. 14, figs 1–9; M)
Trigonirhynchia (Alekseeva *et al.* 1981, p. 75, pl. 18, figs 5–7; M)
Leptostrophia (Hamada 1971, p. 48, pl. 9, figs 1–6; pl. 10, figs 1–4; LH; Alekseeva *et al.* 1981, p. 60, pl. 11, fig. 9; pl. 14, figs 1–9, pl. 15, figs 1–2; M)
Latonotoechia (Su 1976, p. 190, pl. 101, figs 1–3; SH; Alekseeva *et al.* 1981, p. 71, pl. 18, figs 11–13; M)
Spinatrypa (Su 1976, p. 201, pl. 102, figs 5–9; SH; Alekseeva *et al.* 1981, p. 87, pl. 16, figs 15–20; M; Zhang 1983, p. 334, pl. 95, fig. 8; EJ)
Nucleospira (Hamada 1971, p. 72, pl. 25, figs 1–4; LH; Alekseeva *et al.* 1981, p. 102, pl. 21, fig. 16; pl. 22, figs 1–11; M)
Chonetes (Su 1971, p. 182, pl. 100, fig. 12; SH; Alekseeva *et al.* 1981, p. 69, pl. 16, figs 11–14; M)
Howellella (Su 1976, p. 217, pl. 105, figs 4–12; SH; Hamada 1971, p. 76, pl. 25, figs 11–17; pl. 26, figs 1–7; pl. 26, fig. 1; LH)
Cyrtina (Su 1976, p. 211, pl. 114, figs 9–11; SH; Alekseeva *et al.* 1981, p. 99, pl. 21, figs 9–10; M)
"*Stropheodonta*" (Kaplun & Modzalevskaya 1978, p. 99, pl. 3, figs 14–16; FE; Hou 1960, p. 57, pl. 1, figs 3–6; EJ)
Rhynchospirina (Alekseeva *et al.* 1981, p. 107, pl. 22, figs 12–14; M)
Mongolella (Alekseeva *et al.* 1981, p. 111, pl. 23, figs 8–11; M)
Badainjarania (Zhang 1981, p. 388, pl. 1, figs 12–16; BJ)
Howittia (Alekseeva *et al* 1981, p. 96, pl. 20, figs 10–17; pl. 21, figs 1–8; M)
Megastrophia (Hamada 1971, p. 47, pl. 8, figs 1–5; LH; Su 1976, p. 171, pl. 97, figs 1–19; SH)
Meristella (Zhang 1983, p. 375, pl. 88, figs 7–8; WJ)
Borealirhynchia (Su 1976, p. 195, pl. 101, figs 4–10; SH)
Aesopomum (Hamada 1971, p. 59, pl. 19, figs 1–7; pl. 20, figs 1–2; LH)
Resserella (Su 1976, p. 168, pl. 94, figs 2–4; SH; Zhang 1983, p. 262, pl. 88, figs 1–3; WJ)
Eodevonaria (Zhang 1983, p. 281, pl. 87, figs 11–14; WJ, EJ)
Cymostrophia (Zhang 1983, p. 271, pl. 87, figs 9–10; EJ)
Cyrtonopsis (Hou 1960, p. 61, pl. 3, figs 1, 3; EJ)
Neimongolella (Zhang 1981, p. 385, pl. 1, fig. 11; BJ)
Megaplectatrypa (Zhang 1981, p. 386, pl. 2, figs 5–6; BJ)
Rariella (Zhang 1981, p. 386, pl. 2, figs 10–16; BJ)
Pararhynchospirina (Zhang 1981, p. 387, pl. 2, fig. 9; BJ)

SH, Southern Greater Hinggan Mountains; LH, Lesser Hinggan Mountains; BJ, Badan Jaran Desert; FE, Far East of USSR; M, Mongolia (Lake Achitnoor); B, Pribalkhash; EJ, Eastern Junggar; WJ, Western Junggar.

Schischkina, *Xingjiangospirifer* Hou and Zhang, *Neimongolella* Zhang, *Megaplectatrypa* Zhang, *Rariella* Zhang, *Pararhynchospirina* Zhang, *Barainjarania* Zhang, and *Septoparmella* Su. The percentage of endemics may reach >30%, although it is somewhat lower than in other regions. A third feature should be emphasized, namely forms of purely Uralian and South China Regions type are completely lacking in the vast area from Lake Balkhash to the Far East.

All of these features, the biogeographic mixture, the presence of a relatively high percentage and number of endemics, the absence of diagnostic taxa of Uralian and other adjacent Regions, indicate the uniqueness of the Balkhash–Mongolia–Okhotsk Region as contrasted with other biogeographic units.

The question about biogeographic rank is difficult. The percentage of endemics would place the area somewhat below the region level, but the unique admixture of Eastern Americas Realm and Tasman Region taxa makes us decide that region level rank is appropriate. Keep in mind that we still lack any very precise rules for deciding just how biogeographic ranks should be apportioned for the Devonian or any other time interval.

Differences between Altai-Sayan Province and the Balkhash–Mongolia–Okhotsk Region

Rzhonsnitskaya (*in* Nalivkin *et al.* 1973) and Gratsianova (*in* Yolkin *et al.* 1982) systematically summarized the distribution of brachiopods from the Salair, Gorny and Rudny Altai. More than 35 genera were recorded in the Emsian equivalent Salairka, Belova and Shanda Horizons. Of these *Carinatina*, *Punctatrypa*, *Zdimir*, *Vagrania*, *Sieberella*, *Ivdelinia* are widespread in the Old World Realm. It is clear that the presence of *Karpinskia* indicates that the Sayan-Altai should be assigned to the Uralian Region (also occurs in the Kuznetsk, Bublichenko 1928) as pointed out by Rzhonsnitskaya (1978) and Boucot (1975). The faunal similarity may be consistent with Devonian palaeogeography. Russian geologists have suggested a possible marine connection between the Urals and the Altai-Sayan through the Siberian Lowland. From the viewpoint of the tectonics of the continental block, Talent *et al.* (1987) considered that the Altai-Sayan was a part of the continental margin of the Siberian Block. In view of the high level of Emsian age generic similarity between the Urals and the

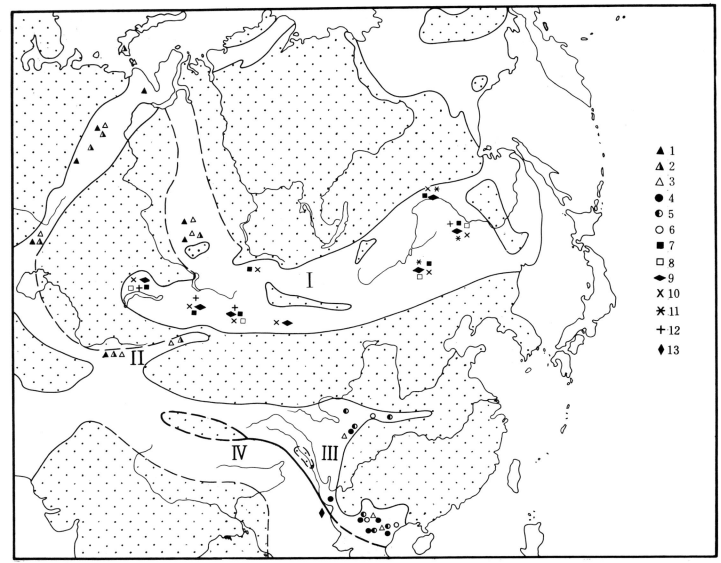

Fig. 3. The 'fixist' biogeography of the Emsian Brachiopoda of Asia. I, Junggar–Hinggan region; II, South Tianshan region; III, South China region; IV, West Yunnan–Tibet region. 1, *Karpinskia*; 2, *Ivdelinia*; 3, *Zdimir*; 4, *Dicoelostrophia*; 5, *Athyrisina*; 6, Euryspirifer; 7, *Sinostrophia*; 8, *Coelospira*; 9, *Rhytistrophia*; 10, '*Paraspirifer*'; 11, *Discomyorthis*; 12, *Xinjiangospirifer*; 13, *Bifida*.

Altai-Sayan, why could they not be placed on the western and southern sides of the Siberian Platform, with the Kazakhstan Block on the south side (see Boucot 1969, for a Silurian view of this possibility)? The brachiopod differences between the Altai-Sayan and the Balkhash–Mongolia–Okhotsk Regions are apparent and are unlikely to be due to ecologic controls as mentioned above. However, it should be noted that the Altai-Sayan fauna does contain a few Eastern Americas Realm and Tasman genera such as *Rhytistrophia* and *Maoristrophia*. Some genera, such as *Wilsoniella*, are first recognized in the Altai-Sayan but occur later in the Balkhash–Mongolia–Okhotsk Region. This suggests Altai-Sayan region boundary mixing with the adjacent Balkhash–Mongolia–Okhotsk Region. It is consistent with the presence of a seaway connecting the Ural Region with the Balkhash–Mongolia–Okhotsk region.

Some geologists have investigated the faunal differences between different zones on the North China Block and on the Siberian Block. From the north margin of the Tianshan Mountain, the Sarimu Lake area south to the Darbut Ophiolite Zone, and the north margin of Yinshan Mountain, and the Aohen Ai area south to the Suolon-Hegen Ophiolite Zone, the Devonian fauna has been recorded and showed relatively low diversity. Unfortunately, it is impossible to make detailed comparisons due to the small sample. In the northern Bedain Jaran desert by the western border of Inner Mongolia, there is a series of normal arenaceous deposits disconformably lying on Silurian, where the Emsian brachiopods are very unique, containing many small size, endemic genera (Zhang Yan 1981). It is difficult to interpret the difference as due to facies control.

Region or Realm: Biogeographic Rank?

Boucot and others (1969; Johnson & Boucot 1973; Boucot 1974) have divided the earlier Devonian into three realms; Old World, Eastern Americas and Malvinokaffric. Alekseeva (1986) is in favor of two units of realm rank. For central East Asia Johnson & Boucot (1973) considered the Mongolo–Okhotsk as a Region within the Old World Realm, and the Junggar–Balkhash as a community unit within the Uralian Region. However, in later publications (Wang Yu *et al.* 1984; Boucot 1985), the Mongolo–Okhotsk unit was not considered a region. Rzhonsnitskaya *et al.* (1981) combined the region with what Boucot termed the Eastern

Americas Realm and eastern Australia Tasman Region into a single unit termed the Pacific Realm, based mainly on the common genera present in these areas. Kauffman (1973) set up a quantitative scale involving percentages of endemic genera as a means for defining various biogeographic ranks. In fact, besides level of endemism, the presence or absence of particular taxa for certain regions, and the overall diversity are important factors for recognition of biogeographic rank. The Balkhash–Mongolia–Okhotsk Region has a well known Emsian brachiopod fauna. Although many of the fossils have not yet been systematically described, the number of endemic brachiopods, excluding those with problems in identification, has reached one-third. This is similar to the number of endemic taxa in the South China Region. Considering the unique mixture feature, which is unknown in other regions of the Old World Realm, one might view the Balkhash–Mongolia–Okhotsk unit as transitional from the Old World Realm to the Eastern Americas Realm, particularly as far as brachiopods are concerned. Therefore we must pay great attention to its importance. Palaeogeographically the region was far removed from the Eastern Americas Realm area and from the Tasman Region of eastern Australia, regardless of whether or not one prefers a plate tectonic or fixist point of view. How then should one logically interpret the faunal rank of these areas? We prefer, as discussed earlier, to consider that a trans-Panthalassic Ocean (a larger Pacific) current might explain the faunal migration whose results are preserved in the Balkhash–Mongolia–Okhotsk Region strata.

From all of the above facts we conclude that the Balkhash–Mongolia–Okhotsk is at least of regional rank. For reasons of both palaeogeography and biogeography (relative numbers of Old World cosmopolitan versus Eastern Americas Realm genera) we prefer to regard the Region as a part of the Old World Realm for the moment. After other groups of important megafossils, such as the corals and trilobites, have been carefully considered biogeographically we may be able to strengthen our present conclusion, modify it seriously, or reject it.

References

ALEKSEEVA, R. E. 1986. On the paleozoogeography of the Early Devonian of Central and East Asia. In: ROYANOV, A. U. (ed.) *Paleobiogeographical problem of Asia*, Scientific Press, Moscow.

——, MENDBAYAR, B. & ERLANGER, O. A. 1981. Brachiopods and biostratigraphy of Lower Devonian of Mongolia. *Scientific Press*, Moscow, 176 pp.

BOUCOT, A. J. 1969. The Soviet Silurian: Recent impressions. *Geological Society of America Bulletin*, 80, 1155–1162.

—— 1974. Silurian and Devonian biogeography. In Ross, C. A. (ed.) *Paleogeographic provinces and provinciality*, SEPM Special Publication, 21, 165–176.

—— 1975. *Evolution and extinction Rate Controls*. Elsevier, Amsterdam.

—— 1985. Late Silurian–Early Devonian biogeography, provincialism, evolution and extinction. *Philosophical Transactions of the Royal Society*, London, B309, 323–339.

——, JOHNSON, J. G. & TALENT, J. A. 1969. Early Devonian brachiopod zoogeography. *Geological Society of America Special Paper*, 119.

BUBLICHENKO, N. L. 1928. Fauna brakhiopod nizhnego paleozoya okrestostei c. Sara-Chumishskogo (Kuznetskii bassein. *Izvestii Geol. Kom.*, XLVI, 8, 979–1008.

—— 1958. Terrigene Fazies des Devons in Kasachstan. *Prager Arbeitstagung uber die Stratigraphie des Silurs und des Devons*, 411–424.

HAMADA, T. 1971. Early Devonian brachiopods from the Lesser Khingan District of Northeast China. *Paleontological Society of Japan Special Paper*, 15.

HOU, H. F. 1981. Devonian brachiopod biostratigraphy of China. *Geological Magazine*, 118(4), 185–192.

—— & WANG, S. T. 1985. Devonian paleogeography of China. *Acta Palaeontologica Sinica*, 24, 186–197.

——, —— & ZHAO, X. W. 1985. Paleogeographic map of the Devonian. In: WANG, H. (ed.) *Atlas of the paleogeography of China*, Cartographic Publishing House, Beijing.

——, XIANG, L. W., LAI, C. G. & LIN, B. Y. 1979. Advances in the Palaeozoic stratigraphy of Tianshan–Xingan region. *Journal of Stratigraphy*, 3, 175–187.

JOHNSON, J. G. & BOUCOT, A. J. 1973. Devonian brachiopods. In: HALLAM, A. (ed.) *Atlas of Palaeobiogeography*, Elsevier, Amsterdam, 89–96.

KAPLUN, L. I. 1961. Brachiopods of Lower Devonian of North Pribalkhash. *Materials on Geology and Mineral Resources of Kazakhstan*, 1(26), stratigraphy and paleontology, 64–114.

—— & MODZALEVSKAYA, E. A. 1978. Brachiopoda. In: RZHONSNITSKAYA, M. A., (ed.) *Stage of Lower Devonian of Pacific Region in territory of U.S.S.R.*, "Nedra", Moskva, 79–94.

KAUFFMAN, E. G. 1973. Cretaceous bivalvia. In: HALLAM, A. (ed.) *Atlas of Palaeobiogeography*, Elsevier, Amsterdam, 353–384.

NALIVKIN, D. V., RZHONSNITSKAYA, M. A. & MARKOVSKI, B. P. (eds) 1973. *Devonian System: Stratigraphy of the U.S.S.R.*, 1–2, "Nedra", Leningrad.

NILOVA, N. V. 1973. On the age of the Ainasu and Karaespe horizons of Central Kazakhstan on Brachiopods. In: NALIVKIN, D. V. (ed.) *Stratigraphy of Lower and Middle Devonian*, "Nauka", Leningrad, 2, 183–188.

RZHONSNITSKAYA, M. A. 1968. Devonian of the USSR. In: OSWALD, D. H. (ed.), *International Symposium on the Devonian System*, Calgary, 1, 331–348.

—— (ed.) 1978 *Stages of Lower Devonian of Pacific Region in territory of USSR*. "Nedra", Moskva.

RZHONSNITSKAYA, M. A., KULIKOVA, V. F. & PETROSYAN, N. M. 1982. Type section of the Lower Devonian and Eifelian of Salair. In: SOKOLOV, B. S. & RZHONSNITSKAYA, M. A. (eds) *Biostratigraphy of Lower and Middle Devonian boundary deposits*, "Nauka", Leningrad, 116–122.

SU, Y. Z. 1976. Brachiopoda. In: *Atlas of Paleontology of North China*, part 1, Nei Mongol, Geological Publishing House, Beijing, 155–277.

—— 1980. Brachiopoda. In: *Atlas of Paleontology of Northeast China*, part 1, Palaeozoic, Geological Publishing House, 254–428.

TALENT, J. A., GRATSIANOVA, R. T., SHISHKINA, G. R. & YOLKIN, E. A. 1987. Devonian faunas in relation to crustal blocks: Kazakhstan, Mongolia, Northern China. *Courier Forschungsinstitut Senckenberg*, 92, 225–233.

USHATINSKAYA, G. T. 1983. Morphogenesis of some stropheodontids from the Upper Silurian and Lower Devonian of Central Kazakhstan. *Paleontologischeskii Zhurnal*, 4, 42–54.

—— & NILOVA, N. V. 1975. Brakhiopodi. In: MENNER, V. V. (ed.) *Materiali po geologii Tsentralnogo Kazakhstana*, T, XII, "Nedra", Moskva, 93–119.

WANG, Y. & RONG, J. Y. 1983. Brachiopod fauna of Yukiang formation and its paleozoogeographical features. In: LU, Y. H. (ed.) *Paleobiogeography of China*, Scientific Press, Beijing, 53–63.

——, BOUCOT, A. J., RONG, J. Y. & YANG, X. C. 1984. Silurian and Devonian biogeography of China. *Bulletin of the Geological Society of America*, 95, 265–279.

YANG, S. P., PAN, J. & HOU, H. F. 1981. The Devonian System in China. *Geological Magazine*, 118 (2), 113–138.

YOLKIN, E. A., GRATSIANOVA, R. T., ZHELTONOGOVA, V. A., & KIM, A. I. 1982. The main biostratigraphic boundaries and subdivisions of the Lower/Middle Devonian in the western Altai-Sayan region. In: SOKOLOV, B. A. & RZHONSNITSKAYA, M. A. (eds) *Biostratigraphy of Lower and Middle Devonian boundary deposits*, "Nauka", Leningrad, 65–80.

ZHANG, F. M. 1983. Brachiopods. In: *Atlas of Paleontology of Northwest China*, Xinjiang Uigur Autonomous Region, 2, Late Paleozoic, Geological Publishing House, Beijing, 262–386.

ZHANG, Y. 1981. Early Devonian brachiopods from Zhusilenghaizhan region, western Neimongol. *Acta Palaontologica Sinica*, 20, 384–392.

Carboniferous—Permian Biogeography

Late Palaeozoic provinciality in the Marine Realm

RICHARD K. BAMBACH

Department of Geological Sciences, Virginia Polytechnic Institute and State University, Blacksburg, Virginia 24061, USA

Abstract: Biogeographic units commonly recognized in large-scale palaeontological studies are usually related to realms, not provinces as recognized in modern zoogeography. In this study the geographic distribution of the Rugosa, Tabulata, Bivalvia, Ammonoidea, Strophomenida, pedunculate Articulata, Bryozoa, and Crinoidea are used to recognize a set of realms and provinces (based on 30% or more generic endemism in the combined fauna within a province) for the Early and Late Carboniferous and Early and Late Permian. Between 12 and 18 provinces in four or five realms are identified in each interval of the Late Palaeozoic. The recently stated view than the Late Palaeozoic was a time of increasing provinciality is related to a decrease in the average range of regionally distributed genera, not an increase in either endemism or in the number of recognizable biogeographic units.

The problem of provinciality and diversity in the Late Palaeozoic

Several recent studies of biogeographic provinciality in the Late Palaeozoic argue that the pattern of change is one of cosmopolitan faunas in the Early Carboniferous giving way to provincial conditions in the Permian. Ross (1979 p. A283) states that, 'Faunas of Tournaisian and Visean [Early Carboniferous] times were relatively widespread; many were nearly cosmopolitan and the remainder had low levels of endemism. Ross & Ross (1979 p. A341) then state that, 'By Artinskian [late Early Permian] time many shallow benthic marine faunas showed marked provinciality.' Ziegler *et al* (1981) summarized a variety of biogeographic studies and concluded that there were six major faunal regions in the Visean (Early Carboniferous) but that as many as twelve could be found in the Kazanian (early Late Permian). Increasing climatic gradients and the elaboration of tectonically produced barriers to dispersal were suggested as possible causes for the apparent increase in provinciality during the Late Palaeozoic.

All data compilations on world diversity through the Phanerozoic indicate that Late Palaeozoic diversity was fairly constant and, in fact, declined slightly from the Early Carboniferous to the Early Permian. For example, the number of families in marine taxa with high preservation potential (robust mineralized skeletons) declined from 405 in the Visean to 350 in the Artinskian (data derived from Sepkoski, 1982). The total number of genera in the eight groups used in this study decreased from 778 in the Early Carboniferous to 505 by Early Permian. Although genera are relatively short lived compared to families, so that summed diversity of genera for epoch-long intervals may be correlated with epoch length, some decline in generic diversity is still revealed by these data. The length of the Early Permian may be 20% less than the length of the Early Carboniferous but generic diversity in the Early Permian is 35% less than in the Early Carboniferous. A further, sharper drop in diversity is recorded for the Late Permian (but prior to the extinction at the end of the Permian). However, the impression of diversity loss in the later Permian has almost certainly been increased artificially by the poor geological record of Late Permian times which limits our knowledge of the geographic distribution and occurrence of taxa. Valentine (1973) and Valentine *et al.* (1978) argued that world diversity is significantly affected by provinciality; higher provinciality resulting in an increased number of endemic taxa and, consequently, higher diversity. If provinciality did increase during the Late Palaeozoic, then why didn't world diversity increase as provinciality increased?

In modern zoogeography, recognition of biogeographic units is based on differences between regions. Definition of provinces is commonly related to some particular level of endemism. For example, in his compilation of world marine zoogeography, Briggs (1974, p. 16) uses the criterion that, 'if there is evidence that 10 per cent or more of the species are endemic to a given area, it is designated as a separate province.' Although there is no fixed modern provincial scheme, about 30 to 35 widely recognized modern marine shelf provinces are of sufficient scope that they might be identifiable in the fossil record at some future time. Briggs recognizes 53 modern shallow water marine provinces. However, 15 of them are isolated oceanic island areas and another five are small basins or large islands, few of which would have a good chance of being recognized as entities in the fossil record. Valentine (1973, p. 355–356) specified 30 large modern marine shelf provinces.

The level of endemism that permits recognition of faunal provinces is caused by climatic and geographic barriers to dispersal, causing at least partial evolutionary isolation of regional faunas. Temperature is the major climatic influence controlling the distribution of marine shelf organisms and land barriers and distance across deep oceanic basins are the major geographic barriers to their dispersal. Significant climatic gradients existed in ancient times, and palaeogeographic reconstructions also reveal widely separated continental blocks at many times. Assuming comparable ecological and evolutionary behavior for ancient and modern organisms, provinces in the past should have formed in areas separated by climatic and geographic differences comparable to those that form barriers to dispersal between provinces today. The degree of taxonomic difference between such ancient provinces should also have been similar to that found between modern provinces.

Are the commonly recognized palaeobiogeographic provinces of the Palaeozoic similar to modern provinces? The number of provinces reported for most time intervals in the past resembles the number of large-scale realms or regions recognized in the modern world rather that the number of provinces. Valentine *et al.* (1978) tabulated the number of provinces reported in the literature for all times in the Phanerozoic and found no more than six for any time interval prior to the Holocene. A few higher totals have been suggested in more recent studies (9 Tournaisian provinces based on the Rugosa (Federowski 1981) and possibly 12 Kazanian provinces (Ziegler *et al.* 1981)) but the number of provinces recognized by palaeontologists continues to be fairly low. Is it possible that many palaeobiogeographic 'provinces' are not defined in the way modern provinces are and, therefore, that they may not be comparable units?

No consistent methodology is used to erect palaeoprovinces. Often only a single higher taxon is the subject of study. The total range of regionally distributed taxa and similarities between regions are usually emphasized rather than endemism and dissimilarities. In palaeontology this is understandable because biostratigraphic correlation is always a concern, and past geographic affinities are concealed and must be deduced from scattered evidence. But when palaeontologists erect 'provinces' based on the total geographic range of higher taxa such as families, as Hill has done for the Rugosa (in Robison 1981) and as Waterhouse &

Bonham-Carter (1975) did for Permian brachiopods, or on the distribution of widespread planktonic organisms, as in the case of conodonts (Charpentier 1984), it is no wonder that we do not find more numerous palaeoprovinces reflecting relatively localized endemic centres. Only regional studies, in which the authors are concerned about differentiating between localities within relatively circumscribed areas, regularly emphasize dissimilarity rather than similarity of faunas (Johnson 1971; Sando et al. 1975; Luo 1984).

For time correlation and other important palaeontological purposes it is necessary to discover faunal linkages between regions, even if they are revealed by relatively small parts of the total fauna, but this is not the focus of attention of zoogeographers in defining modern provinces. If we are to compare the biogeography of the Late Palaeozoic, or any other past time interval, with the modern world, we must deal with units that are in some way comparable to modern biogeographic units. This paper constructs a biogeographic analysis of the Late Palaeozoic, utilizing levels of endemism and patterns of diversity of the whole fauna, in a way that identifies biogeographic units at least crudely comparable to modern units.

Methodology

The fossil record is too imperfect to use species for the analysis of palaeobiogeography over large areas for long intervals of time. Beyond that, the task of compilation of a complete record of species occurrences is just not feasible. Fortunately, Campbell & Valentine (1977), investigating the possibility of recognizing provinces using generic similarity and dissimilarity, demonstrated that modern provinces can be differentiated on generic records alone. For genera, adjacent provinces have similarity coefficients as great as 0.93 when Simpson's coefficient is used, but no greater than 0.70 using the Jaccard coefficient. With greater separation, both similarity coefficients decrease to lower levels. It should be possible to recognize provincial units comparable to modern provinces on the basis of generic distributions.

This paper establishes a set of biogeographic units for the Early Carboniferous, Late Carboniferous, Early Permian, and Late Permian on the basis of 30% generic endemism in each regional fauna. The data on the generic distributions of eight major taxa were compiled from a set of standard taxonomic references. Extremely widespread (cosmopolitan) genera are omitted from the provincial analysis because they contribute no information on geographic differentiation (see also the argument in Savage et al. 1979). A 30% level of generic endemism in the regional fauna is conservative in relation to Campbell & Valentine's comparison of adjacent modern provinces and insures that the provinces recognized are robust. To get a picture of the fauna as a whole and avoid creating subdivisions based exclusively on the vagaries in the distribution of a single group, data for six of the eight major taxa studied were combined for the provincial analysis. These data were supplemented with the diversity data from the other two groups for further palaeoclimatological analysis.

The goals of this study are to apply a hierarchical method of categorizing biogeographic areas that resembles the methods used by recent zoogeographers and to compare the units defined by this method to those recognized in past biogeographic studies of the Late Palaeozoic. There is no claim that the regions defined in this paper represent the actual species-based provinciality of a moment in time, only that a consistent world-wide scheme of definition has been applied in which less similarity exists between the areas recognized at each level of the hierarchy than within areas, at any time scale.

Because of limitations in stratigraphic resolution in the comprehensive taxonomic summaries utilized for this study, it was possible to make these whole fauna analyses only at the epoch level. Dividing the entire Late Palaeozoic into only four intervals obviously can not produce the resolution needed to see the pattern of species-based provinces that actually existed during any one short time interval. For instance, examination of the generic range data for Bryozoa (Ross 1978, 1981) and the discussion of Carboniferous coral distributions by Federowski (1981) both reveal changes in geographic ranges of some genera during the intervals used in this study. However, any time interval that permits world-wide correlation is probably too long to capture a picture of provinciality exactly comparable to the provinces of modern zoogeography. This is an unavoidable difference between biogeography and palaeobiogeography. For example, climatic fluctuations driven by Milankovitch cycles, which rapidly altered features such as the continental glaciation during the Late Carboniferous, can produce concommitant shifts in biogeographic distributions on a scale of 50 000 to 100 000 years, far below the level of world-wide time resolution in the Palaeozoic.

Because taxa are often found in different places at different times, the likelihood of a genus remaining endemic is reduced because epoch length time intervals are used. Thus, in this compilation it is less likely to identify separate regions on the basis of endemic taxa, not more.

The only factor counter to this conclusion is the chance that if there are numerous short-lived genera scattered through the long time intervals used, they could create the impression of considerable endemism based on comparisons of the distributions of non-correlative taxa. The data in the range charts for the Bryozoa (Ross 1978, 1981) permit evaluation of the potential 'oversplitting' in recognizing endemic centers from the use of non-correlative genera. Although it is not possible to test the other major taxa in detail (because of the limited stratigraphic resolution available), the corals, brachiopods, and bivalves have distributions of generic longevities comparable to the Bryozoa (as determined from local studies) and can be presumed to be similar on a global scale as well.

Globally, in the Late Palaeozoic, the Bryozoa have only one quarter of their genera restricted to single stages (roughly one third of an epoch, on average) and 70% of those occur in either the Visean (middle Early Carboniferous) or Kazanian (early Late Permian). Because most short-lived genera occur together in the same time interval (the Visean in the Early Carboniferous and the Kazanian in the Late Permian) they can not create an artifact of non-correlative areas of endemism. Only 7% of all the genera are short-lived and scattered at various other points through the four large time intervals used in this study. The danger of recognizing additional endemic centres (at the 30% level of endemism used in this study to specify a province) by miscorrelation is probably minimal. Although the use of long time intervals gives a 'time-averaged', rather than a 'snapshot', view of provinciality, it is almost certainly a conservative, not an oversplit, representation. On top of that, these are the same time intervals used in most older studies; using them permits direct comparison to the older work with which this study is to be compared.

The data base was compiled from standard taxonomic works, rather than from a variety of more specialized but variable and scattered sources. This sacrificed some geographic detail and also meant that half the data base came from sources over twenty years old. However, the advantages of using a consistent, expert taxonomy avoided the common criticism that a data base synthesized from the literature is internally inconsistent (Johnson 1971, p. 259 and Savage et al. 1979, p. 184). Also, the older taxonomy is often more generalized and therefore the unsplit genera often are more widely distributed than those listed in a recently revised, more precisely defined taxonomy. This is not desirable when searching for geographic resolution but it is likely that the inadequacies of the data base are conservative and, if anything, bias against, rather than in favour of recognizing endemic genera and separate provinces.

The taxa studied are the Rugosa and Tabulata in the Coelenterata, the Bivalvia and Ammonoidea in the Mollusca, the Strophomenida and (as a group together) the pedunculate

Articulata (Orthida, Rhynchonellida, Spiriferida, and Terebratulida) in the Brachiopoda, the ectoproct Bryozoa, and the Crinoidea in the Echinodermata. The *Treatise on Invertebrate Paleontology* served as the data source for the Rugosa and Tabulata (Robison 1981), Bivalvia (Moore 1969), Ammonoidea (Moore 1957), Brachiopoda (Moore 1965), and Crinoidea (Moore & Teichert 1978). Data on the Bryozoa are from J. R. P. Ross (1978, 1981). The data for the Crinoidea are further refined from data in G. D. Webster's *Bibliography of Palaeozoic Crinoids* (Webster 1973, 1975, 1984).

Others have also advocated quantitative methods for defining palaeobiogeographic units and some have developed more sophisticated schemes than that used here. Williams (1969, 1973), working on Ordovician brachiopods, pioneered in quantitative palaeobiogeographic analyses. He used similarity coefficients to cluster localities, emphasizing association of regions rather than their levels of dissimilarity. Johnson (1971) and Savage *et al.* (1979), in studies of Devonian brachiopods, advocated detailed evaluations using indices of provinciality that utilize endemism as well as acknowledging similarities. Kauffman (1973) first advocated the scheme of using particular levels of endemism for defining categories of palaeobiogeographic units in an effort to make Cretaceous units comparable to modern units, and Boucot (1975) followed this example in a study of Silurian and Devonian brachiopods.

Because of the differences in geographic precision and taxonomy between the various data sources used in this study the more complex formulations of some of these earlier studies have not been utilized (especially the calculation of similarity coefficients or provincial indices). However, this study uses a consistent methodology that leads to the recognition of biogeographic entities at least crudely comparable to those seen in the Recent. It is the first such synthesis of provinciality for the entire Late Palaeozoic and the first palaeobiogeographic study to pool data from several taxa to evaluate faunal endemism.

Patterns of distribution of individual higher taxa

The data for seven of the eight major groups studied are plotted on palaeogeographic reconstructions by Scotese (1986) and shown as Figs 1 (Rugosa), 2 (Tabulata), 3 (Bivalvia), 4 (Ammonoidea), 5 (Strophomenida), 6 (pedunculate Articulata), and 7 (Bryozoa). Data for the Crinoidea will be illustrated in a paper by J. B. Bennington & R. K. Bambach on a variety of echinoderm groups to be published elsewhere. Each figure shows the number of genera in each area (enclosed boldface number) and the number of genera endemic to the area (normal type following a slash if endemic genera are present) for up to 30 different areas of the world in each of four time intervals (Early and Late Carboniferous and Early and Late Permian). The number of regionally restricted but not precisely located genera is shown in various places by partly enclosed numbers over an abbreviated title for the area in which they occur. Lines connecting different areas (*ligations*) link areas that share genera. An unnumbered line indicates only one genus ligates the two areas. A small, unenclosed number on a ligation line indicates the number of genera shared if there are multiple ligations between the two areas. The total number of genera tallied for the time interval and the number of genera so widespread that they have no regional significance are listed at the lower right of each map. Each of the eight groups has some interesting biogeographic relationships.

Rugosa (Fig. 1)

The Rugosa consistently show a latitudinal diversity gradient (high diversity in low (tropical) latitudes and low diversity in higher latitudes) during each interval of the Late Palaeozoic. The density of ligations between areas decreases with time, with many ligations in the Early Carboniferous and fewer in the Permian. Connections throughout the Tethyan region become more prominent in the Permian than they were in the Carboniferous. The diversity of the Rugosa decreases steadily throughout the Late Palaeozoic.

Tabulata (Fig. 2)

Very few tabulate corals are found outside the tropical belt (beyond 30 degress north or south latitude). Only one genus ever occurred south of 45° south latitude. The Tabulata dwindle in diversity throughout the Late Palaeozoic and share very few genera between areas after the Early Carboniferous.

Bivalvia (Fig. 3)

The Bivalvia have a high proportion of widespread (cosmopolitan) genera. Those that are restricted in range are mostly tropical in the Early Carboniferous, but genera tolerant of temperate conditions appear in the Late Carboniferous (even as a glacial maximum is reached), and bivalves become common in high southern latitudes in the Early Permian. Note that one of the temperate genera in the Late Carboniferous is found in both Siberia and in three areas in south temperate latitudes but not in the tropics. Ligations between tropical Laurentia and Europe are very high for the Bivalvia throughout the Carboniferous and into the Permian. Unlike many other groups, diversity in the Bivalvia increases during the Carboniferous and into the Early Permian, as do the number of ligations between scattered areas.

Ammonoidea (Fig. 4)

As might be predicted for a pelagic, rather than benthic, group, there are very few endemic genera in the ammonites (except for the central USA in the Late Carboniferous). The variety of ammonite genera shared between areas produces a 'spider-web' of many ligations at all times. As in the Tabulata, very few genera of ammonites are found outside the 30 degree latitude belt. This is particularly noteworthy because of the lack of restriction of their generic ranges within the tropical belt.

Brachiopoda (Figs 5 and 6)

The Strophomenida (Fig. 5) contrast with the pedunculate Articulata (Orthida, Rhynchonellida, Spiriferida, and Terebratulida) (Fig. 6) in having a distinct latitudinal diversity gradient whereas the pedunculate Articulata do not. Interestingly, the surviving pedunculate Articulata (Rhynchonellida and Terebratulida) are relatively abundant and diverse in cooler waters today (Washington State to Alaska, New Zealand). Neither the Strophomenida or the pedunculate Articulata show any significant trend in diversity change during the Late Palaeozoic, but both groups increase in ligations in the Tethyan region in the Permian.

Bryozoa (Fig. 7)

The Bryozoa show a latitudinal diversity gradient throughout the Late Palaeozoic, and it is strongly accentuated in the Late Carboniferous. The general distribution of regional ligations decreases with time but the Tethyan region becomes more interconnected. Bryozoa were more dispersed in the Tethyan region than is easily seen from the maps. The symbols in the centre and in the southwestern and southeastern parts of the Tethys during the Permian represent the three subdivisions of the regional Tethyan fauna recognized by Ross (1978).

Crinoidea

The crinoids have very high tropical diversity during the Late Palaeozoic and appear to be restricted to the tropics exclusively

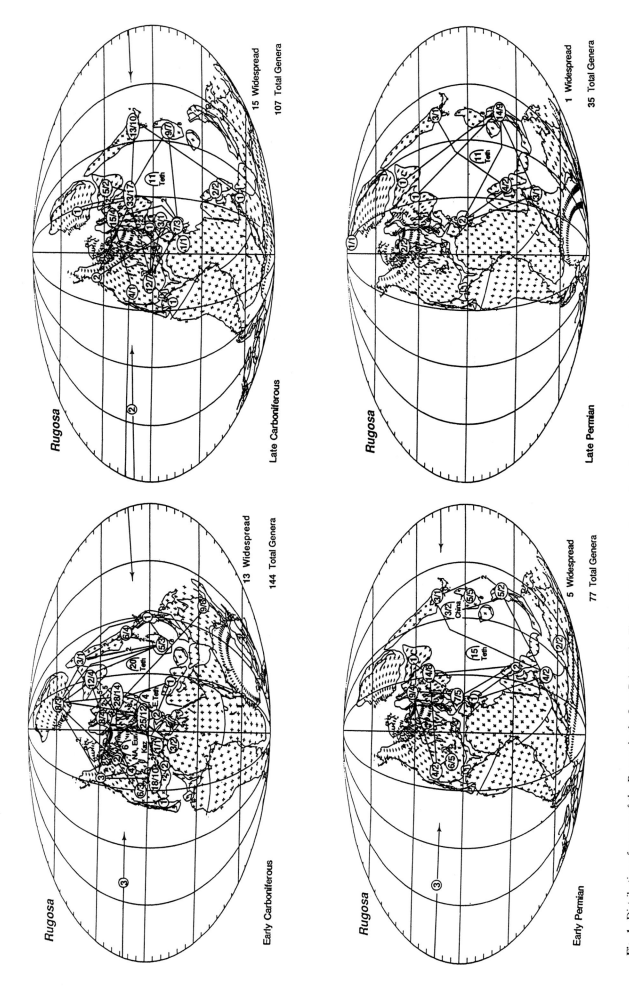

Fig. 1. Distribution of genera of the Rugosa in the Late Palaeozoic. The circled figures are the number of genera/number of endemic genera (when present). Shared genera (ligations) between areas indicated by lines connecting areas. Numbers on ligation lines indicate number of genera shared between areas. Single ligations carry no number.

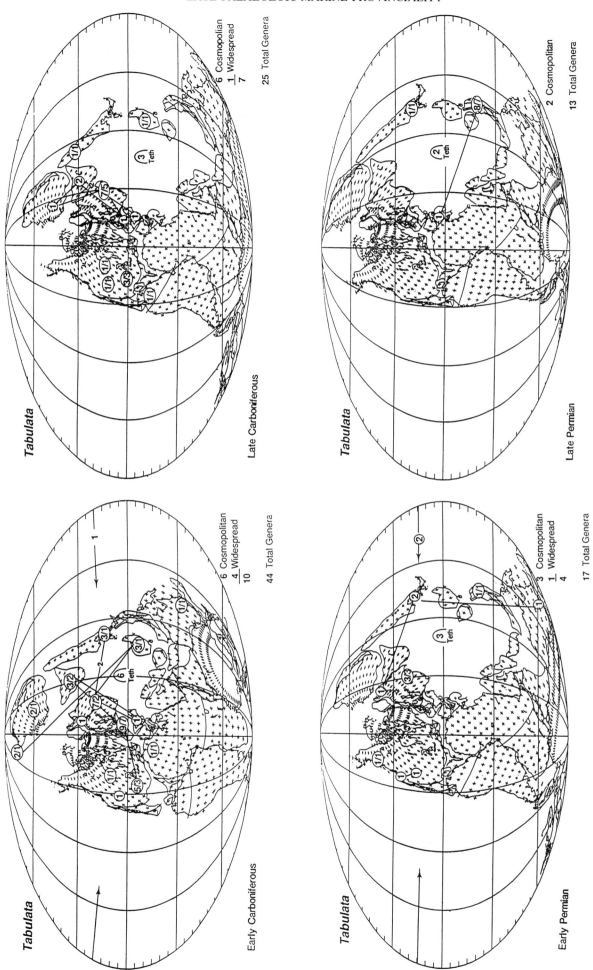

Fig. 2. Distribution of genera of the Tabulata in the Late Palaeozoic. Symbols as in Fig. 1.

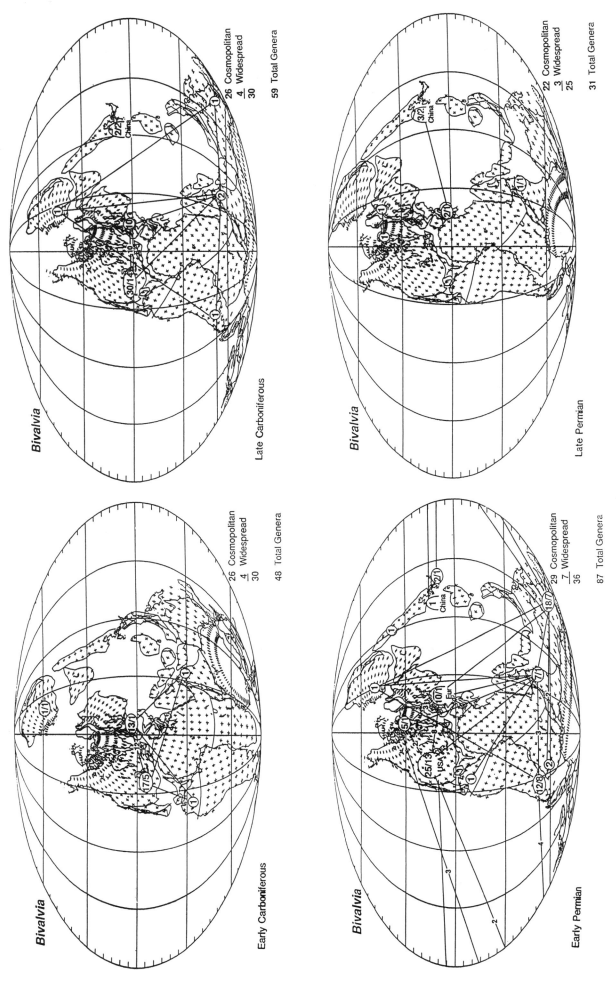

Fig. 3. Distribution of the genera of Bivalvia in the Late Palaeozoic. Symbols as in Fig. 1.

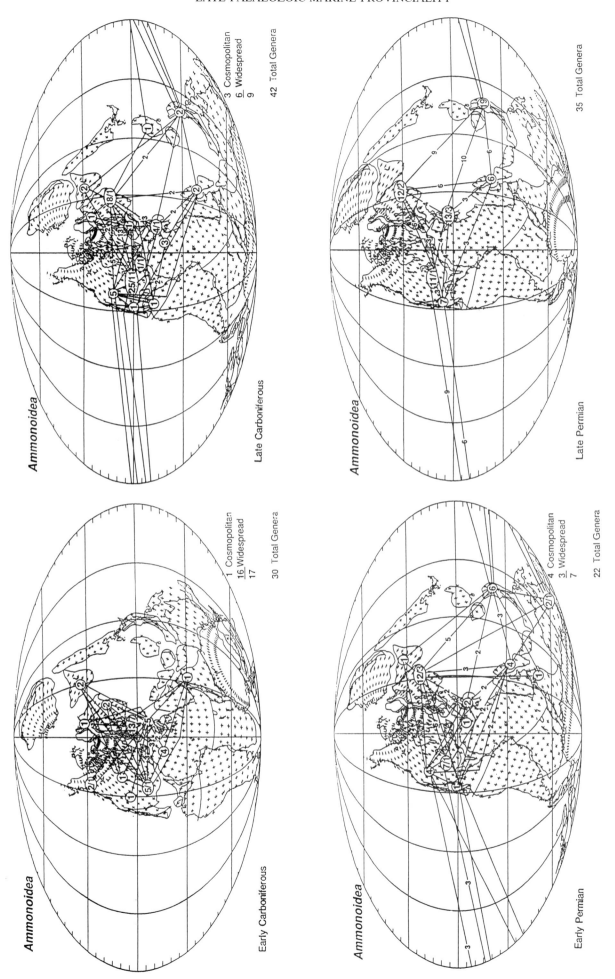

Fig. 4. Distribution of the genera of Ammonoidea in the Late Palaeozoic. Symbols as in Fig. 1.

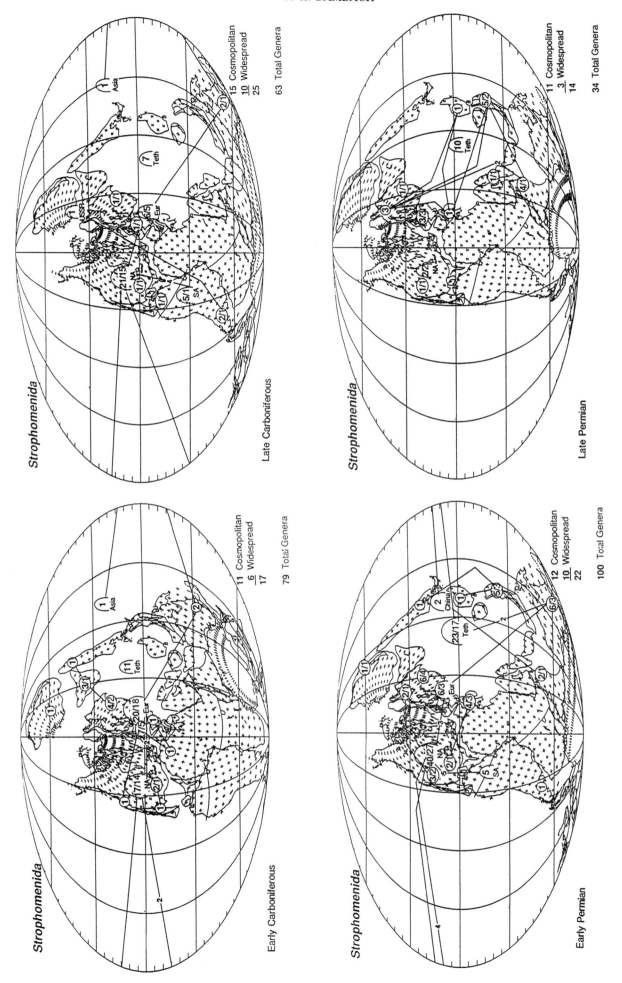

Fig. 5. Distribution of the genera of Strophomenida in the Late Palaeozoic. Symbols as in Fig. 1.

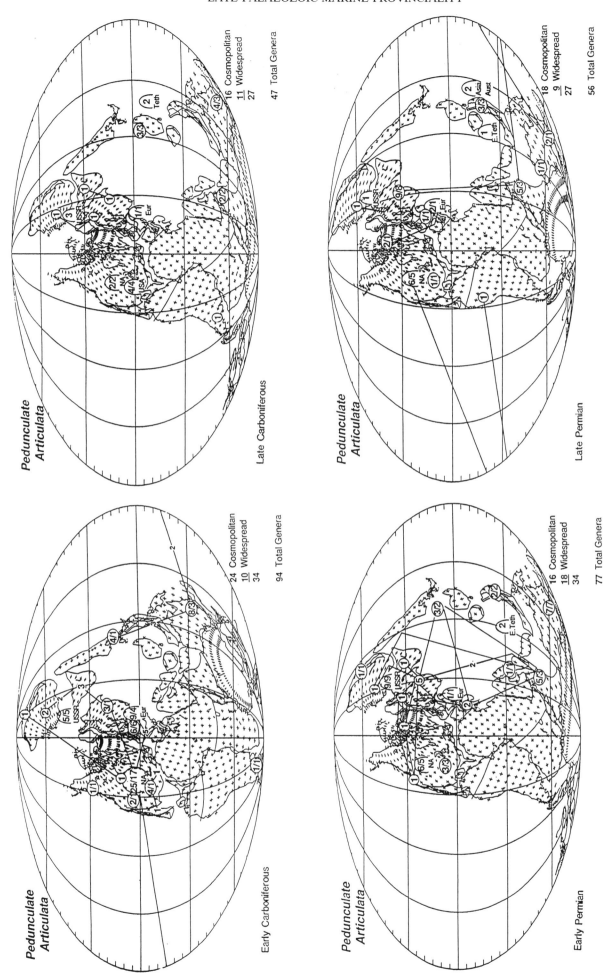

Fig. 6. Distribution of genera of the pedunculate Articulata in the Late Palaeozoic. Symbols as in Fig. 1.

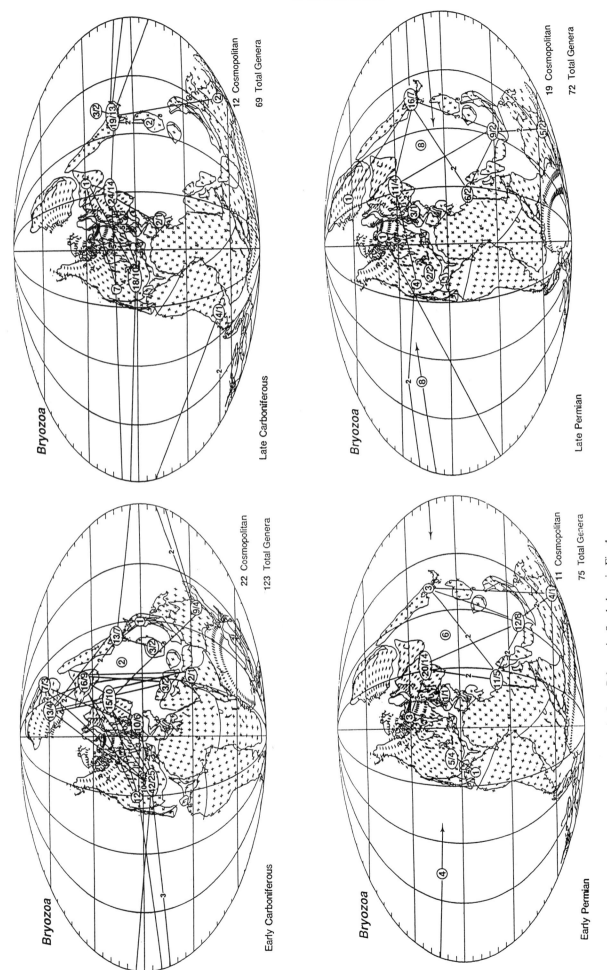

Fig. 7. Distribution of the genera of Bryozoa in the Late Palaeozoic. Symbols as in Fig. 1.

during the Late Carboniferous. As with several other groups, the number of ligated genera in the Crinoidea is high in the Early Carboniferous and decreases over time. The Crinoidea decrease in diversity during the Carboniferous but their diversity stabilizes during the Permian. There are 216 crinoid genera in the Early Carboniferous and only 50 to 60 genera at any time in the Permian. The full geographic range of most crinoid genera is unlikely to be preserved because of their poor individual preservation potential. None of the 458 crinoid genera in the Late Palaeozoic is fully cosmopolitan and 91% are endemic (occur in only one area). These extreme proportions of low cosmopolitanism and high endemism are probably artifacts of poor general preservation, although the overall diversity and ligation patterns of the group apparently reflect real biological relationships, judged by comparison with similar patterns for other taxa with excellent preservation potential.

The fauna as a whole during the Late Palaeozoic

As noted near the beginning of the paper, world diversity declined through the Late Palaeozoic. Familial diversity dropped in six of the eight stages from the Visean to Kazanian, increasingly only once (from the Bashkirian to Moscovian) and remaining constant only once (from the Sakmarian to the Artinskian). The total decrease in familial diversity between the Visean and Kazanian was 15%. For the eight groups studied, the number of genera decreased from 778 in the Early Carboniferous to 554 in the Late Carboniferous, 505 in the Early Permian and 336 in the Late Permian. Although one cannot take the 35% decrease in generic diversity between the Early Carboniferous and Early Permian at face value, because the Early Permian is 20% shorter than the Early Carboniferous, the generic decrease is 15% greater than the length differential between the two intervals. Also, the Early Permian is 32% longer than the Late Carboniferous, yet has 7% fewer genera. Despite difficulties in estimating change in standing generic diversity because of the varied lengths of the time intervals considered, it is clear that generic, as well as familial, diversity was slowly decreasing, not increasing, during most of the Late Palaeozoic. Late Permian diversity is lower still, with a drop in recorded generic diversity of 33% from the Early Permian, but lower sea levels also sharply reduce the area of preserved Late Permian rocks. An accurate measure of Late Permian diversity for the whole marine biosphere is not really possible.

To assess biogeographic relationships between areas and trends in such relationships the proportions of cosmopolitan and endemic taxa in the fauna must be known. In compiling the data it became clear that genera could be assigned to three different range groups; cosmopolitan or widespread, regionally ligated or shared, and endemic. Genera that range into four or more palaeocontinental blocks are defined here as cosmopolitan or widespread. Genera that occur in a variety of areas but are found on only three or fewer palaeocontinental blocks are defined as regionally ligated. Endemic genera are those confined to a single area.

One of the purposes of this study is to evaluate the biogeographic properties of the world fauna as a whole rather than just focussing on the distribution patterns of individual groups. The groups used in this analysis vary in their proportions of cosmopolitan, regional, and endemic genera. In several instances these proportions may be enhanced through taphonomic rather than biological factors. As noted in the review of the eight groups studied, the Ammonoidea are strikingly broadly ligated (although not unusually cosmopolitan), and the Crinoidea are exceptionally endemic. In both cases post-mortem events may be involved; floating of dead ammonite shells spreading them beyond their life range and rapid disarticulation of crinoids before burial degrading the completeness of their record. For these reasons the ammonites and crinoids were not used in the further analyses of endemism and cosmopolitanism, although their occurrences are used in the total diversity analyses.

Table 1. *Percentage of forms in three biogeographic categories*

	Cosmopolitan	Regionally ligated	Endemic
Bivalvia	53.3	20.5	26.2
Pedunculate Articulata	44.5	16.1	39.4
Strophomenida	28.3	27.5	44.2
Tabulata	23.2	29.3	47.5
Bryozoa	19.5	31.2	49.3
Rugosa	9.4	41.6	49
Average	28.1	28.7	43.2

The six remaining groups still have a wide range in their proportions of cosmopolitan, regionally ligating, and endemic genera (Table 1). Cosmopolitanism varies from 9.4% in the Rugosa to 53.3% in the Bivalvia and endemism varies from 26.2% in the Bivalvia to 49.3% in the Bryozoa. Each group also has large fluctuations in each of these proportions at various times, varying by as much as a factor of two from one time interval to the next. This variability within individual higher taxa contributed to the conclusions students of single taxa have reached about changing provinciality during the Late Palaeozoic.

When the data for the six groups are added together, however, and the same proportions calculated for the fauna as a whole, one sees far less variation over time (compare Table 2 with the Average line in Table 1). Cosmopolitanism of the total fauna in any one time interval varied by no more than 6% from the average value of 28.1% for the entire Late Palaeozoic. The proportion of regionally distributed genera varied by just 4% from its average of 28.7%. Endemism varied by only 2% from its average of 43.2% except for the Later Permian, which also is influenced by artifacts introduced by the geographic restriction of the preserved area of rock of this age. Cosmopolitanism is also at its highest value in the Late Permian, but this also may be influenced by the available record. Another possibility is that a higher proportion of widespread taxa accumulated as diversity declined in the Permian (akin to Jablonski's (1986) observations about taxa surviving mass extinctions). In any case, there is no marked change in *either* endemism or cosmopolitan of the total fauna from the Early Carboniferous through the Early Permian. This is an unexpected result, given the impression noted at the beginning of this paper, that cosmopolitanism decreased and provinciality (implying endemism) increased during that interval.

Table 2. *Changing percentage of whole fauna through time in three biogeographic groups*

	Early Carboniferous	Late Carboniferous	Early Permian	Late Permian
Cosmopolitan	23.7	31.4	27.5	34
Regionally Ligated	30.8	24.5	28.4	30.7
Endemic	45.5	44.1	44.1	35.3

Biogeographic units during the Late Palaeozoic

Tables 3 to 6 list the biogeographic units recognized from the analysis of the pooled faunal data using the criteria mentioned above (section on methodology). The locations of these units are shown in Fig. 8.

Realms are designated for large regions that show (a) major differences in diversity from other areas and/or (b) markedly fewer ligations between areas in different realms than between areas within a single realm. The realms (areas outlined by blank bars in Fig. 8) are obviously large-scale regions bounded by the same type of climatic and geographic barriers that separate tropical and boreal, Atlantic and Pacific, and New and Old World realms in the world today. The major change in realms in the Late Palaeozoic was the loss of the distinction between the tropical European and Chinese Realms of the Carboniferous and the emergence of the Tethyan Realm in the Permian. This far-reaching tropical realm (see preliminary reconstruction of oceanic circulation pattern, using an older, less accurate palaeogeographic reconstruction, in Ziegler *et al.* 1981) did maintain sufficient

Table 3. *Early Carboniferous provinciality*

Area	Recorded genera/endemic genera	Percent endemism
I. Siberian Realm		
A. Northeastern USSR	2/1	?
B. Mongolian Province	8/3	37.5
C. Siberian Platform Province	27/9	33
II. American Realm		
A. North Laurentia	2,4/1,6/1	?
B. South Laurentian Province	125/76	60.8
C. Cordillera	22/4	(18.2)
D. Maritime Canada	1/1	?
III. European Realm		
A. Kazakhstanian Province	37/16	43.2
B. West Baltican Province	9/4	44.4
C. East Baltican Province	73/33	45.2
D. Anglo–North German Province	77/55	71.4
E. North African Province	5/3	60
F. Iranian Province	4/2	50
G. Afghanistan–India	3/1	?
IV. Chinese Realm		
A. North China Province	26/12	46.2
B. South China Province	9/4	44.4
V. Austral Realm		
A. East Australian Province	29/16	55.2
B. Southern South America	1/1	?

Table 4. *Late Carboniferous provinciality.*

Area	Recorded genera/endemic genera	Percent endemism
I. Siberian Realm		
A. Northeastern USSR	3/2	?
B. Siberian Platform	4	?
II. American Realm		
A. Central Canada	1/1	?
B. South Laurentian Province	90/55	61.1
C. Cordillera	13/3	(23.1)
D. Northwest S. America Province	8/3	37.5
III. European Realm		
A. Alaska–Yukon	2	?
B. Kazakhstan	9/2	(22.2)
C. West Baltica	16/4	(25)
D. East Baltican Province	73/43	58.9
E. Franco–Czech Province	6/2	33.3
F. South European Province	7/3	42.9
G. North Africa	1/1	?
H. Iran	2/2	?
IV. Chinese Realm		
A. North China Province	32/23	71.9
B. South China Province	18/14	77.8
V. Austral Realm		
A. East Australian Province	9/4	44.4
B. Afghanistan–Indian Province	5/2	40
C. Southern South America	8/2	(25)

Table 5. *Early Permian provinciality*

Area	Recorded genera/endemic genera	Percent endemism
I. Siberian Realm		
1. Mongolia	2/2	?
2. Siberian Platform	2	?
II. American Realm		
1. Alaska–Yukon	1/1	?
2. South Laurentian Province	86/58	67.4
3. Cordilleran Province	8/4	50
III. Tethyan Realm		
A. European Region		
1. Arctic Laurentia	9/1	(11.1)
2. West Baltican Province	13/5	38.5
3. East Baltican Province	53/33	62.3
4. Franco–Czech Province	7/5	71.4
B. Southwest Tethyan Region (11 genera)		
1. South European Province	7/3	42.9
2. Iran	3/1	?
C. North China Region		
1. North China Province	9/3	33.3
D. Southeast Tethyan Region (14 genera)		
1. South China Province	8/6	75
2. Southeast Asia–Timor Province	13/8	61.5
IV. Austral Realm		
1. Australian Province	37/19	51.4
2. Afghanistan–Indian Province	17/7	41.2
3. Southern S.–American Province	13/8	61.5

Table 6. *Late Permian provinciality.*

Area	Recorded genera/endemic genera	Percent endemism
I. Siberian Realm		
1. Northeast USSR	1/1	?
2. Siberian Platform	2/1	?
II. American Realm		
1. South Laurentian Province	10/10	100
2. Cordillera	5/1	(20)
III. Tethyan Realm (32 genera)		
A. European Region		
1. Arctic Canada–Greenland Province	6/4	66.7
2. East Baltican Province	23/14	60.8
3. Anglo–North German Province	9/4	66.7
4. France–Czechoslovakia	1/1	?
B. Southwest Tethyan Region (6 genera)		
1. Southern Europe	7	?
2. Iranian Province	5/3	60
3. Afghanistan–Indian Province	13/6	46.2
C. North China Region		
1. North China Province	20/9	45
D. Southeast Tethyan Region (9 genera)		
1. Southeast Asia–Timor Province	30/21	70
IV. Austral Realm		
1. Australo–New Zealand Province	7/3	42.9
2. West Australia	1/1	?

differences within in it, however, to justify noting regional divisions above the provincial level (shown by dashed blank bars in Fig. 8).

Provinces are designated for areas that have at least 30% endemic genera in the pooled data, more than one endemic genus, and a palaeogeographic location that supports the interpretation of likely barriers to dispersal for taxa in the area. Tables 3 to 6 list the total generic diversities, number of endemic genera, and percentage of endemism for the major locations in each provincial area. In Fig. 8 the areas meeting provincial criteria are outlined, stippled, and identified by the symbol code for the outline in the appropriate Table for each time interval.

Areas not designated as provinces but listed in the Tables are also outlined and labelled with their symbol codes on Fig. 8 but are left unshaded. These are areas that have both some faunal peculiarity and a geographic setting that suggest they are different from the designated provinces in their realm, but, using the data available in this analysis, these areas either do not have a sufficient

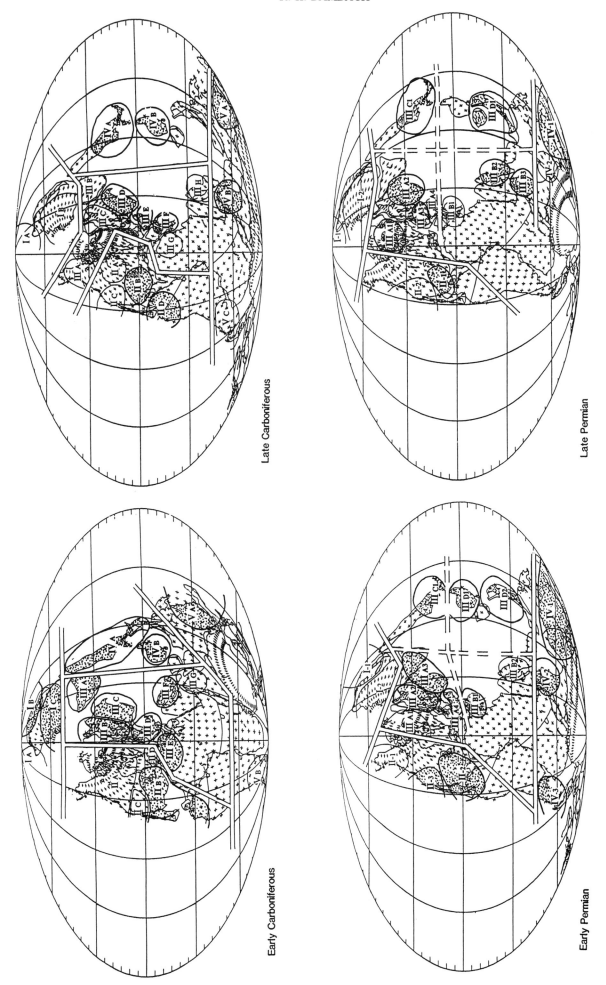

Fig. 8. Distribution of biogeographic regions in the Late Palaeozoic. See Tables 3–6 and text for symbols and discussion.

diversity or a sufficient level of endemism to meet the criteria used here for provincial designation. These areas are named (but not labelled as provinces) in the Tables and their per cent endemism is either in parentheses (for more diverse areas) or noted by a question mark. The combination of location and hints of faunal distinction suggest that these areas, too, might be found to be provinces upon more intensive investigation.

For the Early Carboniferous, 12 areas qualify as provinces, and six others are potential provinces. For example, area II-C is generally recognized as a province and II-A may contain another (Sando et al. 1975). For the Late Carboniferous, 9 areas qualify as provinces, 4 others would qualify at a 20% cutoff on generic endemism, and 6 others are potential provinces. For the Early Permian, 12 areas qualify as provinces and 5 others are potential provinces. In the Late Permian, despite the decrease in preserved geographic area, 9 areas qualify as provinces and 6 others are potential provinces.

This system recognizes biogeographic units that have some comparability with modern biogeographic realms and provinces. It recognizes 12, 11, 12, and 9 almost certain provinces for the four time intervals studied, and if all possible provinces are tallied, a total of 18, 19, 18, and 15 potential provinces for the four time intervals of the Late Palaeozoic. In all cases, the use of pooled faunal data, with an emphasis on endemism and geographic setting as criteria for designating biogeographic units, identifies more provincial units comparable to those recognized in the modern world than past practices, which emphasized single taxa and the similarities between areas. The discovery of a near constant number of provincial units through the Late Palaeozoic also differs from the recently expressed view of early cosmopolitanism and increasing provinciality through this interval.

Diversity gradients and climate change during the late Palaeozoic

Several groups in this study display latitudinal diversity gradients at all times during the Late Palaeozoic, and some appear to have reduced latitudinal range during the Late Carboniferous. The pooled data show both phenomena very clearly (Fig. 9). Very high diversities (over 100 genera), are found only in areas on continental blocks within 15 degrees of the equator. In the Early Carboniferous and Early Permian this high diversity belt is flanked by bands of areas of intermediate diversity (between 20 and 90 genera) that extend as far as 30 to 60 degrees north and 30 to 70 degrees south of the equator. During these intervals, low diversity areas (less than 20 genera) are found predominantly at high latitudes. During the Late Carboniferous, the high diversity tropical belt persisted but the intermediate diversity belts were eliminated and low diversity faunas characterized all latitudes higher than 15 degrees south and 20 degrees north of the equator. This was the time of the most extensive glaciation during the Palaeozoic (Caputo & Crowell 1985, and references therein). These latitudinal diversity gradients and change in gradients probably reflected the climatic gradient of the earth and its changing severity during the Late Palaeozoic, just as the modern diversity gradient does today.

Individual continental blocks, tracked as they change latitude during the Late Palaeozoic, indicate that the diversity gradient is not just a record of more completely reported faunas from particular regions. Diversity rose in East Baltica as it moved southward deeper into the tropics from Early to Late Carboniferous time, despite the cooling trend and steepening of the world diversity gradient that occurred at the same time. The Southeast Asia−Timor block increased markedly in diversity as it moved northward into the tropics in the Permian, even as total world diversity appeared to begin to decline sharply. These conclusions would hold even if the majority of the Timor fauna should prove to be Early, rather than Late Permian in age, as recent study indicates.

In the Early Carboniferous and Late Permian an asymmetry

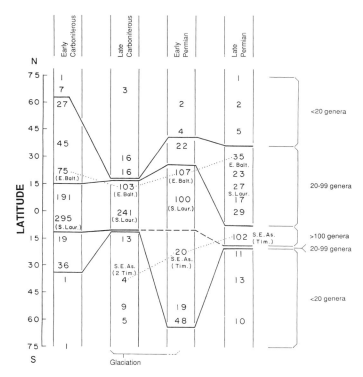

Fig. 9. Diversity of genera of the eight groups studied in this paper in the most diverse region at each latitude in the four time intervals of the Late Palaeozoic. The diversity numbers are placed at the middle latitude for each area.

also appears in the latitude range of intermediate diversities in the Northern as compared to the Southern Hemisphere. This asymmetry also may be represented by the apparent northern displacement of high tropical diversities in the Late Carboniferous and Early Permian, although this is less certain. A similar asymmetry is observed in echinoderm distributions throughout the Palaeozoic (Bennington & Bambach, in prep.). This asymmetry probably is related to a less severe climatic gradient in the more oceanic (and therefore more equable) Northern Hemisphere in contrast to the more continental Southern Hemisphere, in which polar land areas could generate severe continental climates even when extensive ice sheets were not present.

Resolving several conflicts

Three questions need to be resolved. (1) How did the idea of increasing provinciality in the Late Palaeozoic develop, if provinciality actually remained nearly constant during that interval of time? (2) How do the previously recognized 'provinces' of the Late Palaeozoic relate to the biogeographic units defined in this paper? (3) Is there a relationship between provinciality and world diversity?

The answer to the first question lies not in the number of biogeographic units recognized but in the changing pattern of regionally ligating genera during the Late Palaeozoic. Because different, inconsistent, and non-comparable methods were used to recognize biogeographic units by various workers studying various time intervals, the units previously defined for one time are not directly comparable with those defined for other intervals. Therefore, the numbers of biogeographic units previously recognized for different times cannot be used to demonstrate changing levels of provinciality.

In the Early Carboniferous, however, ligating genera in many taxa are shared, on average, among more areas than they are at later times. The decrease in the number of areas connected by ligating genera through the Late Palaeozoic can be seen in the decreasing density of the 'spider-web' of ligation lines over time on many of the maps in the figures showing the distribution of individual taxa (Figs 1–7). So, although neither the proportion of endemic genera or the number of provinces rose during the Late Palaeozoic, the interconnectedness that was the property of an extensive network of regional ligations did decrease. Note that this was not a change in the *proportion* of genera that were shared among areas (Table 2), but was a result of a drop in the number of *areas* among which individual ligating genera were shared. Because ligating genera tended to ligate to fewer different areas in the Permian than in the Early Carboniferous, the impression was created that provinciality was increasing, even though the levels of cosmopolitanism and endemism both remained about the same (Table 2).

Recognizing at least two levels of biogeographic units (provinces and realms) in the Late Palaeozoic fossil record helps explain the relationship between the units recognized by traditional analyses using similarities between areas based on single taxa and the units recognized in this paper. The number of biogeographic units (often called provinces) previously recognized in the fossil record commonly varies from two to six for various times (Valentine *et al.* 1978, table 1, p. 57). These units generally appear to represent the realm level recognized in this paper. This is reasonable since they commonly were defined as large areas that shared some particular faunal similarity, and not as provinces based on some proportion of endemism of their total fauna. For example, of the six Early Carboniferous units noted in Ziegler *et al.* (1981), five correspond to the realms noted here and one is based on a coral province noted by Sando *et al.* (1975) (and designated here as Early Carboniferous unit II-C with 18.2% total generic endemism). The four units noted for the Late Carboniferous in Ziegler *et al.* (1981) represent four of the five realms recognized here. The twelve Kazanian (early Late Permian) units mentioned in Ziegler *et al.* (1981) were compiled from several regionally based studies in which endemic taxa serve more often as a basis for recognizing units. Only in regional studies, where considerably more detailed data may be involved and emphasis placed on differences with other areas in the same general region, have units comparable to modern zoogeographic provincial units commonly been recognized in palaeobiogeographic analyses. Most 'provinces' recognized in large-scale studies of palaeobiogeography, especially those studies that emphasize particular similarities as a criterion for grouping areas together, relate to the realm level.

The view of Ross (1979) and Ross & Ross (1979) and of Ziegler *et al.* (1981) of increasing provinciality during the Late Palaeozoic as world diversity decreased slightly conflicts with the idea that provinciality and diversity are positively correlated (Valentine 1973; Valentine *et al.* 1978). But the distribution in Late Palaeozoic times of provinces that *are* somewhat comparable to modern provinces does not conflict with the idea of a relationship between diversity and provinciality. Levels of endemism remained fairly constant until the Late Permian, and the number of provinces also remained stable, with an apparent slight decrease in the Late Permian. Diversity also remained relatively constant, although decreasing somewhat, and it, too, appeared to decline more markedly in the Late Permian.

Yet this now raises a question about the conclusion of Valentine *et al.* (1978) that provinciality in the world has increased five-fold since the Late Palaeozoic. Valentine *et al.* also argued that such a large increase in provinciality, coupled with the doubling of within habitat diversity between the Late Palaeozoic and the Neogene documented by Bambach (1977), would have caused a ten-fold (order of magnitude) increase in world diversity between the Palaeozoic and the Recent. However, the palaeobiogeographic 'provinces' tabulated by Valentine *et al.* appear to be more comparable to modern realms than modern provinces. The five-fold increase in provinciality they believed they discovered is probably only the difference between the number of non-comparable units,

realms in the past and provinces in the modern world. If one accepts the conclusions of this paper that the provinces recognized here are comparable to modern provinces, then there were, on average, 11 or 12 well established provinces from the Early Carboniferous through the Early Permian. If only half of the other possible provinces noted here for each time interval should prove to be good provinces, this would still mean that there were at least 15 provinces through much of the Late Palaeozoic. In that case, provinciality between then and now would have doubled at most. Consequently, world diversity has probably increased only about four-fold since the Late Palaeozoic and not by the order of magnitude modelled by Valentine *et al.* (1978).

References

BAMBACH, R. K. 1977. Species richness in marine benthic habitats through the Phanerozoic, *Paleobiology*, **3**, 152–167.

BOUCOT, A. J. 1975. *Evolution and Extinction Rate Controls*, Elsevier, New York.

BRIGGS, J. C. 1974. *Marine Zoogeography*, McGraw-Hill, New York.

CAMPBELL, C. A. & VALENTINE, J. W. 1977. Comparability of modern and ancient marine faunal provinces, *Paleobiology*, **3**, 49–57.

CAPUTO, M. V. & CROWELL, J. C. 1985. Migration of glacial centers across Gondwana during Palaeozoic Era, *Geological Society of America Bulletin*, **96**, 1020–1036.

CHARPENTIER, R. R. 1984. Conodonts through time and space: Studies in conodont provincialism, *In*: CLARKE, D. L. (ed.) *Conodont Biofacies and Provincialism* Geological Society of America Special Paper, **196**, 11–32.

FEDEROWSKI, J. 1981. Carboniferous corals: Distribution and sequence, *Acta Palaeontologica Polonica*, **26**, 87–157.

JABLONSKI, D. 1986. Background and mass extinctions: The alternation of macroevolutionary regimes, *Science*, **231**, 129–133.

JOHNSON, J. G. 1971. A quantitative approach to faunal province analysis, *American Journal of Science*, **270**, 257–280.

KAUFFMAN, E. G. 1973. Cretaceous Bivalvia, *In*: HALLAM, A. (ed.) *Atlas of Palaeobiogeography*, Elsevier, New York, 353–383.

LUO, J. 1984. Early Carboniferous coral assemblages and paleobiogeography of China, *Palaeontographica Americana*, **54**, 473–432.

MOORE, R. C. (ed.) 1957. *Treatise on Invertebrate Paleontology*, Part L. Mollusca 4, University of Kansas Press, Lawrence, Kansas.

MOORE, R. C. 1965. *Treatise on Invertebrate Paleontology*, Part H. Brachiopoda, University of Kansas Press, Lawrence, Kansas.

MOORE, R. C. 1969. *Treatise on Invertebrate Paleontology*, Part N. Mollusca 6, University of Kansas Press, Lawrence, Kansas.

—— & TEICHERT, C. (eds) 1978. *Treatise on Invertebrate Paleontology: Part T, Echinodermata 2*, University of Kansas Press, Lawrence, Kansas.

ROBISON, R. A. (ed.) 1981. *Treatise on Invertebrate Paleontology, Part F. Coelenterata, Supplement 1*, The University of Kansas Press, Lawrence, Kansas.

ROSS, C. A. 1979. Carboniferous, *In*: ROBISON, R. A. & TEICHERT, C. (eds), *Treatise on Invertebrate Paleontology, Part A. Introduction*, University of Kansas Press, Lawrence Kansas, A254–A290.

—— & ROSS, J. R. P. 1979. Permian. *In*: ROBISON, R. A. & TEICHERT, C. (eds), *Treatise on Invertebrate Paleontology, Part A. Introduction*, University of Kansas Press, Lawrence, Kansas, A291–A350.

ROSS, J. R. P. 1978. Biogeography of Permian Ectoproct Bryozoa, *Palaeontology*, **21**, 341–356.

—— 1981. Biogeography of Carboniferous Ectoproct Bryozoa, *Palaeontology*, **24**, 313–341.

SANDO, W. J., BAMBER, E. W. & ARMSTRONG, A. K. 1975. Endemism and similarity indices: Clues to the zoogeography of North American Mississippian corals, *Geology*, **2**, 661–664.

SAVAGE, N. M., PERRY, D. G. & BOUCOT, A. J. 1979. A Quantitative Analysis of Lower Devonian Brachiopod Distribution, *In*: GRAY, J. & BOUCOT, A. J. (eds), *Historical Biogeography, Plate Tectonics, and the Changing Environment*, Oregon State University Press, Corvallis, 169–200.

SCOTESE, C. R. 1986. *Phanerozoic Reconstructions: A New Look at the Assembly of Asia*, University of Texas Institute for Geophysics Technical Report N. 66. Austin.

SEPKOSKI, J. J., JR. 1982. *A Compendium of Fossil Marine Families*, Milwaukee Public Museum, Contributions in Biology, Geology, No. 51, Milwaukee.

VALENTINE, J. W. 1973. *Evolutionary Paleoecology of the Marine Biosphere*, Prentice-Hall, Inc., Englewood Cliffs, New Jersey.

—— FOIN, T. C. & PEART, D. 1978. A provincial model of Phanerozoic marine diversity, *Paleobiology*, **4**, 55–66.

WEBSTER, G. D. 1973. Bibliography and Index of Palaeozoic Crinoids 1942–1968, *Geological Society of America, Memoir*, **137**.

—— 1975. Bibliography and Index of Palaeozoic Crinoids 1969–1973, *Geological Society of America, Microfilm Publication*, **8**.

—— 1984. Bibliography and Index of Palaeozoic Crinoids 1974–1980, *Geological Society of America, Microfilm Publication*, **16**.

WILLIAMS, A. 1969. Ordovician faunal provinces with reference to brachiopod distribution, *In*: WOOD, A. (ed.), *The Pre-Cambrian and lower Palaeozoic rocks of Wales*, University of Wales, 117–154.

—— 1973. Distribution of brachiopod assemblages in relation to Ordovician palaeogeography. *In*: HUGHES, N. F. (ed.), *Organisms and Continents Through Time*, Palaeontological Association Special Paper, **12**, 241–269.

WATERHOUSE, J. B. & BONHAM-CARTER, G. F. 1975. Global distribution and character of Permian biomes based on brachiopod assemblages, *Canadian Journal of Earth Sciences*, **12**, 1085–1146.

ZIEGLER, A. M., BAMBACH, R. K., TOTMAN PARRISH, J., BARRETT, S. F., GIERLOWSKI, E. H., PARKER, W. C., RAYMOND, A. & SEPKOSKI, J. J. JR. 1981. Palaeozoic biogeography and climatology, *In*: NIKLAS, K. J. (ed.) *Paleobotany, Paleoecology, and Evolution* (2 vol.), Praeger, New York, **2**, 231–266.

Carboniferous brachiopod migration and latitudinal diversity: a new palaeoclimatic method

P. H. KELLEY[1], A. RAYMOND[2] & C. B. LUTKEN[3]

[1] *Department of Geology and Geological Engineering, University of Mississippi, University, MS 38677, USA*
[2] *Department of Geology, Texas A & M University, College Station, TX 77843-3115, USA*
[3] *Route 1; Box 207, St Joseph, LA 71366, USA*

Abstract: Patterns of migration of organisms, coupled with changes in latitudinal diversity, provide useful information for reconstructing palaeoclimate. We apply these tools to the record of 344 articulate brachiopod genera from the Tournaisian, Visean, and Namurian stages of the Carboniferous. Localities from the Northern Hemisphere were assigned to four palaeolatitudinal zones: the palaeoequatorial zone and the low, middle and high latitudes. Generic migrations were tabulated among latitudinal zones for each pair of successive stratigraphic intervals. Latitudinal diversity gradients were calculated based on the number of genera present within each zone.

Two intervals of climatic change were identified by these methods. Between the middle and late Visean, 42% of the genera moved the northern boundary of their range northward, and nonequatorial diversity rose dramatically. These patterns indicate high-latitude warming, which may have been caused by deflection of the circumequatorial current as the collision of Laurussia and Gondwana progressed. Migration and latitudinal diversity patterns indicate both high-latitude cooling and equatorial warming between the Namurian A and B. The onset of Gondwanan glaciation may have been responsible; similar patterns occurred during glacial onset in the Miocene.

Problems encountered in reconstructing palaeoclimate and palaeogeography are exacerbated for the Palaeozoic Era. No coherent isotopic temperature record exists for most of the Palaeozoic. Interpretations of palaeoclimate are possible based on sedimentological indicators, such as coals, evaporites, and reefs. However, the restricted occurrence of these sediments limits their usefulness. In addition, problems of reconstructing palaeogeography increase with geological age. Palaeomagnetic data are much less reliable prior to the Mesozoic, and the resolution of Palaeozoic radiometric dates is low.

Use of palaeontological data may resolve some of these problems of Palaeozoic studies. Fossils are climatically-sensitive, occur in a wide variety of sediments, and their stratigraphic resolution exceeds that provided by radiometric dating. Patterns of distribution of organisms provide useful information for reconstructing palaeoclimate and palaeogeography.

In this study we present a method for interpreting palaeoclimate and palaeogeography using patterns of migration of organisms, coupled with changes in latitudinal diversity. We apply these tools to the record of articulate brachiopods from the Tournaisian, Visean, and Namurian stages of the Carboniferous. The techniques, however, are applicable to other time intervals and taxonomic groups and yield otherwise unavailable evidence on palaeoclimate. Although palaeotemperatures are not determinable, relative palaeoclimate can be compared throughout the Phanerozoic, using these methods.

Palaeoclimatic analysis using these methods suggests two major events in the early and middle Carboniferous: a period of high-latitude warming in the late Visean, followed by a period of high-latitude cooling and equatorial warming in the middle Namurian. Possible causes for these climatic events include plate collision and glaciation. The collision between Laurussia and Gondwana may have affected global palaeoclimate by disrupting ocean current patterns. Before the collision, warm equatorial currents circled the globe; after the collision, these currents would have been deflected to the north and south along the east coast of Pangaea. Two contrasting interpretations of the climatic consequences of this deflection have been offered. Raymond *et al.* (1985) suggested that high-latitude warming would result from the presence of warm water at higher palaeolatitudes. Conversely, Ross & Ross (1985) and Copper (1986) argued that refrigeration of equatorial currents at the poles would cause global cooling.

Another factor that would have affected global climate is the onset of widespread continental glaciation on Gondwana, which probably occurred in the Namurian A (Veevers & Powell 1987) or near the Namurian A/B boundary (Hambrey & Harland 1981; Bouroz *et al.* 1978). As in the case of plate collision, the climatic consequences of glaciation are not entirely understood. However, a growing body of evidence indicates that polar glaciers increase the equability of equatorial climates, creating continuously warm, wet terrestrial habitats (Ziegler *et al.* 1987) and warmer, nonseasonal marine habitats (Valentine 1984*a*, *b*; Savin 1982). Palaeoclimatic evidence deduced from changes in the distributional patterns of brachiopods can be used to assess the utility of contrasting palaeoclimatic models.

Methods

Data base

The brachiopod data base was compiled by Blanton (1984) and has been deposited with the Society Library and the British Library Document Supply Centre at Boston Spa, West Yorkshire, UK, as Supplementary Publication No. SUP 18060. Included in this Supplementary Publication are a correlation chart, a locality map, and, for each time interval, a list of brachiopod genera occurring at each locality studied. A list of sources from which data were tabulated is also included in the Supplementary Publication.

The data base consists of occurrences of 344 articulate brachiopod genera, which were tabulated from the published literature and from unpublished information made available by J. L. Carter, T. W. Henry, and J. Waters. J. L. Carter also assisted with the taxonomic evaluations. Nine stratigraphic intervals are represented in the data set: lower, middle and upper Tournaisian; lower, middle and upper Visean; and Namurian A, B, and C. The nine intervals vary in duration. Assuming that each subdivision of a stage equals a third of the total duration of that stage, each Tournaisian interval equals about 3 Ma while each Visean interval equals about 6 Ma (Palmer 1983). The entire Namurian is 18 Ma in duration. If the Namurian A is equivalent to the Serpukhovian, this interval lasted 13 Ma (Palmer 1983). Assuming the Namurian B and C were equal in duration, each of these intervals probably lasted 2.5 Ma.

The data base includes information from 59 localities worldwide. However, the present study focuses on the Northern Hemisphere and palaeoequatorial zone because relatively little information is available from localities that were south of the palaeoequatorial zone in the Carboniferous. We also omitted Chinese localities from these analyses; although palaeolatitudes for China are probably known to within 10 to 20 degrees (C. R. Scotese, pers. comm.), this study required more detailed knowledge. Any locality known, or strongly suspected, to include accreted terranes was omitted from the present study.

All usable localities were plotted on the Visean and Namurian palaeogeographic reconstructions of Scotese (1986; Figs 1 and 2). Then, on each reconstruction, we grouped localities into 'zones,' the boundaries of which approximately parallel palaeolatitude. Four zones were defined on each coast of Pangaea: a palaeoequatorial, low latitude, middle latitude, and high latitude zone.

Although the palaeolatitudinal boundaries of zones differ for the two coasts, and for the Visean and Namurian along each coast, the following palaeolatitudes correspond roughly to the zones: palaeoequatorial zone, $10°S-10°N$; low latitudes, $10°N-30°N$; middle latitudes, $30°N-45°N$; high latitudes, $45°N-70°N$. Zones were delineated such that localities did not change zone assignment between two intervals, and localities near one another on the reconstructions were assigned to the same zone. Although

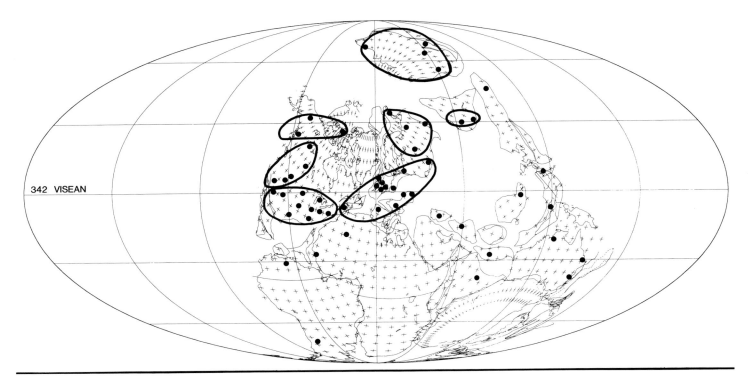

Fig. 1. Articulate brachiopod localities and palaeolatitudinal zones on the Visean reconstruction of Scotese (1986).

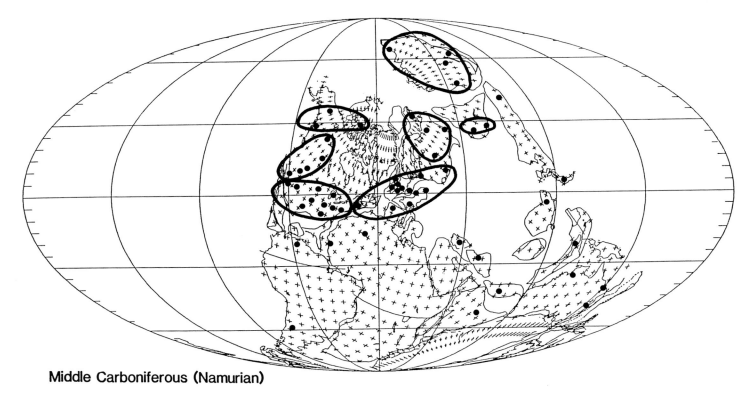

Fig. 2. Articulate brachiopod localities and palaeolatitudinal zones on the Namurian reconstruction of Scotese (1986).

palaeogeography changed somewhat between the Visean and Namurian, the same relative palaeolatitudinal positions were maintained among zones along each coast of Pangaea.

Migration patterns

In order to compensate for biases caused by uneven sampling of localities, brachiopod migration patterns were analysed among palaeolatitudinal zones, rather than among individual assemblages. For each pair of successive intervals, we tabulated: 1, the number of genera that expanded north into a new zone; 2, the number of genera that evacuated the southern portion of their range and expanded north into a new zone; 3, the number of genera that evacuated the northern part of their range; 4, the number of genera that evacuated the northern portion of their range and migrated south into a new zone; 5, the number of genera that expanded the southern boundary of their range southward into a new zone; 6, the number of genera that contracted the southern boundary of their range to the north; and 7, the number of genera that did not change range with respect to the defined zones. Figure 3 illustrates the seven patterns of brachiopod migration. The seven patterns indicate five palaeoclimatic events. If, between any two intervals, many genera expanded the northern boundary of their ranges to the north (patterns 1 and 2), this indicates high-latitude warming. Conversely, if many genera contracted the northern boundary of their ranges to the south (patterns 3 and 4), this suggests high-latitude cooling. If a large percentage of genera moved the southern boundary of their ranges northward, evacuating the southern part of their ranges (patterns 2 and 6), this could indicate equatorial warming (Valentine 1984b). A large percentage of genera expanding their ranges toward the equator (patterns 4 and 5) suggests equatorial cooling (Stanley 1984). If most genera did not change range (pattern 7), or if northward and southward migrations were balanced, we concluded that climate did not change between two intervals. We recorded range changes separately for taxa migrating along the eastern and western coasts of the megacontinent.

All of the Tethyan localities in the mid-latitude zone lie on the palaeocontinent of Kazakhstan, which moved north during the Early Carboniferous and collided with Siberia during the Namurian (Rowley et al. 1985). This movement complicates palaeoclimatic interpretation, because genera that moved their northern range boundary southward may have done so because Kazakhstan moved north, not because high-latitude climate cooled. Therefore, we tabulated migrations from lower latitudes north into Kazakhstan, because these genera migrated north onto a northward-drifting continent. However, we did not include in our tallies genera that migrated south out of Kazakhstan, because such migrations may not have resulted from palaeoclimatic change. The exclusion of these migration events did not affect the results except between the late Visean and the Namurian A. Between these two intervals, migration of genera southward out of Kazakhstan balanced northward migration, whereas without the Kazakhstan-southward migrations, net migration, was northward.

Some apparent migration events may have been the result of uneven sampling of intervals. For instance, if a genus temporarily disappeared from a locality or zone, the cause may have been poor sampling of the interval from which the genus was not reported. Such range fluctuations ('false migrations') were tallied separately from true migration events. Exclusion of these range fluctuations did not affect the observed migration patterns, although it did decrease the percentage of genera whose range changed between intervals.

If our data had included a pre-Tournaisian record, some of the range expansions noted between the early and middle Tournaisian would have been recognized as false migrations resulting from poor sampling of the early Tournaisian. Likewise, the range contractions observed between the Namurian B and C

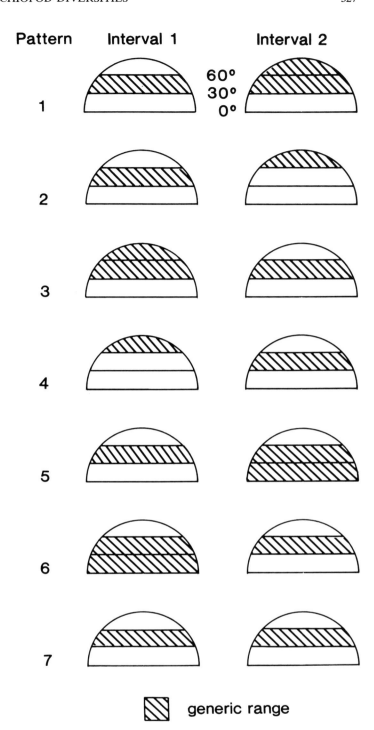

Fig. 3. Patterns of brachiopod migration. Patterns 1 and 2 indicate high-latitude warming. Patterns 3 and 4 indicate high-latitude cooling. Patterns 2 and 6 indicate equatorial warming. Patterns 4 and 5 indicate equatorial cooling. Pattern 7 indicates no change in paleoclimate.

two coasts, and for Visean and Namurian along each coast, the sampling that were undetected because of the lack of a post-Namurian data base. In order to estimate the number of true migrations in each of those cases, we calculated the ratio of true migrations to apparent migrations (true migrations plus fluctuations) for intervals for which both values were known. The average ratio was used to estimate the number of true range expansions between the early and middle Tournaisian and the number of true range contractions between the Namurian A and B (Table 1).

In order to recognize significant pulses of migration, we used a chi-squared test (Simpson et al. 1960) to compare observed

Table 1. *Frequency of true migrations in the Tournaisian through Namurian*

INTERVAL	NORTH WARMS			NORTH COOLS			SOUTH WARMS			SOUTH COOLS		
	TRUE	TOTAL	TRUE/TOT.	TRUE	TOTAL	TRUE/TOT.	TRUE	TOTAL	TRUE/TOT.	TRUE	TOTAL	TRUE/TOT.
T1–T2	17*	24		10	14	0.71	5	5	1.00	6*	9	
T2–T3	17	22	0.77	8	13	0.61	3	4	0.75	5	6	0.83
T3–V1	6	6	1.00	9	25	0.36	5	7	0.71	0	0	—
V1–V2	14	21	0.67	11	22	0.50	2	2	1.00	2	3	0.67
V2–V3	40	56	0.71	3	7	0.43	1	1	1.00	1	1	1.00
V3–NA	13	24	0.54	3	8	0.37	2	4	0.50	1	4	0.25
NA–NB	6	10	0.60	23	27	0.85	9	9	1.00	4	5	0.80
NB–NC	10	16	0.62	12*	21		3*	3		4	6	0.67
Average			0.70			0.55			0.86			0.70

True migrations represent genera that migrated in only one or two directions. Total migrations are true migrations plus genera with range boundaries that fluctuated, possibly due to inconsistent sampling. Values marked with an asterisk were corrected to remove the effect of undetected fluctuating genera using the average ratio of true to total migrations (see text). Migration patterns classifies according to Fig. 3: North warms, patterns 1 and 2; North cools, patterns 3 and 4; South warms, patterns 2 and 6; South cools, patterns 4 and 5. T1, early Tournaisian; T2, middle Tournaisian; T3, late Tournaisian; VI, early Visean; V2, middle Visean; V3, late Visean; NA, Namurian A; NB, Namurian B; NC, Namurian C.

migration frequencies to the corresponding average (background) migration rates for our data set. Migration frequencies estimated for the early to middle Tournaisian and Namurian A and B (described above) were not used to calculate background rates. For each interval, we completed two three-way chi-squared tests. The first compared the frequency of northward migration of the northern range boundary, southward migration of the northern boundary, and non-migrating genera for each interval to the average frequency of these migrations during the Tournaisian through Namurian. The second compared the frequency of northward migration of the southern boundary, southward migration of the southern boundary and non-migrating genera to the average frequency of these migrations. This method allowed us to identify intervals during which the amount of migration differed significantly from the average migration observed in the data.

We interpret intervals with significant changes of generic range as times of climatic change. This method carries the assumption that only species migration causes generic ranges to change. However, generic ranges can also change when species with new environmental tolerances evolve on the margin of the range, or when species on the margin become extinct. Changes in generic range due ot origination and extinction may be unrelated to climate, but are virtually impossible to exclude from the data. However, if these originations and extinctions are unrelated to climatic changes, events that result in apparent northward and southward migrations should be equally common. Significant undirectional pulses of migration should be reliable indicators of climatic change.

Latitudinal diversity gradients

In order to check the climatic interpretations based on migration patterns, we analysed latitudinal diversity gradients for the Northern Hemisphere and palaeoequatorial zone. We first calculated latitudinal diversity gradients by tallying the generic diversity of articulate brachiopods within each zone for each stratigraphic interval. Latitudinal diversity gradients were computed separately for the Cordilleran and Tethyan coasts. Total diversity varied during the Carboniferous; thus to facilitate comparison of diversity gradients we standardised the diversity of each latitudinal zone to diversity within the palaeoequatorial zone for each stratigraphic interval.

Latitudinal diversity gradients calculated in this manner do not provide a completely independent check of the migration pattern interpretations, because the compilations include immigrant genera. Therefore, to produce results independent from those of the migration study, we subtracted the number of genera that were new immigrants into a zone from the zonal tally. Temporal changes in these corrected latitudinal diversity gradients thus are due to origination or extinction within zones, or to earlier migration, and are not simply a reflection of immediate migration patterns. (These corrected latitudinal diversity gradients were also computed for each coast and standardised relative to palaeoequatorial diversity.)

Results

Migrational patterns

Table 2 shows the migrational patterns of brachiopod genera from the Northern Hemisphere and palaeoequatorial zone; these results are corrected for false migrations for the early to middle Tournaisian and Namurian B to Namurian C. Most of the genera that changed their range modified the northern boundary, suggesting that most climatic changes occurred at higher latitudes. Comparison of the observed frequency of northern boundary changes for each interval to their average frequency using a chi-squared test suggests that two pairs of intervals differ significantly from average. Between the middle and late Visean, 42% of the genera present during both intervals migrated north into a new zone, and only 3% evacuated a northern zone and moved south. This pulse of northward migration coincides with a peak in global diversity of articulate brachiopods (Kelley *et al.* 1989) and may indicate high-latitude warming at that time.

This migration trend reversed between the Namurian A and B, when 33% of the genera present during both intervals evacuated a northern zone, moving their northern range boundary south. At the same time, only 9% of the genera migrated in the opposite direction, moving their northern boundary north into a new zone. This pattern of net southward migration suggests an episode of high-latitude cooling during the Namurian.

Namurian climate change was not simple, however, as evidenced by migrations along the southern boundaries of generic ranges. Between the Namurian A and B, 13% of the genera present in both intervals migrated out of the equatorial and low-latitude zones into more northern zones (Table 2). At the same time, only 6% of the genera expanded the southern boundary of their range into equatorial and low-latitude zones. A chi-squared test shows that this episode of migration from the equatorial and low-latitude zones into higher latitudes is statistically greater than

Table 2. *Generic migration patterns along the east and west coasts of Laurussia or northern Pangaea*

INTERVAL		NORTH WARMS			NORTH COOLS			SOUTH WARMS			SOUTH COOLS			DIVERSITY		
		Total	East	West	Total	East	West	Total	East	West	Total	East	West	Total	East	West
T1–T2	#	17*	8*	11*	10	5	5	5	4	1	6*	2*	3*	107	55	78
	%	16%	15%	14%	9%	9%	6%	5%	7%	1%	6%	4%	4%			
T2–T3	#	17	15	4	8	1	7	3	1	2	5	2	3	117	57	97
	%	15%	26%	4%	7%	2%	7%	3%	2%	2%	4%	3%	3%			
T3–V1	#	6	3	4	9	2	7	5	1	4	–	–	–	100	51	81
	%	6%	6%	5%	9%	4%	9%	5%	2%	5%	–	–	–			
V1–V2	#	14	8	6	10	2	9	2	2	–	1	1	–	103	52	73
	%	14%	15%	8%	10%	4%	12%	2%	4%	–	1%	4%	–			
V2–V3	#	40	22	20	3	1	2	4	2	2	1	1	–	95	61	58
	%	42%	36%	34%	3%	2%	3%	4%	3%	3%	1%	2%	–			
V3–NA	#	13	10	5	3	2	2	2	2	–	1	–	1	104	62	73
	%	13%	16%	7%	3%	3%	3%	2%	3%	–	1%	–	1%			
NA–NB	#	6	5	1	23	12	14	9	5	5	4	3	1	70	37	52
	%	9%	13%	2%	33%	32%	27%	13%	14%	10%	6%	8%	2%			
NB–NC	#	10%	4	7	12*	5*	8*	3	2	1	4	–	4	85	49	63
	%	12%	8%	11%	14%	10%	13%	3%	4%	2%	5%	–	6%			

Abbreviations as in Table 1. Values marked with an asterisk and the percentages given beneath them are estimated using the average ratio of true migrations to apparent migrations (see text). Migration patterns 1 and 2 constitute northern warming; patterns 3 and 4 constitute northern cooling. Migration patterns 2 and 6 constitute southern warming; patterns 4 and 5 constitute southern cooling (see Fig. 3).

average. This is the expected pattern during times of equatorial warming (Valentine 1984b). It indicates the simultaneous occurrence of equatorial warming and high-latitude cooling between the Namurian A and B.

Table 2 lists migration patterns separately for the Cordilleran and Tethyan coasts of Laurussia. Migration pulses in the middle to late Visean and the Namurian A and B occurred along both coasts, which suggests that they represent global rather than regional climatic changes.

Latitudinal diversity gradients

The results of the latitudinal diversity study (Fig. 4) corroborate the high-latitude warming and cooling trends revealed by migration patterns. A steep latitudinal diversity gradient existed from the early Tournaisian through the middle Visean along both coasts. Low-latitude diversity represented about one-third to one-half of equatorial diversity, while mid- and high-latitude diversity was generally very low. For most of the Tournaisian through middle Visean, the two methods of computing diversity gradients yielded very similar results, due to low levels of migration during these intervals.

In the late Visean, diversity outside the equatorial zone increased markedly; low-latitude diversity reached two-thirds of equatorial diversity. Diversity was increased at all latitudes relative to the equatorial zone, along both coasts of Pangaea, during the late Visean to Namurian B. In the Namurian A, low-latitude diversity exceeded equatorial diversity on both coasts. (These results occur for both methods of calculating diversity gradients. However, subtracting the numerous northward-migrating genera from the late Visean results somewhat reduces extratropical diversity.)

These patterns suggest a gentle latitudinal temperature gradient from the late Visean through the Namurian B. This climate change could be accomplished by either equatorial cooling or high-latitude warming. However, the high percentage of northward migrations between the middle and late Visean suggests that the high latitudes became warmer; many of the genera present at higher latitudes in the late Visean were immigrants from lower latitudes (Fig. 4). In addition, the relatively high percentage of migrations out of the equatorial zone between the Namurian A and B may contribute to the gentle latitudinal diversity gradient during the Namurian B by depressing equatorial diversity relative to mid-latitude diversity.

By the Namurian C, the trend reversed. Relative diversity declined at all latitudes outside the palaeoequatorial zone, although Cordilleran low-latitude diversity remained fairly high. The decrease in diversity at higher latitudes, compared to that at the equator, in the Namurian C suggests that the pole-to-equator temperature gradient became steeper due to cooling at high latitudes.

Discussion

Tournaisian through Namurian palaeoclimate

The palaeoclimatic history of the Northern Hemisphere during the interval studied is one of high-latitude warming between the middle and late Visean, followed by high-latitude cooling and equatorial warming by the Namurian B. The evidence for these changes in climate comes from migration patterns and changes in latitudinal diversity of articulate brachiopods.

Between the middle and late Visean, a large number of brachiopod genera (42%) extended the northern boundary of their range northward into higher latitudes. At the same time, the diversity of brachiopods at higher latitudes increased relative to their diversity in the equatorial zone, producing a shallow latitudinal diversity gradient. Coupled with poleward migration of equatorial and low-latitude forms, this shallow gradient indicates high-latitude warming. Although the biostratigraphic resolution is not as precise, the phytogeographic history of Early Carboniferous land plants corroborates this trend, suggesting that terrestrial climate warmed between the middle Visean and the late Visean–earliest Namurian A (Raymond 1985). In addition, Waters (1990) has noted the dispersion of species of the blastoid *Nymphaeoblastus* into Kazakhstan from lower latitudes at about this time.

Brachiopod diversity gradients were shallow from the late Visean through the Namurian B. Northern Hemisphere high-latitude climate probably remained warm into the Namurian A. The situation for the Namurian B is less clear. A large number of

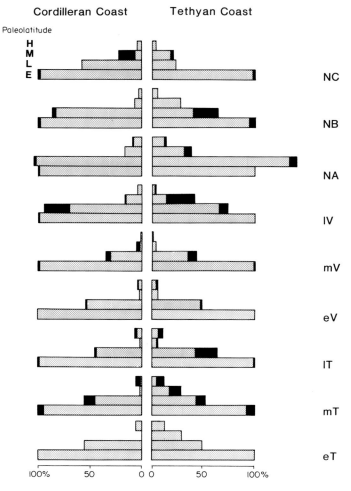

Fig. 4. Latitudinal diversity gradients of Tournaisian through Namurian articulate brachiopods along both coasts of Pangaea. Black pattern indicates new immigrants present in each zone; stippled pattern indicates nonimmigrants. Diversity is standardised to that of the palaeoequatorial zone: diversity of nonimmigrants in each zone is shown as a percentage of the number of nonimmigrants in the palaeoequatorial zone, and diversity including immigrants is shown as a percentage of the total number of genera in the palaeoequatorial zone, including immigrants. Abbreviations: eT, early Tournaisian; mT, middle Tournaisian; lT, late Tournaisian; eV, early Visean; mV, middle Visean; lV, late Visean; NA, Namurian A; NB, Namurian B; NC, Namurian C; E, palaeoequatorial zone; L, low latitudes, M, middle latitudes; H, high latitudes.

genera (33%) evacuated the northern extremities of their ranges between the Namurian A and B, suggesting that high-latitude cooling began between the Namurian A and B. However, the latitudinal diversity gradient of the Namurian B was still shallow. The pulse of southward migration continued into the Namurian C. By the Namurian C, the latitudinal diversity gradient was steep again, due to the loss of taxa in the high latitudes.

Concommitant with high-latitude cooling there is evidence for equatorial warming between the Namurian A and B. At this time 13% of the brachiopod genera common to both intervals evacuated the equatorial zone, contracting the southern boundary of their range into the low latitudes. Evacuation of the equatorial zone could contribute to the shallow diversity gradient of the Namurian B by decreasing relative equatorial diversity as brachiopods migrated into the low-latitude zone. (Over a third of the genera in the Tethyan low–latitudes of the Namurian B were new immigrants; Fig. 4.) Waters (1990) recognised the migration of several groups of blastoids from the equator to Kazakhstan during the Namurian; we suggest that these migrations may be related to equatorial warming.

Geological causes of palaeoclimate change

Two major geological events that could have affected global climate occurred between the Late Devonian and the Late Carboniferous: the collision of Laurussia and Gondwana, and the onset of continental glaciation in the Southern Hemisphere. However, these events, particularly the collision, are difficult to date accurately. In addition, two global regressions occurred in the Namurian A (Saunders & Ramsbottom 1986). Can any of these events be identified as the cause of climatic changes during the Tournaisian through Namurian?

Although palaeogeographic reconstructions differ in details, most agree that the palaeocontinents of Laurussia and Gondwana were separated by the east–west trending Tethys Ocean at the beginning of the Early Carboniferous (Fig. 1). By the end of the Early Carboniferous or during the first part of the Late Carboniferous (Ross & Ross 1985; Hurley & Van der Voo 1987; Veevers 1988), Laurussia and Gondwana collided to form the megacontinent, Pangaea (Fig. 2). Prior to the Laurussia–Gondwana collision, a warm equatorial current encircled the globe (Ziegler et al. 1981; Ross & Ross 1985; Copper 1986). This unobstructed current confined warm water to the equatorial region. After the collision, the current was deflected to the north and south along the eastern coast of the megacontinent.

The climatic consequences of the Laurussia–Gondwana collision and the deflection of the equatorial current are not certain. A number of workers (Benson 1976; Ross & Ross 1985; Copper 1986) have suggested that closing east–west equatorial oceans causes global cooling due to the refrigeration of warm equatorial water at the poles. According to Benson (1976), the closure of Tethys in the late Eocene–Oligocene formed the psychrosphere (deep cold water) in the equatorial basins of the opening Atlantic Ocean. Benson suggested that happened because the closure of Tethys greatly reduced the supply of warm (thermospheric) water to the narrow Atlantic Ocean, while the warm equatorial waters of the Atlantic were refrigerated at the poles. Current deflection may have played a role in the formation of the Atlantic psychrosphere in the late Eocene (although see Matthews & Poore 1980 for a glacial model). However, Benson's model does not apply to the Early Carboniferous. Unlike the Tertiary closure of Tethys, blocking the west side of the Tethys in the Carboniferous probably did not affect the amount of warm water available globally. Even with Laurussia joined to Gondwana, ocean or epicontinental seas covered most of the equatorial latitudes, and these waters are likely to have continued warm.

In contrast to Ross & Ross (1985) and Copper (1986), we think that the Laurussia–Gondwana collision caused warming at high latitudes. Ziegler et al. (1981) invoked deflection of warm equatorial currents to explain the broad palaeolatitudinal extent of very diverse, tropical marine communities in the Permian. This broad distribution of tropical faunas during the Permian, when the west side of the Tethys was certainly blocked, points to high-latitude warming, rather than cooling, as a result of the collision. Hurley & Van der Voo (1987) estimated that the Laurussia–Gondwana collision began in the Visean. High-latitude warming, suggested by brachiopod migrations and latitudinal diversity gradients, as well as by land plants (Raymond 1985), occurred between the middle and late Visean. We suggest that the effective collision of Laurussia and Gondwana occurred at that time, confirming the Hurley & Van der Voo estimate.

The onset of continental glaciation in the Southern Hemisphere, which probably occurred in the Namurian A (Veevers & Powell 1987) or near the Namurian A/B boundary (Hambrey & Harland 1981; Bouroz et al. 1978), may have caused the Namurian B high-latitude cooling and equatorial warming episode. If so, brachiopod migration patterns in the Northern Hemisphere suggest that high-latitude palaeoclimate changed in both hemispheres, even though evidence of polar ice in the Northern Hemisphere is lacking (with the exception of one tillite which is Namurian B to Sakmarian in age; Hambrey & Harland 1981).

During glaciations, the seasonality of equatorial climates may decrease because the intertropical convergence zone is confined to a narrower latitudinal range than during non-glacial intervals. Ziegler et al. (1987) suggested that, during glacial intervals, the terrestrial tropics become less seasonal and wetter. Valentine (1984a, b) and Savin (1982) observed an analogous trend involving decreased equatorial seasonality and warming in shallow marine equatorial environments during the Neogene glaciation.

Southern Hemisphere glaciation also could have caused the world-wide regression observed in the middle Namurian A (Veevers & Powell 1987). If so, the long Namurian A interval contains a palaeoclimatic record of both pre-regression (non-glacial) and post-regression (glacial) climates. Because of this situation, climatic change is not evident in our data set until the Namurian B; the record of migrations between the Namurian A and B reflects the difference between the climate of the early Namurian A, when the high latitudes were still warm, and the Namurian B, when high-latitude cooling and equatorial warming had certainly begun.

Could either the collision or the regressions have caused these Namurian climatic changes in the absence of Southern Hemisphere glaciation? Copper (1986) and Ross & Ross (1985) have argued that the collision caused high-latitude cooling, despite the empirical evidence for widespread tropics in the Late Palaeozoic. However, it seems unlikely that current deflection in the Tethys could decrease equatorial seasonality and increase equatorial temperatures along both coasts of Pangaea. Likewise, though regressions may affect global climate, it seems unlikely that regression alone could cause both high-latitude cooling and equatorial warming. Southern Hemisphere glaciation appears to be the most likely cause of the Namurian climatic change.

Parallels to the onset of Tertiary glaciation

The history of Tournaisian through Namurian climate indicates that high-latitude cooling in the Northern Hemisphere coincided with the onset of Southern Hemisphere glaciation. Although only the south pole became glaciated, both poles became cooler at that time. In addition, equatorial warming and decreased equatorial seasonality accompanied the growth of continental glaciers. Although the co-occurrence of high-latitude cooling and equatorial warming seems contradictory, evidence exists for both equatorial warming and high-latitude cooling during the onset of polar glaciation in the Tertiary (Savin 1982; Valentine 1984a, b).

During the middle Miocene, the surface temperature of equatorial water warmed as high-latitude climate deteriorated (Savin 1982). In response to these climatic changes, migration of high-latitude molluscs toward the equator was coupled with the migration of seasonally-adapted molluscs away from the warming equator and with high rates of equatorial origination (Valentine 1984a, b). The same patterns of migration were observed among Namurian brachiopods; in addition, high rates of equatorial origination also occurred in the Namurian B (Raymond, Kelley, & Lutken, in prep.). The high rate of generic origination among scleractinian corals during the middle Miocene (Rosen 1984) and increases in the diversity of equatorial terrestrial floras between the early and late Tertiary (Bande & Prakash 1986) may be related to a middle Miocene decrease in the seasonality of equatorial climate.

Conclusion

Patterns of migration and latitudinal diversity of organisms provide valuable tools for testing contrasting models of palaeoclimate and palaeogeography. Such methods are applicable throughout the Phanerozoic, and will prove useful in determining the climatic consequences of events such as continental collision and glaciation.

Applying such methods to the record of articulate brachiopods indicates that high-latitude warming occurred in the middle to late Visean, followed by high-latitude cooling and equatorial warming in the Namurian. While strong evidence supports these climatic interpretations, the mechanisms of these changes are less certain. We suggest that the collision of Laurussia and Gondwana produced Visean high-latitude warming, and that Gondwanan glaciation caused high-latitude cooling and equatorial warming in the Namurian. Both of these conclusions are controversial; more widespread use of the techniques described here may help resolve such controversies for the Carboniferous and other time intervals.

We gratefully acknowledge the assistance and advice of J. L. Carter, T. W. Henry, and J. Waters in compiling the data base for this study. Discussions with A. M. Ziegler and P.M. Sheehan were helpful in developing the ideas presented. W. C. Parker and N. Gilinsky rendered statistical counsel. J. Aschberger assisted with manuscript preparation; C. A. Thornton prepared the data for the Supplementary Publication. Finally, we acknowledge the financial support, enthusiasm, and indulgence of the Mary Ingraham Bunting Institute of Radcliffe College, G. Brunton, J. Spang, J. Kelley, and T. Lutken, without which this study would have been impossible to pursue. Partial support for this work was provided by a University of Mississippi Summer Faculty Research Grant.

References

BANDE, M. B. & PRAKASH, U. 1986. The Tertiary flora of southeast Asia with remarks on its paleoenvironment and phytogeography of the Indo-Malayan region. *Reviews of Paleobotany and Palynology*, **49**, 203–234.

BENSON, R. H. 1976. In search of lost oceans: a paradox in discovery. *In*: GRAY, J. & BOUCOT, A. J. (eds) *Historical Biogeography, Plate Tectonics, and the Changing Environment*. Oregon State University Press, Corvallis, 379–390.

BLANTON, C. L. 1984. *Quantitative analyses of Tournaisian, Visean, and Namurian Carboniferous brachiopod biogeography: Comparison of paleobiogeography with paleogeography*. MS thesis, University of Mississippi.

BOUROZ, A., EINOR, O. L., GORDON, M., MEYEN, S. V., & WAGNER, R. H. 1978. Proposals for an international chronostratigraphic classification of the Carboniferous. *Huitieme Congres International de Stratigraphie et de Geologie Carbonifere. Compte Rendu*, **1**, 36–52.

COPPER, P. 1986. Frasnian/Famennian mass extinction and cold-water oceans. *Geology*, **14**, 835–839.

HAMBREY, M. J. & HARLAND, W. B. (eds) 1981. *Earth's Pre-Pleistocene Glacial Record*. Cambridge University Press, Cambridge.

HURLEY, N. F. & VAN DER VOO, R. 1987. Paleomagnetism of Upper Devonian reefal limestones, Canning Basin, Western Australia. *Bulletin of the Geological Society of America*, **98**, 138–146.

KELLEY, P. H., RAYMOND, A. & LUTKEN, C. B. 1989. Paleoclimatic effects of continental collision and glaciation: evidence from Carboniferous articulate brachiopods. *28th International Geological Congress Abstracts*, **2**, 171.

MATTHEWS, R. K. & POORE, R. Z. 1980. Tertiary O^{18} record and glacio-eustatic sea level fluctuations. *Geology*, **8**, 501–504.

PALMER, A. R. 1983. The decade of North American Geology, 1983 Geologic Time Scale. *Geology*, **11**, 503–504.

RAYMOND, A. 1985. Floral diversity, phytogeography, and climatic amelioration during the Early Carboniferous (Dinantian). *Paleobiology*, **11**, 293–309.

——, PARKER, W. C. & PARRISH, J. T. 1985. Phytogeography and paleoclimate of the Early Carboniferous. *In*: TIFFNEY, B. H. (ed.). *Geological Factors and the Evolution of Plants*. Yale University Press, New Haven, 169–222.

——, KELLEY, P. H. & LUTKEN, C. B. In preparation. Articulate brachiopods, paleoclimate, and the mid-Carboniferous extinction event. *Palaios*.

ROSEN, B. R. 1984. Reef coral biogeography and climate through the Late Cainozoic: just islands in the sun or a critical pattern of islands? *In*: BRENCHLEY, P. (ed.). *Fossils and Climate*. John Wiley and Sons

Ltd., New York, 201–262.
Ross, C. A. & Ross, J. R. P. 1985. Carboniferous and early Permian biogeography. *Geology*, **13**, 27–30.
Rowley, D. B., Raymond, A., Parrish, J. T., Lottes, A. L., Scotese, C. R. & Ziegler, A. M. 1985. Carboniferous paleogeographic, phytogeographic and paleoclimatic reconstructions. *International Journal of Coal Geology*, **5**, 7–42.
Saunders, W. B. & Ramsbottom, W. H. C. 1986. The mid-Carboniferous eustatic event. *Geology*, **14**, 208–212.
Savin, S. M. 1982. Stable isotopes in climatic reconstructions. *In*: Geophysics Study Committee, National Academy of Sciences (eds). *Studies in Geophysics: Climate in Earth History*. National Academy Press, Washington, D.C., 164–171.
Scotese, C. R. 1986. *Phanerozoic reconstructions: a new look at the assembly of Asia*. University of Texas Institute for Geophysics Technical Report No. 66.
Simpson, G. G., Roe, A. & Lewontin, R. C. 1960. *Quantitative Zoology*. Harcourt Brace and World, Inc., New York.
Stanley, S. M. 1984. Temperature and biotic crises in the marine realm. *Geology*, **12**, 205–208.
Valentine, J. M. 1984*a*. Neogene marine climate trends: implications for biogeography and evolution of the shallow-sea biota. *Geology*, **12**, 647–650.
—— 1984*b*. Climate and evolution in the shallow sea. *In*: Brenchley, P. (ed.). *Fossils and Climate*. John Wiley and Sons, Ltd., New York, 265–277.
Veevers, J. J. 1988. Gondwana facies started when Gondwanaland merged in Pangea. *Geology*, **16**, 732–734.
—— & Powell, C. MCA. 1987. Late Palaeozoic glacial episodes in Gondwanaland reflected in transgressive–regressive depositional sequences in Euramerica. *Bulletin of the Geological Society of America*, **98**, 475–487.
Waters, J. A. 1990. Palaeobiogeography of the Blastoidea. *In*: McKerrow, W. S. & Scotese, C. R. (eds) *Palaeozoic Palaeogeography and Biogeography*. Geological Society, London, Memoir, **12**, 339–379.
Ziegler, A. M., Bambach, R. K., Parrish, J. T., Barrett, S. F., Gierlowski, E. H., Parker, W. C., Raymond, A. & Sepkoski, J. J. Jr. 1981. Palaeozoic Biogeography and climatology. *In*: Niklas, K. J. (ed.). *Paleobotany, Paleoecology, and Evolution*, Vol. 2, Praeger Press, New York, 231–266.
——, Raymond, A., Gierlowski, T. C., Horrell, M. A., Rowley, D. B. & Lottes, A. L. 1987. Coal, climate and terrestrial productivity: the present and the Early Cretaceous compared. *In*: Scott, A. C. (ed.). *Coal and Coal-bearing Strata: Recent Advances*. Geological Society, London, Special Publication, **32**, 25–49.

Biogeography of Lower Carboniferous crinoids

N. GARY LANE & G. D. SEVASTOPULO

Department of Geology, Indiana University, Bloomington, IN 47405
Department of Geology, Trinity College, Dublin 2, Ireland

Abstract: The stratigraphic are ranges of all known crinoid genera of Mississippian and Lower Carboniferous age in western Europe and North America compared. Camerate crinoid distribution was published in 1987 (Lane and Sevastopulo). Inadunate and flexible crinoid ranges are presented here. A total of 197 genera are involved, 157 in North America and 93 in Europe, with 53 joint occurrences. Conspicuous endemism occurs at both the familial and generic levels in both Europe and North America. In early Mississippian time camerate crinoid genera are especially endemic in North America and many become extinct. In later Mississippian time many poteriocrinoid and flexible crinoid genera and some families are confined to either North America or Europe. Both environmental and physical barriers contributed to isolation and endemism. Genera tend to originate first, and become extinct first, at about equal rates in North America and in Europe. European genera tend to be somewhat longer ranging than North American genera, perhaps due to differences in breadth of definition on the two continents. Primitive groups like disparid and cyathocrinoid inadunates range up from the Late Devonian. Advanced poteriocrinoid inadunates commonly range up into younger Pennsylvanian–Upper Carboniferous rocks. New families originated in both North America and western Europe with about equal frequency. New Japanese occurrences of Pennsylvanian inadunates and the age relations of Permian Timor crinoids are discussed briefly.

The time interval from about 350 to 320 million years ago, roughly the Lower Carboniferous or Mississippian period of geological time, witnessed the acme of the Class Crinoidea. All three major subclasses of Palaeozoic crinoids, the camerates, flexibles, and inadunates, exhibit an expanded taxonomic rate of evolution at the familial, generic, and specific levels. In addition, specimens of many species are abundant and many limestones are constructed of crinoidal skeletal remains.

The record of this tremendous expansion of crinoids is largely known from North America and western Europe, where most intensive research has taken place and where fossiliferous rocks of this age outcrop widely. While contemporaneous crinoids are known from Asia, Australia and northern Africa these latter occurrences are isolated local faunas without continuity of sequence or wide diversity.

For these reasons, the fossil record of the former two areas provides the best basis for discussion of biogeographical aspects of crinoid distribution. This report focuses largely, but not completely, on the crinoids of this age and from these two areas. Brief discussion of other biogeographically significant crinoid occurrences of Late Palaeozoic age concludes this report.

A total of 197 genera in 62 families has been tabulated and their stratigraphic ranges plotted separately both for North America and for Europe (Lane & Sevastopulo 1987; Figs 1–7). These range charts provide the factual basis for this discussion. Of the 197 genera, 157 occur in North America and 93 occur in Europe. A total of 53 genera co-occur on the two continents. There are 104 endemic genera in North America, but only 40 endemic genera in Europe. The greater total number of crinoid genera in North America may partly be due to the much greater area of undeformed, unmetamorphosed rock outcrops available for fossil collection and study.

Crinoid endemism

Camerate crinoids exhibit a high degree of endemism between Europe and North America at the generic level (Lane & Sevastopulo 1987). 66% of Mississippian genera are confined to North America, and 43% of European genera are endemic. The endemism is especially striking in the lower part of the Mississippian, before many camerate genera became extinct. Camerate crinoids typically were more abundant and diversified on carbonate bottoms than they were on terrigenous clastic seafloors. This tendency had been present in camerate crinoids since the Silurian Period (Lane 1971). The land barrier of the rising Appalachians served to isolate European and North American crinoids, but the widespread deltaic muds of shallow marine areas surrounding the Appalachians were also effective barriers to migration of carbonate loving camerates. This clastic barrier was an important element inducing the endemic development of many camerate genera and families.

That these crinoids were able to migrate across areas where suitable bottoms prevailed is demonstrated clearly by the occurrence of many genera both in the central and western United States where carbonate rocks prevailed. With the central United States as a centre of evolution during early Mississippian time, as denoted by maximum diversity of genera and species, westward migration by a few species into New Mexico, Arizona and Nevada occurred. Eastward migration was prevented by spreading clastic wedges of the rising Appalachians.

An outstanding example of endemism at the family level is provided by the camerate family Batocrinidae, consisting of 10 genera, all confined to North America. An example of endemism at the generic level is provided by the camerate family Actinocrinitidae, consisting of 17 genera. Of these, only four genera occur in both Europe and North America. Ten genera are confined to North America and three to Europe.

A second wave of crinoid endemism at both the generic and familial levels occurred during the latter part of the Mississippian-Lower Carboniferous, essentially during the Asbian and Brigantian stages. This interval witnessed the development of largely separate faunas of advanced poteriocrinoids in Europe and North America. While it is true that a few genera occur on both continents and provide a basis for intercontinental correlation (Horowitz & Strimple 1974), the great majority of genera are confined to one area or the other.

Of 50 poteriocrinoid genera confined to this interval, only nine, or 18%, are found on both continents. 26 genera are confined to the United States and 15 genera to western Europe. In this instance environment seemingly had little effect on distribution as these crinoids are found in a wide variety of lithologies. Physical isolation by the Appalachian land barrier must have been important.

Virtually all of the North American genera are confined to the central United States, specifically to the Illinois Basin and adjacent surrounding areas. This region was effectively isolated by land barriers to the east, north, and west. Migration was possible only to the south and southwest, where deep water trenches impeded dispersal. Thus, interchange between North America and Europe was reduced to a minimum for many crinoids during the close of the period.

Flexible crinoids also show a similar endemism of both genera

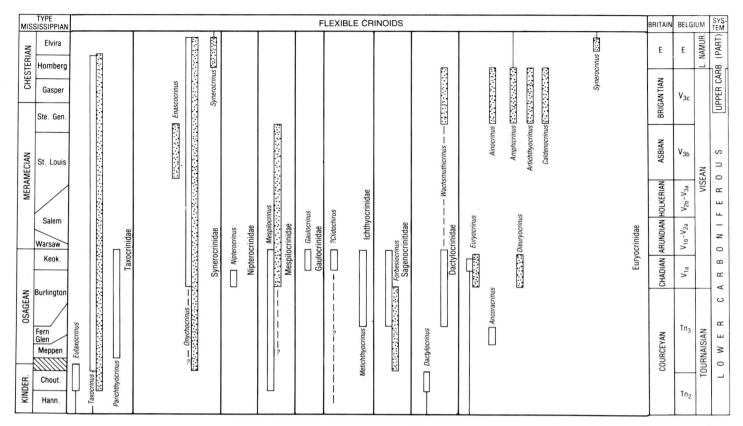

Fig. 1. Range chart for genera of Mississippian–Lower Carboniferous flexible crinoids. Open bars are for North American ranges, stippled bars for European ranges. Families are separated by heavy vertical lines. Solid lines above or below bars indicate that a genus ranges up from the Devonian or up into Pennsylvanian–Upper Carboniferous rocks. Dashed lines indicate that a taxon is either unknown or uncertainly identified from the specified time interval.

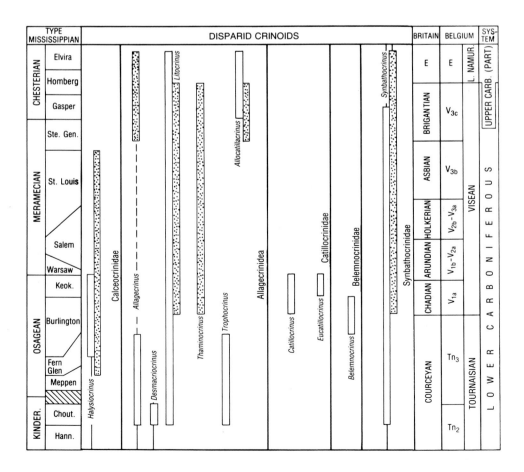

Fig. 2. Range chart for genera of disparid inadunate crinoids. See Fig. 1 for further explanation.

CARBONIFEROUS CRINOIDS

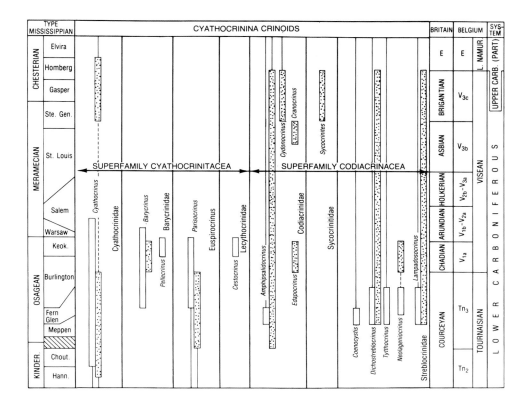

Fig. 3. Range chart for genera of cyathocrinoid inadunate crinoids. See Fig. 1 for further explanation.

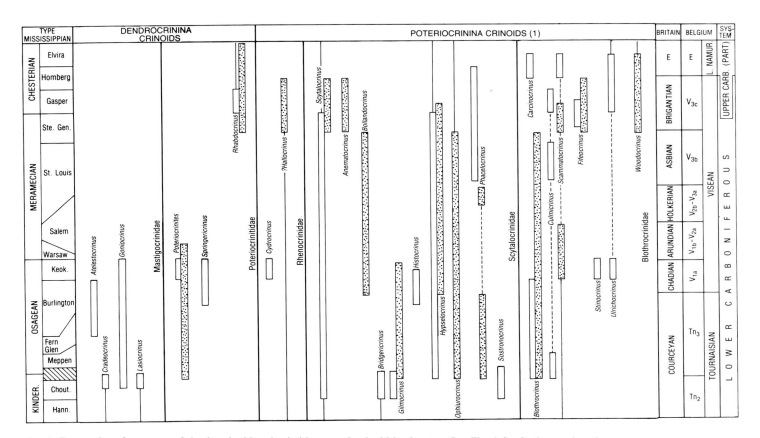

Fig. 4. Range chart for genera of dendrocrinoid and primitive poteriocrinoid inadunates. See Fig. 1 for further explanation.

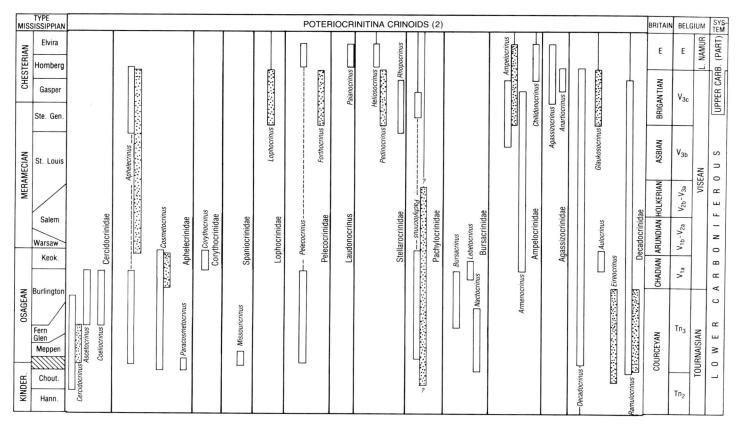

Fig. 5. Range chart for genera of poteriocrinoid inadunates. See Fig. 1 for further explanation.

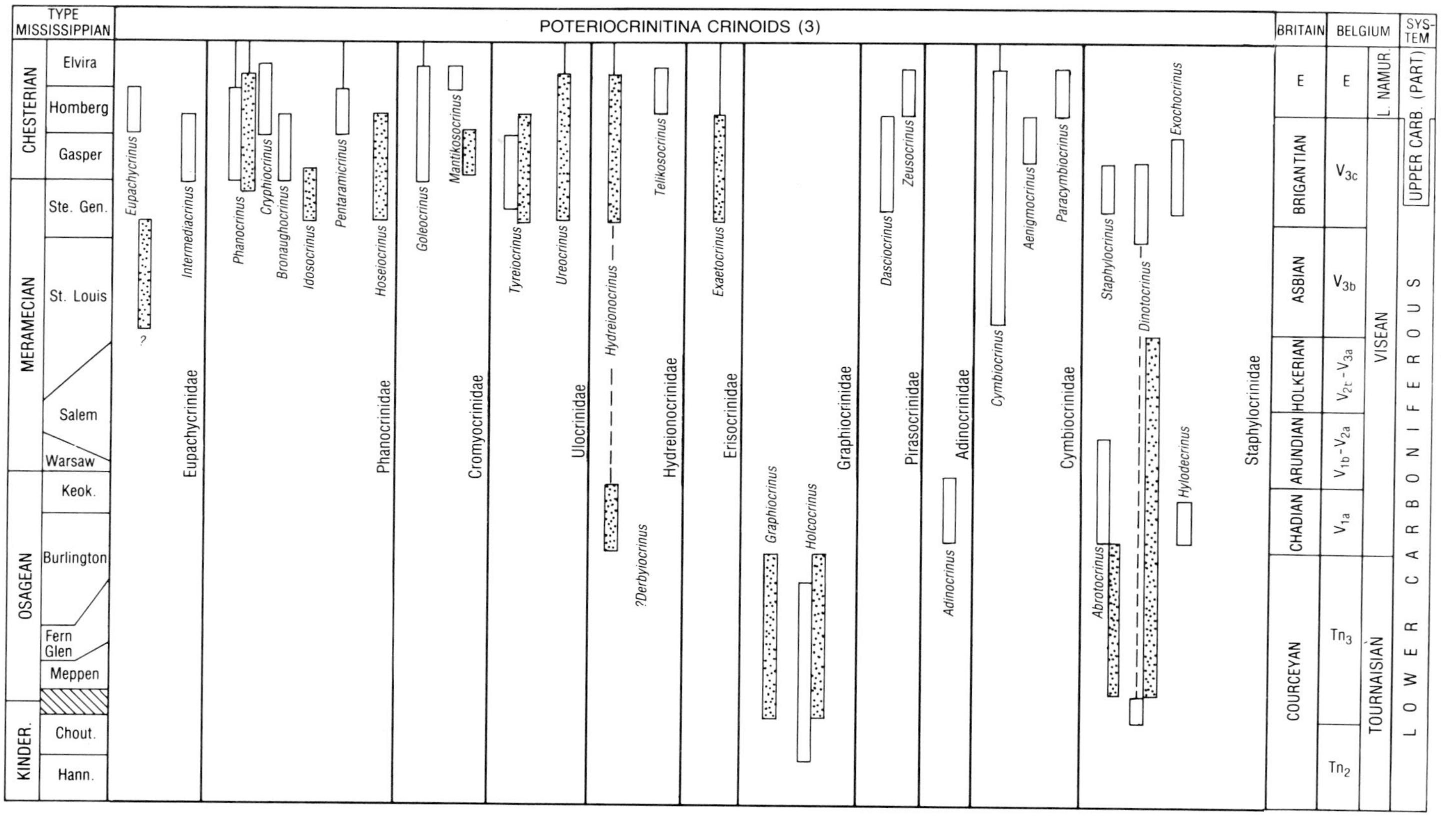

Fig. 6. Range chart for genera of poteriocrinoid inadunates. See Fig. 1 for further explanation.

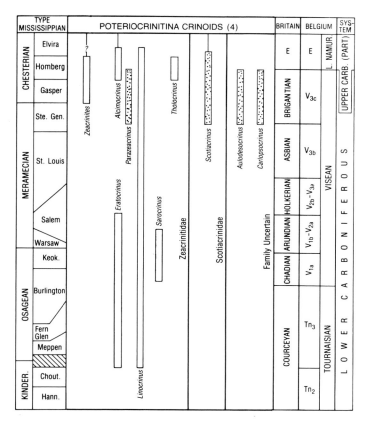

Fig. 7. Range chart for genera of poteriocrinoid inadunates. See Fig. 1 for further explanation.

and families at the close of the Lower Carboniferous. There are seven endemic genera of this small subclass of crinoids that are confined to western Europe. In sharp contrast North American faunas exhibit only two long-ranging and ubiquitous genera, *Taxocrinus* and *Onychocrinus*.

Origination of genera

Among those crinoids that do occur on both continents, the question can be asked as to whether they appear first, therefore apparently evolve first, on one continental area or the other. Obviously vagaries of preservation and collection can strongly bias the record of first appearances. Inspection of the range charts in this paper show that many genera first occur in rocks of closely equivalent age. Slight differences in age are here ignored and only widely different first appearances are taken as indicative of different times and places of origination. Of 40 genera that occur in both Europe and North America about 14 seemingly occur first in North America, 10 start in Europe and 16 are essentially ties (Table 1). Among different higher categories of

Table 1. *Origination of genera that co-occur in Europe and North America*

	Ca	Fl	Di	Cy	Po	Total
Start in North America	1	4	2	1	6	14
Start in Europe	1	1	1	2	5	10
Start, Tie	5	1	0	2	8	16
Totals	7	6	3	5	19	40

Ca, camerates; Fl, flexibles; Di, disparids; Cy, cyathocrinoids; Po, poteriocrinoids.

Table 2. *Generic extinctions of genera that co-occur in Europe and North America*

	Ca	Fl	Di	Cy	Po	Total
End first in North America	3	3	2	2	6	16
End first in Europe	1	1	0	3	0	5
Extinctions, Tie	3	2	1	0	8	14
Totals	7	6	3	5	14	35

See Table 1 for abbreviations.

crinoids no systematic differences between the two continental areas can be noted. Apparently those genera that could migrate did so in relatively short periods of geologic time. Neither area was an overall preferred area for evolution, indicating that opportunities for evolutionary innovation were about the same in both areas.

Generic extinctions

There are many ties in the disappearance of genera in Europe and North America. Numbers of genera that first disappear on one continent are about equal to the number that disappear first on the other continent (Table 2). A major time of extinction in North America that was not nearly so obvious in Europe was close to the Osagean–Meramecian boundary. Genera of camerates, flexibles and disparid crinoids became extinct in North America at or close to this boundary. Near simultaneous disappearances in both areas seem to indicate some widespread change in living conditions, but different times of extermination would seem to argue that local conditions must also have been important. As so commonly happens is considering extinction events a single, simple cause is not obvious.

Persistence of genera in Europe and North America

Generally speaking, crinoid genera survive longer in Europe than in North America. A total of 252 separate generic ranges on each continent were tabulated with respect to their occurrence in one or more series in North America and one or more stages in Europe. In North America 67% of all genera are confined to a single series, and 19% are confined to two series. In contrast 50% of European genera are confined to one stage and 32% occur in two stages. A total of 45% of European genera occur in either two or three stages.

Shorter ranges in North America accord with the greater total number of genera and perhaps with more rapid rates of evolution and extinction. More finely partitioned habitats may have played a role as well as the fact that North America genera may be more narrowly defined than many European genera, especially the many endemic forms. The shorter average range of North American crinoids could also be a sampling artifact. With better sampling more of the rare taxa might be found. The better the sampling the greater the number of taxa known from a single locality. Thus, differences in range may be due, at least in part, to sampling bias or to differing taxonomic philosophies.

Carryover of genera and famlies

By carryover we mean families or genera that persist from the Devonian into younger rocks or that carry over from the Mississippian–Lower Carboniferous into younger Pennsylvanian or Upper Carboniferous rocks. The pattern of carryover depends significantly on the major group of crinoids considered (Table 3). Disparid and cyathocrinoid families have conspicuous carryover from the Devonian. Four of five Mississippian disparid families

Table 3. *Familial and generic carryover into the Devonian and Upper Carboniferous*

Family Carryover	Total Families	Range into Devonian	Range into Up. Carb.
CYATHOCRINOIDS	8	7 (88%)	4 (50%)
DISPARIDS	5	4 (80%)	4 (80%)
DENDROCRINOIDS	1	1 (100%)	0 (0%)
POTERIOCRINOIDS	30	4 (13%)	22 (73%)
FLEXIBLES	9	6 (67%)	5 (56%)
CAMERATES	9	4 (44%)	4 (44%)
TOTAL FAMILIES	62	26 (42%)	39 (63%)

Generic Carryover	Total Genera	Range into Devonian	Range into Up. Carb.
CYATHOCRINOIDS	15	2 (13%)	7 (47%)
DISPARIDS	11	4 (36%)	1 (9%)
DENDROCRINOIDS	4	2 (50%)	0 (0%)
POTERIOCRINOIDS	91	4 (4%)	23 (25%)
FLEXIBLES	22	3 (14%)	3 (14%)
CAMERATES	56	4 (7%)	3 (5%)
TOTAL GENERA	199	19 (10%)	37 (19%)

began in the Devonian, and five of seven Mississippian cyathocrinoid families range up from the Devonian. However, these groups dwindle in importance and become rare or extinct during the Mississippian.

The poteriocrinoids have very few carryovers from the Devonian, only four of 31 families, but expand significantly during the Mississippian with 20 families continuing into Pennsylvanian rocks. The extraordinary evolutionary diversification of poteriocrinoids that characterizes Pennsylvanian and Permian rocks thus began during the Mississippian–Lower Carboniferous.

Centres of evolution

Among those crinoids that occur in both North America and Europe, certain groups clearly originated on one continent and underwent a series of evolutionary events on that continent, with later dispersal to the other continental area. An excellent example of this pattern is afforded by the camerate family Amphoracrinidae. This family clearly originated in western Europe and diversified there into three genera. Two of these are European endemics, the third, *Amphoracrinus*, made a brief incursion into North America during Osagean time but shortly became exterminated there. The genus persisted into the Brigantian Stage in Great Britain. A converse example concerns the poteriocrinoid family Zeacrinitidae. This group originated early in North America, at the base of the Osagean, and six endemic genera evolved. By Brigantian time a single migrant form that was generically distinct, *Parazeacrinus*, appeared in Europe. This short-lived form became extinct at the end of the Brigantian.

Other faunas

In addition to the extensive relationship between European and North American crinoid faunas discussed above, there are other, scattered, Late Palaeozoic occurrences that are of biogeographical interest. The Midcontinent Pennsylvanian of the United States has yielded a rich fauna of advanced poteriocrinoids. Until recently this fauna was known almost exclusively from this area. However, in recent years Hashimoto (1984) has identified and illustrated poteriocrinoids of Middle Pennsylvanian (Atokan) age from Japan that have strong resemblances to North American faunas. The geographical relationships between these two areas during Pennsylvanian (Namurian) time is remote and the biogeographical significance of these new Japanese occurrences is obscure.

For many years it has been clear that the famous crinoid-yielding beds of the island of Timor in Indonesia included two groups of crinoids with very different affinities. One group consists of evolutionarily advanced poteriocrinoids that share families and a few genera with other Permian crinoid faunas from Russia, Australia, and West Texas. In addition, rocks on Timor have yielded other crinoids, especially advanced camerates, that have their closest affinities at both the familial and generic levels with Mississippian–Lower Carboniferous crinoids of North America and Europe. It remains unclear whether these latter crinoids co-occur with truly Permian forms and are thus relics held over from earlier times, or whether they may have been collected from separate limestone lenses of different ages. Lack of modern field studies by trained palaeontologists contributes to this dilemma.

References

HASHIMOTO, K. 1984. Preliminary study of Carboniferous crinoid calyces from the Akiyoshi Limestone Group, Southwest Japan. *Bulletin Yamaguchi Prefectural Yamaguchi Museum*, **10**, 1–30.

HOROWITZ, A. S. & STRIMPLE, H. L. 1974. Chesterian Echinoderm Zonation in Eastern United States. *International Congress of Stratigraphy and Geology of the Carboniferous VII. Compte Rendu*, **3**, 207–220.

LANE, N. G. 1971. Crinoids and Reefs. *Proceedings of the North American Paleontological Convention*, **J**, 1430–1443.

—— & SEVASTOPULO, G. D. 1987. Stratigraphic distribution of Mississippian camerate crinoid genera from North America and western Europe. *Cour. Forsch. Inst. Senckenberg*, **98**, 199–206.

The palaeobiogeography of the Blastoidea (Echinodermata)

JOHNNY A. WATERS

Department of Geology, West Georgia College, Carrollton, GA 30118 USA

Abstract: The 100 blastoid genera are known from some 1500 localities on every continent except Antarctica. Although fissiculates were more geographically widespread than spiraculates, the spiraculates tended to dominate most faunas numerically. The typical blastoid genus is monospecific, relatively short lived (range limited to some part of a single stage), relatively rare in terms of abundance, and geographically restricted to one depositional basin. Throughout their evolutionary history, blastoids were a component (sometimes minor, sometimes important) of echinoderm communities dominated by crinoids.

The palaeobiogeographic history of blastoids can be viewed in three phases. Phase one consisted of the initial radiation from eastern North America in the Late Ordovician, followed by an increase in diversity and geographic range in the Silurian and Devonian and culminating in a Late Devonian extinction. Phase two began with a Tournaisian (Lower Carboniferous) radiation, primarily in North America and Europe and ended in the sudden decline in blastoid diversity and abundance in the Upper Carboniferous. In phase three the biogeographic centre of the blastoids shifted eastward as the re-radiation of Upper Palaeozoic blastoids was concentrated in the Tethyan seaway not in North America and Europe. Upper Carboniferous–Permian blastoids were widespread but the most diversed faunas are found in southeast Asia and Australia.

Blastoids are an extinct class of echinoderms ranging in age from Caradoc (Upper Ordovician) to Kazanian (Upper Permian). They have been reported from every continent except Antarctica, although the North American and European faunas are much better known than faunas from other parts of the world. This distribution undoubtedly reflects collection bias, but after 150 years of collecting, should also reflect the actual biogeographic distribution of the class to some extent. One hundred genera of blastoids are currently recognized although not all have been formally described. New genera have recently been described from North America (Ausich & Meyer 1988) and others await description from North America, Britain and Ireland. It is reasonable to expect that additional fieldwork in South America, Africa and Asia will enhance the fragmentary blastoid record from those continents. Recent unpublished reports of a Fammenian blastoid from China and Upper Carboniferous specimens from Japan only serve to underscore this statement. The Upper Carboniferous is a promising interval for the discovery of new blastoids. Macurda & Mapes (1982) reported a fauna of very small Pennsylvanian blastoids from the United States found by washing very large quantities of shale and picking the residue. The potential for finding similar faunas in other localities is quite good.

Blastoids are divided into two orders, the Fissiculata and the Spiraculata, by the presence or absence of exposed hydrospire slits and spiracles. Monographs by Breimer & Macurda (1972) and Macurda (1983) have revised the taxonomy of the Fissiculata and have increased our understanding of their functional morphology, ontogeny, and phylogeny. In contrast, the Spiraculata have not received comparable treatment and the taxonomic foundation of this order is less firm. The Spiraculata are polyphyletic in origin (Horowitz *et al.* 1986) having originated from several fissiculate ancestors by bridging open hydrospire slits with the side plates and forming spiracular pores. Although the Spiraculata represents a grade in evolution rather than a clade, the concept is retained herein until an alternative classification is defined.

Waters (1988) has summarized the evolutionary palaeoecology of the blastoids. The typical blastoid genus is monospecific, relatively short lived (range limited to some part of a single stage), relatively rare in terms of abundance, and geographically restricted to one depositional basin. Many are predominantly known from a single locality. Blastoids have a fluctuating evolutionary history of local outbursts and declines that can be grouped into three phases of diversification and extinction. Blastoids are typically rare when compared with the most abundant crinoid genera found in a given fauna. However, blastoids did become the numerically dominant echinoderm in faunas from the Middle Devonian and Lower Carboniferous of North America, the Lower Carboniferous of Britain and Ireland, and the Upper Permian of Timor. In each case, blastoids become more diverse taxonomically and morphologically as well as the numerically dominant echinoderm after a major pertubation in the crinoid community.

Palaeobiogeography

Recent studies on the palaeobiogeography of the blastoids include Macurda (1967), Breimer & Macurda (1972, 1973) and Waters (1988). Previous plots of the geographic distribution of blastoids by Breimer & Macurda used modern continental distributions rather than palaeogeographic reconstructions and plotted data in periods of geological time rather stages. Waters (1988) used palaeogeographic reconstructions but only plotted the distribution of fissiculates and spiraculates rather than families as plotted herein. Compilations of biogeographic data are subject to error and the vagaries of correlating rock units on a world-wide basis. I have attempted to minimize errors by cataloging all references to blastoids and all localities where blastoids have been collected. Such catalogues are never complete, but the database currently lists over 500 publications and 1500 localities. The ranges of blastoid families and genera are given in Figs 1–7. The time scale used herein is the one utilized by the Decade of North American Geology Project by the Geological Society of America. The palaeogeography of the blastoids will be discussed within the framework of the three phases of blastoid evolution discussed in Waters (1988).

Phase 1

The oldest blastoid genus, *Macurdablastus*, was described by Broadhead (1984) from the Benbolt Formation (Caradoc; Upper Ordovician) in Tennessee (Fig. 8). Prior to this occurrence, blastoids were the only class of blastozoan echinoderms not reported from the Ordovician. The specimens are not well preserved and cannot be classified to order or family. Blastoids are noticeably absent from other diverse Ordovician echinoderm communities such as the Bromide Formation of Oklahoma. The Bromide is slightly older than the Benbolt of Tennessee, so perhaps the blastoids had not evolved by Bromide time. The echinoderm fauna from the Bromide contains over 11000 complete specimens belonging to 61 genera and 13 classes occurring in a variety of depositional environments (Sprinkle 1982). Both

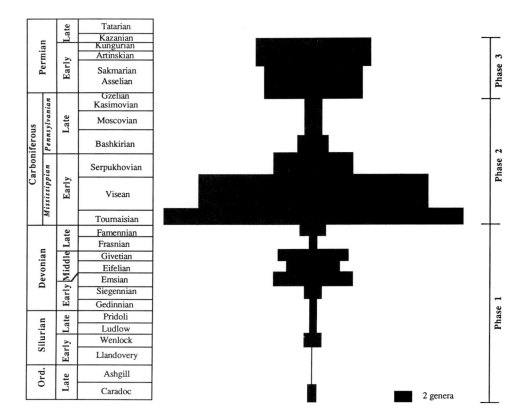

Fig. 1. Spindle diagram showing the generic diversity of blastoids in the Palaeozoic. Blastoids show three phases of diversification and extinction.

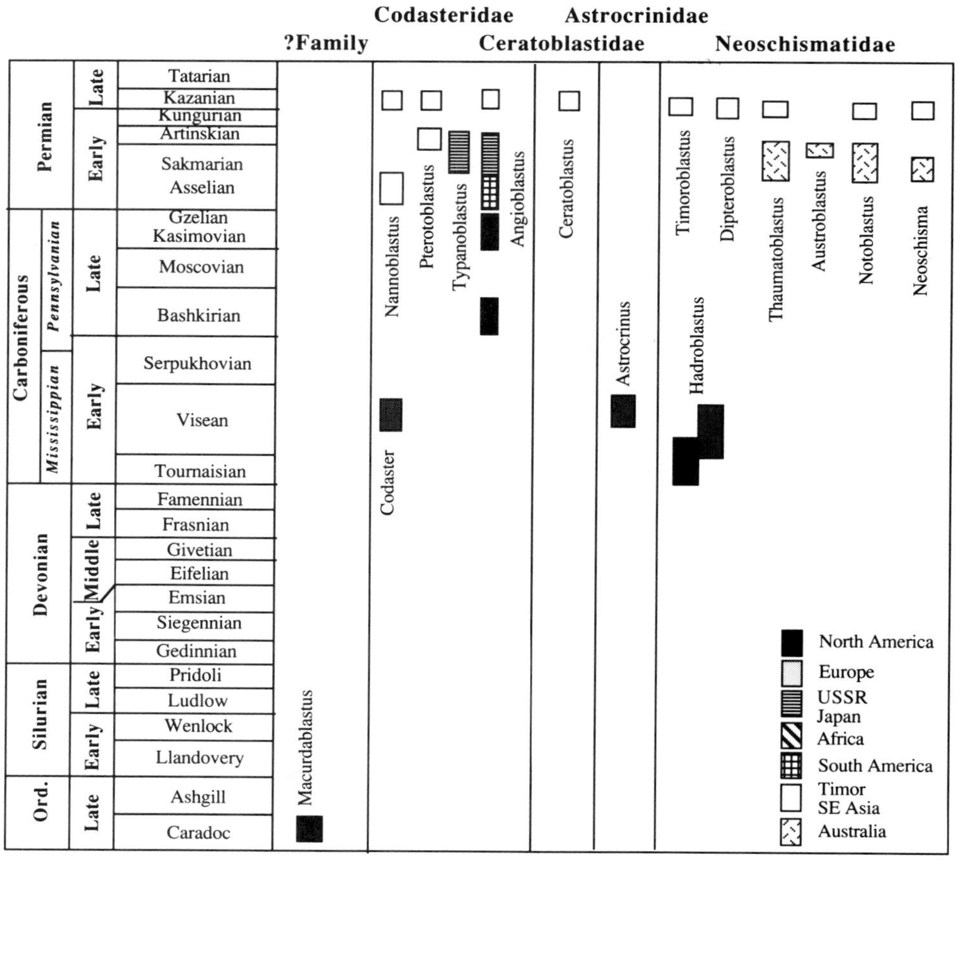

Fig. 2. Ranges and geographic distribution of genera in the Codasteridae, Ceratoblastidae, and Astrocrinidae, and Neoschismatidae well as *Macurdablastus* which cannot be assigned to family.

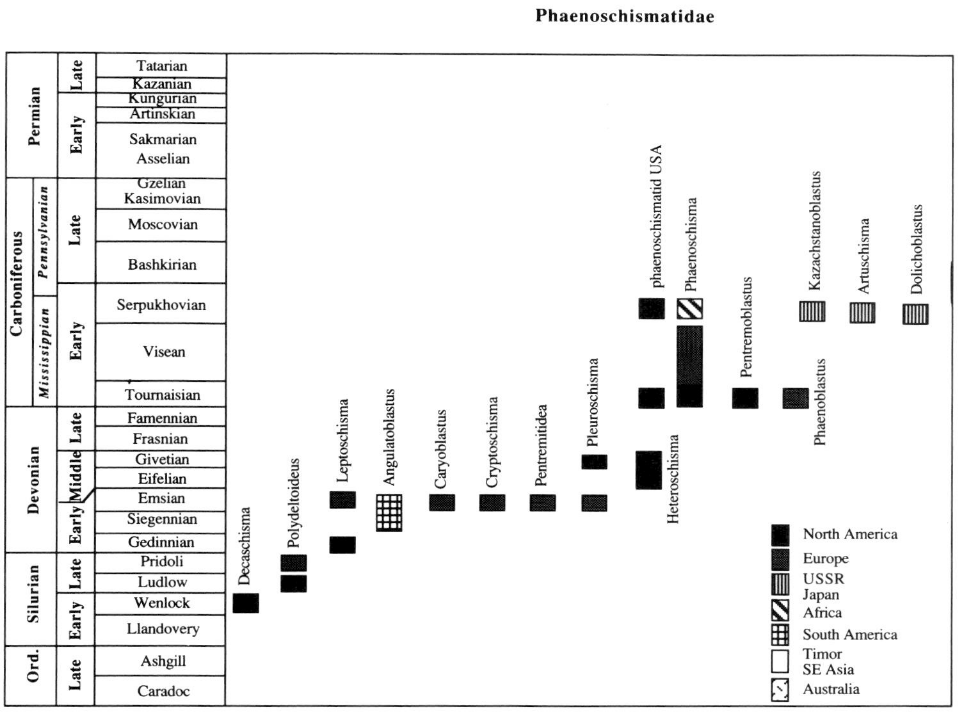

BLASTOIDS 341

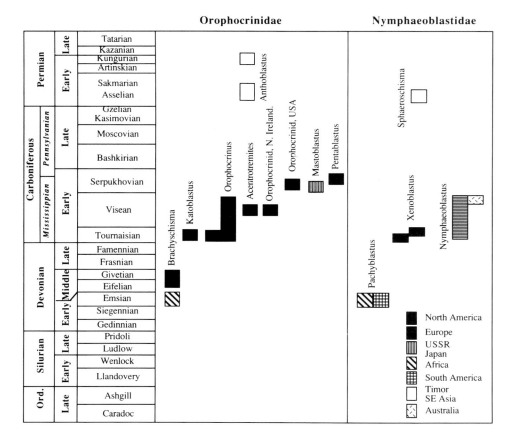

Fig. 3. Ranges and geographic distribution of genera of the Phaenoschismatidae.

Fig. 4. Ranges and geographic distribution of genera of the Orophocrinidae and Nymphaeoblastidae.

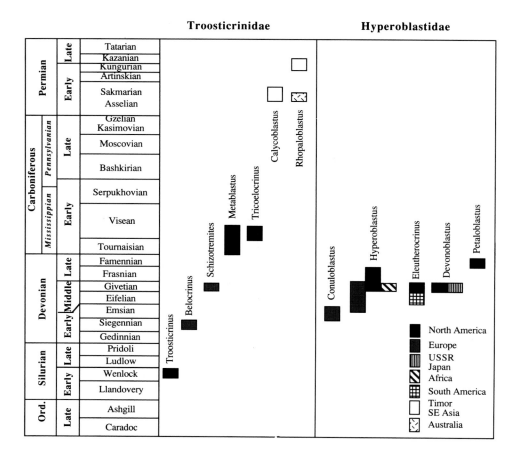

Fig. 5. Ranges and geographic distribution of genera of the Troosticrinidae and Hyperoblastidae.

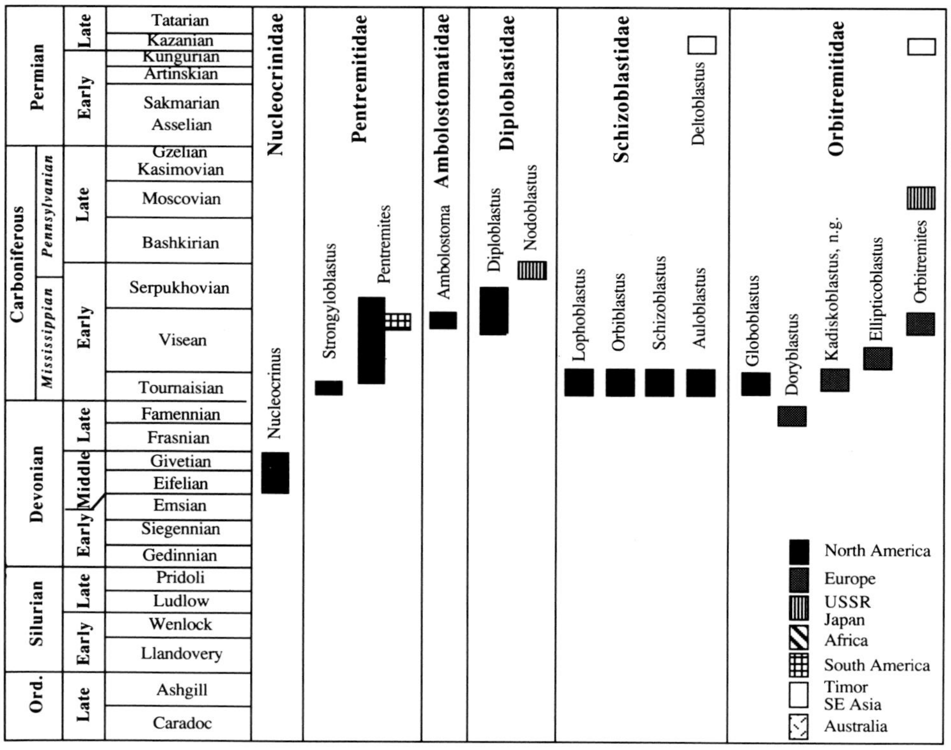

Fig. 6. Ranges and geographic distribution of genera of the Nucleocrinidae, Pentremitidae, Ambolostomatidae, Diploblastidae, Schizoblastidae, and Orbitremitidae.

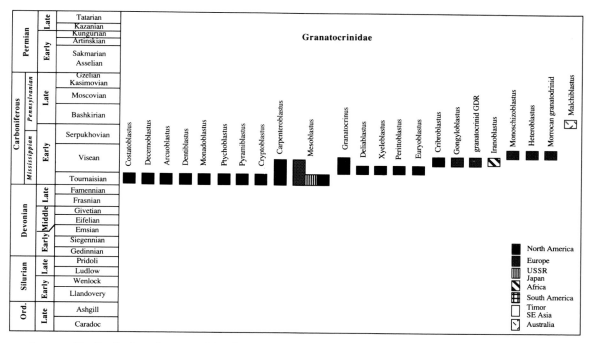

Fig. 7. Ranges and geographic distribution of genera of the Granatocrinidae.

the Bromide and Benbolt formations contain many blastozoan taxa or ecological correlatives such as cystoids, paracrinoids, parablastoids, and hybocrinid inadunate crinoids which occupied the lower tiers of the communities. Blastoids did not become diverse or abundant until many of these morphologically similar taxa declined diversity and abundance.

Two genera of blastoids (*Decaschisma*, a phaenoschismatid, and *Troosticrinus*, a troosticrinid) are known from the Wenlock (Fig. 8). Both of these genera are the oldest members of their respective families and orders, and are well known in eastern North America. The only documented Ludlow blastoid is *Polydeltoideus*, a phaenoschismatid which is known from Oklahoma (Figs. 3 and 9). *Polydeltoideus* is also known from the Pridoli of Czechoslovakia and is the only Silurian blastoid reported from Europe. It is the oldest blastoid to have a transcontinental distribution and range spanning more than one stage.

The eastward migration is repeated in the Gedinnian (Figs 3 and 9) as *Leptoschisma*, another phaenoschismatid from eastern

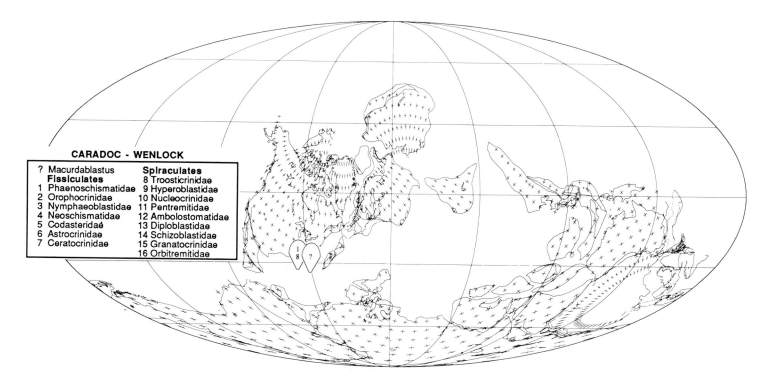

Fig. 8. Distribution of Cardoc and Wenlock blastoids. *Macurdablastus* is the lone blastoid known from the Caradoc. Only North American genera are known from the Wenlock. Numbers correspond to blastoid families as outlined in the legend. This numbering system will be used on all the palaeogeographic maps.

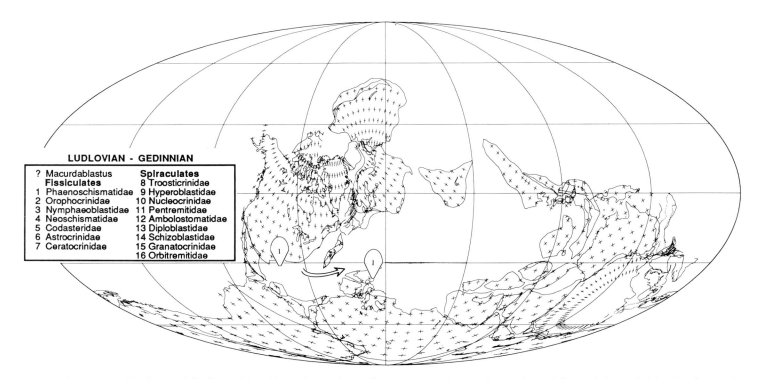

Fig. 9. Distribution of Ludlow and Gedinnian blastoids. Only North American representatives are known for each interval, but *Polydeltoideus* (Ludlow) migrated to Czechoslovakia in the Pridoli as did *Leptoschisma* (Gedinnian) in the Emsian.

North America, is found in the Emsian of Czechoslovakia. Siegennian blastoids are known only from France (*Belocrinus*, a troosticrinid) and South America (*Angulatoblastus*, a phaenoschismatid). Emsian blastoids are well known from a variety of localities (Figs 10 and 11) and generally yield faunas with several co-existing genera for the first time in the evolutionary history of blastoids. Spanish Lower Devonian blastoids represent the best Emsian fauna known. Summarized by Breimer & Dop (1974),

this fauna consists of species of *Cryptoschisma, Pentremitidea, Pleuroschisma, Caryoblastus* (all phaenoschismatids), *Hyperoblastus* (a hyperoblastid) and *Conuloblastus* (a hyperoblastid with features intermediate between the two orders). *Brachyschisma*, the oldest orophocrinid, is also known from South Africa. *Angulatoblastus* ranges into the Emsian in Bolivia. *Pachyblastus*, the oldest nymphaeoblastid, is known from conspecific occurrences in Emsian rocks from Bolivia and South Africa, a rare

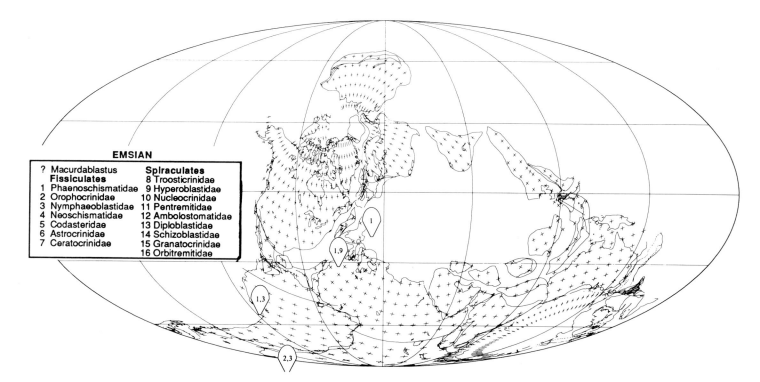

Fig. 10. Distribution of Emsian blastoids. European faunas are best known but fissiculates are reported from South America and South Africa.

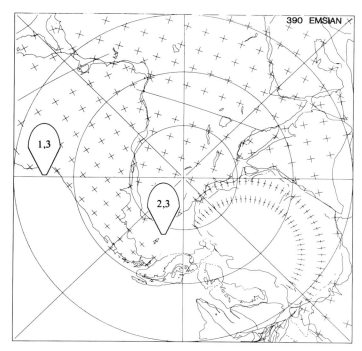

Fig. 11. Distribution of Emsian blastoids from South America and South Africa plotted on a south polar reconstruction. The South African faunas, which include the earliest orophocrinids and nymphaeoblastids, occur close to the Emsian south pole.

case in the blastoids. The South African occurrences are noteworthy because the Emsian continental reconstruction place them near the Devonian south pole (Fig. 11).

The Givetian in the northeastern United States and southern Ontario (Fig. 12) has yielded an abundant fauna of *Brachyschisma*, *Heteroschisma*, *Pleuroschisma*, *Nucleocrinus* (a nucleocrinid), *Hyperoblastus*, *Devonoblastus* and *Eleutherocrinus* (all hyperoblastids). Most elements of this fauna probably migrated into North America from the Emsian of Spain and South America following a southerly route around the emerging Appalachian highlands in eastern North America. *Eleutherocrinus* is first known from the Eifelian of Bolivia and also migrated into this fauna from South America. These occurrences of *Eleutherocrinus* are conspecific. Among the Givetian blastoids, only *Nucleocrinus* (Eifelian–Givetian) and *Devonoblastus* originated in North America. In addition to these faunas, *Hyperoblastus* is also known from the Givetian of New Mexico, and *Devonoblastus* is reported from China. Most Givetian blastoids did not survive into the Frasnian (Fig. 1) and only *Hyperoblastus* survived into the Fammenian. The Late Devonian extinction event in the blastoids apparently occurred at the Givetian–Frasnian boundary rather than the Frasnian–Fammenian as is the case in many marine invertebrates. Although they are not well known as a group, Fammenian blastoids undergo a diversification and geographic expansion (Fig. 13). *Doryblastus* (an orbitremitid) and *Petaloblastus* are known Germany; an unidentified phaenoschimatid is known from Great Britain; and an unidentified spiraculate has been collected in China.

Phase 2

Major changes in dominance and diversity of blastoids, particularly in North America and Europe, accompanied fundamental shifts in the sedimentation patterns and major reorganizations in the echinoderm communities during the Lower Carboniferous. In North America, Kinderhookian (Tournaisian) faunas are known from shallow water carbonates and include *Orophocrinus* and *Pyramiblastus* (a granatocrinid) from the Hampton Formation of Iowa and *Pentremoblastus* (a phaenoschismatid) and a new genus of spiraculate from the McCraney Formation of Missouri. Deeper water occurrences are found in facies associated with the Waulsortian buildups in the Lodgepole Formation in Montana and include an abundant, diverse (6 genera) undescribed fauna dominated by granatocrinids (Sprinkle & Gutschick 1983). Kinderhookian faunas are significant because they illustrate

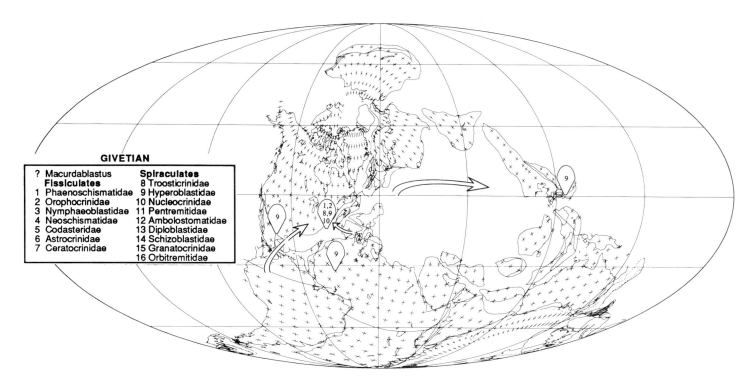

Fig. 12. Distribution of Givetian blastoids. The North American fauna is best known and is composed of genera migrating in from Europe and South America and well as endemic genera.

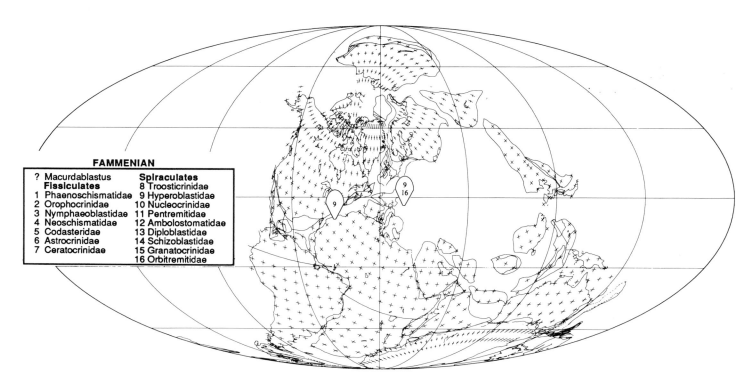

Fig. 13. Distribution of Fammenian blastoids showing the contraction in geographic range and diversity associated with Late Devonian extinction.

blastoid diversification into a variety of environments after the Late Devonian extinction and contain the oldest elements of many important Late Palaeozoic blastoid lineages. The oldest granatocrinids, which are largely responsible for the taxonomic diversity that dominates the Tournaisian–middle Visean blastoid faunas from North America, date from this interval.

During the Osagean (late Tournaisian; early Visean) of North America (Fig. 14), blastoids continued to occur in deep water facies, but became very diverse in shallow water facies. Blastoids associated with the Waulsortian buildups in the Lake Valley Formation in New Mexico include *Phaenoschisma*, *Orophocrinus*, *Hadroblastus*, *Cryptoblastus*, *Schizoblastus* and *Mesoblastus*. This fauna is equally divided into fissiculates and spiraculates, and is significant in that it contains the oldest neoschismatid (*Hadro-*

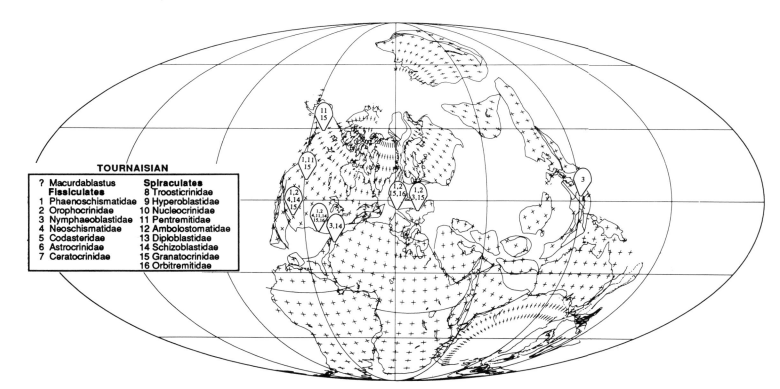

Fig. 14. Distribution of Tournaisian blastoids showing the enormous diversity found in North American and European faunas. This map includes faunas from the Burlington Limestone and correlatives in North America that persisted into the Lower Visean. This interval ended in an extinction event within the Visean.

blastus), a significant Upper Palaeozoic fissiculate family, and contains the only record of *Mesoblastus* in North America.

During the Osagean, monobathrid camerate crinoid communities reached a zenith on the broad shallow water carbonate platforms such as the Burlington Shelf and correlatives. Blastoids were an integral part of this community and attained their maximum diversity (17 genera) in the Burlington. These faunas are dominated by granatocrinids (*Decemoblastus*, *Arcuoblastus*, *Dentiblastus*, *Monadoblastus*, *Ptychoblastus*, and *Cryptoblastus*) which are more diverse than in the Kinderhookian, schizoblastids (*Lophoblastus*, *Orbiblastus*, *Schizoblastus*, and *Auloblastus*), pentremitids (*Pentremites*), and orbitremitids (*Globoblastus*), all globose spiraculates. Sixteen Osagean granatocrinid genera are known from North America (Fig. 7), but most were endemic; only *Mesoblastus* had a cosmopolitan range. *Globoblastus* is the lone North American representative of the orbitremitids which dominate European Visean blastoid communities (Fig. 6). Other occurrences of this approximate age are the faunas from the Banff Formation in Alberta and from the Brooks range of Alaska. These faunas are dominated by granatocrinid spiraculates including *Cryptoblastus*. The oldest occurrences of *Strongyloblastus* and *Pentremites* (both pentremitids) are found in these faunas. *Pentremites* migrated eastward into the Eastern Interior Basin in the later Tournaisian and Visean and later dominated North American Visean echinoderm communities. (Figs 15 and 16).

Elements of Kinderhookian fissiculate families (the phaenoschismatids and orophocrinids) are both well known from the Burlington and equivalents. Blastoid genera found in coeval deltaic deposits are mostly fissiculates not associated with the platform carbonate communities. A notable component of these deltaic faunas is *Xenoblastus* (a nymphaeoblastid) also known from the Tournaisian of Belgium. It represents the first occurrence of a nymphaeoblastid in the northern hemisphere. European late Tournaisian (Courceyan) blastoids are best known from Tournai, Belgium (Macurda 1967). The fauna consists of *Mesoblastus* and four fissiculates, *Katoblastus*, *Orophocrinus* (both orophocrinids), *Phaenoblastus* (a phaenoschismatid), and *Xenoblastus*. Additional coeval faunas consisting of *Orophocrinus*, a new genus of orbitremitid, and two fragmentary genera are known from southern Ireland and Britain (Waters & Sevastopulo 1984). Most of the blastoids are endemic even though they are found in the same facies; only *Orophocrinus* is found in Belgium and the British Isles. The new orbitremitid from Ireland is ancestral to the orbitremitid lineage that dominates Visean faunas in the British Isles.

During the latest Tournaisian (upper Courceyan) and lower Visean (Chadian), buildups of the Walsortian facies were widespread across Europe. Blastoids are commonly found in the flank facies associated with diverse crinoid communities. Faunas include fissiculates, *Orophocrinus* and *Phaenoschisma*, and the more abundant spiraculates, *Ellipticoblastus* (an orbitremitid), and *Mesoblastus*. Blastoids are typically rare compared to crinoids and were a minor element of the crinoid communities as they were in the Waulsortian buildups in North America. *Mesoblastus* is the most geographically widespread of the genera associated with the Waulsortian and is also found in Waulsortian facies in France (Macurda & Racheboeuf 1975), New Mexico and unknown lithologies in China. Curiously, *Mesoblastus* has not been found in Ireland despite the widespread occurrence of the Waulsortian facies there. *Nymphaeoblastus* (a nymphaeoblastid) has been described from the Tournaisian of northern Japan.

A major blastoid extinction event (Ausich *et al.* 1988) occurred in the early Visean after the deposition of the Burlington Limestone in North America. Diplobathrid camerates largely became extinct, and only six families of monobathrids and less than 25% of the blastoids present survived into the middle Visean (Fig. 1).

North American middle Visean granatocrinids are less significant than their precursors (Figs 14 and 15). Although six genera are known (*Granatocrinus*, *Deliablastus*, *Xyeleblastus*, *Perittoblastus*, *Euryoblastus*, and *Cribroblastus*), they are relatively rare and endemic. The most abundant Visean blastoid in North America is *Pentremites*. Having originated in the Tournaisian of western North America, *Pentremites* had migrated throughout the United States and into South America by the Visean (Fig. 16). Waters *et al.* (1982, 1985) have summarized the evolution

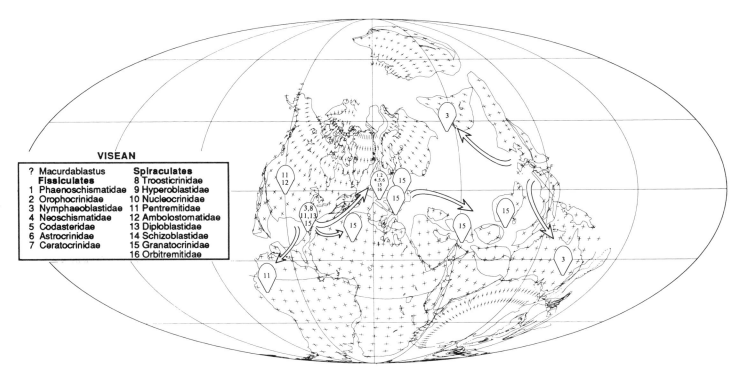

Fig. 15. Distribution of Visean blastoids showing the decrease in diversity but dramatic increase in geographic distribution of the faunas. Most Visean blastoids migrated from centres of origin in North America and Europe, but the Nymphaeoblastids diversified from a centre in the eastern Tethys.

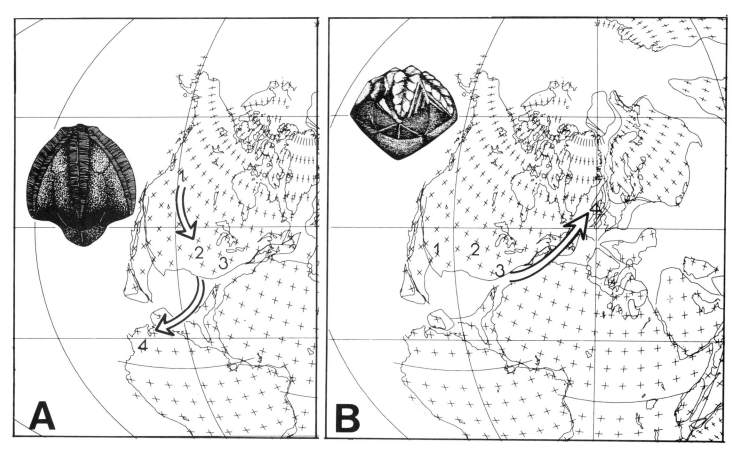

Fig. 16. Distribution of *Pentremites* and *Hadroblastus* showing the eastward migration of the genera (**A**) From Tournaisian localities in Alaska and Alberta (1), *Pentremites* migrated into the central United States by the early Visean (2; Burlington Limestone) and then became very widespread throughout the US, particularly the eastern half, in the middle and late Visean. *Pentremites* is also known from a Visean occurrence in Colombia, South America (4). (**B**) The oldest occurrences of *Hadroblastus* are the Lake Valley Formation (1) in New Mexico and the Burlington Limestone (2) in the midcontinent (late Tournaisian–Early Visean). Hadroblastus has also been found in the middle–late Visean Floyd Shale in Georgia (3) and late Visean localities in Ireland and Scotland (4).

of *Pentremites* and concluded that major speciation events and changes in diversity in this genus were related to changes in Mississippian crinoid communities and were triggered by the ability of the genus to adapt to changing sedimentological regimes. *Pentremites* was a minor member of Osagean carbonate platform communities that survived and appeared to benefit from extinctions among the camerates and other blastoids. *Pentremites* became significantly more abundant in Meramecian carbonate platform communities, invaded deltaic communities for the first time, and successfully competed with advanced Chesterian inadunate crinoids. It has been collected from over 1000 localities in North America. With the further demise of camerate crinoids at the Meramec–Chester boundary (late Visean), *Pentremites* again benefitted from decreased competition and became one of the most abundant and diverse echinoderms of the Chesterian (late Visean–Serpuhkovian).

Although blastoids dominated many echinoderm faunas at this time, middle to late Visean generic diversity in North America was substantially lower than in the Tournaisian–lower Visean. The granatocrinids, nymphaeoblastids, neoschismatids, diploblastids and schizoblastids disappeared from North American faunas after this interval although they migrated into Europe and the Tethyan seaway. The distribution of *Hadroblastus* is a good example of this migration (Fig. 16). From its apparent origins in the lower Visean Lake Valley Formation in New Mexico, *Hadroblastus* migrated through the Eastern Interior Basin in Burlington age deposits, into the Middle Visean Fort Payne Chert of Kentucky, and into the Late Visean Floyd Shale of Georgia as well as late Visean sediments in northwestern Ireland and Scotland. Six genera of neoschismatids descended from *Hadroblastus* have been found in the Permian of Australia and Timor. Tethyan descendants of the North American diploblastids and schizoblastids later dominated their respective echinoderm faunas in Kazachstan and Timor.

Middle Visean blastoids are rare in Europe. They are known only from fragments in the Bundoran Shale, a prodeltaic shale in western Ireland and *Gongyloblastus* (a granatoacrinid) from France. Late Visean blastoids are well known from Ireland and Britain. These blastoids exhibit extremes in environmental setting, range and abundance. *Monoschizoblastus* is a very abundant granatocrinid endemic to northwestern Ireland. It is primarily associated with shallow water reefs in the Dartry Limestone but is also found in non-reef facies. *Orbitremites*, *Astrocrinus* (a diminutive stemless fissiculate) and *Codaster* (ancestral to the very successful Late Palaeozoic codasterids) numerically dominate echinoderm faunas from late Visean reefs in Britain. *Acentrotremites* is a rare orophocrinid endemic to Britain.

Late Visean blastoids in the British Isles also migrated into very shallow environments, a precursor of Late Carboniferous faunas. *Heteroblastus* is a granatocrinid known from a single locality near Hexham, Northumberland, occurring in a shale with siderite nodules and a molluscan fauna. Irish faunas indicate the same trend. The Menymore Formation at Gleniff, County Sligo, contains *Hadroblastus*, *Orbitremites* and *Astrocrinus*. All these blastoids are very small, probably paedomorphic and are associated with a diverse echinoderm fauna including numerous inadunate crinoids, microcrinoids, asteroids, cyclocystoids and ophiuroids (Waters & Sevastopulo 1984). A second Menymore locality at Cashel, County Fermanagh, contains *Orbitremites*, *Phaenoschisma*, and a new orophocrinid genus in a fauna dominated by cephalopods and rugose corals. Collections from Lisdowney, County Kilkenny, include *Orbitremites* (very abundant), *Codaster*, and *Astrocrinus* and occur with an atypical crinoid community.

An undescribed granatocrinid blastoid has been collected from the German Democratic Republic (Macurda, personal communication). Additional Visean blastoids are known from Morocco (an undescribed granatocrinid), Iran (*Iranoblastus*, a granatocrinid), China (*Mesoblastus*), and *Nymphaeoblastus* which occurs in Kazakhstan and Australia. The distribution of Visean species of *Nymphaeoblastus* represent an eastern Tethyan dispersion from the Tournaisian ancestor found in Japan.

Phase 3

Blastoids also underwent another extinction event at the Visean–Serpukhovian boundary (middle Chester in North America). Less than one-third of the Visean genera survived into the Serpukhovian (Fig. 1). In the Early Serpukhovian, *Pentremites* totally dominated echinoderm faunas in North America (Fig. 17), but declined to a single species in the early Pennsylvanian (mid-Serpukhovian). The decline and extinction of *Pentremites* at this time was probably the result of the continuing expansion of poteriocrine inadunate crinoids and the resulting extinction of the majority of Chesterian echinoderms. Arendt *et al.* (1968) described a fauna of *Nodoblastus* (a diploblastid), three phaenoschismatids (*Dolichoblastus*, *Kazachstanoblastus*, and *Artuschisma*), and the orophocrinid *Mastoblastus* from the Serpukhovian of Kazachstan, USSR. *Nodoblastus* dominated the echinoderm community in a fashion similar to *Pentremites* or *Orbitremites*. *Pentablastus* has been collected from the Serpukhovian of Spain and *Phaenoschisma* is known from Algeria at this time. These faunas were derived from lineages migrating into the Tethyan seaway from North America and Europe and contain the last occurrences of the phaenoschismatids and the diploblastids.

Blastoids are extremely rare in Late Carboniferous rocks of North America; the few specimens described are fissiculates (see Breimer & Macurda 1972 and Macurda & Mapes 1982). Additional undescribed blastoid faunas from the mid continent of the United States include a new genus of orophocrinid, a neoschismatid related to *Hadroblastus*, and perhaps a phaenoschismatid. These blastoids are very small (probably paedomorphic), fragmentary and poorly preserved. Nonetheless, they do significantly increase the diversity of Late Carboniferous blastoids in North America. Some of these faunas are associated with reefal deposits but others continue the palaeoecological shift to very shallow water molluscan dominated deposits similar to those of the late Visean cited above. The most widespread late Palaeozoic blastoid is *Angioblastus* (a codasterid) which occurs on Ellesmere Island in the Canadian Arctic, Oklahoma and Texas. The successful European orbitremitid lineage (which continues into the Permian) is represented by *Orbitremites derbiensis moscovi* Arendt from the Moscovian of the USSR (Fig. 18). *Angioblastus* and *Orbitremites* are the only known blastoids with both Carboniferous and Permian representatives. The only other described Late Carboniferous blastoid is *Malchiblastus*, a granatocrinid from the Bashkirian of Queensland, Australia.

Permian blastoids are widespread (Figs 19 and 20) but the most diverse faunas are found in southeast Asia and Australia. The fissiculates are the most cosmopolitan, with *Angioblastus* being known from Bolivia, the Urals, and Indonesia. Poorly known (spiraculate or fissiculate) blastoids have been found in the Canadian archipelago and Sicily (spiraculate). Faunas from Timor are known from the Sakmarian–Asselian and the Kazanian (Fig. 20). Although these faunas are very well known from museum collections, their palaeoecology and stratigraphic relationships are very poorly understood. Australian faunas are intermediate in age (Sakmarian–Artinskian) to the Timor faunas and occur in both eastern and western regions of the continent. The Timor fauna is the most diverse (15 genera) and abundant of faunas. Fissiculates are the most diverse (13 genera) but spiraculates dominate in terms of abundance. *Deltoblastus* is known from tens of thousands of specimens. Australia and Timor were part of the same continental block in the Permian, and several of the genera are found in both faunas (*Notoblastus*, *Thaumatoblastus*,

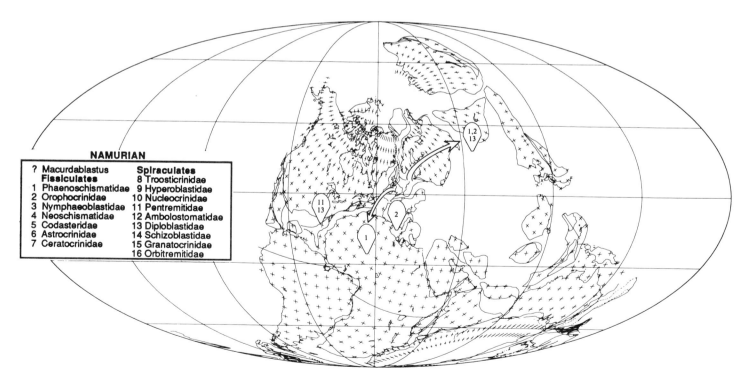

Fig. 17. Distribution of Namurian (Serpukhovian) blastoids showing the decrease in diversity in faunas and retraction of geographic range. The most significant faunas are from Kazakhstan; North America faunas are largely remnants of Visean lineages.

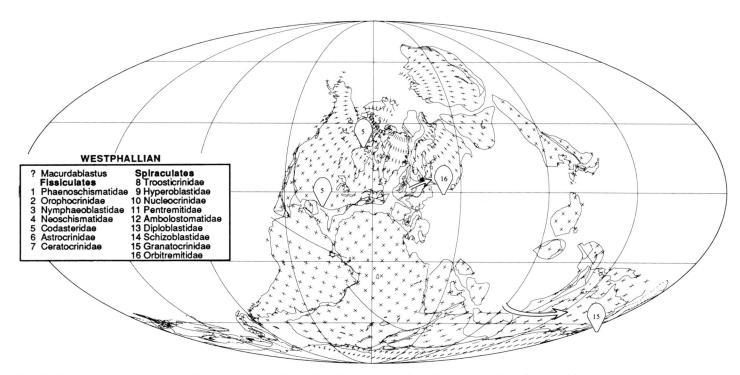

Fig. 18. Distribution of Westphalian blastoids showing the continued decrease in diversity and retraction of geographic range.

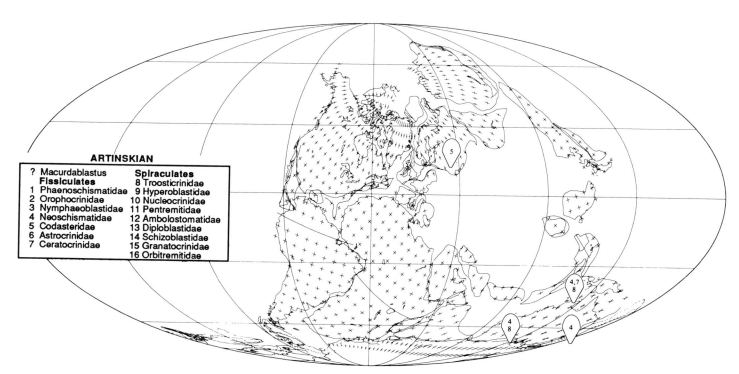

Fig. 19. Distribution of Artinskian blastoids showing the shift in geographic distribution to Tethys, primarily Australia and Timor. Although Timor and Australia were on the same continental block blastoid faunas show considerable endemism.

Neoschisma, Rhopaloblastus), but others are conspicuously absent from one fauna or the other. *Angioblastus* and *Deltoblastus*, among other genera, have not been reported from Australia, and *Australoblastus* have not been found in Timor. These reasons for this local endemism are unclear. *Deltoblastus* has also reported from Permian deposits in Kashmir (Gupta & Webster 1976), but Webster (personal communication, 1987) now believes the occurrence is suspect. Blastoids are unknown from rocks of Tatarian age.

Summary

Blastoids apparently originated in North America in the Caradoc and were largely restricted to North America through the Silurian. They show an increase in diversity and geographic expansion through Lower Carboniferous time and subsequent decline in the Upper Carboniferous and Permian. The rise in blastoid diversity is punctuated by extinction events (and contraction of geographic range) in the Late Devonian, Lower Carboniferous, and Upper

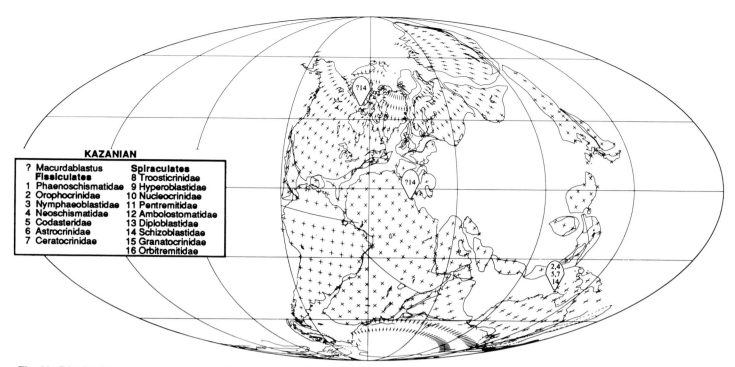

Fig. 20. Distribution of Kazanian blastoids. The fauna from Timor was very abundant and diverse and represents the last successful blastoid community before the class went extinct. The report from Sicily is probably *Deltoblastus*, but this is questionable. The occurrence from the Canadian Arctic is a blastoid unassignable to order or family.

Carboniferous. The history of blastoids can be viewed in three phases. After the initial occurrence, blastoids were largely restricted to North America through the Silurian. Both spiraculate (troosticrinid) and fissiculate (phaenoschismatid) families are found in the Silurian of North America, but only the phaenoschismatids migrated to Europe. This interval was followed by an increase in diversity and geographic range in the Devonian and culminated in the Late Devonian extinction. Phase two included the widespread diversification of blastoids in the Lower Carboniferous. In some cases they became the dominant echinoderm in these communities. The extinction within the lower Visean is particularly striking because it occurred during the interval of maximum blastoid diversification. Ausich *et al.* (1988) have studied this extinction event and have suggested that it was a part of a global reorganization of Visean pelmatozoan echinoderm communities. This phase ended in the sudden decline in blastoid diversity and abundance in the Upper Carboniferous. Phase three involved the re-radiation of blastoids in the Tethyan seaway during the Permian before their final extinction in the Upper Permian.

Among blastoid genera, fissiculates are more geographically widespread and diverse within a fauna, but spiraculates tend to dominate in terms of abundance. The Mississippian of the United States was the zenith of blastoid diversity and abundance. Globose spiraculates (granatocrinids, schizoblastids, and pentremitids) living in platform carbonate communities account for the bulk of this diversity increase. Fissiculates lived in clastic-dominated communities and were less diverse and abundant. One spiraculate, *Pentremites*, successfully invaded clastic communities and became the most successful of all the blastoids. In the Lower Carboniferous of Europe, blastoids successfully adapted to a variety of environments: shallow marine shelf, Waulsortian facies (deep water build-ups), shallow reefal facies and clastic dominated facies. Fissiculates are generally more diverse in a fauna, but spiraculates numerically dominate. The most successful group of European blastoids are the Orbitremitidae (Orbitremitid, n.g., *Ellipticoblastus*, *Monoschizoblastus*, and *Orbitremites*). The evolutionary history of this family is strikingly similar to that of the pentremitids in North America.

Blastoids suffered substantial decrease in diversity in the Upper Carboniferous as advanced cladid inadunates increased their domination of echinoderm communities. The majority of known genera are fissiculates, but the only truly abundant blastoid of this interval is *Nodoblastus*, a spiraculate. Permian blastoids from the Tethyan seaway represent a resurgence of blastoid diversity. The best known faunas from Australia and Timor are numerically dominated by *Deltoblastus*, a spiraculate, although fissiculates account for the bulk of the generic diversity.

Most blastoid families orginated either in North America or Europe and migrated away from these centres of origin into tropical Palaeozoic seaways (Iapetus and Tethys). Two exceptions to this trend can be documented. The oldest representatives of the Orophocrinidae (*Brachyschisma*) and the Nymphaeoblastidae (*Pachyblastus*) are found in Emsian deposits in South Africa. Continental reconstructions place this occurrence near the Emsian south pole and suggest a cold water origin for these families. The orophocrinids, including *Brachyschisma*, later migrated into North America and Europe where they became important members of Carboniferous blastoid communities. A lone representative, *Anthoblastus*, survived into the Permian.

Pachyblastus is also known from Emsian deposits in Bolivia. Later members of the Nymphaeoblastidae migrated into North America and Europe (*Xenoblastus*), and Japan (*Nymphaeoblastus*) by the Tournaisian. From its origin in Japan, *Nymphaeoblastus* migrated north and south in the eastern Tethys in the Visean and is known from Kazakhstan and Australia. *Sphaeroschisma*, a Permian descendant, is also known from Australia. These two families are the only blastoid lineages whose centres of origin were outside the margins of the Iapetus Ocean, although both families are represented in North America and Europe.

My blastoid studies have been made possible by grants from the National Science Foundation (DEB 77–23375), the Petroleum Research Fund, the American Philosophical Society, the Eppley Foundation for Scientific

Research and the West Georgia College learning Resources Committee. The support of these organizations is gratefully acknowledged. The work has benefitted from many discussions with A. S. Horowitz and D. B. Macurda, Jr., G. D. Sevastopulo, A. B. Smith and C. R. C. Paul.

References

ARENDT, Y. A., BREIMER, A. & MACURDA, D. B. 1968. A new blastoid fauna from the lower Namurian of North Kazachstan (U.S.S.R.) *Proceedings of Koninkl. Nederl. Akademis Van Wetenschappen-Amsterdam*, **71**, 159–174

AUSICH, W. I. & MEYER, D. L. 1988. Blastoids from the late Osagean Fort Payne Formation (Kentucky and Tennessee). *Journal of Palaeontology*. **62(2)**, 269–282.

——, —— & WATERS, J. A. 1988. Middle Mississippian blastoid extinction event. *Science*, **240**, 796–798.

BREIMER, A & DOP, A. J. 1974. An anatomic and taxonomic study of some lower and middle Devonian blastoids from Europe and North America. *Proceedings of Koninkl.Nederl.Akademie Van Wetenschappen-Amsterdam*, **78**, 39–217.

—— & MACURDA, D. B., JR. 1972. The phylogeny of the fissiculate blastoids. *Koninkl. Nederl. Akademie Van Wetenschappen-Amsterdam, Eerste Reeks*, Deel **26**.

—— & —— 1973. Palaeozoic Blastoids. *In*: HALLAM, A. (ed.), *Atlas of Palaeobiogeography*. Elsevier Pub. Co. Amsterdam. 207–212.

BROADHEAD, T. W. 1984. *Macurdablastus*, a middle Ordovician blastoid from the southern Appalachians. *The University of Kansas Paleontological Contributions Paper*, **110**, 1–10.

GUPTA, V. J. & WEBSTER, G. D. 1976. *Deltoblastus batheri* from the Kashmir Himalaya. *Rivista Italiana di Paleontologia*, **82**, 279–284.

HOROWITZ, A. S., MACURDA, D. B., JR. & WATERS, J. A. 1986. Polyphyly in the Pentremitidae (Blastoidea, Echinodermata). *Geological Society of America Bulletin*, **97**, 156–161.

MACURDA, D. B., JR. 1967. The Lower Carboniferous (Tournaisian) blastoids of Belgium. *Journal of Paleontology*, **41**, 455–486.

—— 1983. Systematics of the fissiculate Blastoidea. *University of Michigan Museum of Palaeontology Papers on Paleontology*, **22**.

—— & MAPES, R. 1982. The enigma of Pennsylvanian Blastoids. *Proceedings of the Third North American Paleontological Conference*, **2**, 343–345.

—— & RACHEBOEUF, P. R. 1975. Devonian and Carboniferous spiraculate blastoids from Brittany (France). *Journal of Paleontology*, **49**, 845–855.

SPRINKLE, J. (ed.) 1982. Echinoderm faunas from the Bromide Formation (Middle Ordovician) of Oklahoma. *The University of Kansas Paleontological Contributions Monograph*, **1**.

—— & GUTSCHICK, R. C. 1983. Early Mississippian blastoids from western Montana. *Geological Society of America, Program with Abstracts*, **15(6)**, 693.

WATERS, J. A. 1988. The evolutionary paleoecology of the Blastoidea (Echinodermata) in PAUL, C. R. C. & SMITH, A. B. (eds) *Echinoderm Phylogeny and Evolutionary Biology*. Oxford University Press.

—— & SEVASTOPULO, G. D. 1984. The stratigraphical distribution and palaeoecology of Irish Lower Carboniferous blastoids. *Irish Journal of Earth Science*, **6**, 137–154.

——, BROADHEAD, T. W. & HOROWITZ, A. S. 1981. The evolution of *Pentremites* (Blastoidea) and Carboniferous crinoid community succession. *In*: LAWRENCE, J. M. (ed.) *International Echinoderms Conference, Tampa Bay*. A. A. Balkema, Rotterdam, 133–138.

——, HOROWITZ, A. S. & MACURDA, D. B., JR. (1985). Ontogeny and phylogeny of the Carboniferous blastoid *Pentremites*. *Journal of Paleontology*, **59(3)**, 701–712.

Late Palaeozoic bryozoan biogeography

JUNE R. P. ROSS[1] & CHARLES A. ROSS[2]

[1] *Department of Biology, Western Washington University, Bellingham, WA 98225 USA*
[2] *Chevron U.S.A., Post Office Box 1635, Houston, Texas 77251 USA*

Abstract: Late Palaeozoic (Carboniferous and Permian) distributions and times of dispersal of bryozoans, fusulinids, and other marine shelf invertebrates were greatly influenced by the effects of land masses converging to form a supercontinent. This convergence modified oceanographic circulation patterns as well as environmental, climatic, and temperature patterns. Sea level fluctuations of 75 to 100 metres having about two million years duration also strongly affected faunal dispersal patterns. At the end of the Early Carboniferous (late Visean through early Bashkirian), the assemblage of Lesser Pangaea disrupted the tropical, subtropical and warm temperate marine shelf faunas which included cryptostomes, trepostomes, and cystoporates. In the middle of the Early Permian (Leonardian), the assemblage of Greater Pangaea further disrupted the marine shelf faunas giving rise to biogeographic provinces and finally realms. In the assembling of Pangaea, the northward transport of plates and cratons caused the northern marine shelf of Pangaea to be moved into cooler and cooler waters. Tectonic movements caused cool ocean currents to be redirected toward the equator along the western marine shelves of Pangaea. Faunal diversity greatly increased apparently as a result of the flow of consistently warm equatorial currents along the eastern marine shelves of Pangaea.

The earliest known sparse occurrences of ectoproct bryozoans are in the Tremadoc. Continuing through the early part of the Ordovician, bryozoan faunas show low diversity until the Llandeilo at which time the three main Palaeozoic orders, Cryptostomata, Trepostomata, and Cystoporata, are widely represented by many different species, genera, and families. In the Caradoc, bryozoans were dominant keystone species in extensive and diverse faunas encompassing many marine environments on continental shelves, shelf margins, open platforms and platform edges (Ross 1985). These suspension feeders were among the earliest taxa to form biohermal and reefal communities and to comprise extensive flank reef faunas. Through the remainder of the Early Palaeozoic and in the Middle and Late Palaeozoic, bryozoans continued to be dominant members of faunas in widely different environments. During the Late Palaeozoic, further diversification of bryozoan taxa took place and this paper interrelates major changes in the distribution and dispersal of mainly those faunas on subtropical and tropical marine shelves (Fig. 1) with major changes in the palaeogeography. The analysis of bryozoan assemblages using generic distributions (Figs 2 to 5) has involved examination and interpretation of an immense literature on other marine invertebrates including fusulinids, corals, cephalopods and conodonts. Presently, the known distribution of bryozoan taxa delimits eleven regions in the Carboniferous and ten regions in the Permian (Fig. 1). The reconstruction of the palaeobiogeography in the Late Palaeozoic and the location of faunas in particular regions is based on faunal, tectonic, sedimentary and geophysical data. Using the information on the range charts (Figs 2 to 5), the geographic distribution and dispersal both of individual genera and faunal assemblages may be traced. As an example Fig. 6 shows the changes in the geographical distribution through time of the genus *Streblotrypa*.

Palaeogeographic Changes

In the Late Palaeozoic through 100 million years, the coming together of first Gondwana and Euramerica and later Angara took place in a series of steps to form the large land mass of Pangaea (Fig. 1) (Ross & Ross 1983, 1985*a*). These changes in land mass distribution also modified ocean basin shape and size, ocean current dynamics and ultimately changed climates and the distribution of climatic belts. The patterns of dispersals, evolution and distribution of faunas and the sediments and sedimentary patterns provide invaluable information on Late Palaeozoic palaeogeography (Wanless & Shepard 1936; Ross 1967, 1978, 1981; Ross & Ross 1985*b*; Stevens 1985; Heckel 1986).

Globally, four large cratonic blocks (Gondwana, Euramerica, Angara and Cathaysia) and one large ocean basin (Palaeo-Panthalassa) were the dominant physiographic features of the Late Palaeozoic. Gondwana, the largest cratonic block at the beginning of the Early Carboniferous, was apparently assembled from several smaller cratonic blocks that aggregated during the Late Precambrian and Early Palaeozoic. By the beginning of the Carboniferous (Fig. 1A), portions of Gondwana were moving across the southern pole (Caputo & Crowell 1985). Because of its large expanse, some of its marine shelves extended from warm temperate into subtropical regions.

Euramerica, considerably smaller than Gondwana, extended across the equator and into northern tropical latitudes during the Early Carboniferous. The ocean basin between Euramerica and Gondwana was in the process of being deformed into the Hercynian–Appalachian–Marathon orogenic belt and only the Acadian orogenic phase had been completed by the Early Carboniferous. In the middle portion of the orogenic belt (Ziegler 1988), the Maritime Provinces of eastern Canada, Spain, and central France included several microcratons which were accreted to one or other of the larger cratons during Devonian time.

Angara apparently lay to the north and east of Euramerica in warm temperate to cold temperate to cold latitudes. It was separated from other cratonic blocks during the Early Carboniferous by ocean basins of various widths; the narrowest of these being with eastern Euramerica. The position of Cathaysia (various parts of China and east Asia) with respect to the other three cratonic blocks is less well known. This block was also assembled from smaller blocks (Zhang *et al.* 1984; Klimetz 1985; Lin *et al.* 1985) and ocean basins of varying sizes lay between these blocks.

Characteristics of the Late Palaeozoic ocean basins are much less well known than the cratons and adjoining shelves, margins and platforms. Sea floor spreading during the Mesozoic and Cenozoic resulted in some of the sea-floor sediments of Late Palaeozoic age being accreted as structurally deformed material on to margins of the cratons. These accreted terranes have been transported great distances (Monger & Ross 1971) and provide a fragmentary history about some parts of the ocean basins. In the Late Palaeozoic, Palaeo-Panthalassa increased in size and had a large western seaway, Palaeo-Tethys, that contained numerous small island arcs and small to medium size cratons. Our reconstruction of central and eastern Palaeo-Panthalassa suggests several oceanic plates with island arcs and trenches bounded by subduction zones, similar to those now present in the western Pacific.

From McKerrow, W. S. & Scotese, C. R. (eds), 1990, *Palaeozoic Palaeogeography and Biogeography*, Geological Society Memoir No. 12, pp. 353–362.

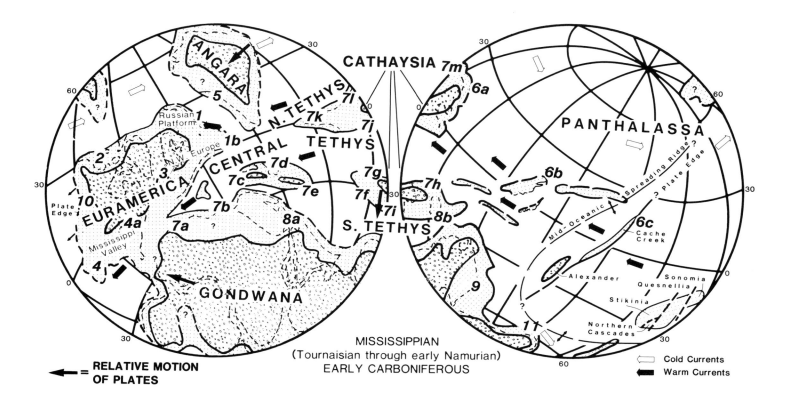

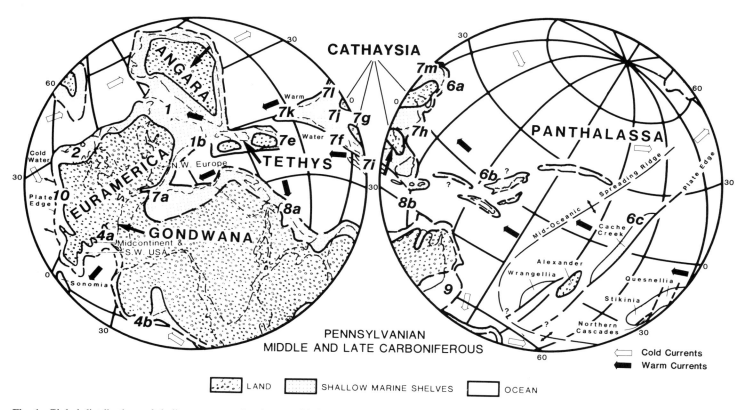

Fig. 1. Global distributions of shallow water marine faunas with bryozoan taxa on continental shelves, shelf margins and platforms. The positions of the major cratonic blocks and seaways are shown, as are the inferred relative motion of tectonic plates and the pattern and direction of cold and warm water currents. (**A**) Lower Carboniferous (Tournaisian through Serpukhovian); cosmopolitan tropical faunas. (**B**) Middle and Late Carboniferous (Pennsylvanian) with two provincial tropical faunas: the Ural–Tethys province and the Midcontinent–Andean province. (**C**) Early Permian (Wolfcampian and Leonardian) with three provincial tropical and subtropical faunas: Tethys province, Franklin–Ural province and Midcontinent–Andean province. Closure of the southern end of the Ural seaway took place in the middle part of the early Permian. (**D**) Early Late Permian (Guadalupian) distribution with two tropical faunal realms, northern and central Tethys, and the Midcontinent–Andean province. Geographic regions are: (1) Russian platform, including Moscow basin, Voronez basin and Ural shelf: lb, Donetz basin; (2) Franklinian shelf and adjacent regions; (3)

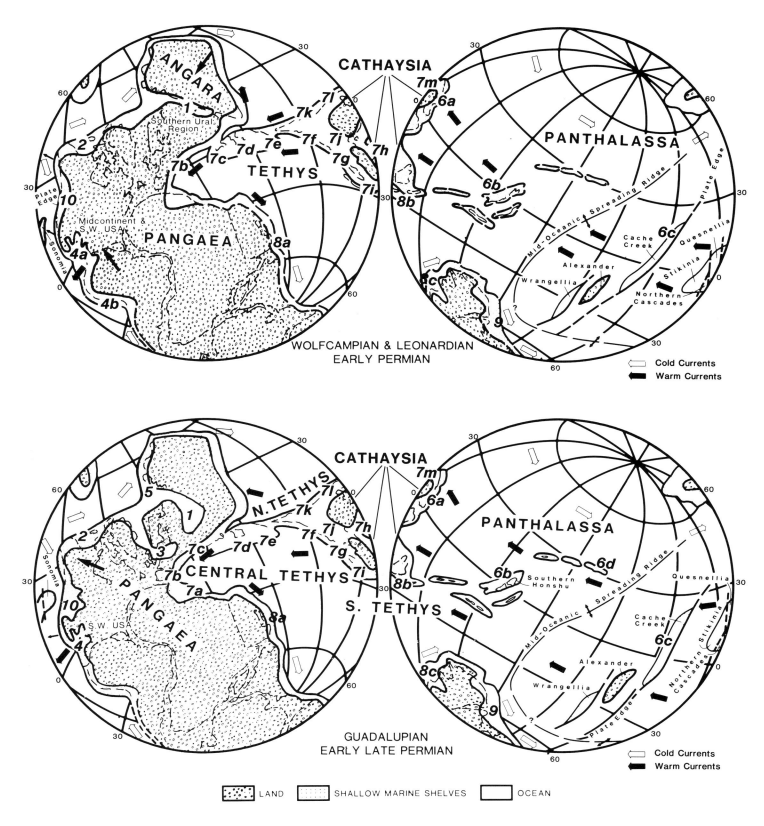

Northwest Europe, including United Kingdom, Eire, Belgium, Germany, Poland, Nova Scotia; (4) American shelves: 4a, Midcontinent and southwestern shelves of the US and west Texas; 4b, Andean sea; (5) Angaran shelves and cratonic basins, including Kuznets basin and central Siberia; (6) Northern Tethys: 6a, Khabarovsk and Transbaikal; 6b, Japan; 6c, western part of North American Cordillera; 6d, northeastern U.S.S.R. (7) Central Tethys: 7a, Morocco; 7b, Carnic Alps; 7c, Transcaucasus; 7d, Uzbekistan, Pamirs, Darvas (Middle Asia); 7e, Afghanistan; 7f, Tibet; 7g, western China; 7h, central China; 7i, southwestern China; 7j, Kazakhstan; 7k, Altai; 7l, Mongolia; 7m, northwest China; (8) Southern Tethys: 8a, parts of southern Afghanistan; 8b, Malaysia and Thailand; (9) Tasman geosyncline and adjacent shelves; (10) North American Cordillera, eastern part; (11) Patagonian shelf, western Argentina.

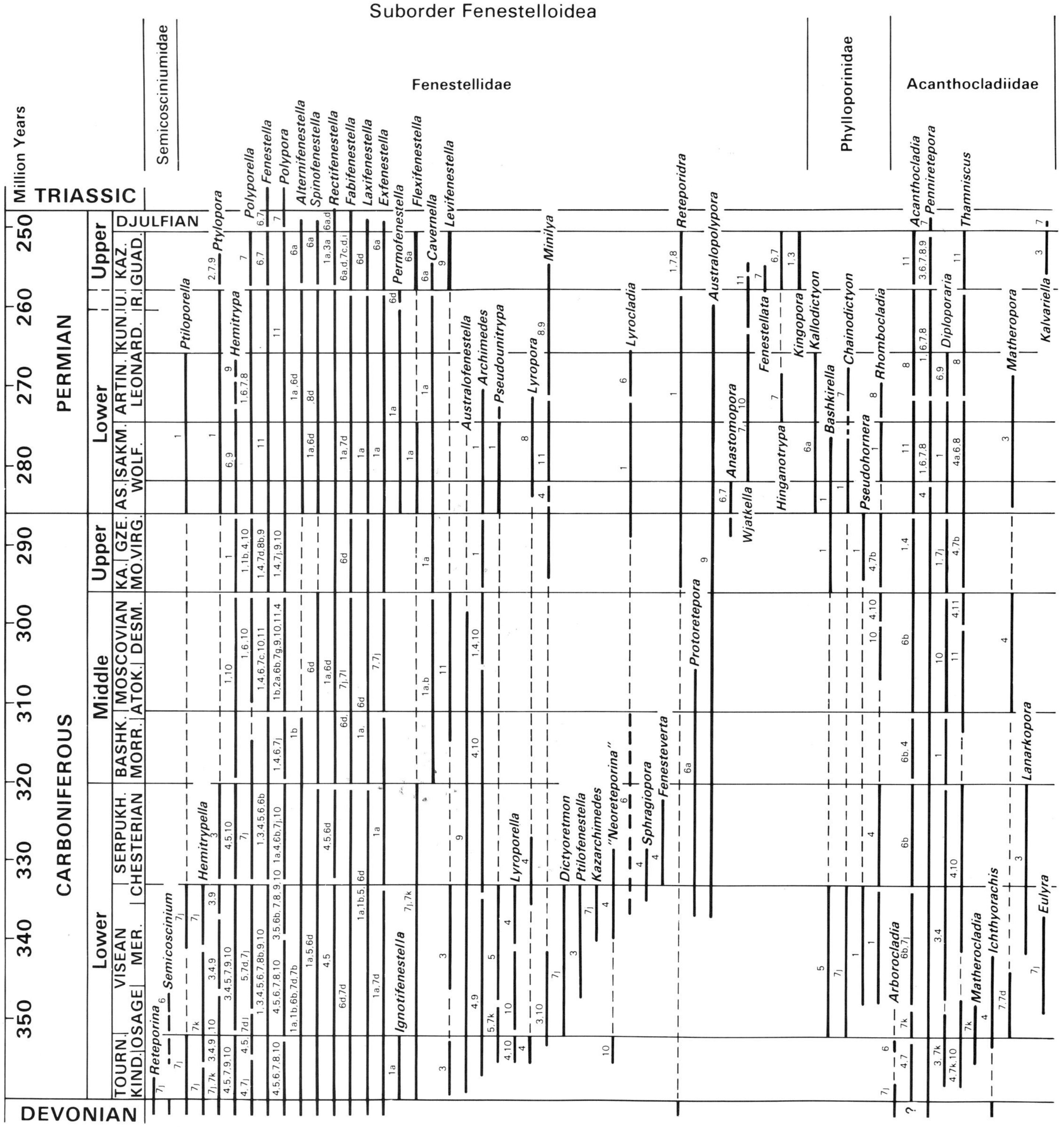

Fig. 2. Stratigraphic and geographic ranges of Carboniferous and Permian ectoproct bryozoan genera: order Cryptostomata, suborder Fenestelloidea, families Semicosciniumidae, Fenestellidae, Phylloporinidae and Acanthocladiidae. See Fig. 1 for key to numbered geographic regions.

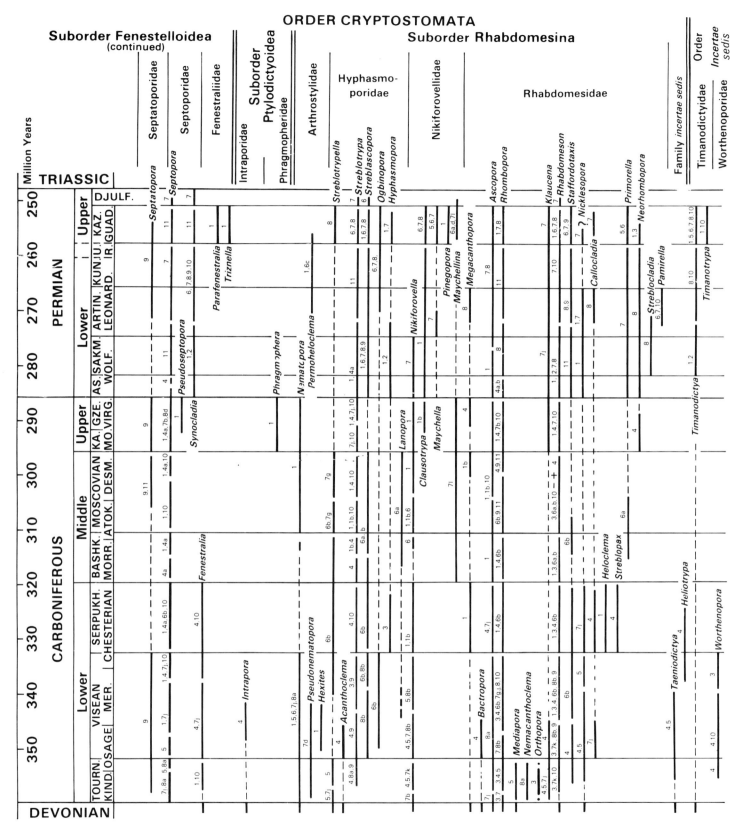

Fig. 3. Stratigraphic and geographic ranges of Carboniferous and Permian ectoproct bryozoan genera: order Cryptostomata, suborder Fenestelloidea, families Septatoporidae, Septoporidae and Fenestraliidae; suborder Ptilodictyoidea, families Intraporidae and Phragmopheridae; suborder Rhabdomesina, families Arthrostylidae, Hyphasmoporidae, Nikiforovellidae, Rhabdomesidae and family *incertae sedis*; order *incertae sedis*, families Timanodictyidae and Worthenoporidae. See Fig. 1 for key to numbered geographic regions.

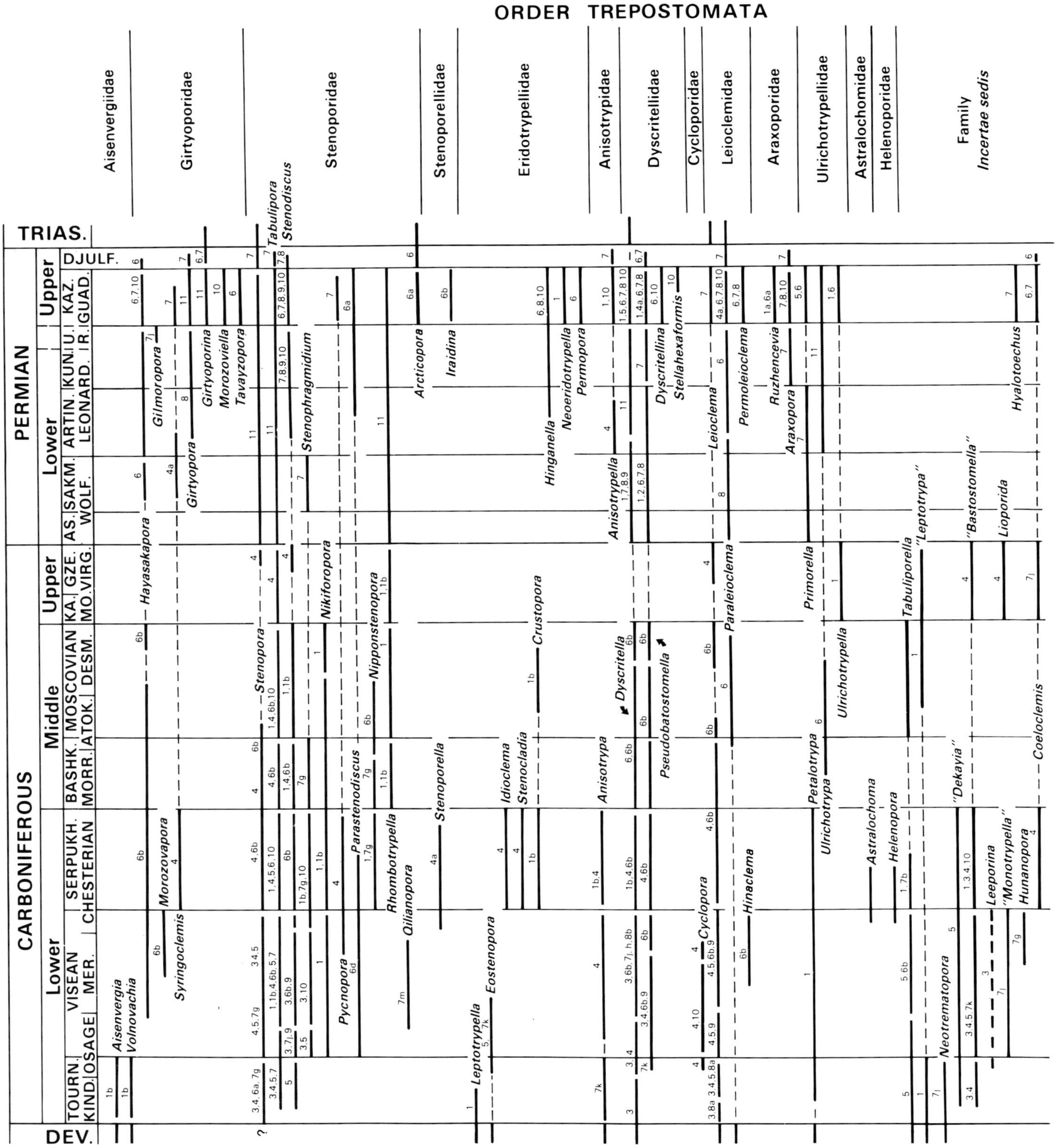

Fig. 4. Stratigraphic and geographic ranges of Carboniferous and Permian ectoproct bryozoan genera; order Trepostomata, families Aisenvergiidae, Girtyoporidae, Stenoporidae, Stenoporellidae, Eridotrypellidae, Anisotrypidae, Dyscritellidae, Cycloporidae, Leioclemidae, Araxoporidae, Ulrichotrypellidae, Astralochomidae, Helenoporidae and family *incertae sedis*. See Fig. 1 for key to numbered geographic regions.

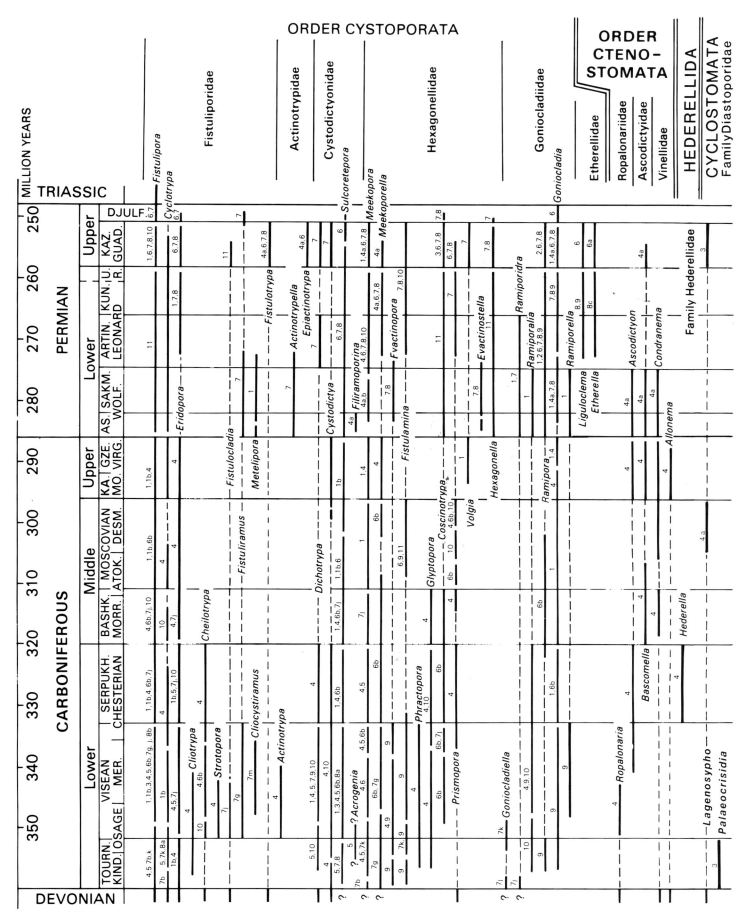

Fig. 5. Stratigraphic and geographic ranges of Carboniferous and Permian ectoproct bryozoan genera: order Cystoporata, families Fistuliporidae; Actinotrypidae, Cystodictyonidae, Hexagonellidae, Goniocladiidae and Etherellidae; order Ctenostomata, families Ropalonariidae, Ascodictyidae and Vinellidae; order Hederellida, family Hederellidae; order Cyclostomata, family Diastoporidae. See Fig. 1 for key to numbered geographic regions.

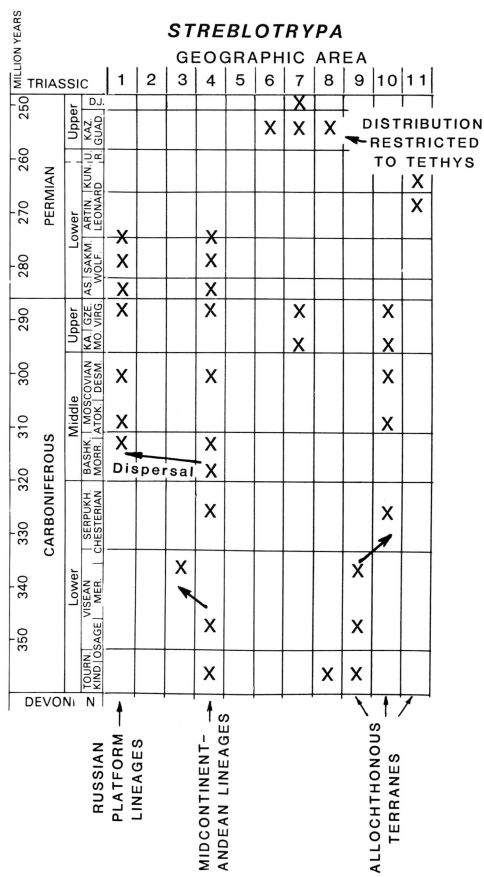

Fig. 6. Changes in the geographical distribution of the genus *Streblotrypa*, using information from Fig. 3.

Distributions and Dispersals of Faunas with Palaeogeographic Changes

The Tournaisian and Visean (Fig. 1A) were apparently times of relatively warm climatic conditions and carbonate-producing organisms show major diversity (Ross & Ross 1987*a*). Sea level fluctuations were apparently of low magnitude and with relatively long frequencies, and, in general, sea level was high (Ross & Ross 1987*b*). Dispersals were common and cosmopolitan faunas had latitudinal gradients. Tournaisian bryozoans commonly had widespread distributions (Figs 2 to 5). Visean bryozoans showed greater generic diversity and broader distributions than those of the Tournaisian. Late in the Early Carboniferous (late Visean through early Bashkirian) Gondwana and Euramerica joined along the Hercynian–Appalachian–Marathon orogenic belt to form Lesser Pangaea. This resulted in elimination of the tropical marine shelf along the southern margin of Euramerica and also the marine connection that joined the shelf faunas of western Euramerica with those of the eastern Palaeo-Tethys. Because parts of the coast of Gondwana remained near the south pole and in cold water, two tropical shelf faunas became isolated, one on either side of Lesser Pangaea, and dispersals between them were infrequent along the warm temperate route on the northern Franklin shelf of Euramerica (Fig. 1B). The closure of the equatorial sea between Euramerica and Gondwana also diverted circumequatorial warm currents into much cooler regions and resulted in increased precipitation and a general climatic cooling. Glaciation in Gondwana resumed during the Serpukhovian and continued with fluctuations until near the end of the Early Permian (Veevers & Powell 1987). With the rapid cooling of world climate during the Serpukhovian, bryozoans showed a marked reduction in number of genera and were less widely distributed as shown by such genera as *Polypora*, *Hemitrypa*, *Ptylopora*, *Nikiforovella*, *Streblotrypa*, *Rhombopora*, *Dyscritella*, *Fistulipora* and *Eridopora*.

Continued cool conditions during the Bashkirian apparently further reduced numbers of genera and species and the dispersal of faunas (Figs 3 & 4). Shelf faunal diversity became very low as a result of many extinctions. Surviving genera and families commonly contained only a few species. Carbonate production was generally limited except for a few equatorial areas. Sea level was generally low, however, sea-level fluctuations were of considerable magnitude and of relatively short frequency. The Moscovian (Fig. 1B) was apparently slightly warmer than the Bashkirian and, although some faunal diversification occurred, additional surviving Early Carboniferous genera became extinct by the end of this stage. Continental shelves and basins such as in regions 1, 4a and 6 were isolated sites of ongoing evolutionary change in bryozoan lineages. In region 6, a group of genera including *Streblascopora*, *Streblotrypella*, *Saffordotaxis*, *Nikiforovella*, *Primorella*, *Hayasakapora*, *Dyscritella*, *Pseudobatostomella*, *Ulrichotrypa*, *Coscinotrypa*, *Meekoporella* and *Ramipora* were part of the newly evolving and dominant Tethyan bryozoan faunas. Conservative, low diversity shelf communities apparently were the rule, faunal dispersals were irregular and probably fortuitous. Although general sea level appeared to rise, sea level fluctuations continued to be of large magnitudes and of short frequencies.

During the Middle and Late Carboniferous and the earliest Early Permian, Lesser Pangaea continued on its northerly track and the north coast of Euramerica was gradually displaced to higher and higher temperate latitudes. Although shelf faunas occasionally dispersed via the northern margin of Euramerica, successful dispersals probably related to times of higher sea levels, which were also times of warmer water temperatures. *Rhombotrypella* from region 1 apparently dispersed via this northern route through the Franklin sea to region 10. Late Carboniferous (Fig. 1B) and earliest Permian (Fig. 1C) were times of gradual warming, few extinctions and modest diversification of bryozoan assemblages, similar to that during the Visean. Dispersals were only slightly more common during the Moscovian. The low diversity carbonate mud bank and mound communities of the Late Carboniferous gradually expanded into more complex biohermal communities during the earliest Permian. For example, regions 1 and 4a were centres of high bryozoan diversity in the Stephanian and several genera gave rise to lineages in the Early Permian.

In the middle Early Permian, a major physiographic change resulted when Lesser Pangaea joined with Angara along the Ural orogenic belt to form Greater Pangaea (Fig. 1C). Angara apparently was displaced from the north or northeast and its coastlines extended well into cold temperate or boreal waters. With the elimination of the warm water currents from the Tethys area that had flowed northward through the Ural marine connexion, the northern shelf of Euramerica lost its physical connexion with the Tethys and also became cooler. Its outpost faunas of warm-adapted, shelf shelly organisms diminished greatly in diversity after this event. By this time, the marine shelves of Greater Pangaea extended into the cold northern seas and effectively completed the isolation of the tropical marine shelf faunas on either side of Pangaea. Each fauna evolved independently with only occasional species dispersal by island 'hopping' across Palaeo-Panthalassa.

The later Early Permian (Leonardian) was warm, perhaps as warm as the Tournaisian, and the shelf carbonate faunas show marked diversification (Figs 2 to 5). Reef-forming communities gradually evolved independently on both the eastern and western tropical margins of Pangaea. This pattern continued into the Guadalupian (Fig. 1D). Dispersals were extremely rare across Palaeo-Panthalassa, giving rise to strongly provincial faunas which became enhanced in the Guadalupian by increased faunal diversity, particularly in the Tethys area. During the later part of the Guadalupian, extinctions again became increasingly common in all bryozoan taxa. Although some families had a few surviving genera, they were composed of only a few species. Sea-level fluctuations in the Leonardian and Guadalupian became longer in duration and less in amplitude.

The latest Permian (Djulfian) saw a burst of diversity in the Tethyan faunal realm (Fig. 1D; regions 6, 7 and 8), and this produced some distinctive and briefly successful lineages. These, and the remaining survivors of the Guadalupian, suffered extensive extinctions before the end of the Permian. The stratigraphic record suggests four rapid sea-level fluctuations of relatively small magnitude which were superimposed on a general lowering of sea level. The shelf faunas in the Tethyan areas included genera and species that had become adapted to warm, perhaps very warm water, however, in other parts of the world, it is difficult to find any faunas or strata that can be identified as being of latest Permian age. The fossil record so far shows only about twenty bryozoan genera surviving into the Triassic.

C. A. Ross thanks Chevron USA Inc. for permission to publish.

References

Caputo, M. V. & Crowell, J. C. 1985. Migration of glacial centers across Gondwana during the Paleozoic era. *Geological Society of America Bulletin*, **96**, 1020–1036.

Heckel, P. H. 1986. Sea-level curve for Pennsylvanian eustatic marine transgressive–regressive depositional cycles along midcontinent out crop belt, North America. *Geology*, **14**, 330–334.

Klimetz, M. P. 1985. An outline of the plate tectonics of China: Discussion and reply. *Geological Society of America Bulletin*, **96**, 407–408.

Lin, Jin-Lu, Fuller, M. & Zhang, Wen-You. 1985. Preliminary Phanerozoic polar wander paths for the North and South China blocks. *Nature*, **313**, 444–449.

Monger, J. W. H. & Ross, C. A. 1971. Distribution of fusulinaceans in the western Canadian Cordillera. *Canadian Journal of Earth Sciences*, **8**, 259–278.

Ross, C. A. 1967. Development of fusulinid (Foraminiferida) faunal realms. *Journal of Paleontology*, **41**, 1341–1354.

—— & Ross, J. R. P. 1983. Late Paleozoic accreted terranes of western North America. *In*: STEVENS, C. H. (ed.) *Pre-Jurassic rocks in western North American suspect terranes*. Pacific Section, Society of Economic Paleontologists and Mineralogists, Los Angeles, 7–22.

—— & —— 1985a. Carboniferous and Early Permian biogeography. *Geology*, **13**, 27–30.

—— & —— 1985b. Late Paleozoic depositional sequences are synchronous and worldwide. *Geology*, **13**, 194–197.

—— & —— 1987a. Biostratigraphic zonation of Late Palaeozoic depositional sequences. *Cushman Foundation for Foraminiferal Research, Special Publication*, **24**, 151–168.

—— & —— 1987b. Late Paleozoic sea levels and depositional sequences. *Cushman Foundation for Foraminiferal Research, Special Publication*, **24**, 137–149.

Ross, J. R. P. 1978. Biogeography of Permian ectoproct Bryozoa. *Palaeontology*, **21**, 341–356.

—— 1981. Biogeography of Carboniferous ectoproct Bryozoa. *Palaeontology*, **24**, 313–341.

—— 1985. Biogeography of Ordovician ectoproct (bryozoan) faunas. *In*: NIELSEN, C. & LARWOOD, G. P. (eds) *Bryozoa: Ordovician to Recent*. Olsen & Olsen, Fredensborg, Denmark, 265–271.

STEVENS, C. H. 1985. Reconstruction of Permian paleogeography based on distribution of Tethyan faunal elements. *9e Congres International Stratigraphie Geologie Carbonifere, Compte Rendu*, **5**, 383–393.

VEEVERS, J. J. & POWELL, C. McA. 1987. Late Palaeozoic glacial episodes in Gondwanaland reflected in transgressive–regressive deposition sequences in Euramerica. *Geological Society of America Bulletin*, **98**, 475–487.

WANLESS, H. R. & SHEPARD, F. P. 1936. Sea level and climatic changes related to late Paleozoic cycles. *Geological Society of America Bulletin*, **47**, 1177–1206.

ZHANG, Z. M., LIOU, J. G. & COLEMAN, R. G. 1984. An outline of the plate tectonics of China. *Geological Society of America Bulletin*, **95**, 295–312.

ZIEGLER, P. A. 1988. Late Silurian to Permian stepwise assembly of Pangaea. *Palaeozoic biogeography and palaeogeography symposium*. Abstracts volume compiled by C. R. Scotese and W. S. McKerrow, Oxford, 13.

Phytogeographic patterns and continental configurations during the Permian Period

A. M. ZIEGLER

Paleogeographic Atlas Project, Department of Geophysical Sciences, 5734 South Ellis Avenue, Chicago, Illinois 60637, USA

Abstract: The climatic influences on vegetation in the present world can serve as an excellent model for the Permian, there being approximately the same level of floral differentiation and the same relatively cold polar regions. Accordingly, a system of present day climatically defined biomes, developed by Walter, is adapted herein for palaeophytogeographic purposes. There are ten biomes altogether which range from the tropical 'everwet' (biome 1) to the polar 'glacial' (biome 10).

Biome 1 is represented by the tropical rainforests of the Cathaysian province *sensu stricto* which was populated by the arborescent lycopods and sphenophytes, and the Gigantopterids which are interpreted as lianas (vines). This is flanked by the lower diversity 'summerwet' biome 2, the Atlantic Province, a zone which today is characterized by savanna vegetation, and in the Permian was represented by the pteridosperm *Callipteris* and the primitive conifers. Biome 3 was effectively 'abiotic' and is expressed geologically by evaporite belts.

The 'winterwet' or Mediterranean climate of today (biome 4) was not well developed in the Permian, or perhaps just not well preserved, but I attribute some low diversity floras of Kazakhstan to this biome. Biome 5, the 'warm temperate' biome is well developed in both hemispheres, and, like biome 1, is characterized by high diversity floras and abundant swamp deposits. Not surprisingly, the biome 5 floras have been mistaken for biome 1 with which they are transitional in the modern world and with which they share similar climatic conditions. In the Permian Angaran Realm, the Pechora Province represents this biome, while the Austrafroamerican Province is the Gondwanan equivalent. Often, biome 5 floras have been simply termed 'mixed'. Proceeding poleward across the 'hard frost line', the cool temperate floras of biome 6 were populated by diverse herbaceous sphenophytes and deciduous trees (the cordaitids of Angara and the similarly diverse glossopterids of Gondwana). Areas in the temperate zone that were remote from moisture sources are assigned to biome 7 and are only known in the southern hemisphere by some aeolian deposits in Argentina. The 'cold temperate' biome 8 is a low diversity equivalent of biome 6, and has been given the name *Gangamopteris* Flora in the southern hemisphere. The 'tundra' environment, biome 9, and the truly glacial deposits, biome 10, are known only from the southern hemisphere.

Pangaea was in an advanced state of assembly in the Permian as the result of a number of Late Carboniferous collisions (including Gondwana and Laurussia along the equator, and Laurussia, Kazakhstania and Siberia in the northern hemisphere), with the exception of a number of independent south Asian microcontinents which existed along an equatorial Tethyan seaway. The three great floral realms of the Permian were subequal in extent and comprised the south temperate Gondwanan Realm of central Gondwana, the tropical Cathaysian Realm of equatorial Gondwana, Laurussia, and the south Asian microcontinents, and the north temperate Angaran Realm of Siberia and Kazakhstania. These realms became distinct in the Carboniferous as a result of increasing equator-to-pole temperature gradients, and the major barriers to floral interchange were the two great subtropical deserts. The southern desert may have extended across Gondwana and is represented by evaporites in Brazil and Arabia, and the northern desert is known from the evaporites of the western United States and northern Europe. True geographic barriers to floral migrations, such as wide seaways, evidently did not play a major role in floral provinciality as there seems to be just one realm per broad latitudinal zone.

Climatic gradients, geographic barriers, and local environmental variations can all result in the differentiation of faunal and floral units, and, of course, it is critical to separate these effects before using fossils to test the various palaeogeographic models. The Late Palaeozoic is excellent for this purpose because: (1) climates varied from very uniform in the Early Carboniferous to highly differentiated in the Permian; (2) the oceans surrounding the various microcontinental elements of southern Asia represent potential geographic barriers; and (3) the dominantly maritime environments were accompanied progressively by more severe continental conditions during this long interval. Of course, these factors are interrelated to a degree, but it is nonetheless theoretically possible to separate the biogeographic effects of climate, geography and environment, and a first attempt at this is included in this paper for two separate stages of the Permian.

The Late Palaeozoic is relatively well known palaeogeographically because most of the continental elements were arrayed in their Pangaean configurations (Ziegler *et al.* 1979), but a number of microcontinents of southern Asia were not, and these range presently from Iran through China to southeast Asia (Rowley *et al.* 1985). Accordingly, the relationships of faunas and floras can be readily discerned for Pangaea and these patterns can then provide constraints on the positions of the less well known microcontinents. Terrestrial floras are used for this review because the effects of the strong climatic differentiation of the Late Palaeozoic are most directly felt in the continental realm, and because the floras have been the subject of a great many studies of broad regional scope (Chaloner & Meyen 1973; Vakhremeev *et al.* 1978; Li & Yao 1979; Asama 1976; Raymond *et al.* 1985). This paper begins, however, with a discussion of the vegetation regions of the present world and the climatic parameters that influence them. I feel that the steep latitudinal gradients typical of the Present world were also developed during the Late Palaeozoic and that the uniformitarian approach is therefore appropriate.

Climate and vegetation in the present world

As a general rule, vegetation in the terrestrial realm is limited by moisture availability, but moderate precipitation that is distributed throughout the year is much more effective than high annual amounts that are subject to pronounced seasonal variations. The lushest forest development, and the greatest preservation potential of vegetation in the form of peat bogs, occurs where precipitation exceeds 20 mm per month for 10 to 12 months of the year (Ziegler *et al.* 1987, p. 30). These conditions are met along the equator, due primarily to the seasonal overlap of the Intertropical Confluence Zone (ITCZ), and in the temperate zones where summer convective rains are replaced during other seasons by frontal systems. In fact, the coals of the geological past are correlated with equatorial and temperate zones (Ziegler *et al.* 1982; Parrish *et al.* 1982). These general patterns have been well established, but it is also clear that palaeobotanists are beginning to interpret some low diversity fossil floras as representing the stress of seasonal drought or cold (Parrish & Spicer 1988).

From McKerrow, W. S. & Scotese, C. R. (eds), 1990, *Palaeozoic Palaeogeography and Biogeography*, Geological Society Memoir No. 12, pp. 363–379.

Although diversity is often low in stressed situations, plant cuticle thickness tends to be greater, leading to enhanced preservation potential. Accordingly, a review of such climatic settings in the modern world is needed before proceeding with the interpretation of Permian phytogeography.

In the present world, vegetation responds to a bewildering array of climatic, edaphic (soil-related) and biotic factors, and maps based on vegetation alone (see Takhtajan 1986) are of limited use for interpreting climate. This is because of the historical, or evolutionary, factors in which floras, evolving in isolation, may or may not respond to similar climates in similar ways. Eyre has warned '... that all the attempts that have been made to explain the distribution of forest types in simple climatic terms are premature and ultimately probably completely fruitless' (1968, p. 87). A reasonably objective and straightforward system has been developed by Walter (1985) and reduces the 'macroclimate' of the land surface to nine major biomes (or 'zonobiomes'). This was affected by preparing about 8000 'ecological climate diagrams', which show monthly rainfall, temperature and other statistics, for the world's meteorological stations (Walter & Lieth 1967) and classifying these into 'monoclimatic' associations. Each association was apparently defined to encompass a homogeneous vegetation system for a particular geographic region. In other words, the choices of climatic boundary conditions were influenced by the natural breaks in the vegetation. The attractive aspect of Walter's scheme is that it is simple and therefore applicable in the geological past, but it also retains information on seasonality in both precipitation and temperature that is crucial in controlling vegetation patterns.

The biomes, as defined by Walter, are illustrated herein by a map (Fig. 1), a table (Table 1), and a highly idealized model (Fig. 2). Note that we add zone 10 (represented on the map by the number 0) for the nearly abiotic glacial regions. The map is redrawn with slight alterations to try to balance the variations between the original map and the text. Also, by a more liberal definition of mountainous regions, we exclude more areas from consideration. Such areas are unlikely sites for fossil preservation, but it is important to realize that mountain chains, like the Andes, do provide migration routes for temperate zone plants and animals across regions that would otherwise be intolerable for them (Raven 1973, p. 219). Table 1 contains data on the present day biomes that may or may not be applicable in kind or degree to times in the geological past. Certainly most of the present plant taxa have evolved since the Palaeozoic, but the hope is that the biomes represent a natural response of vegetation to important climatic subdivisions. They should be recognizable in the past together with some of their more general features such as diversity, influence on soil development, and adaptations to hard frosts. The tropical zone and some temperate zone biome boundaries are based on precipitation characteristics (Table 1), while the higher latitude biomes are based on temperature and the related growing season. This is simply because temperature conditions are extremely uniform in low latitudes, while, in cooler climates, precipitation is variable as in the tropics, but evaporation is low so moisture availability is not limiting and the plants respond mainly to temperature gradients.

The model (Fig. 2) is idealized from the above map and table and shows the relationships of the biomes on a lowland continent the width of North America that extends from pole to pole. It is based on the present but can be applied to the past, with caution, by respecting the nature of the boundaries and understanding the way rainfall patterns would change with different temperature gradients. It is notable that in the model most of the biome boundaries transgress latitudinal lines, indicating that temperature may be of more importance than the length of day or insolation patterns. From this, it could be assumed that in warmer times the temperate biome boundaries would be displaced poleward. In the Cretaceous, for instance, the Arctic and glacial biomes (9 and

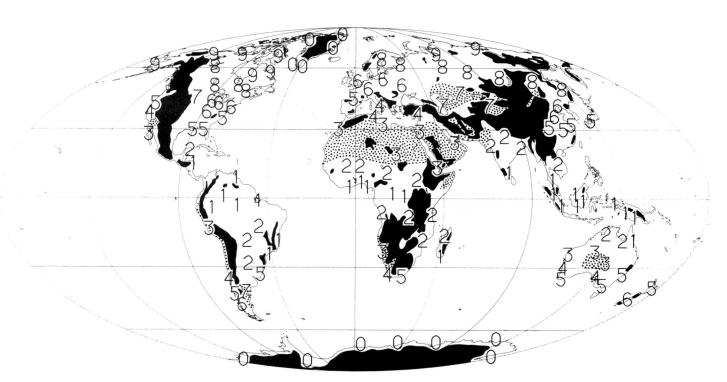

Fig. 1. Present day biome distribution (from Walter 1985, pp. 36, 39–45, 181–182). The biomes are represented by the digits 1 through 9 and 0, with 0 representing the number 10. The digits have been located mainly in obvious depositional settings such as swamps or deltas, so that the map would be as comparable as possible to the Permian fossil-based biome maps (Figs 3 & 4). The black areas represent mountains, uplands, and glaciers, and are defined by the 1000 m contour. The dot pattern represents the modern desert areas, which are defined as those areas that receive 20 mm or less of precipitation during all but 3 or fewer months per year (Ziegler et al. 1987, fig. 5). The digits 3 and 7 represent the desert biomes and are located on well known evaporite basins and the placement of the digit 0 is based on ice sheet margins.

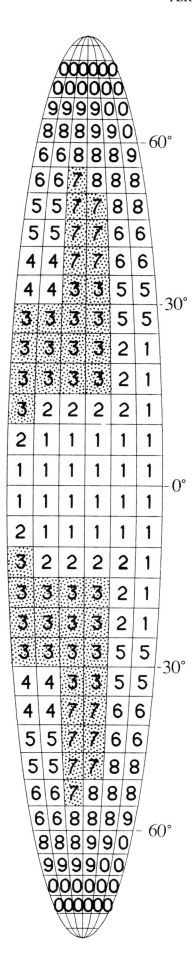

0) were not present, the temperate biomes (4 through 8) were expanded poleward to replace them, while in the tropics, the summerwet biome (2) evidently expanded at the expense of the everwet and desert biomes (1 and 3, respectively). All of this would occur in response to the decrease in the strength of the polar fronts, but the east-to-west asymmetry in the patterns would presumably remain because this is imposed by wind and wind-related ocean currents which are guided by the Coriolis Effect (Parrish *et al.* 1982). In the Late Palaeozoic, as will be seen, equator-to-pole temperature gradients were probably about the same as today, so the model is applicable without modification. However, any applications of the model should be based on the internal properties of the preserved floras, soils, and climatically sensitive sediments, rather than on some theoretical concept of the climate. The following paragraphs describe these properties for each of the biomes. This information has been abstracted for palaeogeographic purposes from Walter (1985) with supplementary information from Strahler & Strahler (1987, Chapter 26) and Eyre (1968) and will not be referenced in detail.

Biome 1: equatorial and tropical everwet biome

The 'tropical rainforest' is developed in areas of constant hot temperature (T mean = $26°-29°C$) and constant heavy rainfall (10–20 mm per month). These conditions occur along the equator due to the Intertropical Confluence Zone, and elsewhere in the tropics where trade winds bring warm moist air to narrow (20–30 km wide) and somewhat discontinuous belts mainly along the east-facing coasts (Barry & Chorley 1982, p. 276–8). It should be mentioned that the forests aid in the constancy of conditions by trapping air, retaining moisture and providing a substantial moisture source through transpiration. The most productive and diverse forests on Earth thrive in this biome which, unfortunately for palaeogeographers, spans 50° of latitude, and is therefore the least restricted environment on Earth. Diversity does decrease away from the equatorial zone but the wide and relatively old Atlantic and Indian Ocean basins constitute the only effective geographic barriers at differentiating floras at the generic and family level. Phanerotypes (trees) dominate the biome and are generally multistoried, but lianas (woody vines) are characteristic with 90% of liana species confined to the tropics. Characteristic adaptations that might be preserved include leaves with drip-tips and thick cuticles, and logs with enormous plank-buttresses. Soils are described mainly as ferralitic or latosols, but peat soils form in sandy substrates where water levels are constant (Ziegler *et al.* 1987, p. 27), and such lowland and coastal settings would be preferentially preserved in the geological record.

Biome 2: tropical and subtropical summerwet biome

This biome is characterized by a variety of vegetative communities ranging from deciduous forests to savannas with varying degrees of woodland and grassland, all of which become active during the rains of the sun-high seasons. They, therefore, occur along the margins of the Intertropical Confluence Zone (ITCZ) and in monsoonal areas where more extreme excursions of the ITCZ occur, such as in India. As such, they are transitional between the tropical Everwet and Desert biomes, and while these floras are quite productive and reasonably differentiated, their preservation potential is fair to poor because of the potential for oxidation during unusually dry seasons. Where depressions get waterlogged the trees of this biome become restricted to the higher surfaces,

Fig. 2. Idealized biome distribution based on an average of all present continents (see Fig. 1 and Table 1). The diagram represents a continent about the width of the Americas, extending from pole to pole. The digits represent the biomes in 5° latitude x 5° longitude squares. The shaded biomes 3 and 7 are the desert regions.

Table 1. *Climatic and latitudinal parameters for the present vegetational biomes*

Biome No.	Present Vegetation	Climate Descriptor	Sub-division	Precipitation Systems	Poleward Boundary	Poleward Latitude
10	none	glacial	none	orographic	none	90°
9	tundra vegetation (treeless)	arctic	none	some summer frontal	0 month $\overline{T} > 0°$ C	c. 70°–60°
8	boreal coniferous forests (taiga)	cold temperate	none	summer frontal	<1 month $\overline{T} \geq 10°$ C	c. 66°–58°
7	steppe to desert with cold winters	midlatitude desert	none	none due to distance or rain shadows	<4 month $\overline{T} \geq 10°$ C	51° ± 2°
6	nemoral broadleaf-deciduous forests	cool temperate	western	winter frontal, summer convective	<4 month $\overline{T} \geq 10°$ C	c. 58°
			eastern	as above	as above	44° ± 2°
5	temperate evergreen forests	warm temperate	western	winter frontal, summer convective	winter $\overline{T} < 0°$ C	46° ± 14°
			eastern	as above	as above	36° ± 2°
4	sclerophyllous woody plants	winterwet	none	winter frontal	≥10 month precip. >20 mm	38° ± 5°
3	subtropical desert vegetation	subtropical desert	coastal	none, due to cold upwelling offshore	limit of winter rains	32° ± 4°
			inland	none, descending limb of Hadley Cell	winter $\overline{T} < 0°$ C	c. 33°
2	tropical deciduous forests or savannas	summerwet	subtropical	summer monsoon	<3 month precip. >20 mm	up to 25°
			tropical	summer extension of ITCZ	as above	15° ± 5°
1	evergreen tropical rain forest	everwet	tropical	coastal diurnal	<11 month precip. >20 mm.	up to 25°
			equatorial	ITCZ	as above	10° ± 7°

Columns one and two are quoted directly from Walter (1985, p. 3) and column five is after Ziegler *et al.* (1987, p. 30–33). Column six is derived from Walter (1985, text) and Strahler & Strahler (1987, Chapter 26) and column seven is determined from Walter (1985, p. 36, 39–45) and Eyre (1968, Appendix 1).

and this could adversely affect their preservation. Sufficient soil moisture normally remains through the dry season and this is why the forests survive, but exceptionally dry years are not uncommon in such regions. The trees and or bushes can have small xeromorphic (drought resistant) leaves and thorns, but large thin deciduous leaves are characteristic and trees may have drought related 'false rings' (Creber & Chaloner 1984). Generalizations on community structure are difficult because there are variations from continent to continent, as well as in relation to soil differences. It is interesting that transitions to biome 1 are usually sharp, and perhaps made distinct by the activity of fire or grazing in biome 2.

Biome 3: coastal and inland tropical desert biome

There is a water deficit during every month in this biome and this happens where the dry adiabatically heated area of the Hadley Cell descends in the zones between 15° and 32° N and S or where cold oceanic upwelling currents inhibit precipitation. On north–south trending coasts, as shown in the model (Fig. 2), these two situations nearly coincide along west facing coasts but elsewhere, such as Venezuela and Somalia, coastal deserts occur in proximity to upwelling zones in configurations not covered by the model, and could, potentially, cause migration discontinuities for plants and animals of biome 2. Vegetation of the deserts is of course not very productive or diverse, and the plants that do occur are adapted for survival rather than competition. Xerophytes and succulents dominate and community structure is variably developed in the deserts of the world. In general, large flat root systems have evolved to gather moisture and transpiring surfaces have been reduced to avoid dessication. Interestingly, sandy dunes or rivercourses are preferred by the plants to clayey substrates because the clay remains hard and unabsorbent, a situation that contrasts with biome 2 and could have taphonomic implications. Soils in the usual sense are not developed. In general, however, desert plants do not get preserved except perhaps as microfloras and the biome is best recognized geologically by the products of the negative precipitation balance, the coastal and inland evaporite deposits and by aeolian sand dunes.

Biome 4: winterwet biome

This biome is at present developed in five geographically distinct regions: Mediterranean, Californian, Chilean, Cape (S. African) and Australian, which are consistent in latitude (32° to 38°N and S) and orientation (west margins of continents). The winterwet aspect of this biome is unique and related to the development of the polar front and associated extratropical cyclones; in the northern hemisphere this occurs because of the radiational cooling of the large continents and in the southern hemisphere because of the extension of sea-ice around Antarctica to about 60°S in the winter (see maps in Zwally et al. 1983). These so-called 'Mediterranean climates' constitute only 1% of the land surface today (Aschmann 1973) and there is evidence that the floras associated with these climates developed in the Late Tertiary (Axelrod 1973). This is not surprising in view of the facts that the Antarctic ice sheet developed and the northern hemisphere continentality increased about the middle of the Tertiary. The vulnerability of this environment may, in part, explain the apparently recent differentiation of these five regions. The five floras from these regions have had distinct origins and each have evolved from relatives in adjacent biomes; north temperate in the case of the Mediterranean and Californian, tropical in the Chilean and Cape, and south temperate in the Australian (Raven 1973, p. 215–220). Despite these separate derivations there is remarkable convergence in the development of the sclerophyllous evergreen forest vegetation which is characterized by low trees with leathery leaves (Kummerow 1973). Diversity is surprisingly high, but in regions of climate stress, small fluctuations can cause population crashes and the resulting 'catastrophic selection', and also, local habitat differences, such as rain shadows, may be enhanced (Raven 1973, p. 221). About half of the species are annuals, a greater proportion than any other biome, and short breeding periods correlate with rapid evolution (Raven 1973, p. 214). However, productivity and preservation potential is low and soils are described as brown earths with virtually no peat accumulation.

Biome 5: western and eastern warm temperate biome

Climates of this biome support the 'temperate evergreen forests' at the present time but this generality covers angiosperm and gymnosperm dominated forests and mixtures of the two, and the latitudinal boundaries of this biome are vague and variously located on the world map even by the same author (cf. Walter 1985, p. 36, 41, 181–182). The rainfall is well distributed throughout the year due to the alternating of winter frontal systems and the summer convective systems, both of which carry precipitation inland about 1000 to 1500 km if unimpeded by orographic effects. The daily temperature minimum of the coldest month is above freezing, but the vegetation spends the winter in a resting state. Productivity is about half that of the tropical rainforest (Lieth 1975, p. 205), due in part to the seasonal slow down, and diversity is quite high. Yellow or red podzolic soils are characteristic and peat swamps are locally developed (Ziegler et al. 1987, p.26). Part of the problem in defining this biome is that it may share boundaries with all other biomes from 1 to 7, a greater number than any other biome, but it is typically developed in climatically isolated sites in the same continent. Also, the east coast sites are the only likely exchange regions for tropical and temperate forest dwelling species.

Biome 6: western and eastern cool temperate biome

This biome is similar in many respects to biome 5, but it does experience hard frosts and cold winters, and the angiosperm dominated forests respond by shedding their leaves. Deciduous angiosperms require a growing season of at least four months with average daily temperatures above 10°C because of the time necessary to generate the foliage. There is an annual change in community structure as spring geophytes take advantage of the leafless trees, whereas a dense canopy is formed in the summer. Forest brown earths and grey forest soils are characteristic, and peat bogs are occasionally developed. Diversity is variable, evidently because some areas, like Europe, were adversely affected by Pleistocene ice sheets. Also, the vegetational boundaries of this biome are gradational to biomes 5, 7, and 8 over distances of hundreds of kilometers.

Biome 7: midlatitude desert biome

This category includes a range of current vegetative types from tall grass and short grass prairies to true desert environments and all experience an annual soil-water deficiency. This biome is simply too remote from moisture sources or isolated by rainshadows for adequate rainfall. The trees and shrubs that do survive occur in patches in valley bottoms. The soils are chernozems or sierozems and evaporites are quite common despite the fact that temperatures are not as warm as in biome 3. Productivity is of course moderate to low and the preservation potential of the plants that do occur is very low. This environment is developed in a coastal situation only in Argentina and this demands an explanation. Here the strong westerlies have dropped their moisture on the windward slopes of the Andes and are heated adiabatically as they descend over Argentina. The continent at this point is narrow enough that these dry winds extend to the east coast. The seasonal fluctuation in the margins of this westerly belt is small in comparison to its width, so the biome is extensively developed. This damped seasonal response is typical of narrow land masses with ocean dominated climates, an important point, since a major focus of this paper is on the effect of seasons on the vegetative patterns.

Biome 8: cold temperate biome

This biome is absent from the southern hemisphere today, but is continuously developed from coast to coast in North America and Eurasia where species richness is variable but community structure is similar. These great boreal coniferous tracts range from forests to taiga (widely spaced trees with a ground cover mostly of mosses and lichens). Precipitation is low to moderate, but this is not limiting because of the low evaporation rates and the short growing season of one to four months with average daily temperatures of 10°C or above. The conifers, except for the larch, are evergreen and so are fully functional throughout the growing season, and they are also apparently less demanding on soil for nutrients than angiosperms. Spruce forests develop within 50 to 100 years in areas where glaciers have retreated, such as Glacier Bay, Alaska. Soils tend to be podzols, that is, raw humus-bleached earths, and peat bogs are well developed, especially in areas where glaciation has upset drainage patterns or in subsiding areas like the West Siberian Lowlands. There is every reason to believe that forests such as this would occur at higher latitudes during times when the Arctic was free of ice, since temperature and not the solar angle is responsible for the short growing season. In fact, reasonably diverse forests are known from palaeolatitudes of 80° to 85° N in the mid-Cretaceous (Parrish & Spicer 1988).

Biome 9: Arctic biome

This biome is devoid of trees because of the extremely short growing season of one month, but like the Cold Temperate biome it forms a band across each of the northern continents. Tundra is the word applied to the modern vegetation which is composed of sedges, grasses, mosses, lichens and forbs, with some dwarf shrubs.

Biome 10: glacial biome

This 'biome' is abiotic for practical purposes, but it is areally significant and well represented in the geological record in the form of tillite. An important point is that this biome can effectively expand greatly in the winter due to the formation of sea-ice, and this enhances the seasonality in temperate latitudes. As has been seen, the precipitation associated with the polar front directly affects all biomes except the tropical ones, and in many parts of the world the polar front is formed by the sea-ice margin.

In summary, climatic differentiation, as represented by the biome boundaries, is rarely sharp, but nonetheless very important in controlling plant life in the terrestrial realm. Even during times of reduced equator-to-pole temperature gradients, floras must respond to rainfall occurrence and seasonality, insolation patterns and some temperature effects. Too often, palaeobiogeographers, pay lip service to climate, but jump to the concept of geographic separation to explain floral discontinuities. Adjacent biomes can be regarded as 'climatically differentiated', but there is an effect that could be termed 'climatically separated' for those biomes that do not share boundaries. For instance, no geographic barrier exists between Africa and Eurasia, but biome 3, in the form of the Sahara Desert, very effectively separates the floras of Europe and tropical Africa, and indeed between tropical Asia and Africa. Also, biome 5 in North America is climatically split by biome 7, and although the climates are by definition, broadly similar between coastal British Columbia and the Carolinas, the vegetation is very different in composition, structure and history. Indeed the separation in such cases is geographic and could be considered as such, but the barriers are climatic, so we prefer the term 'climatic separation' to describe the effect, and we will reserve 'geographic separation' for areas where major ocean basins, deserts or mountains present migration barriers for floras.

Boundaries characterized as 'climatically differentiated', that is, normal biome boundaries, should be regarded as provincial level biogeographic boundaries because the stress encountered by a species moving from one to another can result in new adaptations and hence endemism. The development of the five regions of Mediterranean floras, biome 4, is an example. The Mediterranean and Californian versions of this biome have had completely different evolutionary histories and are 'climatically separated' from each other, but each shares a biome boundary with biome 5 and in turn 6 and 8, and biome 8 is virtually continuous across the northern hemisphere. The geometry is such that hierarchial schemes will prove very difficult to construct in such a two-dimensional array, unless the focus is on the nature of the boundary rather than the province. For the purpose of this paper, the word 'province' will be applied to all biome level associations and the word 'realm' will be used to group these provinces into their tropical versus north and south temperate counterparts. The implication is that biome 3, the subtropical deserts, exerts the primary climatic separation of the Earth's floral regions; but it must be remembered that migration paths often are present along east-facing coasts or at higher elevations that are not well preserved in the geological record. Surprise or puzzlement is often expressed when 'mixed floras' are found in the geological record, as if geographic barriers were ubiquitous. The lesson of the Recent is that gradations are to be expected and this will be especially so in the Pangaean world where potential geographic barriers are limited to the microcontinents of Tethys.

Vegetation and climate in the Permian world

The biome concept is readily applied to the Permian and, as will be seen, Permian vegetation patterns are remarkably similar to the Present when determined on the basis of such general properties as diversity and community structure. However, many of the Permian floral groups have long since been extinct, while most of the rest have survived in reduced or altered states. Accordingly, this section begins with a brief introduction to the thirteen more common Permian floral groups, and some deductions concerning their general adaptations. Few of the Palaeozoic plant genera have been reconstructed in detail, so generalizations based on the ones that have been reconstructed may be misleading. It should also be noted that palaeobotanists vary considerably in their phyletic classifications and I have arbitrarily followed Taylor (1981) except where otherwise noted.

The two most primitive groups considered, the Lycophyta (club mosses and their relatives) and the Sphenophyta (horsetails), are extant but only in the herbaceous form. Arborescent forms of both groups were important elements in tropical and warm temperate environments of the Late Palaeozoic, but the implication that these forms were extinct by the Permian along with the coal swamp environment (Thomas & Spicer 1987, p. 109) is certainly incorrect, as a reading of the Chinese literature will show. The tropical Lycophytes, with their highly distinctive and commonly preserved root form, called *Stigmaria* (Stewart 1983, p. 110−112), seem to have been coal swamp dwellers, but this root form has never been found in the north temperate (Meyen 1982, p. 10) or south temperate occurrences (Cuneo & Andreis 1983, p. 132), nor, apparently, were the temperate forms swamp dwellers. The Sphenophytes were much more widespread and diverse, and are represented in nearly every Permian floral list. Most of them were herbaceous, but some, to judge by the stem or trunk genus *Calamites*, grew to 30 m high and 40 cm in diameter (Thomas & Spicer 1987, p. 83, 89). The arborescent forms were limited to the tropics as they have never been found in the Gondwana (Plumstead 1973, p. 19) or Angara floras (Meyen 1982, p. 19). Unfortunately, the names *Calamites* and *Paracalamites* have been applied to stems of all sizes so it is not possible from a taxonomic list to tell whether an individual was arborescent or herbaceous. Most were probably the latter, and may have constituted the majority of the Palaeozoic herbaceous plants.

The fern orders, Filicales and Marattiales are well represented in the Permian and today by forms evidently similar in structure. However, uniformitarian principles must be applied with caution as modern tree ferns seem to prefer mountainside 'cloud forests' (Troll 1970) while their Late Palaeozoic counterparts were not so limited and are, for instance, encountered in coal swamp floras. The genus *Psaronius* is identified as the trunk of a tree fern (Taylor 1981, p. 234) and was probably restricted, like other manoxylic (literally 'loose wood') trunks already discussed, to warm temperate and tropical climates (Tom L. Phillips, pers. comm.). A problem is that wood fossils are not encountered as often as leaf fossils, and moreover, leaf forms like the common *Pecopteris* are known to have been associated with a number of fern and seed fern orders. Taxonomic lists therefore have their limits in determining so simple a matter as the size of a plant.

The Pteridospermophyte (seed fern) orders are all extinct and the ones that did occur commonly in the Permian, the Medullosales, Callistophytales, Peltaspermales and Gigantonomiales, must have been obvious members of many communities.

'Some pteridosperms are known to have had the general appearance of modern tree ferns, while others had a more prostrate, scrambling, or vine-like habit' (Stewart 1983, p. 242). Again, manoxylic wood is known, and, as will be seen, these groups, with the exception of the Callistophytales, are mainly restricted to the warmer climes, with the Gigantonomiales (a name given by Meyen 1987, p. 157 for the gigantopterids) being confined to the tropical rain forests. Yao (1983, p. 80–81) has concluded that these gigantopterids were 'climbers of the scrambling type' because hook-like structures are present, the roots were superficial and the main stem is unbranched. Reconstructions are available for *Callistophyton*, type genus of the Callistophytales, and show it to have been a scrambling type of plant with adventitious roots extending from a prostrate stem, but this group evidently was in decline and became extinct by the middle of the Permian (Meyen 1987, p. 56, 143). It can be surmised from the examples given that the pteridosperms were structurally as well as taxonomically diverse in the Permian, but much taphonomic and palaeoecologic work on this group will be necessary before the vegetation of this period can be reconstructed in any detail.

The cycads and gingkos, by contrast, were just diversifying in the Permian, and of course, they are represented in modern floras. However, there is disagreement among palaeobotanists as to which of the several purported genera of each group really belong to them, and the fossil reports are so scattered as to be useless at this stage in phytogeographic studies. The one relatively widespread form genus is *Taeniopteris*, a supposed cycad, and this is characteristically found in low to mid-latitude floras.

Very important and diverse in high latitude floras are the cordaitids, dominating in the Angaran Realm, and the glossopterids, with a parallel domination in the Gondwanan Realm. (It should be noted that the Glossopteridales are normally classed with the Pteridosperms but we depart from current taxonomic practice and follow Plumstead 1973, p. 200, in using the 'Division Glossopteridophyta' in our lists). The cordaitids and glossopterids are extinct and are unrelated but possess many similar adaptations, evidently the result of parallel evolution under the stress of cold winter conditions. Both were arborescent and had pycnoxylic (dense wood) trunks (Meyen 1987, p. 185; Gould & Delevoryas 1977, p. 384), and probably ranged in size from shrubs to large trees. Both cordaitids and glossopterids had simple tongue-shaped leaves which were probably seasonally shed. Evidence for this, in the case of *Glossopteris*, comes from '... numerous leaf impressions in the autumn-winter, but not spring-summer, layers of a varved sediment ...' in Australia (Gould & Delevoryas 1977, p. 387). In fact, the leaves of the two groups are similar enough to have been confused, and Meyen (1987, p. 185) has opined that Gondwanan species reported as *Cordaites* are really glossopterids. Interestingly, *Glossopteris* and *Gangamopteris* are not uncommonly reported from various areas in Angara (Meyen 1982, p. 91, 100) but again, these identifications have been doubted (Chaloner & Creber 1988, p. 204). Finally the cordaitids and glossopterids show a striking parallelism in female fructifications (Meyen 1982, p. 78), and Plumstead (1973, p. 189) has emphasized that the seeds of glossopterids 'are enclosed, with varying degrees of completeness inside purse-like two-sided cupules', a 'protection of embryonic life from exposure to extremes of temperature and humidity'.

The final Permian group, the Voltziales, is transitional to the modern conifers and like them, foliage types range from needlelike to scaly, with gross structure being arborescent (Thomas & Spicer 1987, p. 176). The reduced leaf form must have been an adaptation to drought conditions as these Permian conifers are mainly found in regions transitional to evaporite basins in situations where the rainfall that did occur was probably highly seasonal in nature. The great boreal conifer forests of the Present may have been mirrored in the southern hemisphere Permian by the conifer genus *Walkomiella* which possessed scaly shoots (Retallack 1980, p. 404; Plumstead, 1973, p. 200), but conifers are virtually absent in Siberia in the Upper Palaeozoic (Meyen 1982, p. 56).

In the following sections, the phytogeographic units that have been defined in the literature will be assigned biome numbers corresponding to their modern counterparts, and each biome will be characterized in terms of the floral taxa that compose it. A diversity of terms have been used for the phytogeographic units including realm, province, phytochoria, region, area, flora, etc., and these have been prefixed by terms ranging from taxonomic (e.g. *Glossopteris*) to regional (e.g. Kazakhstanian) to climatic (e.g. boreal). Confusion results when a region moved from low to higher latitudes and the geographic term is retained for floras which changed in response to climate by migration rather than evolution. It is the thesis of this paper that the palaeogeographic configuration, climatic setting, and floral response is well enough understood in the Permian to avoid these problems. Moreover, the major floral discontinuities in the Permian, like the Present, are caused by the subtropical deserts (biome 3) and these serve to separate three major realms, the tropical Cathaysian Realm from both the north temperate Angaran Realm and the south temperate Gondwanan Realm. The term, Euramerian Realm or Province, has been applied to the tropical rainforests of the Carboniferous and to floras peripheral to the rainforests in the Permian. It is simply an accident of palaeogeography that Europe and North America moved north out of the equatorial belt about the period boundary, and the rainforests became restricted to Cathaysia. My recommendation is to restrict the term, Euramerian, to the Carboniferous tropical floras and to apply the available term, Atlantic, to the Permian Euramerian floras. I also recommend that the term province be restricted to biome level divisions.

The distribution of the three major realms (Cathaysian, Angaran and Gondwanan) and of the biomes of the Sakmarian and Kazanian stages of the Permian is shown on Figs 3 and 4, and the floral constituents of each biome are given in Tables 2 and 3. These maps are based on Lottes & Rowley (1990) and Nie et al. (1990). The references for all of this information are given in Appendices 1 and 2. It will be seen from the appendices that floral lists are generally available from a number of areas for each biome and these lists range from 'composite lists' from well known and thoroughly collected basins, to outlying localities which are less well known, but attributable, nonetheless, to the appropriate biome. It should be emphasized that the grouping together of the various floras into what I term biomes is taken, with few changes, from the literature, and indeed, many of the climatic implications have been well described as will be seen in the pages that follow. Floral lists for the major realms have also been published (Chaloner & Lacey 1973, p. 280), but rigorous, biome by biome, lists have not, hence Tables 2 and 3. These tables were constructed from the lists given in Appendices 1 and 2, and employ the synonymies of Appendix 3. The lists assigned to each biome were compared and the genera common to two or more basins were included in the tables. Using this methodology, approximately one half of the genera reported were too rare to include. Care was taken to exclude wood, root and seed genera, so that the lists would more closely reflect original diversity. Undoubtedly, many endemic genera, potentially important phytogeographically, were excluded, and in some biomes, there were few lists for even this rather quasi-objective treatment. Another problem is that some lists represent biome transitions, either geographic or temporal, and this is often noted by the original author, so care was taken not to give these lists undue weight. Despite these difficulties, the tables do serve as the crude but adequate basis for assigning the described floras their appropriate biome numbers.

Before proceeding, mention must be made of some problems that exist in correlating Permian floras. Many occur in areas far afield from marine rocks, or in temperate to polar areas where the fusilinid zonation is not applicable. Considerable progress

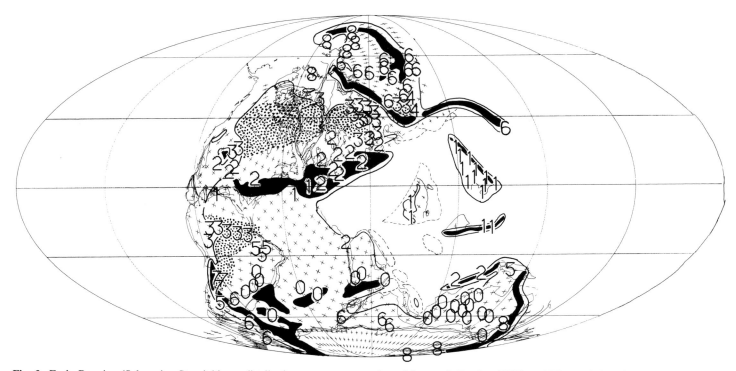

Fig. 3. Early Permian (Sakmarian Stage) biome distribution on a reconstruction of Lottes & Rowley (1990) and Nie *et al.* (1990). The localities and references are listed in Appendix 2, and the symbols and shading are explained in Fig. 1.

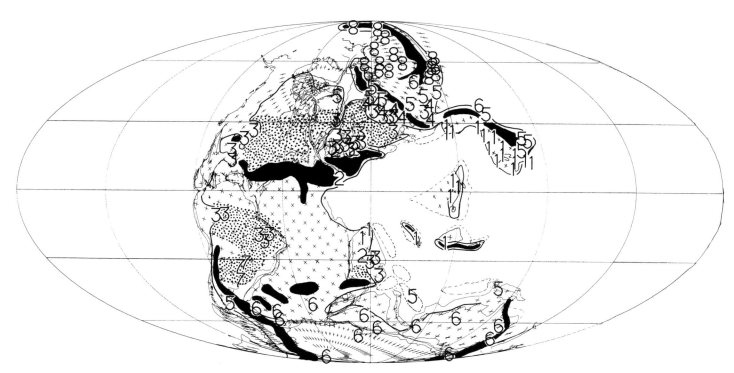

Fig. 4. Late Permian (Kazanian Stage) biome distribution on a reconstruction of Lottes & Rowley (1990) and Nie *et al.* (this volume). The localities and references are listed in Appendix 2, and the symbols and shading are explained in Fig. 1.

has been made recently in establishing floral and microfloral correlations, especially in Gondwanan regions where we have relied on Anderson (1980), Kemp (1975) and Kemp *et al.* (1977). In the Angaran region, Meyen (1982) provides the standard, but by his admission, correlations in the higher northern palaeolatitudes are largely guesswork. In the lower latitudes we have relied on Ross & Ross (1979) and Harland *et al.* (1982). The state of knowledge regarding the correlation of the various Chinese Permian sequences is provided by Yang *et al.* (1986), and here it must be emphasized that the Chinese assignment of the Carboniferous–Permian boundary, as well as the Lower–Upper Permian boundary, to the world standard differs from general practice. Rocks the Chinese regard as Upper Carboniferous are treated here as Lower Permian and the Late Permian Kazanian

Table 2. *Sakmarian floral genera common to two or more basins within each biome*

Divisions	Orders	Genera	Biomes South–North										
			9	8	7	6	5	2	1	2	4	6	8
Lycophyta	not specified	*Brasilodendron*					X						
		Lepidodendron							X				
Sphenophyta	not specified	*Annularia**							X	X		X	
		Annulina										X	
		*Asterophyllites**							X				
		Barakaria			X								
		Lobatannularia						X	X				
		Neokoretrophyllites										X	
		Phyllopitys										X	
		Phyllotheca				X	X				X	X	
		Schizoneura				X							
		Sphenophyllum				X	X	X	X	X	X		
		Tchernovia										X	
Pteridophyta	Filicales	*Sphenopteris**				X	X		X	X		X	
Pteridophyta	Marattiales	*Asterotheca*					X	X		X			
		*Pecopteris**					X		X	X	X	X	
Pteridospermophyta	Medullosales	*Alethopteris**							X				
		*Callipteridium**							X				
		*Neuropteris**							X	X		X	
		*Odontopteris**							X	X			
Pteridospermophyta	Callistophytales	*Angaridium*										X	
		Paragondwanidium										X	
Pteridospermophyta	Peltaspermales	*Callipteris*						X		X	X		
		Compsopteris							X				
Pteridospermophyta	Gigantonomiales	*Emplectopteridium*							X				
		Emplectopteris							X				
		Gigantopteris							X				
Cycadophyta	not specified	*Taeniopteris**					X		X	X	X		
Glossopteridophyta	Glossopteridales	*Gangamopteris*		X		X							
		*Glossopteris**				X	X						
		Ottokaria				X							
		Pursongia										X	
		Rubidgea				X							
Coniferophyta	Cordaitanthales	*Cordaites**				X	X	X	X	X	X	X	X
		Crassinervia										X	
		Dolianita				X							
		Lepeophyllum										X	
		Nephropsis								X		X	X
		Rufloria										X	
		Zamiopteris											X
Coniferophyta	Voltziales	*Buriadia*				X							
		Ernestiodendron								X	X		
		Lebachia						X		X	X		
		Ullmannia									X		
		*Walchia**				X			X	X			
		Walkomiella				X							
Incertae Sedis		*Angaropteridium*										X	
		Botrychiopsis	X	X									
		Tingia							X				
Diversity of 'common' genera			1	2	2	12	9	7	18	13	10	18	3
Maximum generic diversity per locality			?	2	2	13	13	7	22	29	12	24	3

* = form genus

Table 3. *Kazanian floral genera common to two or more basins within each biome*

Divisions	Orders	Genera	Biomes South–North								
			6	5	2	1	2	4	5	6	8
Lycophyta	not specified	Lepidodendron				X					
		Paikhoia							X		
		Signacularia						X			
		Viatcheslavia							X		
Sphenophyta	not specified	Annularia*				X		X	X	X	X
		Annulina	X								
		Lobatannularia				X			X		
		Phyllotheca	X						X	X	X
		Raniganjia	X								
		Schizoneura	X							X	
		Sphenophyllum		X	X	X			X		
		Tchernovia							X		
		Trizygia	X								
Pteridophyta	Filicales	Cladophlebis				X			X		
		Sphenopteris*	X			X	X		X	X	
		Todites								X	
Pteridophyta	Marattiales	Asterotheca		X							
		Dizeugotheca		X							
		Fascipteris		X		X					
		Marattiopsis							X		
		Pecopteris*		X	X	X		X	X	X	
		Prynadaeopteris							X		
		Rajahia				X					
Pteridospermophyta	Medullosales	Alethopteris*	X			X					
		Neuropteris*				X					
		Odontopteris*				X		X			
Pteridospermophyta	Peltaspermales	Callipteris						X	X	X	
		Comia							X	X	
		Compsopteris				X			X		
		Lepidopteris						X			
		Phylladoderma							X	X	
		Rhaphidopteris							X		
		Rhipidopsis				X			X		
Pteridospermophyta	Gigantonomiales	Emplectopteridium				X					
		Emplectopteris				X					
		Gigantonoclea				X					
		Gigantopteris				X					
Cycadophyta	not specified	Pseudoctenis				X					
		Taeniopteris*	X	X	X	X			X		
Ginkgophyta	not specified	Sphenobaiera						X			
Glossopteridophyta	Glossopteridales	Eretmonia	X								
		Dictyopteridium	X								
		Glossopteris*	X	X					X	X	
		Lidgettonia	X								
		Noeggerathiopsis	X								
		Rhabdotaenia	X								
Coniferophyta	Cordaitanthales	Cordaites*				X	X	X	X	X	X
		Crassinervia							X	X	X
		Glottophyllum								X	
		Lepeophyllum							X	X	X
		Nephropsis							X	X	
		Rufloria							X	X	X
		Zamiopteris							X	X	

Table 3. (Continued)

Divisions	Orders	Genera	Biomes South–North								
			6	5	2	1	2	4	5	6	8
Coniferophyta	Voltziales	Culmitzschia			X						
		Entsovia							X		
		Pseudovoltzia			X			X	X		
		Quadrocladus						X	X		
		Ullmannia			X			X			
		Walkomiella	X								
Incertae Sedis		Psygmophyllum							X	X	
		Tingia				X					
		Tychtopteris							X	X	
Diversity of 'common' genera			14	7	7	21	8	13	26	17	6
Maximum generic diversity per locality			23	6	9	21	8	18	30	20	6

* form genera

map is based on sequences the Chinese assign to the Lower Permian. Biostratigraphers are well aware of these varying practices and there is general agreement on the finer stage-level correlations. See Appendices 1 and 2 for the detailed information on formations we have correlated to build our Sakmarian and Kazanian floral and palaeogeographic data sets.

Tropical or Cathaysian Realm: biomes 1, 2 and 3

The heart of the tropical realm in the Permian is the Cathaysian or *Gigantopteris* flora (Li & Yao 1979, 1982; Asama 1984, Zhang & He 1985) and from its equatorial palaeolatitude, common association with coal forming environments (Wang 1985), and generic and ordinal diversity, an assignment of biome 1 (the tropical rainforest) is an obvious one. The interpretation of gigantopterids as lianas, and the association of phanerotypes with coal swamps strengthens this conclusion. This biome is most extensively developed in North and South China but is also known in the Indochina, Qiantang, and Tarim microcontinents, and outliers are known in Arabia, Spain, Venezuela, and Texas, and to judge by microfloras, in Morocco and Iran. Some endemism has been noted in North versus South China (Li & Yao 1979, p. 3), but the amount is slight and individual genera and even species are shared in common, for instance, between China and North America (Yao pers. comm., 1988). In fact, it is Yao's opinion, based on the seed-bearing nature of the gigantopterids, that direct migration routes are implied between the diverse areas mentioned. Certainly, the tropical realm could not have been as divided as today by barriers the width of the Atlantic or Indian oceans, and this assures us that the many continental elements of South Asia, though tectonically isolated, were nonetheless geographically proximal to Pangaea.

Floras assigned to the 'summerwet' biome 2 have been variously termed the Euramerian, Atlantic, Zechstein, or *Callipteris* floras, but all localities from western North America, Europe, and Arabia are routinely linked together by palaeobotanists and I propose that the *Walchia* sites of the Shanthai block be assigned to biome 2 as well. In addition, this flora is known in North China at the very end of the Permian (Wang 1985), but is too young for inclusion on our Kazanian map. Characteristic of these floras is a domination by the Voltziales, the transition conifers, whose association with dry or seasonally dry climates is well understood (Scott 1980, p. 102). In fact, in northern Europe, the transition from the coal swamps of the Late Carboniferous (biome 1) to the Zechstein evaporites of the Late Permian (biome 3) is reflected in a range of floral communities (Barthel 1976, p. 168–9). Generally the diversity of biome 2 flora is quite low, and the European lists, which seem to belie this statement, can be regarded as skewed by a long tradition of collection and description. The north and south hemisphere lists of the biome are similar, and the fact that they are separated on Tables 1 and 2 should not be taken to imply that they are separate provinces. In fact, there are gaps in the tropical rainy zone (biome 1) at present that are not acknowledged in the idealized model (Fig. 1) and therefore biome 2 can be regarded as a single province occurring throughout the tropics in the Permian wherever seasonal drought prevented rainforest development.

Biome 3 is well represented by evaporites in Euramerica and Gondwana in the Permian (Figs 3 and 4, Appendices 1 and 2) and these indicate severe desert conditions that could have been continuous across the northern and southern hemisphere continents, respectively. In addition, the Uralian epeiric seaway must also have been a major obstacle to floral migration in the northern hemisphere. Soviet palaeogeographic maps indicate that this seaway connected the Arctic and Tethyan oceans but that sea level fluctuations were adequate to drain it from time to time in the Late Permian (Vinogradov 1969; Ronov *et al.* 1984).

North temperate or Angaran Realm: biomes 4, 5, 6 and 8

The province assigned to the cool, temperate biome 6 can be considered the heart of the Angaran Realm and has been variously termed the Rufloria Assemblage, Angara Flora, and Tunguska Realm in the Sakmarian, and Rufloria–Cordaitean Assemblage, Taymyr–Kuznets District, and Kuznetsk Province of the Tunguska Realm in the Kazanian (Meyen 1982; Vakhrameev *et al.* 1978; Vinogradov 1969). This province is restricted to Siberia and the Mongolian arcs including the far southeast USSR and indicates that these areas, as today, were contiguous in the Permian. The ordinal diversity is restricted but the generic diversity of cordaitids and sphenophytes is high, and an assignment to the cool temperate biome 6 seems appropriate. This is based on the assumption that the cordaitids were deciduous and the knowledge that the sphenophytes of this region were herbaceous. Palaeogeographic maps (Vinogradov 1969) show that coal formed in the area of biome 6.

Biome 7, the mid-latitude desert climate, apparently was not developed in the northern hemisphere in the Permian because moisture sources were proximal to the rather narrow Siberian landmass and orographic barriers did not exist on the windward (western) side of the continent. The floras assigned to the cold temperate biome 8 are a low diversity version of biome 6. They have never been previously designated as a separate province in the Early Permian, but have been called the Tungusso–Verkhoyan district in the Late Permian (Meyen 1982, p. 68–9). This area was at latitudes of 60° or greater in the Permian and the above mentioned palaeogeographic maps indicate little or no coal and none in the highest latitude areas of biome 8. Meyen realized these low diversity floras were indicative of cold conditions, and all along the northern coast, dropstones in marine strata are thought to have been transported by 'seasonal shore or floe ice' from beach gravels (Epshteyn 1981). Mountain glaciers may have occurred locally, but there is no indication that the glaciers reached sea level.

Biome 5, the warm temperate zone, is developed over a considerable distance in the Late Permian from northern Greenland through the Pechora District at the northern end of the Urals, to the Mongolian arcs and the extreme southeast of the Soviet Union where these floras have been referred to as the Far Eastern Province. Continuity of this flora over this great distance is acknowledged by the authors who have described it recently (See Appendix 1 and 2), and this unity is obvious from the individual floral lists. Coal occurs in the type Pechora area, and the floras are impressive for their diversity of types and numbers of plants which are among the highest known in the Permian. A number of the more typical Angaran forms occur, but arborescent lycophytes and pteridosperms are also present, and typical Cathaysian forms have even been reported in the Far East (Meyen 1982, p. 97–104). Obviously, transitions between biomes 5 and 1 operated in the Permian, just as today, while restrictions to biomes 6 and 8 to the north are related to progressively colder conditions and not to geographic barriers. The junction of biome 5 and 6 probably was coincident with the severe frost line in the Permian. Finally in the Angaran Realm are floras tentatively referred to biome 4, the 'Mediterranean climate'. These are referred to as the Kazakhstan Province of the Early Permian (Vakhrameev et al. 1978, p. 85), and the Kama–Ural Realm, East European Region or Subangara Area in the Late Permian. These floras are similar in many respects to biome 2 on the opposite side of the evaporite belt and are assigned a separate biome number simply on the basis of geographic position. We know of little in the floral compositions that would indicate an adaptation to the 'winter' conditions of biome 4. Moreover we suspect, from the relatively small extent of land in the high northern latitudes, that the polar front and associated winter storms were weakly developed in the Permian. This would mean that the floras we have assigned to biome 4 may simply reflect the transition to desert conditions without the distinctive winter rains of the Mediterranean biome.

South temperate or Gondwanan Realm: biomes 5, 6, 7, 8, 9 and 0

As in the case of the northern hemisphere, the cool temperate biome 6 can be considered the standard Gondwanan, or glossopterid flora of authors. Again, deciduous trees (glossopterids) and herbaceous sphenophytes are characteristic. As will be seen from the maps and tables, the flora is more widely distributed than its northern counterpart. Retallack has termed this biome the *Glossopteris* deciduous swamp forest community and proposed for it a cool temperate climate (1980, p. 398). The well defined growth rings of these trees that were up to 40 m high and 120 years old (ibid, p. 400) attest to the seasonality of the climate. As has been mentioned, the conifer *Walkomiella* occurs in association with *Glossopteris* and this could have been an important constituent of upland floras at this time.

The floras assigned to the warm temperate biome 5 have often been termed 'mixed floras' (Lacey 1975; Li 1986), but the term Austroafroamerican has also been proposed for floras that reflect an amelioration of climate due to the displacement of Gondwana toward the equator (Archangelsky & Arrondo 1975, p. 479). I rather arbitrarily group together some palaeogeographically diverse localities that simply share in common the presence of *Glossopteris* elements with arboresecent lycophytes or tree ferns. Note that these floras are not especially diverse and it might be predicted that with further collecting and study, a higher diversity flora would emerge, and the biome 4 might also be recognized and differentiated. Excellent taphonomic studies are emerging from Argentina which demonstrate a range of communities associated with biome 5 (Cuneo 1983; Cuneo & Andreis 1983). In this area, biome 5 attains a higher latitude, presumably because of the west coast setting.

Biome 7, the midlatitude desert environment, has been identified in Argentina and Brazil where aeolian sands are thought to have formed in the temperate rainshadow of the ancestral Andes (Limarino & Spalletti 1986). Interestingly, the conifer *Walchia* has been collected from one of these sandstones and gives a rare hint of midlatitude desert vegetation in the Permian.

The cold temperate biome 8 has been assigned to low diversity occurrences usually referred to as assemblages of *Gangamopteris* leaves and little else. Retallack called this the 'Gangamopteris Taiga' and it is restricted to the Early Permian and succeeds the *Botrychiopsis* Flora and glacial deposits in time (1980, p. 396). A tundra environment was therein proposed for *Botrychiopsis*, which would make it part of biome 9, but we have no instance of this flora that is appropriate for one of our map intervals.

Finally, biome 0, the glacial environment, is well represented in the Early Permian of Gondwana. By the Late Permian there is no glacial environment indicated, though Banks & Clarke (1987, p. 11) have interpreted the presence of pack ice from dropstones up to 2 m across which are known from Tasmania. For our localities, we have relied on the recent synthesis of Veevers & Powell (1987) and the references therein. Fortunately much progress has been made recently in dating these deposits using microfloral zonations. Veevers & Powell show the glaciers to be on the decline during the Sakmarian but, within this context, there were many cycles within this stage. In fact, we have shown the glaciers during their greatest extent on our Sakmarian map, but have also incorporated floral lists from the very same areas in our tables 2 and 3. It will be noticed that the common assumption that past glaciers were polar is not born out; some northern Pleistocene glaciers were in temperate latitudes. It should also be mentioned that a number of the lower latitude Permian glaciers may have developed on the high ground generated during rifting.

Conclusions

Permian climates of both hemispheres were similar in detail to Quaternary climates in the northern hemisphere. This statement is based on the dual assumptions that the average of the rather dispersed paleomagnetic poles gives the true poles for the Early and Late Permian palaeogeographic maps and that the assignment of the Permian floras to their supposed modern equivalent biomes is correct. The justifications for these assumptions come from the symmetry of the biomes about the equator, and the fact that the biome boundaries are essentially at the same latitudes as their modern equivalents. There is some assymmetry in that glaciers were confined to the southern hemisphere and there was somewhat more land area in the southern hemisphere. Another significant point is the size of evaporite basins in the Permian which encompassed areas that were hundreds by thousands of kilometres in extent. Perhaps this was due to coincidence of epeiric seas with the subtropical belt, and it may not indicate anything significant about climate. It is sobering to realize, however, that modern coastal evaporite basins rarely exceed a hundred kilometres in diameter.

If the above assumptions about the reconstructions and the phytogeographic interpretations are accepted then certain statements about the climate can be made.

(1) The tropical rainforests of biome 1, the Cathaysian Province, were very well developed and interconnected along the equator in the Late Carboniferous and Permian. This situation was not achieved at any other time in geological history until the Late Tertiary and Quaternary (Ziegler et al. 1987). Coal swamp deposits representing this environment are best developed in China but the floras are known from all along the equatorial portions of Pangaea. Lower diversity floras of biome 2 are especially well known from the Permian of Europe, but occur on either side of biome 1 and represent the transition to the desert belts (biome 3). The deserts ranged from about 12° to 25° on the west of Pangaea to 25° to 35° on the east, a situation very comparable to today.

(2) The temperate belts, the Angaran on the north and the Gondwanan on the south, also contain climate information, but this is related more to temperature than to rainfall. The hard frost line, the division between biomes 5 and 6, was at about 45° to 50°N and S as today and the vegetation poleward of this boundary was composed of herbaceous plants and deciduous trees as today. Coal-forming swamps characterized both biomes, again as today. Diversity decreases are seen on the desert side (biome 4), the poleward side (biome 8), and on the interior desert side (biome 7), though not all of these floras have yet been recognized in both hemispheres.

(3) The glaciers (biome 10) of Gondwana were on the wane by the Early Permian and probably were confined to mountains or uplands. As in the case of the Pleistocene, the glaciers were not centred at the pole and reached about 40° from the equator.

This work is supported by the National Science Foundation Grant EAR-86-13412 and is part of the 'Paleogeographic Atlas Project' at the University of Chicago. General support for the atlas project is contributed by a number of companies: Amoco Production Research, Arco Exploration and Production Research, British Petroleum, Chevron Oil Field Research Company, Conoco Exploration Research, Digital Equipment Corporation, Societe Nationale Elf Aquitaine, Exxon Production Research, Mobil Exploration and Producing Services, Shell Development Company, Sun Exploration and Production Company, Texaco USA, Unocal International Oil and Gas, and BHP-Utah International. This support is gratefully acknowledged. We also thank the following for discussions concerning this paper: W. S. McKerrow, S. F. Barrett, M. A. Horrell, R. Langford, T. L. Phillips, A. Raymond, R. A. Spicer and Yao Zhao-qi. Finally, T. C. Gierlowski and M. L. Hulver helped with the preparation of the manuscript.

References

ANDERSON, J. M. 1980. World Permo-Triassic correlations: their biostratigraphic basis. In: CRESSWELL, M. M. & VELLA, P. (eds) *Gondwana Five*, A. A. Balkema, Rotterdam, 3–10.

—— & ANDERSON, H. M. 1985. *Palaeoflora of Southern Africa*. A. A. Balkema, Rotterdam.

APPERT, O. 1977. Die Glossopterisflora der Sakoa in Südwest-Madagaskar. *Palaeontographica Abteilung B*, 162, 1–50.

ARCHANGELSKY, S. & ARRONDO, O. G. 1975. Paleogeographia y plantas fosiles en el Permico inferior Austrosudamericano. In: *Actas del Primer Congreso Argentino de Paleontologia y Bioestratigraphia*, Tomo I, Asociacion Paleontologica Argentina, Tucuman, Argentina, 479–496.

—— & CUNEO, R. 1984. Zonacion del Permico continental de Argentina sobre la base de sus plantas fosiles. In: DEL CARMEN PERRILLIAT, M. (ed.) *Memoria III Congreso Latinoamericano de Paleontologia*, Mexico, 143–153.

—— & WAGNER, R. H. 1983. *Glossopteris anatolica* sp. nov. from uppermost Permian strata in south-east Turkey. *British Museum of Natural History Bulletin*, 37, 81–91.

ASAMA, K. 1976. *Gigantopteris* flora in southeast Asia and its phytopaleogeographic significance. In: KOBAYASHI, T. & HASHIMOTO, W. (eds) *Geology and Palaeontology of Southeast Asia*, 17, 191–207. University of Tokyo Press, Tokyo.

—— 1984. *Gigantopteris* flora in China and Southeast Asia. In: KOBAYASHI, T., TORIYAMA, R. & HASHIMOTO, W. (eds) *Geology and Palaeontology of Southeast Asia*, 25, 311–323. University of Tokyo Press, Tokyo.

——, HONGNUSONTHI, A., IWAI, J., KON'NO, E., RAJAH, S. & VEERABURAS, M. 1975. Summary of the Carboniferous and Permian plants from Thailand, Malaysia and adjacent areas. In: KOBAYASHI, T. & TORIYAMA, R. (eds) *Geology and Palaeontology of Southeast Asia*, 15, 77–101. University of Tokyo Press, Tokyo.

ASCHMANN, H. 1973. Distribution and peculiarity of Mediterranean ecosystems. In: DI CASTRI, F. & MOONEY, H. A. (eds) *Mediterranean Type Ecosystems. Ecological Studies*, 7, 11–19. Springer-Verlag, New York.

AXELROD, D. I. 1973. History of the Mediterranean ecosystem in California. In: DI CASTRI, F. & MOONEY, H. A. (eds) *Mediterranean Type Ecosystems. Ecological Studies*, 7, 225–277, Springer-Verlag, New York.

BANKS, M. R. & CLARKE, M. J. 1987. Changes in the geography of the Tasmania Basin in the Late Palaeozoic. In: MCKENZIE, G. D. (ed.) *Gondwana Six: Stratigraphy, Sedimentology and Paleontology*. Geophysical Monograph, 41, 1–14, American Geophysical Union, Washington, D.C.

BARRETT, P. J. & KYLE, R. A. 1975. The Early Permian glacial beds of South Victoria Land and the Darwin Mountains, Antarctica. In: CAMPBELL, K. S. W. (ed.) *Gondwana Geology*, Australian National University Press, Canberra, 333–346.

BARRY, R. G. & CHORLEY, R. J. 1982. *Atmosphere, Weather and Climate*, 4th edition. Methuen, London.

BARTHEL, M. 1976. Die Rotliegendflora Sachsens. *Abhandlungen des Staatliches für Mineralogie und Geologie zu Dresden*, 24, 1–190.

BESEMS, R. E. & SCHUURMAN, W. M. L. 1987. Palynostratigraphy of Late Palaeozoic glacial deposits of the Arabian Peninsula with special reference to Oman. *Palynology*, 11, 37–53.

BIGARELLA, J. J. 1973. Geology of the Amazon and Parnaiba Basins. In: NAIRN, A. E. M. & STEHLI, F. G. (eds) *The Ocean Basins and Margins Volume 1 The South Atlantic*. Plenum Press, New York, 25–86.

BIRKENMAJER, K. 1981. The geology of Svalbard, the western part of the Barents Sea, and the continental margin of Scandinavia. In: NAIRN, A. E. M., CHURKIN, M. & STEHLI, F. G. (eds) *The Ocean Basins and Margins Volume 5 The Arctic Ocean*. Plenum Press, New York, 265–329.

BROUTIN, J. & GISBERT, J. 1985. Entorno paleoclimatico y ambiental de la flora Stephano-Autuniente del Pirineo Catalan. In: *Dixième Congrès International de Stratigraphie et de Géologie du Carbonifère Compte Rendu*, 3, 53–66. Madrid.

BUNOPAS, S. 1981. *Paleogeographic History of Western Thailand and Adjacent Parts of South East Asia — a Plate Tectonic Interpretation*. PhD Thesis, Victoria University of Wellington, New Zealand.

CAHEN, L. & LEPERSONNE, J. 1981. Late Palaeozoic tillites of the Congo Basin in Zaire. In: HAMBREY, M. J. & HARLAND, W. B. (eds) *Earth's pre-Pleistocene Glacial Record*. Cambridge University Press, Cambridge, 43–47.

CAMINOS, R. 1979. Sierras pampeanas noroccidentales salta, Tucuman, Catamarca, La Rioja y San Juan. In: *Segundo Simposio de Geologia Regional Argentina Volumen I*, Cordoba, 225–291.

CAZZULO-KLEPZIG, M. & GUERRA-SOMMER, M. 1985. Relationship between the taphoflora of the Itarare Group, Parana Basin, south Brazil and the Permocarboniferous boundary. In: *Dixième Congrès International de Stratigraphie et de Géologie du Carbonifère Compte Rendu*, 2, 395–408. Madrid

CHALONER, W. G. & CREBER, G. T. 1988. Fossil plants as indicators of late Palaeozoic plate positions. In: AUDLEY-CHARLES, M. G. & HALLAM, A. (eds) *Gondwana and Tethys*. Geological Society, Special Publication, 37, 201–210.

—— & LACEY, W. S. 1973. The distribution of Late Palaeozoic flora. In: HUGHES, N. F. (ed.) *Organisms and Continents Through Time*. Special Papers in Palaeontology, 12, 271–289.

—— & MEYEN, S. V. 1973. Carboniferous and Permian floras of the northern continents. *In*: HALLAM, A. (ed.) *Atlas of Paleobiogeography*. Elsevier, New York, 169–186.

CHATEAUNEUF, J. J. & STAMPFLI, G. 1978. Palynoflore permo-triasique de l'Elburz oriëntal. *Note du Laboratoire de Paleontologie de l'Universite de Geneve*, **8**, 45–51.

CLARKE, M. J. & BANKS, M. R. 1975. The stratigraphy of the Lower (Permo-Carboniferous) parts of the Parmeener Super-Group, Tasmania. *In*: CAMPBELL, K. S. W. (ed.) *Gondwana Geology*. Australian National University Press, Canberra.

COOK, T. D. & BALLY, A. W. (eds) 1975. *Stratigraphic Atlas of North and Central America*. Princeton University Press.

CREBER, G. T. & CHALONER, W. G. 1984. Climatic indications from growth rings in fossil wood. *In*: BRENCHLEY, P. J. (ed.) *Fossils and Climate*. John Wiley & Sons Ltd, Chichester, 49–74.

CTYROKY, P. 1973. Permian flora from the Ǵa'ara region (western Iraq). Neues Jahrbuch für *Geologie und Paläontologie Monatshefte*, **1**, 383–388.

CUNEO, R. 1983. Paleoecologia de microsecuencias plantiferas del Grupo Rio Genoa, Permico de Chubut, Argentina. *Ameghiniana*, **20**, 111–131.

—— & ANDREIS, R. R. 1983. Estudio de un bosque de licofitas en la Formacion Nueva Lubecka, Permico de Chubut, Argentina. Implicancias paleoclimaticas y paleogeograficas. *Ameghiniana*, **20**, 132–140.

DA SILVA, O. B. 1988. Revisão estratigráfica da Bacia do Solimões. *In*: *Anais do XXXV Congresso Brasileiro de Geologia*, **6**, 2428–2438. Belém-Pará-Brasil.

DOUBINGER, J. & FABRE, J. 1983. Mise en évidence d'un autunien a 'flore mixte' dans le bassin de Béchar-Abadla (Sahara occidental algérien). *Pollen et Spores*, **25**, 91–116.

DURANTE, M. V. 1971. Later Permian flora of Mongolia and the southern boundary of the Angaran floristic region of that time. *Paleontological Journal*, **5**, 511–522.

—— 1976. The Carboniferous and Permian Stratigraphy of Mongolia on the Basis of Palaeobotanical Data. Nauka Publishing House, Moscow.

EL-KHAYAL, A. A. & WAGNER, R. H. 1985. Upper Permian stratigraphy and megafloras of Saudi Arabia: palaeogeographic and climate implications. *In*: *Dixième Congrès International de Stratigraphie et de Géologie du Carbonifère Compte Rendu*, **3**, 371–374. Madrid.

ELLIOT, D. H. 1975. Gondwana Basins of Antarctica. *In*: CAMPBELL, K. S. W. (ed.). *Gondwana Geology*. Australian National University Press, Canberra, 493–536.

EPSHTEYN, O. G. 1981. Late Permian ice-marine deposits of the Atkan Formation in the Kolyma river headwaters region, U.S.S.R. *In*: HAMBREY, M. J. & HARLAND, W. B. (eds) *Earth's pre-Pleistocene Glacial Record*. Cambridge University Press, Cambridge, 270–273.

ESCHER, A. & WATT, W. S. (eds) 1976. *Geology of Greenland*. Geological Survey of Greenland, Denmark.

EYRE, S. R. 1968. Vegetation and Soils; A World Picture 2nd edition. Aldine Publishing Company, Chicago.

Geological Survey of Western Australia 1975. *The Geology of Western Australia*: Western Australia Geological Survey Memoir, **2**.

GOULD, R. E. 1975. The succession of Australian pre-Tertiary megafossil floras. *The Botanical Review*, **41**, 453–483.

—— & DELEVORYAS, T. 1977. The biology of *Glossopteris*: evidence from petrified seed-bearing and pollen-bearing organs. *Alcheringa*, **1**, 387–399.

HARLAND, W. B., COX, A. V., LLEWELLYN, P. G., PICKTON, C. A. G., SMITH, A. G. & WALTERS, R. 1982. *A Geologic Time Scale*. Cambridge University Press, Cambridge.

KEMP, E. M. 1975. The palynology of Late Palaeozoic glacial deposits of Gondwanaland. *In*: CAMPBELL, K. S. W. (ed.) *Gondwana Geology*. Australian National University Press, Canberra, 397–413.

——, BALME, B. E., HELBY, R. J., KYLE, R. A., PLAYFORD, G. & PRICE, P. L. 1977. Carboniferous and Permian palynostratigraphy in Australia and Antarctica: a review. *BMR Journal of Australian Geology and Geophysics*, **2**, 177–208.

KUMMEROW, J. 1973. Comparative anatomy of sclerophylls of Mediterranean climatic areas. *In*: DI CASTRI, F. & MOONEY, H. A. (eds) Mediterranean Type Ecosystems. *Ecological Studies*, **7**, 157–167. Springer-Verlag, New York.

LACEY, W. S. 1974. Some new African Gondwana plants. *In*: *Symposium of Morphological and Stratigraphical Paleobotany: Papers*. Birbal Sahni Institute of Paleobotany Special Publication, **2**, 34–41.

—— 1975. Some problems of 'mixed' floras in the Permian of Gondwanaland. *In*: CAMPBELL, K. S. W. (ed.) *Gondwana Geology*, Australian National University Press, Canberra, 125–134.

LAUDON, T. S., LIDKE, D. J., DELEVORYAS, T. & GEE, C. T. 1987. Sedimentary rocks of the English Coast, eastern Ellsworth Land, Antarctica. *In*: MCKENZIE, G. D. (ed.) *Gondwana Six: Structure, Tectonics, and Geophysics*. Geophysical Monograph, **40**, 183–189. American Geophysical Union.

LEJAL-NICOL, A. 1987. Flores nouvelles du Paleozoique et du Mesozoique d'Egypte et du Soudan septentrional. *In*: KLITZSCH, E. & SCHRANK, E. (eds) *Research in Egypt and Sudan*. Berliner Geowissenschaftliche Abhandlungen Reihe A, **75.1**, 151–248.

LI, X. X. 1986. The mixed Permian Cathaysia-Gondwana flora. *The Palaeobotanist*, **35**, 211–222.

—— & YAO, Z. Q. 1979. Carboniferous and Permian floral provinces in east Asia. *In*: *Neuvième Congrès International de Stratigraphie et de Géologie du Carbonifère Compte Rendu*, **5**, 95–101. Washington and Champaign-Urbana.

—— & —— 1982. A review of recent research on the Cathaysia flora in Asia. *American Journal of Botany*, **69**, 479–486.

——, WU, Y. M. & FU, Z. B. 1985. Preliminary study on a mixed Permian flora from Xiagangjiang of Gerze District, Xizang and its palaeobiogeographic significance. *Acta Palaeontologica Sinica*, **24**, 150–170.

——, YAO, Z. Q. & DENG, L. H. 1982. An Early Late Permian flora from Toba, Qamdo District, eastern Xizang. *Paleontology of Tibet*, **5**, 17–40.

LIETH, H. 1975. Primary production of the major vegetation units of the world. *In*: LIETH, H. & WHITTAKER, R. H. (eds) *Primary Productivity of the Biosphere*, Springer-Verlag, New York, 203–215.

LIMARINO, C. O. & SPALLETTI, L. A. 1986. Eolian Permian deposits in west and northwest Argentina. *Sedimentary Geology*, **49**, 109–127.

LOTTES, A. L. & ROWLEY, D. B. 1990. Early and Late Permian reconstructions of Pangaea. *In*: MCKERROW, W. S. & SCOTESE, C. R. (eds) *Palaeozoic Palaeogeography and Biogeography*. Geological Society, London, Memoir, **12**, 383–395.

MABESOONE, J. M. 1977. Palaeozoic-Mesozoic deposits of the Piauí-Maranhão syneclise (Brazil): geological history of a sedimentary basin. *Sedimentary Geology*, **19**, 7–38.

MENNER, V. V., SARYCHEVA, T. G. & CHERNYAK, G. E. (eds) 1970. Stratigraphy of the Carboniferous and Permian deposits of Verkhoyanie. *Trudy Nauchno-Issledovatel'skogo Instituta Geologii Arktiki*, **154**.

MEYEN, S. V. 1982. The Carboniferous and Permian floras of Angaraland (a synthesis). (*Biological Memoirs*, **7**, 1–110.

—— 1987. *Fundamentals of Palaeobotany*. Chapman & Hall, London.

Nanjing Institute of Geology and Paleontology Academia Sinica (nigpas) 1974. *Plant Fossils of China Volume 1 Palaeozoic*. Science Press, Beijing.

NIE, S., ROWLEY, D. B. & ZIEGLER, A. M. 1990. Constraint on the location of the Asian micro continents in Palaeo-Tethys during the Late Palaeozoic. *In*: MCKERROW, W. S. & SCOTESE, C. R. (eds) *Palaeozoic Palaeogeography and Biogeography*. Geological Society, London, Memoir, **12**, 397–409.

ODREMAN-RIVAS, O. & WAGNER, R. H. 1979. Precisiones sobre algunas floras Carboniferas Permicas de Los Andes Venezolanos. *Boletin de Geologia*, **13**, 77–79.

PARRISH, J. T. & SPICER, R. A. 1988. Late Cretaceous terrestrial vegetation: a near-polar temperature curve. *Geology*, **16**, 22–25.

——, ZIEGLER, A. M. & SCOTESE, C. R. 1982. Rainfall patterns and the distribution of coals and evaporites in the Mesozoic and Cenozoic. *Palaeogeography, Palaeoclimatology, Palaeoecology*, **40**, 67–101.

PETRI, S. & FULFARO, V. J. 1983. *Geologia do Brasil*. Editoria da Universidade de São Paulo, São Paulo.

PLUMSTEAD, E. P. 1973. The Late Palaeozoic Glossopteris flora. *In*: HALLAM, A. (ed.) *Atlas of Palaeobiogeography*, Elsevier, New York, 187–205.

RAVEN, P. H. 1973. The evolution of Mediterranean floras. *In*: DI CASTRI, F. & MOONEY, H. A. (eds) Mediterranean Type Ecosystems. Eco-

logical Studies, 7, 213-224. Springer-Verlag, New York.
RAYMOND, A., PARKER, W. C. & PARRISH, J. T. 1985. Phytogeography and paleoclimate of the Early Carboniferous. *In*: TIFFNEY, B. H. (ed.) *Geological Factors and the Evolution of Plants*, Yale University Press, 169-222.
READ, C. B. & MAMAY, S. H. 1964. Upper Palaeozoic floral zones and floral provinces of the United States. U.S.G.S. Professional Paper, **454-K**.
RETALLACK, G. J. 1980. Late Carboniferous to Middle Triassic megafossil floras from the Sydney Basin. *In*: HERBERT, C. & HELBY, R. (eds) *A Guide to the Sydney Basin*. Geological Survey of New South Wales Bulletin, **26**, 384-430.
RONOV, A., KHAIN, V. & SESLAVINSKY, K. 1984. *Atlas of Lithological-Paleogeographical Maps of the World — Late Precambrian and Palaeozoic of Continents*. USSR Academy of Sciences, Leningrad.
ROSS, C. A. & ROSS, J. R. P. 1979. Permian. *In*: ROBISON, R. A. & TEICHERT, C. (eds) *Treatise on Invertebrate Paleontology, Part A, Introduction*. Geological Society of American & University of Kansas, Boulder & Lawrence, A291-A350.
ROWLEY, D. B., RAYMOND, A., PARRISH, J. T., LOTTES, A. L., SCOTESE, C. R. & ZIEGLER, A. M. 1985. Carboniferous paleogeographic, phytogeographic, and paleoclimatic reconstructions. *International Journal of Coal Geology*, **5**, 7-42.
SCOTT, A. C. 1980. The ecology of some Upper Palaeozoic floras. *In*: PANCHEN, A. L. (ed.) *The Terrestrial Environment and the Origin of Land Vertebrates*. Systematics Association Special Volume, **15**, 87-115. Academic Press, London.
SHARIEF, F. A. 1982. Lithofacies distribution of the Permian-Triassic rocks in the Middle East. *Journal of Petroleum Geology*, **4**, 299-310.
STEWART, W. N. 1983. *Paleobotany and the Evolution of Plants*. Cambridge University Press, Cambridge.
STRAHLER, A. N. & STRAHLER, A. H. 1987. *Modern Physical Geography*. John Wiley & Sons, New York.
SURANGE, K. R. 1975. Indian Lower Gondwana floras: a review. *In*: CAMPBELL, K. S. W. (ed.) *Gondwana Geology*, Australian National University Press, Canberra, 135-147.
SZATMARI, P., CARVALHO, R. S. & SIMOES, I. A. 1979. A comparison of evaporite facies in the Late Palaeozoic Amazon and the Middle Cretaceous South Atlantic Salt Basins. *Economic Geology*, **74**, 432-477.
TAKHTAJAN, A. 1986. *Floristic regions of the World*. University of California Press, Berkeley.
TAYLOR, T. N. 1981. *Paleobotany*. McGraw-Hill Book Company, New York.
THOMAS, B. A. & SPICER, R. A. 1987. *The Evolution and Palaeobiology and Land Plants*. Croom Helm, London.
TROLL, C. 1970. Das 'Baumfarnklima' und die Verbreitung der Baumfarne auf der Erde. *In*: WILHELMY, H., BLUME, H., SCHRÖDER, K. H. & KARGER, A. (eds) *Beiträge zur Geographie der Tropen und Subtropen*. Tübinger Geographische Studien, **34**, 179-189.
TOUZETT, P. J. & SANZ, V. R. 1985. Present and future of petroleum exploration in the Subandean Basins of Peru. *In*: *II Bolivarian Symposium Petroleum Exploration in the Subandean Basins*. Petroperu, Petroleos del Peru.
VAKHREMEEV, V. A., DOBRUSKINA, I. A., MEYEN, S. V. & ZAKLINSKAYA, E. D. 1978. *Paläeozoische und Mesozoische Floren Eurasiens und die Phytogeographie dieser Zeit*. VEB Gustav Fischer Verlag Jena.
VEEVERS, J. J. (ed.) 1984. *Phanerozoic Earth History of Australia*. Clarendon Press, Oxford.

—— & POWELL, C. M. 1987. Late Palaeozoic glacial episodes in Gondwanaland reflected in transgressive-regressive depositional sequences in Euramerica. *Geological Society of America Bulletin*, **98**, 475-487.
VINOGRADOV, A. P. (ed.) 1969. *Atlas of the Lithological-Paleogeographical Maps of the USSR, Volume 2 — Devonian, Carboniferous and Permian*. Akademia Nauk SSSR, Moscow.
VISSER, J. N. J. 1987. The influence of topography on the Permo-Carboniferous glaciation in the Karoo Basin and adjoining areas, Southern Africa. *In*: MCKENZIE, G. D. (ed.) *Gondwana Six: Stratigraphy, Sedimentology and Paleontology*. Geophysical Monograph, **41**, 123-129. American Geophysical Union.
WAGNER, R. H. & MARTINEZ-GARCIA, E. 1982. Description of an Early Permian flora from Asturiasand comments on similar occurrences in the Iberian Peninsula. *Trabajos de Geologia*, **12**, 273-287.
——, SOPER, N. J. & HIGGINS, A. K. 1982. A Late Permian flora of Pechora affinity in North Greenland. *Grønlands Geologiske Undersøgelse*, **108**, 5-13.
WALTER, H. 1985. *Vegetation of the Earth 3rd edition*. Springer-Verlag, Berlin.
—— & LIETH, H. 1967. *Klimadiagramm-Weltatlas*. VEB Gustav Fischer, Jena.
WANG, Z. Q. 1985. Palaeovegetation and plate tectonics: palaeophytogeography of north China during Permian and Triassic times. *Palaeogeography, Palaeoclimatology, Palaeoecology*, **49**, 24-45.
WU, S. Z. 1983. Preliminary study on the Permian phytogeography of Xinjiang. *In*: *Paleobiogeographic Provinces of China*, Science Press, Beijing, 91-99.
YANG, Z. Y., CHENG, Y. Q. & WANG, H. Z. 1986. *The Geology of China*. Clarendon Press, Oxford.
YAO, Z. Q. 1983. Ecology and taphonomy of Gigantopterids. *Bulletin of Nanjing Institute of Geology and Palaeontology*, **6**, 63-84.
Yunnan Stratigraphic Group 1978. *Regional Stratigraphic Data on Southwestern China (Yunnan)*. Geological Publishing House.
ZHANG, S. Z. & HE, Y. L. 1985. Late Palaeozoic palaeophytogeographic provinces in China and their relationships with plate tectonics. *In*: LU, Y. H. (ed.) *Palaeontologia Cathayana*, **2**, 77-86. Science Press, Beijing.
ZHAO, X. H., LIU, L. J. & HOU, J. H. 1987. Carboniferous and Permian flora from the coal-bearing strata of southeastern Shanxi, North China. *In*: *Late Palaeozoic Coal-Bearing Strata and Biota from Southeastern Shanxi, China*, Nanjing University Press, 61-138.
ZIEGLER, A. M., RAYMOND, A. L., GIERLOWSKI, T. C., HORRELL, M. A., ROWLEY, D. B. & LOTTES, A. L. 1987. Coal, climate and terrestrial productivity: the present and early Cretaceous compared. *In*: SCOTT, A. C. (ed.) *Coal and Coal-bearing Strata: Recent Advances*. Geological Society, Special Publication, **32**, 25-49.
——, SCOTESE, C. R. & BARRETT, S. F. 1982. Mesozoic and Cenozoic paleogeographic maps. *In*: BROSCHE, P. & SUNDERMANN, J. (eds) *Tidal Friction and the Earth's Rotation II*, Springer-Verlag, Berlin, 240-252.
——, SCOTESE, C. R., MCKERROW, W. S., JOHNSON, M. E. & BAMBACH, R. K. 1979. Palaeozoic paleogeography. *Annual Review of Earth and Planetary Sciences*, **7**, 473-502.
ZIEGLER, P. A. 1982. *Geological Atlas of Western and Central Europe*. Shell Internationale Petroleum Maatschappij B. V.
ZWALLY, H. J., COMISO, J. C., PARKINSON, C. L., CAMPBELL, W. J., CARSEY, F. D. & GLOERSEN, P. 1983. *Antarctic Sea Ice, 1973-1976: Satellite Passive-Microwave Observations*. NASA SP-459. Washington, D.C.

Appendix 1. *References for Sakmarian lithological and floral data*

Biome	Region	Formation	Type of Information	Reference
8	NE USSR	Khaldanskaya	1 floral list	Menner et al. 1970, 40, 141
8	NE USSR	1st half Lr. Permian	5 map localities	Vakhrameev et al. 1978, 77
6	Siberia	Promezhtochny	1 composite floral list	Meyen 1982, 87, col. 5
6	Siberia	Up. Burguklinsky	1 composite floral list	Meyen 1982, 87, col. 15
6	SE USSR	Dunaiskaya	1 floral list	Meyen 1982, 97, col. 1
6	SE USSR	Dunaiskaya	1 floral list	Meyen 1982, 97, col. 13
6	N Asia	1st half Lr. Permian	13 map localities	Vakhrameev et al. 1978, 77
4	Kazakhstan	1st half Lr. Permian	1 list, 2 localities	Vakhrameev et al. 1978, 81
3	Europe–USSR	Asselian-Sakmarian	14 evaporite localities	Vinogradov 1969, map 24
3	Svalbard	Gipshuken Formation	1 evaporite locality	Birkenmajer 1981, 288
3	W USA	Wolfcampian	2 evaporite localities	Cook & Bally 1975
2	N Europe	Autun	1 list, 10 localities	Vakhrameev et al. 1978, 76
2	N Spain	Un. Roja Inf.	1 list	Broutin & Gispert 1985, 56
2	USA	Wolfcamp	1 list, 4 localities	Read & Mamay 1964, K12
1	N China	Shansi	1 composite list	Zhao et al. 1987, 107
1	China/Korea	Shansi	11 map localities	Li & Yao 1979, 3
1	S China	Lr. Qixia	1 floral list	Yunnan 1978, 93
1	S China	Shansi Equiv.	2 map localities	Li & Yao 1979, 5
1	Sumatra	Unknown	1 list, 1 locality	Asama et al. 1975, 80
1	S Spain	Early Permian	1 list, 1 locality	Wagner & Martinez-G. 1982, 284
1	Venezuela	Palmarito	1 list, 1 locality	Odreman-R. & Wagner 1979, 78
1	Texas	Belle Plains	1 composite floral list	Read & Mamay 1964, K14
1	Algeria	Autunian	1 microfloral locality	Doubinger & Fabre 1983, 113
2	Egypt	Qiseib	1 list, 1 locality	Lejal-Nicol 1987, 170
2	Thailand	Phuket	2 lists, 2 localities	Bunopas 1981, 196
3	Brazil	Evap. Cycle V	4 evaporite localities	Szatmari et al. 1979, fig. 5
3	Brazil	Carauri	1 evaporite locality	da Silva 1988, 2431
3	Brazil	Copacabana	1 evaporite locality	Touzett & Sanz 1985, 83
5	Argentina	N. Lubecka	1 list, 1 locality	Archangelsky & Cuneo 1984, 146
5	Brazil	Pedro de Fogo	3 localities	Bigarella 1973, 64
5	Irian Jaya	Unknown	1 list, 1 locality	Asama et al. 1975, 90
6	India	Karharbari	1 list, 2 localities	Surange 1975, 139
6	Argentina	Bonete	1 list, 1 locality	Archangelsky & Cuneo 1984, 145
6	Falkland Islands	B. Choiseul	1 list, 1 locality	Archangelsky & Cuneo 1984, 145
6	Brazil	Itarare	1 floral list	Cazzulo-K. & Guerra-S. 1985, 399
6	Madagascar	Karharbari Equiv.	1 list, 1 locality	Appert 1977, 1
6	Antarctica	Unknown	1 locality	Laudon et al. 1987, 185
7	Argentina	La Colina	1 floral list	Caminos 1979, 258
7	Argentina	La Colina, etc.	3 localities	Limarino & Spalletti 1986, 109
8	Australia	Shoalhaven-Dalwood	1 locality	Retallack 1980, 396
8	Antarctica	Lr. Weller	1 locality	Barrett & Kyle 1975, 344
8	Antarctica	Fairchild	1 locality	Elliot 1975, 512
9	Australia	Seaham	1 floral list	Retallack 1980, 392
0	Australia	Various	13 glacial localities	Veevers 1984, 236
0	Arabia	Sakmarian	3 glacial localities	Besems & Schuurman 1987, 50
0	Zaire	Lukaga	2 glacial localities	Cahen & Lepersonne 1981, 45
0	S Africa	Up. Dwyka	1 glacial locality	Visser 1987, 128
0	Brazil	Itarare	4 glacial localities	Petri & Fulfaro 1983, 138

Appendix 2. *References for Kazanian lithological and floral data*

Biome	Region	Formation	Type of Information	Reference
8	N Asia	Dugalakhsk, etc.	2 lists, 14 localities	Vakhrameev *et al.* 1978, 94
6	Siberia	Uskatsky-Kaz.	1 composite list	Meyen 1982, 87 col. 9
6	Siberia	Pelyatkinsky	1 composite list	Meyen 1982, 87 col. 17
6	Siberia	1st half Up. Permian	3 map localities	Vakhrameev *et al.* 1978, 88
6	Mongolia	Unknown	1 map locality	Durante 1976, 110
5	N USSR	Pechora	1 composite list	Meyen 1982, 73
5	SE USSR	Vladivostok	3 floral lists	Meyen 1982, 97, cols. 5, 11, 17
5	N Asia	1st half Up. Permian	6 map localities	Vakhrameev *et al.* 1978, 88
5	Mongolia	Tabun-Tologoy	1 list, 1 locality	Durante 1971, 513
5	N Greenland	Unknown	1 list, 1 locality	Wagner *et al.* 1982, 7
4	USSR	Unknown	1 composite list	Meyen 1982, 74
4	USSR	1st half Up. Permian	7 map localities	Vakhrameev *et al.* 1978, 88
3	Europe	Zechstein	9 map localities	Ziegler 1982, encl. 14
3	USSR	Ufimian-Kazanian	6 map localities	Vinogradov 1969, map 28
3	E Greenland	Foldvrik Creek	1 map locality	Escher & Watt 1976, 306
3	W USA	Guadalupian	4 map localities	Cook & Bally 1975
2	Europe	Zechstein	1 list, 4 localities	Vakhrameev *et al.* 1978, 90
1	N China	Lr. Shihhotse	1 composite list	Zhao *et al.* 1987, 107
1	China/Korea	Lr. Shihhotse	14 map localities	Li & Yao 1979, 2
1	Japan	Kashewadaira	1 list, 1 locality	Asama 1984, 239
1	W China	Unknown	1 list, 3 localities	Wu 1983, 96
1	Tibet	Toba	1 list, 1 locality	Li *et al.* 1982, 20
1	Thailand	Dan Sai	1 list, 1 locality	Asama *et al.* 1975, 85
1	Malaysia	Unknown	1 list, 1 locality	Asama *et al.* 1975, 86
1	Iran	Gheshlagh	1 microfloral list	Chateauneuf & Stampfli 1978, 49
1	Iraq	Gaara	1 list, 1 locality	Ctyroky 1973, 386
1	Turkey	Unknown	1 list, 1 locality	Archangelsky & Wagner 1983, 87
1	Texas	Up. Vale	1 list, 1 locality	Read & Mamay 1964, K15
2	Arabia	Khuff	1 list, 1 locality	El-Khayal & Wagner 1985, 22, 24
3	Arabia	Up. Permian	3 evaporite localities	Sharief 1982, 303
3	Brazil	Motuca	3 evaporite localities	Mabesoone 1977, 10
3	Brazil	Fonte-Boa	2 evaporite localities	da Silva 1988, 2432, 2438
5	Argentina	La Golondrina Sup.	1 list, 1 locality	Archangelsky & Cuneo 1984, 146
5	Irian Jaya	Unknown	1 list, 1 locality	Asama *et al.* 1975, 90
5	Tibet	Tianzhanong	1 list, 1 locality	Li *et al.* 1985, 161
6	India	Raniganj	1 list, 2 localities	Surange 1975, 139
6	S Africa	Up. Ecca	1 list, 3 localities	Anderson & Anderson 1985, 25
6	Zambia	Up Madumabisa	1 list, 1 locality	Lacey 1974, 36
6	Australia	Various	1 composite list	Gould 1975, 460
6	E Australia	Tomago	2 map localities	Retallack 1980, 398
6	W Australia	Wagtiina	2 map localities	G.S.W. Australia 1975, 235, 342
6	Tasmania	Cygnet	1 map locality	Clarke & Banks 1975, 465
6	Antarctica	Queen Maud	2 map localities	Elliot 1975, 506, 510
7	S America	Passa Dois, etc.	3 aeolian sandstones	Limarino & Spalletti 1986, 122

Appendix 3. *Synonymies employed on Tables 2 and 3*

Older usage	Usage herein	Justification
Bicoemplectopteris	*Gigantonoclea*	NIGPAS 1974, 127
Tricoemplectopteris	*Gigantonoclea*	NIGPAS 1974, 129
Protoblechnum	*Compsopteris*	NIGPAS 1974, 115
Validopteris	*Fascipteris*	NIGPAS 1974, 99
Shirakiopteris	*Alethopteris*	NIGPAS 1974, 113
Vertebraria	*Glossopteris*	Gould & Delevoryas 1977, 388
Lepidophloios	*Lepidodendron*	Taylor 1981, 134
Umbellaphyllites	*Annulina*	Meyen 1982, 14
Dorycordaites	*Cordaites*	Taylor 1981, 427
Poacordaites	*Cordaites*	Taylor 1981, 427
some *Noeggerathiopsis**	*Cordaites*	Meyen 1982, 68

* identified by Shvedov in Menner *et al.* 1970

Palaeozoic Geography

Reconstruction of the Laurasian and Gondwanan segments of Permian Pangaea

ANN L. LOTTES & DAVID B. ROWLEY

Palaeogeographic Atlas Project, Department of Geophysical Sciences, 5734 S. Ellis Avenue, The University of Chicago, Chicago, IL 60637, USA

Abstract: A reappraisal of the geological and geometrical constraints on the fits of the continents around the Atlantic, Indian and circum-Antarctic oceans is used as a base upon which to re-examine Early and Late Permian palaeomagnetic data. The palaeomagnetic poles from each of the three main Late Palaeozoic palaeo-continents: Eurasia, North America, and Gondwana define statistically well determined means for the Early and Late Permian. When these poles are reconstructed, the means, and their associated A_{95} cones of confidence do not superimpose. The lack of superposition of Late Palaeozoic palaeomagnetic poles on Wegener-style Pangaea reconstructions based on Mesozoic and younger seafloor spreading geometries has been noted for some time. Most solutions to this problem have involved alternative 'tight fits' of the Pangaean continents for the Late Palaeozoic. Solutions invoking the non-dipole behaviour of the magnetic field have also been proposed. We examine the geometric consequences of the 'tight fit' Mesozoic-based Pangaean reconstructions that allow better fit of the palaeomagnetic data and conclude that the magnitudes of overlap (650–950 km) required by these modifications are not compatible with the geological and tectonic evidence from these regions. We prefer to use a 'looser' Pangaea fit that does not necessarily result in the superposition of palaeomagnetic poles. However, when the individual poles from Europe, North America, and Gondwana are combined into a global mean pole, it is observed that both the 'tight' and the 'loose' fit yield virtually indistinguishable mean pole positions. Considering the strong geological and geometrical arguments against the palaeomagnetically derived fits and the lack of a clear statistical difference between the two we use our revised Pangaean reconstruction to determine palaeo-latitudinal framework of the Eurasian, North American, and Gondwanan segments of Pangaea for the Early and Late Permian. The validity of these palaeo-latitudinal reconstructions is tested with climatically sensitive floristic and lithological data by Ziegler, and found to be satisfactory.

We present revised Early and Late Permian reconstructions of the major continents that comprise the western portion of Pangaea (i.e. Gondwana and Laurasia) incorporating important and in some cases newly recognized geological and geometrical constraints on the fits of the continents around the Atlantic, Indian (including Somalian and Mozambique Basins), and circum-Antarctic ocean basins. Recent summaries of the Early and Late Permian palaeomagnetic pole data from North America (Van der Voo, in press), Eurasia (Frei & Cox 1987; Khramov *et al.* 1981), and Gondwana (Van der Voo *et al.* 1984; Brock 1981; Klootwijk & Radhakrishnamurty 1981) are compared and it is shown that the palaeomagnetic data from these continents constitute distinct populations that are not superimposed when continents are re-assembled in a Wegener-style Pangaea based on reconstruction of the post-Palaeozoic ocean basins. This observation is not new (see Morel & Irving 1981*a*, *b*; Smith *et al.* 1981; Van der Voo *et al.* 1984). The lack of superposition of Permian palaeomagnetic poles has been used to argue in favour of modifications to the Wegener-style fits of Bullard *et al.* (1965) such as those suggested by Van der Voo & French (1974), Morel & Irving (1981*a*), and Frei & Cox (1987), as well as alternative possibilities such as long-term non-dipole fields (Briden *et al.* 1970). In this paper we examine the geometric consequences of two of the proposed modifications to the standard fits that yield better superposition of Late Palaeozoic APW paths. In particular we focus on the proposed North America–Eurasia fit of Frei & Cox (1987), and a modified Van der Voo & French (1974) reconstruction of the Gondwana–North America fit and show that these fits yield unacceptable overlaps along the continental margins of the Atlantic. Geological relations (Hallam 1983) have long been used to discount the Morel & Irving (1981*a*, *b*) reconstruction that has generally been labelled as Pangaea B by most workers (Van der Voo *et al.* 1984). The statistics of the palaeomagnetically determined fits are compared with those based on our preferred fit of the Atlantic-bordering continents. The results indicate that the positions of the global mean poles calculated using the two fits are not statistically different.

The orientations and palaeo-latitudes of the Early and Late Permian reconstructions that we favour are derived by averaging individual palaeomagnetic poles from all continents into a global mean based on our geometrically/geologically-derived fits. As discussed elsewhere in this volume (Ziegler 1990), this approach yields satisfactory results when compared with phytogeographically/palaeoclimatically determined palaeolatitudes. The reconstruction of Asian and Middle Eastern blocks is treated separately (Nie *et al.* 1990), and these blocks are not incorporated in the reconstructions presented within this paper.

Fits

Any discussion of Late Palaeozoic palaeogeography must begin with the reconstruction of Mesozoic and Cenozoic plate motions. Particular emphasis is placed on the opening of the various Jurassic and younger ocean basins related to the break-up of Pangaea, because of our inability, as yet, to produce full palinspastic restorations of all subsequent intraplate deformation. The reconstruction of traditional Laurasia, which includes North America*, Eurasia (Europe*, Iberia, Kazakhstania, and Siberian Platform*), and reconstructed Gondwana (comprising South America*, Patagonia*, South Africa*, North Africa, Morocco*, Northeast Africa, East Africa, Arabia, Madagascar*, India*, East Antarctica, Australia, Antarctic Peninsula, Marie Byrd Land, Campbell Plateau/South Island New Zealand, and North Island New Zealand) is shown in Fig. 1, in fixed South American coordinates, and the reconstruction parameters are listed in Table 1. The reconstructions incorporate some intraplate deformation, including deformation within the circum-North Atlantic region (Rowley & Lottes 1988), circum-Gulf of Mexico region (Pindell 1985; Pindell *et al.* 1988; Rowley & Pindell, in press), northern South America (Pindell 1985; Pindell *et al.* 1988; Rowley & Pindell, in press), southern South America (Rowley 1988), Africa (Pindell & Dewey 1982; Pindell 1985; Pindell *et al.* 1988; Rowley & Pindell, in press; Rowley 1988), and Antarctica (modified from Dalziel *et al.* 1987; Storey *et al.* 1988; Lawver & Scotese 1987). The intraplate deformation that has been incorporated has been independently constrained using an iterative combi-

---

* Continents for which Early and/or Late Permian palaeomagnetic data are available and have been used in our analysis.

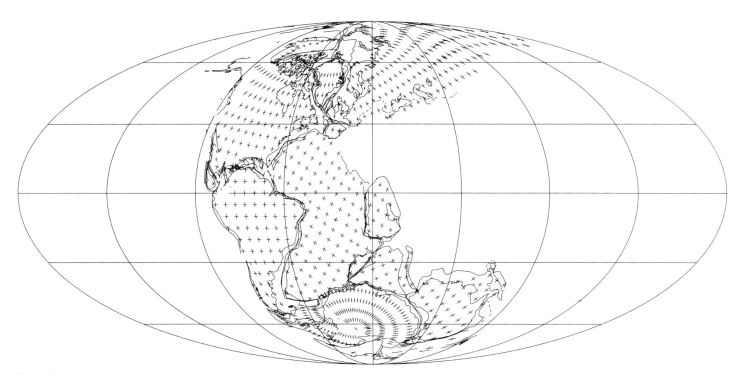

Fig. 1. Reconstruction of Pangaea with South America fixed. The rotation parameters used to produce this reconstruction are listed in Table 1.

Table 1. *Rotation Parameters for Permian Pangaean reconstruction*

Plate	Pole Position		Rotation Angle
	Latitude (°N)	Longitude (°E)	(all rotations are clockwise)
South Africa	0.00	0.00	0.00*
North America	61.30	−16.82	79.54
Greenland	58.87	−4.44	72.28
Ellesmere	59.43	−3.41	66.29
N Mexico	64.28	8.32	67.42
Yucatan	64.26	76.95	62.12
S Florida	59.92	−18.19	77.05
Chortis	49.19	73.78	62.18
S Mexico	62.68	40.79	65.95
South America	46.82	−30.54	55.88
Ecuador	40.64	−33.78	59.53
Venezuela	44.20	−30.13	56.97
Patagonia	43.78	−31.92	57.80
Eurasia	47.04	−2.59	64.58
Iberia	41.10	−2.92	30.17
Hatton Bank	49.68	−4.54	71.63
Rockall Bank	48.74	−3.55	68.05
Spitzbergen	47.85	−4.12	65.98
North Slope	30.66	31.60	53.51
Lomonosov Ridge	59.81	−12.15	74.64
Aldan Shield	47.04	−2.59	64.58
India	−30.27	−140.18	63.06
Arabia	−45.23	−163.72	7.12
Himalayas	−29.72	−139.16	60.25
NW Africa	9.34	5.70	7.82
Morocco	11.06	4.99	7.38
NE Africa	−16.30	41.71	2.53
E Africa	−22.79	40.96	3.44
Australia	−29.16	−57.19	54.02
Antarctica	−12.25	−28.08	59.07
Marie Byrdland	−16.81	−29.24	64.00
N New Zealand	−7.12	−57.44	76.50
S New Zealand	24.27	−60.16	99.06
Antarctic Peninsula	−36.47	−20.16	81.98
Madagascar	−12.53	−54.96	17.46

* Fixed in present coordinates

nation of geometrical and geological arguments that are discussed in these papers, and are not repeated here. All of this intraplate deformation relates to the Late Triassic and younger extensional and seafloor spreading history of the various ocean basins. Our reconstructions do not incorporate Late Palaeozoic and Early Mesozoic extensional or contractional deformation. Important domains of unincorporated Late Palaeozoic and Early Mesozoic extensional deformation include the North Sea (Ziegler 1982, 1988) and its continuation to the north along East Greenland (Surlyk et al. 1986; Ziegler 1988), Barents Shelf (Rønnevik & Jacobsen 1984), West Siberian Lowland (Rudkevich 1970), and Karoo rifts of Africa (Baker 1971; Pallister 1971; Dingle et al. 1983), Madagascar (Boast & Nairn 1982), India (Sastri et al. 1982; Basu & Shrivastava 1981; Mitra et al. 1982), and East Antarctica (Fedorov et al. 1982). Important domains of unincorporated Late Palaeozoic and Early Mesozoic contractional deformation include southern South America (Kilmurray 1975), Cape Foldbelt of South Africa (Dingle et al. 1983), Ellesworth–Whitmore Mountains block (Dalziel et al. 1987), Pensicola Mountains (Dalziel et al. 1987), and eastern Australia (Veevers 1984), all components of Du Toit's (1937) Samfrau Geosyncline, as well as shortening within the Uralian, northern Kazakhstanian, and Taymyr regions (Khain 1986). In addition, we have not attempted a reconstruction of the allochthonous and/or parautochthonous assemblages of the circum-Pacific region (see Howell 1985). Finally, the reconstructions do not palinspastically restore regions that have experienced Mesozoic and Cenozoic contractional and/ or extensional deformation that is particularly important in the Alpine–Himalayan belt and around the circum-Pacific rim. So, for example, the western margins of North and South America are depicted on the reconstructions with their present outlines, even though this clearly does not reflect their Permian configuration.

In order to produce Early and Late Permian Pangaean reconstructions, several assumptions are necessary. One is that the intraplate deformation mentioned above is sufficiently small to allow the use of what are essentially Late Triassic/Early Jurassic reconstructions of the post-Pangaean ocean basins as a foundation for the Permian reconstruction. This assumption is probably reasonable for the majority of Gondwana since contractional deformation is restricted to marginal portions of the continent, and the Karoo rift basins tend to be quite narrow (<100 km) and therefore unlikely to represent substantial amounts of distortion. A second assumption is that regions that have been traditionally included within Laurasia, and in particular, Eurasia [i.e. the European and Siberian platforms, as well as Kazakhstania (see Scotese et al. 1979)] were already amalgamated prior to the mid-Early Permian. As will be discussed below, the European and Siberian platforms, apparently also including northern Kazakhstan appear to have remained fixed with respect to each other within the resolution of the palaeomagnetic data since the Early Permian. One possibility is that Permian shortening within the Uralian orogen is matched within the resolution of the palaeomagnetic data by later Permo-Triassic extension in the West Siberian lowlands. A third area of potential concern is in using a Late Triassic/Early Jurassic fit of Gondwana and North America. Although some deformation within the Appalachian/Ouachita/ Marathon orogen extends into the Early Permian, such as that documented by the folding of the Dunkard Group in the central Appalachians (Rodgers 1970), the vast majority of the shortening, as demonstrated by subsidence histories of the various foredeep basins, and stratigraphic overlap was completed by the end of the Carboniferous (Kluth & Coney 1981; Dewey 1982; Ross 1986). Thus it appears reasonable to assume that the Late Triassic/Early Jurassic reconstruction of the major Pangaean continents should provide an appropriate base upon which to examine Permian palaeomagnetic data and we have used the Late Triassic/Early Jurassic reconstruction of Pangaea (Fig. 1) as the foundation for our analysis.

Palaeomagnetic data

Palaeomagnetic data restricted to the Early and Late Permian, respectively, from the traditional Laurasian and Gondwanan continents have been examined with the goal of producing palaeomagnetically consistent reconstructions of Pangaea. We have used existing compilations for various regions of the traditional Laurasian and Gondwanan continents and have therefore accepted the reliability criteria employed by these workers in their respective papers, with the exception of the Khramov et al. (1981) compilation, in which we have rejected some (<10%) poles because they clearly lie well outside the well defined populations, presumably reflecting structural reorientation or remagnetization.

Permian poles

Eurasia. We have used existing compilations of European (Frei & Cox 1987) and Siberian (Khramov et al. 1981) palaeomagnetic poles as a basis for determining mean poles for the Early and Late Permian. The poles used and the resulting means are listed in Table 2, and are plotted in Fig. 2. Early and Late Permian poles from Europe and Siberia, treated independently and combined yield statistically identical results. This suggests that at least within the resolution of the palaeomagnetic data Europe and Siberia can be assumed to have been rigidly attached since the Early Permian. As mentioned above, interpretation of subsidence history of the Uralian foredeep and folding within the Urals suggests that some Permian motion between Siberia and the European platform should be expected, but there is no geological evidence upon which to suggest substantial (>5–10°) relative motion since the Early Permian. The Early and Late Permian mean poles determined here are similar (i.e. they are essentially encompassed within the A_{95} determined here) to those previously determined by Irving & Irving (1982), Westphal et al. (1986), and Frei & Cox (1987).

North America. Table 3 lists the palaeomagnetic poles used for North America and resulting means for the Early and Late Permian based on the compilation of Van der Voo (in press). The Early and Late Permian pole data for North America are plotted in Fig. 3. Both Early and Late Permian sets of poles yield tight clusters and statistically well defined mean poles, although the number of studies is still small. The means determined here are similar to those previously determined by Van der Voo et al. (1984), Irving & Irving (1982), and Gordon et al. (1984).

Gondwana. Early and Late Permian palaeomagnetic poles from Gondwana are listed in Table 4 and depicted in Fig. 4. The poles used in this analysis are based on compilations by Van der Voo et al. (1984), Brock (1981), and Klootwijk & Radhakrishnamurty (1981) for South America and Africa, Madagascar, and India, respectively. The mean poles are calculated using rotation parameters listed in Table 1. Although the number of poles that have been used from the Gondwanan continents is small in comparison with Eurasia, the poles, when examined in reconstructed coordinates, define clusters and statistically reasonable (i.e. $A_{95} < 15°$, and $\kappa > 15$) Early and Late Permian mean poles. Note, however, that the rate of apparent polar wander between the Early and Late Permian implied by the Gondwanan data is approximately 3 times faster than that implied for Eurasia and almost 10 times faster than that of North America. This requires either substantial Permian motion of Gondwana with respect to Laurasia or some error in the Permian palaeomagnetic data of Gondwana or Laurasia. We follow Van der Voo et al. (1984) in emphasizing the Early Permian over the Late Permian due to the larger number of studies available for the former.

Paleomagnetic poles and Pangaean reconstructions

When the Early and Late Permian palaeo-continental (Eurasia, North America, and Gondwana) mean poles and their associated

Table 2. *Eurasian palaeomagnetic pole positions*

Late Permian European palaeomagnetic North Poles from the compilation of Frei & Cox (1987)

Pole Position		Reference and/or Catalog Number	Pole Age
Lat. (°N)	Long. (°E)		
55.0	142.0	Zijderveld (1975)	248–258
46.7	148.4	Van den Ende (1977)	248–258
48.7	175.2	Konrad & Nairn (1972)	P
50.7	161.3	Thorning & Abrahamsen (1980)	P
47.0	157.0	Van Everdingen (1960)	P
50.2	157.0	European Mean Pole, $N=5$, $A_{95}=8.3$, $\kappa=85.0$	

Late Permian Siberian palaeomagnetic North Poles from the compilation of Khramov et al. (1981)

Pole Position		Reference and/or Catalog Number	Pole Age
Lat. (°N)	Long. (°E)		
42.00	164.00	7–11	Pu
46.00	170.00	7–01	Pu
44.00	164.00	7–02	Pu
46.00	173.00	7–49	Pu
57.00	170.00	7–31	Pu
47.00	161.00	7–32	Pu
43.00	161.00	7–33	Pu
48.00	161.00	7–48	Pu
40.00	160.00	7–23	Pu
43.00	171.00	7–24	Pu
43.00	168.00	7–25	Pu
56.00	165.00	7–26	Pu
55.00	173.00	7–27	Pu
50.00	169.00	7–28	Pu
45.00	180.00	7–29	Pu
42.00	168.00	7–15	Pu
39.00	168.00	7–16	Pu
54.00	168.00	7–06	Pu
42.00	168.00	7–13	Pu
41.00	173.00	7–14	Pu
50.00	160.00	7–35	Pu
50.00	160.00	7–36	Pu
44.00	167.00	7–37	Pu
58.00	157.00	7–07	Pu
50.00	160.00	7–47	Pu
46.00	160.00	7–09	Pu
45.00	147.00	7–45	Pu
60.00	152.00	7–40	Pu
47.00	135.00	7–10	Pu
47.7	164.1	Siberian Mean Pole, $N=29$, $A_{95}=2.7$, $\kappa=97.3$	
48.1	163.1	Eurasian Mean Pole, $N=34$, $A_{95}=2.6$, $\kappa=93.6$	

Early Permian European palaeomagnetic North Poles from the compilation of Frei and Cox (1987)

Pole Position		Reference and/or Catalog Number	Pole Age
Lat. (°N)	Long. (°E)		
46.20	175.20	Nijenhuis (1961) and Berthold et al. (1975)	Pl-m
46.10	156.20	Kruseman (1962)	Saxonian
41.90	168.60	Krs (1968)	Pl-m
37.80	169.70	Krs (1968)	Pl
44.20	175.50	Birkenmajer et al. (1968)	Pl
42.50	168.30	Birkenmajer & Nairn (1964)	Pl
46.60	167.20	Roche et al. (1962)	Pl
43.30	168.40	Kruseman (1962)	Pl
40.30	168.10	Krs (1968)	Pl
41.30	159.30	Krs (1968)	Pl
40.10	157.70	Krs (1968)	Pl
40.10	162.30	Krs (1968)	Pl
45.60	167.10	Rother (1971) and Mauritsch & Rother (1983)	Stef.-Sax.
37.30	173.00	Rother (1971) and Mauritsch & Rother (1983)	Stef.-Sax.
38.80	160.40	Halvorsen (1970)	276
41.60	168.10	Westphal (1972)	Pl-Cu
48.00	163.00	Cornwell (1967)	286
38.00	167.00	Bylund & Patchett (1977)	287
33.20	170.60	Priem et al. (1968) and Mulder (1971)	288
48.70	175.20	Konrad & Nairn (1972)	P
50.70	161.30	Thorning & Abrahamsen (1980)	P
47.00	157.00	Van Everdingen (1960)	P
42.8	166.4	Eurasian Mean Pole, $N=22$, $A^{95}=2.4$, $\kappa=174.9$	

Early Permian Siberian palaeomagnetic North Poles from the compilation of Khramov et al. (1981)

Pole Position		Reference and/or Catalog Number	Pole Age
Lat. (°N)	Long. (°E)		
40.00	162.00	7–19	Pl
51.00	160.00	7–22	Pl
37.00	167.00	7–20	Pl-Cu
41.00	165.00	7–21	Pl-Cu
41.00	171.00	8–55	Pl-Cu
42.1	165.2	Siberian Mean Pole, $N=5$, $A_{95}=5.8$, $\kappa=173.9$	
42.7	166.2	Eurasian Mean Pole, $N=27$, $A_{95}=2.1$, $\kappa=180.6$	

Table 3. *North American palaeomagnetic pole positions*

Late Permian North American palaeomagnetic North Poles from the compilation of Van der Voo (in press)

Pole Position		Reference and/or Catalog Number	Pole Age
Lat. (°N)	Long. (°E)		
51.00	111.00	Larochelle (1967)	Pu
55.00	119.00	Peterson & Nairn (1971)	Pu
51.00	125.00	Peterson & Nairn (1971)	Pu
47.00	103.00	Farrell & May (1969)	Pu
50.00	121.00	Farrell & May (1969)	P
51.1	115.6	North American Mean Pole, $N=5$, $A_{95}=6.0$, $\lambda=164.1$	

Early Permian North American palaeomagnetic North Poles from the compilation of Van der Voo (in press)

Pole Position		Reference and/or Catalog Number	Pole Age
Lat. (°N)	Long. (°E)		
44.00	120.00	Gose & Helsley (1972)	Pl
46.00	122.00	Diehl & Shive (1979)	Pl
44.00	120.00	Helsley (1971)	Pl
44.00	116.00	Gose & Helsley (1972)	Pl
48.00	119.00	McMahon & Strangway (1968)	Pl
43.00	115.00	Miller & Opdyke (1985)	Pl
45.00	115.00	Peterson & Nairn (1971)	Pl
51.00	123.00	Diehl & Shive (1981)	Pl
41.00	118.00	Peterson & Nairn (1971)	Pl
49.00	115.00	Scotese (1985)	Pl
45.4	118.2	North American Mean Pole, $N=10$, $A_{95}=2.2$, $\kappa=489.7$	

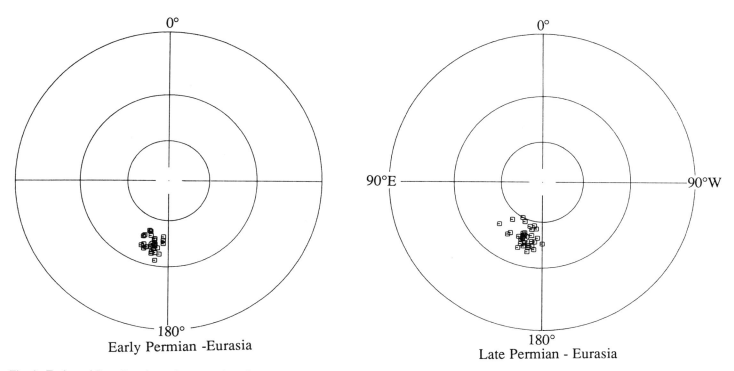

Fig. 2. Early and Late Permian paleomagnetic poles and A_{95} cones of confidence for Eurasia (Europe + Siberia combined) plotted on northern hemisphere polar stereographic projections. The poles that have been used and the pole statistics are listed in Table 2.

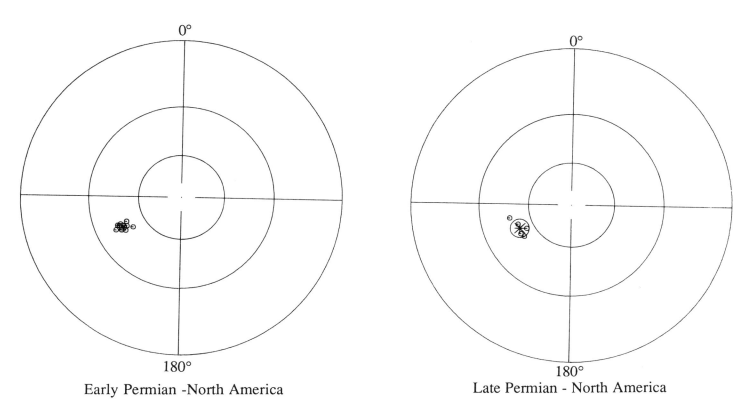

Fig. 3. Early and Late Permian paleomagnetic poles and A_{95} cones of confidence from North America as in Fig. 2. The poles that have been used and the pole statistics are listed in Table 3.

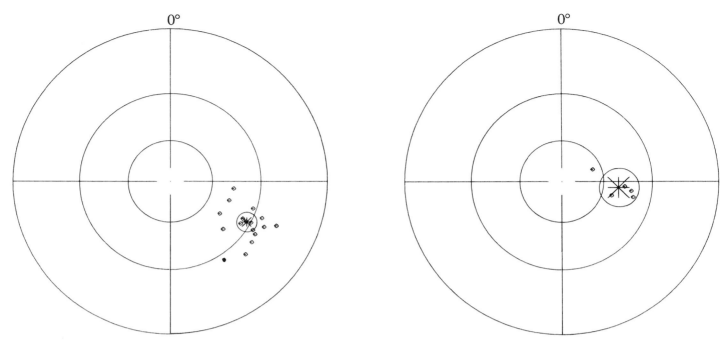

Fig. 4. Early and Late Permian paleomagnetic poles and A_{95} cones of confidence from Gondwana (Africa, India, Madagascar, South America) plotted as in Fig. 2 and rotated into South African coordinates. The poles that have been used and the pole statistics are listed in Table 4.

Table 4. *Gondwanan palaeomagnetic pole positions*

Late Permian Gondwanan palaeomagnetic North Poles

Pole Position Lat. (°N)	Long. (°E)	Reference and/or Catalog Number		Pole Age	Continent
86.00	114.00	Valencio et al. (1975)	M15.104	Guadelupian	S. America
66.00	292.00	McElhinny et al. (1976)	M15.97	Trl-Pu	Madagascar
−4.00	308.50	Klootwijk (1975)	20	Trl-Pu	India
−7.00	304.50	Wensink (1968)	19	Trl-Pu	India
−7.00	300.50	Klootwijk (1974)	18	Trl-Pu	India
47.8	−100.1	Gondwanan Mean Pole, $N=5$, $A_{95}=7.6$, $\kappa=102.7$			

Early Permian Gondwanan palaeomagnetic North Poles

Pole Position Lat.	Long.	Reference and/or Catalog Number		Pole Age	
83.00	234.00	Pascholati et al. (1976)	M15.106	Leonardian	S. America
57.00	177.00	Valencio et al. (1975)	M15.119	280	S. America
82.00	169.00	Thompson (1972)	M14.309	280?	S. America
74.00	133.00	Thompson (1972)	M14.333	263	S. America
59.50	177.50	Embleton (1970)	M12.116	270	S. America
66.00	168.00	Thompson (1972)	M14.345	266	S. America
68.00	174.00	Embleton (1970)	M12.117	266	S. America
49.00	163.00	Sinito et al. (1979)		266	S. America
54.00	171.00	Pascholati & Pacca (1976)		280	S. America
38.80	247.60	Opdyke (1964)	M8.91	266	NW. Africa
38.70	236.80	Daly & Pozzi (1976)	M15.115	272	NW. Africa
36.00	238.00	Westphal et al. (1979)		272	NW. Africa
32.20	244.10	Daly & Pozzi (1976)	M15.114	272	NW. Africa
24.00	243.80	Martin et al. (1978)	M16.142	272	NW. Africa
45.50	220.00	Opdyke (1964)	M8.92	271	NW. Africa
29.00	240.00	Morel et al. (1981)		272	NW. Africa
32.1	−118.8	Gondwanan Mean Pole, $N=16$, $A_{95}=5.7$, $\kappa=43.0$			

S. America and Africa, Van der Voo et al. (1984); Madagascar, Brock (1981); India, Klootwijk & Radhakrishnamurty (1981).

cones of 95% confidence are plotted using the rotation parameters given in Table 1 there is no overlap for either the Early or Late Permian (Fig. 5). This observation has been made previously by many others, as reviewed by Van der Voo et al. (1984), including Van der Voo & French (1974), Van der Voo et al. (1976), and Morel & Irving (1981a, b) for Laurasia and Gondwana, and Frei & Cox (1987) for Eurasia and North America. In each of these publications this discordance of the means of Early and Late Permian poles, as well as Late Carboniferous and Early Triassic poles has been attributed to an unsatisfactory fit of the continents around the Atlantic, and revisions to existing fits were proposed. The Morel & Irving (1981a) (Pangaea B) or Smith et al. (1981) modifications are extreme, placing the northwestern margin of South America against the east cost of North America in a fashion that is completely unsatisfactory on geological grounds, as summarized by Hallam (1983) among others. The Van der Voo & French (1974) modification of the Bullard et al. (1965) Atlantic reconstruction used palaeomagnetic data from western Gondwana (South America and Africa) and North America to demonstrate that a statistically acceptable fit of the continents and apparently of the geology (Ross 1986), could be achieved by rotating Gondwana anti-clockwise with respect to the Bullard et al. (1965) fit, placing the northern margin of South America tight against the Gulf coast of the United States. This fit has subsequently been used by Van der Voo et al. (1976, 1984), and more recently by Lawver & Scotese (1987) to place Florida within a revised Gondwanan framework. Van der Voo et al. (1984) argue that this fit negates the need to consider the more extreme suggestions of Pangaea B (Morel & Irving 1981a) as the Van der Voo & French (1974) reconstruction is slightly better statistically than the Pangaea B, when comparisons are based on only the well determined pole data.

Frei & Cox (1987) noted that recent modifications of the Bullard et al. (1965) reconstruction of North America and Europe based on fits of conjugate continental margins (Srivastava 1978) do not yield a fit of the pre-Jurassic apparent polar wander paths of these continents. They suggested that intracontinental extension prior to seafloor spreading was responsible for the misfit, and derived a best fit pole to superpose the North American and European Late Palaeozoic APW paths.

Figure 6a portrays the fit of the Central and North Atlantic in fixed North American coordinates resulting from the rotation parameters listed in Table 1. In Fig. 6b we plot, in fixed North American coordinates, a modified Van der Voo & French (1974) Gondwana–North America reconstruction, and the Frei & Cox (1987) reconstruction of Europe–North America. In both figures the margins of the blocks have been restored using estimates of extension as discussed in Pindell (1985), Ziegler et al. (1985) and Rowley & Lottes (1988). The Van der Voo & French (1974) reconstruction was modified to incorporate revisions of the Bullard et al. (1965) reconstruction of South Atlantic discussed by Pindell & Dewey (1982), Pindell (1985), Pindell et al. (1988), and Rowley & Pindell (in press) that requires an c. 10° clockwise rotation of South America with respect to NW Africa from its position in the Bullard et al. (1965) reconstruction. The Gondwana–North America fit shown in Fig. 6b preserves the Van der Voo & French fit of South America to North America thereby modifying the Africa–North America segment of their reconstruction. This fit minimizes the implied overlap of continental crust while placing the Early Permian Gondwanan A_{95} cone of confidence adjacent to that of North America (Fig. 7). As noted above, the vast difference in implied rates of apparent polar motion of North America and Gondwana precludes overlap of both Early and Late Permian A_{95} cones of confidence with a single reconstruction pole. Because the Early Permian data are more abundant and statistically better for both North America and Gondwana we have emphasized a fit of these over the data of the Late Permian (Fig. 7). It is clear from Fig. 6b that, even

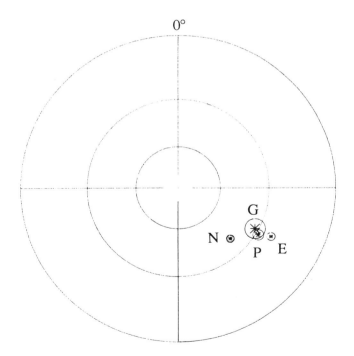

Early Permian - Continental Means
unmodified

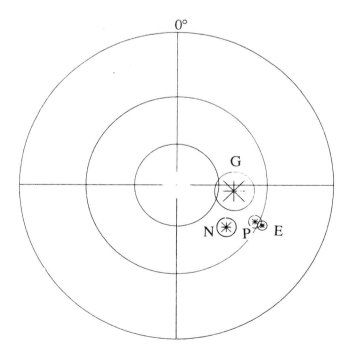

Late Permian - Continental Means
unmodified

Fig. 5. Early and Late Permian mean poles and A_{95} cones of confidence from Eurasia (E), North America (N), and Gondwana (G) and Pangaea (P) plotted on polar stereographic projections rotated into South African coordinates based on fits shown in Fig. 1 and Fig. 6a. The pole statistics are listed in Table 5.

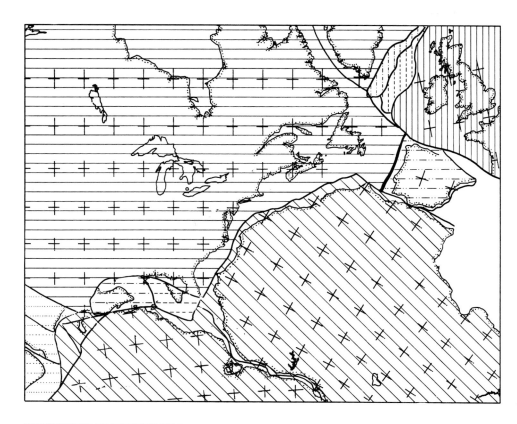

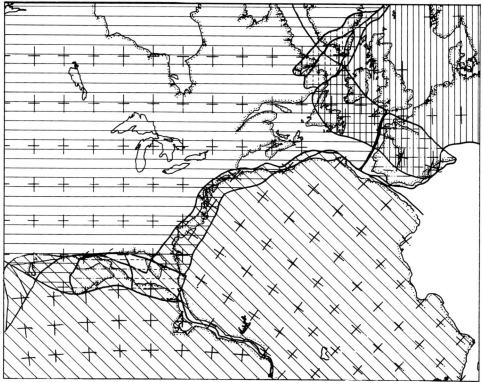

Fig. 6. Reconstructions of the circum-Atlantic continents in fixed North American coordinates. (**6a**) Reconstruction based on rotation parameters listed in Table 1, with continental margins restored to approximate pre–extension geometries. (**6b**) Reconstruction predicated on better fit of APW paths of circum–Atlantic continents including the Frei & Cox (1987) fit of Europe and North America (pole for Eurasia with respect to N. America at lat. 88.0°, long. −145.0°, angle −42.0°), and a modified Van der Voo & French (1974) reconstruction of Gondwana to North America (pole for Gondwana with respect to N. America at lat. 63.3°, long. −16.6°, angle −83.7°). Horizontal rule, North America and Greenland; horizontal dashes, South Florida; horizontal dash-dot, Yucatan; horizontal dots; N. Mexico, S. Mexico, and Chortis; horizontal dash-dot-dot, Iberia. Vertical rule, Eurasia; vertical dashes, Rockall Bank; vertical dash-dot, Hatton-Edoras Bank; slant rule, Africa and South America, including displaced northwestern margin of South America. Heavy lines, reconstructed continental margins of the blocks.

though the overlap of Gondwana and North America was minimized, substantial overlap (this fit is approximately 935 km tighter along the east coast of the United States than Fig. 6a) of both reconstructed continental margins and present day coastlines is required. In fact the northern portion (c. 500 km wide) of the Guyana shield of South America occupies the same area as that of the Coahuila platform (approximately the southern edge of North America), Yucatan block, and South Florida block. Yucatan and South Florida blocks are depicted in their pre-Jurassic positions derived from a detailed analysis of the kinematics of the Gulf of Mexico (Pindell 1985; Pindell & Barrett, in press) and Precambrian and Palaeozoic geology of the circum-Gulf of Mexico region (Rowley & Pindell, in press). This overlap is clearly unacceptable, and any modification of this fit that reduces overlap will not allow the Early Permian A_{95} cones of confidence to be superimposed.

Farther north, the difference between the fits of North America and Europe based on reconstruction of the ocean basins and restoration of conjugate margins versus a palaeomagnetically

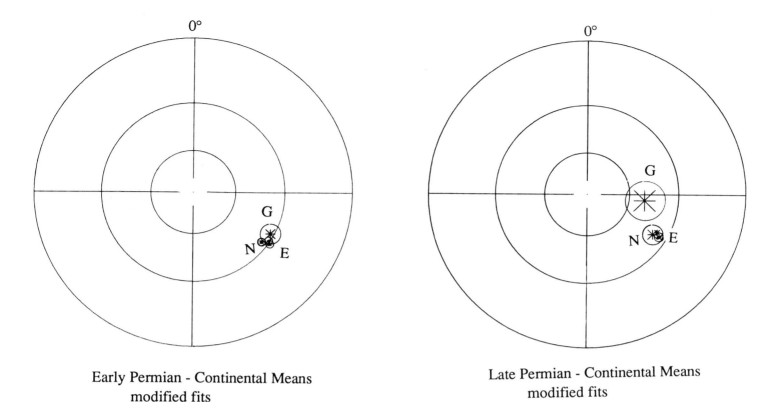

Fig. 7. Early and Late Permian mean poles and A_{95} cones of confidence from Eurasia (E), Gondwana (G), and North America (N) rotated into South African coordinates based on the palaeomagnetic fit shown in Fig. 6b.

determined fit (Fig. 7) can be seen in Fig. 6a and 6b. In Fig. 6a Greenland and Iberia are reconstructed with respect to North America, and Hatton–Edoras and Rockall Banks are reconstructed with respect to Europe according to poles listed in Rowley & Lottes (1988). The Frei & Cox (1987) reconstruction Fig. 6b requires 630 to 850 km of overlap of the restored North American and European margins. Although one might suspect reconstructions of the extension within the Newfoundland and Goban Spur margins, or the North Sea, the nearly complete superposition of southern France and Iberia is clearly unacceptable. Once again, any modification of the Frei & Cox fit (1987) to minimize continental overlap will result in deterioration of the fit of the APW data and closer approach to the fit shown in Fig. 6a.

The statistics based on the Early and Late Permian palaeomagnetic pole data of Eurasia (Table 2), North America (Table 3), and Gondwana (Table 4), can also be compared for the two reconstructions. Figure 7 shows the reconstructed palaeocontinental means for the Early Permian based on the palaeomagnetically determined fit (Fig. 6b). It is characterized by a tight clustering of poles and near overlap of A_{95} cones of confidence. Figure 5a shows the same data on our preferred fit. Table 5 lists the statistics of the Early Permian data treating each pole determination as an equally contributing element. There is little difference in the statistics and in the position of the means of the Early Permian when the poles from individual studies are weighted equally. When the palaeo-continental means are used (Table 5) there is again no difference in the position of the resulting Early Permian mean determined using either fit, however, the palaeomagnetically determined reconstruction yields better values of A_{95} and κ. A similar comparison of Late Permian palaeomagnetic data for both fits is shown in Fig. 5b and 7b. There is virtually no statistical difference and when individual poles are combined to yield a global mean pole when palaeo-continental means are combined. For both the Early and Late Permian comparisons, the global mean poles determined using our preferred Pangaea fit yield results that are not significantly different from those of the palaeomagnetically determined fits, and that are characterized by values of $A_{95} < 10°$ and $\kappa > 15$, which are within the acceptable range used to determined palaeomagnetic reliability. We therefore prefer to utilize the 'looser' fit, since the geologic and geometric problems resulting from fitting the palaeomagnetic data are judged to be unacceptable. In turn, the lack of a clear statistical difference of the Early and Late Permian Pangaean means when compared in either fit supports our preference for using the reconstruction presented in Fig. 1 as a base for reconstructing the palaeolatitudes for the Early and Late Permian of traditional Pangaea.

The reconstructions of Laurasian and Gondwanan components of Pangaea in terms of palaeolatitudes are presented in Fig. 8 for the Early and Late Permian. These reconstructions use the mean poles derived from treating all of the studies as independent, equally weighted elements in the analysis. The 'Pangaean' mean poles used are those listed in Table 5, in South African coordinates. As discussed separately by Ziegler (1990) reconstructions of western Pangaea based on these poles yield results that are compatible with interpretations based on an analysis of palaeolatitudinally sensitive sediments and floras. This framework of Gondwana and Laurasia is used as well by Nie et al. (1990) as a base for discussing the positions and amalgamation history of Asian continental blocks.

Summary

In summary, we use our revised reconstructions of the major continental blocks that comprise Pangaea (i.e. Eurasia, North America, and Gondwana) to examine palaeomagnetic constraints on predicted palaeo-latitudes of the Early and Late Permian palaeogeographic elements. The palaeomagnetic data from the individual major continental blocks generally define well clustered means, with acceptably good statistics. When reconstructed

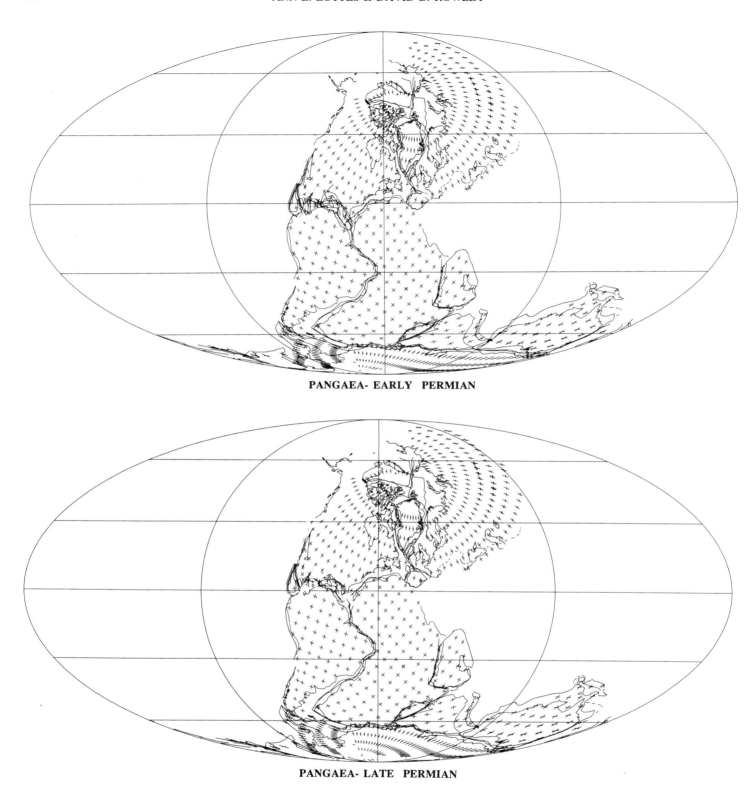

Fig. 8. Pangaea reconstructed with respect to paleolatitude based on the Early and Late Permian mean poles for Pangaea (Table 5).

using a Wegener-style Pangaea reconstruction the means do not overlap, but instead define statistically separate populations. Modifications of this reconstruction that yield better clustering of poles require unacceptably large overlap of palinspastically restored continental margins of both the North and Central Atlantic. However, the statistics for either fit are virtually identical. Considering the continental overlap problems that are required to superpose the palaeomagnetic data we judge that these modifications produce more problems than they resolve and we therefore have chosen not to incorporate these modifications in our reconstructions. Our preferred reconstructions of the Early and Late Permian are based on averaging all of the palaeomagnetic data. These reconstructions yield predicted palaeolatitudes that are in accord with independent estimates derived from an analysis of Sakmarian and Kazanian phytogeography and palaeoclimatically sensitive sediments (Ziegler 1990). These reconstructions are

Table 5. *Early and Late Permian mean pole positions*

Late Permian mean pole positions

Continent	No. of Studies	Kappa	Alpha$_{95}$	Pole Position		(Coordinates)
				Lat. (°N)	Long. (°E)	
Eurasia	34	93.6	2.6	48.1	163.1	(Eurasia)
North America	5	164.1	6.0	51.1	115.6	(North America)
Gondwana	5	37.0	12.7	50.0	−96.2	(S. Africa)
Pangaea (this paper)*	44	35.8	3.6	32.3	−115.5	(S. Africa)
Pangaea (pmag fit, Fig.6b)	44	49.5	3.1	36.5	−118.9	(S. Africa)
Pangaea (this paper)**	3	22.9	26.4	41.6	−114.9	(S. Africa)
Pangaea (pmag fit, Fig.6b)**	3	36.9	20.6	41.3	114.3	(S. Africa)

Early Permian mean pole positions

Continent	No. of Studies	Kappa	Alpha$_{95}$	Pole Position		(Coordinates)
				Lat. (°N)	Long. (°E)	
Eurasia	27	180.6	2.1	42.7	166.2	(Eurasia)
North America	10	489.7	2.2	45.5	118.2	(North America)
Gondwana	16	43.0	5.7	32.1	−118.8	(S. Africa)
Pangaea (this paper)*	53	46.2	2.9	28.8	−121.3	(S. Africa)
Pangaea (pmag fit, Fig. 6b)	53	88.3	2.1	30.9	−123.2	(S. Africa)
Pangaea (this paper)**	3	46.9	18.2	31.7	−123.6	(S. Africa)
Pangaea (pmag fit, Fig. 6b)**	3	392.9	6.2	31.6	−123.3	(S. Africa)

* Poles used for the reconstruction of palaeo-latitudes in Figs 8a and 8b.
** Based on averaging continental means.

used to address some questions related to the Late Palaeozoic palaeogeographic evolution of Asian 'micro-continents' (Nie *et al.* 1990) and their relations to Pangaea.

This work is supported by the National Science Foundation Grant EAR-86−13412 and is part of the 'Paleogeographic Atlas Project' at the University of Chicago. General support for the atlas project is contributed by a number of companies: Amoco Production Research, Arco Exploration and Production Research, British Petroleum, Chevron Oil Field Research Company, Conoco Exploration Research, Digital Equipment Corporation, Societe Nationale Elf Aquitaine, Exxon Production Research, Mobil Exploration and Producing Services, Shell Development Company, Sun Exploration and Production Company, Texaco U.S.A., Unocal International Oil and Gas, and BHP-Utah International. This support is gratefully acknowledged. Three anonymous reviews of a previous version of this manuscript are gratefully acknowledged, and helped improve this paper. We also thank the following for discussions concerning this paper: C. R. Scotese, Nie Shangyou, and A. M. Ziegler. T. Gierlowski and M. Hulver kindly assisted in the preparation of this manuscript.

References

BAKER, B. H. 1971. Explanatory note on the structure of the southern part of the African rift system. *In*: *Tectonique de l'Afrique (Sciences de la terre, 6)*, UNESCO, Paris, 543−548.

BASU, T. N. & SHRIVASTAVA, B. B. P. 1981. Structure and tectonics of Gondwana basins of peninsular India. *In*: *Fifth International Gondwana Symposium*, Wellington, New Zealand, 1980, 177−182.

BERTHOLD, G., NAIRN, A. E. M. & NEGENDANK, J. F. W. 1975. A paleomagnetic investigation of some igneous rocks of the Saar-Nahe basin. *Neues Jahrbuch fur Paleontologie und Geologie*, **3**, 134−150.

BIRKENMAJER, K. & NAIRN, A. E. M. 1964. Palaeomagnetic studies of Polish rocks. I. The Permian igneous rocks of the Krakow district and some results from the Holy Cross Mountains. *Annales de la Societe Geologique de Pologne*, **34**, 225−244.

——, GROCHOLOSKI, A., MILEWICZ, J. & NAIRN, A. E. M. 1968. Palaeomagnetic studies of Polish rocks. II. The Upper Carboniferous and Lower Permian of the Sudetes. *Annales de la Societe Geologique de Pologne*, **38**, 435−474.

BOAST, J. & NAIRN, A. E. M., 1982. An Outline of the Geology of Madagascar. *In*: NAIRN, A. E. M. & STEHLI, F. G. (eds) *The Ocean Basins and Margins, Volume 6, the Indian Ocean*, Plenum Press, New York. 649−696.

BRIDEN, J. C., SMITH, A. G. & SALLOMY, J. T. 1970. The geo-magnetic field in Permo-Triassic time. *Geophysical Journal of the Royal Astronomical Society*, **23**, 101−117.

BROCK, A., 1981, Paleomagnetism of Africa and Madagascar. *In*: MCELHINNY, M. W. & VALENCIO, D. A. (eds) Paleoreconstruction of the Continents, Geodynamics Series, American Geophysical Union, **2**, 65−76.

BULLARD, E. C., EVERETT, J. E. & SMITH, A. G. 1965. The fit of the continents around the Atlantic. *In*: BLACKETT, P. M. S., BULLARD, E. C. & RUNCORN, S. K. (eds). *A Symposium on Continental Drift*. Philosophical Transactions of the Royal Society of London, A, **258**, 41−51.

BYLUND, G. & PATCHETT, P. J. 1977. Paleomagnetic and Rb-Sr isotopic evidence for the age of the Sarna alkaline complex, western central Sweden. *Lithos*, **10**, 73−79.

CORNWELL, J. D. 1967. Paleomagnetism of the Exeter lavas, Devonshire. *Geophysical Journal of the Royal Astronomical Society*, **12**, 181−196.

DALY, L. & POZZI, J-P. 1976. Resultats paleomagnetiques du Permien inferieur et du trias morocain: comparison avec les donees Africanes et sud Americaines. *Earth and Planetary Science Letters*, **29**, 71−80.

DALZIEL, I. W. D., GARRET, S. W., GRUNOW, A. M., PANKHURST, R. J., STOREY, B. C. & VENNUM, W. R. 1987. The Ellsworth−Whitmore Mountains crustal block: its role in the tectonic evolution of West Antarctica. *In*: MCKENZIE, G. D. (ed.) *Gondwana Six: Structure, Tectonics, and Geophysics*. Geophysical Monograph, **40**, American Geophysical Union, Washington, DC, 173−182

DEWEY, J. F. 1982. Plate tectonics and the evolution of the British Isles. *Journal of the Geological Society of London*, **139**, 371−412.

DIEHL, J. F. & SHIVE, P.N. 1979. Paleomagnetic studies of the Early Permian Ingelside Formation of northern Colorado. *Geophysical Journal of Royal Astronomical Society*, **56**, 271−282.

—— & —— 1981. Paleomagnetic results from the Late Carboniferous/Early Permian Casper Formation: Implications for northern Appalachian tectonics. *Earth and Planetary Science Letters*, **54**, 281−292.

DINGLE, R. V., SIESSER, W. G. & NEWTON, A. R. 1983. *Mesozoic and Tertiary geology of southern Africa*. Rotterdam: A. A. Bulkema.

DU TOIT, A. L. 1937. *Our Wandering Continents*. Oliver and Boyd,

EMBLETON, B. J. J. 1970. Palaeomagnetic results for the Permian of South America and a comparison with the African and Australian data. *Geophysical Journal of the Royal Astronomical Society*, **21**, 105–118.

FARRELL, W. E. & MAY, B. T. 1969. Paleomagnetism of Permian redbeds from the Colorado Plateau. *Journal of Geophysical Research*, **74**, 1495–1504.

FEDOROV, L. V., RAVICH, M. G. & HOFMAN, J. 1982. Geologic comparison of southeast peninsular India and Sri Lanka with part of East Antarctica (Enderby Land, MacRobertson Land, and Princess Elizabeth Land). *In*: CRADDOCK, C. (ed.) *Antarctic Geoscience*. The University of Wisconsin Press, Madison. 73–78.

FREI, L. S. & COX, A. 1987. Relative displacement between Eurasia and North America prior to the formation of oceanic crust in the north Atlantic. *Tectonophysics*, **142**, 111–136.

GORDON, R. G., COX, A. & O'HARE, S. O. 1984. Paleomagnetic euler poles and the apparent polar wander and absolute motion of North America since the Carboniferous. *Tectonics*, **3**, 499–537.

GOSE, W. A. & HELSLEY, C. E. 1972. Paleomagnetic and rock magnetic studies of the Permian Cutler and Elephant Canyon Formations in Utah. *Journal of Geophysical Research*, **77**, 1534–1548.

HALLAM, 1983. Supposed Permo-Triassic megashear between Laurasia and Gondwana. *Nature*, **301**, 499–502.

HALVORSEN, E. 1970. Paleomagnetism and the age of the younger diabases in the Ny-Hellesund area, S. Norway. *Norsk Geologiske Tidsskrift*, **50**, 157–166.

HELSLEY, C. E. 1971. Remanent magnetization of the Permian Cutler Formation of western Colorado. *Journal of Geophysical Research*, **76**, 2842–2848.

HOWELL, D. G. (ed.) 1985. *Tectonostratigraphic Terranes of the Circum-Pacific Region, Volume 1*. Circum-Pacific Council for Energy and Mineral Resources, Houston, Texas.

IRVING, E. & IRVING, G. A. 1982. Apparent polar wander paths: Carboniferous through Cenozoic and the assembly of Gondwana. *Geophysical Surveys*, **5**, 141–188.

KHAIN, V. E. 1986. *Geology of the USSR, First Part, Old Craton and Paleozoic Fold Belts*. Gebruder Borntraeger. Berlin.

KHRAMOV, A. N., PETROVA, G. N. & PECHERSKY, D. M. 1981. Paleomagnetism of the Soviet Union. *In*: MCELHINNY, M. W. & VALENCIO, D. A. (eds). Paleoreconstruction of the Continents, Geodynamics Series, American Geophysical Union, v. **2**, 177–194.

KILMURRAY, J. O. 1975. Las Sierras Australes de la Provincia de Buenos Aires, Las Fases deformation y interpretacion Estratigraphica. *Revista de la Associacion Geologicas Argentina*, **30**, 331–348.

KLOOTWIJK, C. T. 1974. Paleomagnetic results from some Panchet Clay beds, Raranpura Coalfield, northeastern India. *Tectonophysics*, **21**, 79–92.

—— 1975. Paleomagnetism of upper Permian red beds in the Wardha Valley, central India. *Tectonophysics*, **25** 115–137.

—— & RADHAKRISHNAMURTY, C. 1981. Phanerozoic paleomagnetism of the Indian plate and the India-Asia collision. *In*: MCELHINNY, M. W. & VALENCIO, D. A. (eds) Geodynamics Series, American Geophysical Union, **2**, 93–105.

KLUTH, C. F., & CONEY, P. J. 1981, Plate tectonics of the Ancestral Rocky Mountains. *Geology*, **9**, 10–15.

KONRAD, H. J., & NAIRN, A. E. M. 1972. The paleomagnetism of the Permian rocks of the Black Forest, Germany. *Geophysical Journal of the Royal Astronomical Society*, **27**, 369–382.

KRS, M. 1968. Rheological aspects of paleomagnetism? *23rd International Geological Congress, Prague*, **5**, 87–96.

KRUSEMAN, G. P. 1962. *Etude paleomagnetique et sedimentologique du bassin Permien de Lodeve Hirault, France*. PhD Dissertation, University of Utrecht.

LAROCHELLE, A. 1967. *Paleomagnetic directions of a basic sill in Prince Edward Island*. Geological Survey of Canada Paper 67–39.

LAWVER, L. & SCOTESE, C. R. 1987. A revised reconstruction of Gondwanaland. *In*: MCKENZIE, G. D. (ed.) *Gondwana Six: Structure, Tectonics, and Geophysics, Geophysical Monograph*, **40**, American Geophysical Union, Washington, D. C. 17–23.

MARTIN, D. L., NAIRN, A. E. M., NOLTIMEIER, H. C., PETTY, M. H. & SCHMIDT, T. J. 1978. Paleozoic and Mesozoic Paleomagnetic results from Morocco. *Tectonophysics*, **44**, 91–114.

MAURITSCH, H. J. & ROTHER, K. 1983. Paleomagnetic investigations in the Thuringer Forest (G. D. R.). *Tectonophysics*, **99**, 63–72.

MCELHINNY, M. W., EMBLETON, B. J. J., DALY, L. & POZZI, J-P. 1976. Paleomagnetic evidence for the location of Madagascar in Gondwanaland. *Geology*, **4**, 455–457.

MCMAHON, B. E. & STRANGWAY, D. W. 1968. Investigation of the Kiaman magnetic division in Colorado redbeds. *Geophysical Journal of the Royal Astronomical Society*, **15**, 265–285.

MILLER, J. D. & OPDYKE, N. D. 1985. Magnetostratigraphy of Red Sandstone Creek section near Vail, Colorado. *Geophysical Research Letters*, **12**, 133–136.

MITRA, N. D., LAHA, C., DE, A. K., DUTT, A. B., BUSU, U. K., JAYKUMAR, R. ET AL. 1982. I. Gondwana Basins Damodar and Koel valley basins. *In*: *Stratigraphic Correlation Between Sedimentary Basins of the ESCAP Region*, **8**, 23–27.

MOREL, P. & IRVING, E. 1981a. Paleomagnetism and the evolution of Pangea. *Journal of Geophysical Research*, **86**, 1858–1872.

—— & —— 1981b. The concept of the a mobile Pangea and the continuity of continental drift. *In*: MCELHINNY, M. W. ET AL. (eds) *Global reconstruction and the Geomagnetic Field during the Paleozoic*. Advances in Earth and Planetary Sciences, **10**, 39–45.

——, —— DALY, L. & MOUSSINE-POUCHKINE, A. 1981. Paleomagnetic results from Permian rocks of the northern Saharan craton and motions for the Moroccan Meseta and Pangea. *Earth and Planetary Science Letters*, **55**, 65–74.

MULDER, F. G. 1971. Paleomagnetic research in some parts of central and southern Sweden. *Svergies Geologiska Undersokning, Arsb.*, **653**, 1–56.

NIE, S-Y, ZIEGLER, A. M. & ROWLEY, D. B. 1990. Constraints on the locations of Asian microcontinents in Palaeo-Tethys during the Late Palaeozoic. *In*: MCKERROW, W. S. & SCOTESE, C. R. (eds) *Palaeozoic Palaeogeography and Biogeography*. Geological Society, London, Memoir, **12**, 397–409.

NIJENHUIS, G. H. W. 1961. A palaeomagnetic study of the Permian volcanics in the Nahe region (SW. Germany). *Geologie en Mijnbouw*, **40**, 26–38.

OPDYKE, N. D. 1964. The paleomagnetism of the Permian redbeds of southwest Tanganyika. *Journal of Geophysical Research*, **69**, 2477–2487.

PALLISTER, J. W. 1971. The tectonics of East Africa. *In*: *Tectonique de l'Afrique (Sciences de la terre*, 6. UNESCO, Paris 511–542.

PASCHOLATI, E. M. & PACCA, I. G. 1976. Estudo paleomagnetico de Seccoes do Subgrupo Itarare. *29th Congress Brasileiro de Geologia, Belo Horizonte*, November 1976.

——, —— & VILAS, J. F. 1976. Paleomagnetism of sedimentary rocks from the Permian Irati Formation, southern Brazil. *Revista Brasileira de Geociencias*, **6**, 156–163.

PETERSON, D. N. & NAIRN, A. E. M. 1971. Paleomagnetism of Permian redbeds from southwestern United States. *Geophysical Journal of the Royal Astronomical Society*, **23**, 191–205.

PINDELL, J. L. 1985. Alleghenian reconstruction and subsequent evolution of the Gulf of Mexico, Bahamas and Proto-Caribbean. *Tectonics*, **4**, 1–39.

—— & Barrett, S. F. in press. Geological evolution of the Caribbean region: a plate tectonic perspective. *In*: DENGO G. & CASE J. E. (eds) *The Caribbean Region*. The Geology of North America: Decade of North America Geological Society America Boulder Co. Geology, **H**.

—— & DEWEY, J. F. 1982. Permo-Triassic reconstruction of western Pangea and the evolution of the Gulf of Mexico/Caribbean region. *Tectonics*, **1**, 179–211.

——, CANDE, S. C., PITMAN, W. C., ROWLEY, D. B., DEWEY, J. F., LABRECQUE, J. & HAXBY, W. 1988. A paleo-kinematic framework for models of Caribbean evolution. *Tectonophysics*, **155**, 121–138.

PRIEM, H. N. A., MULDER, F. G., BOETRIJK, A. A. I. M., HEBEDA, E. H., VERSCHURE, R. H. & VERDURMEN, E. A. Th. 1968. Geochronological and palaeomagnetic reconnaissance survey in parts of central and southern Sweden. *Physics of the Earth and Planetary Interiors*, **1**, 373–380.

ROCHE, A., SAUCIER, H. & LACAZE, J. 1962. Etude paleomagnetique der roches volcaniques Permiennes de la region Nidech-Donon. *Bulletin du Service de la Carte Geologique d' Alsace-Lorraine*, **15**, 59–68.

RODGERS, J. 1970. *Tectonics of the Appalachians*. John Wiley & Sons, Inc.

RONNEVIK, H. & JACOBSEN, H. P. 1984. Chapter 2: Structural highs and basins in the western Barents Sea. *In*: SPENCER, A. M. ET AL. (eds) *Petroleum Geology of the North European Margin*. Norwegian Petroleum Society, London. 19–32.

ROSS, C. 1986. Paleozoic evolution of southern margin of Permian basin. *Geological Society of America Bulletin*, **97**, 536–554.

ROTHER, K. 1971. *Gesteins und Paleomagnetische untersuchungen an Gesteinsproben vom Terretorium der DDR aus dem Präkambrium bis zum Tertiär und Folgerungen für die Veranderungen des geomagnetischen Hauptfeldes sowie für geologisch-geotektonische interpretation smölichkeiten*. PhD Dissertation, University of Berlin.

ROWLEY, D. B. 1988. Revised reconstruction of the South Atlantic incorporating intraplate deformation within South America and Africa. *Geological Society of America Abstracts with Programs*, **20**, A126.

—— & LOTTES, A. L. 1988. Plate kinematic reconstructions of the Arctic and North Atlantic, Jurassic to present. *Tectonophysics*, **155**, 73–120.

—— & PINDELL, J. L. in press. End Paleozoic-Early Mesozoic Western Pangean reconstruction and implications for the distribution of Precambrian and Palaeozoic rocks around Meso-America. *Precambrian Research* (in press).

RUDKEVICH, M. Y. 1970. *Atlas structure, paleotectonic maps, and geological time slice maps of the territory of the West Siberian Lowlands*. Ministry of Geology SSSR (in Russian)

SASTRI, V. V., RAJUU, A. T. R., SHUKLA, S. N. & VENKATACHALA, B. S. 1982. India, Part II: Inland graben. *In*: *Stratigraphic Correlation between Sedimentary Basins of the ESCAP Region*, **8**, 36–49. erals Resources Development Service 48, U.N., New York. 36–49.

SCOTESE, C. R. 1985. *The assembly of Pangea: Middle and Late Palaeozoic paleomagnetic results from North America* PhD Dissertation, University of Chicago.

——, BAMBACH, R. K., BARTON, C., VAN DER VOO, R. & ZIEGLER, A. M. 1979. Paleozoic base maps. *Journal of Geology*, **87**, 217–277.

SINITO A. M., VALENCIO, D. A. & VILAS, J. F. A. 1979. Paleomagnetism of a sequence of Upper Palaeozoic-Lower Mesozoic red beds from Argentina. *Geophysical Journal of the Royal Astronomical Society*, **58**, 237–247.

SMITH, A. G., HURLEY, A. M. & BRIDEN, J. C. 1981. *Phanerozoic World Maps*. Cambridge University Press.

SRIVASTAVA, S. P. 1978. Evolution of the Labrador Sea and its bearing on the early evolution of the North Atlantic. *Geophysical Journal of the Royal Astronomical Society*, **52**, 313–357.

STOREY, B. C., DALZIEL, I. W. D., GARRETT, S. W., GRUNOW, A. N., PANKHURST, R. J. & VENNUM, W. R. 1988. West Antarctica in Gondwanaland: Crustal blocks reconstruction and breakup process, *Tectonophysics*, **155**, 381–390.

SURLYK, F., HURST, J. M., PIASECKI, S., ROLLE, F., SCHOLLE, P. A., STEMMERIK, L. ET AL. 1986. The Permian of the western margin of the Greenland Sea – a future exploration target. *In*: HALBOUTY, M. T. (ed.) *Future Petroleum Provinces of the World*. American Association of Petroleum Geologists Memoir, **40**, 629–659.

THOMPSON, R. 1972. Paleomagnetic results from the Paganzo Basin of northwest Argentina. *Earth and Planetary Science Letters*, **15**, 145–156.

THORNING, L. & ABRAHAMSEN, N. 1980. Paleomagnetism of Permian multiple intrusion dykes in Bohslan, SW Sweden. *Geophysical Journal of the Royal Astronomical Society*, **60**, 163–185.

VALENCIO, D. A., ROCHA-CAMPOS, A. C. & PACCA, I. G. 1975. Paleomagnetism of some sedimentary rocks of the Late Paleozoic Tubarao and Passa Dois Groups, from the Permian Basin, Brazil. *Revista Brasileira de Geociencias*, **5**, 186–197.

VAN DEN ENDE, C. 1977. *Paleomagnetism of Permian redbeds of the Dome de Barrot (S. France)*. PhD Dissertation, University of Utrecht.

VAN DER VOO, R., 1981. Paleomagnetism of North America: A brief review. *In*: MCELHINNY, M. W. & VALENCIO, D. A. (eds) *Paleoreconstruction of the Continents*. Geodynamics Series, American Geophysical Union, v.2, 159–176.

—— in press. Paleomagnetism of continental North America: The craton, its margins and Appalachian belt. *In*: PAKISER, L. C. & MOONEY, W. D. (eds) *Geophysical Framework of the Continental United States*, Geological Society of America Memoir, (in press).

—— & FRENCH, R. B. 1974. Apparent polar wandering for the Atlantic-bordering continents: Late Carboniferous to Eocene. *Earth Sciences Reviews*, **10**, 99–119.

——, MAUK, F. J. & FRENCH, R. B. 1976. Permian-Triassic continental configurations and the origin of the Gulf of Mexico. *Geology*, **4**, 177–180.

——, PEINADO, J. & SCOTESE, C. R. 1984. A paleomagnetic reevaluation of Pangea reconstructions. *In*: Plate Reconstructions from Palaeozoic Paleomagnetism, VAN DER VOO, R., SCOTESE, C. R. & BONHOMMET, (eds) Geodynamics Series, 12, 11–26, American Geophysical Union, Washington.

VAN EVERDINGEN, R. O. 1960. Studies on the igneous rock complex of the Oslo region. XVII. Paleomagnetic analysis of Permian extrusives in the Oslo region, Norway. *Skr Nor Vidensk Akad. Mat.-Natur. Kl*, **1**, 1–80.

VEEVERS, J. J. (ed.) 1984. *Phanerozoic Earth History of Australia*. Clarendon Press, Oxford.

WENSINK, H. 1968. Paleomagnetism of some Gondwana redbeds from central India. *Palaeogeography, Palaeoclimatology and Palaeoecology*, **5**, 323–343.

WESTPHAL, M. 1972. Etude paleomagnetique de certaines formations du Paleozoique superieur. *Memoires du Bureau de Recherches Geologiques et Mineralogique, France*, **77**, 857–860.

——, MONTIGNY, R., THUIZAT, R., BARDON, C., BOSSERT, A., HAMZEH, R. & ROLLEY, J. P. 1979. Paleomagnetisme et datation du volcanisme Permien, Triassique et Cretace du Maroc. *Canadian Journal of Earth Sciences*, **16**, 2150–2164.

——, BAZHENOV, L., LAUER, J. P., PECHERSKY, D. M. & SIBUET, J-C. 1986. Paleomagnetic implications on the evolution of the Tethys belt from the Atlantic Ocean to the Pamirs since the Triassic. *Tectonophysics*, **123**, 37–82.

ZIEGLER, A. M. 1990. Phytogeographic patterns and continental configurations during the Permian Period. *In*: MCKERROW, W. S. & SCOTESE, C. R. (eds) *Palaeozoic Palaeogeography and Biogeography*. Geological Society, London, Memoir, **12**, 363–379.

——, ROWLEY, D. B., LOTTES, A. L., SAHAGIAN, D. L., HULVER, M. L. & GIERLOWSKI, T. C. 1985. Paleogeographic interpretation: with an example from the mid–Cretaceous. *Annual Reviews in Earth and Planetary Science*, **13**: 385–425.

ZIEGLER, P. A. 1982. *Geological Atlas of Western and Central Europe*. Shell Internationale Petroleum Maatschappij B. V.

—— 1988. *Evolution of the Arctic North Atlantic and the Western Tethys*. American Association of Petroleum Geologists Memoir 43.

ZIJDERVELD, J. D. A. 1975. *Paleomagnetism of the Estrel Rocks*. PhD Dissertation, University of Utrecht.

Constraints on the locations of Asian microcontinents in Palaeo-Tethys during the Late Palaeozoic

NIE SHANGYOU, D. B. ROWLEY & A. M. ZIEGLER

Paleogeographic Atlas Project, Department of Geophysical Sciences, The University of Chicago 5734 South Ellis Avenue Chicago, Illinois 60637 USA

Abstract: Useful constraints for Late Palaeozoic reconstructions of the Palaeo-Tethys and the development of Asia come from tectonic, palaeomagnetic, biogeographic and palaeoclimatic data. Tectonic constraints include the timing of collisions of the microcontinents which traditionally have been regarded as proceeding from north to south, and ranging in time from Late Palaeozoic to the Tertiary. Such a view gives a good account of the palaeobiogeographic connections, but the Mongolo–Okhotsk suture of northern Mongolia continued to close by counterclockwise rotation as late as the Jurassic. The biogeographic provinces are well developed in the Late Palaeozoic due to the relatively steep equator-to-pole gradients, thus as continents rifted from the southern margin of Palaeo-Tethys, they lost their south temperate Gondwanan affinities and acquired sub-tropical to tropical floras and faunas. Eventually, the north temperate Angaran floras and faunas inhabiting the northern margin of Palaeo-Tethys invaded some of the Cathaysian microcontinents by the end of the Palaeozoic. Unfortunately, the tropical Cathaysian floras could have occurred over a considerable latitudinal range (25° N to 25° S) and thus do not provide precise constraints. These floras, however, do contain seed-dispersed plants which implies that most of these separate microcontinents must have been geographically connected while apparently tectonically distinct. The climatically sensitive sediments include tillites and glacio-marine deposits associated with some of the terranes, and with Gondwana, but they are clearly temperate in origin, not polar, and are overlain by carbonates in Gondwana and the terranes. We interpret this as due to a climatic amelioration, rather than to a latitudinal plate motion. Palaeomagnetic data are now available for a number of south Asian microcontinents and enough determinations have now been made in North and South China to show consistency with the tectonic, biogeographic and climatic information.

In detail, a southern belt of terranes, from the Helmand block in Iran and Afghanistan, through the Western Qiangtang and Lhasa blocks of Tibet and to the Sibumasu block of Thailand and Malaya, all rifted off the margins of Gondwana in the Permian. Few palaeomagnetic data are available to support this, but the tillites, floras and faunas are shared with the midlatitude portions of Gondwana from Iran, India and Australia. In the Late Permian the Cathaysian flora is known from at least the Helmand and Western Qiangtang blocks, suggesting that they reached lower latitudes, and were in physical contact with other Cathaysian floras, although the nature of this connection is not understood. The major Cathaysian microcontinents of Yangtze, Indochina, Eastern Qiangtang, Sino-Korea and Tarim were tropical throughout the Carboniferous and Permian. Again, a degree of geographic interconnection is implied by their common floras, and this is shared with low latitude portions of Gondwana, including Arabia and North Africa, but the location and nature of the 'land bridges' are unknown. Collision of Tarim with Asia in the Early Permian and of Sino–Korea with the Mongolian arcs in the Late Permian is indicated from tectonic and biogeographic data, but, as stated above, the Mongolo–Okhotsk suture did not close until the Late Jurassic, thus rotation of the combined Mongolia and Sino-Korean block continued until that time. The other Cathaysian microcentinents collided with Asia about the Late Triassic giving rise to the Indosinian Orogeny.

It is now generally accepted that the final assembly of Asian Pangaea elements post-dated the Palaeozoic, and that a number of substantial continental elements that now comprise most of China, southeast Asia, Afghanistan and Iran were located in the Palaeo-Tethyan oceanic realm during the Late Palaeozoic (Ziegler *et al.* 1982; Rowley *et al.* 1985; Sengör 1987*a*). An understanding of the collisional history of these microcontinents in Permian and Mesozoic times is also emerging, though a critical point that is not generally realized is that the northern margin of Palaeo-Tethys was defined by a complex string to marginal arcs that were not fully welded to the Angaran shield until the Late Jurassic to Early Cretaceous (Kosygin & Parfenov 1981). These areas were biogeographically tied to Angara throughout the Palaeozoic, but have rotated as much as 100° counterclockwise, together with Sino–Korea, in the Early Mesozoic. Precise palaeogeographic solutions to these complicated problems are not yet possible, but we can begin to integrate the new palaeomagnetic data with climatic constraints provided by the fossil floras (Ziegler 1990) and our own observations on the deformation history of the region.

We assume that Gondwana and Laurasia had amalgamated by the Permian and that the reconstructions of Lottes & Rowley (1990) provide a good estimate of their Permian palaeolatitudinal configurations. The Palaeo-Tethyan margins of this supercontinental assembly were just outside the tropics except in the regions of Europe and North Africa, and their floras and faunas reflect their temperate origin (Wang & Mu 1981). By contrast, the south Asian microcontinents were dominated mainly by carbonates and thermophylic floras. Accordingly, there is a uniformity in Tethyan biotas which limits their usefulness for palaeogeographic reconstructions. From the numerous reviews now available (Smith 1988 on corals; Nakamura *et al.* 1985 on brachiopods; Ingavat *et al.* 1980 on foraminifera; Ross 1978 and Sakagami 1985 on bryozoa) it appears that a southern belt of microcontinents stretching from Iran and Afghanistan, through southern Tibet (Lhasa and W Qiangtang) to western Thailand (Shan-Thai or Sibumasu block), Malaysia, and Sumatra, had temperate faunas and floras and even tillites in the Early Permian and that these terranes probably were derived by rifting from northern Gondwana in the Permian. Furthermore, free migration through the tropics, including the epeiric Uralian seaway until it began to fill in the Late Permian, is indicated by the foraminifera, but not by the corals, brachiopods, or bryozoans. Prior to the Late Permian, the Uralian seaway contained a range of tropical environments from reef to deeper (thrust-loaded) basin (Ziegler 1988), thus it must be assumed that the comparable habitats of the Tethyan microcontinents must have been isolated from them by a relatively wide expanse of open ocean to explain the faunal discontinuities. A location to the east and south within Palaeo-Tethys is suggested by this rather indirect line of reasoning, and by the fact that the Cathaysian floras, common to many of the south Asian microcontinents, are also known from the northern margin of Gondwana in Arabia.

Asia and the Middle East comprise no less than ten substantial

From McKerrow, W. S. & Scotese, C. R. (eds), 1990, *Palaeozoic Palaeogeography and Biogeography*, Geological Society Memoir No. 12, pp. 397–409

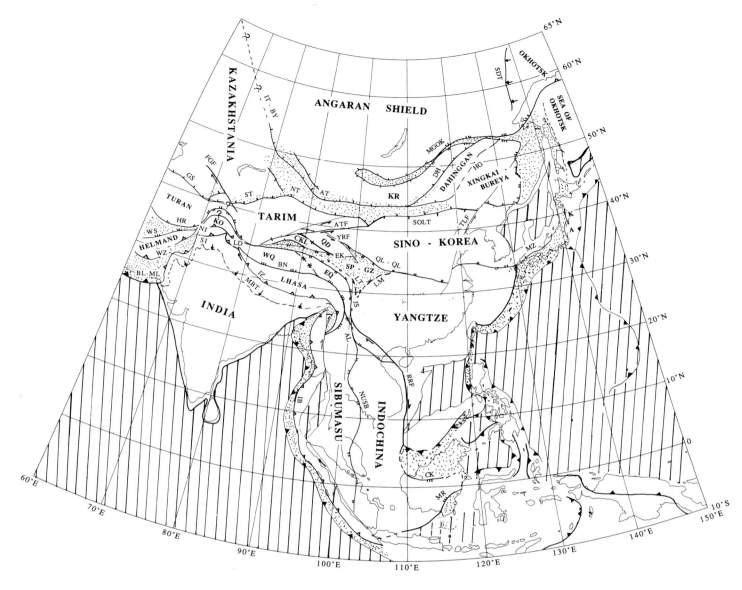

Fig. 1. Tectonic subdivisions and sutures for the eastern Asian blocks. *Microcontinents*: A, Abukuma; CKL, Central Kunlun; DH, Dahinggan; EQ, Eastern Qiangtang; K, Kitikami; KO, Kohistan; QD, Qaidam; WQ Western Qiangtang. *Sutures*: AL, Ailaoshan; AT, Altai; BL-ML, Bela-Muslimbagh; BN, Banggong Co-Nujiang; CK, Crocker; DH, Dahinggan; EK, East Kunlun; GS, Ghissar; HG, Heganshan; HR, Herat; IB, Indo-Burma; IT-BY, Irtysh-Bayitik; IZ, Indus-Zangbo; JS, Jinshajiang; KR, Kerulen; LD, Ladakh; LM, Longmenshan; LT, Litang; MGOK, Mongolo–Okhotsk; MR, Meratus; MZ, Maizuru; NI, North Indus; NT, North Tianshan; NUSB, Nan-Uttaradit-Sra-Kaeo-Bentong; QL-QL, Qilian-Qinling; SI, South Indus; SOLT, Solon Obo-Linxi-Tumen; ST, South Tianshan; WS, Waser; WZ, Waziristan. *Major Strike-slip Faults*: ATF, Altyn Tagh; FGF, Fergana; RRF, Red River; TLF, Tan-Lu; YRF, Yellow River. *Thrust Belts*: MBT, Main Boundary Thrust; SDT, Sette-Daban Thrust. *Accretionary complex*: SP-GZ, Songpan-Ganzi.

continental blocks and numerous continental fragments (not including Kazakhstania, Siberia Platform, India, and Arabia) that have coalesced since the mid-Palaeozoic by the closure of various branches of Palaeo- and Neo-Tethys (Sengör 1987a). Included within this diverse collage are larger domains characterized by generally platformal and shelf sedimentation of Palaeozoic and/or Mesozoic age, including Tarim, Sino−Korea, Yangtze (South China), Eastern Qiangtang, Western Qiangtang, Lhasa, Indochina, and Sibumasu (Shan Thai, Malaya) in Asia, and the Central Iranian micro-continent and Helmand (Sistan) block in the Middle East (Sengör 1984, 1987a) (Fig. 1). In general, we agree with the distribution of sutures and timing of suturing outlined by Sengör et al. (1988), although we have simplified some parts, and for others, such as in Pamirs, we were unable to confidently map all of their positions and have therefore left these off the map. In this account we make only a few comments on the sutures and their timing. The Mongolo−Okhotsk suture zone separates the Kerulen, Dahinggan, and Xingkai−Bureya (Fig. 1) blocks from the Angaran Shield. It is characterized by progressive west to east closure from the Carboniferous and Permian (Marinov et al. 1973) to the latest Jurassic along the Dzhagdi suture (Kosygin & Parfenov 1981, = Shilka suture of Klimetz 1983; Sengör 1987a). Farther east, in the northern Sea of Okhotsk, closure between the Sea of Okhotsk block and the parautochthonous Okhotsk block did not occur until latest Cretaceous or Early Palaeocene (Kosygin & Parfenov, 1981). The Mongolo−Okhotsk suture appears to have terminated in the west in the Early Mesozoic near Lake Har Us Nur, with the area stippled in Fig. 1 representing Palaeozoic and younger turbiditic filling of this basin. We envision this as a Sea of Japan type back-arc basin. The Irtysh−Bayitik segment of the Altai belt separates marginal arcs of Siberia from the Kazakhstanian accretionary complex. Closure along this suture occurred during the Late Carboniferous (Rowley et al. 1985). This suture and its eastward extension in Gobi Altai define the southern limit of the distribution of the Silurian brachiopod *Tuvaella* (Rong & Zhang 1982), which appears to be geographically restricted to Siberia and its marginal arc but not necessarily to a cold environment. The Tianshan−Solon Obo−Linxi−Tumen suture separates the already amalgamated Kazakhstan−Altai from the Tarim block in the west and Sino−Korea in the east, and defines the boundaries between the Angaran and Cathaysian floras and between the Arctic and Tethyan faunas (Marinov et al. 1973). Whether or not Tarim and Sino-Korea were two separate blocks in the late Palaeozoic has been a matter of dispute (Smith et al. 1981; Scotese 1984; Rowley et al. 1985; Lin & Watts 1988). As will be discussed later, the existing palaeomagnetic data seem to suggest they were two different blocks. However, a well defined suture does not exist between them, but this may reflect in part the superposition of Altyn Tagh related strike-slip deformation. Yang et al. (1986) interpret the Altyn Tagh as a suture zone, but where this suture is best defined by ophiolitic rocks is toward the south between Qaidam and Tarim, where two belts of Late Palaeozoic ophiolites have been recognized (Zhang 1981). Biogeographically, Tarim and Sino-Korea are quite similar, but the timing of closure between Tarim and Kazakhstania and between Sino-Korea and the Mongolia arcs is somewhat different. Tarim collided during the Early Permian (end C_3 of the Chinese), as shown by the termination of andesitic volcanism in the Tianshan arc, the development of unconformities between late Early Permian and underlying sequences, and intrusion of widespread granites on both sides of the suture (Xinjiang Stratigraphic Group 1981). Closure along the Solon Obo−Linxi−Tumen suture occurred in the Late Permian (Wang & Liu 1986). The existence of a south-dipping subduction zone north of the Sino-Korean block before the Late Permian is supported by an intercalated assemblage of andesitic rocks with *Gigantopteris*-bearing clastic sequences along its northern margin (Huang 1986; Nei Mongol Stratigraphic Group 1981).

The Kunlun-Qilian suture separating the Songpan-Ganzi accretionary complex (Sengör 1984) from Central Kunlun, Qaidam, and Sino−Korean blocks is herein interpreted as marking a south-dipping, rather than north-dipping subduction zone (Wang 1985; Sengör et al. 1988). We base this on the absence of end Palaeozoic−Triassic arc type magmatic rocks to the north of the ophiolites that mark the suture, where shelf carbonates occur that lack volcanic-derived sediments (Xinjiang Stratigraphic Group 1981; Qinghai Stratigraphic Group 1980; Gansu Stratigraphic Group 1980; Yang et al. 1983). This interpretation accords with observations by the Royal Society−Academia Sinica Traverse south of the Kunlun Pass (Kidd pers. comm.), which interpreted the Triassic as continental rise sediments. Farther east, the Qilian suture is continuous with the Qinling suture separating Yangtze and Sino−Korea, but here it is generally interpreted to dip north, although a well defined andesitic belt is not known along the southern margin of the Sino−Korean block. The continuation of the Qinling suture to the east of the Tan-Lu Fault is controversial. For example Scotese et al. (1979) and Sengör et al. (1988) among many others place a suture across the Korean peninsula through the Chugareong zone. We note that the stratigraphy on opposite sides of this zone is identical, and is very similar to north China (Lee 1987). There is therefore no reason to suspect that northern and southern Korea were separate. The implication is that the Qinling suture extends to the south of Korea and into Japan where it is marked by the Maizuru ophiolites (Tanaka & Nozawa 1977). Another difference in interpretation is that we do not recognize Yangtze and Huanan (literally south China) as separate blocks in the Late Palaeozoic, for the suture as drawn by Hsü et al. (1988) is uncomformably covered by Devonian and younger sediments (Rowley et al. 1989).

The derivation of all of these blocks is not yet well known, but at least some, including central Iran, Helmand, Western Qiangtang, Lhasa, and Sibumasu, are reasonably interpreted to have been rifted off the northern margin of Gondwana in post-Early Permian times, and have been interpreted by Sengör (1984) as belonging to a loosely associated continent named Cimmeria. Blocks farther to the north show no Late Palaeozoic affinities to Gondwana, and although some, such as Yangtze and Indochina, have been said to have facies and faunas that are comparable with earlier Palaeozoic assemblages in Australia (Ziegler et al. 1981), no unequivocal ties have yet been established.

In the body of this paper we discuss the Carboniferous and Permian climatic constraints for a number of areas on the margins and within Palaeo-Tethys (Table 1). The climatic interpretations are based on the biome concept as applied to the fossil floras and described in Ziegler (1990). It must be admitted that some of the biomes, especially the tropical rain forest, can occur over a wide latitudinal range, but in view of the paucity of palaeomagnetic data from many of the regions, any information is useful. Actually, the phytogeographic data are perhaps more useful in the ligations they imply. Many of the plant groups, including *Gigantopteris* were seed bearing plants that could not, presumably, have been dispersed across oceans by winds. A major challenge is therefore posed by the fact that the considerable number of apparently separate microcontinents which shared the *Gigantopteris* flora must have had land connections at times when they were tectonically independent. Moreover, the many epeiric and marginal seaways that existed would have been obstacles around which the *Gigantopteris* flora was successfully dispersed, at least to the degree implied by Table 1. It is true that the floral records of some of the microcontinents are very incomplete, and it is impossible to say which of the gaps represent collection failure and which are phytogeographically significant. Table 1 shows the climatically interpretable constraints that we feel should be respected in preparing the Late Palaeozoic palaeogeographic maps of the Palaeo-Tethyan microcontinents. We hasten to add that many island arcs and other tectonic entities may have disappeared with little trace, and we make no attempt to reconstruct land

Table 1. *Carboniferous through Permian floristic and lithological constraints for the margins and microcontinents of Palaeo-Tethys*

									Microcontinents									
	North Gondwana				Peri-Gondwanan					Cathaysian						South Angara		
	N Arabia	S Arabia	Kashmir	N Guinea	N Iran	Helmand	Lhasa	W Qiangtang	Sibumasu	E Qiangtang	Yangtze	Indochina	Tarim	Sino-Korea	Kitakami	Junggaria	S Mongolia	Primorye
	1	2	3	4	5	6	7	8	9	10	11	12	13	14	15	16	17	18
PERMIAN																		
Changxingian							1				1		5	5,2		5		
Longtanian	(1)									1	1	1	1?	2,1		4	6	
Kazanian		(2)		(5)	1		(5)				1	1	1	1	1	4	5	5
Ufimian	(1)	(1)				4					1	1	1			4		5
Kungurian			5								1		1	1		4	6	6
Artinskian			8	(5)							1	1		1	1	4	6	6
Sakmarian	(2)	0					0		2?		1			1		4	6	6
Asselian		0	0		6?	0?	0	0	0?					1		4	6	
CARBONIFEROUS																		
Gzhelian		0	0											1			6	
Kasimovian			0											1			6	
Moscovian			0		3?									1		1	6	
Bashkirian			5										1	1		2?	6	
Serpukhovian			5		5						1		3	3		2?	5	
Visean			5		5					1	1	1		3		1	5	
Tournaisian	5				5						1			3		1?	5	

The Late Permian Chinese stages, Longtanian and Changxingian, have been substituted for the standard Tartarian, simply to allow inclusion of detail for that part of the column. The numbers refer to the climate related biomes (Ziegler 1990) and are defined on the basis of floras except where noted: (1) tropical everwet, (2) tropical summerwet, (3) subtropical desert (defined by evaporites), (4) temperate winterwet, (5) warm temperate, (6) cool temperate, (8) cold temperate, (0 (or 10)) glacial (defined by tillites or glacio-marine deposits). The parentheses indicate uncertainly in age and the queries indicate uncertainty in biome assignment. Numbers other than 1 in columns 1 through 9 refer to southern hemisphere biomes while numbers other than 1 in the other columns represent northern hemisphere biomas, except for the 3's in columns 13 and 14 which are of uncertain polarity. References for the information are given in the text except for the number 5 in column 1 which is interpreted from Legrand-Blain (1985) and Lejal-Nicol (1985).

bridges implied by the floras if there is no geological evidence for them at this time. We will focus our discussions on the Permian and to a lesser extent the Carboniferous.

North Gondwana

By the Permian, the latitude of north Gondwana ranged from equatorial in northern South America and northern Africa, to tropical in Arabia, to temperate in India and Australia (Fig. 2, Table 1). The marine sediments reflect this gradient, and range from carbonates to clastics going eastward. In fact, Early Permian tillites are known from southern Saudi Arabia (Besems & Schuurman 1987) and glacio-marine deposits occur in India (Singh *et al*. 1982) and Australia (Veevers & Powell 1987) and the glaciers reached nearly to the subtropics, specifically to about 40° S (Ziegler 1990). It is commonly assumed in the literature that the Permo-Carboniferous glaciers 'tracked the pole', but justification for our reconstruction comes from the fact that in northern India it can be shown that the tillites are interbedded with carbonates. This also indicates that the ice sheets reached low elevation, but they may have been centered on highlands possibly of rift-shoulder origin. Such a concept would explain the low latitude setting of the glaciers and would be consistent with the idea that the 'Peri-Gondwanan' terranes described in the next section were rifted from Gondwana about this time.

It is commonly assumed that the Late Palaeozoic floral provinces, including Angaran, Gondwanan and Cathaysian, are biogeographically restricted to Siberia and Kazakhstania, Gondwana, and east Asian micro-continents, respectively, and that this distribution reflects geographic isolation. In recent years doubts have been expressed with regard to the validity of this view. For example, the Cathaysian flora is now known from a number of low latitude Gondwanan sites including Venezuela (Odreman-Rivas & Wagner 1979), Turkey (Archangelsky & Wagner 1983, Iraq (Ctyroky 1973) and Saudi Arabia (El-Khayal & Wagner 1985). A few Cathaysian elements have been reported from India (Singh *et al*. 1982) and New Guinea (Asama *et al*. 1975; Li 1986), but these floras are more appropriate assigned to a 'warm temperate' biome 5 climatic setting (Ziegler 1990) than to a tropical setting. Thus, the present view emphasizes the palaeoclimatic controls rather than strict biogeographic controls on these floras. On the basis of the present evidence, it seems more likely that the Cathaysian floras of China were linked to those of Gondwana through Arabia, rather than through Australia or India. Tectonic evidence for rifting along the northern margin of Gondwana in the Permian includes the major unconformity beneath the Permian throughout the Arabian Peninsula (Weissbrod 1976; Sharief 1982) and the Panjal Traps of Kashmir in northern India (Singh *et al*. 1982). However, the Palaeozoic section in the Himalayas to the east seems to be relatively complete and uninterrupted (Xizang Science Expedition 1987). Farther east, the presence of thick, primarily deltaic and glaciomarine sediments of Late Carboniferous and Early Permian age underlying the northwest shelf of Australia are generally interpreted to reflect rifting at that time (Powell 1976). Still farther east, along the eastern margin of Australia, there is clear evidence of convergent margin tectonism through the Permian, including the initiation of overthrusting in the Sydney and Bowen basins in the Late Permian (Jones *et al.*

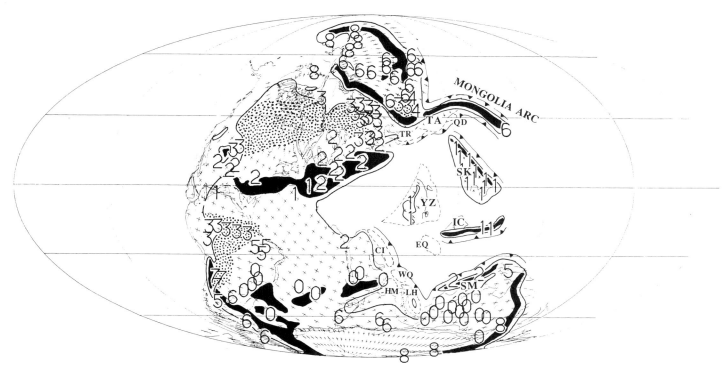

Fig. 2. Early Permian Palaeogeography. Dotted areas are evaporites, black areas mountains. Dashed lines indicate outlines of microcontinents, heavy lines represent shorelines. Subduction zones are schematically shown with arrows indicating dip direction. Where less certain, the subduction zones are represented by the dashed lines along with arrows. Numbers are biome designations for the Sakmarian (Ziegler 1990) as listed in Table 1. Abbreviations for Asian microcontinents: CI, Central Iran; EQ, East Qiangtant; HM, Helmand; IC, Indochina; LH, Lhasa; SK, Sino-Korea; SM, Sibumasu; TA, Tarim; WQ, West Qiangtang; YZ, Yangtze, Reconstruction of Pangaea is after Lottes and Rowley (1990).

1984). The northward extension of this belt is represented by Late Palaeozoic granodiorites that outcrop in the Kubor Ranges of New Guinea (Hamilton 1979), marking the eastward limit for the Late Palaeozoic rifting of the northern Gondwanan margin.

Peri-Gondwanan microcontinents

The microcontinents grouped together in this section seem to be linked in a nearly continuous belt which today, as apparently in the Early Permian, extend along the northern periphery of Gondwana. The present width of this belt is from 200 to 500 km, presumably somewhat less than its original extent, and the lateral limits of the original microcontinents are unknown. This group includes the Helmand, Western Qiangtang, Lhasa, and Sibumasu blocks. As shown in Figs 2 & 3, the Helmand block is recognized as being adjacent to northern Gondwana due to the presence of *Eurydesma* in the Logar synclinorium (Wolfart & Wittekindt 1980), but no glacial or glacio-marine sediments or Gondwanan floras have as yet been recognized. The Western Qiangtang block includes Late Palaeozoic tillites, first recognized by Norin (1946) in the Horpatso area, and recently confirmed by Liang *et al*. (1983), together with the occurrence of *Eurydesma*. The Lhasa block also is characterized by Lower Permian tillites, and by Lower Palaeozoic sediments overlying Late Precambrian metamorphic basement, such as in the Xainza region (Xu *et al*. 1981), which is similar to sections of northern India (Xizang Scientific Expedition 1987). The Sibumasun blocks has long been recognized as a Gondwanan-derived continental fragment (Burrett 1974; Stauffer & Mantajit 1981; Bender 1983; Metcalfe 1988). This block extends northward into western Yunnan, where Cao (1986) has described tillites, *Stepanoviella*, and possible *Eurydesma* in Late Carboniferous−Early Permian sediments. A feature characteristic of all of these blocks is that the sediments immediately above the tillites and/or cool water faunal assemblages are fusulinid-bearing limestones. Note again that this is also typical of much of the northern margin of Gondwana (Nakazawa 1985).

A critical problem is the timing of departure of these blocks from northern Gondwana and their respective sites of derivation. It is clear from the presence of glacial sediments and associated cool-water fauna that these blocks were still adjacent to the northern margin of Gondwanan in the Early Permian. At least in the Sibumasu block, the glacio-marine sediments, including the Phuket Group and Mergui Group, were deposited in fault-bound basins that have been interpreted to be of rift origin (Bender 1983). The presence of coeval active margin facies along the eastern margin of the Sibumasu block clearly suggests that the rifting was along its western margin. As summarized by Metcalfe (1988), there are remarkable correlations that can be made between the Palaeozoic rocks of Sibumasu and Australia. The unconformity between Permian carbonates and underlying Lower Palaeozoic sediments of the central part of the Sibumasu block (Bender 1983) may be interpreted as a 'break-up' unconformity, suggesting that it left Gondwana in the late Early Permian. Sibumasu collided with Indochina by the Late Triassic, as documented by the unconformably overlap of the non-marine lower Khorat Group across the Nan−Uttaradit−Sra−Kreo suture (Bunopas 1981). The Lhasa block is generally reconstructed to a position north of India, essentially identical to their present relative positions. Sengör (1987b) has noted that the Precambrian of the Aravalian Range of India is similar in age to the Precambrian of the Nyanqentanglha Range on the Lhasa block. Andesites ranging in age from Carboniferous to Cretaceous are present in the southeastern part of the Lhasa block (Xu *et al*. 1981), and suggest proximity to an active margin. However, farther west in the Xainza region andesites are not observed in the Palaeozoic units, suggesting that the opening of both the Banggong Co-Nujiang ocean and Indus-Zangbo segment of Neo-Tethys cut across pre-existing tectonic domains. A problem with

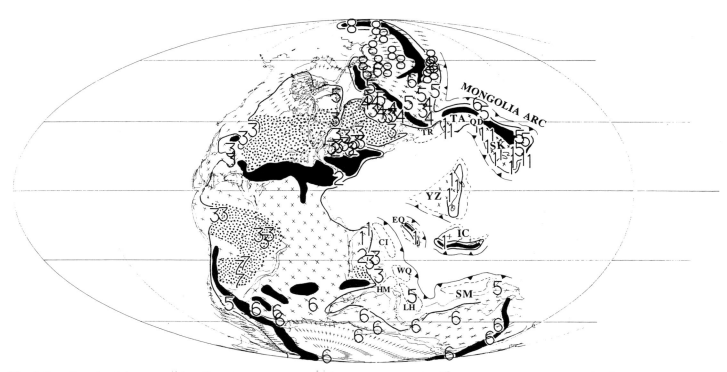

Fig. 3. Late Permian Palaeogeography. (See Fig. 2 for explanations).

placing Lhasa along the northern margin of India is that the andesitic sequences would be in the concave portion of the northern margin of eastern Gondwana. Farther west, the three microcontinents of Helmand, Western Qiangtang and Central Iran are placed to the northeast of Arabia and India. It should be pointed out that little is known about the derivation for these blocks. We place the Western Qiangtang block outside of Helmand simply because it contained typical Cathaysian flora by the end of the Permian (Table 1) and should, presumably, have left Gondwana earlier. These blocks appear to have begun rifting from Gondwana during the Permian, with most recent estimates of the time of spreading along the northern margin of Arabia and India being placed in the Triassic (Sengör et al. 1988).

Late Palaeozoic palaeomagnetic data are still quite limited for these Gondwana-derived fragments and have only been published for the Sibumasu block. The results from four reconnaissance palaeomagnetic studies (McElhinny et al. 1974; Sasajima et al. 1978; Bunopas 1981; and Zhang & Zhang 1986) of the Carboniferous and Permian for the Sibumasu block are listed in Table 2. Because most of Bunopas' (1981) poles are based on only one or two samples, as reflected by their large α_{95} values, they are hardly acceptable for palaeogeographic analysis. The remaining three poles differ considerably from each other and are therefore not used for our reconstructions, although the latitude of 34.1°S of Zhang & Zhang (1986) for western Yunnan part of the Sibumasu block in the Late Carboniferous appears to be in general agreement with our interpretation of Sibumasu being adjacent to northwestern Australia. It is worth noting that a recent palaeomagnetic study (Fang & Van der Voo, pers. comm.) on the Devonian carbonates also suggests a moderately high southern latitude for western Yunnan.

The climatic information summarized for these blocks is also sparse. Data pertaining to northern Iran (Table 1, column 5) is taken from a microfloral analysis of sections in the Alborz Range (Coquel et al. 1977; Chateauneuf & Stampfli 1978). The Early Carboniferous and Early Permian microfloras were matched with Gondwana floras that we assign to biomes 5 and 6 respectively (warm temperate and cool temperate). Cathaysian microfloras, however, occur in Late Permian rocks, and were interpreted by the above mentioned authors to indicate equatorward motion of the central Iranian block. Deposits interpreted as glacio-marine or described as diamictites are known from the early Permian rocks of Lhasa (Jin 1985), western Qiangtang (Liang et al. 1983), and Sibumasu (Stauffer & Mantajit 1981; Cao 1986), and cool water faunas are common to all these areas (Nakamura et al. 1985, Waterhouse 1982), plus the Helmand Block (Wolfart & Witterkindt 1980, tables 14 and 15). Very little can be made of the remaining floral data from the areas. The biome 2 entry of column 9 is based on the occurrence of the conifer *Walchia* (Bunopas 1981) but this fossil is known from other dry climate, or seasonally dry biomes. The biome 4 entry of column 6 is based on a 'Subangaran' flora of Meyen (1982); this would suggest a northern hemisphere setting for the Sistan Block in the Late Permian, though the floral list is short and the locality description vague ('near Kabul'). Another enigmatic flora is the mixed Gondwana−Cathaysian flora from the Lhasa Block (biome 5 in column 7) which contains none of the truly distinctive members of either of these floras (Li et al. 1985). Again, a very late Permian Cathaysian flora (biome 1) is known from the western Qiangtang block. In summary, it appears clear that by the Late Permian some of the Peri-Gondwana terranes were under the influence of the Cathaysian flora.

Cathaysian microcontinents

The remaining blocks, including Yangtze, Indochina, Eastern Qiangtang, Sino−Korea and Tarim, as well as the smaller fragments of Abukuma and Kitikami of Japan were within the tropical Palaeo-Tethyan domain during the Permian. The times of suturing summarized above and in Fig. 1 provide the starting point for placement of these blocks within Palaeo-Tethys. A brief summary is provided below of additional data used in the reconstructions shown in Figs 2 and 3.

Yangtze block

Late Palaeozoic palaeomagnetic data from the Yangtze block are available for only the Late Permian, with one exception (Lin

Table 2. *Palaeomagnetic Pole Data for the Asian Blocks in the Late Palaeozoic*

Formation, Location	Age	Slat. (°N)	Slon. (°E)	Plat. (°N)	Plong. (°E)	α_{95} (°)	K	Reliability 1 2 3 4 5 6 7	Q	Reference
South China block										
Emei Basalt & Xuanwei, Sichuan	P_2	29.6	103.4	54.0*	252.0*	6.2	26.4	x x x x x	5	McElhinny *et al.* 1981
Emei Basalt, Sichuan	P_2	29.6	103.4	54.1	241.8	19.5	10.5	x x x	3	Chan *et al.* 1984
Emei Basalt, Guizhou	P_2	26.4	105.7	−29.3*	55.3*	12.0	41.0	x x x x	4	Lin 1984
Emei Basalt, Sichuan	P_2	29.6	103.4	50.7	248.3	3.3	26.8	x x x	3	Zhang 1984
Emei Basalt, Sichuan	P_2	26.7	102.9	53.5*	241.8*	10.1	37.1	x x x x x x	6	Huang *et al.* 1986
Emei Basalt, Yunnan	P_2	26.1	103.1	52.4*	255.9*	25.0	5.9	x x x x x	5	Huang *et al.* 1986
Emei Basalt, Yunnan	P_2	25.9	100.6	−24.7*	24.3*	24.8	25.9	x x x x	4	Huang *et al.* 1986
Emei Basalt, Sichuan	P_2	28.3	103.0	55.4	251.9	65.1	16.9	x x x	3	Zhou *et al.* 1986
Emei Basalt, Sichuan	P_2	26.8	101.8	25.7*	216.6*	8.7	201.2	x x x x x	5	Zhou *et al.* 1986
Emei Basalt, Sichuan	P_2	29.7	103.5	38.5	231.6	5.1	53.0	x x x	3	Liu *et al.* 1987[a]
Emei Basalt, Sichuan	P_2	28.1	102.9	58.9	246.1	7.0	24.0	x x x	3	Liu *et al.* 1987[a]
Emei Basalt, Sichuan	P_2	25.9	100.6	63.5	264.3	10.7	12.0	x x x	3	Liu *et al.* 1987[a]
Emei Basalt, Sichuan	P_2	29.6	103.4	49.7*	252.0*	3.6	37.4	x x x x	4	Zhao & Coe 1989
For Permian, all Q, N = 13				48.2	236.7	9.0	22.2			
Q ≥ 4, N = 7				42.6*	230.2*	14.8	17.5			
Q ≥ 5, N = 4				47.4	232.0	20.4	21.3			
Q = 6, N = 1				53.5	241.8					
Chuanshan, Zhejiang	C_2–P_1 (C_3)[b]	31.0	119.8	−19.4*	47.1*	16.9	54.0	x x x x	4	Lin 1984
North China block										
U. Shihezi & Shiqianfeng, Shanxi	P_2 (P_2)[b]	37.8	112.3	44.0*	358.0*	6.9	40.3	x x x x x x	6	McElhinny *et al.* 1981
L. & U. Shihezi, Shanxi	P_2 $(P_{1–2})$[b]	37.8	112.4	−49.2*	178.5*	3.9	17.4	x x x x	4	Zhang 1984
U. Shihezi, Shanxi	P_2 (P_2)[b]	37.8	112.3	−38.1*	186.3*	?	?	x x x x x	4	Lin J. 1984
U. Shihezi, Shanxi	P_2 (P_2)[b]	37.9	113.8	56.7*	6.2*	10.9	49.8	x x x x x	5	Lin W. 1984[a]
Dahuanggou, Gansu	P_2 (P_1)[b]	39.8	97.8	−43.9*	171.3*	4.5	35.0	x x x x	4	Coe *et al.* 1987[a]
(?), Nei Monggol	P_2	42.3	110.2	56.7	355.6	?	?	x x x	3	Liu 1987
Shihezi, Hebei	P_2 $(P_{1–2})$[b]	37.5	114.4	55.2*	1.4*	6.9	42.0	x x x x x x	6	Zhao 1987
Shihezi, Shanxi	P_2 $(P_{1–2})$[b]	38.6	112.1	37.9*	2.9*	7.5	14.2	x x x x x x	6	Zhao 1987
Shihezi, Shanxi	P_2 $(P_{1–2})$[b]	40.1	113.2	55.2*	335.7*	9.0	15.2	x x x x x x	6	Zhao 1987
U. Shihezi & Shiqianfeng, Ordos	P_2 (P_2)[b]	35.0	109.0	42.3*	13.7*	9.1	?	x x x x x x	6	Cheng *et al.* 1988
Shuangshiquan, Hebei	P_2 (P_2)[b]	40[c]	116[c]	−35.1*	187.0*	16.6	?	x x x x x x	5	Cheng *et al.* 1988
average for all Q, N = 11				47.1	0.4	5.9	59.4			
Q ≥ 4, N = 10				46.1*	0.8*	6.3	46.2			
Q ≥ 5, N = 7				47.1	1.6	8.9	46.7			
Q = 6, N = 5				47.5	−0.4	11.4	45.8			
Tarim Block										
(red sandstone), Kuqa	P_2	42.1	83.4	73.2*	187.6*	7.3	45.2	x x x x x	5	McFadden *et al.* 1988
(dykes), Keping-Bachu	P_2	39.8[d]	78.8[d]	−65.6*	1.2*	3.9	61	x x x x x	5	Li *et al.* 1988
Kaipaizileike & Kupukuziman, Akesu	P_2 (P_1)[b]	41.0	79.4	−62.8*	11.5*	3.2	87.6	x x x x	4	Cheng *et al.* 1983
Kaipaizileike & Kupukuziman, Akesu	P_1	40.8	79.0	−56.5*	10.1*	4.0	42	x x x x	4	Bai *et al.* 1987
Kangkelin, Akesu	C_2–P_1 (C_3)[b]	40.8	79.0	−52.2*	0.5*	8.6	79	x x x x	4	Bai *et al.* 1987
For P_2: Q ≥ 4, N = 3,				67.3*	186.9*	8.8	196.4			
Q = 5, N = 2				69.4	183.8	17.3	209.6			
For P_1: Q = 4, N = 2				54.5*	184.5*	16.4	232.6			
Mongolia										
Permian,	P_2	48.1	106.0	−11.3	215.2	7.8	13.8	x x x	3	Pruner 1987
Permian,	P_1	49.5	105.1	45.0*	22.9*	6.4	21.3	x x x x	4	Pruner 1987
Carboniferous	C	44.6	108.3	40.9	307.4	3.9	23.2	x x x	3	Pruner 1987
Sibumasu										
Sempah conglomerate & rhyolites, Malaya	C_2–P_1	3.4	101.8	55	164	11.2	13.4	x x x x	4	McElhinny *et al.* 1974
Basalt, Sumatra	P	−1.0	101.0	47.0	182.4	5.1	?	x x	2	Sasajima *et al.* 1978
Permian, N. Thailand	P	15.5[c]	47.1[c]	43.8	155.5	34.5	?	x x	2	Bunopas 1981
Permian, N. Thailand	P	16.7[c]	98.6[c]	87.0	116.9	26.7	?	x x	2	Bunopas 1981
Permian, Malaya	P	8.8[c]	98.3[c]	84.0	22.0	6.7	?	x x	2	Bunopas 1981
Carboniferous, N. Thailand	C	17.7[c]	100.5[c]	64.8	170.1	19.6	?	x x	2	Bunopas 1981
Carboniferous, Malaya	C	7.3[c]	99.6[c]	67.8	163.0	26.2	?	x x	2	Bunopas 1981
Funiushi Basalt, W. Yunnan	C_2	25.1	99.1	20.4	245.9	5.9	18.3	x x x x	4	Zhang & Zhang 1986
Indochina										
Bentong Gp., & Singa Fm., Malaya	C_1 C	3.8 6.2	101.9 99.8	57	182	9.2	37.0	x x	2	McElhinny *et al.* 1974
Pengerang rhyolites, Singapore	P_2–Tr_1	1.4	104.2	57	152	10.7	21.5	x x x x	4	McElhinny *et al.* 1974

Slat./Slong, site latitude/longitude;   Plat./Plong, pole latitude/longitude;
α_{95}, semi-angle of cone of 95% confidence about the mean;   K precision parameter; N number of poles;
Q, reliability index (after Van der Voo, in press); 1, well-dated rocks; 2, sufficient quantity of samples (>25) and reliable statistics (α_{95} > 16) 3, details of demagnetization published; 4, field tests available (including fold-, contact-, and conglomerate test); 5, cratonic coherence (structural control of tilt and tectonic rotation): in this table, Criterion 5 is met if the site locality is not near known plate boundary; 6, reversal test (mixed polarity reported); 7, no suspicion of remagnetization: in this table, Criterion 7 is met for most of the poles simply because no younger poles are yet available for comparison; x, criterion met; *, pole data used for reconstructions; [a], data after Zhao & Coe 1989; [b], original age assignment in parentheses for the formation or group in Chinese usage (see text for details); [c], approximate site latitude and longitude; [d], site locality from which most of the pole data are from, other localities: 40.6° N, 79.5° E and 39.8° N, 78.8° E. Gp, Group; Fm, Formation; P, Permian; C, Carboniferous; Tr, Triassic

1984) from the latest Carboniferous or Early Permian limestone of Chuanshan Formation from Zhejiang Province. As shown in Table 2, almost all of the Late Permian palaeomagnetic results are from the Emei(shan) (also transliterated as Omei) Basalt, a flood-type basaltic sequence that covers much of the western and southwestern part of the Yangtze block, including the eastern part of the Songpan–Ganzi accretionary complex (Wang H. Z. 1985). The Emei Basalt ranges in thickness from several hundred to over 3000 meters (Lin, J. Y. 1985). By and large, it lies stratigraphically between the Maokou Formation (a thick limestone sequence) below, and the Longtan Formation (a terrestrial coal-bearing sequence) above. However, the Emei Basalt is believed to be a diachronous sequence that erupted as early as the Qixian (Artinskian) in the southwest and as late as Changxingian (late Tatarian) in the northeast (Lin, J. Y. 1985). Locally it is interbedded with thin limestone beds, although no pillow structures are known. It is presumably related to back-arc extension behind the Kekexili–Kangdian arc between the Yangtze block and the Songpan-Ganzi accretionary complex.

As listed in Table 2, the palaeomagnetic results from the Emei Basalt are similar, but partly because most of the study areas are near the western edge of the Yangtze block (s.s.) where Mesozoic and Cenozoic deformation is intense, the interpretations of these results have been the subject of some arguments. In their pioneering palaeomagnetic study on China, McElhinny et al. (1981) first interpreted their results to indicate a reversed polarity, correlated with the Late Palaeozoic Kiaman Reversed Interval (from Late Carboniferous to possibly Late Permian, Irving & Parry 1963; Harland et al. 1982), and proposed that the Yangtze block had rotated more than 120° counterclockwise since then. McElhinny (1985) subsequently attributed the rotation to only local tectonic distortion. Lin (1984) argued that McElhinny's Late Permian pole could actually be Jurassic remagnetization. Huang et al. (1986) found mixed polarities in their studies of Sichuan and Yunnan rocks and were the first to point out the possibility that the upper part of the Emei Basalt might postdate the Kiaman Reversed Interval. If this is the case, then only half the amount of rotation (60°, but clockwise) is required to have taken place, a view now also shared by Zhao & Coe (1987). We accept this interpretation, because (1) in addition to its diachronous nature, the Emei Basalt is usually placed at the base of the Upper Permian in Chinese usage, but this corresponds with the late Upper Permian in western usage. It therefore becomes more likely that the upper part of the Emei Basalt could have been deposited after the Kiaman Reversed Interval, and (2) our floral data seem consistent with the reconstructed azimuth of the Yangtze block based on this interpretation. Specifically, the formation of Upper Permian coal on Yangtze is known to have been established earlier (Maokouan) in the east and later (Longtanian) in the west. This can be interpreted to reflect a diachronous passage through the equatorial belt of first the eastern part followed by the western part of the Yangtze block.

In our Late Permian reconstruction (Fig. 3), we have chosen the average pole for which $Q \geq 4$ (quality index, Van der Voo, in press), with its palaeocoordinates and related statistics listed in Table 2. This pole puts the central location of the Yangtze block (28° N, 110° E) astride the equator. For the Early Permian map (Fig. 2), we use Lin's (1984) Late Carboniferous (Chinese usage) pole. The central location of Yangtze block is accordingly placed at about 10° S, with a similar azimuth as in the Late Permian.

Lower Carboniferous and Permian floral data (Zhao & Wu, 1979; Li & Yao, 1979) have been assigned to the Euramerican and Cathaysian Floral Realms, respectively, and they represent the tropical rain forest (biome 1) of the two periods. The absence of any floras of Upper Carboniferous age has been attributed to higher sea level stands, but it could equally well be that the continent moved into slightly drier latitudes that were less favorable for floras. Evaporites are unknown, but dolomites are widespread in the upper Carboniferous of the Yangtze Block (Wang, H. Z. 1985). In summary, the floral data are compatible with available palaeomagnetic information.

Indochina

Since Haile's (1981) summary, little if any advance has been made in regard to the Palaeozoic palaeomagnetism of Indochina. The only pole known to us (McElhinny et al. 1974) is from the Late Permian to Early Triassic Pengerang rhyolites in Singapore (Table 2). It places the Indochina block at about 30° in the northern hemisphere, a result that appears incompatible with the recent understanding of a Triassic closure between the Indochina and South China blocks (Achache & Courtillot 1985; Şengör et al. 1988). Another pole that could apply to Indochina (McElhinny et al. 1974) is from the Bentong Group of Early Carboniferous age (Table 2), but unfortunately, the site locality (3.8° N, 101.9° E) is very close to the Bentong suture. Moreover, the pole is obtained by averaging data from the Bentong Group and those from the Singa Formation (site locality: 6.2° N, 99.8° E) that apparently and the reconnaissance nature of the study, we have not used these poles for our reconstructions.

The floral information (Asama et al. 1975), like that of Yangtze block, indicates the Euramerican Province in the Early Carboniferous and the Cathaysian Province in the Permian. Again, these provinces may be assigned with confidence to the tropical rainforest, or biome 1 climate. No discontinuity in time is implied by the terminology, as these provinces evolved from one another.

Eastern Qiangtang

No palaeomagnetic data are yet available for this region, but the floras (Zhao & Wu 1979; Li et al. 1982) are Euramerican in the Early Carboniferous and Cathaysian in the Late Permian. They have been compared closely to the equivalent provinces in Yangtze. Triassic suturing between Eastern Qiangtang and Yangtze along the Jinshajiang suture zone is suggested by the transition from deep marine to non-marine sedimentation in the Carnian to Norian interval of the Triassic (Sichuan Stratigraphic Group 1978; Yunnan Stratigraphic Group 1978).

Sino–Korea

Suturing along the northern margin of the Sino-Korean block by the Late Permian, preceded by Early Permian subduction below both Sino–Korea and Mongolia arcs (Klimetz 1983; Wang & Liu 1986) suggests progressive northward motion during the Permian. Palaeomagnetic studies have been carried out by eight research groups for the Upper Permian of the Sino–Korean block (Table 2), encompassing five provinces, including, from east to west, Hebei, Shanxi, Shaanxi, Nei Mongol and Gansu. All but the sites in Gansu (Coe et al. 1987) and Nei Mongol (Liu 1987) are well within the North China craton. Coe et al. (1987) provided a fold test for the Dahuanggou redbeds from the North Qilianshan region. Another vague fold test comes from the Shuangshiquan Formation (Cheng et al. 1988) of the Western Mountains (Xishan) near Beijing. It is also a redbed sequence that dips about 60° to the north. After tilt-correction, it yields a direction that is comparable to other results. Both reversed and normal polarities have been found (McElhinny et al. 1981; Zhao 1987; Cheng et al. 1988) suggesting that the magnetization is primary. All the rocks that have been studied are terrestrial clastics, mostly redbeds, belonging to Upper Shihezhi, Shiqianfeng and Dahuanggou Formations. These units are correlated on the basis of their plant fossils with the Longtanian to Changxingian stages of China or the Tartarian of western Europe (Nanjing Institute 1982).

In our Late Permian reconstruction (Fig. 3), we have chosen the average pole for which $Q \geq 4$ (Table 2). It places the center of the Sino–Korean block (38°N, 115°E) at about 10°N, with an orientation about 40° clockwise with respect to the present day.

From the phytogeographic point of view, the Sino–Korean block contains the most continuous development of tropical rainforest type floras in the Late Palaeozoic. The classic area of Shanxi (Li & Yao 1979) and the region to the west in Gansu Province (Wang *et al.* 1986) are comparable in their development of the Euramerican floras in the Upper Carboniferous and the Cathaysian floras through most of the Permian. These are in agreement with the low latitude prediction based on Upper Permian palaeomagnetic directions. Early Carboniferous evaporites in Gansu (Wang, H. Z. 1985) are well developed and dry conditions have been interpreted for the latest Permian of Shanxi (Wang, Z. Q. 1985) where the Zechstein flora is known (biome 2). Also, the latest Permian floras of Gansu show an Angaran influence and we assign them to biome 5 of the Pechora–Far East Province (Ziegler 1990). To summarize, a northward motion of North China from the Early Carboniferous through the Late Permian are compatible with the climatic data that ranges from dry to wet and back to dry again.

Tarim

The northern margin of Tarim appears to have collided with the already amalgamated Kazakhstan–Altai north-dipping arc during the Early Permian as discussed earlier. We place Tarim adjacent to Kazakhstan in both reconstructions, emphasizing its earlier time of suturing when compared with Sino–Korea. Late Palaeozoic palaeomagnetic work has been confined, at present, to the northwestern part of the block. The rocks studied include Carboniferous carbonates and clastic rocks (Bai *et al.* 1987) and Permian basaltic (Li *et al.* 1988) and clastic rocks (Bai *et al.* 1987; Cheng *et al.* 1983; McFadden *et al.* 1988). Li *et al.* (1988) have provided the only fold test by tilt-correcting Devonian redbeds and Ordovician limestones, which are assumed to have been horizontal before being intruded by the Permian mafic dykes, from which the Permian palaeomagnetic results were derived. In another magnetostratigraphic study near the Permo-Triassic boundary (McFadden *et al.* 1988), both normal and reversed polarities are found. It thus appears that the age of magnetization for these directions is properly constrained. Nonetheless, we would like to point out that our own preliminary palaeomagnetic study (Nie *et al.* in progress) of Carboniferous clastic, carbonate and volcanic rocks across the Tianshan Mountain Range along the Dushanzi-Kuqa highway reveals that much later remagnetization should not be completely ruled out. Remagnetization in Carboniferous rocks has also been reported by Sharps *et al.* (1988) for the northwest Tarim and Zhao *et al.* (1988) for the Junggar region, just north of Tianshan, but they have both interpreted the remagnetization to be Permian in age, supposedly corresponding to the collision between the Tarim block with Kazakhstan and Junggar. Better understanding of the geological structures in and near the Tianshan Mountain range will be necessary to unravel this rather interesting problem.

Despite the controversy, we tentatively use these published data in our reconstructions (Figs 2 & 3). The average pole of $Q \geq 4$ places the center of Tarim (37.5°N, 82.5°E) at about 30°N in the Late Permian which seems compatible with the reconstructed position of the southern margin of Kazakhstan.

The climatically interesting data from Tarim (Table 1, column 13) include evaporites in the Early Carboniferous (Wang, H. Z. 1985), Euramerican floras from the Upper Carboniferous (Xinjiang Stratigraphic Group 1981) and Cathaysian to 'mixed Angaran and Cathaysian floras at the very end of the Permian' (Wu 1983). In all these respects, the Tarim block is comparable with the Sino-Korean block.

Kitakami-Abukuma blocks

The Kitikami massif is an outlier of the Cathaysian flora in Japan (Asama 1974; Asama & Murata 1974) that is clearly now displaced with respect to China although it may originally have belonged to the Yangtze block. It apparently arrived adjacent to northern Japan in the Cretaceous (Taira *et al.* 1983).

Northern Margin of Palaeo-Tethys

The northern margin of Palaeo-Tethys is defined by the South Tianshan–North Tianshan–Solon Obo-Linxi-Tumen suture zone. Regions to the north of this suture zone all belong to the previously assembled Laurasia continental block. As briefly described above, the southern margin is characterized by a series of marginal arcs, the eastern parts of which were separated from the Aldan Shield by what we interpret to be a rather large, but closed-ended back-arc basin of south Japan Sea type. Biogeographically, these arcs belong to Angara, and three representative regions are included in Table 1 for comparison with the more southerly regions.

The Junggar area of northwest China was evidently part of the Kazakhstania microcontinent which by the Permian was incorporated in northern Asia. The floral information (Table 1, column 16) is consistent with low latitudes in the Early Carboniferous but increasing to midlatitudes in the Permian (Vakhrameev *et al.* 1978; Wu 1983; Institute of Geology Chinese Academy of Geological Sciences 1986). This floral interpretation is based on Chinese localities and areas immediately across the border in the USSR, south of Lake Balkash.

Pruner (1987) reported the first palaeomagnetic data from Mongolia for Carboniferous, Permian and Cretaceous rocks. Although preliminary, this study provides some rather noteworthy results. It suggests an APW path that shows overall similarity with that of Sino–Korea (Lin *et al.* 1985) since the Carboniferous, yet apparently differs from that of Siberia (Khramov *et al.* 1981). Also, the palaeolatitude of about 20° and azimuth of Mongolia in the Permian appear close to those of the North China block (see Pruner 1987, Fig. 10). If true, this could provide additional support for the disjunct history of amalgamation between Sino–Korea and Siberia in the Late Palaeozoic to Early Mesozoic. It should be noted, however, that the ages of the rocks and magnetization are not well constrained. For example, the age of the Permian andesites is established 'with reference to granite intrusions dated as Jurassic by analogy to the age of the other granite massifs of Mongolia' (Pruner 1987). It is also suspicious that both normal and reversed polarities are found from rocks of Early Permian age, which is supposedly exclusively within the Late Palaeozoic Kiaman Reversed Interval (Irving & Parry 1963; Harland *et al.* 1982).

The floral information for south Mongolia is consistently warm (biome 5) to cool temperate (biome 6) and entirely within the Angaran Realm (Durante 1971, 1976; Vakhrameev *et al.* 1978; Meyen, 1982). North latitudes of 40° and above can be safely inferred. The southeastern portion of the Soviet Union (Primorye, column 18 of Table 1) is similar to Mongolia in its floras (Burago 1976; Zimina 1977, Vakhrameev *et al.* 1978), although the record does not include the Carboniferous.

Conclusions

Precise palaeogeographic reconstructions for the Asian microcontinents during the Late Palaeozoic are not yet possible, but the following conclusions can be drawn based on the tectonic, palaeomagnetic, biogeographic and palaeoclimatic data presented in this paper:

(1) Due in part to the extensive development of the Late Palaeozoic continental glaciation, a steep temperature gradient from the equator to the pole allowed significant faunal and floral

differentiation. Along with other palaeoclimatic indicators, such as tillites, glacio-marine deposits and evaporites, the biogeographical constraints support the general recognition that there were a string of microcontinents that had resided within the tropical Palaeo-Tethys and subsequently moved northward to join Asia, and this is also supported by the sparse yet expanding palaeomagnetic database. These microcontinents include the Sino–Korea, Tarim, Yangtze, Indochina, and Eastern Qiangtang, all of which shared the Late Palaeozoic Cathaysian flora. It must be emphasized, however, that we do not propose a single united and isolated Cathaysian Continent in the Late Palaeozoic, in part because many of the sutures surrounding these Cathaysian continents did not close until much later times. This general northward motion appears to have also been followed by a number of peri-Gondwanan microcontinents that rifted away from Gondwana in the Late Permian.

(2) It is of particular interest that the Cathaysian floras also existed in northern parts of Gondwana, an important fact not often recognized. Because some of the Cathaysian plants such as *Gigantopteris* had to be dispersed by seeds, land connections were required not only among the Cathaysian continents but also between the Cathaysian microcontinents and those of Gondwana, while many of them appear to have had separate tectonic histories. The nature and location of these 'land-bridges' are not now understood and remain as a major challenge to the Palaeozoic palaeogeographic reconstructions.

(3) The suturing of the Cathaysian microcontinents, along with microcontinents rifted from the northern margin of Gondwana (including the Helmand, Western Qiangtang, Lhasa and Sibumasu blocks) with Asia, seems to have occurred in the Late Palaeozoic and the Early Mesozoic to complete the final assembly of the super-continent Pangaea. To be more specific, the Tarim block collided with Kazakhstania in the Early Permian while the Sino–Korean block joined the Mongolian arcs in the Late Permian. Most of the other sutures were closed in the Triassic, giving rise to the extensive Indonisian orogeny of southeastern Asia. It therefore appears that there is a southward younging of these sutures. However, a significant exception is represented by the Mongolo–Okhotsk suture between the southern Mongolian arcs along with the Sino–Korean block and the Angaran shield, which did not close until the Jurassic. As mentioned above, the Sino–Korean block seems to have collided with the southern Mongolia arcs as early as the Late Permian, and they then jointly moved counterclockwise to gradually close the Mongolo–Okhotsk suture, while the Siberian block was rotating clockwise, presumably together with the rest of Pangaea. Such a late final amalgamation between the Siberian and the Sino–Korean blocks has generally not been appreciated and can account for at least part of the discrepancy between the Late Palaeozoic and Early Mesozoic palaeomagnetic data from Siberian block and those from other Asian blocks as noticed by several recent palaeomagnetic studies (Achache & Courtillot 1985; Lin *et al.* 1985; Opdyke *et al.* 1986).

This work is supported by the National Science Foundation Grant EAR-86–13412 and is part of the 'Paleogeographic Atlas Project' at the University of Chicago. General support for the Atlas Project is contributed by a number of companies: Amoco Production Research, Arco Exploration and Production Research, British Petroleum, Chevron Oil Field Research Company, Conoco Exploration Research, Digital Equipment Corporation, Societe Nationale Elf Aquitaine, Exxon Production Research, Mobil Exploration and Producing Services, Shell Development Company, Sun Exploration and Production Company, Texaco USA, Unocal International Oil and Gas, and BHP-Utah International. This support is gratefully acknowledged. We thank Fang W., C. Scotese, R. Van der Voo and an anonymous reviewer for helpful comments. NSY wants to thank the following for making their preprints/reprints available: Lin J., R. Van der Voo, Zhang Z. and Zhao X. Finally, T. Gierlowski and M. Hulver kindly offered help with the preparation of the manuscript.

References

ACHACHE, J. & COURTILLOT, V. 1985. A preliminary Upper Triassic palaeomagnetic pole for the Khorat plateau (Thailand): consequences for the accretion of Indochina against Eurasia. *Earth and Planetary Science Letters*, **73**, 147–157.

ARCHANGELSKY, S. & WAGNER, R. H. 1983. *Glossopteris anatolica* sp. nov. from uppermost Permian strata in south-east Turkey. *British Museum of Natural History Bulletin*, **37**, 81–91.

ASAMA, K. 1974. Permian plants from Takakurayama, Japan. *Bulletin of the National Science Museum*, Tokyo, **17**, 239–250.

——, HONGNUSONTHI, A., IWAI, J., KON'NO, E., RAJAH, S. & VEERABURAS, M. 1975. Summary of the Carboniferous and Permian plants from Thailand, Malaysia and adjacent areas. *In*: KOBAYASHI, T. & TORIYAMA, R. (eds) *Geology and Palaeontology of Southeast Asia*, **15**, University of Tokyo Press, Tokyo, 77–101.

—— & MURATA, M. 1974. Permian plants from Setamai, Japan. *Bulletin of the National Science Museum*, Tokyo, **17**, 251–256.

BAI, Y. H., CHENG, G. L., SUN, Y., LI, Y. A., DONG, Y. & SUN, D. 1987. Late Palaeozoic polar wander path for the Tarim platform and its tectonic significance: *Tectonophysics*, **139**, 145–153.

BENDER, F. 1983. *Geology of Burma*, Berlin, Gebruder Borntraeger.

BESEMS, R. E. & SCHUURMAN, W. M. 1987. Palynostratigraphy of Late Palaeozoic glacial deposits of the Arabian Peninsula with special reference to Oman. *Palynology*, **11**, 37–53

BUNOPAS, S. 1981. *Paleogeographic History of Western Thailand and Adjacent Parts of South East Asia — a Plate Tectonic Interpretation*. PhD Thesis, Victoria University of Wellington, New Zealand.

BURAGO, V. I. 1976. The floristic links between the western and eastern parts of Angarida in the Permian. *Palaeontological Journal*. **10**(1), 84–93.

BURRETT, C. F. 1974. Plate tectonics and the fusion of Asia. *Earth and Planetary Science Letters*, **21**, 181–189.

CAO, R. G. 1986 Discovery of Late Carboniferous glacio-marine deposits in western Yunnan (in Chinese with English abstract). *Geological Review*, **32**, 236–242.

CHAN, L. S., WANG, C. Y. & WU, X. Y. 1984. Paleomagnetic results from some Permian-Triassic rocks from southwestern China, *Geophysical Research Letters*, **11**, 1157–1160.

CHATEAUNEUF, J. J. & STAMPFLI, G. 1978. Palynoflore permotriasique de l'Elburz oriental. *Note du Laboratoire de Paleontologie de l'Universite de Geneve*, **8**, 45–51.

CHENG, G. L., BAI, Y. H. & LI, Y. A. 1983. Palaeomagnetism of Lower Permian in the Wushi-Aksu area of Xinjiang (in Chinese). *Seismology and Geology*, **5**(4).

——, —— & SUN, Y. H. 1988. Palaeomagnetic study on the tectonic evolution of the Ordos block, North China (in Chinese with English abstract). *Seismology and Geology*, **10**(2), 81–87.

COE, R. S., ZHAO, X. X. & MENG, Z. F. 1987. Tectonic implications of palaeomagnetic results from North and South China. *In*: *Abstracts, International Union of Geodesy and Geophysics*, **1**, 115.

COQUEL, R., LOBOZIAK, S., STAMPFLI, G. & STAMPFLI-VUILLE, G. 1977. Palynologiedu Dévonien Supérieur et du Carbonifère inférieur dans l'Elburz oriental (Iran nord-est). *Revue de Micropaléontologie*, **20**, 59–71.

CTYROKY, P. 1973. Permian flora from the Ga'ara region (western Iraq). *Neues Jahrbuch für Geologie und Paläontologie Monatshefte*, **1**, 383–388.

DURANTE, M. V. 1971. Later Permian flora of Mongolia and the southern boundary of the Angaran floristic region of that time. *Paleontological Journal*, **5**, 511–522.

—— 1976. *The Carboniferous and Permian Stratigraphy of Mongolia on the Basis of Palaeobotanical Data*, Nauka Publishing House, Moscow.

EL-KHAYAL, A. A. & WAGNER, R. H. 1985. Upper Permian stratigraphy and megafloras of Saudi Arabia: palaeogeographic and climate implications. *In*: *Dixième Congrès International de Stratigraphie et de Géologie du Carbonifère Compte Rendu*, **3**, Madrid, 371–374.

GANSU STRATIGRAPHIC GROUP 1980. *Regional stratigraphic tables of northwest China, volume of Gansu* (in Chinese). Geological Publishing House, Beijing.

HAILE, N. S. 1981. Palaeomagnetism of Southeast and East Asia. *In*: MCELHINNY, M. W. & VALENCIO, D. A. (eds) *Palaeoreconstruction of the Continents*, American Geophysical Union, Geodynamics Series, **2**, 129–135.

HAMILTON, W. 1979. Tectonics of the Indonesian region. *Geological Survey Professional Paper 1078, USGS*.
HARLAND, W. B., COX, A. V., LLEWELLYN, P. G., PICKTON, C. A. G., SMITH, A. G. & WALTERS, R. 1982. *A Geological Time Scale*, Cambridge University Press, Cambridge, London.
HSU, K. J., SUN, S., LI, J. L., CHEN, H. H., PEN, H. P. & SENGÖR, A. M. C. 1988. Mesozoic overthrust tectonics in south China. *Geology*, 16, 418–421.
HUANG, B. H. 1986. The fossil plants of Elitu Formation at Xianghuang Qi District, Nei Mongol and its significance (in Chinese with English abstract). In: *Contributions to the Project of the Plate Tectonics in Northern China*, 1, Geological Publishing House, Beijing, 115–135.
HUANG, K. N., OPDYKE, N. D., KENT, D. V., XU, G. Z. & TANG, R. L. 1986. Further palaeomagnetic results from the Permian Emeishan basalt in SW China (in Chinese), *Kexue Tongbao* (Science Newsletter), 7, 1195–1201.
INGAVAT, R., TORIYAMA, R. & PITAKPAIVAN, K. 1980. Fusuline zonation and faunal characteristics of the Ratburi Limestone in Thailand and its equivalents in Malaysia. In: KOBAYASHI, T., TORIYAMA, R., HASHIMOTO, W. & KANNO, S. (eds) *Geology and Palaeontology of Southeast Asia*, 21, University of Tokyo Press, Tokyo, 43–62.
INSTITUTE OF GEOLOGY, CHINESE ACADEMY OF GEOLOGICAL SCIENCES 1986. Permian and Triassic strata and fossil assemblages in the Dalongkou area of Jimsar, Xinjiang. *People's Republic of China Ministry of Geology and Mineral Resources Geological Memoirs* (In Chinese with English summary), 2,(3) Geological Publishing House, Beijing.
IRVING, E. & PARRY, L. G. 1963. The magnetism of some Permian rocks from New South Wales. *Geophysical Journal of the Royal Astronomical Society*, 7, 395–411.
JIN, Y. G. 1985. Permian brachiopoda and palaeogeography of the Qinghai-Xizang (Tibet) Plateau. In: LU, Y. H. (ed.) *Palaeontologia Cathayana*, 2, Science Press, Beijing, 19–71.
JONES, J. G., CONAGHAN, P. J., MCDONNELL, K. L., FLOOD, R. H. & SHAW, S. E. 1984. Papuan basin analogue and a foreland basin model for the Bowen-Sydney Basin. In: VEEVERS, J. J. (ed.) *Phanerozoic earth history of Australia*. Oxford Geological Sciences Series, 2, 243–261.
KHRAMOV, A. N., PETROVA, G. N. & PECHERSKY, D. M. 1981. Palaeomagnetism of the Soviet Union. In: MCELHINNY, M. W. & VALENCIO, D. A. (eds) *Palaeoreconstruction of the Continents*. International Union Commission on Geodynamics Final Reports, 177–194.
KLIMETZ, M. P. 1983. Speculations on the Mesozoic plate tectonic evolution of eastern China. *Tectonics*, 2, 139–166.
KOSYGIN, Y. A. & PARFENOV, L. M. 1981. Tectonics of the Soviet Far East. In: NAIRN, A. E. M., CHURKIN, M. & STEHLI, F. (eds) *The Ocean Basins and Margins, Volume 5, The Arctic Ocean*, Plenum Press, New York, 377–412.
LEE, D. S. (ed.) 1987. *Geology of Korea*, Kyohak-Sa Publishing Company Seoul.
LEGRAND-BLAIN, M. 1985. Egypt. In: MARTINEZ-DIAZ, C. (ed.) The Carboniferous of the World, Volume 2. IUGS Publication, 20, 347–351.
LEJAL-NICOL, A. 1985. Egypt. In: MARTINEZ-DIAZ, C. (ed.) *The Carboniferous of the World, Volume 2*. IUGS Publication, 20, 386–403.
LI, X. X. 1986. The mixed Permian Cathaysian-Gondwana flora. *The Palaeobotanist*, 35(2), 211–222.
——, WU, Y. M. & FU, Z. B. 1985. Preliminary study on a mixed Permian flora from Xiagangjiang of Gerze District, Xizang and its palaeobiogeographic significance (in Chinese with English abstract). *Acta Palaeontologica Sinica*, 24, 150–170.
—— & YAO, Z. Q. 1979. Carboniferous and Permian floral provinces in east Asia. In: Neuvième Congrés International de Stratigraphie et de Géologie du Carbonifere Compte Rendu, 5, Washington and Champaign-Urbana, 95–101.
——, —— & DENG, L. H. 1982. An Early Late Permian flora from Toba, Qambo District, eastern Xizang (in Chinese with English abstract). *Palaeontology of Tibet*, 5, 17–40.
LI, Y. P., MCWILLIAMS, M., COX, A., SHARPS, R., LI, Y. A., GAO, Z. J., ZHANG, Z. K., & ZHAI, Y. J. 1988. Late Permian paleomagnetic pole from dikes of the Tarim craton, China. *Geology*, 16, 275–278.
LIANG, D. Y., NIE, Z. T., GUO, T. Y., ZHANG, Y. H., XU, B. W. & WANG, W. P. 1983. Permo-Carboniferous Gondwana-Tethys facies in southern Karakoran, Ali, Xizang (Tibet) (in Chinese with English abstract). *Journal of the Wuhan College of Geology*, 19, 9–27.

LIN, J. L. 1984. *The Apparent Polar Wander Paths for the North and South China blocks*, PhD Thesis, University of California, Santa Barbara.
—— 1987. The apparent polar wander path for the South China block and its geological significance (in Chinese with English abstract). *Scientia Geologica Sinica*, 4, 306–315.
——, FULLER, M. & ZHANG, W. Y. 1985. Preliminary Phanerozoic polar wander paths for the North and South China blocks. *Nature*, 313, 444–449.
——, WATTS, R. 1988. Palaeomagnetic constraints on Himalayan-Tibetan tectonic evolution. *Philosophical Transactions, The Royal Society, London*, 326(A), 177–188.
LIN, J. Y. 1985. Spatial and Temperal Distribution of Permian Basaltic Rocks in the Three Southwestern Provinces (Sichuan, Yunnan and Guizhou) of China (in Chinese). *Kexue Tongbao* (Science Newsletter), 12, 929–932.
LIU, H. S. 1987. Preliminary results of palaeomagnetic study of Permian rocks in Inner Mongolia. In: *Abstracts for the Third National Symposium on Palaeomagnetism*, Guangzhou, China.
LOTTES, A. L. & ROWLEY, D. B. 1990. Reconstruction of the Laurasian and Gondwanan Segments of Permian Pangaea, In: MCKERROW, W. S. & SCOTESE, C. R. (eds) *Palaeozoic Palaeogeography and Biography*. Geological Society, London, Memoir, 12, 383–395.
MARINOV, N. A., ZONENSHAIN, L. & BLAGONRAVOV, V. (eds) 1973. *Geology of the Mongolia People's Republic, Volume 1, Stratigraphy*. Nedra Publishers, Moscow.
MCELHINNY, M. W. 1985. Permian paleomagnetism of the Western Yangtze Block, China: a reinterpretation. *Journal of Geodynamics*, 2, 115–117.
——, EMBLETON, B. J. J., MA, X. H. & ZHANG, Z. K. 1981. Fragmentation of Asia in the Permian. *Nature*, 293, 212–216.
——, HAILE, N. S. & CRAWFORD, A. R. 1974. Palaeomagnetic evidence shows Malay Peninsula was not a part of Gondwanaland. *Nature*, 252, 641–645.
MCFADDEN, P. L., MA, X. H., MCELHINNY, M. W. & ZHANG, Z. K. 1988. Permo-Triassic Magnetostratigraphy in China: northern Tarim. *Earth and Planetary Science Letters*, 87, 152–160.
METCALFE, I. 1988. Origin and assembly of south-east Asian continental terranes. In: AUDLEY-CHARLES, M. & HALLAM, A. (eds) *Gondwana and Tethys*, Geological Society, London, Special Publication, 37, 101–118.
MEYEN, S. V. 1982. The Carboniferous and Permian floras of Angaraland (a synthesis). *Biological Memoirs*, 7, 1–110.
NAKAMURA, K., SHIMUZU, D. & LIAO, Z. T. 1985. Permian palaeobiogeography of brachiopods based on the faunal provinces. In: NAKAZAWA, K. & DICKINS, J. M. (eds) *The Tethys*, Tokai University Press, Tokyo, 185–198.
NAKAZAWA, K. 1985. The Permian and Triassic systems in the Tethys — their palaeogeography. In: NAKAZAWA, K. & DICKINS, J. M. (eds) *The Tethys: Her Palaeobiogeography from Palaeozoic to Mesozoic*, Tokai University Press, Tokyo, 3–111.
NANJING INSTITUTE OF GEOLOGY AND PALAEONTOLOGY, CHINESE ACADEMY OF SCIENCES 1982. *Stratigraphic Correlation Charts in China with Explanatory Text* (in Chinese). Science Press, Beijing.
NEI MONGOL STRATIGRAPHIC GROUP 1978. *Regional Stratigraphic Tables of Northern China, Volume of Nei Mongol* (in Chinese), Geological Publishing House, Beijing.
NORIN, E. 1946. Geological Explorations in Western Tibet. *Sino-Swedish Expedition Publication*, 29, Stockholm.
ODREMAN-RIVAS, O. & WAGNER, R. H. 1979. Precisions sobre alguna floras Carboniferas Permicas de los Andes Venezolanos. *Boletin de Geologia*, 13, 77–79.
OPDYKE, N. D., HUANG, K., XU, G., ZHANG, W. Y. & KENT, D. 1986. Palaeomagnetic results from the Triassic of the Yangtze Platform. *Journal of Geophysical Research*, 91, 9553–9568.
POWELL, D. E. 1976. The Geological Evolution of the continental margin off northwest Australia. *Journal of Australia Petroleum Exploration Association*, 16, 13–23.
PRUNER, P. 1987. Palaeomagnetism and palaeogeography of Mongolia in the Cretaceous, Permian and Carboniferous — preliminary data. *Tectonophysics*, 139, 155–167.
QINGHAI STRATIGRAPHIC GROUP 1980. *Regional Stratigraphic Tables of Northwest China, Volume on Qinghai* (in Chinese). Geological Publishing House, Beijing.

RONG, J. Y. & ZHANG, Z. X. 1982. A southward extension of the Silurian *Tuvaella* brachiopod *Lethaia*, **15**, 133–147.

ROSS, J. R. P. 1978. Biogeography of Permian ectoproct Bryozoa. *Palaeontology*, **21**, 341–356.

ROWLEY, D. B., RAYMOND, A., PARRISH, J. T., LOTTES, A. L., SCOTESE, C. R. & ZIEGLER, A. M. 1985. Carboniferous palaeogeographic, phytogeographic and palaeoclimatic reconstructions. *International Journal of Coal Geology*, **5**, 7–42.

——, ZIEGLER, A. M. & NIE, S. Y. 1989. Mesozoic overthrust tectonics in south China: HSÜture or suture? – a discussion, *Geology*, **17**, 384–386.

SAKAGAMI, S. 1985. Palaeogeographic distribution of Permian and Triassic Ectoprocta (Bryozoa). *In*: NAKAZAWA, K. & DICKINS, J. M. (eds) *The Tethys*, Tokai University Press, Tokyo, 171–183.

SASAJIMA, S., OTOFUJI, Y. HIROOKA, K. SUPARKA, S. & HEHUWAT, F. 1978. Palaeomagnetic studies on Sumatra Island: on the possibility of Sumatra being part of Gondwanaland. *Rock Magnetism and Palaeogeophysics*, **5**, 104–110.

SCOTESE, C. R. 1984. An introduction to this volume: Palaeozoic palaeomagnetism and the assembly of Pangaea. *In*: VAN DER VOO, R. & SCOTESE, C. R. (eds) *Plate reconstruction from Palaeozoic palaeomagnetism*, Geodynamic Series, **12**, American Geophysical Union, Washington DC 1–10.

——, BAMBACH, R. K., BARTON, C., VAN DER VOO, R. & ZIEGLER, A. M. 1979. Palaeozoic base maps, *Journal of Geology*, **87**, 217–277.

SENGÖR, A. M. C. 1984. The Cimmeride orogenic system and the tectonics of Eurasia. *Geological Society of America Special Paper*, **195**.

—— 1987a. Tectonics of the Tethysides: orogenic collage development in a collisional setting. *Annual Review of Earth and Planetary Sciences*, **15**, 213–244.

—— 1987b. Tectonic subdivisions and evolution of Asia. *Bull. Istanbul Tech. Univ*, **40**, 355–435.

——, ALTINER, D., CIN, A., USTAOMER, T. & HSÜ, K. 1988. Origin and assembly of the Tethyside orogenic collage at the expense of Gondwana Land. *In*: AUDLEY-CHARLES, M. & HALLAM, A. (eds) *Gondwana and Tethys*, Geological Society, London, Special Publication, **37**, 119–181.

SHARIEF, F. A. 1982. Lithofacies distribution of the Permian–Triassic rocks in the Middle East. *Journal of Petroleum Geology*, **4**, 299–310.

SHARPS, R., LI, Y. P., MCWILLIAMS, M., LI, Y. A., LI, Q & MCKNIGHT, K. 1988. Age of folding along the northwestern margin of the Tarim Craton: Evidence from remagnetizated Late Palaeozoic strata. *Eos Transactions*, **69** (44), American Geophysical Union, 1171.

SICHUAN STRATIGRAPHIC GROUP 1978. *Regional Stratigraphic Tables of Southwest China, Volume on Sichuan* (in Chinese), Geological Publishing House, Beijing.

SINGH, G., MAITHY, P. K. & BOSE, M. N. 1982. Upper Palaeozoic flora of Kashmir Himalaya. *The Palaeobotanist*, **30**, 185–232.

SMITH, A. B. 1988. Late Palaeozoic biogeography of East Asia and Palaeontological constraints on plate tectonic reconstructions. *Philosophical Transactions, The Royal Society, London*, Series **A**, **326**, 189–227.

SMITH, A. G., HURLEY, A. M. & BRIDEN, J. C. 1981. *Phanerozoic palaeocontinental world maps*, Cambridge University Press, London.

STAUFFER, P. H. & MANTAJIT, N. 1981. Late Palaeozoic tilloids of Malaya, Thailand and Burma. *In*: HAMBREY, M. J. & HARLAND, W. B. (eds) *Earth's pre-Pleostocene Glacial Record*, Cambridge University Press, Cambridge, 331–335.

TAIRA, A., SAITO, Y. HASHIMOTO, M. 1983. The role of oblique subduction and strike-slip tectonics in the evolution of Japan. *In*: HILDE, T. W. C. & UYEDA, S. (eds) *Geodynamics of the western Pacific-Indonesian region*, Geodynamics Series, **11**, American Geophysical Union, Washington DC 303–316.

TANAKA, K. & NOZAWA, T. 1977. Geology and mineral resources of Japan. *In*: TAKANA, K & NOZAWA, T. (eds), *Geology*, **1**, Geological Survey of Japan.

VAKHRAMEEV, V. A., DOBRUSKINA, I. A., MEYEN, S. V. & ZAKLINSKAYA, E. D. 1978. *Paläozoische und Mesozoische Floren Eurasiens und die Phytogeographie dieser Zeit*, VEB Gustav Fischer Verlag, Jena.

VAN DER VOO, R. (in press). Palaeomagnetism of Continental North America: the Craton, its Margins, and the Appalachian Belt. *In*: PAKISER, L. C. & MOONEY, W. D. (eds), *Monograph on 'Geophysical Framework of the Continental United States'*. Geological Society of America Memoir.

VEEVERS, J. J. & POWELL, C. M. 1987. Late Palaeozoic glacial episodes in Gondwanaland reflected in transgressive–regressive depositional sequences in Euramerica. *Geological Society of America Bulletin*, **98**, 475–487.

WANG, D. X., HE, B. and ZHANG, S. L. 1986. Characteristics of Permian flora in Qilian Mountain region (in Chinese with English abstract). *Gansu Geology*, **6**, 37–60.

WANG, H. Z. 1985. *Atlas of the Palaeogeography of China* (in Chinese and English), Cartographic Publishing House, Beijing.

WANG, Q. & LIU, X. Y. 1986. Palaeoplate tectonics between Cathaysia and Angaraland in Inner Mongolia of China. *Tectonics*, **5**(7), 1073–1088.

WANG, Y. J. & MU, X. N. 1981. Nature of the Permian biotas in Xizang and the northern boundary of the Indian Plate. *In*: *Geological and Ecological Studies of Qinghai-Xizang Plateau*, **1**, Gordon and Breach, Science Publishers, Inc., New York, 179–185.

WANG, Z. Q. 1985. Palaeovegetation and plate tectonics: palaeophytogeography of north China during Permian and Triassic times. *Palaeogeography, Palaeoclimatology, Palaeoecology*, **49**, 24–45.

WATERHOUSE, J. B. 1982. An early Permian cold-water fauna from pebbly mudstones in South Thailand. *Geological Magazine*, **119**, 337–354.

WEISSBROD, T. 1976. The Permian in the Near East. *In*: FALKE, H. (ed) *The Continental Permian in Central, West and South Europe*, Reidel Publishing Company, Dordrecht, Holland, 200–214.

WOLFART, R. & WITTEKINDT, H. 1980. *Geologie von Afghanistan*, Gebrüder Borntraeger, Berlin.

WU, S. Z. 1983. Preliminary study on the Permian phytogeography of Xinjiang (in Chinese). *In*: *Paleobiogeographic Provinces of China*, Science Press, Beijing, 91–99.

XINJIANG STRATIGRAPHIC GROUP. 1981. *Regional stratigraphic tables of northwestern China, volume on Xinjiang* (in Chinese). Geological Publishing House, Beijing.

XIZANG (TIBET) SCIENTIFIC EXPEDITION, CHINESE ACADEMY OF SCIENCES. 1987. *Stratigraphy of the Mount Qomolangma region*, Science Press, Beijing.

XU, X., WEI, Z. S., CHENG, G. A. & JIAO, S. R. 1981. *Simplified Regional Stratigraphic Table of Qinghai-Tibet Plateau* (in Chinese). Geological Publishing House, Beijing.

YANG, Z. Y., CHENG, Y. Q. & WANG, H. Z. 1986. *The Geology of China*. Oxford Monographs on Geology and Geophysics No. 3, Clarendon Press, Oxford.

——, YIN, H. F., XU, G. R., WU, S. B. HE, Y. L., LIU, G. C., YIN, J. L. 1983. *Triassic of the south Qilian Mountains* (in Chinese with English summary). Geological Publishing House, Beijing.

YUNNAN STRATIGRAPHIC GROUP. 1978. *Regional Stratigraphic Tables of Southwest China, volume on Yunnan* (in Chinese). Geological Publishing House, Beijing.

ZHANG, C. 1981. Some geological features of the ophiolites in Xinjiang (in Chinese with English abstract). *Geological Review*, **27**(4), 307–314.

ZHANG, Z. K. 1984. Sino–Korean blocks and Yangtze block as part of the Pacifica continent in the late Palaeozoic (in Chinese with English abst.). *Bulletin of the Chinese Academy of Geological Sciences*, **9**, 45–54.

—— & ZHANG, J. X. 1986. Palaeomagnetic research on the upper Carboniferous basalt in Baoshan Block, Yunnan and the tectonic belonging of the block (in Chinese with English abstract). *Bulletin of the Institute of Geology, Chinese Academy of Geological Sciences*, **15**, 183–189.

ZHAO, X. H. & WU, X. Y. 1979. Carboniferous macrofloras of south China. *In*: *Neuvième Congrès International de Stratigraphie et de Géologie du Carbonifère Compte Rendu*, **5**, Washington and Champaign-Urbana, 109–114.

ZHAO, X. X. 1987. *A Paleomagnetic Study of Phanerozoic Rock Units from Eastern China*, PhD Thesis, University of California, Santa Cruz.

—— & COE, R. S. 1987. Paleomagnetic constraints on the collision and rotation of North and South China. *Nature*, **327**, 141–144.

—— & —— 1989. Tectonic implications of Permo-Triassic palaeomagnetic results from North and South China, *In*: HILLHOUSE, J. W. (ed.) *Deep Structure and Past Kinématies of Accreted Terranes*. American Geophysical Union Monograph, **5**, 267–283.

——, ——, WU, H. & ZHANG, C. 1988. Evidence for remagnetization of Palaeozoic rock units from Junggar Basin of Xinjiang Province, China. *Eos Transactions*, **69**,(44), 1170.

ZHOU, Y. X., LU, L. Z. & ZHENG, B. M. 1986. Paleomagnetic polarity of the Permian Emeishan Basalt in Sichuan (in Chinese with English abstract), *Geological Review*, **32**, 465–469.

ZIEGLER, A. M. (1990). Phytogeographic patterns and continental configurations during the Permian Period. *In*: MCKERROW, W. S. & SCOTESE, C. R. (eds) *Palaeozoic Palaeogeography and Biogeography*. Geological Society, London, Memoir, **12**, 363–379.

——, BAMBACH, R. K., PARRISH, J. T., BARRETT, S. F., GIERLOWSKI, E. H., PARKER, W. C., RAYMOND, A. & SEPKOSKI, J. J. JR. 1981. Palaeozoic biogeography and Climatology. *In*: NIKLAS, K. J. (ed.) *Palaeobotany, Palaeoecology and Evolution*, **2**, Praeger, New York, 231–266.

——, SCOTESE, C. R. & BARRETT, S. F. 1982. Mesozoic and Cenozoic palaeogeographic maps. *In*: BROSCHE, F. & SUNDERMANN, J. (eds) *Tidal Friction and Earth's Rotation II*, Springer-Verlag, Berlin, 240–252.

ZIEGLER, P. 1988. Evolution of the Arctic–North Atlantic and the western Tethys. *American Association of Petroleum Geologists Memoir*, **43**.

ZIMINA, V. G. 1977. *Flora of Early and Early Late Permian of Southern Primor'ya*, Akademia Nauk SSSR, Moscow.

Palaeogeographic evolution of southwestern Europe during Early Palaeozoic times

M. ROBARDET[1], F. PARIS[1] & P. R. RACHEBOEUF[2]

[1] *Laboratoire de Paléontologie et de Stratigraphie, Université de Rennes I, 35042 Rennes Cedex, France*
[2] *Centre des Sciences de la Terre, Université Claude-Bernard Lyon I, 69622 Villeurbanne Cedex, France*

Abstract: In the Late Ordovician, both SW Europe and Africa were situated in high latitudes. Sedimentary facies and biogeography suggest that these regions were separated by the Rheic (Mid-European) Ocean from Baltica, with the suture lying between the Armorican Massif and the Ardennes. During the Silurian and the Devonian, palaeobiogeographic affinities persisted between S Europe and N Africa. It is concluded that S Europe consisted of several discrete blocks or microplates prior to the Hercynian Orogeny.

In southern Europe, pre-Carboniferous biogeographic data have only been used rarely even though they can place important constraints on the tectonic evolution of the region. When taken together with sedimentary and structural data, the consideration of shelf benthic faunas at the species level indicates that, while the central and northern regions of the Armorican Massif show close similarities with central Iberia, other parts of southern Europe represent distinct 'domains' which maintain their identity over long periods of time.

Global palaeobiogeographic pattern: Gondwana–southern Europe–Baltica relationships

Ordovician

The Mediterranean Province (Spjeldnaes 1961, 1967) also named 'Tethyan' (Dean 1967) or 'Selenopeltis Province' (Whittington & Hughes 1972) is characterized by an almost entirely terrigenous sequence. The most notable sandy formation is the 'Grès armoricain' (Armorican Quartzite Formation of the Armorican Massif and the Central Iberian regions) and related formations which have been deposited over a wide area (Blaise & Bouyx 1980; Paris *et al.* 1982). In the latest Ordovician (Hirnantian), clast-bearing glaciomarine sediments have been deposited upon the marine shelf that fringed the African ice-sheet to the north and northwest. The geographic distribution of these deposits shows the close connection between Gondwana and the Mediterranean Province at the end of Ordovician times (Robardet & Doré 1988).

As early as 1961, Spjeldnaes recognized the homogeneity of the Ordovician benthic faunas from southern Europe and north Gondwana (Mediterranean Province) and their independence from those of northern Europe (Anglo-Baltic Province). The study of different groups of the marine benthos (trilobites, brachiopods, etc.) has corroborated these views and has lent support to the concept that a mid-European Ocean may have acted as a barrier to migration resulting in the development of two distinct faunal provinces (Spjeldnaes 1961, 1967; Whittington 1966, 1973; Whittington & Hughes 1972; Dean 1967, 1977; Williams 1969, 1973; Burrett 1972; Hughes *et al.* 1975; Cocks & Fortey 1982). The sediments and the faunas of southern Europe and north Gondwana are regarded as corresponding to the high latitude regions (Spjeldnaes 1961).

Northern Europe (Baltica) is characterized by platform deposits mainly composed of nodular detrital limestones and shales. The benthic faunas (Asaphid Trilobite Province of Whittington & Hughes 1972) differ both from those of Laurentia and those of the Mediterranean Province. This type of Ordovician has frequently been regarded as tropical but several authors (Jaanusson 1973; Cocks & Fortey 1982) rather favour a more temperate position of Baltica.

Whatever may be the precise Ordovician palaeolatitude of Baltica, the lithological and faunal characteristics of these regions are clearly distinct from those of the Mediterranean Province: consequently a Mid-European Ocean (or Rheic Ocean) separating Gondwana and southern Europe from Baltica has been postulated (Whittington & Hughes 1972; Burrett 1972; McKerrow & Ziegler 1972).

Silurian

During the Silurian, Baltica and Laurentia had much more in common: reefs or biohermal limestones occur in the Wenlock and Ludlow Series in Baltica (Gotland) and the Silurian of Laurentia mainly consists of low latitude shelf carbonate facies with some evaporitic sequences in restricted areas: such converging features most probably resulted from the narrowing of the Iapetus Ocean. The euxinic black shale facies of the Mediterranean Province strongly differs from the Baltica–Laurentia sequences and it can be postulated that the relative position of the northern continents and the north Gondwanan regions during the Silurian was similar to that of the Ordovician.

Devonian

During the Late Silurian–Early Devonian time period, the global palaeogeographic pattern was marked by the existence of the Old Red Sandstone Continent which included northern Europe. The continental, fluvial and near-shore facies and faunas of these regions thus cannot be compared with those of the south European and north African marine platforms.

Global reconstructions (Dineley 1979; House 1979) have emphasized the general affinities of faunas, floras and sediments from southern Europe and north Africa. The faunal affinities can be more precisely illustrated by different fossil groups. The lower Devonian successions from the Armorican Massif, the Iberian Peninsula and Morocco (Hercynian Meseta and Anti-Atlas) have yielded brachiopod assemblages. The study of the Pragian–Emsian Chonetaceans (Racheboeuf *et al.* 1988) shows (Fig. 1) that: several genera are known only in the Ibero-Armorican and Moroccan regions (*Celtanoplia*, *Ctenochonetes*, *Renaudia*, *Davoustia*).

Most Moroccan species have already been described from the south European regions of France and Spain (*Ctenochonetes ibericus*, *C. aremoricensis*, *Davoustia mezquitensis*, *Philippotia belairensis*, *Plicanoplia carlsi*, *Renaudia mainensis*). Similar faunal affinities between southwestern Europe (Armorican Massif, Iberian Peninsula) and northern Africa (Moroccan Meseta and Anti-Atlas, Algerian Ougarta Chains) are also known at the species level for other fossil groups such as: rhynchonellid brachiopods, *Lanceomyonia occidentalis*, *Glossinotoechia princeps*, *Eucharitina oehlerti*, *Camarotoechia paretiformis* (Drot & L'Hotellier 1976); Leptaeninae as *Hollardina plana* (Racheboeuf *et al.* 1981); tabulate corals as *Cleistopora geometrica* (Plusquellec 1976), *Procterodictyum polentinoi* and the genus *Praemichelina*

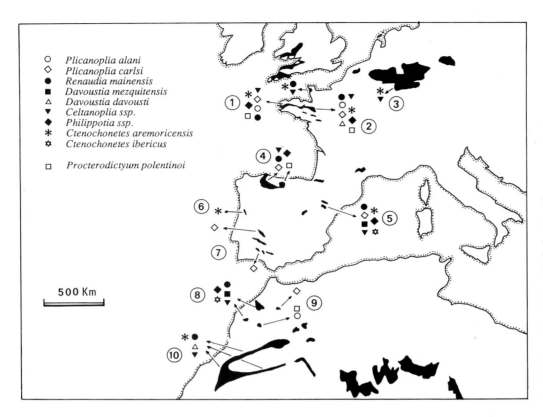

Fig. 1. Geographical distribution of selected Lower Devonian taxa from the north Gondwana margin. The occurrence of the same benthic species of chonetids and corals, both in southern Europe and northern Africa strongly argues against the hypothesis of a Prototethys Ocean. (black areas, Devonian outcrops). 1, western Armorican Massif; 2, eastern Armorican Massif; 3, Ardenno–Rhenish area; 4, Cantabrian Zone; 5, Iberian Chains; 6, northwest Portugal; 7, South Iberian Domain; 8, western Moroccan Meseta; 9, eastern Moroccan Meseta; 10, Anti-Atlas.

which characterizes the Ibero–Armorican and north African regions (Plusquellec 1986).

These strong affinities indicate that the relations between southern Europe and north Africa during the Early Devonian were similar to those that had existed during the Ordovician; it would therefore be impossible to imagine between the two areas any geographical barrier which could have induced reproductive isolation.

During the earliest Devonian there were stronger affinities between the Armorican Massif, the Iberian Peninsula and north Africa than between the Ibero-Armorican area and the European regions located to the north (Artois, Ardennes, Germany) which constituted the southern margin of the Old Red Sandstone Continent (see Racheboeuf & Babin 1986, p. 41).

However in the late Early Devonian (Emsian) the affinities between north Gondwanan benthic faunas (southern Europe and north Africa) and those of other European regions increased (Morzadec et al. 1981, p. 11–18; Morzadec et al. 1988), indicating that the distance separating southern and northern Europe had reduced, foreshadowing the assembly and collision of these plates during the Late Palaeozoic.

Lower Palaeozoic domains in southwestern Europe

All the regions which compose southern Europe (Fig. 2) present sedimentary and faunal similarities and are parts of the Mediterranean Province. However detailed studies of the lithological successions, the litho-facies and the benthic faunas (at the species level) allow the division of this province into several domains, each characterized by a distinctive pattern of sedimentation and fauna. (Henry et al. 1974, 1976; Robardet 1976; Paris & Robardet 1977; Dubreuil 1987; Paris et al. 1987).

Medio-North Armorican–Central Iberian Domain

In the northern and central regions of the Armorican Massif (to the north of the Angers syncline) and in the Central Iberian Zone (Fig. 2), the similarities of the Ordovician successions and, at the species level, of the trilobite (Hammann & Henry 1978), brachiopod (Villas 1985; Young 1985), ostracode (Vannier 1986a, b) and bivalve faunas indicate similar ecological conditions and the absence of any barrier which might have isolated inner shelf marine populations. The occurrence of the same three subdomains (Fig. 3) with slightly differing characteristics, both in the Armorican and the Iberian parts (Clarkson & Henry 1969; Hammann & Henry 1978), and the astonishing identity (Fig. 4) of the whole Ordovician succession and fauna (Henry et al. 1974, 1976; Young 1985, 1988) in Crozon (NW Brittany) and Buçaco

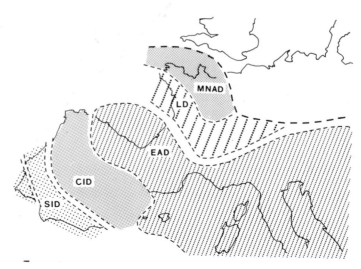

Fig. 2. Present geographical disposition of the Lower Palaeozoic domains in southwestern Europe. MNAD, Medio-North Armorican Domain; LD, Ligerian Domain; EAD, Ebro–Aquitanian Domain; CID, Central Iberian Domain; SID, South Iberian Domain. All the dotted areas are parts of north Gondwana margin (Mediterranean Province); in white, southern part of Baltica.

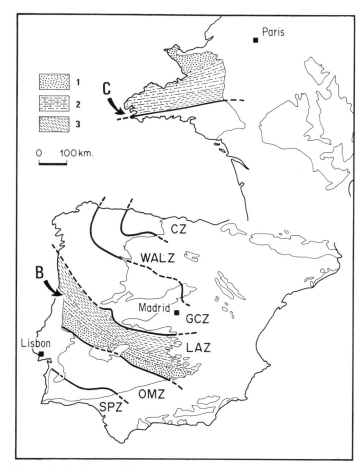

Fig. 3. Sedimentary and faunal 'subdomains' in the Medio-North Armorican area and the Luso-Alcudian Zone (LAZ) of the Iberian Peninsula during Llandeilo times. 1, Sandstone lithofacies; faunal assemblages including *Eohomalonotus brongniarti, E. vicaryi, E. sdzuyi, Kerfornella brevicaudata, Tafilaltia valpyana*.
2, Shale, siltstone and sandstone lithofacies; faunal assemblages including, *Plaesiacomia oehlerti, Morgatia hupei, Crozonaspis struvei, Eorhipidomella musculosa, Appolonorthis bussacensis, Heterorthina morgatensis, H. kerfornei*.
3, Shale-siltstone lithofacies; faunal assemblages including *Colpocoryphe rouaulti, Salterocoryphe salteri, Eccoptochile mariana, Aegiromena mariana*.
C, Crozon peninsula; B, Buçaco syncline (see Fig. 4); CZ, Cantabrian Zone; WALZ, West-Asturian-Leonese Zone; GCZ, Galaico-Castillan Zone; LAZ, Luso-Alcudian Zone; OMZ, Ossa Morena Zone; SPZ, South Portuguese Zone.
Faunal data after Hammann & Henry (1978); Henry (1980); Melou (1974, 1975, 1976); Young (1985).

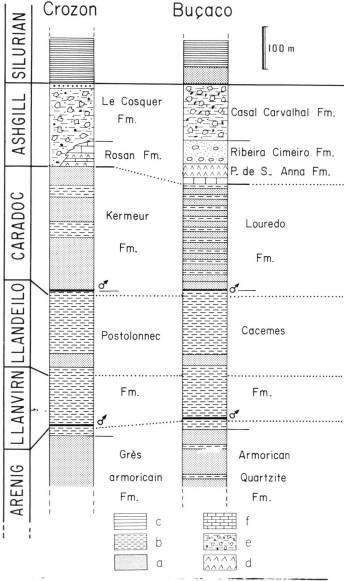

Fig. 4. Ordovician successions in Crozon peninsula, Armorican Massif (C Fig. 3) and Buçaco syncline, Portugal (B Fig. 3); Upper Ordovician of Buçaco after Young (1985, 1988). Both successions display identical lithofacies and sedimentary evolution and identical (at the species level) benthic assemblages. (Henry *et al.* 1974, 1976). or a, sandstones; b, siltstones and shales; c, black shales; d, volcanic and volcaniclastic rocks; e, clastbearing glaciomarine sediments; f, limestones.

(Central Portugal) strongly argues for geographic continuity between the Medio-North Armorican and the Central Iberian regions. These regions were much closer during the Ordovician than they are at present (even after closing the Bay of Biscay) and almost certainly were parts of the same domain.

During the Silurian, the pelagic or epipelagic character of the faunas does not permit the recognition of particular affinities between these two areas.

In the Early Devonian, benthic faunas are again represented in the Armorican Massif and the Iberian Peninsula and are rather similar in both areas (Garcia-Alcalde *et al.* 1979; Lardeux & Morzadec 1979; Racheboeuf 1981; Julivert *et al*, 1983). The best documented affinities are those between the Medio-North Armorican regions and the Iberian Chains in north-east Spain (Carls 1971, 1988; Carls & Gandl 1967).

Ligerian Domain

It has recently been established that the Lower Palaeozoic (Ordovician to Devonian) in Ligeria (south Armorican regions) consists of reworked blocks (olistoliths and sedimentary klippes) incorporated into Lower Carboniferous pull-apart basins (Dubreuil 1980, 1987).

Previous publications had already pointed out the striking originality of the Ligerian faunas (Pillet 1972; Lardeux *et al.* 1977; Henry 1980; Kriz & Paris 1982). However these Bohemian affinities (at the species level) had not been unequivocally interpreted and this area was regarded either as a 'transitional' or 'intermediate' domain (Lardeux *et al.* 1977) or as a totally different palaeogeographic unit (Paris & Robardet 1977). Recent careful regional studies (Dubreuil 1986, 1987) have strongly supported the second interpretation of a distinct Ligerian Domain. The Lower Palaeozoic sequence of this domain was dismembered and reworked blocks (including Famennian rocks) were resedimented most probably during Early Carboniferous times. Oblique motion along strike-slip faults can explain the present juxtaposition of exotic terranes to the south of the Medio-North Armorican Domain.

South Iberian Domain

In the southern part of the Iberian Peninsula, the Ossa Morena Zone extends between the Badajoz-Cordoba blastomylonitic shear zone and the Ficalho thrust. Recent studies have provided numerous palaeontological and stratigraphical data concerning the Ordovician, Silurian and Devonian sequences in this area. The Ordovician succession differs significantly from that of the Central Iberian Zone. The Armorican Quartzite Formation and the overlying dark shales are totally absent, the lower part of the Ordovician sequence being composed of grey-green shales. The distinctness of the Ossa Morena Zone is also illustrated by the benthic faunas though their general characteristics are of Mediterranean type (Gutierrez-Marco et al. 1984). The Silurian succession consists of a condensed sequence (about 140 m) of euxinic graptolitic black shales extending from the true base of the Llandovery up to the lowermost Devonian (Lochkovian): the upper Silurian sandstone formations which occur in the Central Iberian regions do not exist in the Ossa Morena (Jaeger & Robardet 1979; Robardet 1982). The Pragian-Emsian lithofacies and benthic faunas (shales and siltstones with 'Hercynian' fossil assemblages) differ notably from those of the Central Iberian regions but have pronounced affinities with those of the Moroccan Meseta (Racheboeuf & Robardet 1986; Brice et al. 1984). It can therefore be concluded that the Ossa Morena and the Central Iberian Zones belonged to two distinct palaeogeographic domains of the Mediterranean Province (Robardet 1976, 1986).

The north African affinities of the Ossa Morena Zone would rather favour an extension of the domain as far as Morocco; however the occurrence of ophiolitic rocks along the boundary between the Ossa Morena and the South Portuguese Zone (Oliveira 1986) may indicate a greater complexity and the existence of several palaeogeographic units within the South Iberian Domain.

Ebro-Aquitanian Domain

The study of the Palaeozoic rocks covered by the Mesozoic and Cenozoic formations of the Aquitaine Basin has indicated that there are no precise affinities with the contemporaneous rocks of the Armorican Massif (Paris & Robardet 1985; Paris et al. 1987, 1988).

The Lower Ordovician (Tremadoc to Llanvirn) shows strong similarities with that of the Montagne Noire because of the considerable thickness (about 1000 m), the monotonous terrigenous facies with rhythmic alternations and the presence of a *Taihungshania* fauna in the lower Arenig. Moreover Upper Ordovician volcaniclastic rocks occur in both regions. The Ordovician formations of the Pyrénées are most probably of the same type though they are poorly known, due to deformation and metamorphism. The Devonian succession is represented by Lochkovian to Givetian fossiliferous limestones and dolomites. The Tournaisian sequence includes black and grey-green cherts ('lydiennes') similar to the contemporaneous rocks of the Montagne Noire and the Pyrénées. Diamictic deposits probably of lower Carboniferous age are reminiscent of the late Visean flysch sediments which occur in the Montagne Noire, the Mouthoumet and the eastern Pyrénées (Engel et al. 1978).

The exact limits of the Ebro-Aquitanian Domain and the relations with the surrounding units cannot be defined precisely. To the north, the metamorphic rocks of South Brittany, the South Armorican Shear Zone and the relatively poor knowledge of the Palaeozoic series in Vendée create additional difficulties. The northern limit would probably coincide with or run parallel to the so-called South Armorican Suture (Lefort & Ribeiro 1980). The southern limit should be located in Spain as the Pyrénées and most probably the eastern part of the Cantabrian regions were included in this domain. However there is no clear evidence for this limit which might be covered by Cantabrian nappes or Mesozoic series (cf. De Poulpiquet 1986). Finally we must note that the Ebro-Aquitanian Domain is approximately the equivalent of the 'North Spain' continental microplate of Riding (1974).

Palaeobiogeographic pattern of southwestern Europe

Considering the sedimentary and faunal evidence mentioned above, we consider the Lower Palaeozoic palaeobiogeographic pattern of southwestern Europe as characterized by the following.

(1) Two major units, namely the Ebro-Aquitanian and Médio-North Armorican-Central Iberian domains, at present arranged in an arcuate structure (the Ibero-Armorican Arc *sensu stricto*).

(2) The South Iberian domain with Moroccan affinities, most probably comprising at least two distinct units (Ossa-Morena and South Portuguese Zones) and which has no counterpart in the northern branch of the Ibero-Armorican Arc.

(3) The Ligerian domain composed of several displaced terranes of southeastern origin accreted along the southern limit of the Medio-North Armorican Domain.

The position of these units in the Hercynian structures and their sedimentary and faunal characteristics and affinities impose constraints upon any geodynamical model of the Hercynian Belt (Paris & Robardet 1977).

Discussion

The palaeogeographic reconstructions we have proposed above on a global scale and for southwestern Europe can now be compared with the conclusions resulting from other branches of the earth sciences and can be used to test the models that have been proposed.

Palaeocontinental reconstructions

The Rheic Ocean (Mid-European Ocean). There is good agreement between the biogeographic pattern and the plate reconstructions based upon palaeomagnetic results for the Lower Palaeozoic, up to the end of Ordovician times. The palaeomagnetic data define a palaeogeographic entity, the Armorica Plate (Van der Voo 1979), located in high latitudes, close to Gondwana, and corresponding to the Mediterranean Province. This is consistent with the presence of a Mid-European Rheic Ocean which separated southwestern Europe from Baltica and acted as a barrier to migration resulting in the development of two distinct faunal provinces.

These two lines of evidence have led many authors to accept the existence of such a Rheic Ocean. The Mid-European suture should be located between the Armorican Massif which was part of the north Gondwanan area and the Ardenne which may represent the southern border of the Baltica plate. Towards the east, this suture would lie to the southwest of the Holy Cross Mountains and Krakovian areas in Poland since both regions contain Ordovician Baltic faunas (Oliver 1986). Several authors have challenged the existence of a Rheic Ocean because the suture is not clearly identified and illustrated by structural and petrological data. On this point, we can only notice that the suture may have been hidden by major thrusts as those revealed and documented by the deep seismic profiling of the crust in northern France (ECORS project: Bois et al. 1988) and in the South Western Approaches (SWAT project: BIRPS & ECORS 1986).

Another model has been proposed by Cocks & Fortey (1982)

who envisage two successive oceans separating northern and southern Europe during the Palaeozoic: the 'Tornquist's Sea' which closed in the Upper Ordovician; the 'Rheic Ocean' which opened during the Middle Silurian and persisted until Carboniferous times.

This model is based upon the evolution of the faunal characteristics of southern Britain. In the Early Ordovician, the affinities of the British faunas were with those of southern Europe–Gondwana. In the Late Ordovician and the Silurian these affinities had disappeared and Baltic affinities were predominant. According to Cocks & Fortey these data indicate that in Early Ordovician times southern Britain was attached to Gondwana and separated from Baltica by the Tornquist's Sea. After the closure of this ocean by Late Ordovician times, a Rheic Ocean opened to the south of the British Isles, separating the British and Baltic faunas from those of southern Europe which remained attached to Gondwana.

This plate reconstruction is mainly based upon the Gondwanan affinities of the southern Britain faunas during the Arenig and the early Llanvirn when some trilobites and brachiopods (generally at the genus level) are common to both areas (Cocks & Fortey 1982; Fortey & Owens 1987). However, the conclusion that the southern part of the British Isles was located at the northern edge of Gondwanaland and separated from Baltica by a wide ocean does not seem unequivocable. For instance the distribution of the Trinucleidae (Hughes *et al.* 1975) would rather indicate that southern Britain formed 'part of a volcanic island extending from the Baltic plate', a reconstruction that had been previously proposed for the Silurian period by McKerrow and Ziegler (1972) and that can also been regarded as possible for the Ordovician (McKerrow & Cocks 1986, fig. 1).

Considering the Cambrian trilobite palaeogeography, the Mid-European Ocean most probably opened in the late Middle Cambrian or in the early Upper Cambrian, when faunal distinction appeared between a Baltic province and a south European–Gondwanan province (Feist 1984, 1986, 1988).

The Early Ordovician affinities with southern Europe and Gondwana might only indicate that the oceanic separation between these regions and the British Isles was not, at that time, wide enough to prevent faunal exchanges.

Therefore we rather favour a major mid-European oceanic separation situated to the south of the British Isles and of the Ardenne area as it appears in McKerrow & Cocks (1986) reconstruction.

The Prototethys Ocean? We have already noted above that biogeographic and lithologic data indicate that the links between southern Europe and Gondwana were maintained through Silurian and Devonian times. However opposite views have been proposed by palaeomagnetists. The existence of an 'Armorica' plate (Van der Voo 1979, 1982) mainly consisting of the Armorican Massif and the adjacent Hercynian areas of southern and central Europe has been admitted by most palaeomagnetists (Bachtadse *et al.* 1983; Perroud & Bonhommet 1984; Perroud *et al.* 1984a, 1985; Scotese *et al.* 1985, model B). The proposed model was that, after having separated from Gondwana (at the end of the Ordovician), Armorica underwent a rapid drift towards the north and collided Laurussia by Middle or Late Devonian times. This model (Fig. 5 a, b, c) implies that, during the Silurian and the Early Devonian, a wide Prototethys Ocean (25° to 30° in latitude) separated southern Europe (Armorica) from northern Africa (Gondwana). However a few papers presented alternative reconstructions for the relative positions of Gondwana and Armorica on account of conflicting palaeomagnetic data from Gondwana (Kent *et al.* 1984, Livermore *et al.* 1985, Scotese *et al.* 1985, model A). Another model (Van der Voo 1988) also implies that a Prototethys Ocean separated southern Europe and north Africa with this difference that the maximum width of the ocean would have occurred during Middle to Late Devonian times. Sedimentary and faunal evidence does not support these models at all. Such a wide Prototethys Ocean should have acted as a barrier for the benthic faunas and a separation of Gondwana and Armorica

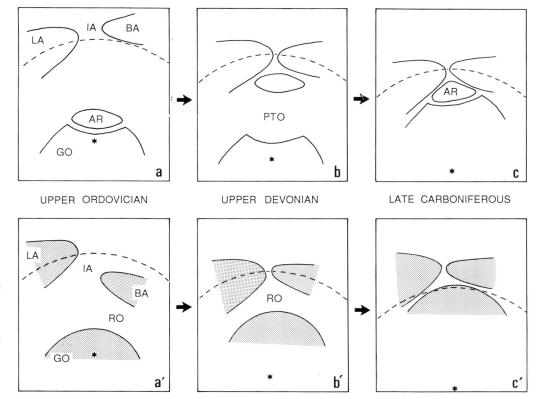

Fig. 5. Two fairly different interpretations of the relative motion of Gondwana (GO), Baltica (BA) and Laurentia (LA) during Palaeozoic times. (a, b, c) diagrams based upon palaeomagnetic data (Van der Voo 1979; Perroud & Bonhommet 1981; Perroud *et al.* 1984). (a', b', c') Alternative model based upon sedimentary and faunal evidence.
AR, Armorica plate sensu Van der Voo (1979); PTO, Prototethys Ocean; RO, Rheic Ocean; IA, Iapetus Ocean; dashed line, equator; black star, south pole.

for several tens of millions years should have induced diverging evolutionary lineages. Such phenomena cannot be observed and the affinities between the south European and Gondwanan faunas persisted during the Silurian and the Devonian, even at the species level. Moreover the assumption that Gondwana was in high palaeolatitudes during Devonian times is based upon rather uncertain or even unreliable palaeomagnetic results.

The Msissi norite, in Morocco, has been for a long time the only reference for the Devonian palaeopole for Africa (Hailwood 1974), upon which most models have been based. This reference recently turned out to be totally unfit for the Devonian period since the dating of the rock (136–139 Ma) denied the supposed Devonian age (Salmon et al. 1986). Other Devonian palaeomagnetic references from Gondwana are rather complex or uncertain and do not permit unquestionable conclusions (see discussion in Kent et al. 1984). Palaeomagnetic data (Perroud et al. 1985) from the Beja gabbro (South Portugal) are not sufficient by themselves to be the bases of a reconstruction. Another discussion might concern the Devonian palaeomagnetic references for the European Hercynides (Bachtadse et al. 1983) as we know that Hercynian (i.e. Carboniferous) remagnetizations have occurred in this mobile area and have sometimes totally overprinted previous magnetizations (Perroud et al. 1984b). Consequently, whatever it may be about these European references, the uncertainties concerning Gondwana are important enough to consider that the existence of a Prototethys Ocean cannot be, at present, demonstrated on palaeomagnetic evidence.

The sedimentary and faunal evidences cited above clearly indicate that a wide ocean cannot have separated northern Gondwana and southern Europe during the Early Devonian. The sedimentary evolution during the Devonian is characterized by shelf carbonates and reefs which are palaeoclimatically significant rock types (Heckel & Witzke 1979) supporting the idea that both areas were located in the warm temperate belt (Morzadec et al. 1988). We therefore favour an alternative reconstruction of the positioning of the Devonian continents with southern Europe in close connection with north Gondwana (b' Fig. 5).

Models of the Hercynian belt in southwestern Europe

When discussing the pre-Carboniferous palaeogeographic pattern of southwestern Europe, we have noted that the existence of the different 'domains', their relative position and their affinities impose constraints that must be considered when proposing geodynamical models for the Hercynian Belt. As yet none of the models that have been proposed satisfactorily conform to these constraints.

Reconstructions are invalid which ignore the close relationships between the Medio-North Armorican and the Central Iberian Domains, and imply that these two units have evolved independently in space and time (Laurent 1972; Ziegler 1985) or have belonged to two different plates (Lefort & Ribeiro 1980, Lefort & Van der Voo 1981, Cogné & Lefort 1985).

In the same way it is not possible to accept any location of oceanic sutures reconstructed only from discontinuous occurrences of mafic and ultramafic rocks. For instance, a continuous suture running from south Iberia (Coimbra–Cordoba shear zone) to south Brittany (parallel to the South Armorican shear zone) dismembers the Medio-North Armorican–Central Iberian Domain and implies that the Medio-North Armorican regions and the Ossa Morena Zone, that show rather distinct characteristics, were homologous areas located on the same margin of the supposed ocean.

Such a proposition (Bard et al. 1980; Matte 1986a, b) cannot be agreed even if most recent publications systematically reproduce this suture in almost identical sketch maps of the Hercynian Belt.

In our opinion this type of reconstruction arises from an excessively 'cylindrical' view of the Hercynian Belt which must be at least partially abandoned.

Conclusion

It has already been pointed out that most of the models of the Hercynian Belt appear overstylized and 'imply a continuity ... that is manifestly not present' (Badham 1982 p. 493). A few alternative interpretations have been proposed (Riding 1974; Badham & Halls 1975; Lorenz 1976; Badham 1982), involving the existence of microplates regarded: (1) either as 'subplates' resulting from the fragmentation of a south-European plate when it collided in Carboniferous times with Gondwana and Laurussia (Lorenz 1976) or (2) as continental fragments separated from the margins of the major plates during Late Silurian–Early Devonian times (Badham & Halls 1975; Badham 1982).

We agree with Badham's general opinion (1982) in considering that most models do not satisfactorily explain the European Hercynides complexity and that a microplate hypothesis may clear up a number of problems. However we cannot accept all his proposals as, for example: the separation between the Armorican Massif and the Iberian Peninsula during Early Devonian times, the suture connecting the Badajoz-Cordoba and the South Armorican shear zone; the denial of a major Rheic Ocean separating southern Europe and northern Europe (Baltica).

We consider that the different sedimentary and faunal domains which existed in southwestern Europe during the Lower Palaeozoic correspond to more or less isolated continental blocks located along the northern margin of Gondwana and separated from northern Europe by a wide Rheic Ocean (Fig. 6).

The existence of numerous occurrences of mafic and ultramafic rocks of ophiolitic character in southern Europe might correspond to the remnants of limited oceanic areas which may have separated the different blocks and which may have acted as minor barriers to the migrations of benthic faunas.

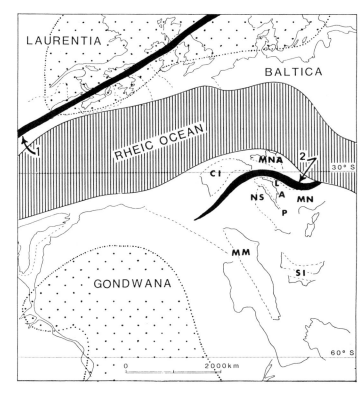

Fig. 6. Tentative palaeogeographical reconstruction for the Early Devonian, based upon sedimentary and faunal evidence. MNA, Medio-North Armorican Domain; CI, Central Iberian Domain; L, Ligerian Domain; A, Aquitaine; P, Pyrénées; MN, Montagne Noire; NS, northern Spain; MM, Moroccan Meseta; SI, South Iberian Domain. 1, Iapetus suture; 2, South Armorican Ocean suture; dotted, land areas.

Considering the palaeobiogeographic characteristics. these blocks would have existed, at least, since the Early Ordovician and would have resulted from the fragmentation of the Panafrican–Cadomian belt.

It would be extremely difficult, not to say almost impossible, to define precisely all the distinct blocks, their exact limits and relative positions. The original positioning of the different north Gondwanan blocks or microplates has been considerably modified during the tectonic evolution of the Hercynian Belt. During the plate convergence period, some of these blocks were subject to important migrations along strike-slip faults. These blocks were also subject to rotations, probably during the compressional episodes, as indicated by the palaeomagnetic declination pattern (Edel & Coulon 1984; Perroud 1986). Further movements along the major faults may have also occurred in latest Palaeozoic times (Arthaud & Matte 1977) and the whole Hercynian Belt has been subsequently involved in Mesozoic and Cenozoic events (i.e. Alpine orogeny) which have finally disturbed the Late Palaeozoic pattern.

However we believe that a microplate model would explain; (1) the discontinuity of the petrological, structural and stratigraphical characteristics along the orogen; (2) the lateral movements and the rotations of blocks; and (3) more generally the complexity of the European Hercynides that 'megaplate' models have failed to explain.

We wish to thank D. Bernard, M. Lautram and M. Le Moigne for assistance. We are grateful to W. S. McKerrow and two anonymous reviewers for constructive criticism and correcting the English manuscript. This work was carried out with the financial support of the Centre National de la Recherche Scientifique.

References

ARTHAUD, F. & MATTE, P. 1977. Late Palaeozoic strike-slip faulting in southern Europe and northern Africa: result of a right-lateral shear zone between the Appalachians and the Urals. *Geological Society of America Bulletin*, **88**, 1305–1320.

BACHTADSE, V., HELLER, F. & KRONER, A. 1983. Paleomagnetic investigations in the Hercynian mountain Belt of Central Europe. *Tectonophysics*, **91**, 285–299.

BADHAM, J. P. N. 1982. Strike-slip orogens-an explanation for the Hercynides. *Journal of the Geological Society, London*, **139**, 493–504.

—— & HALLS, C. 1975. Microplate tectonics, oblique collision, and the evolution of the Hercynian orogenic system. *Geology*, **3**, 373–376.

BARD, J. P., BURG, J. P., MATTE, P. & RIBEIRO, A. 1980. La chaine hercynienne d'Europe occidentale en termes de tectonique des plaques. *26e Congrès géologique International Paris 1980. Colloque C6 Géologie de l'Europe*, 233–246.

BIRPS & ECORS 1986. Deep seismic reflection profiling between England, France and Ireland. *Journal of the Geological, Society London*, **143**, 45–52.

BLAISE, J. & BOUYX, E. 1980. Les séries cambro-ordoviciennes à Cruziana et le problème de l'extension septentrionale des platesformes 'périgondwaniennes' durant le Paléozoique inférieur. *Comptes Rendus de l'Académie des Sciences Paris, D*, **291**, 793–796.

BOIS, C., CAZES, M., HIRN, A., MASCLE, A., MATTE, P., MONTADERT, L. & PINET, B. 1988. Contribution of deep seismic profiling to the knowledge of the lower crust in France and neighbouring areas. *Tectonophysics*, **145**, 253–275. Amsterdam.

BRICE, D., CHARRIERE, A., DROT, J. & REGNAULT, S. 1984. Mise en évidence, par les faunes de Brachiopodes de l'extension des formations dévoniennes dans la Boutonnière d'Immouzer du Kandar (Sud de Fes, Maroc). *Annales de la Société géologique du Nord*, Lille **103**, 445–458, 2 fig., 2 pl.

BURRETT, C. 1972. Plate tectonics and the Hercynian orogeny. *Nature*, **239**, 155–156.

CARLS, P. 1971. Stratigraphische Ubereinstimmungen im hochsten Silur und tieferen Unter-Devon zwischen keltiberien (Spanien) und Bretagne (Frankreich) und das Alter des Grès de Gdoumont (Belgien). *Neues Jahrbuch fur Geologie und Palaontologie, Monatshefte*, **4**, 195–212.

—— 1988. The Devonian of Celtiberia (Spain) and Devonian Palaeogeography of S. W. Europe. *Second International Symposium on the Devonian System. Calgary* 1987. (in press).

CARLS, P. & GANDL, J. 1967. The Lower Devonian of the Eastern Iberian chains (N.E. Spain) and the distribution of its Spiriferacea, Acastavinae and Asteropyginae. *International Symposium on the Devonian System*, Calgary, **2**, 453–464.

CLARKSON, E. N. K. & HENRY, J. L. 1969. Sur une nouvelle espèce du genre Crozonaspis (Trilobite) découverte dans l'Ordovicien de la Mayenne. *Bulletin de la Société géologique de France*, (7), **11**, 116–123. Paris.

COCKS, L. R. M. & FORTEY, R. A. 1982. Faunal evidence for oceanic separations in the Palaeozoic of Britain. *Journal of the Geological Society, London*, **139**, 465–478.

COGNE, J. & LEFORT, J. P. 1985. The Ligerian Orogeny: a proto-variscan event, related to the Siluro-Devonian evolution of the Tethys I ocean. *In*: GEE, D. G. & STURT, B. A. (eds) *The Caledonide Orogen–Scandinavia and related areas*, 1185–1193, John Wiley.

DE POULPIQUET, J. 1986. Etude géophysique d'un marqueur magnétique situé sur la marge continentale sud-armoricaine, arguments en faveur d'un modèle de suture de plaques. *Mémoires et documents du CAESS*, Rennes, n° **6**, 1–159.

DEAN, W. T. 1967. The distribution of Ordovician shelly faunas in the Tethyan region. In ADAMS C. G. & AGER D. V. (eds) *Aspects of Tethyan Biogeography*. Systematics Association Publication, **7**, 11–44.

—— 1977. Some aspects of Ordovician correlation and Trilobite distribution in the Canadian Appalachians. *In*: BASSETT M. G. (ed.) *The Ordovician System*, University of Wales Press, 227–250.

DINELEY, D. L. 1979. Tectonic setting of Devonian sedimentation. *Special Papers in Palaeontology*, **23**, 49–63.

DROT, J. & L'HOTELLIER, J. 1976. Les Brachiopodes Rhychonellida. *In*: LARDEUX, H. (coord.) *Les Schistes et Calcaires éodévoniens de Saint-Céneré (Massif Armoricain, France)*. Mémoire de la Société géologique et minéralogique de Bretagne, Rennes, **19**, 263–270

DUBREUIL, M. 1980. Hypothèse sur la mise en place, au Dinantien, du complexe du Tombeau Leclerc (Bassin d'Ancenis, Sud-Est du Massif Armoricain) sous forme d'un olistostrome. Conséquences géodynamiques. *Comptes Rendus de l' Académie des Sciences Paris, D*, **290**, 1455–1458.

—— 1986. Evolution géodynamique du Paléozoique Ligérien (Massif Armoricain). Thèse doctorat d'Etat, Nantes, 1–278.

—— 1987. Le bassin en décrochement de Saint-Julien-de-Vouvantes – Angers (Carbonifère inférieur du Sud-Est du Massif armoricain). *Bulletin de la Société géologique de France*, (8), III, 2, 215–221. Paris.

EDEL, J. B. & COULON, M. 1984. Mise en évidence par le paléomagnétisme d'une importante rotation anti-horaire des Vosges méridionales entre le Viséen terminal et le Westphalien supérieur. *Tectonophysics*, **106**, 239–257.

ENGEL, W., FEIST, R. D. & FRANKE, W. 1978. Synorogenic gravitational transport in the Carboniferous of the Montagne Noire (S. France). *Zeitschrift der deutschen geologischen Gesellschaft*, Hannover, **129**, 461–472.

FEIST, R. 1984. La Tethys au Cambrien supérieur. *Travaux du Laboratoire de Stratigraphie et Paléoécologie de Marseille, nouvelle série*, **3**, 105–106.

—— 1986. Late Cambrian trilobite biogeography of the western mediterranean area: relationship between Southern Europe and Gondwana. *International Conference on Iberian terranes and their regional correlation, IGCP 233*, Oviedo, September 1986, Abstracts, 50.

—— 1988. Evénement de dispersion faunistique et mouvements des plaques: l'exemple des Trilobites néocambriens. *Colloque C.N.R.S. 'ASP Evolution 1984–1988 bilan et perspectives'*. Résumés, 42.

FORTEY, R. A. & OWENS, R. M. 1987. The Arenig Series in South Wales. *Bulletin of the British Museum (Natural History) Geology*, **41**, 3, 69–307.

GARCIA-ALCALDE, J. L., ARBIZU, M., GARCIA LOPEZ, S. & MENDEZ

BEDIA, I. (eds) 1979. Meeting of the International Subcommission on Devonian Stratigraphy. Guidebook of the field trip. Spain 1979. *Servicio Publicaciones Universidad Oviedo*, 1–41.

GUTIERREZ-MARCO, J. C., RABANO, I. & ROBARDET, M. 1984. Estudio bioestratigrafico del Ordovicico en el Sinclinal del Valle (Provincia de Sevilla, S. O. de Espana). *Memorias e Noticias, Coimbra*, **97**, 12–37, 2 pl.

HAILWOOD, E. A. 1974. Paleomagnetism of the Msissi norite (Morocco) and the Palaeozoic reconstruction of Gondwanaland. *Earth and Planetary Science Letters*, **23**, 376–386.

HAMMANN, W. & HENRY, J. L. 1978. Quelques espèces de *Calymenella*, *Eohomalonotus* et *Kerfornella* (Trilobita, Ptychopariida) de l'Ordovicien de Massif Armoricain et de la Péninsule ibérique. *Senckenbergiana lethaea*, **59**, 401–429.

HECKEL, P. H. & WITZKE, B. J. 1979. Devonian world palaeogeography determined from distribution of carbonates and related palaeoclimatic indicators. *Special Papers in Palaeontology*, **23**, 99–123.

HENRY, J. L. 1980. Trilobites ordoviciens du Massif Armoricain. *Mémoire de la Société géologique et minéralogique de Bretagne, Rennes*, **22**, 1–250, 48 pl.

——, MELOU, M., NION, J., PARIS, F., ROBARDET, M., SKEVINGTON, D. & THADEU, D. 1976. L'apport de Graptolites de la zone à G. teretiusculus dans la datation de faunes benthiques lusitanoarmoricaines. *Annales de la Société Géologique du Nord*, Lille, **96**, 275–281.

HENRY, J. L., NION, J., PARIS, F. & THADEU, D. 1974. Chitinozoaires, Ostracodes et Trilobites de l'Ordovicien du Portugal (Serra de Buçaco) et du Massif Armoricain: essai de comparaison et signification paléogéographique. *Comunicaçoes dos Servicios Geologicos de Portugal*, (1973–1974), **57**, 303–345.

HOUSE, M. R. 1979. Biostratigraphy of early Ammonoidea. *Special Papers in Palaeontology*, **23**, 263–280.

HUGHES, C. P., INGHAM, J. K. & ADDISON, R. 1975. The morphology, classification and evolution of the Trinucleidae (Trilobita). *Philosophical Transactions of the Royal Society of London (Biological Sciences)*, **272**, 920, 537–607.

JAANUSSON, V. 1973. Ordovician Articulate Brachiopods. *In*: HALLAM, A. (ed.) *Atlas of Palaeobiogeography*. Elsevier, 19–25.

JAEGER, H. & ROBARDET, M. 1979. Le Silurien et le Dévonien basal dans le Nord de la Province de Séville (Espagne). *Géobios*, **12**, 5, 687–714.

JULIVERT, M., TRUYOLS, J. & VERGES, J. 1983. El Devonico en el Macizo Iberico. *In: Geologia de Espana, Libro jubilar J. M. Rios*, **1**, 265–311, Instituto Geologico y Minero de Espana.

KENT, D. V., DIA, O. & SOUGY, J. M. A. 1984. Paleomagnetism of Lower-Middle Devonian and Upper-Proterozoic-Cambrian (?) rocks from Mejeria (Mauritania, West Africa). *In*: VAN DER VOO R., SCOTESE C. R. & BONHOMMET N. (eds) *Plate reconstruction from palaeozoic paleomagnetism*, Geodynamics series, American Geophysical Union, **12**, 99–115.

KRIZ, J. & PARIS, F. 1982. Ludlovian, Pridolian and Lochkovian in La Meignanne (Massif Armoricain): Biostratigraphy and correlations based on Bivalvia and Chitinozoa. *Géobios*, **15**, 3, 391–421.

LARDEUX, H. & MORZADEC, P. 1979. *Massif Armoricain 1979. Field excursion guide*. International Subcommission on Devonian Stratigraphy, 1–27.

—— CHAUVEL, J. J., HENRY, J. L., MORZADEC, P., PARIS, F., RACHEBOEUF, P. & ROBARDET, M. 1977. Evolution géologique du Massif Armoricain au cours des temps ordoviciens, siluriens et dévoniens. *In: La chaine varisque d'Europe moyenne et occidentale*. Colloque international du CNRS, **243**, 181–192.

LAURENT, R. 1972. The Hercynides of South Europe, a Model. *24th International Geological Congress, Section 3*, 363–370.

LEFORT, J. P. & RIBEIRO, A. 1980. La faille Porto-Badajoz-Cordoue a-t-elle controlé l'évolution de l'océan paléozoique sud-armoricain? *Bulletin de la Société géologique de France*, Paris (7), **22**, 455–462.

—— & VAN DER VOO, R. 1981. A kinematic model for the collision and complete suturing between Gondwanaland and Laurussia in the Carboniferous. *Journal of Geology*, **89**, 537–550.

LIVERMORE, R. A., SMITH, A. G. & BRIDEN, J. C. 1985. Paleomagnetic constraints on the distribution of continents in the late Silurian and early Devonian. *In*: CHALONER W. G. & LAWSON J. C. (eds) *Evolution and environment in the late Silurian and early Devonian*. Philosophical Transactions of the Royal Society London, series B, **309**, 29–56.

LORENZ, V. 1976. Formation of Hercynian subplates, possible causes and consequences. *Nature*, **262**, 374–377.

MATTE, PH. 1986a. La chaine varisque parmi les chaines paléozoiques péri-atlantiques, modèle d'évolution et position des grands blocs continentaux au Permo-Carbonifère. *Bulletin de la Société géologique de France*, Paris, (**8**), 2, 1, 9–24.

—— 1986b. Tectonics and plate tectonics model for the Variscan belt of Europe. *Tectonophysics*, **126**, 329–374.

MCKERROW, W. S. & COCKS, L. R. M. 1986. Oceans, island arcs and oliststromes: the use of fossils in distinguishing sutures, terranes and environments around the Iapetus Ocean. *Journal of the Geological Society, London*, **143**, 185–191.

—— & ZIEGLER, A. M. 1972. Palaeozoic Oceans. *Nature*, 240, 100, 92–94.

MELOU, M. 1974. Le genre *Aegiromena* (Brachiopode – Strophomenida) dans l'Ordovicien du Massif armoricain (France). *Annales de la Société Géologique du Nord*, Lille **93**, 4, 253–264.

—— 1975. Le genre *Heterorthina* (Brachiopoda, Orthida) dans la Formation des Schistes de Postolonnec (Ordovicien), Finistère, France. *Geobios*, **8**, 3, 191–208.

—— 1976. Orthida (Brachiopoda) de la Formation de Postolonnec (Ordovicien), Finistère, France. *Geobios*, **9**, 6, 693–717.

MORZADEC, P., PARIS, F., PLUSQUELLEC, Y., RACHEBOEUF, P. R. & WEYANT, M. 1988. Devonian stratigraphy and paleogeography of the Armorican Massif (N. W. France). *Second International Symposium on the Devonian System. Calgary 1987*. (in press)

——, —— & RACHEBOEUF, P. 1981. Conclusions stratigraphiques. *In: La tranchée de la Lézais, Emsien supérieur du Massif armoricain. Sédimentologie, Paléontologie, Stratigraphie*. Mémoire de la Société géologique et minéralogique de Bretagne, Rennes, **24**, 11–18.

OLIVEIRA, J. T. 1986. The South-Portuguese terrane: tectono-stratigraphic evolution. *International Conference on Iberian terranes and their regional correlation. IGCP 233*, Oviedo, September 1986, Abstract, 28–29.

OLIVER, G. J. H. 1986. Evidence for the Tornquist Sea: a connection between the Polish and British Caledonides. *International Conference on Iberian Terranes and their regional correlation, IGCP 233*, Oviedo, September 1986, Abstract, 65.

PARIS, F. & ROBARDET, M. 1977. Paléogéographie et relations ibéroarmoricaines au Paléozoique antécarbonifère. *Bulletin de la Société géologique de France*, Paris (7), **19**, 1121–1126.

—— & —— 1985. Evaluation des affinités entre le Paléozoïque caché sous l'Aquitaine et les formations armoricaines contemporaines. In Geologie Profonde de la France, Thème 7. *Documents du Bureau de Recherches Géologiques et Minières*, **95**-7, 11–36.

——, LE POCHAT, G. & PELHATE, A. 1988. Le socle paléozoique nord-aquitain: caractéristiques principales et implications géodynamiques. *Comptes Rendus de l'Académie des Sciences Paris*, **306**, II, 597–603.

——, —— & HOLTZAPFFEL, T. 1987. Synthèse des connaissances sur le Paléozoique Nord-Aquitain. *In: Géologie profonde de la France, thème 7*. Documents du Bureau de Recherches Géologiques et Minières, **144**, 1–57.

——, ROBARDET, M., DURAND, J. & NOBLET, C. 1982. The lower Ordovician transgression on South-West Europe. *Paleontological Contributions from the University of Oslo*, **280**, 41.

PERROUD, H. 1986. Paleomagnetic evidence for tectonic rotations in the Variscan mountain belt. *Tectonics*, **5**, 2, 205–214.

—— & BONHOMMET, N. 1981. Paleomagnetism of the Ibero-Armorican arc and the Hercynian orogeny in Western Europe. *Nature*, **292**, 445–448.

—— & —— 1984. A Devonian palaeomagnetic pole for Armorica. *Geophysical Journal of the Royal astronomical Society*, **77**, 839–845.

——, ——, & RIBEIRO, A. 1985. Paleomagnetism of Late Palaeozoic igneous rocks from Southern Portugal. *Geophysical Research Letters*, **12**, 1, 45–48.

——, VAN DER VOO, R. & BONHOMMET, N., 1984a. Palaeozoic evolution of the Armorica plate on the basis of paleomagnetic data. *Geology*, **12**, 579–582.

——, ROBARDET, M., VAN DER VOO, R., BONHOMMET, N. & PARIS, F. 1984b. Revision of the age of magnetization of the Montmartin red beds, Normandy, France. *Geophysical Journal of the Royal astro-*

nomical Society, **80**, 541–549.

PILLET, J. 1972. Les Trilobites de Dévonien inférieur et Dévonien moyen du Sud-Est du Massif Armoricain. *Mémoires de la Société d' Etudes Scientifiques de l'Anjou, 1972*, **1**, 1–307, 64 pl.

PLUSQUELLEC, Y. 1976. Les Polypiers Tabulata *In*: *Les Schistes et calcaires éodévoniens de Saint-Cénéré (Massif Armoricain, France)*. Mémoire de la Société géologique et minéralogique de Bretagne, Rennes, **19**, 183–226.

—— 1986. Révision de *Michelinia transitoria* KNOD, 1908 (Tabulata, Dévonien de Bolivie). *Annales de la Société Géologique du Nord*, Lille, **105**, 249–252.

RACHEBOEUF, P. 1981. Chonetacés (Brachiopodes) siluriens et dévoniens du Sud-Ouest de l'Europe (Systématique, Phylogénie, Biostratigraphie, Paléobiogéographie). *Mémoire de la Société géologique et minéralogique de Bretagne*, Rennes, **27**, 1–294, 35 pl.

—— & BABIN, C. 1986. Biostratigraphie et corrélations. *In*: RACHEBOEUF, P. R. (ed.) *Le Groupe de Liévin. Pridoli – Lochkovien de l'Artois (N. France)*. Biostratigraphie du Paléozoique, **3**, 31–45.

—— & ROBARDET, M. 1986. Le Pridoli et le Dévonien inférieur de la Zone d'Ossa Morena (Sud-Ouest de la Péninsule Ibérique). Etude des Brachiopodes. *Geologica et Paleontologica*, **20**, 11–37.

——, CARLS, P. & GARCIA-ALCALDE, J. 1981. Hollardina n.g., nouveau Leptaeninae (Brachiopode) du Gedinnien d'Europe occidentale et du Maroc présaharien. *Bulletin de la Société géologique et minéralogique de Bretagne*, C, Rennes, **13**, 2, 45–65.

——, ROBARDET, M. & REGNAULT, S. 1988. Apport des faunes benthiques éodévoniennes à la paléogéographie du Sud-Ouest européen et du Nord-Ouest de l'Afrique. *Colloque C.N.R.S. 'ASP Evolution 1984–1988 bilan et perspectives'. Résumés*, 47.

RIDING, R. 1974. Model of Hercynian Foldbelt. *Earth and Planetary Science Letters*, **24**, 125–135.

ROBARDET, M. 1976. L'originalité du segment hercynien sudibérique au Palaéozoique inférieur: Ordovicien, Silurien et Dévonien dans le Nord de la Province de Séville, Espagne. *Comptes Rendus de l'Académie des Sciences Paris, D*, **283**, 999–1002.

—— 1982. The Silurian-earliest Devonian succession in South Spain (Ossa Morena Zone) and its paleogeographical signification. *IGCP no. 5 Newsletter*, **4**, 72–77.

—— 1986. The Lower Palaeozoic Series of the Ossa-Morena terrane, stratigraphy and paleogeography. *International Conference on Iberian terranes and their regional correlation, IGCP 233, Oviedo September 1986*. Abstract, 47.

ROBARDET, M. & DORE, F. 1988. The Late Ordovician diamictic formations from southwestern Europe: north-Gondwana glaciomarine deposits. *Palaeogeography Palaeoclimatology Palaeoecology*, **66**, 19–31.

SALMON, E., MONTIGNY, R., EDEL, J. B., PIQUE, A., THUIZAT, R. & WESTPHAL, M. 1986. The Msissi norite revisited: K/Ar dating, petrography and paleomagnetism. *Geophysical Research Letters*, **13**, 8, 741–743.

SCOTESE, C. R., BAMBACH, R. K., BARTON, C., VAN DER VOO, R. & ZIEGLER, A. M. 1979. Palaeozoic base maps. *Journal of Geology*, **87**, 217–277.

——, VAN DER VOO, R. & BARRETT, S. F. 1985. Silurian and Devonian base maps. *In*: CHALONER, W. G. & LAWSON, J. D. (eds) *Evolution and environment in the late Silurian and early Devonian*. Philosophical Transactions of the Royal Society London, B **309**, 57–77.

SPJELDNAES, N. 1961. Ordovician climatic zones. *Norsk Geologisk Tidsskrift*, **41**, 45–77.

—— 1967. The palaeogeography of the Tethyan region during the Ordovician. *In*: ADAMS, C. G. & AGER, D. V. (eds) *Aspects of Tethyan biogeography*. Systematics Association Publication, **7**, 45–57.

VAN DER VOO, R. 1979. Palaeozoic assembly of Pangea: a new plate tectonic model for the Taconic, Caledonian and Hercynian orogenies. *EOS, Transactions of the American Geophysical Union*, **60**, 241 (abstract).

—— 1982. Pre-Mesozoic paleomagnetism and plate-tectonics. *Annual Review of Earth and Planetary Sciences*, **10**, 191–220.

—— 1988. Palaeozoic paleogeography of North America, Gondwana and intervening displaced terranes: comparisons of paleomagnetism with paleoclimatology and biogeographical patterns. *Geological Society of America Bulletin*, **100**, 311–324.

VANNIER, J. 1986a. Ostracodes Binodicopa de l'Ordovicien (Arenig-Caradoc) ibéro-armoricain. *Palaeontographica*, (A), **193**, 77–143, 70 fig., 13 pl.

—— 1986b. Ostracodes Palaeocopa de l'Ordovicien (Arenig-Caradoc) ibéro-armoricain. *Palaeontographica (A)*, **193**, 145–218, 74 fig., 13 pl.

VILLAS, E. 1985. Brachiopodos del Ordovicico medio y superior de las Cadenas Ibericas Orientales. *Memorias del Museo Paleontologico Universidad de Zaragoza*, **1**, 1–153, 34 pl.

WHITTINGTON, H. B. 1966. Phylogeny and distribution of Ordovician Trilobites. *Journal of Paleontology*, **40**, 696–737.

—— 1973. Ordovician Trilobites. *In*: HALLAM, A. (ed.) *Atlas of Palaeobiogeography*. Elsevier, 13–18.

—— & HUGHES, C. P. 1972. Ordovician geography and faunal provinces deduced from trilobite distribution. *Philosophical Transactions of the Royal Society London, B*, **263**, 850, 235–278.

WILLIAMS, A. 1969. Ordovician faunal provinces with reference to brachiopod distribution. *In*: WOOD, A. (ed.) *The Pre-Cambrian and Lower Palaeozoic rocks of Wales*, University of Wales Press, Cardiff 117–154.

—— 1973. Distribution of Brachiopod assemblages in relation to Ordovician Palaeogeography. *In: Organisms and Continents through time*. Special Papers in Palaeontology, **12**, 241–269.

YOUNG, T. P. 1985. *The stratigraphy of the Upper Ordovician of Central Portugal*. PhD thesis, University of Sheffield.

—— 1988. The lithostratigraphy of the upper Ordovician of central Portugal. *Journal of the Geological Society, London*, **145**, 377–392.

ZIEGLER, P. A. 1985. Caledonian, Acadian-Ligerian, Bretonian and Variscan orogens: is a clear distinction justified? *In*: GEE D. G. & STURT, B. A. (eds) *The Caledonide Orogen-Scandinavia and related areas*. John Wiley Chicester, 1241–1248.

Ordovician sedimentary facies and faunas of Southwest Europe: palaeogeographic and tectonic implications

T. P. YOUNG

Department of Geology, University of Wales College of Cardiff, PO Box 914, Cardiff CF1 3YE, UK

Abstract: The term West European Platform is introduced for an Ordovician shelf now within the western Variscides. The geometry of the sedimentary facies belts and their faunal variation suggest that it formed a promontory on the Gondwanan margin. The palaeogeography of the platform is discussed with particular reference to the Centro-Iberian/Armorican area, especially central Portugal.

The sedimentary history of the Centro-Iberian/Armorican area was controlled by basement structures. Rapid local subsidence during Early Ordovician extension was followed by later differential subsidence along these early structures. The thin sequences indicate low overall rates of subsidence. The dominantly clastic shallow marine sediments show strong storm influence. Individual sediment packets occur over wide areas indicating little syndepositional relief. The linear nature of facies belts allows reconstruction of the relative orientation of some blocks within the Variscides, and indicates strike-slip faulting during early Variscan events, with subsequent (Carboniferous) rotation of most of Iberia.

Faunas from SW Europe from a homogeneous group related to those of the classic 'Mediterranean Province' areas of Bohemia and Morocco, and more distantly to those of the Avalonian terrane. Faunal migration into the West European Platform from Avalonia accompanied progressive northward drift of the platform, occurring particularly during periods of eustatic sea-level rise in the early Llanvirn, the early Caradoc and the early Ashgill.

Revision of the litho- (Mitchell 1974; Romano & Diggens 1976; Young 1988) and biostratigraphy (Henry *et al*. 1976; Paris 1979, 1981; Romano 1982; Young 1985; Romano & Young research in progress) of the Ordovician of central Portugal has enabled re-evaluation of the Ordovician sequences of the Centro-Iberian/Armorican area. Deposition occurred in shallow marine environments with low subsidence rates, and litle tectonic activity.

The term 'West European Platform' is introduced for the areas bearing thin Ordovician shelf sequences, including the Centro-Iberian/Armorican region, together with the Catalan region, the Ossa Morena Zone, Thuringia, Bohemia and the Carnic Alps. Previous authors have used the term 'Armorica' for part or all of this assemblage, but the new expression is introduced to avoid confusion with the geographical use of Armorica (the region containing the Armorican Massif), and to avoid any connotation that the area formed an independent 'microplate' or 'terrane' during the Ordovician. This contribution uses sedimentological and palaeontological data to examine the extent of the West European Platform and its relationship to other palaeogeographical regions.

Faunal analysis has produced the concepts of the Mediterranean brachiopod province characterized by the Aegiromeninae, Heterorthidae and Orthostrophiinae (Havlíček 1976), and of the *Selenopeltis* trilobite province (Whittington & Hughes 1972) with faunas dominated by dalmanitid and calymenacean trilobites (Cocks & Fortey 1988), particularly in tne Lower and Middle Ordovician by *Neseuretus* (Fortey & Morris 1982). These two faunal provinces occupied the wide shelves of high-latitude Gondwana. Recent studies in SW Europe (Havlíček 1981; Villas Pedruelo 1985; Young 1985) have shown that in the Ashgill much of SW Europe had brachiopod faunas more closely related to those of northern Europe than previously realised.

The present study examines the Llanvirn to Ashgill fauna of central Portugal. The relatively complete sedimentary record has increased understanding of the composition and palaeoecology of the fauna, together with the timing of faunal interchange with adjacent areas.

Previous attempts at producing a palaeogeographic reconstruction for the region have not addressed the details of the pre-Variscan geography. Models of the Mediterranean area as a part of Gondwana, with Avalonia as a separate microplate attached to SW Europe in the Early Ordovician but drifting northward with the creation of the 'Rheic Ocean' in the later Ordovician (Babin *et al*. 1980; Cocks & Fortey 1982, 1988; Fortey & Cocks 1986; Fortey & Morris 1982; Scotese 1986; Soper *et al*. 1987) predominate. Some authors, however, place a 'Mediterranean Ocean' between southern Europe and Gondwana (Smith *et al*. 1981; Whittington & Hughes 1973). All these models use a Permo-Triassic reconstruction within SW Europe; they do not address themselves to the internal palaeogeography of SW Europe before the Variscan orogeny. More detailed palaeogeographic reconstructions, largely based on tectonic evidence, have divided SW Europe into two or more discrete blocks in the Lower Palaeozoic (Bard *et al*. 1980; Lefort & Ribeiro 1980; Matte 1986*a*, 1986*b*; Ziegler 1984, 1986). Very little has been published towards a palaeogeography for the region based on the study of the Ordovician sedimentary sequences themselves.

The nature of the Ordovician sequences of southwest Europe

Centro-Iberian/Armorican region

This includes the Centro-Armorican, Mancellian, Domnonean, Centro-Iberian, and West Asturian-Leonese Zones (Figs 1–3); (the Iberian zones are after Julivert *et al*. (1980)). Early workers regarded the Varsican synclines of this area as following small, individual depositional basins, but current interpretation is that the synclines lie close to depositional strike within a shelf of only very gradual lateral variation, giving each syncline a unique, but persistent character. A further complication is that the synclinal structures often follow basement faults which may have acted both in controlling subsidence in the Ordovician, and as belts of high shear-strain during Variscan deformation.

When the similarities between the isolated exposures of Ordovician sequences are examined, it is apparent that individual lithostratigraphic units often have extremely widespread distribution: the sandstone-dominated 'Armorican Quartzite' and the Monte da Sombadeira Formation, a storm generated sandstone unit of Llandeilo age with a minimum original area of 75 000 km^2 (Brenchley *et al*. 1986), are good examples. Figure 4 shows the great lateral continuity of facies across depositional strike in a transect across the Centro-Iberian/Armorican region in central Portugal. The widespread sedimentary units are affected by areas of differential subsidence, but sequences remain generally rather thin; 1200 m thick Ordovician sequences are rare, 800 m is more typical. The sedimentological and palaeontological evidence

Fig. 1. Varsican Massifs in western Europe; A, Armorican Massif; B, Bohemian Massif; H, Hesperian Massif. Tectonic zones: In the Armorican Massif: 1, Domnonean Domain; 2, Mancellian Domain; 3, Centro-Armorican Domain; 4, Ligerian Domain; 5, Cornouaille Anticline/West Vendée Zone. In the Iberian Peninsula: 6, Cantabrian Zone; 7, West Asturian – Leonese Zone; 8, Centro-Iberian Zone; 9, Ossa Morena Zone; 10, South Portuguese Zone. Other areas referred to in text: CI, Celtiberia; C, Catalonia; S, Sardinia; E, Eastern Pyrenees; M, Mouthoumet Massif; MN, Montagne Noire; CA, Carnic Alps; T, Thuringia; P, Prague Basin.

demonstrates deposition under shallow marine conditions, with strong influence from storm processes. The large areas of shallow, low-topography shelf may have promoted the distribution of storm generated units over large areas. Active tectonism appears to have been very limited. Particularly vigourous subsidence occurred during the Arenig (and Tremadoc?) at the time of the onset of deposition of the Armorican Quartzite and the underlying red beds. Minor volcanism occurred locally at various times during the Ordovician, but the only relatively widespread event was during the early–middle Ashgill when volcanic centres formed along the basin-controlling fault lines in several areas (Crozon, Buçaco, Cabo Peñas), presumably in association with renewed extension.

Several distinct phases can be recognised in the sedimentary evolution of the area.

(1) Tremadoc?–Arenig. Rapid localised subsidence produced deposition of terrestrial and marine red-bed sequences. Acid volcanic rocks are locally related to this extensional phase. For recent discussions see Bonjour & Chauvel (1988), and McDougall *et al.* (1987).

(2) Arenig. The 'Armorican Quartzite': inner shelf sandstone deposition over large areas. The thickest developments occur in the regions of previous red-bed deposition.

(3) Llanvirn–early Ashgill. Storm-dominated shelf environments: the subdued tectonic activity allows strong eustatic control of the depositional system (Young 1989). The major eustatic cycles produced slightly different sediments.

(a) Llanvirn–early Caradoc. Extensive mudstone deposition punctuated by intervals of widespread storm generated sandstone bodies.

(b) Early Caradoc–early Ashgill. Similar to phase 3a, but with the balance shifted in favour of more general sandstone distribution with minor mudstone deposition.

(4) Early–mid Ashgill. Volcanicity occurs locally, with acidic and basic extrusives and sub-volcanic basic intrusions. Carbonate deposition, predominantly of bryozoan bioclastic limestones, occurred over much of the area.

(5) Late Ashgill. The Late Ordovician glaciation locally produced thick sediments in response to glacio-eustatic changes. Diamictites were deposited over almost all of the area (Brenchley *et al.* in press; Robardet & Doré 1988).

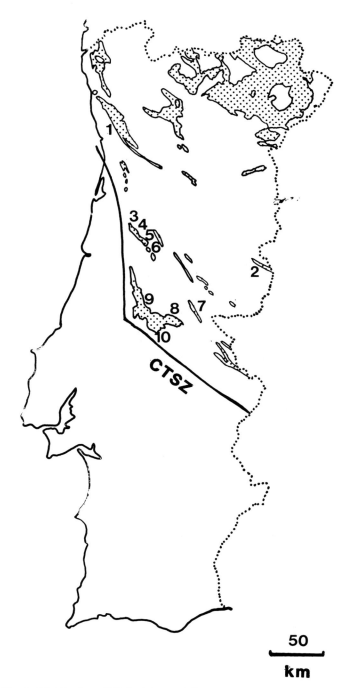

Fig. 2. Distribution of Ordovician and Silurian rocks (stippled) in the Centro-Iberian Zone of Portugal (after Romano 1982). CTSZ Cordoba/Tomar Shear Zone. Location of sections illustrated in Fig. 4: 1, Valongo; 2, Penha Garcia syncline; 3–6, Buçaco syncline; 7, Vila Velha do Rodão syncline; 8, Amêndoa; 9, Dornes; 10, Mação.

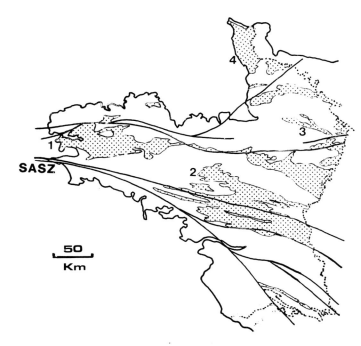

Fig. 3. Distribution of Palaeozoic rocks (stippled) in the Armorican Massif. SASZ South Armorican Shear Zone. Localities referred to in text: 1, Crozon Peninsula; 2, synclines S of Rennes; 3, Domfront; 4, St Germain-sur-Ay.

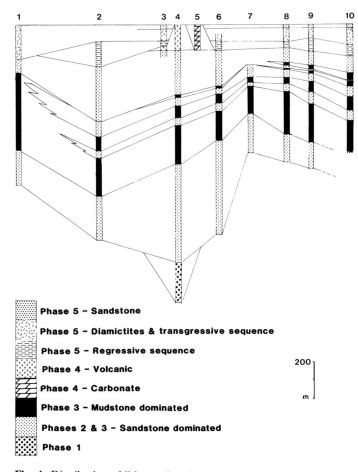

Fig. 4. Distribution of lithostratigraphic units within a schematic cross-section of the southern part of the Centro-Iberian Zone in Portugal. Section drawn perpendicular to depositional strike from NE (1) to SW (10). For location of sections see Fig. 2. The modern length of the section illustrated is 200 km.

Sequences from other related areas of southwest Europe

Catalan region (eastern Pyrenees, Mouthoumet Massif, Montagne Noire, Sardinia and Catalonia). A thick (often greater than 2000 m) Tremadoc/Arenig sequence is seen in some areas (Catalonia?, E. Pyrenees?, Mouthoumet, Montagne Noire), overlain by acid volcanic rocks and conglomerates (Sardinia, Catalonia?, Mouthoumet, Montagne Noire), in turn overlain unconformably by sediments of late Caradoc to Ashgill age (Catalonia, Sardinia, E. Pyrenees, Mouthoumet, Montagne Noire). The ages of the volcanics are often poorly constrained: post-late Arenig, pre-Ashgill (E. Sardinia: Naud 1979), post-early Arenig, pre-late Caradoc (Montagne Noire: Arthaud 1970; Havlicek 1981), interbedded in part with sediments containing a possible Caradoc fauna (Catalonia: Villas *et al.* 1987; SW Sardinia: Arthaud 1970). A major difference between the Catalan region and the Centro-Iberian/Armorican region is that the Llanvirn-middle Caradoc clastic shelf sequence was not developed, or has not been preserved. Arthaud (1970) attributed this break to Upper Ordovician movements.

The Ossa-Morena Zone (southern Iberia) has rather poorly known Ordovician sequences. The 'Armorican Quartzite' is usually absent, and the sequences may include graptolite-bearing Tremadoc strata (Gutierrez-Marco *et al.* 1984). It seems likely that the Ossa Morena Zone represents a part of the West European Platform not originally juxtaposed with the Centro-Iberian Zone in its present position, but which was placed there during Devonian to Carboniferous strike-slip movements on the Cordoba/Tomar Shear Zone.

Thuringia and Prague Basin. Two regions of central Europe have relatively well-known Ordovician sequences, with affinities with the Centro-Iberian/Armorican region. In Thuringia (Sdzuy 1971) a Tremadoc−Arenig sequence grades upwards into a thin sandstone unit of Arenig age, probably the local development of the 'Armorican Quartzite'. The Llanvirn/Llandeilo mudrocks contain ooidal ironstone units very similar to those in the Llanvirn of the Centro-Iberian/Armorican region (Young 1989). The Caradoc sequence is very thin, being represented by a sandstone unit of variable thickness resting on an ooidal ironstone horizon. It is overlain by a further ooidal ironstone and in turn this is overlain by a thin bioclastic limestone yielding conodonts similar to those of the mid-Ashgill bioclastic limestones of the Centro-Iberian/Armorican region. The limestones are overlain by thick glacio-marine deposits (Greiling 1967).

The Prague Basin of Bohemia, Czechoslovakia, has strong faunal affinities with the Centro-Iberian/Armorican region, but the basin shows a rather different sedimentary history (Havlíček 1982), probably controlled more by the tectonic subsidence of the basin than by eustatic sea level change which was the dominant control on sedimentation elsewhere. There is no 'Armorican quartzite' and the Arenig consists of mudrocks with local tuffites. The Llanvirn to Caradoc sequence consists of shelf mudstones with sandstone packets. The mid-Ashgill strata are in a deeper water facies than is developed in the Centro-Iberian/Armorican region. The late Ashgill rocks are generally arenaceous and include only two thin intervals of glaciomarine deposits (Brenchley *et al.* in press).

The Carnic Alps. The sequences of the Carnic Alps (Schonlaub 1971) include sandstones of possible Caradoc age, overlain by calcareous beds bearing brachiopods rather similar to the mid-Ashgill faunas of the bioclastic carbonates of the Centro-Iberian/Armorican region. Shelly faunas of latest Ashgill age referable to the *Hirnantia* Fauna are also seen in this area. The pre-Caradoc Ordovician of this area is unknown.

The nature and affinities of the Ordovician macrofauna of central Portugal

The macrofauna of the Centro-Iberian/Armorican region has often been described as low-diversity, and reflecting the high palaeolatitude and low water temperature of the area (Spjeldnaes 1961). However, relatively little work has been done on the palaeoecology of Mediterranean Province faunas. The palaeogeographical and palaeobiogeographical significance of the fauna of the West European Platform are illustrated here by a discussion of the nature and affinities of the macrofauna from central Portugal.

Phases 1 and 2 of the sedimentary history described above are largely without macrofauna in central Portugal, although a limited shelly fauna, including bivalves and inarticulate brachiopods, has been recovered from the Armorican Quartzite facies (Serra de Brejo Formation of Fig. 5).

During phase 3a faunas show different biofacies in northern and central Portugal (Romano 1982). The Llandeilo faunas are rather similar to the Llanvirn ones, with very little change at generic level (Table 1).

In the northern part of the area (Valongo) a relatively slightly deeper-water facies (deposited just below storm wave base) occurs. Trilobite genera in this facies include *Bathycheilus*, *Dionide*, *Eccoptochile*, *Ectillaenus*, *Eodalmanitina*, *Eoharpes*, *Hungioides*, *Neseuretus*, *Placoparia*, *Prionocheilus*, *Salterocoryphe*, *Selenopeltis*, *Uralichas*, *Valongia* and *Zeliszkella*. The brachiopods are poorly known at present but *Aegiromena* and dalmanellids dominate Llandeilo assemblages, whereas the earlier Llanvirn assemblages are dominated by coarse-ribbed orthaceans.

In the shallower water facies of central Portugal, mudstone/siltstone dominated sequences showing distinct signs of periodic storm activity (coquinas, reworked early diagenetic concretions, thin graded siltstone beds, high disarticulation ratio of shelly fauna) bear a rather different fauna during phase 3a, with abundant ostracodes, brachiopods (*Apollonorthis*, *Cacemia*, *Eorhipidomella*, *Heterorthina*), infaunal bivalves (largely deposit feeders), and trilobites (including *Colpocoryphe*, *Crozonaspis*, *Ectillaenus*, *Eodalmanitina*, *Kerfornella*, *Morgatia*, *Neseuretus*, *Phacopidina*, *Plaesiacomia* and *Selenopeltis*).

During the lower Caradoc transgressive event (Young 1989), at the start of phase 3b, several new groups of shelly fauna arrived in Portugal. The trilobite *Dalmanitina* appears in the earliest Caradoc mudstones, and in the immediately overlying ooidal ironstone *Deanaspis* also occurs, together with the brachiopods *Blyskavomena*?, *Gelidorthis*, *Onniella*, *Tazzarinia* and *Triplesia*.

The 'rise' facies, which developed in a belt of a low subsidence during the lower Caradoc, contains abundant bryozoans, echinoderms, brachiopods (*Aegiromena*, *Blyskavomena*, *Drabovia*, *Gelidorthis*, *Onniella*, *Porambonites* (*Porambonites*), *Rostricellula*, *Saukrodictya*, *Svobodaina*, *Tazzarinia*, *Triplesia*), and trilobites (*Actinopeltis*, *Colpocoryphe*, *Dalmanitina*, *Deanaspis*, *Eccoptochile*, *Ectillaenus*, *Eoharpes*, *Eudolatites*, *Primaspis*, *Prionocheilus*, *Selenopeltis*). The trilobite fauna is strikingly similar to some of the deeper water phase 3a associations.

A bivalve/ostracode/heterorthid brachiopod (*Svobodaina*)/trilobite (*Crozonaspis*, *Dalmanitina*, *Deanaspis*, *Plaesiacomia*) fauna occurs in middle to late Caradoc storm dominated mudstone facies. This fauna is rather similar in general composition to the phase 3a shallower water facies, although very few of the genera are the same. Sandstones in the later Caradoc may contain a fauna of trilobites (?*Scotiella*, *Calymenella*, ?*Deanaspis*) and brachiopods (*Hedstroemina*?, *Horderleyella*, *Rostricellula*, *Svobodaina*).

The onset of the next sea-level rise close to the Caradoc/Ashgill boundary is accompanied by faunal changes similar to those close to the Llandeilo/Caradoc boundary. The faunas from the dominantly carbonate sediments of phase 4 are markedly different from those seen earlier in the Ordovician. Some 31 genera of articulate brachiopods have been found in the phase 4 deposits of central Portugal (Tables 2 and 3), of which only 9 genera occur in earlier strata in Portugal, 4 appear to be Mediterranean endemics, 13 have a pre-Ashgill history in the Welsh basin, and 5 also make their first appearance in the Ashgill of the UK. The 13 genera (*Bicuspina*, *Christiania*, *Dalmanella*, *Dolerorthis*, *Leptaena*, *Leptestiina*, *Nicolella*, *Oxoplecia*, *Platystrophia*, *Parastrophinella*, *Protozyga*, *Rhactorthis*, *Skenidioides*) which appear to expand their range southwards are dominated those particularly abundant in the *Nicolella* Palaeocommunity of Lockley (1983), and include four which have not been recorded in Ashgill rocks of the UK despite being known in the Caradoc. The association of the spread of this fauna with the bioclastic carbonate sediments suggests that the control on the fauna was environmental. The nature of the substrate may have been influenced by climatic change. Webby (1984) has suggested global climatic amelioration as the cause for the increase in carbonate-producing areas during the Ashgill. Alternatively apparent amelioration in SW Europe may have been the result of progressive northward drift of the region during the Ordovician.

The fauna of the top of the carbonate facies in central Portugal is important for it includes elements related to the *Proboscisambon* fauna of the Rawtheyan of Bohemia (Havlíček & Mergl 1982; Cocks & Rong 1988) (*Aegironetes*, *Christiania*, *Eridorthis*, *Jezercia*, *Leptaena*, *Nicolella*, *Ptychopleurella*, *Skenidioides*, *Triplesia*).

Table 1. *Survival and introduction of trilobite faunas at generic level in central Portugal: Llanvirn and Llandeilo (phase 3a).*

	Llanvirn	Llandeilo	
	no. of genera	no. of genera	no. of new gen.
Mudstones (below storm wave base)	9	13	4
Mudstones (Above storm wave base)	11	13	2

Below storm wave base faunas from the Valongo Formation of Valongo, N Portugal. Above storm wave base faunas from the Brejo Fundeiro (Llanvirn) and Fonte da Horta (Llandeilo) Formations of the Buçaco syncline central Portugal. Data from Romano (1982) and author's unpublished data.

Table 2. *Survival and introduction of (1) brachiopod and (2) trilobite faunas at generic level in central Portugal: lower Caradoc to upper Ashgill.*

	A	B	C	D	E	F	G	H	I	J
(1)	2/1	6/5	11/6	1/0	4/3	1/1	18/10	16/6	11/7	3/3
(2)	3/1	3/1	11/1	4/0	3/2	1/0	11/4	4/0	1/1	1/1

First figure total number of genera present, second figure number of new genera.

Phase 3b
    A, L Caradoc: Carregueira Formation (transgressive mudstone)
    B, L Caradoc: Favaçal bed (ooidal ironstone)
    C, L Caradoc: Queixoperra Member (rise facies)
    D, M/U Caradoc: Vale Saido/Zuvinhal members (basin facies: mudstone)
    E, M/U Caradoc: Sandstone below Vaca Member (basin facies: sandstone)
    F, U Caradoc: Galhano Member (transgressive mudstone)
Phase 4
    G, L Ashgill: Leira Má Member (ooidal ironstone)
    H, L Ashgill: Poiares Member (limestone)
    I, M/U Ashgill: Porto de Santa Anna Formation (limestone)
Phase 5
    J, U Ashgill: Ribeira do Braçal Formation (regressive siltstone)

Table 3. *Comparison at generic level of pre-Hirnantian Ashgill brachiopod faunas*

	Crozon	Buçaco	Celtiberia	Montagne Noire	Sardinia
Aegiromena	X	X		X	X
Aegironetes		X	X		
Bicuspina		X			
Christiania		X			
Dalmanella	X	X	X	X	X
Destombesium	X	X			
Dolerorthis	X	X	X	X	
Drabovia	X	X	X	X	
Eoanastrophia	X			X	X
Epitomyonia		X	X		
Eridorthis	X	X	X	X	
Hedstroemina	X	X	X		
Heterorthina		X			
Iberomena	X	X	X	X	X
Jezercia		X			
Kozlowskites			X	X	
Leangella	X	X	X	X	
Leptaena	X	X	X	X	
Leptestiina	X	X	X	X	
Mcewanella	X				
Nicolella	X	X	X	X	X
Oxoplecia	X	X	X		
Palaeostrophomena			X		
Parastrophinella	X	X			
Plaesiomys			X		
Platystrophia		X			
Porambonites	X	X	X	X	X
Portranella	X	X	X	X	X
Protomendacella				X	
Protozyga		X			
Ptychopleurella	X	X	X		
'Rafinesquina'	X	X	X	X	
Rhactorthis	X	X			
Rostricellula	X	X	X		
Saukrodictya	X	X	X		X
Schizophorella			X		
Skenidioides		X	X		
Strophomena				X	
Tafilaltia				X	
Tissintia				X	
Triplesia	X	X	X	X	X

Crozon (author's unpublished data), Buçaco (author's unpublished data), Celtiberia (Villas Pedruelo 1985), Montagne Noire (Havlíček 1981) and Sardinia (Havlíček 1981).

Sediments of phase 5, attributed (Young 1988) to the effects of glacio-eustatic regression, contain a fauna related to the *Hirnantia* fauna. None of the elements of this fauna occurred in the area prior to this event. The faunas show an upward change from trilobite (*Mucronaspis*) dominated faunas to brachiopod dominated faunas ('*Horderleyella*', *Paromalomena*, *Plectothyrella*).

The fauna of phases 3 and 4 in Portugal is very closely related to that of the Armorican Massif and those of other parts of Iberia. The widespread distribution of lithostratigraphic units is paralleled by a similar lateral continuity of the biofacies. In phase 3 generally similar faunas have been described from various parts of the Mediterranean area, including Morocco (Destombes *et al*. 1985; Havlíček 1971), Saudi Arabia (El Khayal & Romano 1988; Fortey & Morris 1982) and Bohemia (Havlíček & Vanek 1966). Many of the genera in phase 3 are related to forms found in the Arenig of the Mediterranean area, and to those of the Arenig of the Anglo-Welsh area. Recent studies of Anglo-Welsh faunas (e.g. Lockley & Williams 1981) reveal that heterorthid-dominated brachiopod associations were also important there in Middle Ordovician shallow water fine-grained clastic environments.

The influx of fauna in the lower Caradoc is also recorded in Morocco and Bohemia, with many trilobites and brachiopods being conspecific. In the later Caradoc the faunas of Portugal can be seen to be very close to those of Armorica (Babin & Melou 1972), Spain (Villas Pedruelo 1985), Montagne Noire (Dreyfuss 1948; Havlíček 1981), Sardinia (Naud 1979; Havlíček 1981) and Morocco (Havlíček 1971; Romano & Young unpublished studies).

The phase 4 faunas of central Portugal show a close relationship at generic and specific level with other western European faunas (Table 3). In the pre-Hirnantian Ashgill deposits of SW Europe some 42 genera of articulate brachiopods have been reported, of which 23 occur in the contemporary rocks of the Welsh Basin, and a further 10 have been found in Caradoc rocks of the Welsh Basin. Very few, if any, of the congeneric stocks are conspecific with the Anglo-Welsh forms. Some elements of this fauna also occur in Morocco in clastic facies (e.g. *Destombesium*).

The major influxes of new elements into the fauna are striking; 11 new brachiopod genera appear in central Portugal in the early Caradoc and 10 in the early Ashgill, compared with three in the mid–late Caradoc. The periods of influx (the early Llanvirn, the early Caradoc and the early Ashgill), correspond to previously described periods of faunal interchange between the more northerly continents. Introduction of North American genera into the Oslo area (Baltica) has been described for each of these periods by Harper (1986) and early Caradoc changes in the North American and Anglo-Welsh has been described by Williams (1965). It is interesting that the dominant sense of influx is southwards; American genera into Baltica, Baltic genera into Avalonia and Avalonian genera into the Mediterranean Province. This strongly suggests that the northward drift of these continents during the Ordovician enabled the apparent southward dispersal of climatically controlled faunas.

The palaeogeography of Southwest Europe during the Ordovician

The relationship between Iberia and Armorica

The correlation of the facies belts of Brittany with those of central Portugal has been discussed by Hammann & Henry (1978), Henry & Thadeu (1971), Henry 1980, Henry *et al.* (1974, 1976), Paris (1981), Paris & Robardet (1977) and Young (1988). These authors have demonstrated the close similarity of the faunas and facies of the Ordovician of Valongo with the synclines S of Rennes, and of the Buçaco Syncline (central Portugal) with the Crozon Peninsula (W. Brittany) (see Fig. 5). In both Brittany and Portugal estimates may be made of the orientation of the facies belts (N80° E in the Crozon Peninsula and approximately N120° E in central Portugal). If the two areas represent parts of an originally linear facies system, then a relative post-Ordovician rotation of Portugal of 140° anticlockwise is indicated.

Palaeomagnetic evidence for such a rotation is sparse, for there is little reliable Ordovician data from Iberia or Armorica. However, Perroud & Bonhommet (1981) suggest a relative rotation of 80° anticlockwise since the Carboniferous. Previous rotation may already have occurred in the Devonian to account for the other 60°. A rotation of this sort may have involved movement on the South Armorican Shear Zone and on the Cordoba/Tomar Shear Zone. Figure 6 shows the postulated configuration of the central Portuguese and Armorican sectors of the Centro-Iberian/Armorican region during the Ordovician (see also Paris & Robardet (1977) fig. 3 and Hammann & Henry (1978) textfig. 7).

As the facies belts are aligned very close to the structural trend in the Centro-Iberian/Armorican region and in the Bohemian Massif (Havlicek 1982), it would seem likely that the facies belts were aligned broadly parallel to the axis of the Centro-Iberian/Armorican to Bohemian crustal block.

The relationship between the Centro-Iberian/Armorican Region and the Catalan Region

The two major facies developments (Centro-Iberian/Armorican, Catalan) of the SW European Ordovician described above have many features in common. The close relationship of the Ashgill shelly fauna between the two areas is particularly striking. The two areas were part of the same faunal province throughout the Ordovician. Some areas bear lithological successions which suggest a gradual transition between the two: in the northern part of the Centro-Iberian/Armorican region in Spain there may be thick Cambrian deposits below the Ordovician, in Celtiberia the Caradoc is reduced to a very thin sandstone between the Middle Ordovician mudrocks and the Ashgill limestones, while in the western Pyrenees a Lower Ordovician sandstone unit resembling the Armorican Quartzite facies has been described (Klaar and Palacios, quoted in Hammann 1976). There seems no reason to doubt that the two regions represent different facies developments on parts of the same crustal block, the West European Platform. Their separate tectonic evolutions probably reflect differing base-

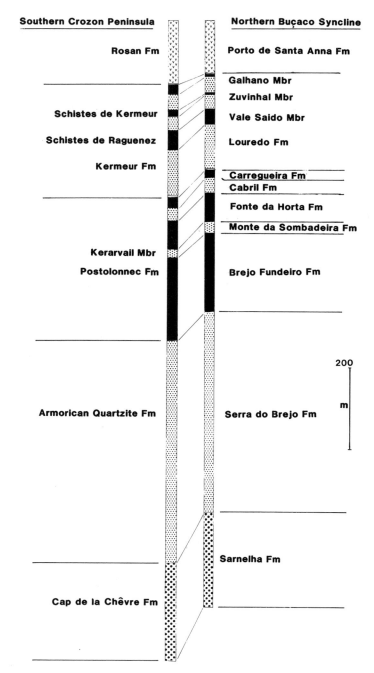

Fig. 5. Comparison of the Ordovician succession of the Crozon Peninsula (W. Brittany) with that of the Buçaco syncline (central Portugal). For key to ornament see Fig. 4.

ment, and the various tectonic settings of different parts of the block.

The relationship between the West European Platform and North Africa

The facies developments of the West European Platform cannot be exactly matched with any Ordovician sequences exposed in North Africa. However, the sequences are similar in many respects to those of the Moroccan Anti-Atlas, particular the western end of the Anti-Atlas where an upper Arenig 'Armorican Quartzite'-like sandstone is developed (the 'Zini sandstone'). Bryozoan rich horizons have been described (Destombes *et al.* 1985) from the Anti-Atlas in deposits of a similar age to the 'Bryozoa Beds' of the Queixoperra Member of central Portugal (Young 1988), and

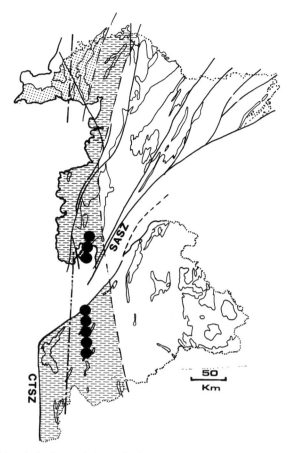

Fig. 6. Illustration of the continuity of depositional facies between western Brittany and Portugal. The distance separating the two blocks is purely schematic. The strike skip displacement on the lesser shear zones within each block has not been restored.

Legend:
- Llandeilo in mudstone facies
- Llandeilo in mudstone facies with sandstone packets
- Llandeilo in sandstone facies
- Line of maximum condensation of Ordovician
- Rosan/Porto de Santa Anna volcanics

bryozoan limestones have been described from the lower Ashgill (Destombes et al. 1985, with revised dating by Elaouad-Debbaj 1984). The most likely position for the part of the West European Platform which abutted the North African area was, therefore, near the western Anti-Atlas.

The extent of the West European Platform

The margins of the West European Platform are largely obscure. The present northwestern limit corresponds to the Rhenohercynian Zone of the Variscides, which separates it from the 'Avalonian' sequences of the southern UK and Belgium. The most external West European Platform sediments farther west lie beneath the English Channel, shown by the clasts with Ordovician Mediterranean Province faunas in the Triassic Budleigh Salterton conglomerates (Cocks & Lockley 1981), and by the blocks in the Roseland Breccia Formation within the Carrick Nappe of south Cornwall (Bassett 1981; Holder & Leveridge 1986; Sadler 1974).

The southeastern limit of the platform is obscured by the Alpine belt, but the acid volcanic rocks of pre-late Caradoc age of the Catalan region (see above) may possibly reflect proximity to a margin.

The northeast termination of the platform lay NE of the Bohemian Massif, where the platform now abuts the southwestern margin of Baltica, at or near the Elbe/Dobrogea line.

The southwestern margin is even more obscure, but there seems little reason to doubt that the platform directly adjoined what is now North Africa, where the Ordovician sequences and faunas have much in common with those of the Ibero-Armorican area at the western end of the platform.

Palaeogeography

Various factors can help constrain the palaeogeographic reconstruction of the West European Platform.

(a) The close faunal similarity between the West European Platform and Morocco suggests that they were associated throughout the Ordovician.

(b) No true margins to the platform have been identified, even in the relatively well-known belt of the Rhenohercynian Zone where the platform now abuts parts of Eastern Avalonia.

(c) During periods of eustatic sea-level rise significant faunal interchange was possible between the platform and Avalonia. Direct evidence of faunal interchange between Baltica and the platform is less certain.

(d) There is some faunal variability within the platform. The faunal differences between Bohemia and Armorica/Iberia (e.g. the absence of *Neseuretus* and *Crozonaspis* from apparently suitable facies in Bohemia; brachiopod faunas rarely being conspecific) suggests some geographical distance between these areas. The common occurrence of some animals (e.g. *Neseuretus* and *Crozonaspis*) in North African, the Iberian Peninsula and the Armorican Massif, as well as in other Gondwanan areas such as Saudi Arabia (El-Khayal & Romano 1988), but not in Bohemia, suggests that this geographical separation may have been a latitudinal difference.

(e) The Centro-Iberian/Armorican region shows little sign of tectonic activity during the Ordovician. The low topography shelf sediments show almost no sign of proximity to a shoreline. Individual sandbodies can be of enormous extent, and may have been sourced from outside the immediate area. In North Africa there are large Cambro-Ordovician marginal marine/non-marine sandstone deposits. This enormous output of sediment from the northern part of Africa could have been the sediment source for the adjacent areas of the West European Platform. If this is so, then it implies that there was no intervening major sedimentary basin to act as a sediment trap.

(f) The limited palaeomagnetic data from the West European Platform (Perroud & Bonhemmet 1981) shows palaeomagnetic declinations interpreted as being of Ordovician age lying close to, but slightly anticlockwise from the structural trend. The depositional strike in these areas is very close to the structural trend, but at Buçaco at least is slightly anticlockwise from it. This suggests that the depositional strike lay close to lines of palaeolongitude. This provides tentative evidence that the facies belts seen in Portugal originally lay at a high angle to the inferred margin of N Africa (Figure 7).

The evidence indicates that the West European Platform was part of Ordovician Gondwana. The Centro-Iberian/Armorican Region formed the core of the platform and lay close to Morocco. The platform extended away from N Africa and geometrical constraints would indicate that the northern tip of the platform (Bohemia) was probably at 20°, and possibly as much as 30°, lower latitude than the southernmost part (S central Spain).

The fragments of the West European Platform seen in the Catalan region probably lay to the east of the Centro-Iberian/Armorican region, and may in part have formed an eastern margin to the platform.

Bands of differential subsidence (e.g. the development of

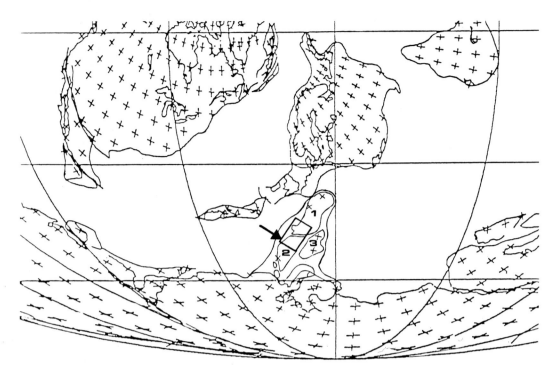

Fig. 7. Tentative reconstruction of the palaeogeography and position of the Central European Platform, illustrated on an ammended Scotese (1986) reconstruction for the Ashgill. The relationship between the Central European Platform and Avalonia/Baltica is not known. Arrowed box indicates area illustrated in Fig. 6. 1, Armorican Massif/Central Europe; 2, Central and northern Iberia; 3, Southern France.

'basins' and 'rises' in central Portugal and Brittany (Young 1988) and the Prague Basin (Havlíček 1982)) were parallel to the axis of the platform, and were probably largely inherited from structures in the Cadomian basement.

The post-Ordovician history of the West European Platform

The Silurian and Devonian sequences of SW Europe show that the platform continued as a relatively stable entity until the Late Devonian. During the Devonian the distribution of the Rhenic clastic- and Hercynian carbonate-dominated facies closely parallel those of the Ordovician Centro-Iberian/Armorican and Catalan developments respectively. In the latest Devonian and earliest Carboniferous the Breton Phase (Guillocheau & Rolet 1982; Rolet 1982) involved disruption of the West European Platform, in the Armorican sector at least, by dextral strike-slip movements and the formation of rapidly subsiding pull-apart basins (e.g. the Châteaulin Basin.) Folding and metamorphism of pre-late Famennian strata accompanied these events. This dextral movement may have caused the separation of the W Britanny and Portuguese sectors of the platform. A second phase of deformation, the Erzgebirge Phase, during the Westphalian, produced further metamorphism, folding and strike-slip faulting. This second phase of deformation has been interpreted (Burg et al. 1981) as demonstrating 'intender' tectonics, with the tightening of the Ibero-Armorican arc around the already displaced northern part of Iberia.

Implications of the model

The widespread distribution of lithostratigraphic units, the low subsidence rates, and the nature of the sediments accumulated in the region all suggest most strongly that the West European Platform behaved as a stable cratonic block during the Ordovician. The sedimentary record would argue against any major tectonic divisions of the region, or any major tectonically active margins to the block within the Ordovician sequences recorded.

The reconstruction of the block as a "promontory" aligned approximately N/S away from the margin of Gondwana allows the most distal portion of the block (Bohemia) to lie at significantly lower latitude than the part closest to N. Africa. It is proposed that this explains the significant faunal differences between Bohemia and other parts of the block, and the reduced significance of the Hirnantian glaciomarine deposits in Bohemia compared to other areas.

The N/S alignment also allows a major reduction in the size of any putative Rheic Ocean. Instead, the West European Platform and Avalonia might have been in relatively close proximity, even if they were not directly in contact across the present Rhenohercynian/Saxothuringian boundary. If the evidence suggesting a high latitude for the Anglo-Welsh area in the Early Ordovician, but of a relatively low latitude by the Late Ordovician (e.g. Soper et al. 1987) is accepted, then it is possible for this movement to have taken place on a zone of lateral displacement along the promontory formed by the West European Platform rather than by the rapid opening of a 'Rheic Ocean' during the Ordovician, for which there is little geological evidence.

The repeated influxes of brachiopod stocks from outside the region suggest that when appropriate conditions were available brachiopods of Baltic or Anglo-Welsh affinites could colonize SW Europe. The promontory form of the block allowed a path for faunal migration between Avalonia and Gondwana. Latitudinal (climatic) constraints are seen as the barriers to faunal dispersion rather than geographical isolation.

Previous tectonic models involving tectonic and/or palaeogeographic separation within SW Europe are militated against by the demonstration that Iberia, Armorica and other blocks within the western European Variscides were part of single crustal block during the Ordovician. Such models include those of Ziegler (1984, 1986), proposing sequential accretion of Armorica and Iberia to Laurasia during the Lower and Middle Palaeozoic, of Matte (1986a, b), who placed a major Lower Palaeozoic ocean along his Galician/Massif Central suture, and of Lefort & Ribeiro (1980), who interpreted the blueschists of S Britanny as a product of closure of a Lower Palaeozoic ocean between northern Iberia and Armorica. The high pressure metamorphic belt of southern Britanny with its blueschists must either be post-Ordovician, or have been subsequently tectonically emplaced between two regions (central Britanny, northern Iberia) originally part of the West European Platform.

The disruption of the block is earlier than the major mid-Carboniferous compression, and probably relates to the Late Devonian/Early Carboniferous strike-slip orogenic event, the Breton Phase. The 'indenter' model of Burg et al. (1981) for the Erzgebirge Phase of deformation can only hold if the central Portuguese and western Breton sectors of the West European Platform have already been separated by dextral strike slip movements in the Breton Phase. This would imply an earlier dextral sense of displacement on the Cordoba/Tomar Shear Zone, prior to the observed sinistral displacement produced by the 'indenter' tectonics.

References

ARTHAUD, F. 1970. Étude tectonique et microtectonique comparée de deux domaines hercyniens: les nappes de la Montagne Noire (France) et l'anticlinorium de l'Eglesiente (Sardaigne). *Publications de la Université de Science et Technologie du Languedoc, Géologie*, **1**.

BABIN, C. & MELOU, M. 1972. Mollusques bivalves et brachiopodes des 'Schistes de Raguenez' (Ordovicien supérieur du Finistère) conséquences stratigraphiques et paléobiogéographiques. *Annales de la Société géologique du Nord*, **92**, 79–94, pls. 7–10.

——, COCKS, L. R. M. & WALLISER, O. H. 1980. Faciès, faunes et paléogéographie antécarbonifère de l'Europe. *In*: COGNÉ, J. & SLANSKY, M. (eds) *Géologie de l'Europe du Précambrien aux bassins sédimentaires post-hercyniens*. Mémoire du Bureau de Recherches Géologiques et Minières, **108**, 191–202.

BARD, J. P., BURG, J. P., MATTE, P. & RIBEIRO, A. 1980. La Chaîne hercynienne d'Europe occidentale en termes de tectonique des plaques. *In*: COGNÉ, J. & SLANSKY, M. (eds) *Géologie de l'Europe du Précambrien aux bassins sédimentaires post-hercyniens*. Mémoire du Bureau de Recherches Géologiques et Minières, **108**, 233–246.

BASSETT, M. G. 1981. The Ordovician brachiopods of Cornwall. *Geological Magazine*, **118**, 647–664.

BONJOUR, J.-L. & CHAUVEL, J.-J. 1988. Un exemple de sédimentation initiale dans un bassin paléozoique: étude pétrographique et géochimique de l'Ordovicien inférieur de la presqu'île de Crozon (Finistère). *Bulletin de la Société géologique de France*, (8), **4**, 81–91.

BRENCHLEY, P. J., ROMANO, M. & GUTIERREZ-MARCO, J. C. 1986. Proximal and distal hummocky cross-stratified facies on a wide Ordovician shelf in Iberia. *In*: KNIGHT, R. J. & MACLEAN, S. R. (eds) *Shelf Sands and Sandstones*. Canadian Society of Petroleum Geologists, Memoir, **11**, 241–255.

——, ——, YOUNG, T. P. & STORCH, P. in press. The glaciomarine sediments of the Upper Ordovician in Europe. The Ordovician Symposium, 1988.

BURG, J.-P., IGLESIAS, M., LAURENT, P., MATTE, P. & RIBEIRO, A. 1981. Variscan intracontinental deformation: The Coimbra-Cordoba shear zone (SW Iberian Peninsula). *Tectonophysics*, **78**, 161–177.

COCKS, L. R. M. & FORTEY, R. A. 1982. Faunal evidence for oceanic separations in the Palaeozoic of Britain. *Journal of the Geological Society, London*, **139**, 465–478.

—— & —— 1988. Lower Palaeozoic facies and faunas around Gondwana. *In*: AUDLEY-CHARLES, M. G. & HALLAM, A. (eds) *Gondwana and Tethys*, Geological Society of London, Special Publication, **37**, 183–200.

—— & LOCKLEY, M. G. 1981. Reassessment of the Ordovician brachiopods from the Budleigh Salterton Pebble Bed. Devon. *Bulletin of the British Museum (Natural History), Geology*, **35**, 111–124.

—— & RONG, J.-Y. 1988. A review of the late Ordovician *Foliomena* brachiopod fauna with new data from China, Wales and Poland. *Palaeontology*, **31**, 53–67.

DESTOMBES, J., HOLLARD, H. & WILLEFERT, S. 1985. Lower Palaeozoic rocks of Morocco. *In*: HOLLAND, C. H. (ed.) *Lower Palaeozoic Rocks of the world*, **4**, 91–336, J. Wiley & Sons.

DREYFUSS, M. 1948. Contribution à l'étude géologique et paléontologique de l'Ordovicien Supérieur de la Montagne Noire. *Mémoire de la Société géologique de France*, **58**.

ELAOUAD-DEBBAJ, Z. 1984. Chitinozoaires Ashgilliens de l'Anti-Atlas (Maroc). *Geobios*, **17**, 45–68.

EL-KHAYAL, A. A. & ROMANO, M. 1988. A revision of the upper part of the Saq Formation and Hanadir Shale (lower Ordovician) of Saudi Arabia. *Geological Magazine*, **125**, 161–174.

FORTEY, R. A. & COCKS, L. R. M. 1986. Marginal faunal belts and their structural implications, with examples from the Lower Palaeozoic. *Journal of the Geological Society, London*, **143**, 151–160.

—— & MORRIS, S. F. 1982. The Ordovician trilobite *Neseuretus* from Saudi Arabia, and the palaeogeography of the *Neseuretus* fauna related to Gondwana in the earlier Ordovician. *Bulletin of the British Museum (Natural History), Geology*, **36**, 63–75.

GREILING, L. G. 1967. Der Thüringische Lederschiefer. *Geologica et Palaeontologica*, **1**, 3–11.

GUILLOCHEAU, F. & ROLET, J. 1982. La sédimentation paléozoïc ouest-armoricaine. Histoire sédimentaire; relations tectonique-sédimentation. *Bulletin de la Société géologique et minéralogique de Bretagne*, **14**, 45–62.

GUTIERREZ-MARCO, J. C., RABANO, I. & ROBARDET, M. 1984. Estudio bioestratigráfico del Ordovícico en el sinclinal del Valle (Provincia de Sevilla, SO de Espana). *Memórias e Noticias, Coimbra*, **97**, 11–37.

HAMMANN, W. 1976. The Ordovician of the Iberian Peninsula – A review. *In*: BASSETT, M. G. (ed.) *The Ordovician System: Proceedings of a Palaeontological Association symposium, Birmingham, September 1974*. University of Wales Press and National Museum of Wales, 387–409.

—— & HENRY, J.-L. 1978. Quelques espèces de *Calymenella, Eohomalonotus* et *Kerfornella* (Trilobita, Ptychopariida) de l'Ordovicien du Massif Armoricain et de la Péninsula Ibérique *Senckenbergiana lethaea*, **59**, 401–429.

HARPER, D. T. 1986. Distributional trends within the Ordovician brachiopod faunas of the Oslo Region, South Norway. *In*: RACHEBOEUF, P. R. & EMIG, C. C. (eds) *Les brachiopodes fossiles et actuels, Biostratigraphie du Paleozoique*, **4**, 465–475.

HAVLÍČEK, V. 1971. Brachiopodes de l'Ordovicien du Maroc. *Notes et Mémoires du Service Géologique du Maroc*, **230**.

—— 1976. Evolution of Ordovician brachiopod communities in the Mediterranean Province. *In*: BASSETT, M. G. (ed.) *The Ordovician System: proceedings of a Palaeontological Association symposium, Birmingham, September 1974*, University of Wales Press and National Museum of Wales, 349–358.

—— 1981. Upper Ordovician Brachiopods from the Montagne Noire. *Palaeontographica*, A, **176**, 1–34, pls. 1–9.

—— 1982. Ordovician in Bohemia: development of the Prague Basin and its benthic communities. *Sborník geologických Věd, geologie*, **37**.

—— & MERGL, M. 1982. Deep water shelly fauna in the latest Kralodvorian (Upper Ordovician, Bohemia). *Vestník Ustredniho ustavu geologickeho*, **57**, 37–46.

—— & VANEK, J. 1966. The biostratigraphy of the Ordovician of Bohemia. *Sborník geolických Věd, paleontologie*, **8**, 7–69.

HENRY, J.-L. 1980. Trilobites ordoviciens du Massif Armoricain. *Mémoires de la Société géologique et minéralogique de Bretagne*, **22**.

——, & THADEU, D. 1971. Interêt stratigraphique et paléogéographique d'un microplancton à Acritarches découvert dans l'Ordovicien de la Serra de Buçaco (Portugal). *Comptes Rendus de l'Academie des Sciences, Paris*, **272**, 1343–1346.

——, MELOU, M., NION, J., PARIS, F., ROBARDET, M., SKEVINGTON, D. & THADEU, D. 1976. L'apport de Graptolites de la zone à *G. teretiusculus* dans la datation de faunes benthiques lusitano-armoricaines. *Annales de la Société géologique du Nord*, **96**, 275–281, 3 figs.

——, NION, J., PARIS, F. & THADEU, D. 1974. Chitinozoaires, Ostracodes et Trilobites de l'Ordovicien du Portugal (Serra de Buçaco) et du Massif Armoricain: essai de comparaison et signification paléogéographique. *Communicações dos Serviços geológicos de Portugal*, **57**, 303–345, pls 1–10.

HOLDER, M. T. & LEVERIDGE, B. E. 1986. A model for the tectonic evolution of south Cornwall. *Journal of the Geological Society of London*, **143**, 125–134.

JULIVERT, M., MARTINEZ, F. J. & RIBEIRO, A. 1980. The Iberian segment of the European Hercynian foldbelt. *In*: COGNÉ, J. & SLANSKY, M. (eds) *Géologie de l'Europe du Précambrien aux bassins sédimentaires post-hercyniens*. Mémoire du Bureau de Recherches Géologiques et Minières, **108**, 132–158.

LEFORT, J.-P. & RIBEIRO, A. 1980. La faille Porto-Badajoz-Cordoue a-t-elle controlé l'évolution de l'ocean paléozoique sud-armoricain? *Bulletin de la Société géologique de France*, (7), 22, 455–462.

LOCKLEY, M. G. 1983. A review of brachiopod dominated palaeocommunities from the type Ordovician. *Palaeontology*, **26**, 111–145.

—— & WILLIAMS, A. 1981. Lower Ordovician Brachiopoda from mid and southwest Wales. *Bulletin of the British Museum (Natural History) Geology*, **35**, 1, 1–78.

MATTE, P. 1986a. La chaîne varisque parmi les chaînes paléozoïques péri atlantiques, modèle d'évolution et position des grands blocs continentaux au Permo-Carbonifère. *Bulletin de la Société géologique de France*, (8), 2, 9–24.

—— 1986b. Tectonics and plate tectonics model for the Variscan belt of Europe. *Tectonophysics*, **126**, 329–374.

McDOUGALL, N., BRENCHLEY, P. J., REBELO, J. A. & ROMANO, M. 1987. Fans and fan deltas – precursors to the Armorican Quartzite (Ordovician) in western Iberia. *Geological Magazine*, **124**, 347–359.

MITCHELL, W. I. 1974. An outline of the stratigraphy and palaeontology of the Ordovician rocks of central Portugal. *Geological Magazine*, **111**, 385–396, 1pl.

NAUD, G. 1979. Les shales de Rio Canoni, formation-repère fossilfère dans l'Ordovicien supérieur de Sardaigne orientale. Conséquences stratigraphiques et structurales. *Bulletin de la Société géologique de France*, 21, 155–159.

PARIS, F. 1979. Les chitinozoaires de la Formation de Louredo, Ordovicien Supérieur du synclinal de Buçaco (Portugal). *Palaeontographica A*, **164**, 24–51.

—— 1981. Les Chitinozoaires dans le Paléozoique du sud-ouest de l'Europe. *Mémoire de la Société géologique et minéralogique de Bretagne*, **26**.

—— & ROBARDET, M. 1977. Paléogéographie et relations ibéro-armoricaines au Paléozoique anté-Carbonifère. *Bulletin de la Société géologique de France*, **19**, 1121–1126.

PERROUD, H. & BONHOMMET, N. 1981. Palaeomagnetism of the Ibero-Armorican arc and the Hercynian orogeny in Western Europe. *Nature*, **292**, 445–448.

ROBARDET, M. & DORÉ, F. 1988. The late Ordovician diamictic formations from southwestern Europe: north-Gondwana glaciomarine deposits. *Palaeogeography, Palaeoclimatology, Palaeoecology*, **66**, 19–31.

ROLET, J. 1982. La "Phase Bretonne" en Bretagne: état des connaissances. *Bulletin de la Société géologique et minéralogique de Bretagne*, **14**, 63–72.

ROMANO, M. 1982. The Ordovician biostratigraphy of Portugal – A review with new data and re-appraisal. *Geological Journal*, **17**, 89–110.

—— & DIGGENS, J. 1976. The stratigraphy and structure of the Ordovician and associated rocks around Valongo, north Potugal. *Communicaç̃oes dos Servicos Geológicos de Portugal*, **57**, 22–50, pls. 1–2.

SADLER, P. M. 1974. Trilobites from the Gorran Quartzites, Ordovician of south Cornwall. *Palaeontology*, **17**, 71–93.

SCHONLAUB, H. P. 1971. Palaeo-environmental studies at the Ordovician/Silurian boundary in the Carnic Alps. *In: Colloque Ordovicien-Silurien, Brest, Sept. 1971*. Mémoire du Bureau de Recherches géologiques et minières, **73**, 379–390.

SCOTESE, C. R. 1986. *Phanerozoic Reconstructions: A new look at the assembly of Asia*. University of Texas Institute for Geophysics Technical Report, No. 66.

SDZUY, K. 1971. The Ordovician in Bavaria. *In: Colloque Ordovicien-Silurien, Brest, Sept. 1971*. Mémoire du Bureau de Recherches géologiques et minières, **73**, 379–390.

SMITH, A. G., HURLEY, A. M. & BRIDEN, J. C. 1981. *Phanerozoic paleocontinental world maps*, Cambridge Earth Science Series, Cambridge Univerity Press.

SOPER, N. J., WEBB, B. C. & WOODCOCK, N. H. 1987. Late Caledonian (Acadian) transpression in north-west England: timing, geometry and geotectonic significance. *Proceedings of the Yorkshire Geological Society*, **46**, 175–192.

SPJELDNAES, N. 1961. Ordovician climatic zones. *Norsk Geologisk Tidsskrift*, **41**, 45–77.

VILLAS PEDRUELO, E. 1985. Braquiopodos del Ordovicico Medio y Superior de las Cadenas Ibericas Orientales. *Memorias del Museo Paleontologico de la Universidad de Zaragoza*, **1**.

VILLAS, E., DURÁN, H. & JULIVERT, M. 1987. The upper Ordovician clastic Sequence of the Catalonian Coastal Ranges and its Brachiopod Fauna. *Neues Jahrbuch für Geologie und Paläontologie Abhandlungen*, **174**, 55–74.

WEBBY, B. D. 1984. Ordovician reefs and climate: a review. *In*: BRUTON, D. L. (ed.) *Aspects of the Ordovician System, Palaeontological Contributions from the Univerity of Oslo*, **295**, 89–100.

WHITTINGTON, H. B. & HUGHES, C. P. 1972. Ordovician geography and faunal provinces deduced from trilobite distribution. *Philosophical Transactions of the Royal Society, Series B*, **263**, 235–278.

—— & —— 1973. Ordovician trilobite distribution and geography. *Special Papers in Palaeontology*, **12**, 235–240.

WILLIAMS, A. 1965. Stratigraphic distribution. *In*: MOORE, R. C. (ed.) *Treatise on Invertebrate Palaeontology, H. Brachiopoda*, H237–250.

YOUNG, T. P. 1985. *The stratigraphy of the upper Ordovician of central Portugal*. PhD thesis, University of Sheffield.

—— 1988. The lithostratigraphy of the upper Ordovician of central Portugal. *Journal of the Geological Society of London*, **145**, 377–392.

—— 1989. Eustatically controlled ooidal ironstone deposition: facies relationships of the Ordovician open-shelf ironstones of western Europe. *In*: YOUNG, T. P. & TAYLOR, W. E. G. (eds) *Phanerozoic Ironstones*, Geological Society of London, Special Publication, **46**, 51–63.

ZIEGLER P. A. 1984. Caledonian and Hercynian crustal consolidation of Western and Central Europe – A working hypothesis. *Geologie en Mijnbouw*, **63**, 93–108.

—— 1986. Geodynamic model for the Palaeozoic crustal consolidation of western and central Europe. *Tectonophysics*, **126**, 303–328.

Devonian palaeogeography and palaeobiogeography of the central Andes

P. E. ISAACSON & P. E. SABLOCK

Department of Geology, University of Idaho, Moscow, Idaho 83843, USA

Abstract: During Devonian time, western Gondwana (northern Chile, Bolivia, and Peru) witnessed a period of thick clastic rock deposition. The sialic Arequipa Massif, apparently a stable terrane since Precambrian time, supplied the bulk of the Bolivian intracratonic basinal sediments. Late Ordovician orogeny in northern Chile–northwestern Argentina included the emplacement of the Puna magmatic arc, with attendant plutonism. The resulting highland supplied coarse-grained sediments to northern Chile. Late Devonian(?) orogenesis deformed Early to Middle Devonian rocks in northern to central Peru. It is suggested that mid-Palaeozoic convergence was 'buffered' by the Arequipa Massif, giving little evidence of orogeny in the Bolivian Devonian–Carboniferous sequence. A highly endemic (Malvinokaffric) shelly fauna, indicative of high palaeolatitudes, occupied the region in Early Devonian (Emsian) time, while allochthonous (Eastern Americas Realm) faunas entered southern Peru. During Middle Devonian time a much lower diversity 'post-Malvinokaffric' fauna (consisting of the brachiopods *Tropidoleptus*, *Globithyris*, and other taxa) arrived, although a Malvinokaffric stock remained. These taxa arrived via the Amazonas and other eastern basins during a rotation of the region to lower latitudes and presumably warmer waters in Middle Devonian time.

Several recent syntheses of Devonian stratigraphy, biostratigraphy, tectonics, and igneous events in Bolivia, Peru, and northern Chile (Fig. 1) permit a tentative reconstruction of mid-Palaeozoic events in that region. Earlier workers suggested (e.g. Ahlfeld & Branisa 1960) that the thick Devonian clastic sediments in the region were deposited in a 'geosynclinal' setting. Other work has supported a passive margin during Devonian time (Zeil 1979), with Late Ordovician and Late Carboniferous–Permian orogeny bracketing the otherwise quiescent sequence.

Devonian stratigraphic data

Basic Bolivian Devonian stratigraphic data essential for palaeogeographic reconstructions are rather well-established (Isaacson 1975, 1977). A thick clastic sequence, the best studied of which is in the Bolivian intracratonic basin, is primarily late Early (Emsian) and early Middle (Eifelian) Devonian in age. Timing of the upper part of the sequence remains tentative; many authors suggest that late Middle Devonian (Givetian) and early Late Devonian (Frasnian) strata exist. Frasnian units are present in the Subandean and subsurface regions of eastern Bolivia and Peru (Isaacson & Sablock 1988; Barrett & Isaacson 1988). Primary lithologies in the northwestern and central Bolivian sequence are, in descending rank abundance, siltstone, quartz arenite, mudstone, and shale. Feldspathic sandstones are found in southern Bolivia. Lithofacies and isopach modelling of the Devonian points to coarser and much thicker lithologies in northwestern Bolivia than those found in the east and south (Isaacson 1975).

In northern Chile (Atacama Desert region) Middle Devonian and Early Carboniferous shallow water conglomerates and litharenites contain andesite and diorite clasts (Davidson *et al.* 1981; Isaacson *et al.* 1985; Isaacson & Sablock 1988). Also, Late(?) Devonian turbidites in northern Chile show a northern to northwestern (Arequipa Massif) source (Bahlburg 1985; Breitkreuz & Bahlburg 1985; Bahlburg *et al.* 1986).

Source areas (Figs 2 & 3) for the Bolivian sequence are an ancestral Arequipa Massif (Isaacson 1975; Boucot *et al.* 1980; Laubacher *et al.* 1982), the Puna Highland of northwest Argentina, and probably the Brazilian Shield (Isaacson & Sablock 1988). These areas supplied sediments to northwestern and central, southern, and eastern Bolivia, respectively. Principal compositions of the strata appear to reflect compositional differences of these source terranes. The Arequipa Massif was largely sialic (Bellido & Guevara 1963; Cobbing *et al.* 1977; Dalmayrac *et al.* 1980) and may have been part of Miller's (1970) 'Pacific Continent'. The Puna Highland is composed of Late Ordovician age magmatic igneous rocks (Davidson *et al.* 1981; Coira *et al.* 1982; Palma *et al.* 1987). It also influenced Atacama (Chilean) depositional events, including westward-propagating, shallow-water, clastic deposits with a nearby magmatic terrane source (Davidson *et al.* 1981; Isaacson *et al.* 1985). The Brazilian Shield is compositionally similar to the Arequipa Massif (Dalmayrac *et al.* 1980).

Orogeny folded and metamorphosed Devonian and early Palaeozoic flysch in central Peru, although these effects are not evident in the Bolivian and northern Chilean Devonian sequences. Also, part of the Devonian sequence may have been eroded in central Peru, and Early Carboniferous (Tournaisian) clastic rocks unconformably overlie it (Megard 1973). Typical of many orogenic zones, evidence of orogeny decreases away (eastward and southeastward) from central Peru, such that only a slight angularity exists between Devonian and Carboniferous rocks in Bolivia (Laubacher 1974, 1977; Isaacson 1975).

Palaeogeography

Current mid-Palaeozoic (Devonian and Early Carboniferous) palaeogeographic reconstructions of the western continental margin of Gondwana in Peru and Chile (Figs 2 & 3), demonstrate that sedimentation on this margin was influenced by at least two different tectonic terranes. The first, the Arequipa Massif, supplied a large portion of clastic sediments in the Peruvian and Bolivian intracratonic basins (Isaacson 1975; Zeil 1979; Dalmayrac *et al.* 1980), as well as part of the northern Chilean 'forearc basin' (Bahlburg *et al.* 1986; Isaacson & Sablock 1988). The second, which is a western Argentine magmatic arc terrane of Late Ordovician age (Coira *et al.* 1982; Forsythe 1982; Mpodozis & Forsythe 1983; Palma *et al.* 1987), influenced northern Chilean and Bolivian depositional events.

Preliminary assessment of Devonian and Early Carboniferous palaeogeography of this complex region suggests that the Arequipa Massif extended beyond the present west coast of South America and had a southern boundary in northern Chile (Isaacson 1975; Isaacson & Sablock 1988). Between northern Chile and central Peru is an intracratonic basin, in which epeirogeny or amagmatic marginal tectonics appear to have been the dominant factor in regional sea level fluctuations and attitudinal differences between Devonian and Carboniferous strata.

During Late Devonian (Fig. 4) and Early Carboniferous time the southern Peruvian sub-basin was uplifted as a result of compression between the Arequipa landmass and the Brazilian Shield. Interestingly, the fold axes in Peru and indicate that the compressional forces were acting in a SW–NE direction (Megard 1973; Dalmayrac *et al.* 1980). The Chanaral melange of northern Chile indicates the same sense of motion (Bell 1987). Some

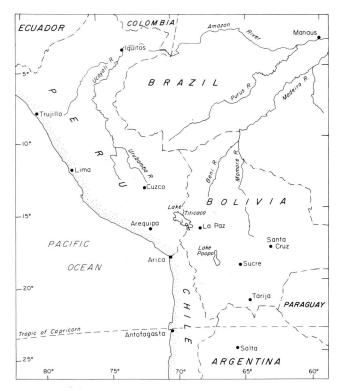

Fig. 1. Present geographic base of central Andes mountains and adjacent regions, discussed herein.

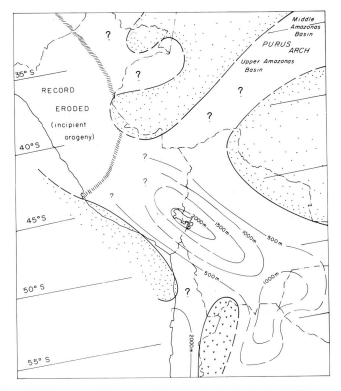

Fig. 3. Palaeogeographic map of the central Andes, during Middle Devonian (late Eifelian–Givetian) time. Maximum transgression, subsidence, and deposition within the Central Andean intracratonic basin is at this time.

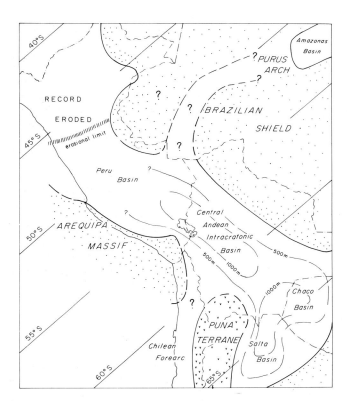

Fig. 2. Palaeogeographic map of central Andes, during late Early Devonian (Emsian) time. Major structural elements responsible for the Central Andean intracratonic basin include the Late Ordovician accretionary volcanic-magmatic Puna Terrane, sialic Arequipa Massif, and Brazilian Shield. Chilean forearc is inactive during Devonian time. Suggested palaeolatitudes from C. Scotese.

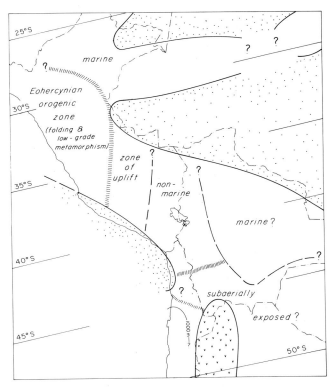

Fig. 4. Palaeogeographic map of the central Andes, during Late Devonian (Frasnian) time. Increased orogeny in northern Peru creates the Devonian–Early Carboniferous unconformity. Adjacent to deformed strata are non-marine sediments. Apparently, marine strata exist in the northern Bolivia–eastern Peru subsurface.

evidence indicates that subduction began in the north (Peru) and propagated south (Bahlburg *et al.* 1986). In central Chile this new convergence emplaced various island arcs and micro continents (Bell 1987; Ramos *et al.* 1986).

Palaeogeographically, these events would restrict formerly 'open' seaways along eastern Peru, western Bolivia, and northern Argentina/Chile. Therefore, possible Late Devonian or Early Carboniferous glaciation along the upper Amazon basin (Rocha Campos 1983) and the Peninsula de Copacabana, Lake Titicaca (Isaacson 1975) may have occurred on substantially uplifted orogenic terranes in northern and western Peru. There were marine seaways adjacent to these features, with apparent black shale deposition in the Upper Amazon region (Hunicken *et al.* 1987).

Palaeobiogeography

Palaeobiogeography of the region is influenced by three major factors, the palaeogeographic setting discussed above, the high latitudinal position of the region during Early Devonian time (Fig. 5), with apparent influx of slightly warmer waters during Middle Devonian time (Fig. 6). The highly endemic Malvinokaffric faunas (Richter & Richter 1942) of the southern South American continent have long been assumed to have lived in cold water, with evolutionary ties to Eastern Americas Realm organisms of the northern Appalachian Basin, USA (Boucot 1971; Isaacson 1977). We suggest, moreover, that Early Devonian Eastern Americas Realm brachiopods in interior southern Peru (Boucot *et al.* 1980) entered the region by means of a warmer current. Higher in the Andean sequence, however, is the much lower diversity fauna, which we tentatively identify as the 'post-Malvinokaffric' fauna. It consists of the circum-'Atlantic'

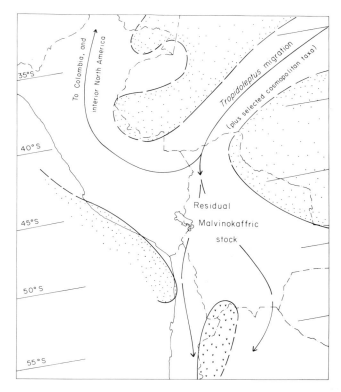

Fig. 6. Palaeobiogeographic map of the central Andes, during Middle Devonian time. Lower latitudes (Gondwana rotation?) and transgression permitted entry of *Tropidoleptus* and probably other taxa from Europe and North Africa. It is suggested that a residual Malvinokaffric stock supported a limited diversity in Bolivia and adjacent regions.

brachiopod genus, *Tropidoleptus* (Isaacson & Perry 1977) and other taxa. It now appears that *Globithyris*, above *Tropidoleptus* in the Devonian sequence (Isaacson 1974), may have its origins in Germany (Boucot 1963) or Libya (Havlicek & Rohrlich 1987). New work on the Bolivian hyoliths, moreover, suggests a probable Bohemian–Malvinokaffric (and post-Malvinokaffric) connection. Pojeta *et al.* (1976) suggest a similar connection between North American, Bolivian, and Bohemian bivalves. This Middle Devonian fauna, however, does not achieve the diversity of 'Hamilton' and other Givetian fauna in New York and North Africa.

Distinct faunal provinciality exhibited during Early to Middle Devonian time was probably influenced by extensive intracratonic basin development within Peru, Chile and Bolivia. The high latitudinal positioning of this region allowed colder water Malvinokaffric fauna to colonize much of South America. The lowest latitudinal penetration of Malvinokaffric faunas roughly corresponds to a postulated cold water sub-polar gyre which paralleled the western margin of South America up to approximately 40°S (Fig. 7).

Faunal Dispersals

We support the suggestion of Harrington (1967, fig. 2), that there was a gradual northerly penetration by Malvinokaffric faunas during Early Devonian time. Colonization of the lower Amazonas and Parnaibas basins by Malvinokaffric faunas would have occurred from the west as the faunas progressed up the western margin and into the interior basins (Fig. 7). Another dispersal route could have been from the Parana (SE Brazil) and Parnaiba basins into the Amazonas Basin, from the east. We have no data by which to evaluate this possibility. The existence of the Purus Arch in the Amazon basin and its lack of Devonian cover could imply that communication between the upper and lower Amazon

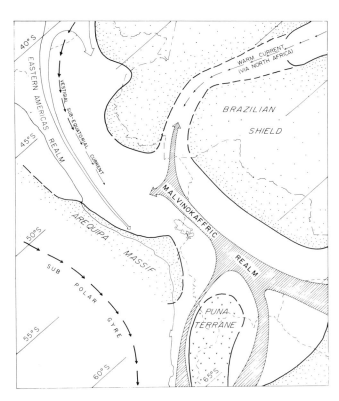

Fig. 5. Palaeobiogeographic map of the central Andes, during Emsian time. High latitude, cold-water currents maintained the low-diversity Malvinokaffric faunas. Vestigial warm-water currents from interior eastern North America and Colombia are responsible for the Eastern Americas Realm faunas of interior southern Peru. Connection(s) to the Amazonas Basin(s) is difficult to determine.

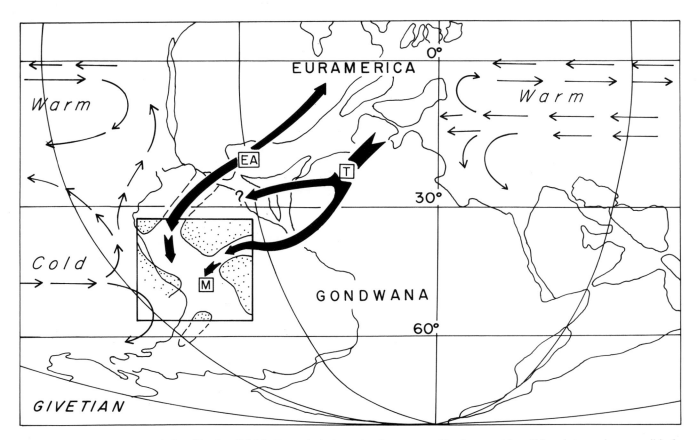

Fig. 7. Gondwana reconstruction during Givetian (Middle Devonian) time, showing western Gondwana setting. Sub-polar gyre is responsible for continued isolation of Malvinokaffric faunas (M), and lower latitude warm currents allow movement of *Tropidoleptus* and Bohemian taxa (T) into South America. Eastern Americas Realm organisms (EA) follow marine connections between North and South America.

basins occurred only sporadically and perhaps only during the Middle or Late Devonian transgression. The northern part of the seaway (into the Amazonas Basin via north-central Peru) was closed by the latest Devonian, when the area underwent uplift in response to the first pulses of the Eohercynian orogeny (Megard 1973).

Eastern Americas Realm faunas are found in South America, to approximately 40°S palaeolatitude (Boucot et al, 1980). This southern penetration into coastal Peru appears to have been established via the interior Peruvian basin. The southwestern Peru brachiopod fauna, with Eastern Americas realm affinities, apparently correlates with the upper Lower Devonian of Colombia.

In our model, the Cocachacra (Peru) Eastern Americas Realm fauna (Boucot et al. 1980) established itself on the western side of the Peruvian interior basin, protected from the immediate effects of the cold water gyre on the distal (western) margin of the emergent and enlarged Arequipa landmass. The high latitudinal positioning of South America in the earliest Devonian allowed emplacement of Malvinokaffric faunas in Bolivia and adjacent areas.

During Middle Devonian time, subsequent northward movement of western Gondwana and marine transgression introduced the warmer water Eastern Americas fauna farther south, into Bolivia. Also, selected brachiopod taxa arrived from North Africa. Isaacson & Perry (1977) suggest that *Tropidoleptus* arrived in Bolivia via the Amazonas Basin. Other taxa (e.g. *Globithyris*) may have taken advantage of this seaway also, although several North African taxa are not found in Bolivia.

The National Geographic Society provided partial support for field data presented herein. The Petroleum Research Fund (American Chemical Society) Grant No. 19464–AC2 provided support for preparation of the manuscript and travel to the symposium. S. F. Barrett offered many constructive comments on an earlier draft, for which we are grateful.

References

AHLFELD, F. & BRANISA, L. 1960. Geologia de Bolivia. La Paz, *Editorial Don Bosco*.

BAHLBURG, H. 1985. Sedimentological aspects of the El Toco Formation (Palaeozoic; Coastal Cordillera) NW of Quillagua, northern Chile. *Actas, Congreso Geologico Chileno No. 4*, **1**, 17–28.

——, BREITKREUZ, C. & ZEIL, W. 1986. Palaozoische Sedimente Nordchiles. *Berliner geowissentschaftliche Abhandlung (A)*, **66**, 147–68.

BARRETT, S. F. & ISAACSON, P. E. 1988. Devonian of South America. *Canadian Society of Petroleum Geologists, Memoir*, **14**, 655–667.

BELL, C. M. 1987. The origin of the upper Palaeozoic Chanaral Melange of Northern Chile. *Journal of the Geological Society of London*, **144**, 599–610.

BELLIDO, E. & GUEVARA, C. 1963. Geologia de los cuadrangulos de Punta de Bombon y Clemesi. Lima, *Comision Carta Geologia Nacional de Peru, Boletin No. 5*.

BOUCOT, A. J. 1963. The globithyrid facies of the Lower Devonian. *Senckenbergiana Lethaia*, **44**, 79–84.

—— 1971. Malvinokaffric Devonian marine community distribution – and implications for Gondwana. *Anais da Academia Brasileira de Ciencias*, **43**, 23–49.

——, ISAACSON, P. E. & LAUBACHER, G. 1980. An Early Devonian, Eastern Americas faunule from the coast of southern Peru. *Journal of Paleontology*, **54**, 359–365.

BREITKREUZ, C. & BAHLBURG, H. 1985. Paleozoic flysch series in the Coastal Cordillera of Northern Chile. *Geologische Rundschau*, **74**, 565–572.

COBBING, E. J., OZARD, J. M. & SNELLING, N. J. 1977. Reconnaissance geochronology of the crystalline basement rocks of the Coastal Cordillera of Peru. *Geological Society of America Bulletin*, **88**, 241–246.

COIRA, B., DAVIDSON, J., MPODOZIS, C. & RAMOS, V. 1982. Tectonic and magmatic evolution of the Andes of northern Argentina and Chile. *Earth Science Reviews*, **18**, 303–332.

DALMAYRAC, B., LAUBACHER, G., MAROCCO, R., MARTINEZ, C. & TOMASI, P. 1980. La chaine hercynienne d'amerique du sud. Structure et evolution d'un orogene intracratonique. *Geologische Rundschau*, **69**, 1–21.

DAVIDSON, J., MPODOZIS, C. & RIVANO, S. 1981. El Paleozoico de Sierra de Almeida, al oeste de Monturaqui, alta cordillera de Antofagasta, Chile. *Revista Geologica de Chile*, **12**, 3–23.

FORSYTHE, R. 1982. The Late Paleozoic to early Mesozoic evolution of southern South America: a plate tectonic interpretation. *Journal of Geological Society of London*, **139**, 671–682.

HARRINGTON, H. J. 1967. Devonian of South America. *In*: OSWALD, D. (ed.) *International Symposium on the Devonian System*. Volume **I**. Alberta Society of Petroleum Geologists, 651–671.

HAVLICEK, V. & ROHLICH, P. 1987. Devonian and Carboniferous brachiopods from the northern flank of the Murzuq Basin (Libya). Prague, *Sbornik Geologickych Ved. Paleontologie*, **28**, 117–177.

HUNICKEN, M. A., LEMOS, V. B. & MELO, J. H. G. 1987. Devonian conodonts from the Upper Amazon Basin, northwestern Brazil. *Second International Symposium on the Devonian System, Calgary Alberta, Program and Abstracts*, 118.

ISAACSON, P. E. 1974. First South American occurrence of *Globithyris*: its ecological and age significance in the Malvinokaffric Realm. *Journal of Paleontology*, **48**, 778–784.

—— 1975. Evidence for a western extracontinental land source during the Devonian Period in the central Andes. *Geological Society America Bulletin*, **86**, 39–46.

ISAACSON, P. E. 1977. Devonian stratigraphy and brachiopod paleontology of Bolivia. Part A: Orthida and Strophomenida. *Palaeontographica, Abteilung A*, **155**, 133–192.

—— & PERRY, D. G. 1977. Biogeography and Morphological Conservatism of *Tropidoleptus* (Brachiopoda, Orthida) during the Devonian. *Journal of Paleontology*, **51**, 1108–1122.

—— & SABLOCK, P. E. 1988. Devonian System in Bolivia, Peru, and Northern Chile. *Canadian Society of Petroleum Geologists Memoir*, **14**, 719–728.

——, FISHER, L. L. & DAVIDSON, J. 1985. Devonian and Carboniferous stratigraphy of Sierra de Almeida, preliminary results. *Revista Geologica de Chile*, **25–26**, 113–121.

LAUBACHER, G. 1974. Le Paleozoique Inferieur de la Cordillere Orientale du SE du Perou. *Cahiers Office de la Recherche Scientifique et Technique Outre-Mer, serie Geologie*, Paris, **VI**, 29–40.

—— 1977. *Geologie des Andes Peruviennes: Geologie de l'Altiplano et de la Cordillere Orientale au nord et nord-ouest du Lac Titicaca (Perou)*. Theses, Academie de Montpellier, Universite des Sciences et techniques du Languedoc.

——, BOUCOT, A. J. & GRAY, J. 1982. Additions to Silurian stratigraphy, lithofacies, biogeography and paleontology of Bolivia and southern Peru. *Journal of Paleontology*, **56**, 1138–1170.

MEGARD, F. 1973. *Etude geologique d'une transversale des Andes au niveau du Perou central*. Theses, Academie, de Montpellier, Universite des sciences et techniques du Languedoc.

MILLER, H. 1970. Das Problem des hypothetischen 'Pazifischen Kontinentes', gesehen von der chilenischen Pazifikkuste. *Geologische Rundschau*, **59**, 927–938.

MPODOZIS, C. & FORSYTHE, R. 1983. Stratigraphy and geochemistry of accreted fragments of the ancestral Pacific floor in southern South America. *Palaeogeography, Palaeoclimatology, Palaeoecology*, **41**, 103–124.

PALMA, M. A., PARICA, P. O., & RAMOS, V. A. 1987. El Granitico Archibarca: su edad y significado tectonico, Provincia de Catamarca. *Revista, Asociacion Geologica Argentina*, **41**, 414–419.

POJETA, J., KRIZ, J. & BERDAN, J. M. 1976. Silurian–Devonian pelecypods and Palaeozoic stratigraphy of subsurface rocks in Florida and Georgia and related Silurian pelecypods from Bolivia and Turkey. *U.S. Geological Survey Professional Paper*, 879.

RAMOS, V. A., JORDAN, T. E., ALLMENDINGER, R. W., MPODOZIS, C., KAY, S. M., CORTES, J. M. & PALMA, M. A. 1986. Palaeozoic terranes of the central Argentine–Chilean Andes. *Tectonics* 5, 855–880.

RICHTER, R. & RICHTER, E. 1942. Die Trilobiten der Weismes-Schichten am Hohen Venn, mit Bemerkungen uber die Malvinocaffrische Provinz. *Senckenbergiana*, **25**, 156–179.

ROCHA CAMPOS, A. C. 1983. North Andean Area. *In*: MARTINEZ DIAZ, C. (ed.) *The Carboniferous of the World*. Volume **II**. Instituto Geologico y Minero de Espana, 180–200.

ZEIL, W., 1979. *The Andes – a geological review*. Gebruder Borntraeger, Berlin.

Origins and Evolution of the Antarctic Biota

Geological Society Special Publication No. 47
Edited by J. A. Crame (British Antarctic Survey)

Within the last 25 years there has been a dramatic increase in our knowledge of the fossil record of Antarctica. Improved access to the remotest parts of the continent, the advent of offshore drilling and intensive study of early expedition collections have all led to the accumulation of a vast amount of data that stretches back nearly 600 Ma to the begining of the Cambrian Period. No longer can Antarctica be dismissed from our view of the history of life on Earth simply because so little is known about it; it is fast becoming another crucial reference point for global palaeontological syntheses.

Outline of Contents

Principal Authors

L.R.M. Cocks(UK)
F. Debrenne (France)
G.F. Webers (USA)
G.C. Young (Australia)
S.L. De Fauw (USA)
W.G. Chaloner (UK)
M.E. Dettmann (Australia)
R.A. Askin (USA)
T.H. Rich (Australia)
R.E. Molnar (Australia)
G.R. Stevens (New Zealand)
P. Doyle (UK)
R.M. Feldmann (USA)
S. Chatterjee (USA)
J.A. Case (USA)
K. Birkenmajer (Poland)
J.T. Eastman (USA)
A. Clarke (UK)
R.E. Fordyce (New Zealand)
E. Thomas (USA)
L. Watling (USA)

Antarctica in Cambrian-Devonian Gondwana • Cambrian archaeocyaths • Cambrian molluscs and cephalopod origins • Devonian fish of Southern Victoria Land • evolution of Dicynodontia • forest growth in Antarctica • Antarctica: cradle of temperate forests • Endemism in Seymour Island palynofloras • Cretaceous terrestrial tetrapods • biotic links between New Zealand and Antarctica • Antarctic belemnite biogeography • evolutionary patterns in crustaceans • Upper Cretaceous Plesiosaurs • origin of Australian marsupials • floras from King George Island • evolution of Antarctic fishes • origin of Southern Ocean marine fauna • Antarctic marine mammals • Cenozoic deep-sea benthic foraminifera • Antarctica as evolutionary incubator.

- Traces palaeontological record of Antarctica over the last 600Ma
- International field of contributors
- 22 papers
- includes review articles
- over 100 illustrations
- Published October 1989
- Price £58.00, US $89.00
- ISBN 0 903317 44 3
- Primary audience: palaeontologists, geologists, biologists.

Please send your orders to:
Geological Society Publishing House, Unit 7 Brassmill Enterprise Centre, Brassmill Lane, Bath BA1 3JN, UK. Tel: 0225 445046 Fax: 0225 442836.
Please add 10% of the invoice total for overseas (surface) mail.

FORTHCOMING TITLE OF RELATED INTEREST
Palaeozoic Palaeogeography and Biogeography
• Publication date: February 1990
• ISBN 0 903317 49 4

Order Slip

Please add 10% of the invoice total for overseas (surface) mail.

☐ Please enter my standing order to this book series.

☐ I enclose a cheque made out to "Geological Society" for £/US $ ....
☐ I wish to pay by Visa / Access / Mastercard / American Express / Diners Club.
Please debit my card no. ☐☐☐☐☐☐☐☐☐☐☐☐☐☐☐☐
with the sum of £/ US $ .... Expiry date ....
Signature .... Date ....
Please give the address at which the card is registered separately, if not registered at the delivery address.
Name ....
Delivery Address ....

Geological Society Publishing House, Unit 7, Brassmill Enterprise Centre, Brassmill Lane, Bath BA1 3JN UK.

GEOLOGICAL APPLICATIONS OF WIRELINE LOGS

Geological Society Special Publication No. 48
Edited by A. Hurst (Statoil), M.A. Lovell (Nottingham University) and A.C. Morton (British Geological Survey)

Oil and gas exploration personnel will find Geological Applications of Wireline Logs of great value as it concentrates on the geological interpretation of downhole subsurface measurements. This volume covers a wide range of topics from conventional wireline log interpretation through to the evaluation of the latest nuclear (geochemical) and Formation Micro Scanner measurements. The volume centres on applications in potential hydrocarbon bearing environments. Also information is presented on engineering, hydrogeology, crystalline basement studies and numerical methods of interpretation. The papers are grouped according to their geological topics as: sedimentology, mineralogy and geochemistry, stratigraphic correlation, fault and fracture identification and physical properties.

Principal Authors
S.M. Luthi (Schlumberger-Doll Research, USA)
S.D. Harker (Occidental Petroleum (Caledonia), UK)
M.H. Rider (Rider-French Consulting, UK)
J.C. Herweijer (Koninklijke Shell, Netherlands)
B.P. Moss (Scientific Software - intercomp, UK)
C.M. Griffiths (Trondheim University, Norway)
R. Nurmi (Schlumberger, United Arab Emirates)
M.H. Dorfman (Texas University, USA)
P.F. Worthington (BP Research, UK)
C.M. Griffiths (Trondheim University, Norway)
D.K. Buckley (British Geological Survey, UK)
M.M. Herron (Schlumberger-Doll Research, USA)
R.N. Anderson (Columbia University, USA)
T.S. Brewer (Nottingham University, UK)
T.J. Primmer (B.P. Research, UK)
A. Hurst (Statoil, Norway)
B. Humphreys (British Geological Survey, UK)
A.E. Stocks (Exploration Consultants, UK)
W.H. Fertl (Atlas Wireline Services, USA)
K.A. Lehne (Statoil, Norway)
T.M. Ronningsland (Norsk Hydro, Norway)
M.C. Devilliers (Total CFP, France)
D. Goldberg (Columbia University, USA)
J.S. Bell (Geological Survey of Canada, Canada)
C.J. Evans (British Geological Survey, UK)
M.A. Lovell (Nottingham University, UK)
D.C. Entwisle (British Geological Survey, UK)
D.E. King (Schlumberger Well Services, USA)

Outline of Contents
Sedimentary structures of clastic rocks • FMS images in North Sea oil fields • gamma-ray log shapes • SHDT dip interpretation • well log interpretation • stochastic reservoir description • the language of rocks • carbonate reservoir heterogeneities: detection & analysis • lithofacies determination: new techniques • sediment cyclicity from well logs • interwell matching • geophysical logging, central India • geochemical logging applications • geochemical well logs • geochemical results: core and log data •mudrock evaluation: in situ analysis • NGS measurements in sandstones • North Sea Jurassic sandstones • source rock indentification • circumferential acoustic logs •fracture detection • Gullfaks field dipmeter results • fault identification using dipmeter data • hydraulic conductivity from pulse tests •stress regimes in sedimentary basins•in situ stress from borehole breakouts•electrical properties of basalt •Christensen's equation

- Presents unique insight into the diverse applications of wireline logs.
- International field of contributors from oil industry, service companies, academia and geological survey organisations.
- 28 papers
- 273 illustrations including colour plates
- Published January 1990
- Price £59.00, US$97.00
- ISBN 0 903317 45 1
- Primary audience: petroleum geologists, reservoir engineers, petrophysicists, sedimentolgists, mineralogists, stratigraphers and geochemists.

Please send your order to:
Geological Society Publishing House
Unit 7, Brassmill Enterprise Centre, Brassmill Lane, Bath BA1 3JN, UK
Telephone: 0225 445046   Fax: 0225 442836

Please add 10% of the invoice total for overseas (surface) mail.

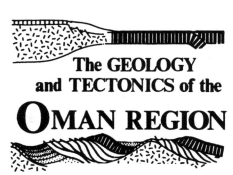

The GEOLOGY and TECTONICS of the OMAN REGION

Geological Society Special Publication No. 49
Edited by A.H.F. Robertson (Edinburgh University, UK)
M.P. Searle (Leicester University, UK)
A.C. Ries (ESRI, Reading, UK)

The Geology and Tectonics of the Oman Region is an extensive collaborative work involving a large number of academic, oil company and government earth scientists. The resulting unique synthesis will be of interest to earth scientists concerned with hydrocarbon habitat, structural geology, sedimentology and palaeoceanography. *Part 1* is concerned with the Oman Mountains which form probably the world's best exposed deformed continental margin and as such the region is an ideal field laboratory for the study of fundamental processes including rifting, passive margin development and thrust tectonics. The contributors discuss the sedimentary and volcanic units underlying the Semail Ophiolite and the post-emplacement sedimentary cover. The sedimentary and structural evolution of the Mesozoic continental margin is particularly relevant to hydrocarbon exploration in Oman. *Part 2* is concerned with the geology and tectonics of that part of Oman that lies outside the Oman Mountains. Attention is focussed on this previously geological unknown area following realisation of the major hydrocarbon potential of the Palaeozoic sequences of southern Oman. *Part 3* sets the Oman area in its Middle Eastern plate tectonic context, including Iran.

- International field of contributors
- 864 pages
- 49 papers
- 492 illustrations
- Published February 1990
- Price £99.00, US $ 160.00
- ISBN 0 903317 46 X
- Primary audience: hydrocarbon geologists, structural geologists, industry and researchers on Middle East continental margin evolution

Principal Authors
A.H.F. Robertson (Edinburgh University, UK)
W. Blendinger (Petroleum Development Oman, Oman)
C.W. Lee (Wales Polytechnic, UK)
D. Rabu (B.R.G.M., France)
B.R. Pratt (University of Saskatchewan, Canada)
R.W. Scott (AMOCO, USA)
E.A. Haan (Petroleum Development, Oman, Oman)
P.D. Wagner (AMOCO, USA)
K.F. Watts (Alaska University, USA)
D.J.W. Cooper (Leicester University, UK)
D. Bernoulli (ETH-Zentram, Switzerland)
E.T. Tozer (Geological Survey of Canada, Canada)
F. Bechennec (B.R.G.M., France)
P. De Wever (C.N.R.S., France)
W. Kickmaier (Berne University, Switzerland)
C.D. Blome (U.S. Geological Survey, USA)
A.E.S. Kemp (Southampton University, UK)
A. Mann (Cambridge University, UK)
J. Le Metour (B.R.G.M., France)
S.S. Hanna (Sultan Qaboos University, Oman)
M.P. Searle (Leicester University, UK)
L.A. Dunne (AMOCO, USA)
P.L. Michaelis (Dallas, USA)
D.R.D. Boote (Occidental Oil & Gas, USA)
J. Warburton (BP Petroleum Development, UK)
P.A. Cawood (Newfoundland University, Canada)
D.Q. Coffield (South Carolina University, USA)
A.W. Shelton (Sultan Qaboos University, Oman)
J. Maizels (Aberdeen University, UK)
I.G. Gass (Open University, UK)
V.P. Wright (P.R.I.S., Reading, UK)
B.W. Mattes (Houston, USA)
A.P. Heward (Shell, UK)
A.C. Ries (ESRI, University of Reading, UK)
F. Moseley (Birmingham University, UK)
R.M. Shackleton (East Hendred, UK)
J.B. Filbrandt (Shell, London)
G.S. Mountain (Columbia University, USA)
W.L. Prell (Brown University, USA)
G.B. Shimmield (Edinburgh University, UK)
K.W. Glennie (Ballater, UK)
R. Stoneley (Imperial College, UK)
A.M.C. Sengor (Istanbul, Turkey)

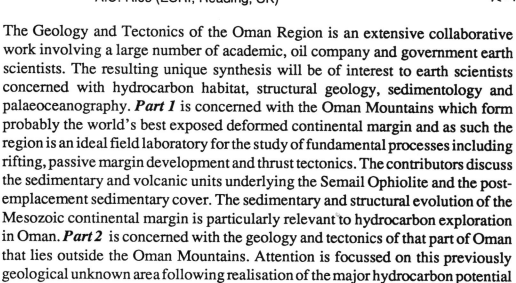

For further information on titles published by the Geological Society and the AAPG contact:

The Geological Society Publishing House, Unit 7
Brassmill Enterprise Centre,
Brassmill Lane, Bath BA1 3JN, UK.
Telephone: (0225) 445046
Fax: (0225) 442836

Contents:

- The northern Oman Tethyan continental margin: stratigraphy, structure concepts and controversies
- Updoming, rifting and continental margin development during the late Palaeozoic in northern Oman
- A review of platform sedimentation in the Early and Late Permian of Oman, with particular reference to the Oman Mountains
- Sedimentary aspects of the Eo-Alpine cycle on the northeast edge of the Arabian Platform (Oman Mountains)
- Jurassic and Early Cretaceous platform margin configuration and evolution, central Oman Mountains
- Chronostratigraphy of the Cretaceous carbonate shelf, southeastern Arabia
- The Lower Kahmah Group of Oman: the carbonate fill of a marginal shelf basin
- Geochemical stratigraphy and porosity controls in Cretaceous carbonates near the Oman Mountains
- Mesozoic carbonate slope facies marking the Arabian platform margin in Oman: depositional history, morphology and palaeogeography
- Sedimentary evolution and palaeogeographical reconstruction of the Mesozoic continental rise in Oman: evidence from the Hamrat Duru Group
- Evolution of the Triassic Hawasina Basin, central Oman Mountains
- Triassic ammonoids from Jabal Safra and Wadi Alwa, Oman, and their significance
- The Hawasina Nappes: stratigraphy palaeogeography and structural evolution of a fragment of the south-Tethyan passive continental margin
- Permian to Cretaceous radiolarian biostratigraphic data from the Hawasina Complex, Oman Mountains
- Manganese occurrences in the Al Hammah Range-Wahrah Formation
- Evolution of the Arabian continental margin in the Dibba Zone, northern Oman Mountains
- Sedimentary and structural evolution of a continental margin transform lineament: the Hatta Zone, northern Oman Mountains
- The tectonic evolution of pre-Permian rocks, Central and southeastern Oman Mountains
- Subduction and obduction: two stages in the Eo-Alpine tectonometamorphic evolution of the Oman Mountains.
- The Alpine deformation of the central Oman Mountains
- Structure of Jebel Sumeini-Jebel Ghawil areas, northern Oman
- Structural style and domains of northern Oman Mountains (Oman and United Arab Emirates). Seismic interpretation of the structure and stratigraphy of the Strait of Hormuz
- Structural evolution of the Suneinah Foreland, central Oman Mountains
- The evolution of the Oman Mountains Foreland Basin
- Origin of culminations within the southeast Oman Mountains at Jebel Ma-Jhool and Ibra Dome
- Structures associated with nappe emplacement and culmination collapse in the central Oman Mountains
- The interpretation of garvity data in Oman: constraints on the ophiolite emplacement mechanism
- Metamorphism in the Oman Mountains in relation to the Semail ophiolite emplacement
- Maastrichtian to early Tertiary stratigraphy and palaeogeography of the central and northern Oman Mountains
- The Maastrichtian transgression onto the northwestern flank of the Proto-Oman Mountains: sequences of rudist-bearing beach to open shelf facies
- The post-Campanian tectonic evolution of the central Oman Mountains: Tertiary extension of the Eastern Arabian Margin
- Cenozoic alluvial fan systems of interior Oman: palaeoenvironmental reconstruction based on discrimination of palaeochannels using remotely sensed data
- Tectonics, geochronology and geochemistry of the Precambrian rocks of Oman
- Infraplatformal basin-fill deposits from the Infra-cambrian Huqf Group, east central Oman
- Carbonate/evaporite deposition in the Late Precambrian-Early Cambrian Ara Formation of Southern Oman
- Salt removal and sedimentation in Southern Oman
- Structures in the Huqf-Haushi Uplift, east central Oman
- The structure of Masirah Island, Oman
- The Batain Melange of NE Oman
- Late Cretaceous and early Tertiary evolution of Jebel Ja' alan and adjacent areas, NE Oman
- Tectonics of the Masirah Fault Zone and eastern Oman
- A multiphase plate tectonic history of the southeast continental margin of Oman
- Neogene tectonics and sedimentation of the SE Oman continental margin: results from ODP Leg 117
- The influence of hydrography, bathymetry and productivity on sediment type and composition of the Oman margin and in the northwest Arabian Sea
- Inter-relationship of the Makran-Oman Mountains belts of convergence
- The Arabian continental margin in Iran during the Late Cretaceous
- A new model for the late Palaeozoic-Mesozoic tectonic evolution of Iran and implications for Oman

Journal of the Geological Society

Chief Editor: M. J. Le Bas

The Journal of the Geological Society is among the world's leading references for significant research in geology. It has been continuously published since 1845 and it enjoys a wide circulation throughout the world.

Only material of the highest quality is accepted for publication. As well as major research papers, the Journal publishes Conference Reports, Discussions and rapid publication Short Papers.

Indexed/Abstracted in: Current Contents, Geological Abstracts, GEOREF, Petroleum Abstracts, Mineralogical Abstracts.

Editors:

J.R. ANDREWS
D. BARR
K.H. BRODIE
B. CHADWICK
H. COLLEY
I.J. FAIRCHILD
N.B.W. HARRIS
R.J. PANKHURST
K.T. PICKERING
R.A. SCRUTTON
D.J. SIVETER
N.J. SOPER
P.J. WILLIAMS

Advisory Editors:
J.D. BELL
F.W. CAMBRAY
L.R.M. COCKS
I.W.D. DALZIEL
L.E. FROSTICK
P.L. HANCOCK
R.T. HAWORTH
J.D. HUDSON
A. NOTHOLT
D. ROBINSON
A. TAIRA
P.A. ZIEGLER

Forthcoming Thematic Sets

Palaeoclimates
Nature and Analysis of the Stratigraphic Record
Tectonics and Sedimentation
The Murchison Silurian Symposium
Geology and Geophysics of the Irish Shelf and Continental Margin
Monitoring Active Volcanoes

Forthcoming Papers

The San Nicolas batholith of coastal Peru: early Palaeozoic continental arc or continental rift magmatism? S.B. MUKASA (USA) & D.J. HENRY (USA)

A buried granite batholith beneath the East Midland Shelf of the Southern North Sea Basin, J.A. DONATO (UK) & J.B. MEGSON (UK)

Rapid thermal recovery of thrust related metamorphism in basement windows of the Scandinavian Caledonides, J.E. LINQVIST (SWEDEN)

Geochemistry and origin of the Archaean Beit Bridge Complex, Limpopo Belt, South Africa, K.C. CONDIE (USA) & M. BORYTA (USA)

Geomorphology and surface tilting in an active extensional basin, SW Montana, USA, J. ALEXANDER (UK) & M.R. LEEDER (UK)

Submission procedure
Manuscripts should be sent to the Publications Secretary: Angharad Hills, The Geological Society Publishing House, Unit 7, Brassmill Enterprise Centre, Brassmill Lane, Bath, Avon BA1 3JN, UK (Tel: 0225 445046 - Fax: 0225 442836) There are no page charges. Advertisement enquiries should also be sent to this address.

Subscription rates 1990 (Volume 147)
(The journal is published in January, March, May, July, September and November)
£198.00 (UK)   £238.00 (overseas)   US $ 400.00 (North America and Japan only)
Single issues: £39.60 (UK) or £47.60 (overseas)   US $ 80.00 (North America and Japan only)
All orders and business correspondence relating to subscriptions should be addressed to: Journals Subscriptions Dept., Room J1, The Geological Society Publishing House, Unit 7, Brassmill Enterprise Centre, Brassmill Lane, Bath BA1 3JN, UK (Tel: 0225 445046 - Fax 0225 442836)
Orders can be placed directly or through your usual agent.